SELECTED PAPERS (1937–1976) OF JULIAN SCHWINGER

MATHEMATICAL PHYSICS AND APPLIED MATHEMATICS


*Editors:*

M. FLATO, *Université de Dijon, Dijon, France*

R. RĄCZKA, *Institute of Nuclear Research, Warsaw, Poland*

*with the collaboration of:*

M. GUENIN, *Institut de Physique Théorique, Geneva, Switzerland*

D. STERNHEIMER, *Collège de France, Paris, France*


VOLUME 4

JULIAN SCHWINGER

# SELECTED PAPERS
## (1937 - 1976)
## OF
# JULIAN SCHWINGER

*Edited by*

M. FLATO, C. FRONSDAL AND K. A. MILTON

D. REIDEL PUBLISHING COMPANY

DORDRECHT : HOLLAND / BOSTON : U.S.A.
LONDON : ENGLAND

**Library of Congress Cataloging in Publication Data**

Schwinger, Julian Seymour, 1918–
Selected Papers (1937–1976) of Julian Schwinger

(Mathematical physics and applied mathematics ; v. 4)
1. Physics–Addresses, Essays, Lectures. 2. Schwinger, Julian Seymour, 1918– – –Bibliography.
I. Flato, Moshe. II. Title. III. Series.
QC71.S32 530'.08 79-11576
ISBN 90-277-0974-2
ISBN 90-277-0975-0 pbk. (Pallas edition)



Published by D. Reidel Publishing Company,
P.O. Box 17, Dordrecht, Holland.

Sold and distributed in the U.S.A., Canada, and Mexico
by D. Reidel Publishing Company, Inc.
Lincoln Building, 160 Old Derby Street, Hingham,
Mass. 02043 U.S.A.



Printed in The Netherlands

# TABLE OF CONTENTS



---

The papers appearing in this volume are numbered according to the comprehensive list of publications of Julian Schwinger which are presented in chronological order beginning on p. xiii.

# PREFACE

Very few people have contributed as much to twentieth-century physics as Julian Schwinger. It is therefore appropriate to offer a retrospective of his work on the occasion of his sixtieth birthday (February 12, 1978). We hope, in offering this selection of his papers, to bring to light ideas and results that may have been partly overlooked at the time of the original publication.

Schwinger has published prodigiously on a great variety of subjects, as is evident from the comprehensive list of publications arranged in chronological order which appears on p. xiii. Needless to say, only a small subset could be included in the present modest volume. In the selection, great weight was assigned to papers that seem to be less widely known or appreciated than they deserve. Many important papers are therefore omitted. (Examples: Paper [64] 'On Gauge Invariance and Vacuum Polarization' and Paper [69] 'On Angular Momentum', both of which have been reprinted elsewhere.)

The collection is a personal one, having been chosen by Schwinger himself, and is therefore of particular interest. It would probably not be interesting to offer an analysis, by the editors, of Schwinger's contributions to physics. However, we are very pleased to be able to include Schwinger's own informal and very personal comments about each article that appears in this volume. These comments indicate his reasons for choosing these particular articles and, in many cases, provide a capsule synopsis of what he considers most valuable.

It would be a grave error, unlikely to befall anybody who knows Schwinger, to interpret the appearance of the collection as a finis to this remarkable physicist's productivity. We are looking forward to many more years of unorthodox and incisive research from Julian Schwinger.

THE EDITORS

# JULIAN SCHWINGER – LIST OF PUBLICATIONS

[ 0] 'On the Interaction of Several Electrons', unpublished (1934).

[ 1] 'On the Polarization of Electrons by Double Scattering' (with O. Halpern), *Phys. Rev.* **48**, 109 (1935).

[ 2] 'On the $\beta$-Radioactivity of Neutrons' (with L. Motz), *Phys. Rev.* **48**, 704 (1935).

[ 3] 'On the Magnetic Scattering of Neutrons', *Phys. Rev.* **51**, 544 (1937).

[ 4] 'On Non-Adiabatic Processes in Inhomogeneous Fields', *Phys. Rev.* **51**, 648 (1937).

[ 5] 'The Scattering of Neutrons by Ortho- and Parahydrogen' (with E. Teller), *Phys. Rev.* **51**, 775 (1937).

[ 6] 'Depolarization by Neutron-Proton Scattering' (with I. I. Rabi), *Phys. Rev.* **51**, 1003 (1937).

[ 7] 'Neutron Energy Levels' (with J. Manley and H. Goldsmith), *Phys. Rev.* **51**, 1022 (1937).

*[ 8] 'The Scattering of Neutrons by Ortho- and Parahydrogen' (with E. Teller), *Phys. Rev.* **52**, 286 (1937).

[ 9] 'On the Spin of the Neutron', *Phys. Rev.* **52**, 1250 (1937).

[10] 'The Widths of Nuclear Energy Levels' (with J. Manley and H. Goldsmith), *Phys. Rev.* **55**, 39 (1939).

*[11] 'The Neutron-Proton Scattering Cross Section' (with V. Cohen and H. Goldsmith), *Phys. Rev.* **55**, 106 (1939).

[12] 'The Resonance Absorption of Slow Neutrons in Indium' (with J. Manley and H. Goldsmith), *Phys. Rev.* **55**, 107 (1939).

[13] 'On the Neutron-Proton Interaction', *Phys. Rev.* **55**, 235 (1939).

[14] 'The Scattering of Neutrons by Hydrogen and Deuterium Molecules' (with M. Hamermesh), *Phys. Rev.* **55**, 679 (1939).

*[15] 'On Pair Emission in the Proton Bombardment of Fluorine' (with J. R. Oppenheimer), *Phys. Rev.* **56**, 1066 (1939).

[16] 'Neutron-Deuteron Scattering Cross Section' (with L. Motz), *Phys. Rev.* **57**, 161 (1940).

[17] 'The Scattering of Thermal Neutrons by Deuterons' (with L. Motz), *Phys. Rev.* **58**, 26 (1940).

---

*The papers reprinted in this volume are marked with an asterisk and appear in chronological order. In the Table of Contents of the present volume these articles retain the numbering as given in this comprehensive list of publications.

[18] 'The Electromagnetic Properties of Mesotrons' (with H. Corben), *Phys. Rev.* **58**, 191 (1940).

[19] 'The Electromagnetic Properties of Mesotrons' (with H. Corben), *Phys. Rev.* **58**, 953 (1940).

[20] 'Neutron Scattering in Ortho- and Parahydrogen and the Range of Nuclear Forces', *Phys. Rev.* **58**, 1004 (1940).

[21] 'The Photodisintegration of the Deuteron' (with W. Rarita), *Phys. Rev.* **59**, 215 (1941).

[22] 'The Photodisintegration of the Deuteron' (with W. Rarita and H. Nye), *Phys. Rev.* **59**, 209 (1941).

[23] 'On the Neutron-Proton Interaction' (with W. Rarita), *Phys. Rev.* **59**, 436 (1941).

[24] 'On the Exchange Properties of the Neutron-Proton Interaction' (with W. Rarita), *Phys. Rev.* **59**, 556 (1941).

*[25] 'On a Theory of Particles with Half-Integral Spin' (with W. Rarita), *Phys. Rev.* **60**, 61 (1941).

*[26] 'On the Interaction of Mesotrons and Nuclei' (with J. R. Oppenheimer), *Phys. Rev.* **60**, 150 (1941).

[27] 'The Theory of Light Nuclei' (with E. Gerjuoy), *Phys. Rev.* **60**, 158 (1941).

[28] 'On the Charged Scalar Mesotron Field', *Phys. Rev.* **60**, 159 (1941).

[29] 'The Quadrupole Moment of the Deuteron and the Range of Nuclear Forces', *Phys. Rev.* **60**, 164 (1941).

[30] 'On Tensor Forces and the Theory of Light Nuclei' (with E. Gerjuoy), *Phys. Rev.* **61**, 138 (1942).

*[31] 'On a Field Theory of Nuclear Forces,' *Phys. Rev.* **61**, 387 (1942).

[32] 'On the Magnetic Moments of $H^3$ and $He^3$' (with R. Sachs), *Phys. Rev.* **61**, 732 (1942).

[33] 'The Scattering of Slow Neutrons by Ortho and Para Deuterium' (with M. Hamermesh), *Phys. Rev.* **69**, 145 (1946).

*[34] 'Polarization of Neutrons by Resonance Scattering in Helium', *Phys. Rev.* **69**, 681 (1946).

[35] 'Electron Orbits in the Synchrotron' (with D. Saxon), *Phys. Rev.* **69**, 702 (1946).

[36] 'The Magnetic Moments of $H^3$ and $He^3$' (with R. Sachs), *Phys. Rev.* **70**, 41 (1946).

[37] 'Electron Radiation in High Energy Accelerators', *Phys. Rev.* **70**, 798 (1946).

[38] 'Neutron Scattering in Ortho- and Parahydrogen' (with M. Hamermesh), *Phys. Rev.* **71**, 678 (1947).

[39] 'On the Radiation of Sound from a Unflanged Circular Pipe' (with H. Levine), *Phys. Rev.* **72**, 742 (1947).

[40] 'A Variational Principle for Scattering Problems', *Phys. Rev.* **72**, 742 (1947).

[41] 'On the Radiation of Sound from an Unflanged Circular Pipe' (with H. Levine), Phys. *Rev.* **73**, 383 (1948).

*[42] 'On the Polarization of Fast Neutrons', *Phys. Rev.* **73**, 407 (1948).

*[43] 'On Quantum-Electrodynamics and the Magnetic Moment of the Electron', *Phys. Rev.* **73**, 416 (1948).

*[44] 'A Note on Saturation in Microwave Spectroscopy' (with R. Karplus), *Phys. Rev.* **73**, 1020 (1948).

[45] 'On the Electromagnetic Shift of Energy Levels' (with V. Weisskopf), *Phys. Rev.* **73**, 1272 (1948).

[46] 'On the Theory of Diffraction by an Aperture in an Infinite Plane Screen. I', (with H. Levine), *Phys. Rev.* **74**, 958 (1948).

[47] 'An Invariant Quantum Electrodynamics', *Phys. Rev.* **74**, 1212 (1948).

[48] 'Variational Principles for Diffraction Problems' (with H. Levine), *Phys. Rev.* **74**, 1212 (1948).

[49] 'On Tensor Forces and the Variation-Iteration Method' (with H. Feshbach and J. Eisenstein), *Phys. Rev.* **74**, 1223 (1948).

[50] 'Quantum Electrodynamics I. A Covariant Formulation', *Phys. Rev.* **74**, 1439 (1948).

[51] 'Radiative Correction to the Klein–Nishina Formula' (with D. Feldman), *Phys. Rev.* **75**, 358 (1949).

[52] 'Quantum Electrodynamics II. Vacuum Polarization and Self Energy', *Phys. Rev.* **75**, 651 (1949).

[53] 'On Radiative Corrections to Electron Scattering', *Phys. Rev.* **75**, 898 (1949).

[54] 'On the Theory of Diffraction by an Aperture in an Infinite Plane Screen. II.' (with H. Levine), *Phys. Rev.* **75**, 1423 (1949).

[55] 'On the Transmission Coefficient of a Circular Aperture' (with H. Levine), *Phys. Rev.* **75**, 1608 (1949).

[56] 'On the Classical Radiation of Accelerated Electrons', *Phys. Rev.* **75**, 1912 (1949).

[57] 'Quantum Electrodynamics III. The Electromagnetic Properties of the Electron-Radiative Corrections to Scattering', *Phys. Rev.* **76**, 790 (1949).

*[58] 'On the Charge Independence of Nuclear Forces', *Phys. Rev.* **78**, 135 (1950).

[59] 'On the Self-Stress of the Electron', (with S. Borowitz and W. Kohn) *Phys. Rev.* **78**, 345 (1950).

[60] 'Variational Principles for Scattering Processes. I.' (with B. Lippman), *Phys. Rev.* **79**, 469 (1950).
[61] 'On the Theory of Electromagnetic Wave Diffraction by an Aperture in an Infinite Plane Conducting Screen' (with H. Levine), *Communications on Pure and Applied Mathematics III,* **4**, 355 (1950).
[62] 'Quantum Dynamics', *Science* **113**, 479 (1951).
[63] 'On the Representation of the Electric and Magnetic Fields Produced by Currents and Discontinuities in Wave Guides. I.' (with N. Marcuvitz), *J. Appl. Phys.* **22**, 806 (1951).
[64] 'On Gauge Invariance and Vacuum Polarization', *Phys. Rev.* **82**, 664 (1951).
[65] 'The Theory of Quantized Fields. I', *Phys. Rev.* **82**, 914 (1951).
*[66] 'On the Green's Functions of Quantized Fields. I, II', *Proc. Natl. Acad. Sci. U.S.A.* **37**, 452, 455 (1951).
[67] 'On a Phenomenological Neutron-Proton Interaction' (with H. Feshbach), *Phys. Rev.* **84**, 194 (1951).
[68] 'Electrodynamic Displacement of Atomic Energy Levels' (with R. Karplus and A. Klein), *Phys. Rev.* **84**, 597 (1951).
[69] 'On Angular Momentum', 1952, later published in *Quantum Theory of Angular Momentum,* Academic Press, New York, 1965, edited by L. C. Biedenharn and H. Van Dam, p. 229.
[70] 'Electrodynamic Displacement of Atomic Energy Levels. II. Lamb Shift' (with R. Karplus and A. Klein), *Phys. Rev.* **86**, 288 (1952).
[71] 'Radiation Force and Torque' (with H. Levine), *Phys. Rev.* **87**, 224 (1952).
[72] 'On High Energy Nucleon Scattering and Isobars', (with R. B. Raphael), *Phys. Rev.* **90**, 373 (1953).
[73] 'The Theory of Quantized Fields. II', *Phys. Rev.* **91**, 713 (1953).
*[74] 'The Theory of Quantized Fields. III', *Phys. Rev.* **91**, 728 (1953).
[75] 'A Note on the Quantum Dynamical Principle', *Philos. Mag.* **44**, 1171 (1953).
*[76] 'The Theory of Quantized Fields. IV', *Phys. Rev.* **92**, 1283 (1953).
[77] 'The Theory of Quantized Fields. V', *Phys. Rev.* **93**, 615 (1954).
*[78] 'The Quantum Correction in the Radiation by Energetic Accelerated Electrons', *Proc. Natl. Acad. Sci. U.S.A.* **40**, 132 (1954).
[79] 'Use of Rotating Coordinates in Magnetic Resonance Problems' (with I. I. Rabi and N. Ramsey), *Rev. Mod. Phys.* **26**, 167 (1954).
[80] 'The Theory of Quantized Fields. VI', *Phys. Rev.* **94**, 1362 (1954).
[81] 'Dynamical Theory of K Mesons', *Phys. Rev.* **104**, 1164 (1956).

*[82] 'A Theory of the Fundamental Interactions', *Ann. Phys. (N.Y.)* **2**, 407 (1957).
[83] *'Quantum Electrodynamics'*, Editor, Dover, New York, 1958.
[84] 'Spin Statistics and the TCP Theorem', *Proc. Natl. Acad. Sci. U.S.A.* **44**, 223 (1958).
[85] 'Addendum to Spin, Statistics and the TCP Theorem', *Proc. Natl. Acad. Sci. U.S.A.* **44**, 617 (1958).
*[86] 'On the Euclidean Structure of Relativistic Field Theory', *Proc. Natl. Acad. Sci. U.S.A.* **44**, 956 (1958).
[87] 'Four-Dimensional Euclidean Formulation of Quantum Field Theory', Proceedings of the 1958 International Conference on High Energy Physics, CERN, CERN, Geneva, 1958, p. 134.
*[88] 'Euclidean Quantum Electrodynamics', *Phys. Rev.* **115**, 721 (1959).
[89] 'Theory of Many-Particle Systems, I', (with P. C. Martin), *Phys. Rev.* **115**, 1342 (1959).
[90] 'Field Theory Commutators', *Phys. Rev. Letters* **3**, 296 (1959).
*[91] 'The Algebra of Microscopic Measurement', *Proc. Natl. Acad. Sci. U.S.A.* **45**, 1542 (1959).
[92] 'Field Theory Methods', 1959 Brandeis University Summer Institute in Theoretical Physics.
[92] 'The Geometry of Quantum States', *Proc. Natl. Acad. Sci. U.S.A.* **46**, 257 (1960).
[94] 'Field Theory of Unstable Particles', *Ann. Phys. (N.Y.)* **9**, 169 (1960).
[95] 'Euclidean Gauge Transformation', *Phys. Rev.* **117**, 1407 (1960).
[96] 'Unitary Operator Bases', *Proc. Natl. Acad. Sci. U.S.A.* **46**, 570 (1960).
[97] 'Unitary Transformations and the Action Principle', *Proc. Natl. Acad. Sci. U.S.A.* **46**, 883 (1960).
*[98] 'The Special Canonical Group', *Proc. Natl. Acad. Sci. U.S.A.* **46**, 1401 (1960).
[99] 'Field Theory Methods in Non-Field Theory Contexts', 1960 Brandeis University Summer Institute in Theoretical Physics, Lecture Notes, p. 223.
*[100] 'On the Bound States of a Given Potential', *Proc. Natl. Acad. Sci. U.S.A.* **47**, 122 (1961).
*[101] 'Brownian Motion of a Quantum Oscillator', *J. Math. Phys.* **2**, 407 (1961).
[102] 'Quantum Variables and the Action Principle', *Proc. Natl. Acad. Sci. U.S.A.* **47**, 1075 (1961).

[103] 'Spin and Statistics', (with L. Brown), *Prog. Theor. Phys. (Kyoto)* **26**, 917 (1961).

*[104] 'Gauge Invariance and Mass', *Phys. Rev.* **125**, 397 (1962).

*[105] 'Non-Abelian Gauge Fields. Commutation Relations', *Phys. Rev.* **125** 1043 (1962).

*[106] 'Exterior Algebra and the Action Principle. I', *Proc. Natl. Acad. Sci. U.S.A.* **48**, 603 (1962).

*[107] 'Non-Abelian Gauge Fields. Relativistic Invariance', *Phys. Rev.* **127**, 324 (1962).

*[108] 'Gauge Invariance and Mass. II', *Phys. Rev.* **128**, 2425 (1962).

*[109] 'Quantum Variables and Group Parameters', *Il Nuovo Cimento* **30**, 278 (1963).

[110] 'Non-Abelian Gauge Fields. Lorentz Gauge Formulation', *Phys. Rev.* **130**, 402 (1963).

*[111] 'Commutation Relations and Conservation Laws', *Phys. Rev.* **130**, 406 (1963).

*[112] 'Energy and Momentum Density in Field Theory', *Phys. Rev.* **130**, 800 (1963).

[113] 'Quantized Gravitational Field', *Phys. Rev.* **130**, 1253 (1963).

*[114] 'Quantized Gravitational Field. II', *Phys. Rev.* **132**, 1317 (1963).

[115] 'Gauge Theories of Vector Particles', Theoretical Physics (Trieste Seminar, 1962), IAEA, Vienna, 1963, p. 89.

*[116] 'Coulomb Green's Function', *J. Math. Phys.* **5**, 1606 (1964).

*[117] 'Non-Abelian Vector Gauge Fields and the Electromagnetic Field', *Rev. Mod. Phys.* **36**, 609 (1964).

*[118] 'Field Theory of Matter', *Phys. Rev.* **135**, B816 (1964).

[119] 'A Ninth Baryon' Coral Gables Conference on Symmetry Principles at High Energy, 1964, edited by B. Korsunoglu and A. Perlmutter, Freeman, San Francisco, 1964, p. 127.

[120] 'A Ninth Baryon? ', *Phys. Rev. Letters* **12**, 237 (1964).

[121] '$\Delta T = 3/2$ Nonleptonic Decay', *Phys. Rev. Letters* **12**, 630 (1964).

[122] 'Broken Symmetries and Weak Interactions', *Phys. Rev. Letters* **13**, 355 (1964).

[123] 'Broken Symmetries and Weak Interactions. II', *Phys. Rev. Letters* **13**, 500 (1964).

*[124] 'Field Theory of Matter. II', *Phys. Rev.* **136**, B1821 (1964).

[125] *Field Theory of Particles*, Lectures on Particles and Field Theory (1964 Brandeis Lectures), edited by S. Deser and K. Ford, Prentice-Hall, Englewood Cliffs, N.J., 1965, p. 145.

[126] *Field Theory of Matter,* Proceedings of the 12th International Conference on High Energy Physics, Dubna, 1964, Atomizdat, Moscow 1966, Vol. 1, p. 771.

[127] 'Field Theory of Matter.. III. Phenomenological Field Theory', Coral Gables Conference on Symmetry Principles at High Energy, 1965, edited by B. Kursunoglu, A. Perlmutter, and I. Sakmar, Freeman, San Francisco, 1965, p. 372.

*[128] 'Field Theory of Matter. IV', *Phys. Rev.* **140**, B158 (1965).

[129] 'Magnetic Charge and Quantum Field Theory', *Phys. Rev.* **144**, 1087 (1966).

[130] 'Magnetic Charge and Quantum Field Theory', Coral Gables Conference on Symmetry Principles at High Energy, 1966, edited by A. Perlmutter, J. Wojtaszek, G. Sudarshan, and B. Kursunoglu, Freeman, San Francisco, 1966, p. 233.

[131] 'Lectures on Quantum Field Theory', Coral Gables, 1966, University of Miami, 1967.

*[132] 'Relativistic Quantum Field Theory', Nobel Lecture, in *Nobel Lectures - Physics, 1963–1970,* Elsevier, Amsterdam, 1972.

[133] 'Electric and Magnetic-Charge Renormalization. I', *Phys. Rev.* **151**, 1048 (1966).

[134] 'Electric and Magnetic-Charge Renormalization. II', *Phys. Rev.* **151**, 1055 (1966).

*[135] 'Particles and Sources', *Phys. Rev.* **152**, 1219 (1966).

[136] 'Sourcery', Coral Gables Conference on Symmetry Principles at High Energy, 1967, edited by A. Perlmutter and B. Kursunoglu, Freeman, San Francisco, 1967, p. 180.

*[137] 'Chiral Dynamics', *Phys. Letters* **24B**, 473 (1967).

[138] 'Mass Empirics', *Phys. Rev. Letters* **18**, 797 (1967).

*[139] 'Partial Symmetry', *Phys. Rev. Letters* **18**, 923 (1967).

[140] 'Photons, Mesons and Form Factors', *Phys. Rev. Letters* **19**, 1154 (1967).

[141] 'Radiative Corrections in $\beta$ Decay', *Phys. Rev. Letters* **19**, 1501 (1967).

[142] 'Sources and Electrodynamics', *Phys. Rev.* **158**, 1391 (1967).

[143] 'Boson Mass Empirics', *Phys. Rev. Letters* **20**, 516 (1968).

*[144] 'Gauge Fields, Sources and Electromagnetic Masses', *Phys. Rev.* **165**, 1714 (1968); *Phys. Rev.* **167**, 1546 (1968).

[145] 'Chiral Transformations', *Phys. Rev.* **167**, 1432 (1968).

[146] 'Sources and Gravitons', *Phys. Rev.* **173**, 1264 (1968).

*[147] 'Sources and Magnetic Charge', *Phys. Rev.* **173**, 1536 (1968).

[148] *Discontinuities in Wave Guides* (with D. Saxon), Gordon and Breach, New York, 1968.

[149] *Particles and Sources,* Gordon and Breach, New York, 1969.

*[150] 'A Magnetic Model of Matter', *Science,* **165**, 757 (1969).

*[151] 'Theory of Sources', *Contemporary Physics* (Trieste Symposium 1968), IAEA, Vienna, 1969, Vol. II, p. 59.

[152] *Quantum Kinematics and Dynamics,* Benjamin, New York, 1970.

[153] *Particles, Sources and Fields, Vol. I,* Addison-Wesley, Reading, Mass., 1970.

[154] 'Unit-Spin Propagation Functions and Form Factors', *Phys. Rev.* D**3**, 1967 (1971).

*[155] 'How Massive is the $W$ Particle? ', *Phys. Rev.* D**7**, 908 (1973).

*[156] 'Classical Radiation of Accelerated Electrons II. A Quantum Viewpoint', *Phys. Rev.* D**7**, 1696 (1973).

*[157] 'How to Avoid $\Delta Y = 1$ Neutral Currents', *Phys. Rev.* D**8**, 960 (1973).

[158] *Particles, Sources and Fields, Vol. II,* Addison-Wesley, Reading, Mass., 1973.

[159] 'Radiative Polarization of Electrons', (with W.-y. Tsai), *Phys. Rev.* D**9**, 1843 (1974).

*[160] 'A Report on Quantum Electrodynamics', in *The Physicist's Conception of Nature,* edited by J. Mehra, Reidel, Dordrecht, 1973, p. 413.

[161] 'Spectral Forms for Three-Point Functions', *Phys. Rev.* D**9**, 2477 (1974).

[162] 'Precession Tests of General Relativity - Source Theory Derivations', *Am. J. Phys.* **42**, 507 (1974).

[163] 'Spin Precession– A Dynamical Discussion', *Am. J. Phys.* **42**, 510 (1974).

*[164] 'Photon Propagation Function: Spectral Analysis of Its Asymptotic Form', *Proc. Natl. Acad. Sci. U.S.A.* **71**, 3024 (1974).

[165] 'Photon Propagation Function: A Comparison of Asymptotic Functions', *Proc. Natl. Acad. Sci. U.S.A.* **71**, 5047 (1974).

[166] 'Interpretation of a Narrow Resonance in $e^+ e^-$ Annihilation', *Phys. Rev. Letters* **34**, 37 (1975).

*[167] 'Source Theory Viewpoints in Deep Inelastic Scattering', *Proc. Natl. Acad. Sci. U.S.A.* **72**, 1 (1975).

[168] 'Source Theory Discussion of Deep Inelastic Scattering with Polarized Particles', *Proc. Natl. Acad. Sci. U.S.A.* **72**, 1559 (1975).

[169] 'Psi Particles and Dyons', *Science* **188**, 1300 (1975).

[170] 'Resonance Interpretation of the Decay of $\psi'$ (3.7) into $\psi$ (3.1)', (with K. A. Milton, W.-y. Tsai and L. L. DeRaad, Jr.), *Phys. Rev.* D**12**, 2617 (1975).

[171] 'Pion Spectrum in Decay of $\psi'$ (3.7) to $\psi$ (3.1)', (with K. A. Milton, W.-y. Tsai and L. L. DeRaad, Jr.), *Proc. Natl. Acad. Sci. U.S.A.* **72**, 4216 (1975).

*[172] 'Magnetic Charge and the Charge Quantization Condition', *Phys. Rev.* D**12**, 3105 (1975).

[173] 'Source Theory Analysis of Electron–Positron Annihilation Experiments', *Proc. Nat. Acad. Sci. U.S.A.* **72**, 4725 (1975).

*[174] 'Casimir Effect in Source Theory', *Lett. Math. Phys.* **1**, 43 (1975).

[175] 'Magnetic Charge', in *Gauge Theories and Modern Field Theory*, edited by R. Arnowitt and P. Nath, MIT Press, Cambridge, Mass., 1976, p. 337.

[176] 'Classical and Quantum Theory of Synergic Synchrotron-Cerenkov Radiation', (with W.-y. Tsai and T. Erber), *Ann. Phys. (N.Y.)* **96**, 303 (1976).

[177] 'Gravitons and Photons: The Methodological Unification of Source Theory', *Gen. Relativity and Gravitation* **7**, 251 (1976).

*[178] 'Deep Inelastic Scattering of Leptons', *Proc. Natl. Acad. Sci. U.S.A.* **73**, 3351 (1976).

[179] 'Deep Inelastic Scattering of Charged Leptons', *Proc. Natl. Acad. Sci. U.S.A.* **73**, 3816 (1976).

[180] 'Nonrelativistic Dyon-Dyon Scattering', (with K. A. Milton, W.-y. Tsai, L. L. DeRaad, Jr., and D. C. Clark), *Ann. Phys. (N.Y.)* **101**, 451 (1976).

[181] 'Adler's Sum Rule in Source Theory', *Phys. Rev.* D**15**, 910 (1977).

[182] 'Deep Inelastic Neutrino Scattering and Pion-Nucleon Cross Sections', *Phys. Rev. Letters* **67B**, 89 (1977).

[183] 'Deep Inelastic Sum Rules in Source Theory', *Nucl. Phys.* B**123**, 223 (1977)

# COMMENTS BY JULIAN SCHWINGER
# ON THE PAPERS REPRINTED IN THIS VOLUME

## 1934–1939

[8] Because I, not my distinguished colleague, wrote it.
[11] An experimental paper. The result has remained valid over the years.

## 1940–1949

[15] An interesting application of electrodynamics.
[25] Spin 3/2 and the vector-spinor.
[26] Strong Coupling theory and nucleonic isobars.
[31] The prediction of the $\rho$-meson.
[34] How to polarize fast neutrons.
[42] Another polarization method of some interest.
[43] The first publication of the new electrodynamics: renormalization.
[44] Contains the ordered expansion of exponentials.

## 1950–1959

[58] First published effective range derivation.
[66] Dunctional derivative formulation of field theory; introduction of anticommuting Fermi sources.
[74] Systematic use for Bose systems of states based on non-Hermitian operators (later popularized as 'coherent states').
[76] Systematic use for Fermi systems of totally anticommutative number system (Grassmann algebra); Green's functions for multiparticle states.
[78] It's cute.
[82] A speculative paper that was remarkably on target: VA weak interaction theory, two neutrinos, charged intermediate vector meson, dynamical unification of weak and electromagnetic interactions, scale invariance, chiral transformations, mass generation through vacuum expectation value of scalar field. Concerning the idea of unifying weak and electromagnetic interactions, Rabi once reported to me: "They hate it".

[86] I still recall the utter disbelief this idea engendered.
[88] A detailed example, with incidental remarks about gauge transformations, functional differential and integral methods.
[91] First of a series devoted to the mathematics of quantum mechanics as a symbolic presentation of physics.

1960–1969

[98] Quantum phase space.
[100] Demystifying mathematics.
[101] On the transition between quantum and classical mechanics with temperature and non-Hermitian operators.
[104] Dynamical mass generation.
[105] The action principle at work.
[106] Systematic use of Grassmann algebra.
[107] Energy density commutator equation.
[108] Electrodynamics in one spatial dimension.
[109] Putting together that which man has sundered.
[111] Commutators as the expression of kinematics through dynamics.
[112] Tetrads (vierbein) and the action principle.
[114] Ref. [109] in action.
[116] Too bad it wasn't published in the 40's.
[117] A precursor of photon-vector meson mixing; the unit spin magnetic moment reappears.
[118] Speculative (two intermediate vector bosons, with incomplete lepton polarization) and practical (mass formulas, selection rules).
[124] Dominance of dynamics by short-ranged operator products.
[128] For its remarks on photon-vector meson mixing, spin 2 multiplets, non-linear meson transformations.
[132] The past is prologue.
[135] Now source theory begins, already using generalized sources that unify all spins. It is interesting to see the psychological difficulty in breaking completely with operator field theory. But once accomplished, there could be no turning back.
[137] The first application of source theory to new physics.
[139] This one has not lost its provocativeness.
[144] Photon-vector meson mixing, and mass regularities.
[147] The beginnings of the dyon model.
[150] Nothing is too wonderful to be true.

[151] A passionate, colloquial, overview of Source Theory, two years after its inception.

1970–

[155] Old ideas revisited.
[156] Proper time and errata.
[157] Anti-Cabibbo; it turned out to be a prediction of the $\psi$ particles.
[160] The genesis of source theory in history.
[164] Has anyone ever read this?
[167] A modest beginning.
[172] If only the Price had been right.
[174] Fun.
[178] I like the pictures.

# ACKNOWLEDGEMENTS

The publishers and editors wish to thank the following publishers for granting permission to reprint the papers which appear in this volume (the papers are identified according to the List of Publications appearing on p. xiii):

The American Physical Society (*The Physical Review* and *Physical Review Letters*) for papers [8, 11, 15, 25, 26, 31, 34, 42, 43, 44, 58, 74, 76, 88, 104, 105, 107, 108, 111, 112, 114, 118, 124, 128, 135, 139, 144, 147, 155, 156, 157, 172].

Academic Press (*Annals of Physics*) for paper [82].

The American Institute of Physics (*Journal of Mathematical Physics* and *Reviews of Modern Physics*) for papers [101, 116, 117].

Società Italiana di Fisica (*Il Nuovo Cimento*) for paper [109].

Elsevier Scientific Publishing Company (*Nobel Lectures – Physics 1963–1970*) for paper [132].

North-Holland Publishing Company (*Physics Letters*) for paper [137].

American Association for the Advancement of Science (*Science*) for paper [150].

International Atomic Energy Agency (*Contemporary Physics*) for paper [151].

D. Reidel Publishing Company (*The Physicist's Conception of Nature* and *Letters in Mathematical Physics*) for papers [160, 174].



Unless otherwise stated within this volume, the copyright to the papers listed above is held by the respective publishers cited.



The publishers and editors also wish to acknowledge that papers [66, 78, 86, 91, 98, 100, 106, 164, 167, 178] are reprinted from *Proceedings of the National Academy of Sciences, U.S.A.*

# The Scattering of Neutrons by Ortho- and Parahydrogen

JULIAN SCHWINGER
*Columbia University, New York, New York*

AND

E. TELLER
*George Washington University, Washington, D. C.*

(Received June 12, 1937)

Calculations have been performed which indicate that experiments on the scattering of neutrons by ortho- and parahydrogen would enable one to determine the sign of the singlet state binding energy and the range of the neutron-proton interaction, in addition to providing direct information concerning the spin dependence of the neutron-proton interaction. A dependence of the neutron-proton interaction upon the relative spin orientation of the particles will manifest itself in a marked difference between the slow neutron scattering cross sections of orthohydrogen (parallel proton spins) and parahydrogen (anti-parallel proton spins). Neutrons with energy less than 0.068 ev, incident upon *para*-$H_2$ in its ground state ($J=0$, $v=0$, $S=0$), may be either elastically scattered, or inelastically scattered with excitation of the molecule to the ground state of the *ortho* system ($J=1$, $v=0$, $S=1$). This latter process, requiring 0.023 ev, occurs only if the neutron-proton interaction is spin dependent. When the neutron energy is less than 0.045 ev, the cross section for the scattering of neutrons by *ortho*-$H_2$ in its ground state will be the sum of the elastic scattering cross section and the cross section for the inelastic process in which the molecule is converted to a *para*-$H_2$ molecule in its ground state, with the neutron taking up the excess energy. The cross sections of these four processes have been calculated, assuming an interaction range of $2\times10^{-13}$ cm and a virtual singlet state of the deuteron. For liquid-air temperature neutrons ($3kT/2=0.012$ ev), $\sigma_{\text{para}}(0.012)=0.21\times10^{-24}$ cm$^2$, while $\sigma_{\text{ortho}}(0.012)=65\times10^{-24}$ cm$^2$. The cross sections for neutrons at ordinary temperatures ($3kT/2=0.037$ ev), however, are $\sigma_{\text{para}}(0.037)=19\times10^{-24}$ cm$^2$, and $\sigma_{\text{ortho}}(0.037)=50\times10^{-24}$ cm$^2$. Therefore, if the present concept of the neutron-proton interaction is valid, one would expect the following results: (a) The *ortho*-scattering cross section for liquid air neutrons should be about 300 times the corresponding *para*-scattering cross section. (b) The *para*-scattering cross section for ordinary thermal neutrons should be roughly 100 times the *para*-scattering cross section for liquid air neutrons. For a real singlet state, however, these ratios are of the order of one. The elastic *para*-scattering cross section is quite sensitive to the value of the range of interaction if the singlet state is virtual. For example, the value of this cross section at liquid air neutron temperatures with zero range of interaction is $1.75\times10^{-24}$ cm$^2$, as compared with $0.26\times10^{-24}$ cm$^2$ for an interaction range of $2\times10^{-13}$ cm. Hence, from a measurement of the *para* elastic scattering cross section for homogeneous neutrons at some energy less than 0.023 ev, the range of interaction in the triplet state may be inferred with some degree of accuracy. A discussion of the influence of intermolecular forces on the previous results is given.

## INTRODUCTION

WIGNER and Bethe and Peierls[1] have given a simple theory of the scattering of neutrons by protons, assuming a short range neutron-proton interaction. The resultant scattering cross section:

$$\sigma=\frac{4\pi\hbar^2}{M}\frac{1}{|E_0|+\frac{1}{2}E}, \tag{1}$$

depends only upon the binding energy $|E_0|$ of the deuteron, and the energy $E$ of the neutron measured in the system in which the proton is initially at rest.

For small neutron energies ($E\ll|E_0|$), Eq. (1) predicts a scattering cross section of $2.40\times10^{-24}$ cm$^2$, since $|E_0|=2.2\times10^6$ ev. This is in complete disagreement with the experimental value of $\sigma=13\times10^{-24}$ cm$^2$.[2] This discrepancy cannot be explained by using a finite value of $r_0$, the range of the neutron-proton force. Bethe and Bacher[3] have shown, for a rectangular potential hole, that the Bethe-Peierls formula must be multiplied by $1+\alpha_0r_0$, i.e.,

$$\sigma=\frac{4\pi\hbar^2}{M}\frac{1+\alpha_0r_0}{|E_0|+\frac{1}{2}E}, \tag{2}$$

where

$$\alpha_0=(M|E_0|/\hbar^2)^{\frac{1}{2}}=2.29\times10^{12}\ \text{cm}^{-1}. \tag{3}$$

It is easily shown that this formula is practically

[1] E. Wigner, Zeits. f. Physik **83**, 253 (1933); H. A. Bethe and R. Peierls, Proc. Roy. Soc. **A149**, 176 (1935).

[2] E. Amaldi and E. Fermi, Phys. Rev. **50**, 899 (1936).

[3] H. A. Bethe and R. F. Bacher, Rev. Mod. Phys. **8**, 82 (1936).



Reprinted from *The Physical Review* **52**, 286–295 (1937).

independent of the shape of the potential well. With a reasonable value of $r_0$ obtained from the theory of the light nuclei, namely $r_0=2\times10^{-13}$ cm, this modified Bethe-Peierls formula gives a cross section of $3.50\times10^{-24}$ cm², for slow neutrons.

To explain this marked lack of agreement, Wigner[4] advanced the suggestion that the neutron-proton interaction in the singlet state differs from that in the triplet state. From this assumption, it follows that

$$\sigma=\frac{4\pi\hbar^2}{M}\left(\frac{3}{4}\frac{1+\alpha_0 r_0}{|E_0|+\frac{1}{2}E}+\frac{1}{4}\frac{1+\alpha_1 r_0}{|E_1|+\frac{1}{2}E}\right), \quad (4)$$

where $|E_1|$ is the energy of the singlet state, and

$$\alpha_1=\pm(M|E_1|/\hbar^2)^{\frac{1}{2}}, \quad (5)$$

the plus or minus sign obtaining according as the singlet state is real or virtual, respectively. Since the ground state of the deuteron gives no information about $|E_1|$, it can be chosen to give agreement with the experimental results. The large slow neutron scattering will be explained, therefore, if $|E_1|=142{,}000$ ev for a real level, or $|E_1|=114{,}000$ ev for a virtual level.

Fermi[5] has shown that one can determine which of these two possibilities is realized by a measurement of the cross section for radiative capture of neutrons by protons. Such measurements have been performed,[2] and indicate the existence of a virtual singlet state of the deuteron It should be pointed out, however, that on the basis of the $\beta$-ray theory of magnetic moments, it is questionable whether Fermi's mechanism of magnetic dipole capture can be correctly treated by simply assuming the additivity of magnetic moments.

It is the purpose of this paper to point out that experiments on the scattering of neutrons by ortho- and parahydrogen would afford a direct means of testing Wigner's hypothesis of a spin-dependent interaction between neutrons and protons. Furthermore, it will be shown that, from the results of such experiments, one can deduce the energy of the singlet state and the range of the forces.

[4] E. Feenberg and J. K. Knipp, Phys. Rev. **48**, 906 (1935).
[5] E. Fermi, Phys. Rev. **48**, 570 (1935); Ricerca Scient. VII–II, 13 (1936).

# I

Neutrons of thermal energy have a wave-length which is of the same order as the distance between the protons in $H_2$. One should therefore observe interference between the neutron waves scattered from the two protons in the hydrogen molecule. If the neutron-proton interaction depends upon the relative spin orientation of the particles, this should manifest itself in a difference between the interference effects in ortho- and parahydrogen, i.e., parallel and antiparallel proton spins, respectively. In addition to demonstrating directly the existence or nonexistence of spin forces, an experimental determination of these scattering cross sections would give the amplitudes of the singlet and the triplet scattered waves, save for a possible common factor of $(-1)$. It should be emphasized that these results can be obtained without a detailed theory of the mechanism of interaction. Making the usual assumptions about the interaction, one can then determine the energy of the singlet level, and the range of the neutron-proton force.

The problem of calculating scattering cross sections is complicated by the presence of inelastic scattering processes. In general, all possible transitions consistent with conservation of energy and momentum will occur. However, a neutron-proton interaction which is independent of spin will only induce transitions between states of the molecule with the same total spin angular momentum, that is, *ortho-ortho* and *para-para* transitions, while spin-dependent interactions will also cause *ortho-para* and *para-ortho* transitions.

It is interesting to note that, for neutrons with energy comparable to the excitation energy of the first few rotational levels, the *ortho*-scattering cross section differs from the *para*-scattering cross section, even if no spin-dependent forces act between neutrons and protons. This is a consequence of the different rotational energy levels of ortho- and parahydrogen. Neutrons with energies large compared with the lowest rotational levels, however, will have the same scattering cross sections in ortho- and parahydrogen. If, on the other hand, the neutron wave-length is large compared with the internuclear separation in hydrogen, the elastic scattering in ortho- and parahydrogen will differ if spin-dependent forces are present, but will be the same if no such spin forces are operative.

At sufficiently low temperatures, the molecules of $H_2$ are in their respective ground states, namely $J=0$, $v=0$ for *para*-molecules and $J=1$, $v=0$ for *ortho*-molecules. In the absence of spin-dependent neutron-proton forces, slow neutrons will suffer only elastic collisions. The scattering cross sections of slow neutrons in ortho- and parahydrogen will, therefore, be the same.

If, however, the neutron-proton interaction involves the relative spin orientation of the particles, the cross section corresponding to the conversion of an *ortho*-molecule into a *para*-molecule will also contribute to the *ortho*-scattering cross section, whereas the *para* cross section will still consist of only the elastic cross section. Therefore, at sufficiently low neutron energies, a difference between the scattering of ortho- and parahydrogen should manifest itself.

## II

The mathematical treatment of the scattering of neutrons by hydrogen molecules may be given with the aid of a theorem which Fermi[6] utilized in his considerations on the effect of molecular binding forces on the neutron-proton scattering cross section. With neglect of the dependence of the neutron-proton interaction upon the relative spin orientation of the particles, this theorem states that the scattering of a neutron by a bound proton may be calculated with the Born approximation, by using

$$-(4\pi\hbar^2/M)a\delta(\mathbf{r}_n-\mathbf{r}_p) \qquad (6)$$

as the effective neutron-proton interaction. Here $\delta(\mathbf{r})$ is the three-dimensional Dirac $\delta$-function, and $a$ is defined by the solution corresponding to zero energy of the wave equation for the relative motion of the particles, namely:

$$1+a/r \quad \text{for} \quad r>r_0. \qquad (7)$$

The condition of validity of this result is that there exist a quantity $R$ which is large compared with $r_0$ and $a$, but small compared with the de Broglie wave-lengths of the neutron and the proton. This condition is easily satisfied for slow neutrons and the protons in the hydrogen molecule.

The extension of this theorem to an interaction which differs in the singlet state from that in the triplet state is made without difficulty. We need merely replace $a$ in formula (6) by an operator whose eigenvalue for a triplet state is $a_1$, the value of $a$ calculated for the triplet interaction, and whose eigenvalue for a singlet state is $a_0$, the value of $a$ calculated for the singlet interaction. Such an operator can be written:

$$\tfrac{1}{2}a_1(1+Q)+\tfrac{1}{2}a_0(1-Q), \qquad (8)$$

where $Q$ is an operator with the eigenvalues $+1$ and $-1$ for triplet and singlet states, respectively. It has been shown by Dirac[7] that $Q$ can be expressed algebraically in terms of $\boldsymbol{\sigma}_n$ and $\boldsymbol{\sigma}_p$, the Pauli matrices of the neutron and the proton, *viz.:*

$$Q=\tfrac{1}{2}(1+\boldsymbol{\sigma}_n\cdot\boldsymbol{\sigma}_p). \qquad (9)$$

Therefore, (6) must be replaced by

$$-(\pi\hbar^2/M)(3a_1+a_0 + (a_1-a_0)\boldsymbol{\sigma}_n\cdot\boldsymbol{\sigma}_p)\delta(\mathbf{r}_n-\mathbf{r}_p). \qquad (10)$$

The effective interaction between a neutron and the protons in the hydrogen molecule will then be

$$\begin{aligned}&-(\pi\hbar^2/M)(3a_1+a_0 + (a_1-a_0)\boldsymbol{\sigma}_n\cdot\boldsymbol{\sigma}_1)\delta(\mathbf{r}_n-\mathbf{r}_1)\\ &-(\pi\hbar^2/M)(3a_1+a_0 + (a_1-a_0)\boldsymbol{\sigma}_n\cdot\boldsymbol{\sigma}_2)\delta(\mathbf{r}_n-\mathbf{r}_2), \qquad (11)\end{aligned}$$

in consequence of the small probability that the protons approach within a distance of the order of $R$.

It is convenient to separate (11) into two parts which are, respectively, symmetrical and antisymmetrical in the proton spins, namely:

$$\begin{aligned}&-(\pi\hbar^2/M)(3a_1+a_0+(a_1-a_0)\boldsymbol{\sigma}_n\cdot\mathbf{S})\times(\delta(\mathbf{r}_n-\mathbf{r}_1)+\delta(\mathbf{r}_n-\mathbf{r}_2))\\ &-(\pi\hbar^2/2M)(a_1-a_0)\boldsymbol{\sigma}_n\cdot(\boldsymbol{\sigma}_1-\boldsymbol{\sigma}_2)\times(\delta(\mathbf{r}_n-\mathbf{r}_1)-\delta(\mathbf{r}_n-\mathbf{r}_2)), \qquad (12)\end{aligned}$$

where

$$\mathbf{S}=\tfrac{1}{2}(\boldsymbol{\sigma}_1+\boldsymbol{\sigma}_2) \qquad (13)$$

represents the total spin angular momentum of the molecule. It is evident that the symmetrical part of the interaction will be responsible for

[6] E. Fermi, Ricerca Scient. VII–II, 13 (1936).

[7] P. A. M. Dirac, *Quantum Mechanics*, second edition (Oxford, 1935).

those transitions in which the spin symmetry of the molecule does not change, that is, *ortho-ortho* and *para-para* transitions. The antisymmetrical part of the interaction, however, will induce *ortho-para* and *para-ortho* transitions.

Consider a scattering process in which a neutron with momentum $\mathbf{p}^0$ collides with a hydrogen molecule with the momentum $-\mathbf{p}^0$, the vibrational quantum number $v$, the rotational quantum number $J$, and the quantum number $S$ of the resultant proton spin. We wish to calculate the probability that the neutron is scattered with the momentum $\mathbf{p}$, leaving the molecule in a state described by the set of quantum numbers $-\mathbf{p}$, $v'$, $J'$, $S'$. The wave function of the initial state, normalized in a volume $V$, is:

$$\Psi_i = V^{-\frac{1}{2}} \exp((i/\hbar)\mathbf{p}^0\cdot\mathbf{r}_n)\chi_m V^{-\frac{1}{2}} \exp(-(i/\hbar)\mathbf{p}^0\cdot(\mathbf{r}_1+\mathbf{r}_2)/2)\phi_{v,\,J,\,m_J}(\mathbf{r}_1-\mathbf{r}_2)\chi_{S,\,m_S}, \tag{14}$$

where $V^{-\frac{1}{2}} \exp((i/\hbar)\mathbf{p}^0\cdot\mathbf{r}_n)$ is the wave function of a free neutron with momentum $\mathbf{p}^0$, $\chi_m$ is the eigenfunction of the $z$-component of the neutron spin corresponding to the eigenvalue $m$, $V^{-\frac{1}{2}} \exp(-(i/\hbar)\mathbf{p}^0\cdot(\mathbf{r}_1+\mathbf{r}_2)/2)$ is the wave function which describes the motion of the center of gravity of the hydrogen molecule with momentum $-\mathbf{p}^0$, $\phi_{v,\,J,\,m_J}(\mathbf{r}_1-\mathbf{r}_2)$ is the wave function of the relative motion of the two protons in a state with vibrational quantum number $v$, rotational quantum number $J$ and magnetic quantum number $m_J$, and, finally, $\chi_{S,\,m_S}$ is the eigenfunction of the proton spins corresponding to a resultant spin angular momentum $S$ and an eigenvalue of the $z$-component of the total spin equal to $m_S$. Similarly, the wave function of the final state is:

$$\Psi_f = V^{-\frac{1}{2}} \exp((i/\hbar)\mathbf{p}\cdot\mathbf{r}_n)\chi_{m'} V^{-\frac{1}{2}} \exp(-(i/\hbar)\mathbf{p}\cdot(\mathbf{r}_1+\mathbf{r}_2)/2)\phi_{v',\,J',\,m'_J}(\mathbf{r}_1-\mathbf{r}_2)\chi_{S',\,m'_S}. \tag{15}$$

According to the well-known Born formula, the probability of the transition per unit time is given by:

$$(2\pi/\hbar)\,|(\Psi_f,\, U\Psi_i)|^2\rho_E, \tag{16}$$

where $U$ denotes the expression (12) for the interaction, and $\rho_E$ represents the number of neutron momentum states per unit range of the total final energy. Writing $E_{J,\,v}$ for the internal energy of the molecule in a state characterized by the quantum numbers $v$ and $J$, we have

$$E=(3/4M)p^{0^2}+E_{J,\,v}=(3/4M)p^2+E_{J',\,v'}, \tag{17}$$

by the law of conservation of energy. Therefore, if we consider a process in which the neutron is scattered through an angle $\Theta$ into a solid angle $d\Omega$, the number of final states per unit range of the total final energy is

$$\rho_E=\frac{Vp^2}{8\pi^3\hbar^3}\frac{dp}{dE}d\Omega=V\frac{Mp}{12\pi^3\hbar^3}d\Omega. \tag{18}$$

Formula (16) gives the transition probability for definite values of $m$, $m_J$, $m_S$ and $m'$, $m_J'$, $m_S'$. Since we are not interested in the transition probabilites between particular degenerate states, we must sum over all values of $m'$, $m_J'$, $m_S'$ and average with respect to $m$, $m_J$, $m_S$. Therefore, the total transition probability per unit time will be

$$\frac{1}{2(2S+1)(2J+1)}\sum_{m',\,m'_J,\,m'_S}\;\sum_{m,\,m_J,\,m_S}|(\Psi_f,\,U\Psi_i)|^2\frac{VMp}{6\pi^2\hbar^4}d\Omega. \tag{19}$$

Dividing by the neutron flux relative to the hydrogen molecule, namely

$$3p^0/2MV, \tag{20}$$

we obtain the differential cross section $\sigma_{J',\,v',\,S';\,J,\,v,\,S}(\Theta)d\Omega$ for scattering of a neutron through an angle $\Theta$ into the solid angle $d\Omega$ with excitation of the molecule from the state $J$, $v$, $S$ to the state

$J'$, $v'$, $S'$, referred to the system in which the center of gravity of the neutron and the molecule is at rest, *viz.:*

$$\sigma_{J',\,v',\,S';\,J,\,v,\,S}(\Theta)d\Omega=\frac{4}{9}\frac{p}{p^0}\frac{1}{2(2S+1)(2J+1)}\sum_{m',\,m'_J,\,m'_S}\sum_{m,\,m_J,\,m_S}\left|\left(\Psi_f,\frac{MV}{2\pi\hbar^2}U\Psi_i\right)\right|^2 d\Omega. \quad (21)$$

The summation with respect to $m$, $m_S$ and $m'$, $m_S'$ may be performed without difficulty. We consider first those transitions in which the total spin of the molecule remains unchanged, i.e., $S'=S$. Utilizing the expressions (14) and (15) for the initial and final wave functions, we obtain:

$$\left(\Psi_f,\frac{MV}{2\pi\hbar^2}U\Psi_i\right)=-\frac{1}{2V}(m';S,m_S'|3a_1+a_0+(a_1-a_0)\boldsymbol{\sigma}_n\cdot\mathbf{S}|m;S,m_S)$$

$$\cdot\int\exp\left[(i/\hbar)(\mathbf{p}^0-\mathbf{p})\cdot(\mathbf{r}_n-(\mathbf{r}_1+\mathbf{r}_2)/2)\right]\phi^*_{v',\,J',\,m'_J}(\mathbf{r}_1-\mathbf{r}_2)\phi_{v,\,J,\,m_J}(\mathbf{r}_1-\mathbf{r}_2)$$

$$\times(\delta(\mathbf{r}_n-\mathbf{r}_1)+\delta(\mathbf{r}_n-\mathbf{r}_2))d\tau_n d\tau_1 d\tau_2=-(m';S,m_S'|3a_1+a_0+(a_1-a_0)\boldsymbol{\sigma}_n\cdot\mathbf{S}|m;S,m_S)$$

$$\cdot\int\cos\frac{(\mathbf{p}^0-\mathbf{p})\cdot\mathbf{r}}{2\hbar}\phi^*_{v',\,J',\,m'_J}(\mathbf{r})\phi_{v,\,J,\,m_J}(\mathbf{r})d\tau. \quad (22)$$

Therefore,

$$\sum_{m,\,m_S}\sum_{m',\,m'_S}\left|\left(\Psi_f,\frac{MV}{2\pi\hbar^2}U\Psi_i\right)\right|^2=\left|\int\cos\frac{(\mathbf{p}^0-\mathbf{p})\cdot\mathbf{r}}{2\hbar}\phi^*_{v',\,J',\,m'_J}(\mathbf{r})\phi_{v,\,J,\,m_J}(\mathbf{r})d\tau\right|^2$$

$$\cdot\sum_{m,\,m_S}\sum_{m',\,m'_S}|(m';S,m_S'|3a_1+a_0+(a_1-a_0)\boldsymbol{\sigma}_n\cdot\mathbf{S}|m;S,m_S)|^2. \quad (23)$$

Since $3a_1+a_0+(a_1-a_0)\boldsymbol{\sigma}_n\cdot\mathbf{S}$ is diagonal with respect to the total spin of the molecule, the second factor of this expression may be written:

$$\sum_{m,\,m_S}\sum_{m',\,S',\,m'_S}|(m';S',m_S'|3a_1+a_0+(a_1-a_0)\boldsymbol{\sigma}_n\cdot\mathbf{S}|m;S,m_S)|^2, \quad (24)$$

which becomes

$$\sum_{m,\,m_S}(m;S,m_S|(3a_1+a_0+(a_1-a_0)\boldsymbol{\sigma}_n\cdot\mathbf{S})^2|m;S,m_S) \quad (25)$$

by the matrix law of multiplication. Now

$$(3a_1+a_0+(a_1-a_0)\boldsymbol{\sigma}_n\cdot\mathbf{S})^2=(3a_1+a_0)^2+(a_1-a_0)^2\mathbf{S}^2+2(3a_1+a_0)(a_1-a_0)\boldsymbol{\sigma}_n\cdot\mathbf{S}-(a_1-a_0)^2\boldsymbol{\sigma}_n\cdot\mathbf{S}, \quad (26)$$

from which we obtain

$$2(2S+1)((3a_1+a_0)^2+S(S+1)(a_1-a_0)^2) \quad (27)$$

for the sum (25), using the fact that the diagonal sum of an angular momentum matrix is zero.

Collecting our formulae, we obtain finally

$$\sigma_{J',\,v',\,S;\,J,\,v,\,S}(\Theta)d\Omega=(4p/9p^0)((3a_1+a_0)^2+S(S+1)(a_1-a_0)^2)$$

$$\cdot\frac{1}{2J+1}\sum_{m_J,\,m'_J}\left|\int\cos\frac{(\mathbf{p}^0-\mathbf{p})\cdot\mathbf{r}}{2\hbar}\phi^*_{v',\,J',\,m_{J'}}(\mathbf{r})\phi_{v,\,J,\,m_J}(\mathbf{r})d\tau\right|^2 d\Omega. \quad (28)$$

We may note that Eq. (28) gives a nonvanishing transition probability only if $J$ and $J'$ are both even or both odd, since $\phi_{v,\,J,\,m_J}(-\mathbf{r})=(-1)^J\phi_{v,\,J,\,m_J}(\mathbf{r})$. Indeed, this is required by the well-known condition

$$(-1)^J=(-1)^S, \quad (29)$$

imposed by the Pauli exclusion principle. Therefore *ortho-ortho* transitions are associated with transitions between states of odd rotational quantum number, while *para-para* transitions are associated with transitions between states of even $J$.

A similar calculation shows that the scattering cross sections for transitions in which the total spin of the molecule changes, i.e., $S'=1-S$, are given by:

$$\sigma_{J',\,v',\,1-S;\,J,\,v,\,S}(\Theta)d\Omega=(4p/9p^0)(a_1-a_0)^2(3-2S)\cdot\frac{1}{2J+1}\sum_{m_J,\,m'_J}\left|\int \sin\frac{(\mathbf{p}^0-\mathbf{p})\cdot\mathbf{r}}{2\hbar}\phi^*_{v',\,J',\,m'_J}(\mathbf{r})\phi_{v,\,J,\,m_J}(\mathbf{r})d\tau\right|^2 d\Omega. \quad (30)$$

It is evident from (29) that *ortho-para* and *para-ortho* transitions are accompanied by odd-even and even-odd transitions in the rotational quantum number $J$, respectively.

## III

The minimum energy necessary to excite the molecule from the state $J$, $v$ to the state $J'$, $v'$ is $(2/3)(E_{J',\,v'}-E_{J,\,v})$, if the neutron energy is measured in a system moving with the center of gravity of the neutron and the molecule. Measured in the system in which the molecule is initially at rest, the requisite energy is $(3/2)(E_{J',\,v'}-E_{J,\,v})$ since $\mathbf{p}_0$, the initial momentum of the neutron in the latter system of reference, is given by

$$\mathbf{p}_0=3\mathbf{p}^0/2. \quad (31)$$

For small rotational and vibrational quantum numbers, the energy levels of the molecule, in volts, are:[8]

$$E_{J,\,v}=0.015\tfrac{1}{2}J(J+1)+0.533v. \quad (32)$$

Therefore, if the neutron energy is less than 0.045 ev, the energy necessary to excite the $J$ transition $1\rightarrow 2$, the only excitation process which may take place is the $J$ transition $0\rightarrow 1$, requiring 0.023 ev.

The most interesting effects are obtained at low temperatures which are such that practically all the hydrogen molecules will be in their respective ground states, namely $v=0$, $J=0$ for *para*-molecules, and $v=0$, $J=1$ for *ortho*-molecules. With these restrictions upon the neutron energy and the initial states of the molecule, the only processes which can occur are those in which the vibrational states remain unexcited and $J$ undergoes any of the four possible transitions: $0\rightarrow 0$, $0\rightarrow 1$, $1\rightarrow 0$, $1\rightarrow 1$.

To a sufficient approximation, the probability of these transitions may be calculated by treating the $H_2$ molecule as a rigid rotator. The wave function appropriate to this model is:

$$\phi_{0,\,J,\,m_J}(\mathbf{r})=\left(\frac{2J+1}{4\pi}\frac{(J-|m_J|)!}{(J+|m_J|)!}\right)^{\frac{1}{2}}\sin^{|m_J|}\vartheta\left(\frac{d}{d\cos\vartheta}\right)^{J+|m_J|}\frac{(\cos^2\vartheta-1)^J}{2^J J!}e^{im_J\varphi}(\delta(r-r_e))^{\frac{1}{2}}/r_e, \quad (33)$$

where $r_e$, the equilibrium nuclear separation, has the numerical value[8]

$$r_e=0.75\times 10^{-8}\text{ cm}. \quad (34)$$

According to the principle of spectroscopic stability, the sums which occur in Eqs. (28) and (30) have a value which is independent of the direction of quantization. It is evident from (33) that, if the direction of quantization is taken to be the direction of the vector $\mathbf{p}^0-\mathbf{p}$, the only nonvanishing terms in these sums are those for which $m_J=m_J'$. Therefore, the number of terms is equal to the rotational multiplicity of the state with smaller rotational quantum number.

The only matrix element which need be evaluated to calculate the elastic scattering from parahydrogen $(0\rightarrow 0)$, is

$$\int\cos\left(\frac{|\mathbf{p}^0-\mathbf{p}|r\cos\vartheta}{2\hbar}\right)|\phi_{0,\,0,\,0}(r)|^2 2\pi r^2 dr\sin\vartheta d\vartheta=\frac{\sin(|\mathbf{p}^0-\mathbf{p}|r_e/2\hbar)}{|\mathbf{p}^0-\mathbf{p}|r_e/2\hbar}. \quad (35)$$

[8] W. Jevons, *Report on Band-Spectra of Diatomic Molecules* (Cambridge, 1932).

In terms of $\Theta$, the angle of scattering in the system of the center of gravity, this may be written

$$\frac{\sin[(p^0r_e/\hbar)\sin\frac{1}{2}\Theta]}{(p^0r_e/\hbar)\sin\frac{1}{2}\Theta}. \tag{36}$$

Therefore,

$$\sigma_{0,0,0;0,0,0}(\Theta)2\pi\sin\Theta d\Theta=\frac{4}{9}(3a_1+a_0)^2\left(\frac{\sin[(p^0r_e/\hbar)\sin\frac{1}{2}\Theta]}{(p^0r_e/\hbar)\sin\frac{1}{2}\Theta}\right)^2 2\pi\sin\Theta d\Theta. \tag{37}$$

The total elastic scattering cross section is

$$\sigma_{0,0,0;0,0,0}=\frac{32\pi}{9}(3a_1+a_0)^2\left(\frac{\hbar}{p^0r_e}\right)^2\int_0^{p^0r_e/\hbar}\frac{\sin^2 x}{x}dx$$

$$=\frac{16\pi}{9}(3a_1+a_0)^2\left(\frac{\hbar}{p^0r_e}\right)^2\left[\log\frac{2p^0r_e}{\hbar}-Ci\frac{2p^0r_e}{\hbar}+0.5772\right], \tag{38}$$

where

$$Cix=-\int_x^\infty\frac{\cos t}{t}dt. \tag{39}$$

A similar calculation yields the following cross sections for the $0\rightarrow1$, $1\rightarrow0$, $1\rightarrow1$ transitions, respectively:

$$\sigma_{1,0,1;0,0,0}=32\pi(a_1-a_0)^2\left(\frac{\hbar}{p^0r_e}\right)^2\left[\frac{\sin x}{x}+\frac{\cos x-1}{x^2}+\tfrac{1}{2}\log x-\tfrac{1}{2}Cix\right]_{(p^0-p)r_e/\hbar}^{(p^0+p)r_e/\hbar},\quad p=\left(p^{02}-\frac{4M}{3}E_{1,0}\right)^{\frac{1}{2}};$$

$$\sigma_{0,0,0;1,0,1}=\frac{32\pi}{3}(a_1-a_0)^2\left(\frac{\hbar}{p^0r_e}\right)^2\left[\frac{\sin x}{x}+\frac{\cos x-1}{x^2}+\tfrac{1}{2}\log x-\tfrac{1}{2}Cix\right]_{(p-p^0)r_e/\hbar}^{(p+p^0)r_e/\hbar},\quad p=\left(p^{02}+\frac{4M}{3}E_{1,0}\right)^{\frac{1}{2}};$$

$$\sigma_{1,0,1;1,0,1}=\frac{16\pi}{3}((3a_1+a_0)^2+2(a_1-a_0)^2)\frac{1}{\xi^2}\left[-\frac{3}{\xi^2}\left(\frac{\sin\xi}{\xi}-\cos\xi\right)^2+\frac{\sin^2\xi}{\xi^2}+\log 2\xi-Ci2\xi-0.4228\right],$$

$$\xi=\frac{p^0r_e}{\hbar}. \tag{40}$$

These four cross sections, calculated for a virtual singlet state of the deuteron and a range $r_0=2\times10^{-13}$ cm, are plotted in Figs. 1, 2 as a function of the neutron energy.[9] The curves for any other choice of the range of interaction and the sign of the binding energy of the singlet state[10] may be obtained by an appropriate change of scale since only the values of $a_1$ and $a_0$ need be changed. Thus, if the singlet state is taken as real rather than virtual, the cross sections for the elastic scattering by parahydrogen should be multiplied by 244.

It is evident from these graphs that the present concept of the neutron-proton interaction as a spin-dependent force leading to a virtual state of the deuteron, and of range $r_0=2\times10^{-13}$ cm, implies the following qualitative predictions:

(a) The scattering cross section of parahydrogen for ordinary thermal neutrons ($E=0.037$ ev) should be about 100 times the scattering cross section of parahydrogen for liquid-air neutrons ($E=0.012$ ev);

[9] The quantities $a_1$ and $a_0$ are obtained from the free proton triplet and singlet scattering cross sections with the aid of the relations: $\sigma_{\text{triplet}}=4\pi a_1^2$, $\sigma_{\text{singlet}}=4\pi a_0^2$. The sign of $a_0/a_1$ is positive if the singlet state is real, negative if the singlet state is virtual.

[10] We speak of a positive binding energy if the state is real, a negative binding energy if it is virtual.

(b) The scattering cross section of orthohydrogen for liquid-air neutrons should be roughly 300 times the scattering cross section of parahydrogen for liquid-air neutrons;

(c) The scattering cross section of orthohydrogen for ordinary thermal neutrons should be only 2 or 3 times larger than the scattering cross section of parahydrogen for these neutrons;

(d) The *ortho*-scattering cross section for ordinary thermal neutrons should be about three-fourths of the *ortho*-scattering cross section for liquid-air neutrons.

If, however, the singlet state is real, the parahydrogen scattering cross section for ordinary thermal neutrons should be about $\frac{3}{4}$ of the parahydrogen scattering cross section for liquid-air neutrons, and the orthohydrogen cross section should be approximately 7/5 of the *para*-scattering cross section for liquid-air neutrons. A comparison of these two predictions shows clearly that this, rather than the capture of neutrons by protons, is the ideal method for determining the sign of the deuteron singlet state binding energy.

The parahydrogen elastic scattering cross section varies sensitively with the range of the neutron-proton interaction. This is illustrated in Fig. 3 for a virtual singlet state. It is seen, for example, that the scattering cross section for zero range is 7 times the scattering cross section for a range of $r_0=2\times10^{-13}$ cm. An experimental determination of the parahydrogen scattering cross section for neutrons of some definite energy

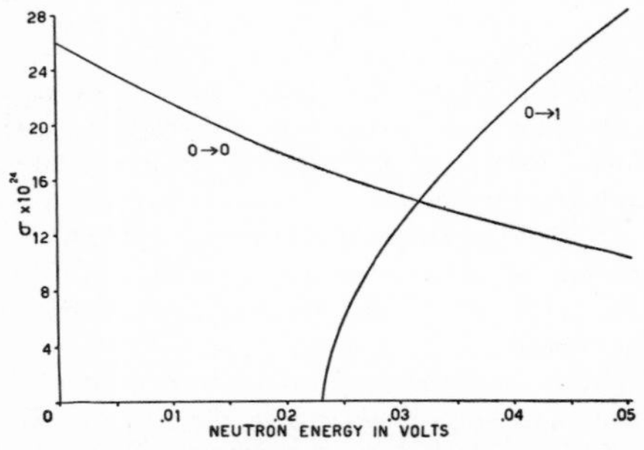


FIG. 1. The elastic (0→0) and inelastic (0→1) scattering cross sections of parahydrogen, calculated for a virtual singlet state of the deuteron and a range of interaction in the triplet state equal to $2\times10^{-13}$ cm. The ordinates of the elastic scattering curve are enlarged 100 fold.

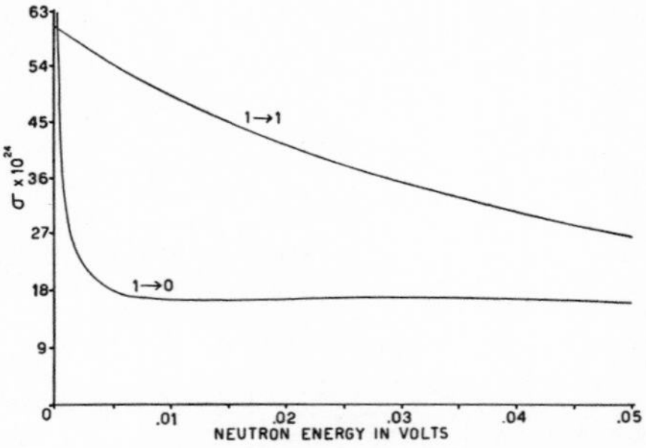


FIG. 2. The elastic (1→1) and inelastic (1→0) scattering cross sections of orthohydrogen, calculated under the same assumptions as Fig. 1.

less than 0.023 ev would, therefore, permit one to calculate the range of the neutron-proton interaction with some degree of accuracy. Neutron beams with a sufficient energy homogeneity can be obtained with a mechanical velocity selector[11] or with the aid of Bragg reflection from crystals.[12]

Unfortunately, this scattering cross section is of the same order as the capture cross section if the singlet state is virtual. This necessitates a theoretical knowledge of the capture cross section if one performs absorption measurements. Since the deuteron formed in a capture process has a recoil energy ($\sim$1000 ev) which is very large compared with the molecular binding energy of a proton, the cross section for this process can be calculated by assuming the protons to be free.[13] The capture cross section per molecule, therefore, is twice the free proton capture cross section, and thus has the numerical magnitude[5] $0.6\times10^{-24}$ cm$^2$ at liquid-air temperatures ($E=0.012$ ev), provided the singlet state is virtual.

## IV

The neglect of intermolecular forces in the preceding discussion limits its applicability to the scattering by $H_2$ in its gaseous state. However, experiments are most easily carried out with liquid or solid hydrogen where intermolecular

[11] G. A. Fink, Phys. Rev. **50**, 738 (1936).
[12] D. P. Mitchell and P. N. Powers, Phys. Rev. **50**, 486 (1936).
[13] An analytical proof has been given by W. E. Lamb, Jr., Phys. Rev. **51**, 187 (1937).

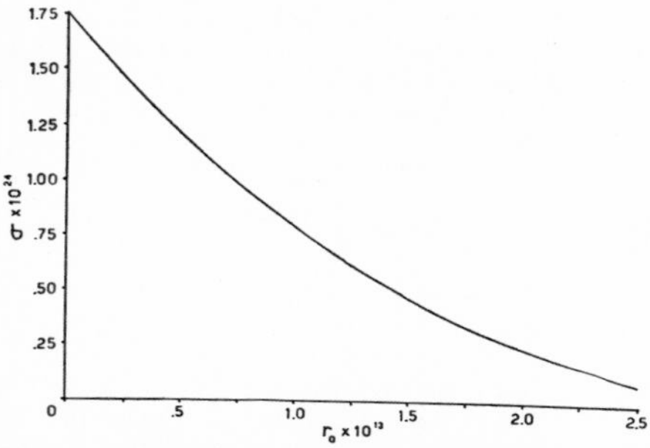


FIG. 3. The elastic scattering cross section of parahydrogen for neutrons of zero energy, as a function of the range of interaction in the triplet state. The singlet state of the deuteron is assumed to be virtual.

forces may have an appreciable effect. The discussion of this effect may be simplified by supposing the molecules to vibrate independently about positions of equilibrium with a frequency $\nu$, given by

$$h\nu = k\Theta_E, \tag{41}$$

where $\Theta_E$, the characteristic Einstein temperature, is related to the characteristic Debye temperature $\Theta_D$ by[14]

$$\Theta_E = 3\Theta_D/4. \tag{42}$$

Experimentally,[15] $\Theta_D = 105°\text{K}$, from which we obtain

$$h\nu = 0.007 \text{ ev}. \tag{43}$$

Consequently, if the neutron energy is large compared with 0.007 ev, the molecules may be treated as if they were free and the preceding results are applicable.

The heat of evaporation and the heat of fusion of normal and parahydrogen agree to within 1 percent,[16] which is good evidence for the free rotation of hydrogen molecules in the liquid and solid phase.[17] This is supported by experiments upon the Raman effect in liquid $H_2$.[18] We may conclude that intermolecular forces act only upon the center of gravity of a hydrogen molecule, and do not affect its internal motion. Therefore, the appropriate description of the effect of intermolecular forces upon the *elastic* scattering cross sections is obtained by multiplying our previous results by a function $F(E/h\nu)$ which approaches 1 rapidly as $E$ becomes greater than $h\nu$. The limiting value of this function for $E/h\nu \ll 1$ may be obtained by an elementary consideration. When $E$ is small compared with $h\nu$, the effective mass of a molecule is the same as the mass of the whole liquid or solid. Now the scattering cross section is proportional to the square of the reduced mass of the system, that is, $4M^2/9$ for a free molecule, and $M^2$ for a strongly bound molecule. Therefore $F(0) = 9/4$ so that the elastic scattering cross sections we have obtained should be multiplied by a function which decreases rapidly from 9/4 to 1 with increasing neutron energy.[19] The effect of intermolecular forces upon the $1 \rightarrow 0$ transition will be much less important, for even with $E = 0$ the molecule has a recoil energy of the order of $h\nu$.

If the *ortho*-scattering cross section consisted only of the elastic cross section, the ratio of the *ortho*- to the *para*-scattering would be independent of intermolecular forces, since the interaction between *ortho*-molecules is practically identical with the interaction between *para*-molecules.[20] It is clear, therefore, that the effect of intermolecular forces will not be such as to alter our qualitative comparison between the *ortho*- and *para*-scattering cross sections. Unfortunately, however, the $1 \rightarrow 0$ cross section, with its different dependence upon the intermolecular forces, is added to the elastic ortho-scattering cross section. It is necessary, therefore, to perform measurements at energies which are large compared with $h\nu$, so that intermolecular forces may be neglected, but yet are less than 0.023 ev. A measurement of the scattering cross sections of both ortho- and parahydrogen at some such energy would enable one to calculate the amplitudes $a_1$ and $a_0$, from which we may obtain $r_0$ and the free proton scattering cross

[14] N. F. Mott and H. Jones, *The Theory of the Properties of Metals and Alloys* (Oxford, 1936).

[15] G. Bartholomé, Zeits. f. physik. Chemie **33**, 387 (1936).

[16] A. Farkas, *Orthohydrogen, Parahydrogen and Heavy Hydrogen* (Cambridge, 1935).

[17] The anomalous specific heat of solid orthohydrogen at very low temperatures indicates that, at these temperatures, rotation ceases to be free.

[18] J. C. McLennan and J. H. McLeod, Nature **123**, 160 (1929).

[19] A similar argument enables one to deduce Fermi's result of an increase in the scattering cross section of a proton bound to a heavy molecule by a factor of 4 over the corresponding value for a free proton, in the limiting case of zero neutron velocity.

[20] R. B. Scott and F. G. Brickwedde, Phys. Rev. **51**, 684 (1937).

section, thus permitting a check of the consistency of the whole theory.

*Note added in proof:* Recently published experiments by J. Halpern, I. Estermann, O. C. Simpson and O. Stern, Phys. Rev. **52**, 142 (1937) indicate that the scattering cross section of *ortho*-$H_2$ for liquid-air neutrons is much larger than the corresponding *para*-$H_2$ scattering cross section. This proves conclusively that the singlet state of the deuteron is virtual.

## The Neutron-Proton Scattering Cross Section*

The scattering of thermal neutrons in paraffin has been extensively investigated.[1] The results of such experiments, however, do not yield directly the cross section for the scattering of slow neutrons by free protons, for, as Fermi has indicated,[2] the protons in the paraffin may be considered free only if the neutron energy, $E$, is large compared with $h\nu$, the energy of the lowest vibrational level of the paraffin molecule. Since this condition is not realized for thermal neutrons ($E\sim0.026$ ev; $h\nu\sim0.3$ ev) the free proton cross section may be inferred from the thermal neutron cross section only by means of a theoretical calculation of the effect of chemical binding. In view of the inaccuracies necessarily attendant upon this calculation, a direct measurement of the neutron-proton scattering cross section, a quantity of great importance in the investigation of nuclear forces, seems highly desirable.

The existence of resonance levels in the region of $1-10$ ev affords a convenient method of employing neutrons which satisfy the condition $E\gg h\nu$. An ideal experiment on the scattering of resonance neutrons would be one which utilizes an initial beam homogeneous in energy, and a detector sensitive only to neutrons of that energy. The fact that a neutron upon suffering a collision in the hydrogenous scatterer would undergo an energy loss and escape detection obviates the necessity of correcting for scattering into the detector. In practice this homogeneous beam may be approximately realized by utilizing those neutrons which are strongly absorbed in a resonance filter, thus introducing an energy in homogeneity of the order of the resonance width. That contribution to the activity of a detector of the same material which arises from scattered neutrons will then be limited to those which have been deviated through such small angles that the associated energy loss is insufficient to remove them from the resonance region.

An experiment of the above type has been performed with filter and detector films of Rh of 0.1 g/cm². As a source of neutrons we used a Rn-Be bulb of 600–300 mC in a cylinder of paraffin. A boron carbide diaphragm with an aperture 5 cm in diameter served to define the effective source of the neutron beam. Both source and detector were shielded with Cd to eliminate any effect of thermal neutrons. The detector, 45 cm² in area, was placed 30 cm from the source with the detector placed midway between source and detector. Each detector consisted of four rectangles of nickel sheet coated with Rh powder mixed with a small amount of Zapon lacquer as a binder. The four segments were jointed so that irradiations could be made with the detector spread out, while the activity could be measured with the detector folded so as to enclose a Geiger-Mueller counter.

From our data we conclude the mean free path of Rh resonance neutrons in paraffin to be 0.56 cm. This corresponds to a cross section of $20\times10^{-24}$ cm², if we take the value of the carbon cross section[3] to be $4.8\times10^{-24}$ cm². The probable error as calculated from the data is about five percent. However, in view of the approximations made in correcting the transmissions for scattering into the detector, we consider the above result to be good to about ten percent.

The experiments on the neutron scattering in ortho- and parahydrogen[4] permit an independent determination of the neutron-proton scattering cross section which is in substantial agreement with the value obtained in these measurements.

Since the scattering cross section obtained by direct measurement is considerably larger than the value of $14\times10^{-24}$ cm² deduced from the thermal neutron cross section,[5] nuclear quantities calculated from the singlet scattering cross section will be altered appreciably. Two instances are: the increase in the magnitude of the singlet neutron-proton interaction, thereby affecting the comparison with the singlet proton-proton interaction; and, numerically more significant, the effect on the calculated cross section for slow neutron capture by protons, which increases approximately in the same ratio as the scattering cross section. A more detailed discussion of these experiments and their consequences will be presented in a forthcoming publication.

VICTOR W. COHEN
H. H. GOLDSMITH
JULIAN SCHWINGER

Columbia University,
New York, New York,
December 2, 1938.

* Publication assisted by the Ernest Kempton Adams Fund for Physical Research.

[1] M. Goldhaber and G. H. Briggs, Proc. Roy. Soc. **162**, 127 (1937); S. Frisch, H. v. Halban, H. Koch, Kgl. Dansk. Vid. Selsk. **15**, 10 (1938); H. Carroll and J. Dunning, Phys. Rev. **54**, 541 (1938).
[2] E. Fermi, Ric. Sci. **7**, 13 (1936).
[3] M. Goldhaber and G. H. Briggs, Proc. Roy. Soc. **162**, 127 (1937); J. Whittaker and O. Beyer, unpublished.
[4] F. G. Brickwedde, J. C. Dunning, H. J. Hoge and J. H. Manley, Phys. Rev. **54**, 266 (1938).
[5] H. Bethe, Rev. Mod. Phys. **9**, 69 (1937).

Reprinted from *The Physical Review* **55**, 106 (1939).

## On Pair Emission in the Proton Bombardment of Fluorine

The gamma-rays emitted when fluorine is bombarded by protons have been studied by Fowler and Lauritsen, who have shown that these gamma-rays are monochromatic, with an energy of $(6.3 \pm 0.1)$ Mev, independent of the proton energy; that they are emitted by an excited oxygen nucleus formed from the compound nucleus $Ne^{20}$ by short range alpha-particle emission; and that they are emitted strongly only at certain well-defined proton energies which correspond to resonance levels in $Ne^{20}$. To account for the sharpness of these resonances it has been suggested that the parity-angular momentum selection rule prohibits the emission from these states of a long range alpha-particle, with the formation of a normal oxygen nucleus.

Recently Fowler and Lauritsen[1] have shown that under proton bombardment fluorine also emits electron pairs of total energy $(5.9 \pm 0.5)$ Mev. The excitation function for the pairs also exhibits resonances, but uncorrelated with those for the gamma-rays—at some energies there are more than 100 times as many gamma-rays as pairs; at others the number of gamma-rays is comparable with, and may be smaller than, the number of pairs. Since, for any multipole order the pair creation internal conversion coefficient for 6-Mev gamma-rays is less than half of one percent, the pair emission must arise from a nuclear transition for which gamma-radiation is strictly forbidden, i.e., either $Ne^{20}$ or $O^{16}$ possesses an excited state with $j=0$ with no lower states of nonvanishing $j$, from which these pairs must originate. Since $Ne^{20}$ is known from the work of Bonner to have a number of low-lying excited states, we are led to ascribe the pairs to a transition from an excited state of oxygen, with $j=0$ and about 6 Mev excitation energy, to the ground state.

If long range alpha-particle emission from the corresponding $Ne^{20}$ is to be forbidden by the parity-angular momentum selection rule, this state in oxygen must be of odd parity $(0^-)$. A $(0^- \rightarrow 0^+)$ transition, however, corresponds to vanishing matrix elements of charge and current density; decay by pair emission can only occur if there is a nonelectromagnetic coupling between nuclear particles and the pair field. Thus, a coupling term of the form suggested by the Gamow-Teller, theory of nuclear forces,

$$(\boldsymbol{\sigma}_{\mathrm{el}} \cdot \boldsymbol{\sigma}_{\mathrm{prot}}) \quad \text{or} \quad (\boldsymbol{\sigma}_{\mathrm{el}} \cdot \boldsymbol{\sigma}_{\mathrm{neut}}),$$

would permit this transition to occur with the emission of pairs. A value of the coupling constant much smaller than that suggested by these authors would be sufficient to give a rate of pair emission greater than the rate of the competitive process of two-quantum emission. This latter process takes place by the successive emission of an electric dipole and a magnetic dipole quantum, through the intermediary of states possessing the character $1^+$ or $1^-$. The lifetime of the $0^-$ state associated with this method of decay may be roughly estimated at $10^{-6}$ sec.

The only alternative to this radical suggestion is to assume the excited state in oxygen to be of even parity $(0^+)$. Pair emission, no longer forbidden, proceeds at the rate:

$$\left(\frac{1}{135\pi}\right)\left(\frac{e^2}{\hbar c}\right)^2\left(\frac{\gamma^5}{\hbar^5 c^4}\right)\left|\left(\psi_{\mathrm{exc}},\ \sum_i r_i^2 \psi_{\mathrm{norm}}\right)\right|^2,$$

which exceeds the rate of two-quantum emission:

$$\left(\frac{2}{7}\right)\left(\frac{\gamma}{\Delta}\right)^2\left(\frac{1}{135\pi}\right)\left(\frac{e^2}{\hbar c}\right)^2\left(\frac{\gamma^5}{\hbar^5 c^4}\right)\left|\left(\psi_{\mathrm{exc}},\ \left[\sum_i \mathbf{r}_i\right]^2 \psi_{\mathrm{norm}}\right)\right|^2$$

since $\gamma$, the energy emitted in the transition, is rather less than $\Delta$ ($\sim$20 Mev), the mean energy difference between the ground state and the excited states of type $1^-$ which act as intermediates in the double electric dipole emission process. (The summation in these formulae is to be extended over all nuclear protons.) The quanta would give a weak continuous spectrum extending up to 6 Mev and would not have been observed.

This second alternative involves abandoning the explanation of the long life of the resonance states of $Ne^{20}$ in terms of the parity-angular momentum selection rule. Neither spin nor isotopic spin conservation can appreciably reduce the probability of long range alpha-particle emission. Indeed, if the first excited state of $O^{16}$ is $0^+$, it would seem likely on the basis of the alpha-particle model, which could make this intelligible, that this state has the same spin and isotopic spin as the ground state. For this reason we must expect the yield of long range alpha-particles to show resonance at the energies which produce pairs. Some indication that this may be so is afforded by the fact that the maximum pair yield found by Fowler and Lauritsen is under a tenth of the maximum gamma-ray yield, which suggests that the emission of long range alpha-particles competes with the alpha-particle emission leading to the excited state. We should in fact expect that at bombarding energies of the order of 1 Mev, where pair emission becomes appreciable, the Coulomb barrier would not materially reduce the emission probability of even the short range alpha-particles. The value of 10 to 1 suggested by the observed yield of pairs for the ratio of the decay rates of long and short range alpha-particles is thus a reasonable one. The fact that the gamma-ray yields are larger, and, near the 330-kev resonance, much larger than the yield of long range alpha-particles, makes it necessary to assume that, at least for this $\gamma$-ray resonance, the long range alpha-emission is either forbidden or reduced.

If then the excited state is even, we should expect the resonance yield of long range alpha-particles to be comparable with, and probably considerably greater than the yield of pairs. If this is not so, the pair emission itself would seem to provide strong evidence for a nonelectromagnetic coupling between electrons and heavy particles.

J. R. Oppenheimer
J. S. Schwinger*

University of California,
Berkeley, California,
October 29, 1939.

* National Research Fellow.

[1] W. A. Fowler and C. C. Lauritsen, Phys. Rev. 56, 840 (1939).

Reprinted from *The Physical Review* **56**, 1066–1067 (1939).

# On a Theory of Particles with Half-Integral Spin

WILLIAM RARITA AND JULIAN SCHWINGER
*Department of Physics, University of California, Berkeley, California*
June 18, 1941

FIERZ and Pauli[1] have developed a general theory of particles with arbitrary spin, both integral and half-integral. The quantities describing particles of integral spin are tensors of suitable rank; half-integral spins are described by spinors of multiple order. It is the purpose of this note to suggest an alternative formulation of the theory on half-integral spins which avoids the complicated spinor formalism of Fierz and Pauli. The fundamental quantities of this theory $\Psi_{\mu_1\ldots\mu_k}$, have the mixed transformation properties of a Dirac four-component wave function, and of a symmetric tensor of rank $k$. (The Dirac index is suppressed.) The equations, which describe a particle of spin $k+\frac{1}{2}$ are:

$$(\gamma_\tau\partial_\tau+\kappa)\Psi_{\mu_1\ldots\mu_k}=0,\quad \gamma_\alpha\Psi_{\alpha\mu_2\ldots\mu_k}=0. \tag{1}$$

The usual supplementary conditions of the integral spin theory:

$$\partial_\alpha\Psi_{\alpha\mu_2\ldots\mu_k}=0,\quad \Psi_{\alpha\alpha\mu_3\ldots\mu_k}=0, \tag{2}$$

appear as consequences of these equations.

The verification that such wave fields are indeed associated with a particle of spin $k+\frac{1}{2}$ may proceed either by demonstrating that the number of independent plane wave solutions of definite energy and momentum is $2(k+\frac{1}{2})+1$, or by a direct proof that the square of the intrinsic angular momentum has the value $(k+\frac{1}{2})(k+\frac{3}{2})$ in the rest system. (It will not have this value in an arbitrary reference system.) The spin multiplicity is an invariant and is easily calculated in the rest system. Since each $\Psi$ obeys the Dirac equation, the "small" components vanish and the two "large" components altogether form $2(k+3)!/k!3!$ quantities. The second equation of (1) provides the information that the $\Psi_{4\mu_2\ldots\mu_k}=0$, which are $2(k+2)!/(k-1)!3!$ in number, and that $\sigma_i\Psi_{i\mu_2\ldots\mu_k}=0$ ($\mu_s\neq 4$) which constitute $2(k+1)!/(k-1)!2!$ relations among the $\Psi$'s. The total number of independent components is, therefore, $2(k+1)$, as desired. The operator of total spin consists of the sum of the $k$ infinitesimal rotation operators associated with the tensor indices in addition to the spin operator $\frac{1}{2}\boldsymbol{\sigma}$ of the Dirac theory. The proof that the spin possesses the correct eigenvalue is an elementary consequence of the condition $\sigma_i\Psi_{i\mu_2\ldots\mu_k}=0$.

The special case of spin $\frac{3}{2}$ has been treated in detail by Fierz and Pauli employing a method that necessitated the adjunction of auxiliary spinors in order to produce the proper subsidiary conditions from a variational principle. It is an advantage of the new formalism that a suitable Lagrangian can be constructed without the intervention of additional fields. One of a class of possible Lagrangians is

$$L=\bar{\Psi}_\mu(\gamma_\tau\partial_\tau+\kappa)\Psi_\mu-\tfrac{1}{3}\bar{\Psi}_\mu(\gamma_\mu\partial_\nu+\gamma_\nu\partial_\mu)\Psi_\nu+\tfrac{1}{3}\bar{\Psi}_\mu\gamma_\mu(\gamma_\tau\partial_\tau-\kappa)\gamma_\nu\Psi_\nu. \tag{3}$$

Although the Lagrangian is not unique in the absence of external fields, the form (3) is the only one which permits a relatively simple expression of the equations of motion in the presence of electromagnetic fields. The four-vector of charge and current, obtained by the usual prescription, is

$$\begin{aligned} j_\nu &= \bar{\Psi}_\mu\gamma_\nu\Psi_\mu-\tfrac{1}{3}(\bar{\Psi}_\mu\gamma_\mu)\Psi_\nu-\tfrac{1}{3}\bar{\Psi}_\nu(\gamma_\mu\Psi_\mu) \\ &\quad+\tfrac{1}{3}(\bar{\Psi}_\sigma\gamma_\sigma)\gamma_\nu(\gamma_\tau\Psi_\tau) \\ &= \bar{\Psi}_\mu\gamma_\nu\Psi_\mu \end{aligned} \tag{4}$$

in the absence of external electromagnetic fields. Although the charge density is not a positive definite form, the total charge is necessarily definite since $\Psi_4=0$ in the rest system. In the exceptional case of zero rest mass the wave function admits a gauge transformation,

$$\Psi_\mu'=\Psi_\mu+\partial_\mu\varphi,\quad \gamma_\tau\partial_\tau\varphi=0,$$

which leaves all physical quantities invariant.

The method here presented for developing the theory of spin $\frac{3}{2}$ thus contains many of the features of both the Proca and the Dirac theory. This is particularly evident in the application to $\beta$-decay which Mr. Kusaka[2] has discussed in an accompanying letter.

[1] M. Fierz and W. Pauli, Proc. Roy. Soc. **A173**, 211 (1939).
[2] S. Kusaka, Phys. Rev. **60**, 61 (1941), this issue.

Reprinted from *The Physical Review* **60**, 61 (1941).

# On the Interaction of Mesotrons and Nuclei

J. R. Oppenheimer and Julian Schwinger
*Department of Physics, University of California, Berkeley, California*
(Received June 19, 1941)

THE formalism of mesotron field theories was first developed in analogy either with the method of classical electrodynamics that involves the determination of fields produced by pregiven sources, or with the perturbation technique of quantum electrodynamics. This program not only has led to predictions in contradiction with experiment (the singular nature of nuclear forces, and large cross sections for mesotron scattering by nuclei) but is intrinsically inconsistent, for the methods of electrodynamics, involving the neglect of the reaction of field on source, and based on the small magnitude of the coupling between charges and the electromagnetic field, are here quite unjustified. This inconsistency is rendered far sharper by the charge- and spin-dependent couplings of the mesotron theory. These difficulties have led to two apparently unrelated suggestions for modifying the formal treatment of mesotron theory.

Thus Heisenberg has attempted a more complete solution of the mesotron equations by including, within the framework of a classical theory, the large reaction of the field on the emitting source. He treated the problem of a neutral vector field interacting with a fixed, spatially extended source to which is attached a classical spin. His calculations showed that the inertia of the spin, arising from the reaction of its proper field, would greatly reduce the spin-dependent mesotron scattering. A relativistic variant of this procedure has been elaborated by Bhabha, who based his work on Dirac's classical electron model. Those terms in the field reaction on the source which become singular as $a$, the spatial extension of the source, vanishes, are discarded and the finite terms are evaluated in the limit of point coupling. However, it is precisely these singular terms that are essential for Heisenberg's explanation of the small mesotron scattering. Furthermore, quite apart from the serious question of whether this method affords a classical description of elementary particles, it is so essentially restricted to the classical domain that it provides no basis for a correspondence treatment of the actual quantum-mechanical problems encountered in a discussion of the scattering of charged mesotrons and of nuclear forces.

For this reason Bhabha himself and Heitler suggested the alternative theory of proton isobars. They observed that, from the point of view of perturbation theory, the large scattering arising from charge- and spin-dependent couplings is a consequence of the prohibition of certain intermediate states in the scattering process by the conservation laws for charge and angular momentum. They therefore proposed a modification of the assumptions of mesotron theory by postulating the existence of slightly more massive nuclear particles with arbitrary integral charge and half-integral spin. By a suitable choice of the excitation energy of the lowest isobars—high enough to make them escape ready detection—the scattering cross section could be reduced sufficiently to avoid conflict with experiment.

The connection between these two sets of ideas lies in the circumstance that a field strongly coupled to a charge or spin-dependent source will itself possess states in which charge or angular momentum is bound to the source, with the energy of the system increasing quadratically in its dependence on the total charge or angular momentum. This has been shown in detail by Wentzel for the charged scalar field. Wentzel used a lattice space to achieve convergence and considered only the limit of strong coupling. Despite the fact that his theory predicted isobars, his calculations indicated an inacceptably large value for the mesotron scattering cross section.

The evidence of nuclear forces shows, however, that the mesotron coupling is spin dependent, and indicates that the coupling constant itself is not very large. In order to see how these facts would modify Wentzel's conclusions, we have investigated a class of problems that throws some light on the earlier work and that also offers promise of a limited but consistent description of



Reprinted from *The Physical Review* **60**, 150–152 (1941).

the mesotron, in agreement with experience: nuclear forces on the one hand, and, on the other, the small scattering, zero spin, and highly multiple production observed for mesotrons in cosmic rays. We have, in part, generalized Heisenberg's treatment, and considered the classical problem of the coupling of neutral and charged, scalar and pseudo-scalar mesotrons to an extended, spatially fixed source. These problems are all rigorously soluble, for all values of the coupling constant and source size. In addition, we have treated Wentzel's quantum problem of the charged scalar field,[1] using an extended source instead of a lattice space, in the limit where the coupling constant $g$ is large, $\gamma=g^2/\hbar c\gg 1$, and have made the analogous calculation for the neutral pseudo-scalar in the corresponding limit $\gamma\gg\kappa a$ (here $\kappa=\mu c/\hbar$ and $\mu$ is the mesotron mass). The problem of the charged pseudo-scalar has not been solved quantum-mechanically even in this limit, nor has any quantum solution been found for intermediate values of $\gamma$, or $\gamma/\kappa a$; nor have the revisions in nuclear forces been calculated in detail.

Nevertheless, we believe our results to be of some interest. The classical solutions for the pseudo-scalar always give states of bound angular momentum if $\gamma/\kappa a$ is not too small; the energy of these states depends on the total angular momentum $J$, as

$$\tfrac{3}{2}(\kappa a/\gamma)\mu c^2(J^2-\tfrac{1}{4}).$$

Here $a$ is defined by

$$a^{-1}=\int d\tau K(\mathbf{r})\,|\mathbf{r}-\mathbf{r}'|^{-1}K(\mathbf{r}')d\tau'$$

and $K(\mathbf{r})$ is the source function. The symmetrical scalar and charged scalar theories give solutions with bound charge, the former only if $2\gamma>1$; and here the mass of the normal state of an isobar, for $\gamma\gg 1$, depends on the charge $Q$ as $(\mu/\gamma)(Q-\tfrac{1}{2})^2$. For the symmetrical pseudo-scalar the condition for the existence of bound spin and charge is $2\gamma>3\kappa a$. The quantum solutions for $\gamma\gg 1$ (charged scalar) and $\gamma\gg\kappa a$ (neutral pseudo-scalar) agree in these conclusions.[2] This agreement confirms the *a priori* expectation that, for large enough coupling, the quantum fluctuations of the source will be negligible compared to the reaction of the source to the field.

Classical scattering formulae, applicable when the proton recoil is negligible, give for all pseudo-scalar theories a scattering vanishing with $a^2$: this is a direct consequence of the fact that, in these theories, only $p$ states are coupled to the source. Thus, for $\gamma\gg\kappa a$, and a mesotron momentum $p<\hbar/a$, the scattering cross section for a charged mesotron on the symmetrical pseudo-scalar theory is

$$d\sigma=\tfrac{3}{8}a^2(pc/E)^4(1+2\cos^2\vartheta)d\Omega, \tag{1}$$

whereas for a charged scalar with $\gamma\gg 1$, $a\to 0$, it is just

$$d\sigma=(\hbar c/E)^2 d\Omega. \tag{2}$$

The results of the quantum solution agree with (2) for the charged scalar, and are of the same form as (1) for the neutral pseudo-scalar.

It is thus clear that pseudo-scalar theories can give a scattering small enough to agree with that observed, but that scalar theories could, at most, do so with a choice of $\gamma$ far too small to account for nuclear forces. In fact, the experimental scattering results demand a value of $a$ of the order of the proton Compton wave-length $\hbar/Mc$, or possibly slightly smaller. Indeed, this length marks the extreme limit of validity of the methods we are using, and of the classical localizability of the source.

With this small value of $a$, and a value $\gamma\sim\tfrac{1}{10}$, which is derived from the magnitude of nuclear forces in singlet states, the spin and charge isobars will have an excitation only somewhat smaller than the rest energy of the mesotron.

This small value of $\gamma$ would at first sight seem to be inconsistent with the high multiplicity of mesotron production in high energy nuclear collisions; but for pseudo-scalar theories this is not so. We have made an admittedly crude estimate of this multiplicity by calculating classically the excitation of the mesotron field when a proton's velocity and spin are suddenly

[1] Julian Schwinger, to be published soon.

[2] G. Wentzel (Helv. Phys. Acta **13**, 269 (1940)) gives $\gamma\kappa a\gg 1$ as the condition for the validity of his solution and, for the isobar separation $(\mu/\gamma\kappa a)(Q-\tfrac{1}{2})^2$. In both results $\gamma\kappa a$ must be replaced by $\gamma$. Wentzel's conclusions were obtained by overlooking the contribution to the isobar energy of order $g^{-1}$ in the expansion of the Hamiltonian in descending powers of $g$. For a detailed discussion see reference 1.

altered, and then limiting the actual spectrum to mesotrons of total energy equal to or lower than the energy loss $\Delta E$ of the proton. In this way one finds a multiplicity $N \sim \gamma^{\frac{1}{4}}(\Delta E/\mu c^2)^{\frac{3}{4}}$, with a value 10 for $\gamma = \frac{1}{10}$, and an energy loss of $10^{10}$ volts. Under these conditions, and with a comparable value of $\gamma$, a scalar theory would give practically no multiple production.

It will be observed that we have used a value of $\gamma$ given by the perturbation theoretic evaluation of nuclear forces. This is because, for singlet states, our theories still give forces of range $\hbar/\mu c$, which have an effective depth of the order $\gamma\mu c^2$, and which behave like $g^2/a$ for $\gamma < a$. The detailed radial dependence, and in particular the form of the tensor forces, remains to be investigated.

In conclusion, we should like again to emphasize that our present methods involve, in their physical content, only a rather more complete effort to take into account the reaction of the source to the mesotron field than either Bhabha's classical methods or the *a priori* postulation of isobars afforded. But it would seem that these methods are sufficient to decide definitely in favor of a pseudo-scalar, rather than a scalar or vector, field and to fix roughly the values of the coupling constant and source size needed to make the model definite.

**8. On a Field Theory of Nuclear Forces.** JULIAN SCHWINGER, *Purdue University.*—Mesotron theories of nuclear forces based on perturbation methods have suffered from the grave defect that the tensor forces which they predict possess inadmissible $1/r^3$ singularities. Although the concept of strong coupling, applied to the pseudo-scalar theory, is remarkably successful in accounting for cosmic-ray phenomena, it does not remove the objectionable singularities. If, however, in addition, a vector mesotron field is postulated which possesses the same nuclear coupling constant as the pseudo-scalar field, but whose particles differ in mass from the pseudo-scalar mesotrons observed in cosmic-rays, the inadmissible singularities are removed. The sign and magnitude of the resultant tensor interaction, which behaves as $1/r$ at small distances, is determined by the mass difference of the two mesotrons. It is satisfactory that the positive quadripole moment of the deuteron corresponds to the vector mesotron being more massive than the pseudo-scalar mesotron, for such particles will be highly unstable against disintegration into a pseudo-scalar mesotron and a $\gamma$-ray; and thus would not have been observed. However, the spontaneous disintegration of the vector mesotrons formed at the top of the atmosphere will provide a source of soft component to be added to the soft component production by the $\beta$ decay of the pseudo-scalar mesotrons.

Reprinted from *The Physical Review* **61**, 387 (1942).

**B12. Polarization of Neutrons by Resonance Scattering in Helium.** JULIAN SCHWINGER, *Harvard University*.—Neutron scattering in helium exhibits an anomaly for neutron energies in the vicinity of 1 Mev, which has been attributed to a $P$ resonance associated with the formation of the unstable $He^5$ nucleus. The energy dependence of the cross section for back-scattering is accounted for qualitatively on the assumption of a splitting of the two $P$ levels associated with $J=1/2$ and $3/2$, with an energy separation comparable with the width of each level, $|E_{1/2}-E_{3/2}|\sim\Gamma\sim0.4$ Mev. In consequence of the large energy difference between the levels, a neutron scattered by He in the resonance region is effectively subjected to a strong spin-orbit interation, which manifests itself in a polarization of neutrons scattered through a definite angle. The polarization effects are quite large; neutrons scattered through 90° in the center of mass system are practically completely polarized at energies corresponding roughly to the positions of the two levels, the direction of polarization being reversed at the two resonances. The polarization effect can be exhibited by a second scattering process, the resultant intensity being asymmetrical relative to the direction of the first scattered neutron. For neutrons scattered twice through 90° in the center of gravity system, the intensity ratio can be as great as ten. An essential factor contributing to this large ratio is that the energy loss on the first collision is comparable with the doublet separation, whence large polarization effects can occur at both collisions. The polarization effect offers the possibility of producing strongly polarized beams of thermal neutrons by slowing down polarized neutrons in a non-depolarizing medium. The investigation of depolarization in various substances is also of interest.

Reprinted from *The Physical Review* **69**, 681 (1946).

# On the Polarization of Fast Neutrons

JULIAN SCHWINGER
*Harvard University, Cambridge, Massachusetts*
(Received January 8, 1948)

ALTHOUGH the production of polarized thermal neutrons has long been an accomplished fact, no such success has been forthcoming with fast neutrons. Only one method for the polarization of fast neutrons has thus far been suggested,[1] of which the essential mechanism is the large, effective nuclear spin-orbit interaction present when neutrons are resonance scattered by helium and similar nuclei. It is the purpose of this note to suggest a second mechanism for polarizing fast neutrons—the spin-orbit interaction arising from the motion of the neutron magnetic moment in the nuclear Coulomb field. Despite the apparent small magnitude of this interaction, the long-range nature of the Coulomb field is such that the use of small scattering angles will produce almost complete polarization under ideal conditions. A closely related phenomenon produced by this electromagnetic interaction is an additional scattering of unpolarized neutrons which increases rapidly with decreasing scattering angle and is comparable with purely nuclear scattering at the small angles effective in producing polarized neutrons.

The energy of a neutron moving in an electric field, $\mathbf{E}=-\nabla\phi$, is described by the following contribution to the neutron Hamiltonian:

$$H'=\mu_n(e\hbar/2M^2c^2)\boldsymbol{\sigma}\cdot\mathbf{E}\times\mathbf{p}, \qquad (1)$$

where $\mu_n=1.91$ is the numerical value of the neutron moment in units of $e\hbar/2Mc$, $\boldsymbol{\sigma}$ is the Pauli spin vector, and $\mathbf{p}$ is the momentum of the neutron. In order that the electric field be fully effective in producing spin-dependent scattering the major portion of this scattering should take place outside of the nucleus ($r\gg R$), but well within the screening radius of the atomic electrons ($r\ll a$). This restricts the range of useful scattering angles, since the waves scattered through an angle $\vartheta$ are primarily generated at a distance $r$ from the nucleus, given by

$$2kr\sin\vartheta/2\sim 1, \qquad (2)$$

where $k=p/\hbar$ is the neutron wave number. Hence, the unscreened Coulomb field of a point nucleus will be effective for scattering in the angular range:

$$1/ka\ll 2\sin\vartheta/2\ll 1/kR. \qquad (3)$$

If the nuclear radius and atomic screening radius are taken to be

$$R=1.5\cdot 10^{-13}A^{\frac{1}{3}}\text{ cm} \quad\text{and}\quad a=0.53\cdot 10^{-8}Z^{-\frac{1}{3}}\text{ cm},$$

the angle restrictions for a 1-Mev neutron scattered in Pb, for example, are

$$4\cdot 10^{-4}\ll 2\sin\vartheta/2\ll\tfrac{1}{2}. \qquad (4)$$

The electromagnetic scattering of a neutron under these conditions can be calculated with the plane wave Born approximation, for the nuclear scattered wave is negligible compared with the incident wave at the significant scattering distances. We denote the incident plane wave by

$$\psi_{\text{inc}}=e^{i\mathbf{k}_0\cdot\mathbf{r}}\chi, \qquad (5)$$

where $\mathbf{k}_0$ is the initial propagation vector and $\chi$ is a spin function. The asymptotic form of the wave scattered in the direction of the propagation vector $\mathbf{k}$ is then

$$\psi_{sc}\sim(e^{ikr}/r)f(\vartheta)\chi, \qquad (6)$$

with

$$f(\vartheta)=f_0(\vartheta)+\tfrac{1}{2}i\boldsymbol{\sigma}\cdot\mathbf{n}\cot\vartheta/2(\hbar/Mc)(Ze^2/\hbar c). \qquad (7)$$

In this formula, $\mathbf{n}$ is the unit vector defined by

$$\mathbf{k}\times\mathbf{k}_0=\mathbf{n}k^2\sin\vartheta, \qquad (8)$$

and $f_0(\vartheta)$ is the amplitude of the wave scattered by specifically nuclear forces. We assume the latter to be spin-independent and, further, ignore any inelastic nuclear scattering. Both of these assumptions are not unreasonable for neutron energies in the vicinity of 1 Mev.

The intensity of the scattered wave is determined by the following spin scalar product:

$$r^2(\psi_{sc},\psi_{sc})=(\chi, f^{\dagger}(\vartheta)f(\vartheta)\chi), \qquad (9)$$

[1] J. Schwinger, Phys. Rev. 69, 681 (1946).



Reprinted from *The Physical Review* 73, 407–409 (1948).

which yields

$$r^2(\psi_{sc}, \psi_{sc}) = |f_0(\vartheta)|^2 + \gamma^2 \cot^2\vartheta/2 + 2\gamma Im\, f_0(\vartheta) \cot\vartheta/2\mathbf{n}\cdot\mathbf{P}_{inc}, \quad (10)$$

in which

$$\gamma = \tfrac{1}{2}\mu_n(\hbar/Mc)(Ze^2/\hbar c) \quad (11)$$

and

$$\mathbf{P}_{inc} = (\chi, \sigma\chi) \quad (12)$$

is a vector describing the polarization state of the incident beam. The corresponding vector for the scattered wave is

$$\mathbf{P}_{sc} = (\psi_{sc}, \sigma\psi_{sc})/(\psi_{sc}, \psi_{sc}), \quad (13)$$

where

$$r^2(\psi_{sc}, \sigma\psi_{sc}) = \mathbf{n}2\gamma Im\, f_0(\vartheta)\cot\vartheta/2 + |f_0(\vartheta)|^2\mathbf{P}_{inc} + \gamma^2\cot^2\vartheta/2(2\mathbf{nn}\cdot\mathbf{P}_{inc} - \mathbf{P}_{inc}) - 2\gamma Re\, f_0(\vartheta)\cot\vartheta/2\mathbf{n}\times\mathbf{P}_{inc}. \quad (14)$$

For an initially unpolarized beam:

$$\mathbf{P}_{sc} = \mathbf{n}\frac{2Im\, f_0(\vartheta)\gamma\cot\vartheta/2}{|f_0(\vartheta)|^2 + \gamma^2\cot^2\vartheta/2} = \mathbf{n}P(\vartheta). \quad (15)$$

The discussion of the latter formula is greatly simplified by noting that, within the restricted angular range (3), the specifically nuclear scattering must be insensitive to angle and can be replaced by the value appropriate to forward scattering. Now, according to a well-known theorem, $Im\, f_0(0)$ is related to the total scattering cross section by

$$Im\, f_0(0) = (k/4\pi)\sigma, \quad (16)$$

while

$$|f_0(0)|^2 = (\sigma/4\pi)G \quad (17)$$

expresses the differential cross section for forward scattering in terms of the "gain," the ratio of the actual forward scattered intensity to that of an isotropic scatterer. It follows from the form of (15) that there is an optimum scattering angle, $\vartheta_0$, for the production of polarized neutrons, namely,

$$\tan\vartheta_0/2 = \frac{\gamma}{|f_0(0)|} = \tfrac{1}{2}\mu_n\frac{Ze^2}{\hbar c}\left[\frac{4\pi(\hbar/Mc)^2}{\sigma G}\right]^{\frac{1}{2}}. \quad (18)$$

The maximum polarization is

$$P(\vartheta_0) = \frac{Im\, f_0(0)}{|f_0(0)|} = \left[\frac{k^2\sigma}{4\pi G}\right]^{\frac{1}{2}}. \quad (19)$$

In order to estimate the magnitudes of these quantities, it is necessary to have some knowledge of the energy dependence of $\sigma$ and $G$. The model of an impenetrable sphere provides the following information in the limits $kR \ll 1$ and $kR \gg 1$:

$$kR \ll 1:\ \sigma = 4\pi R^2,\ G = 1$$
$$kR \gg 1:\ \sigma = 2\pi R^2,\ G = \tfrac{1}{2}(kR)^2. \quad (20)$$

The predicted limiting forms of the quantities (18) and (19) are then

$$\tan\vartheta_0/2 = \begin{cases} \tfrac{1}{2}\mu_n\dfrac{Ze^2}{\hbar c}\dfrac{\hbar/Mc}{R}, & kR \ll 1 \\ \mu_n\dfrac{Ze^2}{\hbar c}\dfrac{\hbar/Mc}{R}\dfrac{1}{kR}, & kR \gg 1 \end{cases} \quad (21)$$

and

$$P(\vartheta_0) = \begin{cases} kR, & kR \ll 1 \\ 1, & kR \gg 1. \end{cases} \quad (22)$$

It appears to be a reasonable interpolation to place

$$\tan\vartheta_0/2 \cong \tfrac{1}{2}\mu_n\frac{Ze^2}{\hbar c}\frac{\hbar/Mc}{R}$$
$$P(\vartheta_0) \cong 1 \quad (23)$$

for $kR \sim 2$, which is the value appropriate to 1-Mev neutrons scattered in Pb. Under these conditions, we expect practically complete polarization for a scattering angle of $\vartheta_0 = 1.5°$. In view of the stationary character of the polarization in the vicinity of this angle, somewhat larger angles can be employed without undue impairment of the degree of polarization. Thus, for $\vartheta = 3°$, $P = 0.80$; $\vartheta = 6°$, $P = 0.47$; $\vartheta = 9°$, $P = 0.32$.

To detect the polarization produced by scattering, it is necessary to subject the polarized neutrons to a second scattering process. If the two scattering angles are $\vartheta_1$ and $\vartheta_2$, with the normals to the two scattering planes being $\mathbf{n}_1$ and $\mathbf{n}_2$, respectively, the intensity after the second deflection is, according to (10) and (15), proportional to

$$1 + \mathbf{n}_1\cdot\mathbf{n}_2 P(\vartheta_1)P(\vartheta_2). \quad (24)$$

When both scattering events occur in the same plane, the intensity for the situation in which the two deflections occur in the same sense exceeds that for deflections in the opposite sense in the ratio:

$$R=\frac{1+P(\vartheta_1)P(\vartheta_2)}{1-P(\vartheta_1)P(\vartheta_2)}. \tag{25}$$

If both scattering angles equal the optimum angle $\vartheta_0$, this ratio can be very large. Thus the experimental difficulties accompanying the small angles involved are ameliorated to some extent by the large effects under investigation. However, somewhat larger angles can be employed without destroying the experimental effect. For example, under the numerical conditions previously employed, $R=2.0$ for $\vartheta_1=1.5°$, $\vartheta_2=9°$, while $R=1.2$ for $\vartheta_1=\vartheta_2=9°$.

Finally, we note that the differential cross section for the scattering of an unpolarized neutron beam is modified at small angles. According to Eq. (10),

$$\sigma(\vartheta)=\sigma_0(\vartheta)+\gamma^2 \cot^2\vartheta/2,$$

where $\sigma_0(\vartheta)$ is the specifically nuclear differential cross section, which assumes the value $(\sigma/4\pi)G$ at the small angles significant for the electromagnetic scattering. Thus the additional contribution to the scattering increases rapidly with diminishing angle and equals the purely nuclear scattering at precisely the angle $\vartheta_0$ that is optimum for polarization. In view of the small angular range over which it is effective, the electromagnetic scattering provides a negligible contribution to the total cross section, namely,

$$\delta\sigma \sim 2\pi\gamma^2 \log a/R = 2\pi\mu_n{}^2(\hbar/Mc)^2 \times (Ze^2/\hbar c)^2 \log 3.5\cdot 10^4 (AZ)^{-\frac{1}{3}} \tag{26}$$

for $kR>1$, which has the value $\delta\sigma=2.6\cdot 10^{-26}$ cm$^2$ for Pb.

# On Quantum-Electrodynamics and the Magnetic Moment of the Electron

JULIAN SCHWINGER
*Harvard University, Cambridge, Massachusetts*
December 30, 1947

ATTEMPTS to evaluate radiative corrections to electron phenomena have heretofore been beset by divergence difficulties, attributable to self-energy and vacuum polarization effects. Electrodynamics unquestionably requires revision at ultra-relativistic energies, but is presumably accurate at moderate relativistic energies. It would be desirable, therefore, to isolate those aspects of the current theory that essentially involve high energies, and are subject to modification by a more satisfactory theory, from aspects that involve only moderate energies and are thus relatively trustworthy. This goal has been achieved by transforming the Hamiltonian of current hole theory electrodynamics to exhibit explicitly the logarithmically divergent self-energy of a free electron, which arises from the virtual emission and absorption of light quanta. The electromagnetic self-energy of a free electron can be ascribed to an electromagnetic mass, which must be added to the mechanical mass of the electron. Indeed, the only meaningful statements of the theory involve this combination of masses, which is the experimental mass of a free electron. It might appear, from this point of view, that the divergence of the electromagnetic mass is unobjectionable, since the individual contributions to the experimental mass are unobservable. However, the transformation of the Hamiltonian is based on the assumption of a weak interaction between matter and radiation, which requires that the electromagnetic mass be a small correction $(\sim(e^2/\hbar c)m_0)$ to the mechanical mass $m_0$.

The new Hamiltonian is superior to the original one in essentially three ways: it involves the experimental electron mass, rather than the unobservable mechanical mass; an electron now interacts with the radiation field only in the presence of an external field, that is, only an accelerated electron can emit or absorb a light quantum;* the interaction energy of an electron with an external field is now subject to a *finite* radiative correction. In connection with the last point, it is important to note that the inclusion of the electromagnetic mass with the mechanical mass does not avoid all divergences; the polarization of the vacuum produces a logarithmically divergent term proportional to the interaction energy of the electron in an external field. However, it has long been recognized that such a term is equivalent to altering the value of the electron charge by a constant factor, only the final value being properly identified with the experimental charge. Thus the interaction between matter and radiation produces a renormalization of the electron charge and mass, all divergences being contained in the renormalization factors.

The simplest example of a radiative correction is that for the energy of an electron in an external magnetic field. The detailed application of the theory shows that the radiative correction to the magnetic interaction energy corresponds to an additional magnetic moment associated with the electron spin, of magnitude $\delta\mu/\mu=(\frac{1}{2}\pi)e^2/\hbar c=0.001162$. It is indeed gratifying that recently acquired experimental data confirm this prediction. Measurements on the hyperfine splitting of the ground states of atomic hydrogen and deuterium[1] have yielded values that are definitely larger than those to be expected from the directly measured nuclear moments and an electron moment of one Bohr magneton. These discrepancies can be accounted for by a small additional electron spin magnetic moment.[2] Recalling that the nuclear moments have been calibrated in terms of the electron moment, we find the additional moment necessary to account for the measured hydrogen and deuterium hyperfine structures to be $\delta\mu/\mu=0.00126\pm0.00019$ and $\delta\mu/\mu=0.00131\pm0.00025$, respectively. These values are not in disagreement with the theoretical prediction. More precise conformation is provided by measurement of the $g$ values for the $^2S_{\frac{1}{2}}$, $^2P_{\frac{1}{2}}$, and $^2P_{3/2}$ states of sodium and gallium.[3] To account for these results, it is necessary to ascribe the following additional spin magnetic moment to the electron, $\delta\mu/\mu=0.00118\pm0.00003$.

The radiative correction to the energy of an electron in a Coulomb field will produce a shift in the energy levels of hydrogen-like atoms, and modify the scattering of electrons in a Coulomb field. Such energy level displacements have recently been observed in the fine structures of hydrogen,[4] deuterium, and ionized helium.[5] The values yielded by our theory differ only slightly from those conjectured by Bethe[6] on the basis of a non-relativistic calculation, and are, thus, in good accord with experiment. Finally, the finite radiative correction to the elastic scattering of electrons by a Coulomb field provides a satisfactory termination to a subject that has been beset with much confusion.

A paper dealing with the details of this theory and its applications is in course of preparation.

* A classical non-relativistic theory of this type was discussed by H. A. Kramers at the Shelter Island Conference, held in June 1947 under the auspices of the National Academy of Sciences.

[1] J. E. Nafe, E. B. Nelson, and I. I. Rabi, Phys. Rev. **71**, 914 (1947); D. E. Nagel, R. S. Julian, and J. R. Zacharias, Phys. Rev. **72**, 971 (1947).

[2] G. Breit, Phys. Rev. **71**, 984 (1947). However, Breit has not correctly drawn the consequences of his empirical hypothesis. The effects of a nuclear magnetic field and a constant magnetic field do not involve different combinations of $\mu$ and $\delta\mu$.

[3] P. Kusch and H. M. Foley, Phys. Rev. **72**, 1256 (1947), and further unpublished work.

[4] W. E. Lamb, Jr. and R. C. Retherford, Phys. Rev. **72**, 241 (1947).

[5] J. E. Mack and N. Austern, Phys. Rev. **72**, 972 (1947).

[6] H. A. Bethe, Phys. Rev. **72**, 339 (1947).

Reprinted from *The Physical Review* 73, 416–417 (1948).

# A Note on Saturation in Microwave Spectroscopy

ROBERT KARPLUS* AND JULIAN SCHWINGER
*Harvard University, Cambridge, Massachusetts*
(Received January 9, 1948)

The investigation of Van Vleck and Weisskopf, on the shape of collision-broadened absorption lines, is extended to high power levels of the exciting radiation. Transitions among the molecular states are then induced at a rate that is not negligible compared with the collision rate, thus invalidating the assumption of thermal equilibrium. The theory is based on a quantum transcription of the previous semiclassical treatment, the essential tool being the density matrix. When the resonant frequencies of the molecule are widely spaced, an absorption line is ultimately broadened as the power level increases. Correspondingly, the peak absorption coefficient decreases, the power absorbed per unit volume approaching saturation with increasing incident power. It is shown that the broadening of an absorption line is not to be attributed to any intrinsic modification of the line shape, but rather to a frequency dependent alteration of the energy level populations. A molecule for which all resonant frequencies coincide, a one-dimensional harmonic oscillator, is treated in Appendix II. No saturation effect occurs here, since the absorption is independent of the molecular distribution among the oscillator energy levels.

THE shape of collision-broadened absorption lines has been investigated by Van Vleck and Weisskopf[1] for weak monochromatic radiation. Their results are of importance for the interpretation of the extensive data recently acquired on microwave absorption spectra. However, the investigation of Van Vleck and Weisskopf does not include a discussion of two aspects of these experimental studies—saturation and frequency modulation. At high power levels of the exciting radiation, transitions among the molecular states are induced at a rate that is not negligible compared with the collision rate, thus invalidating the assumption of thermal equilibrium. This note contains a theoretical analysis of the resultant modification of the absorption coefficient. A discussion of frequency modulation is given in a companion note by R. Karplus.

To facilitate the extension of the theory, we shall first derive the results of Van Vleck and Weisskopf, replacing their semiclassical treatment by a suitable quantum transcription. The linear absorption coefficient for radiation of angular frequency $\omega$, in a gas of molecular density $N$, is

$$\alpha = 4\pi(\omega/c)N Im\chi, \tag{1}$$

where $\chi$ is the molecular susceptibility. The latter is determined by the dipole moment

$$\mathbf{p}(t) = Re(\chi\mathbf{F}e^{-i\omega t}), \tag{2}$$

induced by the field

$$\mathbf{F}(t) = \mathbf{F}\cos\omega t = Re(\mathbf{F}e^{-i\omega t}), \tag{3}$$

which may be either electric or magnetic. The assumptions of Van Vleck and Weisskopf are most compactly expressed in quantum language with the aid of the density matrix.[2] Thus, to calculate the dipole moment at time $t$, we suppose that a strong collision has occurred at a previous time $t_0 = t - \vartheta$. This assumption of complete thermal equilibrium, immediately following the collision, is described by

$$\rho_0(t_0) = Ce^{-H(t_0)/kT}, \quad 1/C = Spe^{-H(t_0)/kT}, \tag{4}$$

where $H(t)$ is the Hamiltonian of the molecule, including the energy in the external field,

$$H(t) = H_0 - \mathbf{p}\cdot\mathbf{F}(t) = H_0 + V\cos\omega t. \tag{5}$$

In the latter formula, $H_0$ is the Hamiltonian of the isolated molecule and $\mathbf{p}$ is the dipole moment operator. The subsequent development of the density matrix in time is determined by

$$i\hbar(\partial/\partial t)\rho = H\rho - \rho H. \tag{6}$$

The final task of the theory lies in the calculation

* U. S. Rubber Company, Predoctoral Fellow.
[1] J. H. Van Vleck and V. F. Weisskopf, Rev. Mod. Physics 17, 227 (1945).
[2] Cf. R. C. Tolman, *Principles of Statistical Mechanics* (Oxford University Press, New York, 1930), Chapter IX



Reprinted from *The Physical Review* 73, 1020–1026 (1948).

of the average dipole moment at time $t$,

$$\mathbf{p}(t) = Sp\mathbf{p}\bar{\rho}(t), \tag{7}$$

where the bar indicates that one must eventually average over $t_0$, the time of the last collision, in accordance with the assumption of random collisions spaced by an average time interval $\tau$.

This program is greatly simplified by first performing the average over the time of the last collision, using the fact that the probability that the last collision occurs in the time interval $t-\vartheta$, $t-\vartheta-d\vartheta$ is $e^{-\vartheta/\tau}d\vartheta/\tau$. The density matrix may be considered to depend parametrically upon the initial time $t_0$, say $\rho(t,t_0)$, and is subject to the equation of motion (6) together with the initial condition (Cf. Eq. (4))

$$\rho(t_0,t_0) = \rho_0(t_0). \tag{8}$$

Now

$$\bar{\rho}(t) = \int_0^\infty \rho(t, t-\vartheta)e^{-\vartheta/\tau}d\vartheta/\tau, \tag{9}$$

and

$$(\partial/\partial t)\bar{\rho}(t) = \int_0^\infty [(\partial/\partial t)\rho(t,t_0)]_{t_0=t-\vartheta}e^{-\vartheta/\tau}d\vartheta/\tau - \int_0^\infty [(\partial/\partial\vartheta)\rho(t, t-\vartheta)]e^{-\vartheta/\tau}d\vartheta/\tau, \tag{10}$$

whence

$$(\partial/\partial t)\bar{\rho}(t) = -(i/\hbar)[H(t)\bar{\rho}(t) - \bar{\rho}(t)H(t)] - (1/\tau)[\bar{\rho}(t) - \rho_0(t)]. \tag{11}$$

We have thus succeeded in constructing an equation of motion for the averaged density matrix $\bar{\rho}(t)$, which is the quantity directly concerned in the evaluation of the average dipole moment. However, it is somewhat more convenient to introduce

$$D(t) = \bar{\rho}(t) - \rho_0(t) \tag{12}$$

which obeys the equation of motion

$$(\partial/\partial t)D(t) = -(i/\hbar)[H(t)D(t) - D(t)H(t)] - (1/\tau)D(t) - (\partial/\partial t)\rho_0(t) \tag{13}$$

and measures the deviation of the density matrix $\bar{\rho}(t)$ from instantaneous thermal equilibrium.

We shall first solve this equation of motion under the conditions of Van Vleck and Weisskopf, namely, a weak monochromatic radiation field. On employing a matrix scheme in which the unperturbed Hamiltonian, $H_0$, is diagonal, the equation of motion (13) becomes

$$(\partial/\partial t + i\omega_{mn} + 1/\tau)D_{mn}(t) = -(\partial/\partial t)[\rho_0(t)]_{mn} - (i/\hbar)\textstyle\sum_k(V_{mk}D_{kn}(t) - D_{mk}(t)V_{kn})\cos\omega t, \tag{14}$$

where

$$\omega_{mn} = (E_m - E_n)/\hbar \tag{15}$$

designates the angular frequency associated with the transition from state $m$ to state $n$ of the unperturbed molecule. The assumption of a weak radiation field implies that the density matrix differs little from that of the isolated molecule at temperature $T$:

$$\rho^{(0)} = C^{(0)}e^{-H_0/kT}, \quad 1/C^{(0)} = Spe^{-H_0/kT}. \tag{16}$$

The latter operator is represented by the diagonal matrix

$$\rho_{mn}{}^{(0)} = \rho_m{}^{(0)}\delta_{mn}, \ \rho_m{}^{(0)} = e^{-E_m/kT}/\textstyle\sum_n e^{-E_n/kT}. \tag{17}$$

Thus, we may neglect the terms in Eq. (14) in which $D$ is multiplied by the magnitude of the field. In addition, the matrix $\rho_0(t)$ may be simplified since the energy of the molecule in the external radiation field will always be small compared to $kT$. It is shown in Appendix I that, to a sufficient approximation,

$$[\rho_0(t)]_{mn} = \rho_m{}^{(0)}\delta_{mn} + (\rho_m{}^{(0)} - \rho_n{}^{(0)})(V_{mn}/\hbar\omega_{mn})\cos\omega t. \tag{18}$$

With these simplifications, Eq. (14) becomes

$$(\partial/\partial t + i\omega_{mn} + 1/\tau)D_{mn}(t) = \omega(\rho_m{}^{(0)} - \rho_n{}^{(0)})(V_{mn}/\hbar\omega_{mn})\sin\omega t. \tag{19}$$

The steady-state solution of the latter equation is

$$D_{mn}(t) = \frac{\omega}{\omega - \omega_{mn} + i/\tau}(\rho_n{}^{(0)} - \rho_m{}^{(0)}) \times (V_{mn}/2\hbar\omega_{mn})e^{-i\omega t} + \frac{\omega}{\omega - \omega_{mn} - i/\tau}(\rho_n{}^{(0)} - \rho_m{}^{(0)}) \times (V_{mn}/2\hbar\omega_{mn})e^{i\omega t}, \tag{20}$$

whence

$$\bar{\rho}_{mn}(t)=\rho_m{}^{(0)}\delta_{mn}+\left(\frac{\omega}{\omega-\omega_{mn}+i/\tau}-1\right)\times(\rho_n{}^{(0)}-\rho_m{}^{(0)})(V_{mn}/2\hbar\omega_{mn})e^{-i\omega t}+\left(\frac{\omega}{\omega-\omega_{mn}-i/\tau}-1\right)\times(\rho_n{}^{(0)}-\rho_m{}^{(0)})(V_{mn}/2\hbar\omega_{mn})e^{i\omega t},\quad(21)$$

and, finally,

$$\mathbf{p}(t)=\sum_{m,n}\mathbf{p}_{nm}\bar{\rho}_{mn}(t)=\sum_{m,n}\mathbf{p}_{nm}\mathbf{p}_{mn}\cdot\mathbf{F}e^{-i\omega t}\times\left(1-\frac{\omega}{\omega-\omega_{mn}+i/\tau}\right)\frac{\rho_n{}^{(0)}-\rho_m{}^{(0)}}{2\hbar\omega_{mn}}+\sum_{m,n}\mathbf{p}_{mn}\mathbf{p}_{nm}\cdot\mathbf{F}e^{i\omega t}\times\left(1-\frac{\omega}{\omega-\omega_{mn}-i/\tau}\right)\frac{\rho_n{}^{(0)}-\rho_m{}^{(0)}}{2\hbar\omega_{mn}}\quad(22)$$

in which the indices $m$ and $n$ have been interchanged in the second term. In virtue of the spherical symmetry of an isolated molecule, it is possible to replace $\mathbf{p}_{nm}\mathbf{p}_{mn}\cdot\mathbf{F}$ with $(1/3)\mathbf{p}_{nm}\cdot\mathbf{p}_{mn}\mathbf{F}$, or $(1/3)|\mathbf{p}_{mn}|^2\mathbf{F}$. We therefore obtain a formula for $\mathbf{p}(t)$ of the form (2), with

$$\chi=\sum_{m,n}\tfrac{1}{3}|\mathbf{p}_{mn}|^2\left(1-\frac{\omega}{\omega-\omega_{mn}+i/\tau}\right)\frac{\rho_n{}^{(0)}-\rho_m{}^{(0)}}{\hbar\omega_{mn}}=\sum_{m,n}\tfrac{1}{6}|\mathbf{p}_{mn}|^2\left(2-\frac{\omega}{\omega-\omega_{mn}+i/\tau}-\frac{\omega}{\omega+\omega_{mn}+i/\tau}\right)\frac{\rho_n{}^{(0)}-\rho_m{}^{(0)}}{\hbar\omega_{mn}}.\quad(23)$$

In particular,

$$Im\chi=\sum_{m,n}\frac{1}{6\hbar}|\mathbf{p}_{mn}|^2\frac{\omega}{\omega_{mn}}\left[\frac{1/\tau}{(\omega-\omega_{mn})^2+1/\tau^2}+\frac{1/\tau}{(\omega+\omega_{mn})^2+1/\tau^2}\right](\rho_n{}^{(0)}-\rho_m{}^{(0)}),\quad(24)$$

and

$$\alpha=\frac{2\pi}{3}\frac{\omega^2}{c}\frac{N}{kT}\sum_{m,n}|\mathbf{p}_{mn}|^2\left[\frac{1/\tau}{(\omega-\omega_{mn})^2+1/\tau^2}+\frac{1/\tau}{(\omega+\omega_{mn})^2+1/\tau^2}\right]\frac{1-e^{-\hbar\omega_{mn}/kT}}{\hbar\omega_{mn}/kT}\rho_n{}^{(0)},\quad(25)$$

which is essentially the final result of Van Vleck and Weisskopf.

An important specialization of this formula refers to the situation in which the resonant frequencies of the molecule are widely spaced, in comparison with the resonance width $\Delta\omega=1/\tau$. If the applied frequency lies in the vicinity of a particular resonance frequency, $\omega_0=(E_a-E_b)/\hbar$, only terms involving $\omega-\omega_0$ will appreciably contribute to the summation in Eq. (25). However, the energy levels $E_a$ and $E_b$ will be degenerate, in general, and we must further specify the states contributing to (25) by a degeneracy index, say $\kappa$ and $\lambda$, respectively. Thus

$$\alpha=\frac{4\pi}{3}\frac{\omega^2}{c}\frac{N}{kT}\Big[\sum_{\kappa,\lambda}|\mathbf{p}_{a\kappa,b\lambda}|^2\Big]\frac{1/\tau}{(\omega-\omega_0)^2+1/\tau^2}\times\frac{1-e^{-\hbar\omega_0/kT}}{\hbar\omega_0/kT}\frac{e^{-E_b/kT}}{\sum_m e^{-E_m/kT}}.\quad(26)$$

## SATURATION

We shall now extend our treatment to include the effects of a strong monochromatic radiation field. The discussion will, however, be limited to the situation embodied in Eq. (26), that is, where $\omega-\omega_0$ is comparable with $1/\tau$, but all other such differences, $\omega-\omega_{mn}$, are large compared to $1/\tau$. In addition, we shall impose an inocuous restriction on the field strength that amounts to requiring that the external radiation field be small compared to the internal molecular field; we ignore any time variation of $\mathbf{p}(t)$ that involves harmonics of the fundamental frequency $\omega$. This restriction is expressed by the assumption

$$D_{mn}(t)=R_{mn}+P_{mn}{}^{(+)}e^{-i\omega t}+P_{mn}{}^{(-)}e^{i\omega t},\quad(27)$$

which is to be inserted in the equation of motion

(14). The resultant set of equations are:

$$P_{mn}^{(+)}=\frac{1}{2\hbar(\omega_{mn}-\omega-i/\tau)}\left[\frac{\omega}{\omega_{mn}}V_{mn}(\rho_m^{(0)}-\rho_n^{(0)})-\sum_k(V_{mk}R_{kn}-R_{mk}V_{kn})\right],$$

$$P_{mn}^{(-)}=-\frac{1}{2\hbar(\omega_{mn}+\omega-i/\tau)}\left[\frac{\omega}{\omega_{mn}}V_{mn}(\rho_m^{(0)}-\rho_n^{(0)})+\sum_k(V_{mk}R_{kn}-R_{mk}V_{kn})\right],\quad (28)$$

$$R_{mn}=-\frac{1}{2\hbar(\omega_{mn}-i/\tau)}\sum_k[V_{mk}(P_{kn}^{(+)}+P_{kn}^{(-)})-(P_{mk}^{(+)}+P_{mk}^{(-)})V_{kn}].$$

With the impressed frequency in approximate resonance with the transition frequency $\omega_0=(E_a-E_b)/\hbar$, only the matrix elements $P_{ab}^{(+)}$, $P_{ba}^{(-)}$, $R_{aa}$, and $R_{bb}$ are of significance, if we suppose the energy levels $a$ and $b$ to be non-degenerate. This statement is accurate to the extent that $\Delta\omega=1/\tau\ll\omega_{mn}$. The simultaneous equations (28) then reduce to

$$P_{ab}^{(+)}=-\frac{V_{ab}}{2\hbar(\omega-\omega_0+i/\tau)}\left[\frac{\omega}{\omega_0}(\rho_a^{(0)}-\rho_b^{(0)})+(R_{aa}-R_{bb})\right],$$

$$P_{ba}^{(-)}=-\frac{V_{ba}}{2\hbar(\omega-\omega_0-i/\tau)}\left[\frac{\omega}{\omega_0}(\rho_a^{(0)}-\rho_b^{(0)})+(R_{aa}-R_{bb})\right],\quad (29)$$

$$R_{aa}-R_{bb}=-i(\tau/\hbar)[V_{ab}P_{ba}^{(-)}-P_{ab}^{(+)}V_{ba}],$$

$$R_{aa}+R_{bb}=0.$$

The solution of these equations is

$$R_{aa}-R_{bb}=-\frac{\omega}{\omega_0}\frac{|V_{ab}|^2/\hbar^2}{(\omega-\omega_0)^2+1/\tau^2+|V_{ab}|^2/\hbar^2}\times(\rho_a^{(0)}-\rho_b^{(0)}),\quad (30)$$

$$P_{ab}^{(+)}=P_{ba}^{(-)*}=-\frac{V_{ab}}{2\hbar}\frac{\omega}{\omega_0}\frac{\omega-\omega_0-i/\tau}{(\omega-\omega_0)^2+1/\tau^2+|V_{ab}|^2/\hbar^2}\times(\rho_a^{(0)}-\rho_b^{(0)}).$$

In computing the average dipole moment,

$$\mathbf{p}(t)=Sp\mathbf{p}\bar{\rho}(t)=Sp\mathbf{p}\rho_0(t)+Sp\mathbf{p}D(t),\quad (31)$$

we may ignore the time-dependent contribution of $\rho_0$, for this is neither resonant nor contributes to the absorption. Hence

$$\mathbf{p}(t)=Sp\mathbf{p}D(t)=2Re\mathbf{p}_{ba}P_{ab}^{(+)}e^{-i\omega t}=Re\frac{1}{\hbar}\mathbf{p}_{ba}\mathbf{p}_{ab}\cdot\mathbf{F}e^{-i\omega t}\times\frac{\omega}{\omega_0}\frac{\omega-\omega_0-i/\tau}{(\omega-\omega_0)^2+1/\tau^2+|V_{ab}|^2/\hbar^2}\times(\rho_a^{(0)}-\rho_b^{(0)}).\quad (32)$$

The assumption of non-degeneracy must be relinquished at this point, for orientation degeneracy, at least, is necessarily present. However, our theory is easily extended if the component of the dipole moment in the direction of the field has vanishing matrix elements between different degenerate states. This is indeed realized in the most important situation of orientation degeneracy since the magnetic quantum number relative to the direction of the field is unaltered in any transition induced by the radiation. Under these circumstances the appropriate generalization of Eq. (32) is

$$\mathbf{p}(t)=Re\frac{1}{\hbar}\sum_\kappa|p_{ab}^\kappa|^2\mathbf{F}e^{-i\omega t}\times\frac{\omega}{\omega_0}\frac{\omega-\omega_0-i/\tau}{(\omega-\omega_0)^2+1/\tau^2+|p_{ab}^\kappa|^2F^2/\hbar^2}\times(\rho_a^{(0)}-\rho_b^{(0)}),\quad (33)$$

where $p_{ab}^\kappa$ designates a non-vanishing matrix element of the component of the dipole moment in the field direction, and $\kappa$ is a degeneracy index. This is of the desired form (2), with

$$Im\chi=\frac{1}{\hbar}\sum_\kappa|p_{ab}^\kappa|^2\times\frac{\omega}{\omega_0}\frac{1/\tau}{(\omega-\omega_0)^2+1/\tau^2+|p_{ab}^\kappa|^2F^2/\hbar^2}\times(\rho_b^{(0)}-\rho_a^{(0)}),\quad (34)$$

whence, finally

$$\alpha = 4\pi \frac{\omega^2}{c} \frac{N}{kT} \sum_\kappa |p_{ab}{}^\kappa|^2 \times \frac{1/\tau}{(\omega-\omega_0)^2 + 1/\tau^2 + |p_{ab}{}^\kappa|^2 F^2/\hbar^2} \times \frac{1-e^{-\hbar\omega_0/kT}}{\hbar\omega_0/kT} \frac{e^{-E_b/kT}}{\sum_m e^{-E_m/kT}}. \quad (35)$$

In the weak field limit, this result reduces to Eq. (26), as it must, since

$$\tfrac{1}{3} \sum_{\kappa,\lambda} |\mathbf{p}_{a\kappa,b\lambda}|^2 = \sum_\kappa |p_{ab}{}^\kappa|^2. \quad (36)$$

When, however, $|p_{ab}{}^\kappa|F$ becomes comparable with $\hbar/\tau$, the absorption line begins to widen, and the peak absorption coefficient correspondingly decreases. Indeed, in the limit $|p_{ab}{}^\kappa|F \gg \hbar/\tau$ and $\hbar(\omega-\omega_0)$, but $|p_{ab}{}^\kappa|F \ll \hbar(\omega-\omega_{mn})$, the power absorbed per unit volume exhibits a saturation phenomenon, approaching a value independent of the field strength, namely,

$$P = \alpha c F^2/8\pi \to \frac{(\hbar\omega)^2}{2\tau} \frac{N}{kT} g \frac{1-e^{-\hbar\omega_0/kT}}{\hbar\omega_0/kT} \frac{e^{-E_b/kT}}{\sum_m e^{-E_m/kT}}. \quad (37)$$

In this formula, $g$ denotes the total number of degenerate transitions contributing to the absorption line.

It is well to note, finally, that our essential results, Eqs. (29) and (30), permit an elementary interpretation. In the immediate vicinity of resonance, $\omega/\omega_0$ may be replaced by unity, and the first of Eq. (29) can be written

$$P_{ab}{}^{(+)} = -\frac{V_{ab}}{2\hbar(\omega-\omega_0+i/\tau)} [\bar{\rho}_{aa} - \bar{\rho}_{bb}]. \quad (38)$$

It is apparent from this form that the broadening of the absorption line at high power levels is not to be attributed to any intrinsic modification of the line shape, but rather to a frequency dependent alteration of the populations of the two levels. The net rate of absorption is the difference between the rates of true absorption and stimulated emission, with the common transition probability for the two processes possessing the characteristic frequency dependence $1/[(\omega-\omega_0)^2 + 1/\tau^2]$. The frequency dependence of the population difference is contained in the first of Eq. (30), which may be written

$$\bar{\rho}_{aa} - \bar{\rho}_{bb} = \frac{1/\tau}{1/\tau + w} (\rho_a{}^{(0)} - \rho_b{}^{(0)}), \quad (39)$$

where

$$w = \frac{1}{\hbar^2} \frac{|V_{ab}|^2/\tau}{(\omega-\omega_0)^2 + 1/\tau^2}. \quad (40)$$

To interpret this result, we note that $1/\tau$ is the rate at which the population difference tends toward its thermal equilibrium value, as induced by collisions, whence $w$ may be considered the rate at which radiative transitions tend to produce equality in population. That this is a self-consistent interpretation may be seen on computing the power absorbed per unit volume in accordance with the detailed radiative processes:

$$P = \tfrac{1}{2} w\hbar\omega(\bar{\rho}_{aa} - \bar{\rho}_{bb})N = \tfrac{1}{2}\hbar\omega \frac{w(1/\tau)}{w+1/\tau} (\rho_a{}^{(0)} - \rho_b{}^{(0)})N \quad (41)$$

which, with appropriate modifications for degeneracy, agrees precisely with Eq. (35). In writing the last equation, the transition probability for absorption or stimulated emission has been placed equal to $\tfrac{1}{2}w$, since both processes contribute equally to the rate at which radiative transitions tend to decrease the population difference.

We have enjoyed several conversations on this subject with J. H. Van Vleck.

### APPENDIX I

To simplify the density matrix $\rho_0(t)$, in accordance with the smallness of $\mathbf{p}\cdot\mathbf{F}/kT$, we consider the problem of expanding the operator $\exp(A+B)$ in powers of $B$. We shall be content with the first two terms of the expansion. It is convenient to introduce

$$F(\lambda) = e^{\lambda(A+B)} \quad (I.1)$$

which satisfies the differential equation

$$d/d\lambda F(\lambda) = (A+B)F(\lambda) \quad (I.2)$$

and the initial condition

$$F(0)=1. \tag{I.3}$$

On performing the transformation

$$F(\lambda)=e^{\lambda A}G(\lambda), \quad G(0)=1, \tag{I.4}$$

we encounter the differential equation

$$d/d\lambda G(\lambda)=e^{-\lambda A}Be^{\lambda A}G(\lambda) \tag{I.5}$$

which is equivalent to the integral equation

$$G(\lambda)=1+\int_0^\lambda e^{-\lambda' A}Be^{\lambda' A}G(\lambda')d\lambda'. \tag{I.6}$$

The latter may be solved by successive substitution, which generates a series of which the first two terms are

$$G(\lambda)=1+\int_0^\lambda e^{-\lambda' A}Be^{\lambda' A}d\lambda'+\cdots. \tag{I.7}$$

Hence

$$e^{(A+B)}=F(1)=e^A+\int_0^1 e^{(1-\lambda)A}Be^{\lambda A}d\lambda+\cdots. \tag{I.8}$$

The matrix elements of $\exp(A+B)$, in a scheme with $A$ diagonal, are

$$[e^{(A+B)}]_{a'a''}=e^{a'}\delta_{a'a''}+\frac{e^{a'}-e^{a''}}{a'-a''}B_{a'a''}+\cdots. \tag{I.9}$$

Thus, with $A=-H_0/kT$, $B=-(V/kT)\cos\omega t$,

$$[e^{-H(t)/kT}]_{mn}=e^{-E_m/kT}\delta_{mn}+\frac{e^{-E_m/kT}-e^{-E_n/kT}}{E_m-E_n}V_{mn}\cos\omega t, \tag{I.10}$$

and

$$1/C=Spe^{-H(t)/kT}=Spe^{-H_0/kT}+\frac{1}{kT}Spe^{-H_0/kT}\mathbf{p}\cdot\mathbf{F}\cos\omega t \tag{I.11}$$
$$=1/C^{(0)}.$$

The latter statement is valid since the average dipole moment of the molecule vanishes, in the absence of an external field. We finally obtain

$$[\rho_0(t)]_{mn}=\rho_m{}^{(0)}\delta_{mn}+(\rho_m{}^{(0)}-\rho_n{}^{(0)})(V_{mn}/\hbar\omega_{mn})\cos\omega t, \tag{I.12}$$

which is Eq. (18) of the text.

## APPENDIX II

The situation treated in the text is that of a molecule with widely spaced resonant frequencies. We now treat the opposite extreme, a one-dimensional harmonic oscillator, and we shall show that no saturation effect occurs. The electric dipole moment of a charged particle confined to the $x$ axis, $p=ex$, has the following non-vanishing matrix elements:

$$p_{n,n-1}=p_{n-1,n}=e\left[\frac{n\hbar}{2m\omega_0}\right]^{\frac{1}{2}}, \quad n=1,2,3,\cdots, \tag{II.1}$$

where $m$ and $\omega_0$ are the mass and natural frequency of the particle. According to the third equation of the set (28), only the diagonal elements, $R_{nn}$, are significant if $\omega_0\gg 1/\tau$. The properties of the dipole moment matrix then restrict the other non-vanishing matrix elements of $D(t)$ to

$$\overset{(+)}{P}_{n,n-1}=\overset{(-)}{P}_{n-1,n}{}^*=\frac{V_{n,n-1}}{2\hbar(\omega_0-\omega-i/\tau)}\left[\frac{\omega}{\omega_0}(\rho_n{}^{(0)}-\rho_{n-1}{}^{(0)})+R_{n,n}-R_{n-1,n-1}\right],$$
$$\overset{(+)}{P}_{n-1,n}=\overset{(-)}{P}_{n,n-1}{}^*=-\frac{V_{n-1,n}}{2\hbar(\omega_0+\omega+i/\tau)}\times\left[\frac{\omega}{\omega_0}(\rho_n{}^{(0)}-\rho_{n-1}{}^{(0)})-R_{n,n}+R_{n-1,n-1}\right]. \tag{II.2}$$

The average dipole moment may now be computed as

$$p(t)=2Re\sum_{n=1}^{\infty}(p_{n-1,n}\overset{(+)}{P}_{n,n-1}+p_{n,n-1}\overset{(+)}{P}_{n-1,n})e^{-i\omega t}$$
$$=Re\sum_{n=1}^{\infty}\frac{|p_{n,n-1}|^2}{\hbar(\omega_0-\omega-i/\tau)}Ee^{-i\omega t}\times\left[\frac{\omega}{\omega_0}(\rho_{n-1}{}^{(0)}-\rho_n{}^{(0)})+R_{n-1,n-1}-R_{n,n}\right]$$
$$-Re\sum_{n=1}^{\infty}\frac{|p_{n,n-1}|^2}{\hbar(\omega_0+\omega+i/\tau)}Ee^{-i\omega t}\times\left[\frac{\omega}{\omega_0}(\rho_{n-1}{}^{(0)}-\rho_n{}^{(0)})-R_{n-1,n-1}+R_{n,n}\right], \tag{II.3}$$

whence

$$Im\chi=\frac{e^2}{2m\omega_0}\frac{1/\tau}{(\omega-\omega_0)^2+1/\tau^2}\times\sum_{n=1}^{\infty}n\left[\frac{\omega}{\omega_0}(\rho_{n-1}{}^{(0)}-\rho_n{}^{(0)})+R_{n-1,n-1}-R_{n,n}\right]+\frac{e^2}{2m\omega_0}\frac{1/\tau}{(\omega+\omega_0)^2+1/\tau^2}\times\sum_{n=1}^{\infty}n\left[\frac{\omega}{\omega_0}(\rho_{n-1}{}^{(0)}-\rho_n{}^{(0)})-R_{n-1,n-1}+R_{n,n}\right]. \quad \text{(II.4)}$$

However

$$\sum_{n=1}^{\infty}n(\rho_{n-1}{}^{(0)}-\rho_n{}^{(0)})=\sum_{n=0}^{\infty}\rho_n{}^{(0)}=Sp\rho^{(0)}=1, \quad \text{(II.5)}$$

and

$$\sum_{n=1}^{\infty}n(R_{n-1,n-1}-R_{n,n})=SpD=Sp\bar{\rho}-Sp\rho_0=0. \quad \text{(II.6)}$$

We thus obtain

$$Im\chi=\frac{e^2}{2m\omega_0}\frac{\omega}{\omega_0}\left[\frac{1/\tau}{(\omega-\omega_0)^2+1/\tau^2}+\frac{1/\tau}{(\omega+\omega_0)^2+1/\tau^2}\right], \quad \text{(II.7)}$$

which leads, finally, to

$$\alpha=2\pi\frac{e^2}{mc}\left(\frac{\omega}{\omega_0}\right)^2N\left[\frac{1/\tau}{(\omega-\omega_0)^2+1/\tau^2}+\frac{1/\tau}{(\omega+\omega_0)^2+1/\tau^2}\right] \quad \text{(II.8)}$$

in complete agreement with Eq. (17) of reference (1). It is evident that the absorption coefficient of an oscillator is independent of the distribution of molecules among the states of the system and is therefore unaffected by the redistribution induced by the radiation field.

# On the Charge Independence of Nuclear Forces

JULIAN SCHWINGER
*Harvard University, Cambridge, Massachusetts*

(Received January 5, 1950)

It is shown that the small difference between the neutron-proton and proton-proton interactions in the $^1S$ state can be accounted for by magnetic forces, provided the nuclear interaction is described by the Yukawa potential. A variational principle is used to facilitate the comparison of the interactions, and for the determination of the effect of magnetic forces. A short discussion of variational principles is presented in the Appendix.

It has long been known that the nuclear potentials operating in the $^1S$ state between a pair of protons, and a neutron and proton, are of approximately equal depth if range and shape are assumed identical.[1] There exists, however, a small but real difference, the neutron-proton potential depth exceeding that of two protons by several percent.[2] It will be shown in this note that magnetic interaction between nucleons adequately accounts for this difference, if the nuclear potential resembles that of Yukawa. The evidence in favor of the fundamental charge independence of nuclear forces is thereby greatly strengthened.

The comparison between neutron-proton and proton-proton interactions inferred from scattering data, as well as the determination of the effect of magnetic forces, is facilitated by a formulation of the scattering problem in terms of a variational principle.[3] A useful variational expression for the $^1S$ phase shift $\delta$, which characterizes the effect of non-Coulomb forces between protons, is the following,

$$\lim_{\substack{R\to\infty \\ \epsilon\to 0}}\left\{\int_0^R [(du/dr)^2-(M/\hbar^2)(E-(e^2/r)-V_{pp})u^2]dr - \int_\epsilon^R [(dv/dr)^2-(M/\hbar^2)(E-(e^2/r))v^2]dr - (1/b_0)\log(\gamma^2(\epsilon/b_0))v^2(0)\right\}$$
$$=\left[k\cot\delta\frac{2\pi\eta}{e^{2\pi\eta}-1}+\frac{1}{b_0}\left(Re\frac{\Gamma'(i\eta)}{\Gamma(i\eta)}-\log\eta\right)\right]v^2(0), \tag{1}$$

where $\log\gamma=C=0.5772$, $b_0=\hbar^2/Me^2$, $\eta=1/(2kb_0)$, and the other symbols have their standard significance. This expression for $\cot\delta$ has the property that it is stationary with respect to independent variations of $u(r)$ and $v(r)$, subject only to the restrictions

$$u(0)=0,\quad u(r)-v(r)\to 0,\quad r\to\infty. \tag{2}$$

The stationary value is attained for functions that satisfy the differential equations

$$[(d^2/dr^2)+(M/\hbar^2)(E-(e^2/r)-V_{pp})]u=0,$$
$$[(d^2/dr^2)+(M/\hbar^2)(E-(e^2/r))]v=0, \tag{3}$$

in addition to (2). Furthermore, this stationary value yields the correct nuclear phase shift. The verification of these assertions will be performed in the Appendix.

The advantage of this formula for the phase shift is that it involves the behavior of the wave function $u$ only within the region of nuclear interaction, since $u\to v$ as $V_{pp}\to 0$, according to (2) and (3). In addition, the replacement of $u$ and $v$ within this region by slightly different functions will produce only a small error in the calculated value of $\cot\delta$, in virtue of the stationary property. A convenient set of functions for this purpose is provided by the solutions of the equations

$$((d^2/dr^2)-(M/\hbar^2)V_{pp})u_0=0;\quad (d^2/dr^2)v_0=0,$$
$$u_0(0)=0;\quad u_0-v_0\to 0,\quad r\to\infty, \tag{4}$$

that is, (2) and (3) with $E=0$, and $e=0$. We are thus going to employ trial functions which describe the zero energy scattering of two protons arising from non-Coulomb interactions alone. That this is a useful approximation stems from the large magnitude of the nuclear potential within the range of nuclear forces, in comparison with the Coulomb potential and the energy of relative motion, provided the latter is suitably restricted. From the comparison of the resulting phase shift formula with the empirical data, we shall be able to infer the hypothetical scattering properties of two protons interacting by non-Coulomb forces, which can be contrasted with the corresponding observed properties of a neutron and proton, thus affording a direct comparison of the two non-Coulomb interactions. We shall, finally, be able to compute the effect of a magnetic interaction addition to the nuclear potential by simply altering that potential as it appears explicitly in the variational expression for the phase shift. The implied alteration in the wave function has no first-order effect, in view of the stationary property.

On inserting the approximate functions $u_0$ and $v_0$ in (1), we obtain

$$\frac{\pi}{e^{2\pi\eta}-1}\cot\delta+Re\frac{\Gamma'(i\eta)}{\Gamma(i\eta)}-\log\eta=b_0\left[\frac{1}{a}+\frac{1}{2}k^2r_e\right], \tag{5}$$

[1] Breit, Condon, and Present, Phys. Rev. **50**, 825 (1936).
[2] Breit, Hoisington, Share, and Thaxton, Phys. Rev. **55**, 1103 (1939).
[3] J. Schwinger, Lectures on nuclear physics, Harvard, 1947; Phys. Rev. **72**, 724 (1947). J. M. Blatt, Phys. Rev. **74**, 92 (1948); J. M. Blatt and J. D. Jackson, Phys. Rev. **76**, 18 (1949).



Reprinted from *The Physical Review* 78, 135–139 (1950).

where

$$r_e=\frac{2}{v_0^2(0)}\int_0^\infty (v_0^2-u_0^2)dr, \tag{6}$$

and

$$\frac{1}{a}=\frac{1}{a_{pp}}-\frac{1}{b_0}\left[\frac{1}{v_0^2(0)}\int_\epsilon^\infty \frac{v_0^2-u_0^2}{r}dr+\log\left(\gamma^2\frac{\epsilon}{b_0}\right)\right]_{\epsilon\to 0}$$
$$=\frac{1}{a_{pp}}+\frac{1}{b_0}\int_0^\infty \log\left(\gamma^2\frac{r}{b_0}\right)\frac{d}{dr}\left(\frac{v_0^2-u_0^2}{v_0^2(0)}\right)dr. \tag{7}$$

Here $a_{pp}$ is the amplitude for the zero energy scattering of two protons by non-Coulomb forces, which replaces the phase shift in the form assumed by (1) with $e$ and $E$ equal to zero:

$$\lim_{R\to\infty}\left\{\int_0^R\left[\left(\frac{du_0}{dr}\right)^2+\frac{M}{\hbar^2}V_{pp}u_0^2\right]dr - \int_0^R\left(\frac{dv_0}{dr}\right)^2 dr\right\}=\frac{1}{a_{pp}}v_0^2(0). \tag{8}$$

Equation (5) represents an expression for the phase shift in terms of two parameters, $a$ and $r_e$, which completely characterize the effect of short range non-Coulomb interactions within a restricted energy domain. That the experimental data permits such a representation is demonstrated by plotting the left side of (5) as a function of proton kinetic energy and observing that the ensuing points form an accurate straight line.[3,4] The result of such a treatment of the data obtained with Van de Graaff generators[5] is expressed by

$$r_e=(2.65\pm 0.07)10^{-13}\text{ cm},\quad b_0/a=3.755\pm 0.024. \tag{9}$$

Although the constant $a$ is obtained directly from experiment, the determination of the scattering amplitude $a_{pp}$ requires some assumption concerning the nuclear potential. This is indicated by the formula

$$1/a_{pp}=1/a-1/b_0[\log(b_0/\bar{r})-2C], \tag{10}$$

since the average distance $\bar{r}$, defined by

$$\int_0^\infty \log\frac{r}{\bar{r}}\frac{d}{dr}(v_0^2-u_0^2)dr=0, \tag{11}$$

involves the detailed behavior of the wave functions. A good initial approximation to $\bar{r}$ is provided by the wave functions associated with $1/a_{pp}=0$, that is, a state of zero binding energy. We shall consider two potentials, a rectangular well and a Yukawa potential. For the rectangular well, of range $r_0$, the functions $u_0$ and $v_0$ are

$$u_0=\sin(\pi/2)r/r_0,\quad r<r_0;\quad u_0=1,\quad r>r_0, \tag{12}$$
$$v_0=1,$$

whence

$$\log\frac{r_e}{\bar{r}}=\int_0^1\frac{\sin^2(\pi/2)x}{x}dx=0.8241, \tag{13}$$

in which we have replaced $r_0$ by $r_e$, since the two are identical, according to (6) and (12). The approximate but accurate function

$$u_0=1-e^{-\beta r},\quad v_0=1, \tag{14}$$

will be used for the Yukawa potential, in which $\beta$ is related to the measured $r_e$ by

$$1/\beta=\tfrac{1}{3}r_e. \tag{15}$$

On performing the integrations in (11), we find

$$\log(r_e/\bar{r})=\log\tfrac{3}{2}+C=0.9827. \tag{16}$$

From these approximate evaluations of $\bar{r}$ we obtain, for the rectangular and Yukawa potentials, respectively,

$$b_0/a_{pp}=3.755-\log(b_0/r_e)+0.330$$
$$=1.699, \tag{17}$$

and

$$b_0/a_{pp}=3.755-\log(b_0/r_e)+0.172$$
$$=1.541. \tag{18}$$

Having computed these approximate values of $b_0/a_{pp}$, we may return to the evaluation of $\bar{r}$, according to (11), and insert wave functions appropriate to a finite scattering amplitude. This is easily done for the rectangular potential, and the result, correct to first order in $r_e/a_{pp}$, is

$$\log\frac{r_e}{\bar{r}}=\left(1+2\frac{r_e}{a_{pp}}\right)\int_0^1\frac{\sin^2(\pi/2)x}{x}dx-2\frac{r_e}{a_{pp}}$$
$$=0.7692. \tag{19}$$

For the Yukawa potential, we employ the approximate wave function

$$u_0=1+(r/a_{pp})-e^{-\beta r},\quad v_0=1+(r/a_{pp}), \tag{20}$$

and obtain

$$\log(r_e/\bar{r})=\log\tfrac{3}{2}+C-(2/9)(r_e/a_{pp})=0.9512. \tag{21}$$

We may now compute corrected values of $b_0/a_{pp}$:

$$b_0/a_{pp}=1.754\pm 0.035\text{ (rectangular potential)}$$
$$=1.573\pm 0.035\text{ (Yukawa potential)}. \tag{22}$$

The probable error is that associated with the inaccuracy of $1/a$ and $r_e$. Our results[6] are, finally, to be compared with the $^1S$ neutron-proton zero energy scattering amplitude which, in these units ($b_0=2.881\times 10^{-12}$

[4] L. Landau and J. Smorodinsky, J. Phys. U.S.S.R. 8, 154 (1944).
[5] J. D. Jackson, thesis, M.I.T., 1949.

[6] The numerical values for $1/a_{pp}$ differ slightly from those obtained by H. A. Bethe, Phys. Rev. 76, 38 (1949).

cm), is given by[7]

$$b_0/a_{np}=1.217\pm 0.003. \tag{23}$$

The discrepancy between (22) and (23) appears to be real and indicates that the neutron-proton interaction is somewhat stronger than the non-Coulomb proton-proton interaction.

We shall now show that the magnetic interaction between nucleons produces just such a difference, thereby implying that specifically nuclear forces are accurately charge independent. A suitable basis for such a discussion is the following approximate Dirac Hamiltonian for two nucleons with charges $e_1$, $e_2$ and anomalous magnetic moments $\mu_1'$, $\mu_2'$,

$$\begin{aligned} H=&c\boldsymbol{\alpha}_1\cdot\mathbf{p}_1+\beta_1Mc^2+c\boldsymbol{\alpha}_2\cdot\mathbf{p}_2+\beta_2Mc^2+V\\ &+\frac{e_1e_2}{r}\left(1-\frac{1}{2}\boldsymbol{\alpha}_1\cdot\boldsymbol{\alpha}_2-\frac{1}{2}\frac{\boldsymbol{\alpha}_1\mathbf{r}\ \boldsymbol{\alpha}_2\cdot\mathbf{r}}{r^2}\right)\\ &-e_1\boldsymbol{\alpha}_1\cdot\frac{\mu_2'\beta_2\boldsymbol{\sigma}_2\times\mathbf{r}}{r^3}-e_2\boldsymbol{\alpha}_2\cdot\frac{\mu_1'\beta_1\boldsymbol{\sigma}_1\times(-\mathbf{r})}{r^3}\\ &-\frac{\mu_1'\mu_2'}{r^3}\left(3\frac{\beta_1\boldsymbol{\sigma}_1\cdot\mathbf{r}\beta_2\boldsymbol{\sigma}_2\cdot\mathbf{r}}{r^2}-\beta_1\boldsymbol{\sigma}_1\cdot\beta_2\boldsymbol{\sigma}_2\right)\\ &-\frac{8\pi}{3}\mu_1'\mu_2'\beta_1\boldsymbol{\sigma}_1\cdot\beta_2\boldsymbol{\sigma}_2\delta(\mathbf{r})+i\mu_1'\beta_1\boldsymbol{\alpha}_1\cdot e_2\frac{\mathbf{r}}{r^3}\\ &+i\mu_2'\beta_2\boldsymbol{\alpha}_2\cdot e_1\frac{(-\mathbf{r})}{r^3}. \end{aligned} \tag{24}$$

Here $\mathbf{r}=\mathbf{r}_1-\mathbf{r}_2$, and the successive terms following the sum of the free particle Hamiltonians describe the nuclear force interaction energy; the mutual energy of the charges, correct to order $(v/c)^2$; the energy of each charge in the magnetic field produced by the anomalous magnetic moment of the other particle; the mutual potential energy of the additional magnetic moments, separated into a tensor and a spherically symmetrical interaction; and finally the energy which the electric dipole of each particle, associated with the anomalous magnetic moment, possesses in the electric field of the other particles.

By eliminating the small components of the Dirac wave function, the Dirac Hamiltonian can be replaced by a Pauli-type Hamiltonian, in which the Coulomb and nuclear potentials are supplemented by a magnetic interaction energy. The latter can be written

$$\begin{aligned} V^{(\mathrm{mag})}=&\frac{e_1e_2}{M^2c^2}\mathbf{p}\cdot\frac{1}{r}\mathbf{p}-4\pi\frac{e_1\hbar}{2Mc}\frac{e_2\hbar}{2Mc}\delta(\mathbf{r})\\ &-4\pi\left(\mu_1'\frac{e_2\hbar}{2Mc}+\mu_2'\frac{e_1\hbar}{2Mc}\right)\delta(\mathbf{r})-\frac{8\pi}{3}\mu_1\mu_2\boldsymbol{\sigma}_1\cdot\boldsymbol{\sigma}_2\delta(\mathbf{r}), \end{aligned} \tag{25}$$

in which various terms, ineffective in $S$ states, have been discarded. The successive contributions to $V^{(\mathrm{mag})}$ have the following significance. The first term represents the magnetic interaction between the convection currents associated with the charges, as specialized to the center of gravity system ($\mathbf{p}_1=-\mathbf{p}_2=\mathbf{p}$); the second term contains the sum of the energies which the electric dipole moment of each particle, accompanying the Dirac magnetic moment, possesses in the electric field of the other particle; the third term is the analogous electric energy attributable to the anomalous magnetic moments, differing fundamentally from the preceding term by the absence of the Thomas factor; and finally we have the interaction energy of the total spin magnetic moments, $\mu=\mu'+e\hbar/2Mc$. We are actually interested in the $^1S$ neutron-proton and proton-proton magnetic interaction energies. These can be written

$$V_{np}{}^{(\mathrm{mag})}=8\pi\mu_n(\mu_p-\tfrac{1}{2})\mu_0{}^2\delta(\mathbf{r}), \tag{26}$$

and

$$V_{pp}{}^{(\mathrm{mag})}=8\pi(\mu_p{}^2-\mu_p+\tfrac{1}{2})\mu_0{}^2\delta(\mathbf{r})+4\mu_0{}^2\frac{1}{\hbar^2}\mathbf{p}\cdot\frac{1}{r}\mathbf{p}, \tag{27}$$

in which $\mu_n=-1.913$, and $\mu_p=2.793$ are the magnetic moments of neutron and proton in units of $\mu_0=e\hbar/2Mc$ the nuclear magneton. It will be noted that the neutron-proton magnetic interaction is attractive, while that for two protons is repulsive.

The magnetic part of the neutron-proton interaction produces a contribution to $1/a_{np}$ which, according to the analog of (8), is given by

$$\begin{aligned} \delta\left(\frac{1}{a_{np}}\right)&=\frac{M}{\hbar^2}\frac{1}{v_0{}^2(0)}\int_0^\infty V_{np}{}^{(\mathrm{mag})}u_0{}^2dr\\ &=\tfrac{1}{2}\mu_n(\mu_p-\tfrac{1}{2})\frac{e^2}{Mc^2}\left[\frac{u_0'(0)}{v_0(0)}\right]^2, \end{aligned} \tag{28}$$

since $\delta(\mathbf{r})=\delta(r)/4\pi r^2$. The ratio $u_0'(0)/v_0(0)$ can be computed rigorously for the rectangular well, and with reasonable accuracy from the wave function (20) for the Yukawa potential. The results, including first-order corrections in $r_e/a_{np}$, are

$$\begin{aligned} \frac{u_0'(0)}{v_0(0)}&=\frac{\pi}{2r_e}\left(1+\frac{r_e}{a_{np}}\right) \quad \text{(rectangular potential)}\\ &=\frac{3}{r_e}\left(1+\frac{7}{9}\frac{r_e}{a_{np}}\right) \quad \text{(Yukawa potential),} \end{aligned} \tag{29}$$

from which we obtain

$$\begin{aligned} \delta\left(\frac{b_0}{a_{np}}\right)&=\frac{\pi^2}{8}\mu_n(\mu_p-\tfrac{1}{2})\left(1+2\frac{r_e}{a_{np}}\right)\left(\frac{\hbar/Mc}{r_e}\right)^2\\ &=-(0.042\pm 0.002), \end{aligned} \tag{30}$$

[7] This value is obtained from the free proton cross-section measurement of E. Melkonian, Phys. Rev. 76, 1744 (1949), and the recent determination of the coherent scattering amplitude by Hughes, Burgy and Ringo, Phys. Rev. 77, 291 (1950).

and

$$\delta\left(\frac{b_0}{a_{np}}\right)=\frac{9}{2}\mu_n(\mu_p-\tfrac{1}{2})\left(1+\frac{14}{9}\frac{r_e}{a_{np}}\right)\left(\frac{\hbar/Mc}{r_e}\right)^2 \tag{31}$$
$$=-(0.145\pm0.007),$$

for the rectangular and Yukawa potentials, respectively. The numerical values are obtained by the assumption that $r_e$ is charge independent, which is consistent with the present inaccurate determinations of the neutron-proton $r_e$ in the $^1S$ state. We may now compute the hypothetical scattering amplitude $a_{np}'$,

$$\frac{b_0}{a_{np}'}=\frac{b_0}{a_{np}}-\delta\left(\frac{b_0}{a_{np}}\right) \tag{32}$$
$$=1.259\pm0.004 \text{ (rect.)}$$
$$=1.362\pm0.008 \text{ (Yukawa)},$$

which describes the neutron-proton scattering that would be produced by specifically nuclear forces alone.

The magnetic contribution to $1/a_{pp}$ is similarly given by

$$\delta\left(\frac{1}{a_{pp}}\right)=\tfrac{1}{2}(\mu_p^2-\mu_p+\tfrac{1}{2})\frac{e^2}{Mc^2}\left[\frac{u_0'(0)}{v_0(0)}\right]^2+\frac{e^2}{Mc^2}\frac{1}{v_0^2(0)}\int_0^\infty\left(u_0'-\frac{u_0}{r}\right)^2\frac{dr}{r}. \tag{33}$$

The integral in this formula can be transformed to

$$-\int_0^\infty u_0u_0''\frac{dr}{r}-\frac{1}{2}(u_0'(0))^2+\frac{1}{2a_{pp}^2}v_0^2(0), \tag{34}$$

which can be evaluated, with sufficient precision, as

$$\left(\int_0^1\frac{\sin^2(\pi/2)x}{x}dx-\frac{1}{2}\right)(u_0'(0))^2=0.324(u_0'(0))^2, \tag{35}$$

and

$$(\log 2-\tfrac{1}{2})(u_0'(0))^2=0.193(u_0'(0))^2, \tag{36}$$

for the two potentials. Hence

$$\delta\left(\frac{b_0}{a_{pp}}\right)=\frac{\pi^2}{8}(\mu_p^2-\mu_p+1.15)\left(1+2\frac{r_e}{a_{pp}}\right)\left(\frac{\hbar/Mc}{r_e}\right)^2 \tag{37}$$
$$=0.056\pm0.003 \text{ (rect.)}$$

and

$$\delta\left(\frac{b_0}{a_{pp}}\right)=\frac{9}{2}(\mu_p^2-\mu_p+0.886)\left(1+\frac{14}{9}\frac{r_e}{a_{pp}}\right)\left(\frac{\hbar/Mc}{r_e}\right)^2 \tag{38}$$
$$=0.195\pm0.010 \text{ (Yukawa)}.$$

The scattering amplitude $a_{pp}'$,

$$\frac{b_0}{a_{pp}'}=\frac{b_0}{a_{pp}}-\delta\left(\frac{b_0}{a_{pp}}\right), \tag{39}$$

which describes zero energy proton-proton scattering by specifically nuclear forces, is now obtained as

$$\frac{b_0}{a_{pp}'}=1.698\pm0.035, \tag{40}$$

for the rectangular potential, and

$$\frac{b_0}{a_{pp}'}=1.378\pm0.040, \tag{41}$$

for the Yukawa potential.

We finally observed that $a_{np}'$ and $a_{pp}'$ are in definite disagreement for the rectangular potential, and in very satisfactory agreement for the Yukawa potential; the hypothesis of charge independent nuclear forces is accurately confirmed by experiment, provided the Yukawa potential is also assumed. It should be noted that a small difference between the neutron-proton and proton-proton nuclear interactions is to be anticipated theoretically, as a consequence of electrical perturbations of the mesons responsible for nuclear forces. Independent evidence concerning the shape of the nuclear potential, as well as a more accurate evaluation of the magnetic interaction, will be required to reveal this effect.

*Note added in proof.*—A more precise calculation for the Yukawa potential has been performed by Mr. B. R. Mottelson. The wave function employed in the text actually corresponds to a potential of the form $e^{-\beta r}/(1-e^{-\beta r})$, which possesses the singularity of the Yukawa potential, but differs at large distances. A corrected wave function leads to a slightly smaller value of $\bar{r}$, and increases somewhat the effect of the magnetic interaction. The results are $b_0/a_{np}'=1.396\pm0.01$, $b_0/a_{pp}'=1.302\pm0.04$. This comparison is not unsatisfactory, particularly in view of the expected, but uncertain, decrease in magnetic interaction produced by the finite spatial extension of the nucleon magnetization.

## APPENDIX

The discussion of the variational principle (1) in the text will be preceded by a treatment of the simpler variational principle appropriate to neutron-proton $S$ scattering:

$$\lim_{R\to\infty}\left\{\int_0^R\left[\left(\frac{du}{dr}\right)^2-\frac{M}{\hbar^2}(E-V_{np})u^2\right]dr-\int_0^R\left[\left(\frac{dv}{dr}\right)^2-\frac{ME}{\hbar^2}v^2\right]dr\right\}=k\cot\delta\, v^2(0). \tag{A1}$$

We shall demonstrate that the quantity called $k\cot\delta$ is stationary[8] with respect to independent variations of $u$ and $v$, subject only to the restriction

$$u(0)=0,\quad u(r)-v(r)\to 0,\quad r\to\infty. \tag{A2}$$

[8] L. Hulthèn has independently remarked on the stationary property of the phase shift. However, he has not introduced an explicit stationary expression for the phase shift, but rather an implicit characterization. See Mott and Massey, *The Theory of Atomic Collisions* (Clarendon Press, Oxford, 1949), second edition, p. 128, and W. Kohn, Phys. Rev. 74, 1763 (1948).

The stationary value is attained for functions that satisfy

$$\left[\frac{d^2}{dr^2}+\frac{M}{\hbar^2}(E-V_{np})\right]u=0, \quad \text{(A3a)}$$

$$\left[\frac{d^2}{dr^2}+\frac{ME}{\hbar^2}\right]v=0, \quad \text{(A3b)}$$

and the stationary value of $\delta$ is the rigorous phase shift. On introducing such a variation in (A1), we obtain

$$u'\delta u-v'\delta v\Big]_0^R-\int_0^R\left[\frac{d^2u}{dr^2}+\frac{M}{\hbar^2}(E-V_{np})u\right]\delta u dr$$

$$+\int_0^R\left[\frac{d^2v}{dr^2}+\frac{ME}{\hbar^2}v\right]\delta v dr=(k\cot\delta)v(0)\delta v(0)+\delta(k\cot\delta)\tfrac{1}{2}v^2(0). \quad \text{(A4)}$$

Now

$$u'\delta u-v'\delta v\Big]_0^R=v'(0)\delta v(0), \quad \text{(A5)}$$

in consequence of the restrictions (A2). Hence the conditions that $k\cot\delta$ be stationary with respect to otherwise unrestricted variations are expressed by the differential Eqs. (A3), and by

$$\frac{v'(0)}{v(0)}=k\cot\delta. \quad \text{(A6)}$$

Now the solution of (A3b) consistent with (A6) has the form

$$v(r)/v(0)=\sin(kr+\delta)/\sin\delta, \quad \text{(A7)}$$

which is indeed the asymptotic form of $u$, thereby identifying the stationary value of $\delta$ with the correct phase shift.

Among the advantages of a variational principle is the ease with which one infers the dependence of the stationary quantity on various parameters, such as kinetic energy and depth of potential well in (A1). In considering a small change of the energy, for example, no corresponding alteration in the wave function need be introduced, in virtue of the stationary property. Thus, we have the rigorous result

$$\frac{d}{dk^2}(k\cot\delta)=\frac{1}{v^2(0)}\int_0^\infty(v^2-u^2)dr. \quad \text{(A8)}$$

The integral in this formula involves the wave functions $u$ and $v$ associated with the energy $E$, and the effective domain of integration is the region of nuclear force. With a short range interaction, however, $u$ and $v$ are only slightly energy dependent, for kinetic energies that are small in comparison with the average nuclear potential. It is therefore useful to introduce an energy expansion for the wave functions within the interaction region,

$$u=u_0+k^2u_1+\cdots,\quad v=v_0+k^2v_1+\cdots, \quad \text{(A9)}$$

subject to the convenient normalization, $v(0)=1$, and obtain

$$d/(dk^2)(k\cot\delta)=\tfrac{1}{2}r_e+2pk^2r_e^3+\cdots. \quad \text{(A10)}$$

Here

$$\tfrac{1}{2}r_e=\int_0^\infty(v_0^2-u_0^2)dr,\quad v_0(0)=1, \quad \text{(A11)}$$

and

$$pr_e^3=\int_0^\infty(v_0v_1-u_0u_1)dr,\quad v_1(0)=0, \quad \text{(A12)}$$

define two energy independent parameters; the length $r_e$, called the effective range, and the numerical coefficient $p$. The resulting expansion[9] for $k\cot\delta$ is

$$k\cot\delta=\frac{1}{a}+\frac{1}{2}k^2r_e+pk^4r_e^3+\cdots, \quad \text{(A13)}$$

in which the length $a$ describes the amplitude of the scattered wave for zero energy.

The determination of the conditions that $\cot\delta$ in (1) be stationary proceeds in the same manner. One easily obtains the differential Eqs. (3) and the requirement

$$\lim_{\epsilon\to\infty}\left[\frac{v'(\epsilon)}{v(0)}-\frac{1}{b_0}\log\left(\gamma^2\frac{\epsilon}{b_0}\right)\right]=k\cot\delta\frac{2\pi\eta}{e^{2\pi\eta}-1}$$

$$+\frac{1}{b_0}\left(Re\frac{\Gamma'(i\eta)}{\Gamma(i\eta)}-\log\eta\right). \quad \text{(A14)}$$

Now the solution of the differential equation for $v$ has the form

$$v(r)=f(r)\cos\delta+g(r)\sin\delta, \quad \text{(A15)}$$

where $\delta$ is the desired phase shift, and $f$, $g$ are the regular and irregular Coulomb wave functions, identified by the asymptotic formulas

$$f\sim\sin\varphi,\quad g\sim\cos\varphi,\quad \varphi=kr-\eta\log 2kr+\arg\Gamma(1+i\eta). \quad \text{(A16)}$$

To compute the left side of (A14) from (A15),

$$\lim_{\epsilon\to 0}\left[\frac{v'(\epsilon)}{v(0)}-\frac{1}{b_0}\log\left(\gamma^2\frac{\epsilon}{b_0}\right)\right]=\frac{f'(0)}{g(0)}\cot\delta$$

$$+\lim_{\epsilon\to 0}\left[\frac{g'(\epsilon)}{g(0)}-\frac{1}{b_0}\log\left(\gamma^2\frac{\epsilon}{b_0}\right)\right], \quad \text{(A17)}$$

we require the functions $f$ and $g$ in the vicinity of the origin. These are known[10] to be given by

$$f(r)\sim C_0kr,\quad C_0^2=\frac{2\pi\eta}{e^{2\pi\eta}-1}, \quad \text{(A18)}$$

$$g(r)\sim\frac{1}{C_0}\left\{1+\frac{r}{b_0}\left[\log\left(\gamma^2\frac{r}{b_0}\right)-1+Re\frac{\Gamma'(i\eta)}{\Gamma(i\eta)}-\log\eta\right]\right\}.$$

Hence

$$\frac{f'(0)}{g(0)}=k\frac{2\pi\eta}{e^{2\pi\eta}-1}, \quad \text{(A19)}$$

and

$$\frac{g'(\epsilon)}{g(0)}=\frac{1}{b_0}\left[\log\left(\gamma^2\frac{\epsilon}{b_0}\right)+Re\frac{\Gamma'(i\eta)}{\Gamma(i\eta)}-\log\eta\right], \quad \text{(A20)}$$

which, inserted in (A17), yields (A14) thereby demonstrating that the stationary value of $\delta$ is the correct phase shift.

[9] Since the original construction of this expansion by somewhat different variational methods (reference 3), a number of derivations have been published in which the stationary property of the phase shift is not explicitly employed: F. C. Barker and R. E. Peierls, Phys. Rev. 75, 312 (1949); Hatcher, Arfken, and Breit, Phys. Rev. 75, 1389 (1949); G. F. Chew and M. L. Goldberger, Phys. Rev. 75, 1637 (1949); H. A. Bethe, Phys. Rev. 76, 38 (1949).

[10] Yost, Wheeler, and Breit, Phys. Rev. 49, 174 (1936).

# *ON THE GREEN'S FUNCTIONS OF QUANTIZED FIELDS. I*

BY JULIAN SCHWINGER

HARVARD UNIVERSITY

Communicated May 22, 1951

The temporal development of quantized fields, in its particle aspect, is described by propagation functions, or Green's functions. The construction of these functions for coupled fields is usually considered from the viewpoint of perturbation theory. Although the latter may be resorted to for detailed calculations, it is desirable to avoid founding the formal theory of the Green's functions on the restricted basis provided by the assumption of expandability in powers of coupling constants. These notes are a preliminary account of a general theory of Green's functions, in which the defining property is taken to be the representation of the fields of prescribed sources.

We employ a quantum dynamical principle for fields which has been described elsewhere.[1] This principle is a differential characterization of the function that produces a transformation from eigenvalues of a complete set of commuting operators on one space-like surface to eigenvalues of another set on a different surface,[2]

$$\delta(\zeta_1', \sigma_1|\zeta_2'', \sigma_2) = i(\zeta_1', \sigma_1|\delta\int_{\sigma_2}^{\sigma_1}(dx)\mathcal{L}|\zeta_2'', \sigma_2). \qquad (1)$$

Here $\mathcal{L}$ is the Lagrange function operator of the system. For the example of co[illegible] Dirac and Maxwell fields, with external sources for each field, the Lagrange function may be taken as

$$\mathcal{L} = -{}^1/_4[\bar{\psi}, \gamma_\mu(-i\partial_\mu - eA_\mu)\psi + m\psi] + {}^1/_2[\bar{\psi}, \eta] + \text{Herm. conj.} + {}^1/_4F_{\mu\nu}^2 - {}^1/_4\{F_{\mu\nu}, \partial_\mu A_\nu - \partial_\nu A_\mu\} + J_\mu A_\mu, \qquad (2)$$

which implies the equations of motion

$$\gamma_\mu(-i\partial_\mu - eA_\mu)\psi + m\psi = \eta,$$
$$F_{\mu\nu} = \partial_\mu A_\nu - \partial_\nu A_\mu, \quad \partial_\nu F_{\mu\nu} = J_\mu + j_\mu, \qquad (3)$$

where

$$j_\mu = e\,{}^1/_2[\bar{\psi}, \gamma_\mu\psi]. \qquad (4)$$

With regard to commutation relations, we need only note the anticommutativity of the source spinors with the Dirac field components.

We shall restrict our attention to changes in the transformation function that arise from variations of the external sources. In terms of the notation

$$(\zeta_1', \sigma_1|\zeta_2'', \sigma_2) = \exp i\mathcal{W},$$
$$(\zeta_1', \sigma_1|F(x)|\zeta_2'', \sigma_2)/(\zeta_1', \sigma_1|\zeta_2'', \sigma_2) = \langle F(x)\rangle, \qquad (5)$$

Reprinted from the *Proceedings of the National Academy of Sciences U.S.A.* **37**, 452–459 (1951).

the dynamical principle can then be written

$$\delta \mathfrak{W} = \int_{\sigma_2}^{\sigma_1} (dx)\langle \delta \mathfrak{L}(x)\rangle, \tag{6}$$

where

$$\langle \delta \mathfrak{L}(x)\rangle = \langle \bar{\psi}(x)\rangle \delta\eta(x) + \delta\bar{\eta}(x)\langle \psi(x)\rangle + \langle A_\mu(x)\rangle \delta J_\mu(x). \tag{7}$$

The effect of a second, independent variation is described by

$$\delta'\langle \delta \mathfrak{L}(x)\rangle = i \int_{\sigma_2}^{\sigma_1} (dx')[\langle(\delta \mathfrak{L}(x)\delta' \mathfrak{L}(x'))_+\rangle - \langle \delta \mathfrak{L}(x)\rangle\langle \delta' \mathfrak{L}(x')\rangle], \tag{8}$$

in which the notation $(\ )_+$ indicates temporal ordering of the operators. As examples we have

$$\delta_\eta\langle \psi(x)\rangle = i \int_{\sigma_2}^{\sigma_1} (dx')[\langle(\psi(x)\bar{\psi}(x')\delta\eta(x'))_+\rangle - \langle \psi(x)\rangle\langle \bar{\psi}(x')\delta\eta(x')\rangle], \tag{9}$$

and

$$\delta_J\langle \psi(x)\rangle = i \int_{\sigma_2}^{\sigma_1} (dx')[\langle(\psi(x)A_\mu(x'))_+\rangle - \langle \psi(x)\rangle\langle A_\mu(x')\rangle]\delta J_\mu(x'). \tag{10}$$

The latter result can be expressed in the notation

$$-i(\delta/\delta J_\mu(x'))\langle \psi(x)\rangle = \langle(\psi(x)A_\mu(x'))_+\rangle - \langle \psi(x)\rangle\langle A_\mu(x')\rangle, \tag{11}$$

although one may supplement the right side with an arbitrary gradient. This consequence of the charge conservation condition, $\partial_\mu J_\mu = 0$, corresponds to the gauge invariance of the theory.

A Green's function for the Dirac field, in the absence of an actual spinor source, is defined by

$$\delta_\eta\langle \psi(x)\rangle]_{\eta=0} = \int_{\sigma_2}^{\sigma_1} (dx')G(x, x')\delta\eta(x'). \tag{12}$$

According to (9), and the anticommutativity of $\delta\eta(x')$ with $\psi(x)$, we have

$$G(x, x') = i\langle(\psi(x)\bar{\psi}(x'))_+\rangle\epsilon(x, x'), \tag{13}$$

where $\epsilon(x, x') = (x_0 - x_0')/|x_0 - x_0'|$. On combining the differential equation for $\langle \psi(x)\rangle$ with (11), we obtain the functional differential equation

$$[\gamma_\mu(-i\partial_\mu - e\langle A_\mu(x)\rangle + ie\delta/\delta J_\mu(x)) + m]G(x, x') = \delta(x - x'). \tag{14}$$

An accompanying equation for $\langle A_\mu(x)\rangle$ is obtained by noting that

$$\langle j_\mu(x)\rangle = ie \operatorname{tr} \gamma_\mu G(x, x')_{x' \to x}, \tag{15}$$

in which the trace refers to the spinor indices, and an average is to be taken of the forms obtained with $x_0' \to x_0 \pm 0$. Thus, with the special choice of gauge, $\partial_\nu\langle A_\nu(x)\rangle = 0$, we have

$$-\partial_\nu^2\langle A_\mu(x)\rangle = J_\mu(x) + ie \operatorname{tr} \gamma_\mu G(x, x). \tag{16}$$

The simultaneous equations (14) and (16) provide a rigorous description of $G(x, x')$ and $\langle A_\mu(x)\rangle$.

A Maxwell field Green's function is defined by

$$\mathcal{G}_{\mu\nu}(x, x') = (\delta/\delta J_\nu(x'))\langle A_\mu(x)\rangle = (\delta/\delta J_\mu(x))\langle A_\nu(x')\rangle = \\ i[\langle(A_\mu(x)A_\nu(x'))_+\rangle - \langle A_\mu(x)\rangle\langle A_\nu(x')\rangle]. \quad (17)$$

The differential equations obtained from (16) and the gauge condition are

$$-\partial_\lambda{}^2\mathcal{G}_{\mu\nu}(x, x') = \delta_{\mu\nu}\delta(x - x') + ie \operatorname{tr} \gamma_\mu(\delta/\delta J_\nu(x'))G(x, x), \\ \partial_\mu\mathcal{G}_{\mu\nu}(x, x') = 0\ (= \partial_\nu\chi). \quad (18)$$

More complicated Green's functions can be discussed in an analogous manner. The Dirac field Green's function defined by

$$\delta_\eta{}^2\langle(\psi(x_1)\psi(x_2))_+\rangle\epsilon(x_1, x_2)_{\eta=0} = \\ \int_{\sigma_2}^{\sigma_1}(dx_1')\int_{\sigma_2}^{\sigma_1}(dx_2')G(x_1, x_2; x_1', x_2')\delta\eta(x_1')\delta\eta(x_2'), \quad (19)$$

may be called a "two-particle" Green's function, as distinguished from the "one-particle" $G(x, x')$. It is given explicitly by

$$G(x_1, x_2; x_1', x_2') = \langle(\psi(x_1)\psi(x_2)\bar{\psi}(x_1')\bar{\psi}(x_2'))_+\rangle\ \epsilon, \\ \epsilon = \epsilon(x_1, x_2)\epsilon(x_1', x_2')\epsilon(x_1, x_1')\epsilon(x_1, x_2')\epsilon(x_2, x_1')\epsilon(x_2, x_2'). \quad (20)$$

This function is antisymmetrical with respect to the interchange of $x_1$ and $x_2$, and of $x_1'$ and $x_2'$ (including the suppressed spinor indices). It obeys the differential equation

$$\mathfrak{F}_1G(x_1, x_2; x_1', x_2') = \delta(x_1 - x_1')G(x_2, x_2') - \delta(x_1 - x_2')G(x_2, x_1'), \quad (21)$$

where $\mathfrak{F}$ is the functional differential operator of (14). More symmetrically written, this equation reads

$$\mathfrak{F}_1\mathfrak{F}_2G(x_1, x_2; x_1', x_2') = \delta(x_1 - x_1')\delta(x_2 - x_2') - \\ \delta(x_1 - x_2')\delta(x_2 - x_1'), \quad (22)$$

in which the two differential operators are commutative.

The replacement of the Dirac field by a Kemmer field involves alterations beyond those implied by the change in statistics. Not all components of the Kemmer field are dynamically independent. Thus, if 0 refers to some arbitrary time-like direction, we have

$$m(1 - \beta_0{}^2)\psi = (1 - \beta_0{}^2)\eta - \beta_k(-i\partial_k - eA_k)\beta_0{}^2\psi, \\ k = 1, 2, 3, \quad (23)$$

which is an equation of constraint expressing $(1 - \beta_0{}^2)\psi$ in terms of the independent field components $\beta_0{}^2\psi$, and of the external source. Accordingly, in computing $\delta_\eta\langle\psi(x)\rangle$ we must take into account the change induced in $(1 - \beta_0{}^2)\psi(x)$, whence

$$G(x, x') = i\langle(\psi(x)\bar{\psi}(x'))_+\rangle + (1/m)(1 - \beta_0{}^2)\delta(x - x'). \quad (24)$$

The temporal ordering is with respect to the arbitrary time-like direction

The Green's function is independent of this direction, however, and satisfies equations which are of the same form as (14) and (16), save for a sign change in the last term of the latter equation which arises from the different statistics associated with the integral spin field.

[1] Schwinger, J., *Phys. Rev.*, June 15, 1951 issue.
[2] We employ units in which $\hbar = c = 1$.

---

# ON THE GREEN'S FUNCTIONS OF QUANTIZED FIELDS. II

BY JULIAN SCHWINGER

HARVARD UNIVERSITY

Communicated May 22, 1951

In all of the work of the preceding note there has been no explicit reference to the particular states on $\sigma_1$ and $\sigma_2$ that enter in the definitions of the Green's functions. This information must be contained in boundary conditions that supplement the differential equations. We shall determine these boundary conditions for the Green's functions associated with vacuum states on both $\sigma_1$ and $\sigma_2$. The vacuum, as the lowest energy state of the system, can be defined only if, in the neighborhood of $\sigma_1$ and $\sigma_2$, the actual external electromagnetic field is constant in some time-like direction (which need not be the same for $\sigma_1$ and $\sigma_2$). In the Dirac one-particle Green's function, for example,

$$\begin{aligned} G(x, x') &= i\langle\psi(x)\bar{\psi}(x')\rangle, \; x_0 > x_0', \\ &= -i\langle\bar{\psi}(x')\,\psi(x)\rangle, \; x_0 < x_0', \end{aligned} \tag{25}$$

the temporal variation of $\psi(x)$ in the vicinity of $\sigma_1$ can then be represented by

$$\psi(x) = \exp\,[iP_0(x_0 - X_0)]\psi(X)\exp\,[-iP_0(x_0 - X_0)], \tag{26}$$

where $P_0$ is the energy operator and $X$ is some fixed point. Therefore,

$$x \sim \sigma_1\colon G(x, x') = i\langle\psi(X)\exp\,[-i(P_0 - P_0^{\text{vac}})(x_0 - X_0)]\bar{\psi}(x')\rangle, \tag{27}$$

in which $P_0^{\text{vac}}$ is the vacuum energy eigenvalue. Now $P_0 - P_0^{\text{vac}}$ has no negative eigenvalues, and accordingly $G(x, x')$, as a function of $x_0$ in the vicinity of $\sigma_1$, contains only positive frequencies, which are energy values for states of unit positive charge. The statement is true of every time-like direction, if the external field vanishes in this neighborhood.

A representation similar to (26) for the vicinity of $\sigma_2$ yields

$$x \sim \sigma_2\colon G(x, x') = -i\langle\bar{\psi}(x')\exp\,[i(P_0 - P_0^{\text{vac}})(x_0 - X_0)]\psi(X)\rangle, \tag{28}$$

which contains only negative frequencies. In absolute value, these are the energies of unit negative charge states. We thus encounter Green's functions that obey the temporal analog of the boundary condition characteristic of a source radiating into space.[1] In keeping with this analogy, such Green's functions can be derived from a retarded proper time Green's function by a Fourier decomposition with respect to the mass.

The boundary condition that characterizes the Green's functions associated with vacuum states on $\sigma_1$ and $\sigma_2$ involves these surfaces only to the extent that they must be in the region of outgoing waves. Accordingly, the domain of these functions may conveniently be taken as the entire four-dimensional space. Thus, if the Green's function $G_+(x, x')$, defined by (14), (16), and the outgoing wave boundary condition, is represented by the integro-differential equation,

$$\gamma_\mu(-i\partial_\mu - eA_{+\mu}(x))G_+(x, x') + \int (dx'')M(x, x'')G_+(x'', x') = \delta(x - x'), \tag{29}$$

the integration is to be extended over all space-time. This equation can be more compactly written as

$$[\gamma(p - eA_+) + M]G_+ = 1, \tag{30}$$

by regarding the space-time coordinates as matrix indices. The mass operator $M$ is then symbolically defined by

$$MG_+ = mG_+ + ie\gamma(\delta/\delta J)G_+. \tag{31}$$

In these formulae, $A_+$ and $\delta/\delta J$ are considered to be diagonal matrices,

$$(x|A_{+\mu}|x') = \delta(x - x')A_{+\mu}(x). \tag{32}$$

There is some advantage, however, in introducing "photon coordinates" explicitly (while continuing to employ matrix notation for the "particle coordinates"). Thus

$$\gamma A_+ \rightarrow \int (d\xi)\gamma(\xi)A_+(\xi), \tag{33}$$

where $\gamma(\xi)$ is defined by

$$(x|\gamma_\mu(\xi)|x') = \gamma_\mu\delta(x - \xi)\delta(x - x'). \tag{34}$$

The differential equation for $A_+(\xi)$ can then be written

$$-\partial_\xi^2 A_+(\xi) = J(\xi) + ie \operatorname{Tr} [\gamma(\xi)G_+], \tag{35}$$

where Tr denotes diagonal summation with respect to spinor indices and particle coordinates. The associated photon Green's function differential equation is

$$-\partial_\xi^2 \mathcal{G}_+(\xi, \xi') = \delta(\xi - \xi') + ie \operatorname{Tr} [\gamma(\xi)(\delta/\delta J(\xi'))G_+]. \tag{36}$$

To express the variational derivatives that occur in (31) and (36) we introduce an auxiliary quantity defined by

$$\Gamma(\xi) = -(\delta/\delta eA_+(\xi))G_+^{-1} \\ = \gamma(\xi) - (\delta/\delta eA_+(\xi))M. \tag{37}$$

Thus

$$(\delta/\delta J(\xi))G_+ = e\int(d\xi')G_+\Gamma(\xi')G_+\mathcal{G}_+(\xi', \xi), \tag{38}$$

from which we obtain

$$M = m + ie^2\int(d\xi)(d\xi')\gamma(\xi)G_+\Gamma(\xi')\mathcal{G}_+(\xi', \xi), \tag{39}$$

and

$$-\partial_\xi^2\mathcal{G}_+(\xi, \xi') + \int(d\xi'')P(\xi, \xi'')\mathcal{G}_+(\xi'', \xi') = \delta(\xi - \xi'), \\ P(\xi, \xi') = -ie^2 \operatorname{Tr}[\gamma(\xi)G_+\Gamma(\xi')G_+] \tag{40}$$

With the introduction of matrix notation for the photon coordinates, this Green's function equation becomes

$$(k^2 + P)\mathcal{G}_+ = 1, \ [\xi_\mu, k_\nu] = i\delta_{\mu\nu}, \tag{41}$$

and the polarization operator $P$ is given by

$$P = -ie^2 \operatorname{Tr}[\gamma G_+\Gamma G_+]. \tag{42}$$

In this notation, the mass operator expression reads

$$M = m + ie^2 \operatorname{T\rho}[\gamma G_+\Gamma\mathcal{G}_+], \tag{43}$$

where $\operatorname{T\rho}$ denotes diagonal summation with respect to the photon coordinates, including the vector indices.

The two-particle Green's function

$$G_+(x_1, x_2; x_1', x_2') = (x_1, x_2|G_{12}|x_1', x_2'), \tag{44}$$

can be represented by the integro-differential equation

$$[(\gamma\pi + M)_1(\gamma\pi + M)_2 - I_{12}]G_{12} = 1_{12}, \\ \pi = p - eA_+, \tag{45}$$

thereby introducing the interaction operator $I_{12}$. The unit operator $1_{12}$ is defined by the matrix representation

$$(x_1, x_2|1_{12}|x_1', x_2') = \delta(x_1 - x_1')\delta(x_2 - x_2') - \\ \delta(x_1 - x_2')\delta(x_2 - x_1'). \tag{46}$$

On comparison with (21) we find that the interaction operator can be characterized symbolically by

$$\begin{aligned} I_{12}G_{12} &= -ie^2\, \mathrm{T}\rho[\gamma_1\Gamma_2\mathcal{G}_+]G_{12} - ie^2\, \mathrm{T}\rho[\gamma_1 G_1\delta/\delta J](I_{12}G_{12}) \\ &= -ie^2\, \mathrm{T}\rho[\gamma_2\Gamma_1\mathcal{G}_+]G_{12} - ie^2\, \mathrm{T}\rho[\gamma_2 G_2\delta/\delta J](I_{12}G_{12}), \end{aligned} \quad (47)$$

where $G_1$ and $G_2$ are the one-particle Green's functions of the indicated particle coordinates.

The various operators that enter in the Green's function equations, the mass operator $M$, the polarization operator $P$, the interaction operator $I_{12}$, can be constructed by successive approximation. Thus, in the first approximation,

$$\begin{aligned} M(x, x') &= m\delta(x - x') + ie^2\gamma_\mu G_+(x, x')\gamma_\mu D_+(x, x'), \\ P_{\mu\nu}(\xi, \xi') &= -ie^2\, \mathrm{tr}[\gamma_\mu G_+(\xi, \xi')\gamma_\nu G_+(\xi', \xi)], \\ I(x_1, x_2; x_1', x_2') &= -ie^2\gamma_{1\mu}\gamma_{2\mu}D_+(x_1, x_2)(x_1, x_2|\, 1_{12}\,|x_1', x_2'), \end{aligned} \quad (48)$$

where

$$\mathcal{G}_{\mu\nu}(\xi, \xi') = \delta_{\mu\nu}D_+(\xi, \xi'), \quad (49)$$

and the Green's functions that appear in these formulae refer to the $0$th approximation ($M = m$, $P = 0$). We also have, in the first approximation,

$$\begin{aligned} \Gamma_\mu(\xi; x, x') &= \gamma_\mu\delta(\xi - x)\delta(x - x') \\ &- ie^2\gamma_\nu G_+(x, \xi)\gamma_\mu G_+(\xi, x')\gamma_\nu D_+(x, x') \end{aligned} \quad (50)$$

Perturbation theory, as applied in this manner, must not be confused with the expansion of the Green's functions in powers of the charge. The latter procedure is restricted to the treatment of scattering problems.

The solutions of the homogeneous Green's function equations constitute the wave functions that describe the various states of the system. Thus, we have the one-particle wave equation

$$(\gamma\pi + M)\psi = 0, \quad (51)$$

and the two particle wave equation

$$[(\gamma\pi + M)_1(\gamma\pi + M)_2 - I_{12}]\psi_{12} = 0, \quad (52)$$

which are applicable equally to the discussion of scattering and to the properties of bound states. In particular, the total energy and momentum eigenfunctions of two particles in isolated interaction are obtained as the solutions of (52) which are eigenfunctions for a common displacement of the two space-time coordinates. It is necessary to recognize, however, that the mass operator, for example, can be largely represented in its effect by an alteration in the mass constant and by a scale change of the Green's function. Similarly, the major effect of the polarization operator is to multiply the photon Green's function by a factor, which everywhere appears associated with the charge. It is only after these renormalizations have been performed that we deal with wave equations that involve the empirical mass and charge, and are thus of immediate physical applicability.

The details of this theory will be published elsewhere, in a series of articles entitled "The Theory of Quantized Fields."

[1] Green's functions of this variety have been discussed by Stueckelberg, E. C. G., *Helv. Phys. Acta*, **19**, 242 (1946), and by Feynman, R. P., *Phys. Rev.*, **76**, 749 (1949).

# The Theory of Quantized Fields. III

JULIAN SCHWINGER
*Harvard University, Cambridge, Massachusetts*

(Received March 31, 1953)

In this paper we discuss the electromagnetic field, as perturbed by a prescribed current. All quantities of physical interest in various situations, eigenvalues, eigenfunctions, and transition probabilities, are derived from a general transformation function which is expressed in a non-Hermitian representation. The problems treated are: the determination of the energy-momentum eigenvalues and eigenfunctions for the isolated electromagnetic field, and the energy eigenvalues and eigenfunctions for the field perturbed by a time-independent current; the evaluation of transition probabilities and photon number expectation values for a time-dependent current that departs from zero only within a finite time interval, and for a time-dependent current that assumes non-vanishing time-independent values initially and finally. The results are applied in a discussion of the infrared catastrophe and of the adiabatic theorem. It is shown how the latter can be exploited to give a uniform formulation for all problems requiring the evaluation of transition probabilities or eigenvalue displacements.

## INTRODUCTION

WE shall approach the general problem of coupled fields through the simpler situation presented by a single field which is externally perturbed. In this paper we illustrate the treatment of a Bose-Einstein system by discussing the Maxwell field with a prescribed electric current. A succeeding paper will be devoted to the Dirac field.

The solution to all dynamical questions is obtained by constructing the transformation function linking two descriptions of the system that are associated with different space-like surfaces. Thus, for a closed system, the general transformation function can be expressed as

$$(\zeta_1'\sigma_1|\zeta_2''\sigma_2)=\sum_{\gamma',\gamma''}(\zeta_1'|\gamma')(\gamma'\sigma_1|\gamma''\sigma_2)(\gamma''|\zeta_2''),$$

where the $\gamma$ are a complete set of compatible constants of the motion, in terms of which the energy-momentum vector $P_\mu$ can be exhibited. In the $\gamma$ representation, the effect of an infinitesimal translation of $\sigma_1$ is given by

$$\delta_\epsilon(\gamma'\sigma_1|\gamma''\sigma_2)=i(\gamma'\sigma_1|P_\mu\delta\epsilon_\mu|\gamma''\sigma_2)=iP_\mu'\delta\epsilon_\mu(\gamma'\sigma_1|\gamma''\sigma_2),$$

where $P_\mu'=P_\mu(\gamma')$. Accordingly, if $\sigma_1$ is parallel to $\sigma_2$, and is generated from the latter by the translation $X_\mu$, we have

$$(\gamma'\sigma_1|\gamma''\sigma_2)=\delta(\gamma',\gamma'')\exp(iP_\mu'X_\mu),$$

and

$$(\zeta_1'\sigma_1|\zeta_2''\sigma_2)=\sum_{\gamma'}(\zeta_1'|\gamma')\exp(iP_\mu'X_\mu)(\gamma'|\zeta_2''). \quad (1)$$

This shows how a knowledge of the transformation function that relates two conveniently chosen representations on parallel surfaces yields all the eigenvalues and eigenfunctions of $P_\mu$.

Another illustration of the utility of transformation functions relates to the situation in which the same system is externally perturbed, in the interior of the space-time region bounded by $\sigma_1$ and $\sigma_2$. The transformation function $(\gamma'\sigma_1|\gamma''\sigma_2)$, inferred from the knowledge of $(\zeta_1'\sigma_1|\zeta_2''\sigma_2)$, then yields the probability of a transition from the initial state $\gamma''$ to the final state $\gamma'$,

$$p(\gamma',\gamma'')=|(\gamma'\sigma_1|\gamma''\sigma_2)|^2. \quad (2)$$

Representations of particular convenience are suggested by the characterization of the vacuum state for a complete system. The vacuum is the state of minimum energy. If this natural origin of energy is adjusted to zero, the vacuum can be described as that state presenting identical properties to all observers, $P_\mu\Psi_0=J_{\mu\nu}\Psi_0=0$, and is therefore independent of the surface $\sigma$. Now, if the general field component $\chi$ is analyzed into contributions of various frequencies, $\chi_{p_0}$, we have $[\chi_{p_0},P_0]=p_0\chi_{p_0}$, or $P_0\chi_{p_0}=\chi_{p_0}(P_0-p_0)$. When this relation, involving a positive frequency, $p_0>0$, is

Reprinted from *The Physical Review* **91**, 728–740 (1953).

applied to the vacuum state vector, we obtain

$$P_0(\chi_{p0}\Psi_0) = -p_0(\chi_{p0}\Psi_0).$$

Hence

$$\chi_{p0}\Psi_0 = 0, \quad p_0 > 0, \tag{3}$$

since this state, of energy less than that of the vacuum, must be nonexistent. A similar discussion yields

$$\Psi_0^\dagger \chi_{p0} = 0, \quad p_0 < 0,$$

which is the statement adjoint to (3). The vector $\Psi_0$ is thus characterized as the right eigenvector of the positive frequency parts of the field components, $\chi^{(+)}$, with zero eigenvalues, and $\Psi_0^\dagger$ appears as the left eigenvector, with zero eigenvalues, of the $\chi^{(-)}$, the negative frequency parts of the field components. It should be noted that the decomposition into positive and negative frequency parts is invariant under orthochronous Lorentz transformations. The complete sets of eigenvectors of these types will evidently be of particular value for the construction of energy eigenstates.

## THE MAXWELL FIELD

Elementary descriptions of the electromagnetic field on a given $\sigma$ are provided by the alternative complete sets of commuting operators, the transverse potential $A_k(x)$, and the transverse electric field[1] $F_{0k}(x)$. Following the suggestion of the preceding section, we employ instead the non-Hermitian operators, $F_{0k}^{(+)}(x)$ and $F_{0k}^{(-)}(x)$, which in the absence of an external current, are the positive and negative frequency parts of $F_{0k}(x)$. The transverse field equations, for zero current, are

$$-\partial_0 A_k = F_{0k}, \quad \partial_0 F_{0k} = -\partial_l^2 A_k = \omega^2 A_k,$$

where $\omega$, as a coordinate operator, is defined by the matrix

$$(\mathbf{x}|\omega|\mathbf{x}') = \int \frac{(d\mathbf{k})}{(2\pi)^3} |\mathbf{k}| \exp(i\mathbf{k}\cdot(\mathbf{x}-\mathbf{x}')),$$

which is symmetrical and positive-definite. On writing

$$F_{0k} = F_{0k}^{(+)} + F_{0k}^{(-)}, \quad A_k = A_k^{(+)} + A_k^{(-)},$$

where

$$F_{0k}^{(\pm)} = \pm i\omega A_k^{(\pm)} = \tfrac{1}{2}(F_{0k} \pm i\omega A_k),$$

the equations of motion assume the form

$$\partial_0 F_{0k}^{(\pm)} = \mp i\omega F_{0k}^{(\pm)},$$

which, in virtue of the positive-definite nature of $\omega$, confirms the interpretation of $F_{0k}^{(\pm)}$.

The canonical form of the infinitesimal generators

$$G_A = \int d\sigma(-F_{0k})\delta A_k, \quad G_F = \int d\sigma A_k \delta F_{0k},$$

[1] When no confusion is likely, we shall not employ the more complete notation $F_{(0)(k)}^{(T)}(x)$, which indicates that these are the transverse field components, relative to a local coordinate system based on $\sigma$.

can be extended to the generators of infinitesimal changes in the non-Hermitian operators $F_{0k}^{(\pm)}$, in the sense of the transformation equation $G - \bar{G} = \delta(W)$. Thus

$$G_F = \int d\sigma (A_k^{(+)} + A_k^{(-)})(\delta F_{0k}^{(+)} + \delta F_{0k}^{(-)})$$

$$= \int d\sigma 2A_k^{(-)}\delta F_{0k}^{(+)} + \delta\left[\int d\sigma(\tfrac{1}{2}A_k^{(+)}F_{0k}^{(+)} + \tfrac{1}{2}A_k^{(-)}F_{0k}^{(-)} + A_k^{(+)}F_{0k}^{(-)})\right],$$

and

$$G_F = \int d\sigma 2A_k^{(+)}\delta F_{0k}^{(-)} + \delta\left[\int d\sigma(\tfrac{1}{2}A_k^{(+)}F_{0k}^{(+)} + \tfrac{1}{2}A_k^{(-)}F_{0k}^{(-)} + A_k^{(-)}F_{0k}^{(+)})\right],$$

which yields

$$G_{F^{(+)}} = \int d\sigma 2A_k^{(-)}\delta F_{0k}^{(+)} = 2i\int d\sigma F_{0k}^{(-)}\omega^{-1}\delta F_{0k}^{(+)},$$

and

$$G_{F^{(-)}} = \int d\sigma 2A_k^{(+)}\delta F_{0k}^{(-)} = -2i\int d\sigma F_{0k}^{(+)}\omega^{-1}\delta F_{0k}^{(-)}.$$

The commutation relations on $\sigma$, implied by these generators, are

$$[F_{0k}^{(+)}(x), F_{0l}^{(+)}(x')] = [F_{0k}^{(-)}(x), F_{0l}^{(-)}(x')] = 0,$$

and

$$[F_{0k}^{(+)}(x), 2A_{0l}^{(-)}(x')] = [F_{0k}^{(-)}(x), 2A_{0l}^{(+)}(x')] = i(\delta_{kl}\delta_\sigma(x-x'))^{(T)}.$$

The latter can also be written

$$[F_{0k}^{(+)}(x), F_{0l}^{(-)}(x')] = \tfrac{1}{2}(\delta_{kl}(\mathbf{x}|\omega|\mathbf{x}'))^{(T)}. \tag{4}$$

These operator properties can be verified directly from those of $F_{0k}$ and $A_k$.

It should be noted that there exists some freedom in choosing the generator for a given set of independent variables. Thus,

$$'G_{F^{(+)}} = \int d\sigma 2A_k \delta F_{0k}^{(+)}$$

is also a generator of infinitesimal changes in $F_{0k}^{(+)}$, since

$$G_{F^{(+)}} - {}'G_{F^{(+)}} = -\int d\sigma 2A_k^{(+)}\delta F_{0k}^{(+)}$$

$$= 2i\int d\sigma F_{0k}^{(+)}\omega^{-1}\delta F_{0k}^{(+)}$$

$$= \delta\left[i\int d\sigma F_{0k}^{(+)}\omega^{-1}F_{0k}^{(+)}\right].$$

Similarly,

$$'G_{F^{(-)}}=\int d\sigma 2A_k\delta F_{0k}{}^{(-)}$$

is an alternative generator of changes in $F_{0k}{}^{(-)}$,

$$G_{F^{(-)}}-'G_{F^{(-)}}=\delta\left[-i\int d\sigma F_{0k}{}^{(-)}\omega^{-1}F_{0k}{}^{(-)}\right].$$

The eigenvector concept can be extended to non-Hermitian operators, with some limitations. We introduce the right eigenvector of the complete set of commuting operators, $F_{0k}{}^{(+)}(x)$ on $\sigma$,

$$F_{0k}{}^{(+)}(x)\Psi(F^{(+)\prime}\sigma)=F_{0k}{}^{(+)\prime}(x)\Psi(F^{(+)\prime}\sigma),$$

and the left eigenvector of the complete set, $F_{0k}{}^{(-)}(x)$ on $\sigma$,

$$\Phi(F^{(-)\prime}\sigma)F_{0k}{}^{(-)}(x)=\Phi(F^{(-)\prime}\sigma)F_{0k}{}^{(-)\prime}(x).$$

In virtue of the relation $F_{0k}{}^{(-)}(x)=F_{0k}{}^{(+)}(x)^\dagger$, these eigenvectors and eigenvalues are connected by

$$\Phi(F^{(-)\prime}\sigma)=\Psi(F^{(+)\prime}\sigma)^\dagger,\quad F_{0k}{}^{(-)\prime}(x)=F_{0k}{}^{(+)\prime}(x)^*. \quad (5)$$

However, the right eigenvector of the $F_{0k}{}^{(-)}$ and the left eigenvector of the $F_{0k}{}^{(+)}$ do not exist. This can be inferred from the commutator (4), in the form

$$['F_{0k}{}^{(-)}(x)^\dagger, 'F_{0l}{}^{(-)}(x')]=\tfrac{1}{2}(\delta_{kl}(\mathbf{x}|\omega|\mathbf{x}'))^{(T)},$$

where

$$'F_{0k}{}^{(-)}(x)=F_{0k}{}^{(-)}(x)-F_{0k}{}^{(-)\prime}(x).$$

When applied to the hypothetical eigenvector $\Psi(F^{(-)\prime}\sigma)$, this relation yields

$$-'F_{0l}{}^{(-)}(x')\,'F_{0k}{}^{(-)}(x)^\dagger\Psi(F^{(-)\prime}\sigma)=\tfrac{1}{2}(\delta_{kl}(\mathbf{x}|\omega|\mathbf{x}'))^{(T)}\Psi(F^{(-)\prime}\sigma).$$

The contradiction between the negative-definite nature of the operator on the left, and the positive-definite character of the numerical quantity on the right establishes the nonexistence[2] of $\Psi(F^{(-)\prime}\sigma)$, and similarly, of $\Phi(F^{(+)\prime}\sigma)$.

Let us consider the significance of the change induced in the eigenvectors $\Psi(F^{(+)\prime}\sigma)$ and $\Phi(F^{(-)\prime}\sigma)$ by the respective generators $G_{F^{(+)}}$ and $G_{F^{(-)}}$, according to the mutually Hermitian conjugate equations,

$$\delta\Psi(F^{(+)\prime}\sigma)=-iG_{F^{(+)}}\Psi(F^{(+)\prime}\sigma),$$

$$\delta\Phi(F^{(-)\prime}\sigma)=i\Phi(F^{(-)\prime}\sigma)G_{F^{(-)}}.$$

Now $\Psi(F^{(+)\prime}\sigma)+\delta\Psi(F^{(+)\prime}\sigma)$ is the eigenvector of the operator set $F_{0k}{}^{(+)}-\delta F_{0k}{}^{(+)}$, with the eigenvalues $F_{0k}{}^{(+)\prime}$. Since the $\delta F_{0k}{}^{(+)}$ are arbitrary infinitesimal numbers, this vector is also the eigenvector of the $F_{0k}{}^{(+)}$ with the eigenvalues $F_{0k}{}^{(+)\prime}+\delta F_{0k}{}^{(+)}$. Hence the alteration of the eigenvector $\Psi(F^{(+)\prime}\sigma)$ is that associated with the change of the eigenvalues by $\delta F_{0k}{}^{(+)}$. A similar statement applies to $\delta\Phi(F^{(-)\prime}\sigma)$.

The relation between the eigenvectors $\Psi(F^{(+)\prime}\sigma)$ and $'\Psi(F^{(+)\prime}\sigma)$, which are affected analogously by the respective generators $G_{F^{(+)}}$ and $'G_{F^{(+)}}$, can be deduced from

$$\delta'\Psi(F^{(+)\prime}\sigma)=-i\,'G_{F^{(+)}}\,'\Psi(F^{(+)\prime}\sigma)$$
$$=-iG_{F^{(+)}}\,'\Psi(F^{(+)\prime}\sigma)-\delta\left(\int d\sigma F_{0k}{}^{(+)\prime}\omega^{-1}F_{0k}{}^{(+)\prime}\right)'\Psi(F^{(+)\prime}\sigma),$$

namely,

$$'\Psi(F^{(+)\prime}\sigma)=\exp\left(-\int d\sigma F_{0k}{}^{(+)\prime}\omega^{-1}F_{0k}{}^{(+)\prime}\right)\Psi(F^{(+)\prime}\sigma)$$
$$=\exp\left(-i\int d\sigma A_k{}^{(+)\prime}F_{0k}{}^{(+)\prime}\right)\Psi(F^{(+)\prime}\sigma). \quad (6)$$

The adjoint equation reads

$$'\Phi(F^{(-)\prime}\sigma)=\exp\left(-\int d\sigma F_{0k}{}^{(-)\prime}\omega^{-1}F_{0k}{}^{(-)\prime}\right)\Phi(F^{(-)\prime}\sigma)$$
$$=\exp\left(i\int d\sigma A_k{}^{(-)\prime}F_{0k}{}^{(-)\prime}\right)\Phi(F^{(-)\prime}\sigma). \quad (7)$$

We shall now discuss the Maxwell field under the influence of a prescribed current distribution $J_\mu(x)$. It is convenient, initially, to describe the relations between states on the two arbitrary plane surfaces, $\sigma_1$ and $\sigma_2$, by means of the transformation function

$$'(F^{(-)\prime}\sigma_1|F^{(+)\prime}\sigma_2)=('\Phi(F^{(-)\prime}\sigma_1)\,'\Psi(F^{(+)\prime}\sigma_2)). \quad (8)$$

The dependence of this transformation function on the eigenvalues $F_{(0)(k)}{}^{(-)\prime}$ and $F_{(0)(k)}{}^{(+)\prime}$ is indicated by

$$\delta_F\,'(F^{(-)\prime}\sigma_1|F^{(+)\prime}\sigma_2)$$
$$=i\,'(F^{(-)\prime}\sigma_1|['G_{F^{(-)}}(\sigma_1)-'G_{F^{(+)}}(\sigma_2)]|F^{(+)\prime}\sigma_2)$$
$$=2i\,'\left(F^{(-)\prime}\sigma_1\middle|\left[\int_{\sigma_1}d\sigma_\mu\delta F_{\mu\nu}{}^{(-)\prime}A_\nu-\int_{\sigma_2}d\sigma_\mu\delta F_{\mu\nu}{}^{(+)\prime}A_\nu\right]\middle|F^{(+)\prime}\sigma_2\right), \quad (9)$$

while an infinitesimal change of the external current produces the alteration

$$\delta_J\,'(F^{(-)\prime}\sigma_1|F^{(+)\prime}\sigma_2)$$
$$=i\,'\left(F^{(-)\prime}\sigma_1\middle|\int_{\sigma_2}^{\sigma_1}(dx)\delta J_\mu A_\mu\middle|F^{(+)\prime}\sigma_2\right). \quad (10)$$

The current variations are subject to the restriction

[2] It is evident from the discussion of the first section, that this is related to the nonexistence of a state with *maximum* energy.

$\partial_\mu \delta J_\mu = 0$. Accordingly, if we rewrite (10) in the notation

$$(\delta/\delta J_\mu(x))\,'(F^{(-)'}\sigma_1|F^{(+)'}\sigma_2) = i\,'(F^{(-)'}\sigma_1|A_\mu(x)|F^{(+)'}\sigma_2), \quad (11)$$

we are at liberty to add an arbitrary gradient to $A_\mu(x)$. Since this coincides with the freedom of gauge transformations, we do not indicate it explicitly.

The advantage provided by the transformation function (8) rests in the possibility of combining (9) and (11) into

$$\delta_F\,'(F^{(-)'}\sigma_1|F^{(+)'}\sigma_2) = \left[2\int_{\sigma_1} d\sigma_\mu \delta F_{\mu\nu}{}^{(-)'}(\delta/\delta J_\nu) - 2\int d\sigma_\mu \delta F_{\mu\nu}{}^{(+)'}(\delta/\delta J_\nu)\right] \times '(F^{(-)'}\sigma_1|F^{(+)'}\sigma_2),$$

which possesses the formal solution

$$'(F^{(-)'}\sigma_1|F^{(+)'}\sigma_2) = \exp\left[2\int_{\sigma_1} d\sigma_\mu F_{\mu\nu}{}^{(-)'}(\delta/\delta J_\nu) - 2\int_{\sigma_2} d\sigma_\mu F_{\mu\nu}{}^{(+)'}(\delta/\delta J_\nu)\right] '(0\sigma_1|0\sigma_2). \quad (12)$$

The problem is thus reduced to the construction of the transformation function referring to null eigenvalues.

We shall write[3]

$$(0\sigma_1|0\sigma_2) = \exp(i\mathcal{W}_0),$$

and

$$(0\sigma_1|A_\mu(x)|0\sigma_2)/(0\sigma_1|0\sigma_2) = \langle A_\mu(x)\rangle.$$

In this notation, the dependence of the null eigenvalue transformation function upon the external current is described by

$$(\delta/\delta J_\mu(x))\mathcal{W}_0 = \langle A_\mu(x)\rangle,$$

or

$$\delta\mathcal{W}_0 = \int_{\infty}^{\infty} (dx)\delta J_\mu(x)\langle A_\mu(x)\rangle, \quad (13)$$

in which we have extended the integration over the entirety of space-time by supposing that the current vanishes externally to the region of interest, the volume bounded by $\sigma_1$ and $\sigma_2$. According to the operator field equation

$$\partial_\nu F_{\mu\nu} = -\partial_\nu{}^2 A_\mu + \partial_\mu \partial_\nu A_\nu = J_\mu,$$

the numerical quantity $\langle A_\mu(x)\rangle$ obeys the differential equation

$$-\partial_\nu{}^2\langle A_\mu\rangle + \partial_\mu\partial_\nu\langle A_\nu\rangle = J_\mu. \quad (14)$$

Now the gauge ambiguity of $\langle A_\mu\rangle$ is completely without effect in (13), since $J_\mu$ vanishes on the boundary of the extended region. Therefore, for the purpose of constructing $\mathcal{W}_0$, we can replace the differential equation (14) with

$$-\partial_\nu{}^2\langle A_\mu\rangle = J_\mu. \quad (15)$$

We are concerned with the solution of this equation that is compatible with the boundary conditions

$$\langle F_{(0)(k)}{}^{(-)}\rangle = 0, \text{ on } \sigma_1, \quad \langle F_{(0)(k)}{}^{(+)}\rangle = 0, \text{ on } \sigma_2,$$

which follow from the nature of the null eigenvalue states on $\sigma_1$ and $\sigma_2$. Since the current vector is zero in the external region, we can rephrase these boundary conditions as the requirement that the field shall contain only positive frequencies in the domain constituting the future of $\sigma_1$, and only negative frequencies in the region prior to $\sigma_2$. This excludes a possible homogeneous solution of (15), whence

$$\langle A_\mu(x)\rangle = \int_{-\infty}^{\infty} (dx')D_+(x-x')J_\mu(x'), \quad (16)$$

in which $D_+(x-x')$ is the Green's function defined by

$$-\partial_\nu{}^2 D_+(x-x') = \delta(x-x'),$$

together with the statement that it contains only positive frequencies for $x_0 > x_0'$, and only negative frequencies for $x_0 < x_0'$. It therefore satisfies the temporal analog of the outgoing wave or radiation condition familiar in the spatial description of a harmonic source.[4]

The expression of (16) provided by

$$(\delta/\delta J_\nu(x'))\langle A_\mu(x)\rangle = (\delta/\delta J_\mu(x))(\delta/\delta J_\nu(x'))\mathcal{W}_0 = \delta_{\mu\nu}D_+(x-x'),$$

indicates that $D_+(x-x')$ is a symmetrical function of $x$ and $x'$. Accordingly, the integral of (13) is

$$\mathcal{W}_0 = \frac{1}{2}\int_{-\infty}^{\infty} (dx)(dx')J_\mu(x)D_+(x-x')J_\mu(x'), \quad (17)$$

apart from the additive constant which is the value of $\mathcal{W}_0$ for the isolated electromagnetic field ($J_\mu = 0$). It is an advantage of the representation we have been employing that this integration constant has the value zero. Indeed, the null eigenvalue states of the complete system provided by the electromagnetic field with no external current are just the $\sigma$-independent vacuum state, whence

$$J_\mu = 0: \quad (0\sigma_1|0\sigma_2) = 1, \quad \mathcal{W}_0 = 0.$$

The differential operator appearing in (12) has the effect of inducing the substitution

$$J_\mu(x) \to J_\mu(x) + 2[\delta_\mu(x, \sigma_1)F_{\mu\nu}{}^{(-)'}(x) - \delta_\mu(x, \sigma_2)F_{\mu\nu}{}^{(+)'}(x)]$$

in $\mathcal{W}_0$. Here $\delta_\mu(x, \sigma)$ represents a one-dimensional delta

[3] The dash is omitted, since there is no distinction between the eigenvectors $\Psi(F^{(+)'}\sigma)$ and $'\Psi(F^{(+)'}\sigma)$, for zero eigenvalues.

[4] Green's functions of this type have been discussed by E. C. G. Stueckelberg, Helv. Phys. Acta 19, 242 (1946); and R. P. Feynman, Phys. Rev. 76, 749 (1949).

function, which is defined by

$$\int (dx)\delta_\mu(x,\sigma)f_\mu(x)=\int_\sigma d\sigma_\mu f_\mu(x).$$

Hence

$$'(F^{(-)'}\sigma_1|F^{(+)'}\sigma_2)=\exp(i'\mathcal{W}),$$

where

$$'\mathcal{W}=\mathcal{W}_0+2\int_{\sigma_1} d\sigma_\mu F_{\mu\nu}{}^{(-)'}(x)\langle A_\nu(x)\rangle$$
$$-2\int_{\sigma_2} d\sigma_\mu F_{\mu\nu}{}^{(+)'}(x)\langle A_\nu(x)\rangle \qquad (18)$$
$$+2\int_{\sigma_1} d\sigma_\mu\int_{\sigma_1} d\sigma_\nu' F_{\mu\lambda}{}^{(-)'}(x)D_+(x-x')F_{\nu\lambda}{}^{(-)'}(x')$$
$$+2\int_{\sigma_2} d\sigma_\mu\int_{\sigma_2} d\sigma_\nu' F_{\mu\lambda}{}^{(+)'}(x)D_+(x-x')F_{\nu\lambda}{}^{(+)'}(x')$$
$$-4\int_{\sigma_1} d\sigma_\mu\int_{\sigma_2} d\sigma_\nu' F_{\mu\lambda}{}^{(-)'}(x)D_+(x-x')F_{\nu\lambda}{}^{(+)'}(x').$$

The following symbolic form of the Green's function $D_+(x-x')$,

$$D_+(x-x')=\tfrac{1}{2}i\omega^{-1}\exp(-i\omega|x_0-x_0'|)\delta(\mathbf{x}-\mathbf{x}'), \qquad (19)$$

shows that

$$x_0=x_0': \quad D_+(x-x')=\tfrac{1}{2}i(\mathbf{x}|\omega^{-1}|\mathbf{x}'),$$

which identifies the double surface integrals, referring to a single surface, in (18), with the factors appearing in (6) and (7), Our result is therefore expressed more simply as

$$(F^{(-)'}\sigma_1|F^{(+)'}\sigma_2)=\exp(i\mathcal{W}), \qquad (20)$$

with

$$\mathcal{W}=\mathcal{W}_0+2\int_{\sigma_1} d\sigma_\mu F_{\mu\nu}{}^{(-)'}(x)\langle A_\nu(x)\rangle$$
$$-2\int_{\sigma_2} d\sigma_\mu F_{\mu\nu}{}^{(+)'}(x)\langle A_\nu(x)\rangle$$
$$-4\int_{\sigma_1} d\sigma_\mu\int_{\sigma_2} d\sigma_\nu' F_{\mu\lambda}{}^{(-)'}(x)D_+(x-x')F_{\nu\lambda}{}^{(+)'}(x').$$

In particular,

$$J_\mu=0: \quad (F^{(-)'}\sigma_1|F^{(+)'}\sigma_2)$$
$$=\exp\left[-4i\int_{\sigma_1} d\sigma_\mu\int_{\sigma_2} d\sigma_\nu' F_{\mu\lambda}{}^{(-)'}(x)\times D_+(x-x')F_{\nu\lambda}{}^{(+)'}(x')\right]. \qquad (21)$$

## APPLICATIONS

Explicit forms of the Green's function $D_+(x-x')$ are required for further work. The Fourier integral version of the three-dimensional delta function in (19) yields

$$D_+(x-x')=\frac{1}{2}i\int\frac{(d\mathbf{k})}{(2\pi)^3}\frac{1}{k_0}\begin{cases}e^{ik(x-x')}, & x_0>x_0'\\ e^{-ik(x-x')}, & x_0<x_0',\end{cases} \qquad (22)$$

where $k_0=|\mathbf{k}|$ is a *positive* frequency. The invariance of this structure is more evident in the four-dimensional transcription

$$D_+(x-x')=i\int\frac{(dk)}{(2\pi)^3}\delta(k^2)e^{ik(x-x')},$$

in which the integration is restricted to positive frequencies for $x_0>x_0'$, and to negative frequencies for $x_0<x_0'$. No conditions on the domain of integration are involved in the alternative four-dimensional form

$$D_+(x-x')=\int\frac{(dk)}{(2\pi)^4}\frac{1}{k^2-i\epsilon}e^{ik(x-x')}, \quad \epsilon\to+0.$$

We shall express the tensor Green's function, $\delta_{\mu\nu}D_+(x-x')$, with the aid of four orthonormal vectors associated with each plane wave,

$$\delta_{\mu\nu}=\sum_{\lambda=1}^{4} e_\mu(\lambda k)e_\nu(\lambda k).$$

We choose the first two vectors to obey the conditions

$$\lambda=1,2: \quad n_\mu e_\mu(\lambda k)=k_\mu e_\mu(\lambda k)=0,$$

in which $n_\mu$ is an arbitrary time-like unit vector,

$$n_\mu{}^2=-1.$$

The remaining two are given explicitly by

$$e_\mu(3k)=n_\mu+k_\mu/(n_\nu k_\nu), \quad n_\mu e_\mu(3k)=0,$$

and

$$e_\mu(4k)=in_\mu.$$

Thus, employing the three-dimensional form (22), we have

$$\delta_{\mu\nu}D_+(x-x')=\frac{1}{2}i\sum_{\lambda=1}^{4}\int\frac{(d\mathbf{k})}{(2\pi)^3}\frac{1}{k_0}e_\mu(\lambda k)\times e^{\pm ikx}e_\nu(\lambda k)e^{\mp ikx'}. \qquad (23)$$

For applications referring to parallel surfaces, there is a useful alternative form of $\mathcal{W}_0$, which corresponds to the construction of $\langle A_\mu\rangle$ in the radiation gauge common to both surfaces,

$$\langle A_0(x)\rangle_{\text{r.g.}}=\int(dx')\mathfrak{D}(x-x')J_0(x'),$$

$$\langle A_k(x)\rangle_{\text{r.g.}}=\int(dx')D_+(x-x')J_k{}^{(T)}(x'),$$

where

$$\mathfrak{D}(x-x')=\delta(x_0-x_0')\mathfrak{D}(\mathbf{x}-\mathbf{x}'),$$

and

$$\mathfrak{D}(\mathbf{x}-\mathbf{x}')=(1/4\pi)|\mathbf{x}-\mathbf{x}'|^{-1}.$$

Thus,

$$\mathfrak{W}_0=\tfrac{1}{2}\int(dx)(dx')[J_k{}^{(T)}(x)D_+(x-x')J_k{}^{(T)}(x') - J_0(x)\mathfrak{D}(x-x')J_0(x')]. \quad (24)$$

The direct proof of equivalence with (17) employs the expression for the longitudinal current,

$$J_k{}^{(L)}(x)=\partial_k\int(dx')\mathfrak{D}(x-x')\partial_0'J_0(x'),$$

and the identity

$$D_+(x-x')=\mathfrak{D}(x-x') +\partial_0\partial_0'\int(dx'')D_+(x-x'')\mathfrak{D}(x''-x'). \quad (25)$$

The latter, incidentally, can be expressed in the symbolic form

$$\tfrac{1}{2}i\omega^{-1}e^{-i\omega|x_0-x_0'|}\delta(\mathbf{x}-\mathbf{x}')=\omega^{-2}\delta(x_0-x_0')\delta(\mathbf{x}-\mathbf{x}') +\partial_0\partial_0'[\tfrac{1}{2}i\omega^{-3}e^{-i\omega|x_0-x_0'|}\delta(\mathbf{x}-\mathbf{x}')]. \quad (26)$$

## Zero Current

We shall use the appropriate form of the transformation function, (21), to illustrate the construction of the eigenvalues and eigenfunctions of $P_\mu$ for a complete system. It is supposed that the surface $\sigma_1$ is obtained from $\sigma_2$ by a translation $X$, which brings the point $x_2$ of $\sigma_2$ into the point $x_1$. Since the surfaces are parallel, and the eigenvalues refer to transverse fields, one can write (21) as

$$(F^{(-)\prime}\sigma_1|F^{(+)\prime}\sigma_2) =\exp\left[-4i\int_{\sigma 1}d\sigma\int_{\sigma 2}d\sigma' F_{0m}{}^{(-)\prime}(x) \times(\delta_{mn}D_+(x-x'))^{(T)}F_{0n}{}^{(+)\prime}(x')\right].$$

With the time-like vector $n_\mu$ identified with the common normal to both surfaces, we see that the $e_\mu(\lambda k)$, $\lambda=1, 2, 3$ are pure space vectors, while the fourth vector possesses only a time component. Furthermore, the first two vectors are orthogonal to $\mathbf{k}$, which the third vector parallels. Hence,

$$x_0>x_0':\quad (\delta_{mn}D_+(x-x'))^{(T)}=\frac{1}{2}i\sum_{\lambda=1,2}\sum_k\left(\frac{(d\mathbf{k})}{(2\pi)^3}\frac{1}{k_0}\right)^{\frac{1}{2}} \times e_m(\lambda k)e^{ikx}\left(\frac{(d\mathbf{k})}{(2\pi)^3}\frac{1}{k_0}\right)^{\frac{1}{2}}e_n(\lambda k)e^{-ikx'},$$

where we have also replaced the integration with respect to $\mathbf{k}$ by a summation over cells of volume $(d\mathbf{k})$.

On defining

$$a_{\lambda k}{}^{(-)}(\sigma_1)=-i\left(\frac{(d\mathbf{k})}{(2\pi)^3}\frac{2}{k_0}\right)^{\frac{1}{2}}\int_{\sigma 1}d\sigma \times F_{0m}{}^{(-)}(x)e_m(\lambda k)e^{ik(x-x_1)}, \quad (27)$$

and

$$a_{\lambda k}{}^{(+)}(\sigma_2)=i\left(\frac{(d\mathbf{k})}{(2\pi)^3}\frac{2}{k_0}\right)^{\frac{1}{2}}\int_{\sigma 2}d\sigma \times e^{-ik(x-x_2)}e_m(\lambda k)F_{0m}{}^{(+)}(x), \quad (28)$$

which are correspondingly constructed linear combinations of the $F_{0m}{}^{(-)}$ on $\sigma_1$, and of the $F_{0m}{}^{(+)}$ on $\sigma_2$, we obtain

$$\begin{aligned}(F^{(-)\prime}\sigma_1|F^{(+)\prime}\sigma_2)&=\exp[\sum_{\lambda k}e^{ikX}a_{\lambda k}{}^{(-)\prime}a_{\lambda k}{}^{(+)\prime}]\\ &=\prod_{\lambda k}\exp[e^{ikX}a_{\lambda k}{}^{(-)\prime}a_{\lambda k}{}^{(+)\prime}]\\ &=\prod_{\lambda k}\sum_{n_{\lambda k}=0}^{\infty}e^{inkX}\frac{(a^{(-)\prime})^n}{(n!)^{\frac{1}{2}}}\frac{(a^{(+)\prime})^n}{(n!)^{\frac{1}{2}}},\end{aligned}$$

or

$$(F^{(-)\prime}\sigma_1|F^{(+)\prime}\sigma_2)=\sum_n\left(\prod_{\lambda k}\frac{(a^{(-)\prime})^n}{(n!)^{\frac{1}{2}}}\right) \times\exp[i(\sum_{\lambda k}nk_\mu)X_\mu]\left(\prod_{\lambda k}\frac{(a^{(+)\prime})^n}{(n!)^{\frac{1}{2}}}\right).$$

A comparison with (1) shows that

$$P_\mu'=P_\mu(n)=\sum_{\lambda k}n_{\lambda k}k_\mu,\quad n_{\lambda k}=0, 1, 2, \cdots, \quad (29)$$

where, in particular,

$$P_0'=\sum_{\lambda k}n_{\lambda k}k_0\geq 0,$$

and that

$$(F^{(-)\prime}|n)=\prod_{\lambda k}\frac{(a^{(-)\prime})^n}{(n!)^{\frac{1}{2}}},\quad (n|F^{(+)\prime})=\prod_{\lambda k}\frac{(a^{(+)\prime})^n}{(n!)^{\frac{1}{2}}}.$$

The occupation numbers $n_{\lambda k}$ provide the complete set of constants of the motion.

Note that if the eigenvalues at corresponding points are in the relation $F_{0m}{}^{(-)\prime}=F_{0m}{}^{(+)\prime *}$, we have $a_{\lambda k}{}^{(-)\prime}=a_{\lambda k}{}^{(+)\prime *}$, and therefore $(F^{(-)\prime}|n)=(n|F^{(+)\prime})^*$, as required by (5). With the knowledge of these simple eigenfunctions, one can construct eigenfunctions for any other representation of interest. We can also present our results without reference to a representation. On remarking that the vacuum state eigenfunction is

$(F^{(+)'}\sigma|0)=1$, we can write

$$(F^{(+)'}\sigma|n\sigma)=\left(\prod_{\lambda k}\frac{(a_{\lambda k}{}^{(+)'})^n}{(n!)^{\frac{1}{2}}}\right)(F^{(+)'}\sigma|0)$$

$$=\left(F^{(+)'}\sigma\left|\prod_{\lambda k}\frac{(a_{\lambda k}{}^{(+)}(\sigma))^n}{(n!)^{\frac{1}{2}}}\right|0\right).$$

Therefore,

$$\Psi(n\sigma)=\left(\prod_{\lambda k}\frac{(a^{(+)}(\sigma))^n}{(n!)^{\frac{1}{2}}}\right)\Psi_0,$$

and

$$\Psi(n\sigma)^\dagger=\Psi_0{}^\dagger\prod_{\lambda k}\frac{(a^{(-)}(\sigma))^n}{(n!)^{\frac{1}{2}}},$$

are the eigenvectors of the state with photon occupation numbers, $n_{\lambda k}$.

### Time Independent Current

In this situation, $J_\mu(x)=J_\mu(\mathbf{x})$, the energy operator $P_0$ is still a constant of the motion, and its eigenvalues and eigenfunctions are obtained from the transformation function that characterizes the time translation

$$T=t_1-t_2,$$

where $t_1$ and $t_2$ are the time coordinates that label $\sigma_1$ and $\sigma_2$.

On employing the form (24) for $\mathfrak{W}_0$ we get

$$\mathfrak{W}_0=-E(0)T+\tfrac{1}{2}\int(d\mathbf{x})(d\mathbf{x}')J_k(\mathbf{x})J_k(\mathbf{x}')$$

$$\times\int_{t_2}^{t_1}dx_0dx_0'[D_+(x-x')-\mathfrak{D}(x-x')],$$

where

$$E(0)=-\tfrac{1}{2}\int(d\mathbf{x})(d\mathbf{x}')J_\mu(\mathbf{x})\mathfrak{D}(\mathbf{x}-\mathbf{x}')J_\mu(\mathbf{x}').$$

According to the symbolic form (26),

$$\int_{t_2}^{t_1}dx_0dx_0'[D_+(x-x')-\mathfrak{D}(x-x')]$$

$$=\int_{t_2}^{t_1}dx_0dx_0'\partial_0\partial_0'[\tfrac{1}{2}i\omega^{-3}e^{-i\omega|x_0-x_0'|}\delta(\mathbf{x}-\mathbf{x}')]$$

$$=i\omega^{-3}(1-e^{-i\omega T})\delta(\mathbf{x}-\mathbf{x}'),$$

so that

$$\mathfrak{W}_0=-E(0)T+\tfrac{1}{2}i\int(d\mathbf{x})J_k\omega^{-3}(1-e^{-i\omega T})J_k.$$

Furthermore,

$$x_0=t_1:\quad\langle A_k(x)\rangle=\int_{t_2}^{t_1}dx_0'\tfrac{1}{2}i\omega^{-1}e^{-i\omega(t_1-x_0')}J_k(\mathbf{x})$$

$$=\tfrac{1}{2}\omega^{-2}(1-e^{-i\omega T})J_k(\mathbf{x}),$$

and

$$x_0=t_2:\quad\langle A_k(x)\rangle=\int_{t_2}^{t_1}dx_0'\tfrac{1}{2}i\omega^{-1}e^{-i\omega(x_0'-t_2)}J_k(\mathbf{x})$$

$$=\tfrac{1}{2}\omega^{-2}(1-e^{-i\omega T})J_k(\mathbf{x}).$$

The transformation function (20) is thus obtained as

$$\frac{(F^{(-)'}\sigma_1|F^{(+)'}\sigma_2)}{(F^{(-)'}|F^{(+)'})}$$

$$=\exp\Big[-iE(0)T-2\int(d\mathbf{x})(F_{0k}{}^{(-)'}+\tfrac{1}{2}i\omega^{-1}J_k)$$

$$\times\frac{1-e^{-i\omega T}}{\omega}(F_{0k}{}^{(+)'}-\tfrac{1}{2}i\omega^{-1}J_k)\Big],\quad(30)$$

in which we have divided by the transformation function referring to a common surface,

$$(F^{(-)'}|F^{(+)'})$$

$$=\exp\left[-4i\int_\sigma d\sigma d\sigma' F_{0k}{}^{(-)'}(x)D_+(x-x')F_{0k}{}^{(+)'}(x')\right]$$

$$=\exp\left[2\int(d\mathbf{x})F_{0k}{}^{(-)'}\omega^{-1}F_{0k}{}^{(+)'}\right].$$

It is evidently desirable to employ a new description, characterized by the eigenvalues $\bar{F}_{0k}{}^{(\pm)'}$, where

$$\bar{F}_{0k}{}^{(\pm)}=F_{0k}{}^{(\pm)}\mp\tfrac{1}{2}i\omega^{-1}J_k,$$

and

$$\bar{A}_k{}^{(\pm)}=A_k{}^{(\pm)}-\tfrac{1}{2}\omega^{-2}J_k.\quad(31)$$

The relation between the eigenvectors $\Psi(\bar{F}^{(+)'}\sigma)$ and $\Psi(F^{(+)'}\sigma)$ can be inferred from the generator

$$G(\bar{F}^{(+)})=\int d\sigma 2\bar{A}_k{}^{(-)}\delta\bar{F}_{0k}{}^{(+)}$$

$$=\int d\sigma(2A_k{}^{(-)}-\omega^{-2}J_k)\delta F_{0k}{}^{(+)}$$

$$=G(F^{(+)})-\delta\left[\int d\sigma J_k\omega^{-2}\tfrac{1}{2}(F_{0k}{}^{(+)}+\bar{F}_{0k}{}^{(+)})\right],$$

where we have maintained the symmetry between $F_{0k}{}^{(+)}$ and $\bar{F}_{0k}{}^{(+)}$ that accompanies the substitution $J_k\rightarrow -J_k$. Thus,

$$\Psi(\bar{F}^{(+)'}\sigma)$$

$$=\exp\left[i\int d\sigma J_k\omega^{-2}\tfrac{1}{2}(F_{0k}{}^{(+)'}+\bar{F}_{0k}{}^{(+)'})\right]\Psi(F^{(+)'}\sigma)$$

$$=\exp\left[i\int d\sigma J_k\omega^{-2}F_{0k}{}^{(+)'}+\tfrac{1}{4}\int d\sigma J_k\omega^{-3}J_k\right]\Psi(F^{(+)'}\sigma),$$

and

$$\Phi(\bar{F}^{(-)\prime}\sigma)=\exp\left[-i\int d\sigma J_k\omega^{-2}F_{0k}{}^{(-)\prime}+\tfrac{1}{4}\int d\sigma J_k\omega^{-3}J_k\right]\Phi(F^{(-)\prime}\sigma).$$

In further confirmation, observe that

$$\begin{aligned}(\bar{F}^{(-)\prime}|\bar{F}^{(+)\prime})&=\exp\left[2\int d\sigma\bar{F}_{0k}{}^{(-)\prime}\omega^{-1}\bar{F}_{0k}{}^{(+)\prime}\right]\\&=\exp\left[-i\int d\sigma J_k\omega^{-2}F_{0k}{}^{(-)\prime}+\tfrac{1}{4}\int d\sigma J_k\omega^{-3}J_k\right]\\&\quad\times(F^{(-)\prime}|F^{(+)\prime})\exp\left[i\int d\sigma J_k\omega^{-2}F_{0k}{}^{(+)\prime}+\tfrac{1}{4}\int d\sigma J_k\omega^{-3}J_k\right],\end{aligned}$$

which results in the same eigenvector transformation properties. Since these conversion factors do not refer explicitly to the surface, the transformation function ratio in (30) preserves its structure on introducing the new representations. Therefore,

$$(\bar{F}^{(-)\prime}\sigma_1|\bar{F}^{(+)\prime}\sigma_2)=\exp(i\bar{\mathfrak{W}}),\qquad(32)$$

where

$$\begin{aligned}\bar{\mathfrak{W}}&=-E(0)T-2i\int(d\mathbf{x})\bar{F}_{0k}{}^{(-)\prime}\omega^{-1}e^{-i\omega T}\bar{F}_{0k}{}^{(+)\prime}\\&=-E(0)T-4\int_{\sigma_1}d\sigma\int_{\sigma_2}d\sigma'\bar{F}_{0k}{}^{(-)\prime}(x)D_+(x-x')\times\bar{F}_{0k}{}^{(+)\prime}(x').\end{aligned}$$

Apart from the factor $\exp(-iE(0)T)$, this transformation function is identical in form with (21). The expanded version of (32) is, therefore,

$$(\bar{F}^{(-)\prime}\sigma_1|\bar{F}^{(+)\prime}\sigma_2)=\sum_n(\bar{F}^{(-)\prime}|n)e^{-iP_0'T}(n|\bar{F}^{(+)\prime}),$$

where

$$P_0'=E(0)+\sum_{\lambda k}n_{\lambda k}k_0,\quad n_{\lambda k}=0,1,2,\cdots,$$

and

$$(n|\bar{F}^{(+)\prime})=\prod_{\lambda k}\frac{(\bar{a}^{(+)\prime})^n}{(n!)^{\frac{1}{2}}},\quad(\bar{F}^{(-)\prime}|n)=\prod_{\lambda k}\frac{(\bar{a}^{(-)\prime})^n}{(n!)^{\frac{1}{2}}}.$$

We see that $E(0)$ is to be interpreted as the energy of the photon vacuum, the state of minimum energy, $n_{\lambda k}=0$. With respect to this displaced origin, the energy eigenvalues are the same as in the absence of a current. One may say that the field $\bar{F}_{0k}$ describes pure radiation, which is without coupling to the external current. Indeed, (31), written as

$$\bar{A}_k{}^{(\pm)}(x)=A_k{}^{(\pm)}(x)-\tfrac{1}{2}\int(dx')\mathfrak{D}(x-x')J_k(x'),$$

represents the removal from $A_k(x)$ of the time-independent potential produced by the static current, which is allocated equally to $A_k{}^{(+)}$ and to $A_k{}^{(-)}$. In view of this uncoupling of the static field and the radiation field, one can assign momentum as well as energy eigenvalues to the radiation quanta, as given by (29).

### Time Dependent Currents

We shall now discuss the class of problems in which the physical information contained in the general transformation function (20) refers to transition probabilities rather than eigenvalues. Let us first suppose that the current is zero on $\sigma_2$, varies in an arbitrary manner in the region between the parallel surfaces $\sigma_1$ and $\sigma_2$, but again reduces to zero on $\sigma_1$. Thus, the physical states on $\sigma_1$ and $\sigma_2$ are those of the isolated electromagnetic field, and we wish to compute the probabilities of the transitions induced by this perturbing current.

Now,

$$\begin{aligned}2\int_{\sigma_1}d\sigma F_{0m}{}^{(-)\prime}(x)\langle A_m(x)\rangle&=2\int_{\sigma_1}d\sigma F_{0m}{}^{(-)\prime}(x)\int_{\sigma_2}^{\sigma_1}(dx')(\delta_{mn}D_+(x-x'))^{(T)}J_n(x')\\&=-\sum_{\lambda k}a_{\lambda k}{}^{(-)\prime}e^{ikx_1}J_{\lambda k},\end{aligned}$$

and

$$\begin{aligned}2\int_{\sigma_2}d\sigma F_{0m}{}^{(+)\prime}(x)\langle A_m(x)\rangle&=2\int_{\sigma_2}d\sigma F_{0m}{}^{(+)\prime}(x)\int_{\sigma_2}^{\sigma_1}(dx')(\delta_{mn}D_+(x-x'))^{(T)}J_n(x')\\&=\sum_{\lambda k}a_{\lambda k}{}^{(+)\prime}e^{-ikx_2}J_{\lambda k}{}^*,\end{aligned}$$

where

$$J_{\lambda k}=\left(\frac{(d\mathbf{k})}{(2\pi)^3}\frac{1}{2k_0}\right)^{\frac{1}{2}}\int_{-\infty}^{\infty}(dx)e_m(\lambda k)e^{-ikx}J_m(x).\qquad(33)$$

The quantity $\mathfrak{W}$, determining the transformation function $(F^{(-)\prime}\sigma_1|F^{(+)\prime}\sigma_2)$, is thus obtained in the form

$$\mathfrak{W}=\mathfrak{W}_0-\sum_{\lambda k}[a_{\lambda k}{}^{(-)\prime}e^{ikx_1}J_{\lambda k}+a_{\lambda k}{}^{(+)\prime}e^{-ikx_2}J_{\lambda k}{}^*+ia_{\lambda k}{}^{(-)\prime}e^{ikx_1}a_{\lambda k}{}^{(+)\prime}e^{ikx_2}].$$

The transformation function then serves, according to

$$(F^{(-)}\sigma_1|F^{(+)\prime}\sigma_2)=\sum_{n,n'}(F^{(-)\prime}|n)(n\sigma_1|n'\sigma_2)(n'|F^{(+)\prime}),$$

as a generating function for $(n\sigma_1|n'\sigma_2)$, from which the transition probabilities are found in the manner of (2).

It is somewhat more convenient to deal with the elements of the matrix

$$(n|S|n')=e^{-iP^{(n)}x_1}(n\sigma_1|n'\sigma_2)e^{iP^{(n')}x_2},$$

$$p(n,n')=|(n|S|n')|^2,$$

since they are independent of $\sigma_1$ and $\sigma_2$, provided the current vanishes on these surfaces. The following substitution, representing a transformation to a common reference surface,

$$a_{\lambda k}{}^{(-)\prime}e^{ikx_1}\to a_{\lambda k}{}^{(-)\prime}, \quad a_{\lambda k}{}^{(+)\prime}e^{-ikx_2}\to a_{\lambda k}{}^{(+)\prime},$$

yields the generating function

$$\exp(i\mathcal{W}_0)\prod_{\lambda k}\exp[a^{(-)\prime}a^{(+)\prime}-ia^{(-)\prime}J-ia^{(+)\prime}J^*] \qquad (34)$$
$$=\sum_{n,n'}(F^{(-)\prime}|n)(n|S|n')(n'|F^{(+)\prime}).$$

On picking out the coefficient of a particular $(F^{(-)\prime}|n)$, we obtain the partial generating function

$$\exp(i\mathcal{W}_0)\prod_{\lambda k}\left[\frac{(a^{(+)\prime}-iJ)^n}{(n!)^{\frac{1}{2}}}\exp(-ia^{(+)\prime}J^*)\right]$$
$$=\sum_{n'}(n|S|n')(n'|F^{(+)\prime}), \quad (35$$

and, similarly,

$$\exp(i\mathcal{W}_0)\prod_{\lambda k}\left[\frac{(a^{(-)\prime}-iJ^*)^{n'}}{(n'!)^{\frac{1}{2}}}\exp(-ia^{(-)\prime}J)\right]$$
$$=\sum_{n}(F^{(-)\prime}|n)(n|S|n'). \quad (36)$$

Preliminary to a direct verification of the unitary property of the operator $S$, we evaluate the imaginary part of $\mathcal{W}_0$. According to (24),

$$2\,\mathrm{Im}\,\mathcal{W}_0=\int(dx)(dx')J_m(x)$$
$$\times\mathrm{Im}(\delta_{mn}D_+(x-x'))^{(T)}J_n(x'),$$

where, referring to (23),

$$\mathrm{Im}(\delta_{mn}D_+(x-x'))^{(T)}$$
$$=\mathrm{Re}\sum_{\lambda=1,2}\int\frac{(d\mathbf{k})}{(2\pi)^3}\frac{1}{2k_0}e_m(\lambda k)e^{ikx}e_n(\lambda k)e^{-ikx'}.$$

This form is valid without restriction on $x_0-x_0'$. Hence,

$$2\,\mathrm{Im}\,\mathcal{W}_0=\sum_{\lambda k}|J_{\lambda k}|^2.$$

Alternatively, the invariant expression (17) yields

$$2\,\mathrm{Im}\,\mathcal{W}_0=\int(dx)(dx')J_\mu(x)\,\mathrm{Im}\,D_+(x-x')J_\mu(x'),$$

with

$$\mathrm{Im}\,D_+(x-x')=\mathrm{Re}\int\frac{(d\mathbf{k})}{(2\pi)^3}\frac{1}{2k_0}e^{ikx}e^{-ikx'},$$

which can be written

$$2\,\mathrm{Im}\,\mathcal{W}_0=\sum_k I_k.$$

Here

$$I_k=\frac{(d\mathbf{k})}{(2\pi)^3}\frac{1}{2k_0}\left|\int(dx)e^{-ikx}J_\mu(x)\right|^2=\sum_{\lambda=1,2}|J_{\lambda k}|^2,$$

in which it must be understood that the complex conjugation does not extend to $J_4=iJ_0$. The necessary equivalence of the two evaluations indicates the complete cancellation of the integrals associated with $\lambda=3$ and 4.

Let us multiply (35) with the complex conjugate equation,

$$\exp(-i\mathcal{W}_0{}^*)\prod_{\lambda k}\left[\frac{(a^{(-)\prime}+iJ^*)^n}{(n!)^{\frac{1}{2}}}\exp(ia^{(-)\prime}J)\right]$$
$$=\sum_{n'}(F^{(-)\prime}|n')(n'|S^\dagger|n), \quad (37)$$

and perform the summation with respect to $n$,

$$(F^{(-)\prime}|S^\dagger S|F^{(+)\prime})$$
$$=\prod_{\lambda k}\exp[-|J|^2+(a^{(-)}+iJ^*)(a^{(+)}-iJ)$$
$$+ia^{(-)}J-ia^{(+)}J^*]$$
$$=\prod_{\lambda k}\exp(a^{(-)\prime}a^{(+)\prime})=(F^{(-)\prime}|F^{(+)\prime}),$$

whence $S^\dagger S=1$. A symmetry property of $S$ may be noted here. The invariance of the generating function (34), under the substitution

$$a_{\lambda k}{}^{(+)\prime}\to a_{\lambda k}{}^{(-)\prime}(J_{\lambda k}/J_{\lambda k}{}^*),$$
$$a_{\lambda k}{}^{(-)\prime}\to a_{\lambda k}{}^{(+)\prime}(J_{\lambda k}{}^*/J_{\lambda k}),$$

shows that

$$(n|S|n')=(n'|S|n)\prod_{\lambda k}(J/J^*)^{n-n'},$$

which has the consequence $p(n,n')=p(n',n)$.

As an elementary application of the generating function, we place $n_{\lambda k}'=0$ in (36), which yields

$$(n|S|0)=\exp(i\mathcal{W}_0)\prod_{\lambda k}\frac{(-iJ)^n}{(n!)^{\frac{1}{2}}},$$

and

$$p(n,0)=\prod_{\lambda k}\left[\frac{(|J|^2)^n}{n!}\exp(-|J|^2)\right], \qquad (38)$$

for the situation in which no quanta are present initially. If we are not concerned with the polarization of the emitted quanta, we can employ the binomial theorem to replace (38) with

$$p(n,0)=\prod_k\left[\frac{(I_k)^n}{n!}\exp(-I_k)\right].$$

The general matrix element of $S$ is obtained as

$$(n|S|n')=\exp(i\mathcal{W}_0)\prod_{\lambda k}\left[\frac{i^{n+n'}}{(n!n'!)^{\frac{1}{2}}}J^nJ^{*n'}f_{n,n'}(|J|^2)\right],$$

where the function $f_{n,n'}(x)$, which is symmetrical in $n$

and $n'$, is given by

$$f_{n,n'}(x)=x^{-n'}e^{x}(d/dx)^{n}(-x)^{n'}e^{-x}$$
$$=x^{-n}e^{x}(d/dx)^{n'}(-x)^{n}e^{-x}$$
$$=(-1)^{n_>}n_<!x^{-n_<}L_{n_<}^{(n_>-n_<)}(x).$$

In the indicated relation to the Laguerre polynomials,[5] $n_>$ and $n_<$ represent the greater and lesser of the integers $n$ and $n'$. The general transition probability is thus obtained as

$$p(n,n')=\prod_{\lambda k}\left[\frac{n_<!}{n_>!}(|J|^2)^{n_>-n_<}(L_{n_<}^{(n_>-n_<)}(|J|^2))^2 \times\exp(-|J|^2)\right]. \quad (39)$$

In particular, the probability that there be no change in the numbers of quanta is

$$p(n,n)=\prod_{\lambda k}[(L_n^{(0)}(|J|^2))^2\exp(-|J|^2)].$$

Should the quantum numbers $n$ and $n'$ be large in comparison with unity and $\Delta n=n-n'\ll n,\ n'$, for a particular mode of the radiation field, we can replace the factor in the transition probability referring to that mode with the Bessel function asymptotic form $[J_{\Delta n}(2n^{\frac{1}{2}}|J|)]^2$.

One can devise another generating function for the transition probabilities which has the further advantage of yielding the expectation values of powers of the final occupation numbers. We first perform the substitution $a_{\lambda k}^{(+)\prime}\to a_{\lambda k}^{(+)\prime}e^{-i\gamma_{\lambda k}}$ in (35), where the $\gamma_{\lambda k}$ are arbitrary constants. The result,

$$\exp(i\mathcal{W}_0)\prod_{\lambda k}\left[\frac{(a^{(+)\prime}-ie^{i\gamma}J)^n}{(n!)^{\frac{1}{2}}}\exp(-ia^{(+)}e^{-i\gamma}J^*)\right]$$
$$=\sum_{n'}[\prod_{\lambda k}\exp(i\gamma(n-n'))](n|S|n')(n'|F^{(+)}),$$

is then multiplied by (37) and the summation with respect to $n$ performed. This gives

$$\sum(F^{(-)\prime}|n')(n'|S^{\dagger}|n)[\prod_{\lambda k}\exp(i\gamma(n-n''))]\times(n|S|n'')(n''|F^{(+)\prime})$$
$$=\prod_{\lambda k}\exp[a^{(-)\prime}a^{(+)\prime}-ia^{(-)\prime}(e^{i\gamma}-1)J -ia^{(+)\prime}(e^{-i\gamma}-1)J^*+(e^{i\gamma}-1)|J|^2],$$

which exhibits the same structure as (34). On confining our attention to the diagonal matrix elements, $n_{\lambda k}'=n_{\lambda k}''$, we find

$$\sum_n[\prod_{\lambda k}\exp(i\gamma(n-n'))]p(n,n')=\langle\prod_{\lambda k}\exp(i\gamma(n-n'))\rangle_{n'}$$
$$=\prod_{\lambda k}[L_{n'}^{(0)}((e^{i\gamma}-1)(e^{-i\gamma}-1)|J|^2)\times\exp((e^{i\gamma}-1)|J|^2)]. \quad (40)$$

The right side thus serves as a generating function for the transition probabilities if developed in positive and negative powers of the $e^{i\gamma_{\lambda k}}$. The expansion in powers of the $\gamma_{\lambda k}$ exhibits it as the generator of expectation values of all powers of the quantities $n_{\lambda k}-n_{\lambda k}'$.

The alternative presentation of this result,

$$\langle\prod_{\lambda k}(1+x_{\lambda k})^{n-n'}\rangle_{n'}=\prod_{\lambda k}\left[L_{n'}^{(0)}\left(-\frac{x^2}{1+x}|J|^2\right)\exp(x|J|^2)\right],$$

supplies the expectation values of products successively decreasing by unity. Thus, in the special example referring to the vacuum as the initial state, where

$$\langle\prod_{\lambda k}(1+x)^n\rangle_0=\prod_{\lambda k}\exp(x|J|^2),$$

we find, for a particular mode,

$$\langle n!/(n-k)!\rangle_0=(|J|^2)^k,$$

which is characteristic of the Poisson distribution. The first two expectation values, derived from the general generating function, are

$$\langle(n_{\lambda k}-n_{\lambda k}')\rangle_{n'}=|J_{\lambda k}|^2 \quad (41)$$

and

$$\langle(n_{\lambda k}-n_{\lambda k}')^2\rangle_{n'}=\langle(n_{\lambda k}-n_{\lambda k}')\rangle_{n'}^2+(2n_{\lambda k}'+1)|J_{\lambda k}|^2.$$

We need hardly remark on the statistical independence of different modes.

If we are not interested in the polarizations of the emitted quanta, it suffices to identify the parameters distinguishing the different polarizations, $\gamma_{\lambda k}=\gamma_k$. To obtain statements referring also to unpolarized incident quanta, we must average, with equal weight, over the various polarized photon numbers that are consistent with a given number of photons in a certain propagation mode, $n_{1k}'+n_{2k}'=n_k'$. This can be accomplished with the aid of the addition theorem for the Laguerre polynomials. The resulting generating function, without reference to polarization, is

$$\sum_n[\prod_k\exp(i\gamma(n-n'))]p(n,n')$$
$$=\langle\prod_k\exp(i\gamma(n-n'))\rangle_{n'}$$
$$=\prod_k[(n'+1)^{-1}L_{n'}^{(1)}((e^{i\gamma}-1)(e^{-i\gamma}-1)I)\times\exp((e^{i\gamma}-1)I)].$$

[5] We employ the definition of W. Magnus and F. Oberhettinger, *Special Functions of Mathematical Physics* (Chelsea Publishing Company, New York, 1949), p. 84.

Some expectation values are $\langle(n_k-n_k')\rangle_{n'}=I_k$ and

$$\langle(n_k-n_k')^2\rangle_{n'}=\langle(n_k-n_k')\rangle_{n'}^2+(n_k'+1)I_k.$$

For the second example that is concerned with the evaluation of transition probabilities, we suppose that the current is time-independent in the vicinity of $\sigma_2$, varies in an arbitrary manner in the region between the parallel surfaces $\sigma_1$ and $\sigma_2$, but again becomes time-independent in the neighborhood of $\sigma_1$. These limiting forms, $J_\mu(\mathbf{x}, 1)$ and $J_\mu(\mathbf{x}, 2)$, need not be the same.

On each surface, we use the description appropriate to the current on that surface,

$$\begin{aligned}(\bar{F}^{(-)\prime}\sigma_1|\bar{F}^{(+)\prime}\sigma_2)&=\exp\Big[-i\int d\sigma J_k(1)\omega^{-2}F_{0k}^{(-)\prime}\\&\quad+\tfrac{1}{4}\int d\sigma J_k(1)\omega^{-3}J_k(1)\Big](F^{(-)\prime}\sigma_1|F^{(+)\prime}\sigma_2)\\&\quad\times\exp\Big[i\int d\sigma J_k(2)\omega^{-2}F_{0k}^{(+)\prime}\\&\quad+\tfrac{1}{4}\int d\sigma J_k(2)\omega^{-3}J_k(2)\Big],\end{aligned}\tag{42}$$

where

$$\bar{F}_{0k}^{(-)}=F_{0k}^{(-)}+\tfrac{1}{2}i\omega^{-1}J_k(1),\ \bar{F}_{0k}^{(+)}=F_{0k}^{(+)}-\tfrac{1}{2}i\omega^{-1}J_k(2).$$

Were the current constant, this transformation function would possess the form (32). Accordingly, it must be possible to express all additional contributions in terms of the time derivative of the current. The manner in which this occurs can be illustrated with the evaluation of

$$\begin{aligned}\int_{\sigma_1} d\sigma F_{0k}^{(-)\prime}(x)\langle A_k(x)\rangle&=-\int_{\sigma_1} d\sigma F_{0k}^{(-)\prime}(x)\int_{t_2}^{t_1}dx_0'\int(d\mathbf{x}')\partial_0'\\&\qquad\times D_+(x-x')i\omega^{-1}J_k(x')\\&=\int_{\sigma_1}d\sigma F_{0k}^{(-)\prime}\tfrac{1}{2}\omega^{-2}J_k(1)+\int_{\sigma_1}d\sigma\int_{\sigma_2}d\sigma' F_{0k}^{(-)\prime}(x)\\&\qquad\times D_+(x-x')i\omega^{-1}J_k(\mathbf{x}',2)\\&\quad+\int_{\sigma_1}d\sigma F_{0k}^{(-)\prime}(x)\int_{\sigma_2}^{\sigma_1}(dx')D_+(x-x')i\omega^{-1}\partial_0'J_k(x').\end{aligned}$$

The terms containing $J_k(1)$ and $J_k(2)$ are canceled on expressing (42) as a function of the variables $\bar{F}_{0k}^{(-)\prime}$ and $\bar{F}_{0k}^{(+)\prime}$. Carrying out a similar reduction of $\mathfrak{W}_0$ with the aid of (25), we obtain

$$\begin{aligned}(\bar{F}^{(-)\prime}\sigma_1|\bar{F}^{(+)\prime}\sigma_2)&=\exp\Big[i\bar{\mathfrak{W}}_0-4i\int_{\sigma_1}d\sigma\int_{\sigma_2}d\sigma'\bar{F}_{0k}^{(-)\prime}(x)D_+(x-x')\\&\quad\times\bar{F}_{0k}^{(+)\prime}(x')+2i\int_{\sigma_1}d\sigma\bar{F}_{0k}^{(-)\prime}(x)\int_{\sigma_2}^{\sigma_1}(dx')\\&\quad\times D_+(x-x')i\omega^{-1}\partial_0'J_k(x')\\&\quad+2i\int_{\sigma_2}d\sigma\bar{F}_{0k}^{(+)\prime}(x)\int_{\sigma_2}^{\sigma_1}\\&\quad\times D_+(x-x')i\omega^{-1}\partial_0'J_k(x')\Big],\end{aligned}$$

where

$$\begin{aligned}\bar{\mathfrak{W}}_0&=\tfrac{1}{2}\int_{\sigma_2}^{\sigma_1}(dx)(dx')[J_k^{(T)}(x)J_k^{(T)}(x')-J_0(x)J_0(x')]\\&\quad\times\mathfrak{D}(x-x')+\tfrac{1}{2}\int_{\sigma_2}^{\sigma_1}(dx)(dx')\partial_0J_k^{(T)}(x)\omega^{-1}\\&\quad\times D_1(x-x')\omega^{-1}\partial_0'J_k^{(T)}(x').\end{aligned}$$

The introduction of the variables $\bar{a}_{\lambda k}^{(-)}$ and $\bar{a}_{\lambda k}^{(+)}$, in the manner of (27) and (28), brings this transformation function into the form

$$\begin{aligned}(\bar{F}^{(-)\prime}\sigma_1|\bar{F}^{(+)\prime}\sigma_2)&=\exp(i\bar{\mathfrak{W}}_0)\prod_{\lambda k}\exp[\bar{a}^{(-)\prime}e^{ikx_1}\bar{a}^{(+)\prime}e^{-ikx_2}\\&\quad-i\bar{a}^{(-)\prime}e^{ikx_1}\bar{J}-i\bar{a}^{(+)\prime}e^{-ikx_2}\bar{J}^*],\end{aligned}\tag{43}$$

where

$$\bar{J}_{\lambda k}=\left(\frac{(d\mathbf{k})}{(2\pi)^3}\frac{1}{2k_0}\right)^{\frac{1}{2}}\int_{-\infty}^{\infty}(dx)e_m(\lambda k)e^{-ikx}(i/k_0)\partial_0J_m(x)\tag{44}$$

is expressed as an integral over all space-time by supposing that the current in the extended domain exhibits the time-independent value appropriate to the nearest surface bounding the region of interest. In order to present this result as a generating function for the surface independent unitary matrix

$$(n|S|n')=e^{-iP(n,1)x_1}(n\sigma_1|n'\sigma_2)e^{iP(n',2)x_2},$$

$$P_0(n)=E(0)+\textstyle\sum_{\lambda k}n_{\lambda k}k_0,$$

a further rearrangement of $\bar{\mathfrak{W}}_0$ is required. Indeed, the first term of this quantity is

$$-\int_{t_2}^{t_1}dx_0E(0,x_0)=-[t_1E(0,1)-t_2E(0,2)]+\int_{-\infty}^{\infty}dx_0x_0\partial_0E(0,x_0).$$

The substitution

$$\bar{a}^{(-)\prime}e^{ikx_1}\to\bar{a}^{(-)\prime},\quad \bar{a}^{(+)\prime}e^{ikx_2}\to\bar{a}^{(+)\prime},$$

now yields the required generating function

$$\exp(iw_0)\prod_{\lambda k}[\bar{a}^{(-)\prime}\bar{a}^{(+)\prime}-i\bar{a}^{(-)\prime}\bar{J}-i\bar{a}^{(+)\prime}\bar{J}^*]$$
$$=\sum_{n,n'}(\bar{F}^{(-)\prime}|n)(n|S|n')(n'|\bar{F}^{(+)\prime}), \quad (45)$$

where

$$w_0=\int_{-\infty}^{\infty}dx_0x_0\partial_0E(0,x_0)+\tfrac{1}{2}\int_{-\infty}^{\infty}(dx)(dx')$$
$$+\partial_0J_k^{(T)}(x)\omega^{-1}D_+(x-x')\omega^{-1}\partial_0'J_k^{(T)}(x') \quad (46)$$

is such that

$$2\,\mathrm{Im}\,w_0=\sum_{\lambda k}|\bar{J}_{\lambda k}|^2.$$

The generating function (45) is identical in structure with (34). Hence the transition probabilities and expectation values are given by (39) and (40), with $\bar{J}_{\lambda k}$ replacing $J_{\lambda k}$. If the current is zero on the boundaries of the region, an integration by parts reduces $\bar{J}_{\lambda k}$ to $J_{\lambda k}$. Notice also that if the current is time-independent, (45) asserts that $(n|S|n')=\delta(n,n')$, attesting to the stationary character of the states labeled by the photon numbers.

*The infrared catastrophe.*—According to (41), the average number of photons emitted into a particular mode is given by

$$|\bar{J}_{\lambda k}|^2=\frac{(d\mathbf{k})}{(2\pi)^3}\frac{1}{2k_0^3}\left|\int_{-\infty}^{\infty}(dx)e_m(\lambda k)e^{-ikx}\partial_0J_m(x)\right|^2.$$

We shall now consider frequencies that are sufficiently low for the wavelength to be large in comparison with the linear dimensions of the spatio-temporal region in which the current changes. The average number of photons, of either polarization, that emerge in a range of such low frequencies is then

$$\int\frac{(d\mathbf{k})}{(2\pi)^3}\frac{1}{2k_0^3}\left|\int_{-\infty}^{\infty}(dx)\partial_0J_\mu(x)\right|^2$$
$$=\frac{1}{4\pi^2}\int\frac{dk_0}{k_0}\left|\int(d\mathbf{x})(J_\mu(\mathbf{x},1)-J_\mu(\mathbf{x},2))\right|^2,$$

which becomes infinite as the lower frequency limit approaches zero. Any time variation of the current thus produces a logarithmically infinite number of zero frequency photons—a fact well-known as the "infrared catastrophe." Accordingly, we find a zero probability for the emission of a finite number of photons. To avoid this type of statement we recognize that in any experimental arrangement there is a minimum detectable frequency, $k_{0,\min}$, such that we have no knowledge of the number of photons emitted into modes with frequencies less than $k_{0,\min}$. If we sum the general transition probability (39) over all final occupation numbers referring to these unobservable modes, we are left with the same expression constructed only from the observable modes. The latter will yield nonvanishing probabilities for the emission of a finite number of photons, each transition probability being dependent upon $k_{0,\min}$ through the factor

$$\exp(-\sum_{\lambda k}|\bar{J}_{\lambda k}|^2)=\exp\left[-\int_{k_{0,\min}}^{\infty}\frac{(d\mathbf{k})}{(2\pi)^3}\frac{1}{2k_0^3}\times\left|\int_{-\infty}^{\infty}(dx)e^{-ikx}\partial_0J_\mu(x)\right|^2\right].$$

This quantity represents the probability that no (observable) photon will be emitted, if none are present initially.

*The adiabatic theorem.*—This important statement refers to the situation in which the current changes from its initial to its final value at a rate determined by the total elapsed time $T=t_1-t_2$. We are particularly interested in the limit in which $T$ becomes very large compared to the periods of all observable modes, $k_{0,\min}T\to\infty$. The above description of the current time variation is expressed quantitatively by

$$\int(d\mathbf{x})e_m(\lambda k)e^{-i\mathbf{k}\cdot\mathbf{x}}J_m(x)=j_{\lambda k}(\zeta),\quad \zeta=(x_0-t_2)/T.$$

Hence the integral occurring in (44) is essentially determined by

$$\int_0^1 d\zeta e^{ik_0T\zeta}j'(\zeta).$$

Now, according to the Riemann-Lebesgue lemma,[6]

$$\lim_{k_0T\to\infty}\left|\int_0^1 d\zeta e^{ik_0T\zeta}j'(\zeta)\right|=0, \quad (47)$$

provided only that

$$\int_0^1 d\zeta|j'(\zeta)|<\infty.$$

This suffices to establish that

$$\lim_{T\to\infty}|\bar{J}_{\lambda k}|^2=0.$$

If $j_{\lambda k}'(\zeta)$ is of limited total fluctuation, the integral in (47) approaches zero as $(k_0T)^{-1}$, which enables us to satisfy

$$\lim_{T\to\infty}\sum_{\lambda k}|\bar{J}_{\lambda k}|^2=0,$$

without essential restriction on the spatial distribution of the current (it must not be as singular as the gradient of a delta function). Under these conditions we obtain the probability zero for any change in the photon numbers, despite the alteration in the current.

[6] E. T. Whittaker and G. N. Watson, *Modern Analysis* (Macmillan Company, New York, 1927), p. 172.

This theorem can be exploited to give a uniform expression for the results of all problems involving transition probabilities. Thus, in the integration over the extended region in (44), it is supposed that the current is constant in the exterior region. If we were to replace these constant currents by currents decreasing adiabatically to zero at infinity, the null contribution from the external region would not be affected. But we would have succeeded in substituting for the original problem one in which the current vanishes on the boundaries of the extended region. Accordingly, we can integrate by parts in (44) and regain the form (33) appropriate to null currents on the boundaries. The most general problem requiring the evaluation of transition probabilities between stationary states, involves initial and final currents that are time-independent with respect to different reference systems. When modified with the aid of the adiabatic device, this situation also falls into the class of problems covered by (34).

The adiabatic device is also applicable to eigenvalue problems. Thus, we can use the transformation function (34), appropriate to zero current on the boundary surfaces, to construct the energy eigenvalues for the situation of a time-independent current. We suppose that the current, which is zero on the surface $\sigma_{-\infty}$, grows adiabatically and maintains a constant value between surfaces $\sigma_2$ and $\sigma_1$, and reduces adiabatically to zero on $\sigma_\infty$. The designations $\sigma_{\pm\infty}$ refer to the fact that the adiabatic theorem involves the limit of infinite temporal separation between $\sigma_\infty$ and $\sigma_1$, and between $\sigma_2$ and $\sigma_{-\infty}$. Then

$$(n\sigma_\infty | n'\sigma_{-\infty}) = \delta(n, n') \exp[i\mathcal{W}_0 + iP(n)(x_\infty - x_{-\infty})],$$

where [reversing the integration by parts in the first term of (46)],

$$\mathcal{W}_0 = -\int_{-\infty}^{\infty} dx_0 E(0, x_0),$$

and

$$\exp(i\mathcal{W}_0) = \exp\left(-i\int_{t_1}^{\infty} dx_0 E(0, x_0)\right)$$
$$\times \exp(-iE(0)(t_1 - t_2)) \exp\left(-i\int_{-\infty}^{t_2} dx_0 E(0, x_0)\right).$$

On recalling the composition property of transformation functions, we recognize immediately that

$$(n\sigma_1 | n'\sigma_2) = \delta(n, n') \times \exp[-iE(0)(t_1 - t_2) + iP(n)(x_1 - x_2)],$$

which shows that, in the presence of a time-independent current, the energy eigenvalues of the radiation field are displaced by $E(0)$.

The methods discussed in this paper and illustrated for the electromagnetic field are equally applicable to other Bose-Einstein systems, such as the symmetrical pseudoscalar meson field.

# The Theory of Quantized Fields. IV

JULIAN SCHWINGER
*Harvard University, Cambridge, Massachusetts*
(Received August 6, 1953)

The principal development in this paper is the extension of the eigenvalue-eigenvector concept to complete sets of anticommuting operators. With the aid of this formalism we construct a transformation function for the Dirac field, as perturbed by an external source. This transformation function is enlarged to describe phase transformations and, when applied to the isolated Dirac field, yields the charge and energy-momentum eigenvalues and eigenfunctions. The transformation function describing the system in the presence of the source is then used as a generating function to construct the matrices of all ordered products of the field operators, for the isolated Dirac field. The matrices in the occupation number representation are exhibited with a classification that effectively employs a time-reversed description for negative frequency modes. The last section supplements III by constructing the matrices of all ordered products of the potential vector, for the isolated electromagnetic field.

## INTRODUCTION

THIS paper and its sequel are continuations of III[1] in their concern with a single externally perturbed field. We shall discuss the Dirac field as perturbed by a second prescribed Dirac field, which appears as an external source, or by a prescribed Bose-Einstein field, as exemplified by a given electromagnetic field. The Lagrange function of this system is

$$\mathcal{L}=-\tfrac{1}{4}[\bar{\psi},\gamma_\mu(-i\partial_\mu-eA_\mu)\psi+m\psi] -\tfrac{1}{4}[(i\partial_\mu-eA_\mu)\bar{\psi}\gamma_\mu+m\bar{\psi},\psi] +\tfrac{1}{2}[\bar{\psi},\eta]+\tfrac{1}{2}[\bar{\eta},\psi]. \quad (1)$$

The resulting field equations are

$$\begin{aligned}\gamma_\mu(-i\partial_\mu-eA_\mu)\psi+m\psi&=\eta,\\ (i\partial_\mu-eA_\mu)\bar{\psi}\gamma_\mu+m\bar{\psi}&=\bar{\eta},\end{aligned} \quad (2)$$

and the generators of infinitesimal changes in $\psi$ or $\bar{\psi}$ on a surface $\sigma$ are given by

$$G(\psi)=i\int d\sigma_\mu\bar{\psi}\gamma_\mu\delta\psi=i\int d\sigma\bar{\psi}\gamma_{(0)}\delta\psi \quad (3)$$

and

$$G(\bar{\psi})=-i\int d\sigma_\mu\delta\bar{\psi}\gamma_\mu\psi=-i\int d\sigma\delta\bar{\psi}\gamma_{(0)}\psi. \quad (4)$$

It was shown in III that the vacuum state of a closed system, $\Psi_0$ can be characterized as the right eigenvector, with zero eigenvalues, of the positive frequency parts of the field components, and that $\Psi_0{}^\dagger$ is the left eigenvector, with zero eigenvalues, of the negative frequency parts of the field components. The inference that the totality of eigenvectors of these types would be of particular utility led us, in discussing a Bose-Einstein system, to introduce eigenvectors and eigenvalues for complete sets

[1] J. Schwinger, Phys. Rev. **91**, 728 (1953).

Reprinted from *The Physical Review* **92**, 1283–1299 (1953).

of commuting non-Hermitian operators. We shall now find it desirable to extend the eigenvalue-eigenvector concept to complete sets of *anticommuting*, non-Hermitian operators.

### THE TRANSFORMATION FUNCTION

The discussion in this paper will be limited to the situation of zero electromagnetic field, as described with the elementary gauge, $A_\mu=0$. Relative to a coordinate system based on a given surface, the field equations in the absence of the sources can be written as the equations of motion,

$$i\partial_0\psi=-i\gamma_0\gamma_k\partial_k\psi+m\gamma_0\psi=H\psi,$$
$$-i\partial_0\bar\psi=i\partial_k\bar\psi\gamma_k\gamma_0+m\bar\psi\gamma_0=\bar\psi\gamma_0H\gamma_0. \qquad (5)$$

We define two Hermitian coordinate operators,

$$P^{(\pm)}=\tfrac{1}{2}(1\pm(H/E)), \qquad (6)$$

where $E$ is the positive-definite quantity

$$E=(H^2)^{\frac{1}{2}}. \qquad (7)$$

These operators have the projection properties

$$P^{(+)}+P^{(-)}=1,\quad P^{(+)}P^{(-)}=P^{(-)}P^{(+)}=0 \qquad (8)$$

and

$$P^{(\pm)}H=\pm EP^{(\pm)}. \qquad (9)$$

Hence the representation of $\psi$ and $\bar\psi$ as

$$\psi=\psi^{(+)}+\psi^{(-)},\quad \bar\psi=\bar\psi^{(+)}+\bar\psi^{(-)}, \qquad (10)$$

where

$$\psi^{(\pm)}=P^{(\pm)}\psi,\quad \bar\psi^{(\pm)}=\bar\psi\gamma_0P^{(\mp)}\gamma_0 \qquad (11)$$

is a decomposition into positive and negative frequency parts, according to the resulting form of the equations of motion,

$$i\partial_0\psi^{(\pm)}=\pm E\psi^{(\pm)},\quad i\partial_0\bar\psi^{(\pm)}=\pm E\bar\psi^{(\pm)}. \qquad (12)$$

It should also be noted that

$$\bar\psi^{(\pm)}=\overline{\psi^{(\mp)}}. \qquad (13)$$

In view of the orthogonality properties expressed by

$$\int d\sigma\bar\psi^{(+)}\gamma_0\psi^{(+)}=\int d\sigma\bar\psi\gamma_0P^{(-)}P^{(+)}\psi=0, \qquad (14)$$

and

$$\int d\sigma\bar\psi^{(-)}\gamma_0\psi^{(-)}=\int d\sigma\bar\psi\gamma_0P^{(+)}P^{(-)}\psi=0, \qquad (15)$$

the generator $G(\psi)$ appears as

$$G(\psi)=i\int d\sigma\bar\psi^{(-)}\gamma_0\delta\psi^{(+)}+i\int d\sigma\bar\psi^{(+)}\gamma_0\delta\psi^{(-)}. \qquad (16)$$

By the addition of suitable variations to this generator, we deduce the generators of infinitesimal changes in the positive frequency, or in the negative frequency parts of $\psi$ and $\bar\psi$,

$$G_+=i\int d\sigma\bar\psi^{(-)}\gamma_0\delta\psi^{(+)}-i\int d\sigma\delta\bar\psi^{(+)}\gamma_0\psi^{(-)}, \qquad (17)$$

$$G_-=i\int d\sigma\bar\psi^{(+)}\gamma_0\delta\psi^{(-)}-i\int d\sigma\delta\bar\psi^{(-)}\gamma_0\psi^{(+)}. \qquad (18)$$

The orthogonality properties contained in (14) and (15) also enable us to write these generators as

$$G_+=i\int d\sigma\bar\psi\gamma_0\delta\psi^{(+)}-i\int d\sigma\delta\bar\psi^{(+)}\gamma_0\psi \qquad (19)$$

and

$$G_-=i\int d\sigma\bar\psi\gamma_0\delta\psi^{(-)}-i\int d\sigma\delta\bar\psi^{(-)}\gamma_0\psi, \qquad (20)$$

or, alternatively, as

$$G_+=i\int d\sigma\bar\psi^{(-)}\gamma_0\delta\psi-i\int d\sigma\delta\bar\psi\gamma_0\psi^{(-)} \qquad (21)$$

and

$$G_-=i\int d\sigma\bar\psi^{(+)}\gamma_0\delta\psi-i\int d\sigma\delta\bar\psi\gamma_0\psi^{(+)}. \qquad (22)$$

The latter forms facilitate the derivation of commutation properties on $\sigma$ for the field components $\psi^{(\pm)}$, $\bar\psi^{(\pm)}$. These are expressed by the vanishing of all anticommutators save

$$\{\psi^{(+)}(x),\bar\psi^{(-)}(x')\}=P^{(+)}\gamma_0\delta_\sigma(x-x') \qquad (23)$$

and

$$\{\psi^{(-)}(x),\bar\psi^{(+)}(x')\}=P^{(-)}\gamma_0\delta_\sigma(x-x'). \qquad (24)$$

In particular, all positive frequency components are anticommutative, as are all negative frequency components.

Complete sets of anticommuting operators on $\sigma$ are thus provided by $\chi^{(+)}(x)=\psi^{(+)}(x)$, $\bar\psi^{(+)}(x)$, and by $\chi^{(-)}(x)=\psi^{(-)}(x)$, $\bar\psi^{(-)}(x)$. The existence of right and left eigenvectors, respectively, with null eigenvalues, follows from the equivalence of these states with the vacuum state. Let us now extend the number system by introducing quantities $\chi^{(+)\prime}(x)=\psi^{(+)\prime}(x)$, $\bar\psi^{(+)\prime}(x)$ and $\chi^{(-)\prime}(x)=\psi^{(-)\prime}(x)$, $\bar\psi^{(-)\prime}(x)$, which anticommute among themselves and with all Dirac field operators. Then the operators

$${}'\chi^{(\pm)}=\chi^{(\pm)}-\chi^{(\pm)\prime}$$

have the same commutation properties as the $\chi^{(\pm)}$, so that there exists a right eigenvector of the complete set ${}'\chi^{(+)}$ with zero eigenvalues,

$$(\chi^{(+)}(x)-\chi^{(+)\prime}(x))\Psi(\chi^{(+)\prime}\sigma)=0,$$

and a left eigenvector of the complete set ${}'\chi^{(-)}$ with zero eigenvalues,

$$\Phi(\chi^{(-)\prime}\sigma)(\chi^{(-)}(x)-\chi^{(-)\prime}(x))=0.$$

As the notation indicates, we thereby obtain a right eigenvector of the complete set $\chi^{(+)}$ with the "eigenvalues" $\chi^{(+)\prime}$, and a left eigenvector of the set $\chi^{(-)}$ with the "eigenvalues" $\chi^{(-)\prime}$. In view of the relation (13), the eigenvectors and eigenvalues are connected by

$$\Phi(\chi^{(-)\prime}\sigma)=\Psi(\chi^{(+)\prime}\sigma)^{\dagger}, \tag{25}$$

$$\bar{\psi}^{(\pm)\prime}=\overline{\psi^{(\mp)\prime}}. \tag{26}$$

The interpretation of the infinitesimal transformation equations

$$\delta\Psi(\chi^{(+)\prime}\sigma)=-iG_{+}\Psi(\chi^{(+)\prime}\sigma) \tag{27}$$

and

$$\delta\Phi(\chi^{(-)\prime}\sigma)=i\Phi(\chi^{(-)\prime}\sigma)G_{-}, \tag{28}$$

employs the identical operator properties of the field variations and of the eigenvalues. Thus $\Psi(\chi^{(+)\prime}\sigma)+\delta\Psi(\chi^{(+)\prime}\sigma)$ is the eigenvector of the $\chi^{(+)}-\delta\chi^{(+)}$ with the eigenvalues $\chi^{(+)\prime}$. But this is also the eigenvector of the $\chi^{(+)}$ with the eigenvalues $\chi^{(+)\prime}+\delta\chi^{(+)}$. Hence the alteration given by (27) is that associated with the change of the eigenvalues by $\delta\chi^{(+)}$. A similar statement applies to (28).

We shall now discuss the Dirac field under the influence of the external sources $\eta$, $\bar{\eta}$ in terms of the transformation function

$$(\chi^{(-)\prime}\sigma_1|\chi^{(+)\prime}\sigma_2)=(\Phi(\chi^{(-)\prime}\sigma_1)\Psi(\chi^{(+)\prime}\sigma_2)).$$

The dependence of this transformation function upon the eigenvalues is expressed by

$$\delta_{\chi^{\prime}}(\chi^{(-)\prime}\sigma_1|\chi^{(+)\prime}\sigma_2)=i(\chi^{(-)\prime}\sigma_1|(G_{-}(\sigma_1)-G_{+}(\sigma_2))|\chi^{(+)\prime}\sigma_2), \tag{29}$$

where

$$\begin{aligned}G_{-}(\sigma_1)-G_{+}(\sigma_2)&=-i\int_{\sigma_1}d\sigma_\mu\delta\bar{\psi}^{(-)\prime}\gamma_\mu\psi+i\int_{\sigma}d\sigma_\mu\bar{\psi}\gamma_\mu\delta\psi^{(-)\prime}\\&\quad+i\int_{\sigma_2}d\sigma_\mu\delta\bar{\psi}^{(+)\prime}\gamma_\mu\psi-i\int_{\sigma_2}d\sigma_\mu\bar{\psi}\gamma_\mu\delta\psi^{(+)\prime}\\&=-i\oint d\sigma_\mu\delta\bar{\psi}^{\prime}\gamma_\mu\psi+i\oint d\sigma_\mu\bar{\psi}\gamma_\mu\delta\psi^{\prime}.\end{aligned} \tag{30}$$

In the latter form, it is understood that negative frequency eigenvalues are employed on $\sigma_1$, and positive frequency eigenvalues on $\sigma_2$.

An infinitesimal change of the external source produces the alteration

$$\delta(\chi^{(-)\prime}\sigma_1|\chi^{(+)\prime}\sigma_2)=i\left(\chi^{(-)\prime}\sigma_1\left|\int_{\sigma_2}^{\sigma_1}(dx)(\delta\bar{\eta}\psi+\bar{\psi}\delta\eta)\right|\chi^{(+)\prime}\sigma_2\right). \tag{31}$$

Accordingly, an infinitesimal change in the eigenvalues can be simulated by the surface distributed source variation

$$\begin{aligned}\delta\bar{\eta}(x)&=-i\delta\bar{\psi}^{(-)\prime}(x)\gamma_\mu\delta_\mu(x,\sigma_1)+i\delta\bar{\psi}^{(+)\prime}(x)\gamma_\mu\delta_\mu(x,\sigma_2),\\ \delta\eta(x)&=i\delta_\mu(x,\sigma_1)\gamma_\mu\delta\psi^{(-)\prime}(x)-i\delta_\mu(x,\sigma_2)\gamma_\mu\delta\psi^{(+)\prime}(x),\end{aligned} \tag{32}$$

where $\delta_\mu(x,\sigma)$ is defined by

$$\int(dx)\delta_\mu(x,\sigma)f_\mu(x)=\int_\sigma d\sigma_\mu f_\mu(x).$$

We conclude that the general transformation function is obtained from the one referring to zero eigenvalues on making the followlng substitution in the latter,

$$\begin{aligned}\bar{\eta}(x)&\rightarrow\bar{\eta}(x)-i\bar{\psi}^{(-)\prime}(x)\gamma_\mu\delta_\mu(x,\sigma_1)+i\bar{\psi}^{(+)\prime}(x)\gamma_\mu\delta_\mu(x,\sigma_2),\\ \eta(x)&\rightarrow\eta(x)+i\delta_\mu(x,\sigma_1)\gamma_\mu\psi^{(-)\prime}(x)-i\delta_\mu(x,\sigma_2)\gamma_\mu\psi^{(+)\prime}(x).\end{aligned} \tag{33}$$

With the notation

$$(0\sigma_1|0\sigma_2)=\exp(i\mathcal{W}_0) \tag{34}$$

and

$$(0\sigma_1|F|0\sigma_2)/(0\sigma_1|0\sigma_2)=\langle F\rangle, \tag{35}$$

the dependence of the null eigenvalue transformation function upon the source is expressed by

$$\begin{aligned}\delta_l\mathcal{W}_0/\delta\bar{\eta}(x)&=\langle\psi(x)\rangle,\\ \delta_r\mathcal{W}_0/\delta\eta(x)&=\langle\bar{\psi}(x)\rangle,\end{aligned} \tag{36}$$

or

$$\delta\mathcal{W}_0=\int_{-\infty}^{\infty}(dx)[\delta\bar{\eta}\langle\psi\rangle+\langle\bar{\psi}\rangle\delta\eta], \tag{37}$$

in which it is supposed that the source vanishes externally to the volume bounded by $\sigma_1$ and $\sigma_2$. According to the field equations (2), we have

$$-i\gamma_\mu\partial_\mu\langle\psi(x)\rangle+m\langle\psi(x)\rangle=\eta(x), \tag{38}$$

and

$$i\partial_\mu\langle\bar{\psi}(x)\rangle\gamma_\mu+m\langle\bar{\psi}(x)\rangle=\bar{\eta}(x), \tag{39}$$

which are to be solved subject to the boundary conditions

$$\langle\psi^{(-)}\rangle=\langle\bar{\psi}^{(-)}\rangle=0,\quad\text{on }\sigma_1, \tag{40}$$

and

$$\langle\psi^{(+)}\rangle=\langle\bar{\psi}^{(+)}\rangle=0,\quad\text{on }\sigma_2, \tag{41}$$

that follow from the nature of the null eigenvalue states on $\sigma_1$ and $\sigma_2$. We can express these as boundary conditions in the extended domain through the requirement that the fields shall contain only positive frequencies in the region constituting the future of $\sigma_1$, and only negative frequencies in the region prior to $\sigma_2$.

The solutions meeting these conditions are

$$\langle\psi(x)\rangle=\int_{-\infty}^{\infty}(dx^{\prime})G_{+}(x,x^{\prime})\eta(x^{\prime}) \tag{42}$$

and

$$\langle\bar{\psi}(x')\rangle=\int_{-\infty}^{\infty}(dx)\bar{\eta}(x)G_+(x,x'), \qquad (43)$$

where $G_+(x,x')$ is the Green's function defined by the differential equations

$$-i\gamma_\mu\partial_\mu G_+(x,x')+mG_+(x,x')$$
$$=i\partial_\mu' G_+(x,x')\gamma_\mu+mG_+(x,x')$$
$$=\delta(x-x'), \qquad (44)$$

and the boundary condition that $G_+$, as a function of $x$, shall contain only positive frequencies for $x_0>x_0'$, and only negative frequencies for $x_0<x_0'$. Since $G_+$ is only dependent upon $x-x'$, the same statement applies with $x$ and $x'$ interchanged. That the identical Green's function is encountered in (42) and (43) follows from the integrability condition deduced from (36),

$$(\delta_r/\delta\eta(x'))\langle\psi(x)\rangle=(\delta_l/\delta\bar{\eta}(x))\langle\bar{\psi}(x')\rangle=G_+(x,x').$$

We thus construct $\mathcal{W}_0$ as

$$\mathcal{W}_0=\int_{-\infty}^{\infty}(dx)(dx')\bar{\eta}(x)G_+(x,x')\eta(x'), \qquad (45)$$

with a constant of integration that has the value zero since, in the absence of sources, the null eigenvalue states have the significance of the $\sigma$-independent vacuum state

$$\eta=\bar{\eta}=0:\quad (0\sigma_1|0\sigma_2)=1,\quad \mathcal{W}_0=0.$$

On performing the substitution (33), the general transformation function is obtained as

$$(\chi^{(-)\prime}\sigma_1|\chi^{(+)\prime}\sigma_2)=\exp(i\mathcal{W}), \qquad (46)$$

where

$$\mathcal{W}=\int(dx)(dx')\bar{\eta}(x)G_+(x,x')\eta(x')$$
$$-i\oint d\sigma_\mu\int(dx')\bar{\psi}'(x)\gamma_\mu G_+(x,x')\eta(x')$$
$$+i\int(dx)\oint d\sigma_\nu'\bar{\eta}(x)G_+(x,x')\gamma_\nu\psi'(x')$$
$$+\oint d\sigma_\mu\oint d\sigma_\nu'\bar{\psi}'(x)\gamma_\mu G_+(x,x')\gamma_\nu\psi'(x'). \qquad (47)$$

In particular,

$$\eta=\bar{\eta}=0:\quad (\chi^{(-)\prime}\sigma_1|\chi^{(+)\prime}\sigma_2)$$
$$=\exp\left[i\oint d\sigma_\mu\oint d\sigma_\nu'\bar{\psi}'(x)\gamma_\mu G_+(x,x')\gamma_\nu\psi'(x')\right]. \qquad (48)$$

The Green's function $G_+(x,x')$ can be exhibited in the three-dimensional symbolic form

$$G_+(x,x')=iP^{(+)}\exp[-iE(x_0-x_0')]\gamma_0\delta(\mathbf{x}-\mathbf{x}'),\quad x_0>x_0'$$
$$=-iP^{(-)}\exp[-iE(x_0'-x_0)]\gamma_0\delta(\mathbf{x}-\mathbf{x}'),\quad x_0<x_0', \qquad (49)$$

or, combining both situations,

$$G_+(x,x')=(i\partial_0+H)\tfrac{1}{2}iE^{-1}$$
$$\times\exp(-iE|x_0-x_0'|)\gamma_0\delta(\mathbf{x}-\mathbf{x}'). \qquad (50)$$

The function of $E$ in the latter equation has the integral representation

$$\tfrac{1}{2}iE^{-1}\exp(-iE|x_0-x_0'|)$$
$$=\frac{1}{2\pi}\int_{-\infty}^{\infty}dp_0\frac{e^{-ip_0(x_0-x_0')}}{E^2-p_0^2-i\epsilon},\quad \epsilon\to+0. \qquad (51)$$

The Fourier integral expression of the three-dimensional delta function in (50), combined with (51), leads to the four-dimensional integral

$$G_+(x,x')=\frac{1}{(2\pi)^4}\int(dp)\frac{m-\gamma p}{p^2+m^2-i\epsilon}e^{ip(x-x')}$$
$$=\frac{1}{(2\pi)^4}\int(dp)\frac{1}{\gamma p+m-i\epsilon}e^{ip(x-x')},$$
$$\epsilon\to+0, \qquad (52)$$

which shows that the mass $m$ must be supplemented with an infinitesimal negative imaginary constant in constructing the Green's function as the reciprocal of the differential operator in (44).

The three-dimensional Fourier integrals derived from (49) are advantageously presented as

$$G_+(x,x')=\tfrac{1}{2}i\int\frac{(d\mathbf{p})}{(2\pi)^3}\frac{1}{p_0}\times\begin{cases}(m-\gamma p)e^{ip(x-x')}, & x_0>x_0'\\(m+\gamma p)e^{-ip(x-x')}, & x_0<x_0',\end{cases} \qquad (53)$$

where $p_0$ is a *positive* frequency

$$p_0=(\mathbf{p}^2+m^2)^{\frac{1}{2}}.$$

The four-rowed square matrix $-\gamma p$ has two distinct eigenvalues, $\pm m$,

$$(-\gamma p)^2=-p^2=m^2,$$

each of which is twofold degenerate. We shall designate the eigenvectors by $u_{\lambda p}$ where $\lambda=1,2$ refers to the eigenvalue $+m$, and $\lambda=-1,-2$ indicates the eigenvalue $m$. Thus

$$-\gamma p u_{\lambda p}=\epsilon(\lambda)m u_{\lambda p}, \qquad (54)$$

and

$$-\bar{u}_{\lambda p}\gamma p=\bar{u}_{\lambda p}\epsilon(\lambda)m,$$

where

$$\epsilon(\lambda)=\lambda/|\lambda|.$$

In view of the indefinite character of the quantities

$$(\bar{u}_{\lambda p}u_{\lambda p})=(u_{\lambda p}{}^*\gamma_0 u_{\lambda p}),$$

the orthonormality and completeness properties of these eigenvectors appear as

$$(\bar{u}_{\lambda p}u_{\lambda' p})=\delta_{\lambda\lambda'}\epsilon(\lambda),$$

and

$$\sum_\lambda \epsilon(\lambda)u_{\lambda p}\bar{u}_{\lambda p}=1.$$

The positive definite quantities

$$(\bar{u}_{\lambda p}\gamma_0 u_{\lambda p})=(u_{\lambda p}{}^*u_{\lambda p})$$

are then given by

$$(\bar{u}_{\lambda p}\gamma_0 u_{\lambda' p})=\delta_{\lambda\lambda'}(p_0/m).$$

The latter result is deduced from

$$\begin{aligned}-(\bar{u}_{\lambda p}\{\gamma_0,\gamma p\}u_{\lambda' p})&=2p_0(\bar{u}_{\lambda p}u_{\lambda' p})\\&=m(\epsilon(\lambda)+\epsilon(\lambda'))(\bar{u}_{\lambda p}\gamma_0 u_{\lambda' p}),\end{aligned}$$

which shows that $(\bar{u}_{\lambda p}u_{\lambda p})$ is indeed negative for $\lambda<0$.

When the eigenvalue equation (54), in the form

$$(1/2m)(m-\gamma p)u_{\lambda p}=\tfrac{1}{2}(1+\epsilon(\lambda))u_{\lambda p},$$

is multiplied by $\epsilon(\lambda)\bar{u}_{\lambda p}$ and summed over all $\lambda$ with the aid of the completeness relation, there results

$$(1/2m)(m-\gamma p)=\sum\nolimits_+ u_{\lambda p}\bar{u}_{\lambda p},$$

where + signifies the eigenvectors with $\lambda>0$. Similarly,

$$(1/2m)(m+\gamma p)=-\sum\nolimits_- u_{\lambda p}\bar{u}_{\lambda p}.$$

We employ these projection operator representations in (53), and replace the integral by a summation over cells of volume $(d\mathbf{p})$. This yields

$$\begin{aligned}G_+(x,x')&=i\sum_{+,p}\psi_{\lambda p}(x)\bar{\psi}_{\lambda p}(x'),\quad x_0>x_0'\\&=-i\sum_{-,p}\psi_{\lambda p}(x)\bar{\psi}_{\lambda p}(x'),\quad x_0<x_0',\end{aligned}\tag{55}$$

in which

$$\psi_{\lambda p}(x)=\left(\frac{(d\mathbf{p})}{(2\pi)^3}\frac{m}{p_0}\right)^{\frac{1}{2}}u_{\lambda p}e^{i\epsilon(\lambda)px}.$$

The completeness of these functions on a given surface is implied by the discontinuity of the Green's function at $x_0=x_0'$, as derived from (44),

$$\sum\nolimits_{\lambda p}\psi_{\lambda p}(x)\bar{\psi}_{\lambda p}(x')=\gamma_0\delta_\sigma(x-x'),\tag{56}$$

The associated *ortho*-normality statement is

$$\int d\sigma\bar{\psi}_{\lambda p}\gamma_0\psi_{\lambda' p'}=\delta_{\lambda p,\lambda' p'}.\tag{57}$$

**EIGENVALUES AND EIGENFUNCTIONS**

We begin our applications of (46) with the isolated Dirac field, as described by (48). The eigenvalues and eigenfunctions of the energy-momentum vector $P_\mu$ are obtained from the transformation function that connects representations associated with parallel surfaces. The effect of infinitesimal translations of $\sigma_1$ and $\sigma_2$ is given by

$$\delta_x\Phi(\chi^{(-)\prime}\sigma_1)=i\Phi(\chi^{(-)\prime}\sigma_1)P_\mu\delta x_{1\mu},$$

$$\delta_x\Psi(\chi^{(+)\prime}\sigma_2)=-iP_\mu\delta x_{2\mu}\Psi(\chi^{(+)\prime}\sigma_2).$$

Accordingly, if $x_1$ and $x_2$ are the finite translations that produce $\sigma_1$ and $\sigma_2$ from a standard surface, we have

$$\Phi(\chi^{(-)\prime}\sigma_1)=\Phi(\chi^{(-)\prime})\exp(iP_\mu x_{1\mu}),$$

$$\Psi(\chi^{(+)\prime}\sigma_2)=\exp(-iP_\mu x_{2\mu})\Psi(\chi^{(+)\prime}),$$

and

$$\begin{aligned}(\chi^{(-)\prime}\sigma_1|\chi^{(+)\prime}\sigma_2)&=(\chi^{(-)\prime}|\exp(iP_\mu X_\mu)|\chi^{(+)\prime})\\&=\sum\nolimits_{\gamma'}(\chi^{(-)\prime}|\gamma')\exp(iP_\mu' X_\mu)(\gamma'|\chi^{(+)\prime}),\end{aligned}$$

where

$$X=x_1-x_2,$$

and the $\gamma$ are a complete set of constants of the motion.

Before employing the transformation function (48) in this manner, we shall extend it to serve also as a generating function for the eigenvalues and eigenfunctions of the charge operator $Q$. An infinitesimal phase transformation on $\sigma$,

$$\begin{aligned}\delta\psi(x)&=-ie\delta\alpha\psi(x),\\\delta\bar{\psi}(x)&=ie\delta\alpha\bar{\psi}(x),\end{aligned}\tag{58}$$

induces the eigenvector transformations generated by

$$G_\alpha=Q\delta\alpha.$$

According to the orthogonality properties (14) and (15), the charge operator can be written

$$\begin{aligned}Q&=e\tfrac{1}{2}\int d\sigma([\bar{\psi}^{(-)},\gamma_0\psi^{(+)}]+[\bar{\psi}^{(+)},\gamma_0\psi^{(-)}])\\&=e\int d\sigma(\bar{\psi}^{(-)}\gamma_0\psi^{(+)}-\psi^{(-)}(\bar{\psi}^{(+)}\gamma_0)).\end{aligned}\tag{59}$$

In arriving at the latter form with the aid of the commutation relations (23) and (24), we have assigned the value zero to the quantity

$$\mathrm{Tr}(P^{(+)}-P^{(-)})=\mathrm{Tr}(H/E),\tag{60}$$

where the trace is applied to spatial coordinates and spinor indices. This evaluation is based upon the time reflection invariance of the theory, which indicates that a one to one correspondence can be established between positive and negative frequency modes. Thus complete cancellation occurs in summing the eigenvalues of $H/E$, which are $\pm 1$.

The infinitesimal eigenvector-transformation equations are

$$\delta_\alpha\Phi(\chi^{(-)'}\sigma\alpha)=i\Phi(\chi^{(-)'}\sigma\alpha)e\delta\alpha \times\int d\sigma(\bar{\psi}^{(-)'}\gamma_0\psi^{(+)}+\bar{\psi}^{(+)}\gamma_0\psi^{(-)'})$$

and

$$\delta_\alpha\Psi(\chi^{(+)'}\sigma\alpha)=-ie\delta\alpha \times\int d\sigma(\bar{\psi}^{(-)}\gamma_0\psi^{(+)'}+\bar{\psi}^{(+)'}\gamma_0\psi^{(-)})\Psi(\chi^{(+)'}\sigma\alpha).$$

A comparison with (17) and (18) shows that the eigenvector transformations are just those produced by the eigenvalue changes

$$\delta\psi^{(\pm)'}=-ie\delta\alpha\psi^{(\pm)'},$$
$$\delta\bar{\psi}^{(\pm)'}=ie\delta\alpha\bar{\psi}^{(\pm)'},$$

as we could anticipate from (58). Finite phase transformations

$$\Phi(\chi^{(-)'}\sigma\alpha)=\Phi(\chi^{(-)'}\sigma)\exp(iQ\alpha),$$
$$\Psi(\chi^{(+)'}\sigma\alpha)=\exp(-iQ\alpha)\Psi(\chi^{(+)'}\sigma),$$

are thus described by

$$\Phi(\psi^{(-)'}\bar{\psi}^{(-)'}\sigma\alpha)=\Phi(e^{-ie\alpha}\psi^{(-)'},e^{ie\alpha}\bar{\psi}^{(-)'},\sigma) \quad (61)$$

and

$$\Psi(\psi^{(+)'}\bar{\psi}^{(+)'}\sigma\alpha)=\Psi(e^{-ie\alpha}\psi^{(+)'},e^{ie\alpha}\bar{\psi}^{(+)'},\sigma). \quad (62)$$

If different phase transformations are performed on $\sigma_1$ and $\sigma_2$, we have

$$(\chi^{(-)'}\sigma_1\alpha_1|\chi^{(+)'}\sigma_2\alpha_2)=(\chi^{(-)'}\sigma_1|\exp(iQ\alpha)|\chi^{(+)'}\sigma_2), \quad (63)$$

where

$$\alpha=\alpha_1-\alpha_2.$$

Hence

$$(\chi^{(-)'}\sigma_1\alpha_1|\chi^{(+)'}\sigma_2\alpha_2)=(\chi^{(-)'}|\exp(iP_\mu X_\mu+iQ\alpha)|\chi^{(+)'})$$
$$=\sum_{\gamma'}(\chi^{(-)'}|\gamma')\exp(iP_\mu'X_\mu+iQ'\alpha)(\gamma'|\chi^{(+)'}) \quad (64)$$

is the generating function of the simultaneous eigenfunctions and eigenvalues of the commuting operators $P_\mu$ and $Q$, and this transformation function is obtained from (48) by applying the substitutions indicated in (61) and (62).

It should be noted that (48) does not contain terms in which both integrals refer to a common surface. Consider, for example,

$$\int_{\sigma_1}d\sigma\int_{\sigma_1}d\sigma'\bar{\psi}^{(-)'}(x)\gamma_0G_+(x,x')\gamma_0\psi^{(-)'}(x')$$
$$=\int_{\sigma_1}d\sigma\int_{\sigma_1}d\sigma'\bar{\psi}'\gamma_0P^{(+)}G_+\gamma_0P^{(-)}\psi'. \quad (65)$$

Now $G_+\gamma_0$ becomes $iP^{(+)}$ or $-iP^{(-)}$ as $x_0-x_0'\to\pm0$. Either limit results in a null value for (65) ($P^{(+)}P^{(-)}=0$). Hence,

$$(\chi^{(-)'}\sigma_1\alpha_1|\chi^{(+)'}\sigma_2\alpha_2)$$
$$=\exp\Big[-i\int_{\sigma_1}d\sigma_\mu\int_{\sigma_2}d\sigma_\nu'\bar{\psi}^{(-)'}(x)\gamma_\mu e^{ie\alpha}\times G_+(x,x')\gamma_\nu\psi^{(+)'}(x')-i\int_{\sigma_2}d\sigma_\mu\times\int_{\sigma_1}d\sigma_\nu'\bar{\psi}^{(+)'}(x)\gamma_\mu G_+(x,x')e^{-ie\alpha}\gamma_\nu\psi^{(-)'}(x')\Big] \quad (66)$$

and we need not have indicated the positive or negative frequency parts of the field eigenvalues since these are automatically selected by the structure of the Green's function.

We insert the expressions (55) for the Green's function and introduce the following linear combinations of eigenvalues, which are not explicit functions of the surface,

$$\lambda>0:\quad \chi_{\lambda p}{}^{(-)'}=\int_{\sigma_1}d\sigma_\mu\bar{\psi}^{(-)'}(x)\gamma_\mu\psi_{\lambda p}(x)e^{-ipx_1}$$
$$\chi_{\lambda p}{}^{(+)'}=\int_{\sigma_2}d\sigma_\mu\bar{\psi}_{\lambda p}(x)e^{ipx_2}\gamma_\mu\psi^{(+)'}(x),$$

$$\lambda<0:\quad \chi_{\lambda p}{}^{(-)'}=\int_{\sigma_1}d\sigma_\mu\bar{\psi}_{\lambda p}(x)e^{-ipx_1}\gamma_\mu\psi^{(-)'}(x)$$
$$\chi_{\lambda p}{}^{(+)'}=\int_{\sigma_2}d\sigma_\mu\bar{\psi}^{(+)'}(x)\gamma_\mu\psi_{\lambda p}(x)e^{ipx_2}.$$

With these definitions, the transformation function (64) becomes[2]

$$(\chi^{(-)'}\sigma_1\alpha_1|\chi^{(+)'}\sigma_2\alpha_2)$$
$$=\exp[\sum_{\lambda p}\chi_{\lambda p}{}^{(-)'}\exp(ipX+ie\epsilon(\lambda)\alpha)\chi_{\lambda p}{}^{(+)'}]$$
$$=\prod_{\lambda p}\exp[\chi_{\lambda p}{}^{(-)'}\exp(ipX+ie\epsilon(\lambda)\alpha)\chi_{\lambda p}{}^{(+)'}]$$
$$=\prod_{\lambda p}\sum_{n_{\lambda p}}\frac{1}{n!}(\chi_{\lambda p}{}^{(-)'}\chi_{\lambda p}{}^{(+)'})^n\times\exp[in(pX+e\epsilon(\lambda)\alpha)]. \quad (67)$$

In view of the anticommutative nature of the eigenvalues, the square of any $\chi_{\lambda p}{}^{(\pm)'}$ is zero, whence the expansion of the exponential terminates after the first two terms, $n_{\lambda p}=0,1$. It will be observed that the distinction between B.E. and F.D. systems is embodied primarily in the nature of the eigenvalues rather than in the formula containing those eigenvalues.

[2] A basic statement in all manipulations with eigenvalues is that the product of two eigenvalues, as a unit, behaves like an ordinary number.

On comparison with (64) we see that

$$P_\mu' = P_\mu(n) = \sum_{\lambda p} n_{\lambda p} p_\mu, \quad n_{\lambda p} = 0, 1,$$

where

$$P_0' = \sum_{\lambda p} n_{\lambda p} p_0 \geq 0,$$

and that

$$Q' = \sum_{\lambda p} n_{\lambda p} e\epsilon(\lambda).$$

The occupation numbers $n_{\lambda p}$ thus form the complete set of constants of the motion. The associated eigenfunctions are

$$(n|\chi^{(+)'}) = \prod_{\lambda p} (\chi_{\lambda p}{}^{(+)'})^{n_{\lambda p}} = (\chi_1{}^{(+)'})^{n_1} (\chi_2{}^{(+)'})^{n_2} \cdots,$$

and

$$(\chi^{(-)'}|n) = {}'\prod_{\lambda p} (\chi_{\lambda p}{}^{(-)'})^{n_{\lambda p}} = \cdots (\chi_2{}^{(-)'})^{n_2} (\chi_1{}^{(-)'})^{n_1},$$

in which we have introduced an arbitrary standard order of the field modes. That order, when read from left to right, is symbolized by $\prod$, and in the reverse sense by $'\prod$. Thus, if only modes 1 and 2 are occupied, we have in (67)

$$\chi_1{}^{(-)'}\chi_1{}^{(+)'}\chi_2{}^{(-)'}\chi_2{}^{(+)'} = (\chi_2{}^{(-)'}\chi_1{}^{(-)'})(\chi_1{}^{(+)'}\chi_2{}^{(+)'})$$
$$= {}'\prod_{\lambda p} (\chi_{\lambda p}{}^{(-)'})^{n_{\lambda p}} \prod_{\lambda p} (\chi_{\lambda p}{}^{(+)'})^{n_{\lambda p}}.$$

With the eigenvalues $\chi^{(\pm)'}(x)$ at corresponding points in the relation (26), we have

$$\chi_{\lambda p}{}^{(-)'} = \chi_{\lambda p}{}^{(+)'\dagger},$$

and therefore

$$(\chi^{(-)'}|n) = (n|\chi^{(+)'})^\dagger,$$

as demanded by the eigenvector connection (25).

The eigenfunction of the vacuum state, referring to an arbitrary surface, is

$$(\chi^{(-)'}\sigma|0) = 1,$$

and therefore

$$(\chi^{(-)'}\sigma|n\sigma) = (\chi^{(-)'}\sigma|0)'\prod_{\lambda p} (\chi_{\lambda p}{}^{(-)'})^{n_{\lambda p}}$$
$$= (\chi^{(-)'}\sigma|{}'\prod_{\lambda p} (\chi_{\lambda p}{}^{(-)})^{n_{\lambda p}}|0),$$

in which we have introduced the operators on $\sigma$ possessing the $\chi_{\lambda p}{}^{(-)'}$ as eigenvalues. Accordingly, the eigenvectors of the state with particle occupation numbers $n_{\lambda p}$ are

$$\Psi(n\sigma) = [{}'\prod_{\lambda p} (\chi_{\lambda p}{}^{(-)})^{n_{\lambda p}}]\Psi_0,$$

and

$$\Psi(n\sigma)^\dagger = \Psi_0{}^\dagger [\prod_{\lambda p} (\chi_{\lambda p}{}^{(+)})^{n_{\lambda p}}].$$

## THE MATRICES OF FIELD OPERATORS

We shall now use the transformation function (46) to obtain the matrix elements, for the isolated Dirac field, of all products of the field operators $\psi(x)$ and $\bar{\psi}(x)$. For this purpose, we remark that the transformation function describing the system, with sources present, is the matrix of a certain time-ordered operator for the isolated system. Indeed,

$$(\chi^{(-)'}\sigma_1|\chi^{(+)'}\sigma_2)$$
$$= \left(\chi^{(-)'}\sigma_1 \left| \left( \exp\left[ i\int (dx)(\bar{\eta}(x)\psi(x) + \bar{\psi}(x)\eta(x)) \right] \right)_+ \right| \chi^{(+)'}\sigma_2 \right)\Big]_0, \quad (68)$$

where $]_0$ indicates that the operators and states correspond to $\eta=\bar{\eta}=0$. To prove this, let us replace $\eta$ and $\bar{\eta}$ with $\lambda\eta$ and $\lambda\bar{\eta}$, where $\lambda$ is a numerical parameter. The effect of an infinitesimal change of the latter is expressed by

$$(\partial/\partial\lambda)(\chi^{(-)'}\sigma_1|\chi^{(+)'}\sigma_2) = \left(\chi^{(-)'}\sigma_1 \left| \int (dx) l(x) \right| \chi^{(+)'}\sigma_2 \right),$$

in which we have temporarily written

$$l(x) = i(\bar{\eta}(x)\psi(x) + \bar{\psi}(x)\eta(x)).$$

On differentiating again, we find[3]

$$(\partial^2/\partial\lambda^2)(\chi^{(-)'}\sigma_1|\chi^{(+)'}\sigma_2)$$
$$= \left(\chi^{(-)'}\sigma_1 \left| \int (dx_1)(dx_2)(l(x_1)l(x_2))_+ \right| \chi^{(+)'}\sigma_2 \right),$$

and, in general,

$$(\partial/\partial\lambda)^n(\chi^{(-)'}\sigma_1|\chi^{(+)'}\sigma_2)$$
$$= \left(\chi^{(-)'}\sigma_1 \left| \int (dx_1)\cdots(dx_n)(l(x_1)\cdots l(x_n))_+ \right| \chi^{(+)'}\sigma_2 \right).$$

If we now construct the transformation function describing the system in the presence of sources ($\lambda=1$) in terms of that for zero sources ($\lambda=0$), as a Taylor series expansion, we obtain an infinite series which is compactly represented by (68).

The transformation function (46) can be expressed as

$$(\chi^{(-)'}\sigma_1|\chi^{(+)'}\sigma_2) = (\chi^{(-)'}\sigma_1|\chi^{(+)'}\sigma_2)]_0$$
$$\times \exp\left[ i\int (dx)(dx')\bar{\eta}(x)G_+(x,x')\eta(x') + i\int (dx)(\bar{\eta}(x)\psi'(x) + \bar{\psi}'(x)\eta(x)) \right], \quad (69)$$

in which we have used the symbols $\psi'(x)$, $\bar{\psi}'(x)$, at points in the interior of the region bounded by $\sigma_1$ and $\sigma_2$,

[3] J. Schwinger, Phys. Rev. 82, 914 (1951), Eq. (2.133).

to mean

$$\psi'(x)=i\oint d\sigma_\mu' G_+(x,x')\gamma_\mu\psi'(x')$$
$$=i\int_{\sigma_1} d\sigma_\mu' G_+(x,x')\gamma_\mu\psi^{(-)\prime}(x')$$
$$-i\int_{\sigma_2} d\sigma_\mu' G_+(x,x')\gamma_\mu\psi^{(+)\prime}(x'),$$

and

$$\bar\psi'(x)=-i\oint d\sigma_\mu'\bar\psi'(x')\gamma_\mu G_+(x',x)$$
$$=-i\int_{\sigma_1} d\sigma_\mu'\bar\psi^{(-)\prime}(x')\gamma_\mu G_+(x',x)$$
$$+i\int_{\sigma_2} d\sigma_\mu'\bar\psi^{(+)\prime}(x')\gamma_\mu G_+(x',x).$$

We see that

$$\psi'(x)=\sum_{+,p}\psi_{\lambda p}(x)e^{-ipx_2}\chi_{\lambda p}{}^{(+)\prime}+\sum_{-,p}\psi_{\lambda p}(x)e^{ipx_1}\chi_{\lambda p}{}^{(-)\prime}$$

is the solution of the Dirac equation with the prescribed positive frequency part $\psi^{(+)\prime}(x)$, on $\sigma_2$, and the prescribed negative frequency part $\psi^{(-)\prime}(x)$, on $\sigma_1$. Similarly,

$$\bar\psi'(x)=\sum_{-,p}\bar\psi_{\lambda p}(x)e^{-ipx_2}\chi_{\lambda p}{}^{(+)\prime}+\sum_{+,p}\bar\psi_{\lambda p}(x)e^{ipx_1}\chi_{\lambda p}{}^{(-)\prime}$$

is the solution of the adjoint Dirac equation which has the positive and negative frequency parts, $\bar\psi^{(+)\prime}(x)$ and $\bar\psi^{(-)\prime}(x)$, on $\sigma_2$ and $\sigma_1$, respectively. Note, however, that $\bar\psi'(x)$ is not the adjoint of $\psi'(x)$. Indeed,

$$\psi'(x)^\dagger\gamma_0=-i\oint d\sigma_\mu'\bar\psi'(x')\gamma_\mu G_-(x',x),$$

where

$$G_-(x,x')=\gamma_0 G_+(x',x)^\dagger\gamma_0 \tag{70}$$

satisfies the same differential equations as $G_+(x,x')$, but obeys ingoing rather than outgoing wave temporal boundary conditions,

$$G_-(x,x')=i\sum_{-,p}\psi_{\lambda p}(x)\bar\psi_{\lambda p}(x'),\qquad x_0>x_0'$$
$$=-i\sum_{+,p}\psi_{\lambda p}(x)\bar\psi_{\lambda p}(x'),\quad x_0<x_0'. \tag{71}$$

In terms of the notation [not to be confused with (35)]

$$(\chi^{(-)\prime}\sigma_1|F|\chi^{(+)\prime}\sigma_2)/(\chi^{(-)\prime}\sigma_1|\chi^{(+)\prime}\sigma_2)]_0=\langle F\rangle, \tag{72}$$

we express (68) and (69) as

$$\left\langle\left(\exp\left[i\int(dx)(\bar\eta\psi+\bar\psi\eta)\right]\right)_+\right\rangle$$
$$=\exp\left[i\int(dx)(dx')\bar\eta G_+\eta+i\int(dx)(\bar\eta\psi'+\bar\psi'\eta)\right], \tag{73}$$

or, more simply,

$$\left\langle\left(\exp\left[i\int(dx)(\bar\eta'\psi+'\bar\psi\eta)\right]\right)_+\right\rangle$$
$$=\exp\left[i\int(dx)(dx')\bar\eta(x)G_+(x,x')\eta(x')\right], \tag{74}$$

in which we have introduced the operators

$$'\psi(x)=\psi(x)-\psi'(x),$$
$$'\bar\psi(x)=\bar\psi(x)-\bar\psi'(x).$$

An expansion of both sides in powers of $\eta$ and $\bar\eta$ will yield the matrix elements of ordered operator products of the $\psi$ and $\bar\psi$.

From the absence of terms that are linear in $\eta$ and $\bar\eta$, on the right of (74), we see that

$$\langle'\psi(x)\rangle=\langle'\bar\psi(x)\rangle=0,$$

or that

$$\langle\psi(x)\rangle=\psi'(x), \tag{75}$$
$$\langle\bar\psi(x)\rangle=\bar\psi'(x). \tag{76}$$

The term on the left of (74) that in bilinear in $\eta$ and $\bar\eta$ is

$$-\left\langle\int(dx)(dx')(\bar\eta(x)'\psi(x)'\bar\psi(x')\eta(x'))_+\right\rangle$$
$$=-\int(dx)(dx')\bar\eta(x)\langle('\psi(x)'\bar\psi(x'))_+\rangle\epsilon(x,x')\eta(x'),$$

where

$$\epsilon(x,x')=\begin{cases}+1, & x_0>x_0'\\ -1, & x_0<x_0'.\end{cases}$$

Hence

$$\langle('\psi(x)'\bar\psi(x'))_+\rangle\epsilon(x,x')=-iG_+(x,x'),$$

or

$$\langle(\psi(x)\bar\psi(x'))_+\rangle\epsilon(x,x')=\psi'(x)\bar\psi'(x')-iG_+(x,x'). \tag{77}$$

The complete expansion of the left side in (74) is

$$\sum_{k,l}\frac{i^{k+l}}{k!l!}\int(dx_1)\cdots(dx_k)(dx_1')\cdots(dx_l')\bar\eta(x_k)\cdots$$
$$\times\bar\eta(x_1)\langle('\psi(x_1)\cdots'\psi(x_k)'\bar\psi(x_l')\cdots'\bar\psi(x_1'))_+\rangle\epsilon_{k,l}$$
$$\times\eta(x_1')\cdots\eta(x_l'), \tag{78}$$

where $\epsilon_{k,l}$ is the alternating symbol expressed by

$$\epsilon_{k,l}=(\prod_{i<j}\epsilon(x_i,x_j))(\prod_{i,j}\epsilon(x_i,x_j'))(\prod_{i>j}\epsilon(x_i',x_j')). \tag{79}$$

This is to be compared with

$$\sum_k\frac{i^k}{k!}\int(dx_1)\cdots(dx_k)(dx_1')\cdots(dx_k')\bar\eta(x_k)\cdots$$
$$\times\bar\eta(x_1)G_+(x_1,x_1')\cdots G_+(x_k,x_k')\eta(x_1')\cdots\eta(x_k')$$
$$=\sum_k\frac{i^k}{(k!)^2}\int(dx_1)\cdots(dx_k')\bar\eta(x_k)\cdots\bar\eta(x_1)$$
$$\times[\det_{(k)}G_+(x_i,x_j')]\eta(x_1')\cdots\eta(x_k'),$$

where the $k$-dimensional determinant constructed from the elements $G_+(x_i, x_j')$ has been introduced by subjecting the variables $x_1'\cdots x_k'$ to the set of $k!$ permutations. The anticommutativity of the $\eta$ provides the algebraic signs to form the alternating combination of terms which constitutes the determinant. Therefore

$$\langle('\psi(x_1)\cdots'\psi(x_k)'\bar\psi(x_l')\cdots'\bar\psi(x_1'))_+\rangle\epsilon_{k,l} = \delta_{k,l}(-i)^k \det{}_{(k)}G_+(x_i, x_j'), \quad (80)$$

in which both sides are completely antisymmetrical in the variables $x_i$, and in the variables $x_j'$.

Straightforward algebraic rearrangement would yield the matrix elements for successive products of the operators $\psi$ and $\bar\psi$, as illustrated by (75) and (77). However, one can obtain an explicit formula from (73). We first consider operator products with $k=l$, and remark that such terms are isolated in (73) by substituting $\bar\eta\to t\bar\eta$, $\eta\to t^{-1}\eta$, and evaluating the integral

$$\frac{1}{2\pi i}\oint\frac{dt}{t}\left\langle\left(\exp\left[i\int(dx)(t\bar\eta\psi+t^{-1}\bar\psi\eta)\right]\right)_+\right\rangle = \frac{1}{2\pi i}\oint\frac{dt}{t}\exp\left[i\int(dx)(dx')\bar\eta G_+\eta + i\int(dx)(t\bar\eta\psi'+t^{-1}\bar\psi'\eta)\right]. \quad (81)$$

The further substitution performed on the right, $it\int(dx)\bar\eta\psi'\to t$, yields

$$\frac{1}{2\pi i}\oint\frac{dt}{t}e^t\exp\left[i\int(dx)(dx')\bar\eta(x)(G_+(x,x') + it^{-1}\psi'(x)\bar\psi'(x'))\eta(x')\right].$$

The known result of expanding the right side of (74), as expressed in (80), now shows that

$$\langle(\psi(x_1)\cdots\psi(x_k)\bar\psi(x_k')\cdots\bar\psi(x_1'))_+\rangle\epsilon_{k,k} = \frac{1}{2\pi i}\oint\frac{dt}{t}e^t\det{}_{(k)}[-iG_+(x_i,x_j')+t^{-1}\psi(x_i)\bar\psi(x_j')]$$
$$=\psi'(x_1)\cdots\psi'(x_k)\bar\psi'(x_k')\cdots\bar\psi'(x_1')+\cdots + (-i)^k\det{}_{(k)}G_+(x_i,x_j'),$$

in which the various terms will be given by the development of the determinant, combined with the theorem

$$\frac{1}{2\pi i}\oint\frac{dt}{t^{n+1}}e^t=\frac{1}{n!}.$$

The effect of the $t$ integration is to compensate the numerical factors that appear on expanding the determinant. An example is

$$\langle(\psi(x_1)\psi(x_2)\bar\psi(x_2')\bar\psi(x_1'))_+\rangle\epsilon_{2,2}=\frac{1}{2\pi i}\oint\frac{dt}{t}e^t\begin{vmatrix}-iG_+(x_1,x_1')+t^{-1}\psi'(x_1)\bar\psi'(x_1'), & -iG_+(x_1,x_2')+t^{-1}\psi'(x_1)\bar\psi'(x_2')\\ -iG_+(x_2,x_1')+t^{-1}\psi'(x_2)\bar\psi'(x_1'), & -iG_+(x_2,x_2')+t^{-1}\psi'(x_2)\bar\psi'(x_2')\end{vmatrix}$$
$$=\psi'(x_1)\psi'(x_2)\bar\psi'(x_2')\bar\psi'(x_1')-i\psi'(x_1)\bar\psi'(x_1')G_+(x_2,x_2')-i\psi'(x_2)\bar\psi'(x_2')G_+(x_1,x_1')$$
$$+i\psi'(x_1)\bar\psi'(x_2')G_+(x_2,x_1')+i\psi'(x_2)\bar\psi'(x_1')G_+(x_1,x_2')$$
$$-G_+(x_1,x_1')G_+(x_2,x_2')+G_+(x_1,x_2')G_+(x_2,x_1').$$

Operator products with $k-l>0$ are isolated by the integral

$$\frac{1}{2\pi i}\oint\frac{dt}{t^{k-l+1}}\left\langle\left(\exp\left[i\int(dx)(t\bar\eta\psi+t^{-1}\bar\psi\eta)\right]\right)_+\right\rangle = \frac{1}{2\pi i}\oint\frac{dt}{t}e^t\left(it^{-1}\int(dx)\bar\eta\psi'\right)^{k-l} \times\exp\left[i\int(dx)(dx')\bar\eta(G_++it^{-1}\psi'\bar\psi')\eta\right].$$

On expanding the right side, it is seen that

$$\langle(\psi(x_1)\cdots\psi(x_k)\bar\psi(x_l')\cdots\bar\psi(x_1'))_+\rangle\epsilon_{k,l} = \frac{1}{2\pi i}\oint\frac{dt}{t}e^t\det{}_{(k)}[t^{-1}\psi'(x_i)^{(k-l)}, -iG_+(x_i,x_j')+t^{-1}\psi'(x_i)\bar\psi'(x_j')],$$

in which the determinant is constructed with $t^{-1}\psi'(x_i)$, $i=1\cdots k$, occupying the first $k-l$ columns, and the rectangular matrix $-iG_+(x_i, x_j')+t^{-1}\psi'(x_i)\bar\psi'(x_j')$, $i=1\cdots k$, $j=1\cdots l$, completing the $k$-dimensional square array. It is understood that the determinant is defined by alternating permutations of the *row* indices, applied to the product of the diagonal elements written successively from left to right. This is illustrated by

$$\langle(\psi(x_1)\psi(x_2)\psi(x_3)\bar{\psi}(x_1'))_+\rangle_{\epsilon_{3,1}}=\frac{1}{2\pi i}\oint\frac{dt}{t}e^t\begin{vmatrix}t^{-1}\psi'(x_1), & t^{-1}\psi'(x_1), & -iG_+(x_1,x_1')+t^{-1}\psi'(x_1)\bar{\psi}'(x_1')\\ t^{-1}\psi'(x_2), & t^{-1}\psi'(x_2), & -iG_+(x_2,x_1')+t^{-1}\psi'(x_2)\bar{\psi}'(x_1')\\ t^{-1}\psi'(x_3), & t^{-1}\psi'(x_3), & -iG_+(x_3,x_1')+t^{-1}\psi'(x_3)\bar{\psi}'(x_1')\end{vmatrix}$$

$$=\psi'(x_1)\psi'(x_2)\psi'(x_3)\bar{\psi}'(x_1')-i\psi'(x_1)\psi'(x_2)G_+(x_3,x_1')$$
$$-i\psi'(x_2)\psi'(x_3)G_+(x_1,x_1')-i\psi'(x_3)\psi'(x_1)G_+(x_2,x_1').$$

A similar treatment for $l-k>0$ yields

$$\langle(\psi(x_1)\cdots\psi(x_k)\bar{\psi}(x_l')\cdots\bar{\psi}(x_1'))_+\rangle_{\epsilon_{k,l}}$$
$$=\frac{1}{2\pi i}\oint\frac{dt}{t}e^t\det{}_{(l)}[-iG_+(x_i,x_j')$$
$$+t^{-1}\psi'(x_i)\bar{\psi}'(x_j'),\ t^{-1}\bar{\psi}'(x_j')^{(l-k)}],$$

where this determinant contains $t^{-1}\bar{\psi}'(x_j')$, $j=1\cdots l$ in the first $l-k$ rows, and the $l$-dimensional array is filled out with the rectangular matrix, $-iG_+(x_i,x_j')+t^{-1}\psi'(x_i)\bar{\psi}'(x_j')$, $i=1\cdots k$, $j=1\cdots l$. Here the determinant is defined by alternating permutations of the *column* indices applied to the product of the diagonal elements, written successively from right to left. Thus, for the example $k=0$, we have

$$\langle(\bar{\psi}(x_l')\cdots\bar{\psi}(x_1'))_+\rangle_{\epsilon_{0l}}=\langle\bar{\psi}(x_l')\cdots\bar{\psi}(x_1')\rangle$$
$$=\bar{\psi}'(x_l')\cdots\bar{\psi}'(x_1').$$

### The Occupation Number Representation

Matrices in the occupation number description can be derived from these results. The simplest examples are the diagonal matrix elements referring to the vacuum state—the vacuum expectation values. Indeed, on placing all eigenvalues equal to zero in (80), we obtain

$$(0|(\psi(x_1)\cdots\psi(x_k)\bar{\psi}(x_l')\cdots\bar{\psi}(x_1'))_+|0)_{\epsilon_{k,l}}$$
$$=\delta_{k,l}(-i)^k\det{}_{(k)}G_+(x_i,x_j'),\quad(82)$$

and, in particular,

$$(0|(\psi(x)\bar{\psi}(x'))_+|0)_{\epsilon(x,x')}=-iG_+(x,x').\quad(83)$$

To obtain the occupation number matrices of $\psi(x)$ and $\bar{\psi}(x)$, we observe that (75), for example, can be written

$$(\chi^{(-)'}\sigma_1|\psi(x)|\chi^{(+)'}\sigma_2)=\psi'(x)(\chi^{(-)'}\sigma_1|\chi^{(+)'}\sigma_2),$$

or

$$\sum_{n,n'}(\chi^{(-)'}|n)(n\sigma_1|\psi(x)|n'\sigma_2)(n'|\chi^{(+)'})$$
$$=[\sum_{+,p}\psi_{\lambda p}(x)e^{-ipx_2}\chi_{\lambda p}{}^{(+)'}+\sum_{-,p}\psi_{\lambda p}(x)e^{ipx_1}\chi_{\lambda p}{}^{(-)'}]$$
$$\times\sum_n(\chi^{(-)'}|n)\exp[iP(n)(x_1-x_2)](n|\chi^{(+)'}),\quad(84)$$

which exhibits it as a generating function for $(n\sigma_1|\psi(x)|n'\sigma_2)$. We shall prefer to construct

$$(n|\psi(x)|n')=\exp(-iP(n)x_1)$$
$$\times(n\sigma_1|\psi(x)|n'\sigma_2)\exp(iP(n')x_2),$$

which is independent of $\sigma_1$ and $\sigma_2$, and refers to the standard surface. On incorporating $e^{-ipx_2}$ into $\chi_{\lambda p}{}^{(+)'}$ and $e^{ipx_1}$ into $\chi_{\lambda p}{}^{(-)'}$, (84) becomes

$$\sum_{n,n'}(\chi^{(-)'}|n)(n|\psi(x)|n')(n'|\chi^{(+)'})$$
$$=[\sum_{+,p}\psi_{\lambda p}(x)\chi_{\lambda p}{}^{(+)'}+\sum_{-,p}\psi_{\lambda p}\chi_{\lambda p}{}^{(-)'}]$$
$$\times\sum_n(\chi^{(-)'}|n)(n|\chi^{(+)'}).$$

Now

$$\dot{\chi}_{\lambda p}{}^{(-)'}(\chi_{\lambda p}{}^{(-)'}|n)=\chi_{\lambda p}{}^{(-)'}\,'\prod(\chi^{(-)'})^n$$
$$=0,\quad n_{\lambda p}=1$$
$$=(-1)^{n_{>\lambda p}}(\chi^{(-)'}|n+1_{\lambda p}),\quad n_{\lambda p}=0,$$

where $n_{>\lambda p}$ is the number of occupied states that follow $\lambda p$ in the standard order. Similarly,

$$(n|\chi^{(+)'})\chi_{\lambda p}{}^{(+)'}=\prod(\chi^{(+)'})^n\chi_{\lambda p}{}^{(+)'}$$
$$=0,\quad n_{\lambda p}=1$$
$$=(-1)^{n_{>\lambda p}}(n+1_{\lambda p}|\chi^{(+)'}),\quad n_{\lambda p}=0.$$

We see that the nonvanishing matrix elements of $\psi(x)$ are of the form

$$(n|\psi(x)|n+1_{\lambda p})=(-1)^{n_{>\lambda p}}\psi_{\lambda p}(x),\quad\lambda>0,\quad(85)$$
$$(n+1_{\lambda p}|\psi(x)|n)=(-1)^{n_{>\lambda p}}\psi_{\lambda p}(x),\quad\lambda<0,\quad(86)$$

where $n_{\lambda p}=0$ in both statements. These exhibit $\psi(x)$ as a unit charge annihilator.

In an analogous way,

$$(\chi^{(-)'}\sigma_1|\bar{\psi}(x)|\chi^{(+)'}\sigma_2)=\bar{\psi}'(x)(\chi^{(-)'}\sigma_1|\chi^{(+)'}\sigma_2)$$

yields

$$\sum_{n,n'}(\chi^{(-)'}|n)(n|\bar{\psi}(x)|n')(n'|\chi^{(+)'})$$
$$=[\sum_{-,p}\bar{\psi}_{\lambda p}(x)\chi_{\lambda p}{}^{(+)'}+\sum_{+,p}\bar{\psi}_{\lambda p}(x)\chi_{\lambda p}{}^{(-)'}]$$
$$\times\sum_n(\chi^{(-)'}|n)(n|\chi^{(+)'}),$$

from which we obtain the nonvanishing matrix elements

$$(n+1_{\lambda p}|\bar\psi(x)|n)=(-1)^{n_{>\lambda p}}\bar\psi_{\lambda p}(x), \quad \lambda>0,$$
$$(n|\bar\psi(x)|n+1_{\lambda p})=(-1)^{n_{>\lambda p}}\bar\psi_{\lambda p}(x), \quad \lambda<0,$$

with $n_{\lambda p}=0$, which display $\bar\psi(x)$ as a unit charge creator.

The matrices of $\psi$ and $\bar\psi$ suggest the utility of a classification of matrix elements that would unify a given change in occupation number for positive frequency modes with that in the reverse sense for negative frequency modes. This is accomplished by transposing the matrices with respect to the occupation numbers of modes with $\lambda<0$, which effectively introduces a time-reversed description for the negative frequency modes.

The generating function for the matrices of all ordered products is (73), written as

$$\sum_{n,n'}(\chi^{(-)\prime}|n)\left(n\left|\left(\exp\left[i\int(dx)(\bar\eta\psi+\bar\psi\eta)\right]\right)_+\right|n'\right)\times(n'|\chi^{(+)\prime})=(\chi^{(-)\prime}|\chi^{(+)\prime})$$
$$\times\exp\left[i\int(dx)(dx')\bar\eta G_+\eta+i\int(dx)(\bar\eta\psi'+\bar\psi'\eta)\right], \tag{87}$$

in which all states refer to the standard surface. The sources $\bar\eta$ and $\eta$ are understood to be placed on the extreme left and right, respectively, as we have done in (78). We now indicate the positive and negative frequency modes separately, placing the negative frequency modes first in the standard order,

$$(\chi^{(-)\prime}|n)=(\chi_+^{(-)\prime}|n_+)(\chi_-^{(-)\prime}|n_-),$$
$$(n'|\chi^{(+)\prime})=(n_-'|\chi_-^{(+)\prime})(n_+'|\chi_+^{(+)\prime}),$$

and define the mixed eigenfunctions

$$[\bar\psi'|N]=(\chi_+^{(-)\prime}|n_+)(n_-'|\chi_-^{(+)\prime})(-1)^{\frac12 n_-'(n_-'-1)},$$
$$[N'|\psi']=(-1)^{\frac12 n_-(n_--1)}(\chi_-^{(-)\prime}|n_-)(n_+'|\chi_+^{(+)\prime}). \tag{88}$$

Here the integers $n_-$ and $n_-'$ indicate the respective number of occupied negative frequency modes, while $N$ and $N'$ symbolize the sets of occupation numbers $\{n_+, n_-'\}$ and $\{n_+', n_-\}$, respectively. We shall also write $N_-=n_-'$, $N_-'=n_-$. The notation $[\bar\psi'|N]$ refers to the fact that the $\chi_+^{(-)\prime}$ and $\chi_-^{(+)\prime}$ together comprise the quantities

$$\bar\psi_{\lambda p}'=\int d\sigma_\mu\bar\psi'(x)\gamma_\mu\psi_{\lambda p}(x) \tag{89}$$

for positive and negative $\lambda$. Thus

$$[\bar\psi'|N]={}'\prod_{\lambda p}(\bar\psi_{\lambda p}')^{N_{\lambda p}}, \tag{90}$$

and this product is in standard order, from right to left, in virtue of the factor $(-1)^{\frac12 n_-'(n_-'-1)}$, which effectively reverses the sense of multiplication for the $n_-'$ anticommuting eigenvalues in $(n_-'|\chi_-^{(+)\prime})$. Similarly, $\chi_+^{(+)\prime}$ and $\chi_-^{(-)\prime}$ are comprised in

$$\psi_{\lambda p}'=\int d\sigma_\mu\bar\psi_{\lambda p}(x)\gamma_\mu\psi(x)', \tag{91}$$

and

$$[N'|\psi']=\prod_{\lambda p}(\psi_{\lambda p}')^{N_{\lambda p}'}. \tag{92}$$

To carry out the transposition, we must take

$$\sum_{n,n'}(\chi_+^{(-)\prime}|n_+)(\chi_-^{(-)\prime}|n_-)(n_+n_-|F|n_+'n_-')\times(n_-'|\chi_-^{(+)\prime})(n_+'|\chi_+^{(+)\prime}), \tag{93}$$

where $F$ is any product of the operators $\psi$, $\bar\psi$, and reverse the positions of the two negative frequency eigenfunctions. This introduces the factor $(-1)^{n_-n_-'}$, so that (93) becomes

$$\sum_{N,N'}(\pm)[\bar\psi'|N][N|F|N'][N'|\psi'], \tag{94}$$

in which we have written

$$[N|F|N']=(n|F|n'),$$

and

$$(\pm)=(-1)^{N_-}(-1)^{\frac12(N_--N_-')(N_-+1-N_-')}$$
$$=(-1)^{N_-'}(-1)^{\frac12(N_--N_-')(N_-'+1-N_-)}. \tag{95}$$

Thus, if $F$ is the unit operator, we have

$$(\chi^{(-)\prime}|\chi^{(+)\prime})=\sum_N[\bar\psi'|N](-1)^{N_-}[N|\psi']$$
$$=\exp\left[\sum_{\lambda p}\bar\psi_{\lambda p}'\epsilon(\lambda)\psi_{\lambda p}'\right]. \tag{96}$$

To complete the re-expression of the generating function (87), we remark that

$$\psi'(x)=\sum_{\lambda p}\psi_{\lambda p}(x)\psi_{\lambda p}', \tag{97}$$

and

$$\bar\psi'(x)=\sum_{\lambda p}\bar\psi_{\lambda p}(x)\bar\psi_{\lambda p}'. \tag{98}$$

Hence,

$$\sum_{N,N'}(\pm)[\bar\psi'|N]\times\left[N\left|\left(\exp\left[i\int(dx)(\bar\eta\psi+\bar\psi\eta)\right]\right)_+\right|N'\right][N'|\psi']$$
$$=\exp\left[i\int(dx)(dx')\bar\eta G_+\eta\right]$$
$$\times\exp\left[\sum_{\lambda p}\bar\psi_{\lambda p}'\epsilon(\lambda)\psi_{\lambda p}'+i\bar\eta_{\lambda p}\psi_{\lambda p}'+i\bar\psi_{\lambda p}'\eta_{\lambda p}\right], \tag{99}$$

in which we have employed the notation,

$$\bar\eta_{\lambda p}=\int(dx)\bar\eta(x)\psi_{\lambda p}(x), \quad \eta_{\lambda p}=\int(dx)\bar\psi_{\lambda p}(x)\eta(x).$$

It should be emphasized again that the sources $\bar{\eta}$ and $\eta$ are written to the left and to the right, respectively, rather than as indicated in the left-hand member of Eq. (99), since the sign factor expressed by (95) is valid only if the matrix element in (93) is a number, rather than a quantity possessing anticommutative properties.

The second exponential factor in (99) can be written

$$\prod_{\lambda p} \exp[\bar{\psi}_{\lambda p}'\epsilon(\lambda)\psi_{\lambda p}'+i\bar{\eta}_{\lambda p}\psi_{\lambda p}'+i\bar{\psi}_{\lambda p}'\eta_{\lambda p}]$$
$$=\prod_{\lambda p}\sum_{N,N'}(\bar{\psi}_{\lambda p}')^N[\delta_{N,N'}\epsilon^N(1+N\epsilon\bar{\eta}\eta)$$
$$+i\bar{\eta}N'(1-N)+i\eta N(1-N')](\psi_{\lambda p}')^{N'}, \quad (100)$$

where the expansion of the exponential referring to a given mode yields only five nonvanishing terms, as represented on the right of (100). We shall first extract the diagonal matrix elements, which constitute the following terms of (100),

$$\prod_{\lambda p}\sum_{N_{\lambda p}}(\bar{\psi}_{\lambda p}')^{N_{\lambda p}}[\epsilon(\lambda)]^{N_{\lambda p}}[1+N_{\lambda p}\epsilon(\lambda)\bar{\eta}_{\lambda p}\eta_{\lambda p}](\psi_{\lambda p}')^{N_{\lambda p}}$$
$$=\sum_N \exp[\sum_{\lambda p} N_{\lambda p}\epsilon(\lambda)\bar{\eta}_{\lambda p}\eta_{\lambda p}]$$
$$\times[\bar{\psi}'|N](-1)^{N-}[N|\psi']. \quad (101)$$

The exponential so obtained combines with the first factor on the right of (99) to form

$$\exp\left[i\int(dx)(dx')\bar{\eta}(x)(G_+(x,x')\right.$$
$$\left.-i\sum_{\lambda p}N_{\lambda p}\epsilon(\lambda)\psi_{\lambda p}(x)\bar{\psi}_{\lambda p}(x'))\eta(x')\right].$$

No minus signs are introduced on moving the $\eta$ to the right of the eigenfunctions in (101). On comparison with (74) and (80), we see that $(N_{\lambda p}=n_{\lambda p})$:

$$(n|(\psi(x_1)\cdots\psi(x_k)\bar{\psi}(x_l')\cdots\bar{\psi}(x_1'))_+|n)\epsilon_{k,l}$$
$$=\delta_{k,l}\det{}_{(k)}[-iG_+(x_i,x_j')$$
$$-\textstyle\sum_{\lambda p}n_{\lambda p}\epsilon(\lambda)\psi_{\lambda p}(x_i)\bar{\psi}_{\lambda p}(x_j')], \quad (102)$$

of which the simplest example is

$$(n|(\psi(x)\bar{\psi}(x'))_+|n)\epsilon(x,x')$$
$$=-iG_+(x,x')-\textstyle\sum_{\lambda p}n_{\lambda p}\epsilon(\lambda)\psi_{\lambda p}(x)\bar{\psi}_{\lambda p}(x'). \quad (103)$$

The modes that occur in the general matrix element can be divided into three classes: class $a$, those for which $N_{\lambda p}=0$, $N_{\lambda p}'=1$; class $b$, modes with $N_{\lambda p}=N_{\lambda p}'$; class $c$, those with $N_{\lambda p}=1$, $N_{\lambda p}'=0$. A typical term in the expansion of the product (100) can then be written

$$\prod_a(i\bar{\eta}\psi')\prod_b[(\bar{\psi}')^N\epsilon^N(1+N\epsilon\bar{\eta}\eta)(\psi')^N]\prod_c(i\bar{\psi}'\eta). \quad (104)$$

The $b$ mode product can be re-arranged as in the discussion of the diagonal matrix elements, which yields

$$\exp[\sum_b N_{\lambda p}\epsilon(\lambda)\bar{\eta}_{\lambda p}\eta_{\lambda p}](-1)^{N-_b}\,{}'\prod_b(\bar{\psi}_{\lambda p}')^N\prod_b(\psi_{\lambda p}')^N.$$

If we now bring the eigenvalues $\psi_{\lambda p}'$ of the $a$ modes and $\bar{\psi}_{\lambda p}'$ of the $c$ modes into standard order, (100) becomes

$$'\prod_a(i\bar{\eta}(-1)^{N>})\exp[\sum_b N\epsilon\bar{\eta}\eta](-1)^{N-_b}(-1)^{N_aN_c}$$
$$\times[\bar{\psi}'|N][N'|\psi']\prod_c(i\eta(-1)^{N>}),$$

in which $N_{>\lambda p}$ represents the number of occupied $b$ modes that follow $\lambda p$ in the standard order, while $N_a$ and $N_c$ are the total number of modes in the $a$ and $c$ classes.

The right side of (99) is thus expressed as

$$\sum_{N,N'}\,{}'\prod_a(i\bar{\eta}(-1)^{N>})\exp[\ ](-1)^{N-_b+N_aN_c}$$
$$\times[\bar{\psi}'|N][N'|\psi']\prod_c(i\eta(-1)^{N>})$$

where

$$\exp[\ ]=\exp\left[i\int(dx)(dx')\bar{\eta}(x)\right.$$
$$\left.\times(G_+(x,x')-i\sum_b N_{\lambda p}\epsilon(\lambda)\psi_{\lambda p}(x)\bar{\psi}_{\lambda p}(x'))\eta(x')\right].$$

We must obey the injunction that all $\bar{\eta}$ appear on the left, and all $\eta$ on the right. The effect of moving an $\eta$ in the above exponential past the product of the eigenfunctions is to introduce a factor of $(-1)^{N-N'}$, where $N$ and $N'$ represent the total number of occupied modes in the respective eigenfunctions. Accordingly,

$$(\pm)\left[N\left|\left(\exp\left[i\int(dx)(\bar{\eta}\psi+\bar{\psi}\eta)\right]\right)_+\right|N'\right]$$
$$=(-1)^{N-_b+N_aN_c}\,{}'\prod_a(i\bar{\eta}(-1)^{N>})\exp\left[i(-1)^{N-N'}\right.$$
$$\left.\times\int(dx)(dx')\bar{\eta}(G_+-i\sum_b N\epsilon\psi\bar{\psi})\eta\right]\prod_c(i\eta(-1)^{N>}).$$

On expansion, we obtain the following result for the general nonvanishing matrix element,

$$[N|(\psi(x_1)\cdots\psi(x_k)\bar{\psi}(x_l')\cdots\bar{\psi}(x_1'))_+|N']\epsilon_{k,l}$$
$$=(-1)^{\frac{1}{2}(N-_a-N-_c)^2-\frac{1}{2}(N-_a+N-_c)}\det{}_{(k+l-r)}\begin{bmatrix}0, & (-1)^{N>}\bar{\psi}_c(x_j')\\ (-1)^{N>}\psi_a(x_i), & -iG_+(x_i,x_j')-\sum_b N_{\lambda p}\epsilon(\lambda)\psi_{\lambda p}(x_i)\bar{\psi}_{\lambda p}(x_j')\end{bmatrix} \quad (105)$$

in which

$$k-N_a=l-N_c=r \tag{106}$$

must be a non-negative integer. Also,

$$N_a=N'-N_b, \quad N_c=N-N_b.$$

In this determinant, 0 stands for the null matrix of $N_c$ rows and $N_a$ columns, and $(-1)^{N>}\psi_a(x_i)$ represents a matrix of $k$ rows and $N_a$ columns in which the eigenfunctions of the various $a$ modes are standardly arrayed in the successive columns. The matrix $(-1)^{N>}\bar{\psi}_c(x_j')$ is one of $N_c$ rows and $l$ columns, with the various $c$ mode adjoint eigenfunctions, in standard order, occupying the successive rows. Finally, we have the matrix $-iG_+(x_i,x_j')-\sum_b N_{\lambda p}\epsilon(\lambda)\psi_{\lambda p}(x_i)\bar{\psi}_{\lambda p}(x_j')$ of $k$ rows and $l$ columns. Thus the dimension of this determinant is

$$k+N_c=l+N_a=k+l-r.$$

Since this can also be written as $N_a+N_c+r$, we see that the integer $r$ is also the maximum number of Green's function factors that appear in the development of the determinant.

For the elementary example $k=1$, $l=0$, we have $N_a=1$, $N_c=r=0$, and the nonvanishing matrix element

$$[N|\psi(x)|N+1_{\lambda p}]=(-1)^{N>\lambda p}\psi_{\lambda p}(x)$$

which unifies (85) and (86) as intended. Similarly,

$$[N+1_{\lambda p}|\bar{\psi}(x)|N]=(-1)^{N>\lambda p}\bar{\psi}_{\lambda p}(x).$$

With this classification, $\psi(x)$ and $\bar{\psi}(x)$ appear as single "particle" annihilators and creators, respectively. The selection rules (106) for the general matrix element can then be described as follows. The operator contains $k$ annihilators and $l$ creators. If $r$ of these operators combine in pairs to produce a null net effect, the remaining $k-r$ annihilators and $l-r$ creators will empty $N_a=k-r$ occupied modes, and fill $N_c=l-r$ occupied modes. Of course $N=N'+l-k$.

The diagonal matrix elements (102) represent the extreme situation in which

$$r=k=l, \quad N_a=N_c=0.$$

At the opposite limit is

$$r=0, \quad N_a=k, \quad N_c=l,$$

where the matrix element (105) becomes

$$\pm\det{}_{(k)}[(-1)^{N>}\psi_a(x_i)]\,\det{}_{(l)}[(-1)^{N>}\bar{\psi}_c(x_j')]$$

and

$$\pm=(-1)^{\frac{1}{2}N_a(N_a-1)}(-1)^{\frac{1}{2}N_c(N_c-1)}(-1)^{N_aN_c-N_aN_c}.$$

## MAXWELL FIELD MATRIX ELEMENTS

This section is a supplement to paper III, in which the transformation function describing the Maxwell field with an external current is used to construct the matrices of all products of the potential vector for the isolated electromagnetic field. The transformation function (III, 20) can be expressed as

$$(F^{(-)\prime}\sigma_1|F^{(+)\prime}\sigma_2)=(F^{(-)\prime}\sigma_1|F^{(+)\prime}\sigma_2)]_0 \times\exp\left[i\mathcal{W}_0+i\int(dx)J_\nu(x)A_\nu'(x)\right], \tag{107}$$

where $]_0$ indicates zero external current, and

$$\begin{aligned}A_\nu'(x')&=2\int_{\sigma_1}d\sigma_\mu F_{\mu\nu}^{(-)\prime}(x)D_+(x-x')\\&\quad-2\int_{\sigma_2}d\sigma_\mu F_{\mu\nu}^{(+)\prime}(x)D_+(x-x')\\&=2\oint d\sigma_\mu F_{\mu\nu}'(x)D_+(x-x').\end{aligned} \tag{108}$$

We also recall that

$$\begin{aligned}\mathcal{W}_0&=\tfrac{1}{2}\int(dx)(dx')J_\mu(x)D_+(x-x')J_\mu(x')\\&=\tfrac{1}{2}\int(dx)(dx')[J_k(x)(\delta_{kl}D_+(x-x'))^{(T)}J_l(x')\\&\qquad-J_0(x)\mathfrak{D}(x-x')J_0(x')],\end{aligned}$$

where the latter form is appropriate to the radiation gauge.

The dependence of the transformation function upon the external current is expressed by

$$\delta_J(F^{(-)\prime}\sigma_1|F^{(+)\prime}\sigma_2)=i\left(F^{(-)\prime}\sigma_1\left|\int(dx)\delta J_\mu(x)A_\mu(x)\right|F^{(+)\prime}\sigma_2\right). \tag{109}$$

In the radiation gauge, $A_0(x)$ is the numerical quantity

$$A_0(x)=\int(dx')\mathfrak{D}(x-x')J_0(x').$$

Hence

$$\delta_{J_0}(F^{(-)\prime}\sigma_1|F^{(+)\prime}\sigma_2)=-i\int(dx)(dx')\delta J_0(x)\mathfrak{D}(x-x')J_0(x')(F^{(-)\prime}\sigma_1|F^{(+)\prime}\sigma_2),$$

and

$$(F^{(-)\prime}\sigma_1|F^{(+)\prime}\sigma_2)=(F^{(-)\prime}\sigma_1|F^{(+)\prime}\sigma_2)]_{J_0=0}\times\exp\left[-i\tfrac{1}{2}\int(dx)(dx')J_0(x)\mathfrak{D}(x-x')J_0(x')\right],$$

which gives the simple dependence of the transformation function upon $J_0$, in the radiation gauge. This factor is evident in the radiation gauge version of $\exp(i\mathcal{W}_0)$. Accordingly, we restrict ourselves to transverse currents, for which $J_0=0$.

On introducing the scale factor $\lambda$, $J_k \to \lambda J_k$ we infer from (109) that

$$(\partial/\partial\lambda)(F^{(-)\prime}\sigma_1|F^{(+)\prime}\sigma_2) = i\left(F^{(-)\prime}\sigma_1\middle|\int (dx)J_k(x)A_k(x)\middle|F^{(+)\prime}\sigma_2\right).$$

Repeated differentiation yields

$$(\partial/\partial\lambda)^n(F^{(-)\prime}\sigma_1|F^{(+)\prime}\sigma_2) = i^n\left(F^{(-)\prime}\sigma_1\middle|\int (dx_1)\cdots(dx_n)J_{k_1}(x_1)\cdots J_{k_n}(x_n) \times (A_{k_1}(x_1)\cdots A_{k_n}(x_n))_+\middle|F^{(+)\prime}\sigma_2\right),$$

and the transformation function appropriate to the external current ($\lambda=1$), is obtained from that of the isolated electromagnetic field ($\lambda=0$) as

$$(F^{(-)\prime}\sigma_1|F^{(+)\prime}\sigma_2) = \left(F^{(-)\prime}\sigma_1\middle|\left(\exp\left[i\int (dx)J_k(x)A_k(x)\right]\right)_+\middle|F^{(+)\prime}\sigma_2\right)\Big]_0.$$

If we employ the notation

$$(F^{(-)\prime}\sigma_1|\quad|F^{(+)\prime}\sigma_2)/(F^{(-)\prime}\sigma_1|F^{(+)\prime}\sigma_2)]_0 = \langle\ \rangle,$$

the transformation function (107) can be expressed by

$$\left\langle\left(\exp\left[i\int (dx)J_k(x)A_k(x)\right]\right)_+\right\rangle = \exp\left[i\tfrac{1}{2}\int (dx)(dx')J_k(x)(\delta_{kl}D_+(x-x'))^{(T)} \times J_l(x') + i\int (dx)J_k(x)A_k'(x)\right]. \quad (110)$$

It will be advantageous to suppress the vector indices. Accordingly, we rewrite (110) as

$$\left\langle\left(\exp\left[i\int (dx)J(x)A(x)\right]\right)_+\right\rangle = \exp\left[i\tfrac{1}{2}\int (dx)(dx')J(x)D_+(x-x')J(x') + i\int (dx)J(x)A'(x)\right]. \quad (111)$$

An alternative version is

$$\left\langle\left(\exp\left[i\int (dx)J(x)'A(x)\right]\right)_+\right\rangle = \exp\left[i\tfrac{1}{2}\int (dx)(dx')J(x)D_+(x-x')J(x')\right], \quad (112)$$

where

$$'A(x) = A(x) - A'(x).$$

An expansion of both sides in (111) or (112) will supply the matrix elements of ordered $A$ products.

The right side of (112) is an even function of $J$. Accordingly,

$$\langle('A(x_1)\cdots 'A(x_{2n-1}))_+\rangle = 0.$$

In particular,

$$\langle 'A(x)\rangle = 0,$$

or

$$\langle A(x)\rangle = A'(x). \quad (113)$$

The general even term in the expansion of the left side in (112) is

$$\frac{i^{2n}}{(2n)!}\int (dx_1)\cdots(dx_{2n})J(x_1)\cdots J(x_{2n}) \times \langle('A(x_1)\cdots 'A(x_{2n}))_+\rangle.$$

This is to be compared with

$$\frac{i^n}{2^n n!}\int (dx_1)\cdots(dx_{2n})J(x_1)\cdots J(x_{2n}) \times D_+(x_1-x_2)\cdots D_+(x_{2n-1}-x_{2n}) = \frac{i^n}{(2n)!}\int (dx_1)\cdots(dx_{2n})J(x_1)\cdots J(x_{2n}) \times \mathrm{sym}_{(n)}D_+(x_i-x_j),$$

in which has been introduced what we shall call the $n$th symmetrant of $D_+$. This is defined by

$$\mathrm{sym}_{(n)}D_+(x_i-x_j) = \sum_{\mathrm{perm}} D_+(x_{i_1}-x_{i_2})\cdots D_+(x_{i_{2n-1}}-x_{i_{2n}}),$$

and the summation is extended over all distinct permutations of the indices $i_1\cdots i_{2n}$, which are some rearrangement of the integers $1\cdots 2n$. Since $D_+$ is an even function of its argument, and the order of the $n$ factors is irrelevant, the number of such permutations is $(2n)!/2^n n! = (2n-1)(2n-3)\cdots$.

The matrix of an even product of the $'A$ is thus expressed by

$$\langle('A(x_1)\cdots 'A(x_{2n}))_+\rangle = (-i)^n\,\mathrm{sym}_{(n)}D_+(x_i-x_j). \quad (114)$$

The first two such products are

$$\langle('A(x)'A(x'))_+\rangle = -iD_+(x-x'),$$

or

$$\langle(A(x)A(x'))_+\rangle = A'(x)A'(x') - iD_+(x-x') \quad (115)$$

and

$$\langle('A(x_1)'A(x_2)'A(x_3)'A(x_4))_+\rangle = -[D_+(x_1-x_2)D_+(x_3-x_4) + D_+(x_1-x_3)D_+(x_2-x_4) + D_+(x_2-x_3)D_+(x_1-x_4)].$$

To obtain corresponding results for products of the operators $A$, as in the simple examples (113) and (115), we first consider the even terms of (111). The even function, $\cos x$, can be obtained from an exponential function of $x^2$ by a suitable operation

$$\cos x = C_t[\exp(-\tfrac{1}{2}tx^2)],$$

where

$$C_t(t^n) = 2^n n!/(2n)!$$

effectively defines the operator $C_t$, although an explicit integral representation can also be exhibited, as in (81). Hence the even part of (111) is

$$\left\langle\left(\cos\left[\int (dx)J(x)A(x)\right]\right)_+\right\rangle$$
$$= C_t \exp\left[-\tfrac{1}{2}\int (dx)(dx')J(x) \times(-iD_+(x-x')+tA'(x)A'(x'))J(x')\right],$$

which, in view of (112) and (114), yields

$$\begin{aligned}\langle(A(x_1)\cdots A(x_{2n}))_+\rangle &= C_t\,\mathrm{sym}_{(n)}[-iD_+(x_i-x_j)+tA'(x_i)A'(x_j)]\\ &= A'(x_1)\cdots A'(x_{2n})+\cdots\\ &\quad+(-i)^n\,\mathrm{sym}_{(n)}D_+(x_i-x_j). \qquad (116)\end{aligned}$$

As the initial term of the developed version indicates, the effect of the $C_t$ operation is to reinstate the unique counting of each distinct permutation in the expansion of the symmetrant (116).

For the construction of matrices describing odd products of the $A$, we remark that

$$(\delta/\delta A'(x))\left\langle\left(\exp\left[i\int (dx)JA\right]\right)_+\right\rangle = iJ(x)\left\langle\left(\exp\left[i\int (dx)JA\right]\right)_+\right\rangle.$$

On expanding both sides, we obtain

$$\begin{aligned}(\delta/\delta A'(x))\langle(A(x_1)\cdots A(x_m))_+\rangle &= \delta(x-x_1)\langle(A(x_2)\cdots A(x_m))_+\rangle+\cdots\\ &\quad+\delta(x-x_m)\langle(A(x_1)\cdots A(x_{m-1}))_+\rangle,\end{aligned}$$

which can also be expressed by

$$(\partial/\partial A'(x_m))\langle(A(x_1)\cdots A(x_m))_+\rangle = \langle(A(x_1)\cdots A(x_{m-1}))_+\rangle.$$

Hence the matrices of odd products can be obtained by differentiation from those of even products,

$$\begin{aligned}\langle(A(x_1)\cdots A(x_{2n-1}))_+\rangle &= (\partial/\partial A'(x_{2n}))\langle(A(x_1)\cdots A(x_{2n}))_+\rangle\\ &= (\partial/\partial A'(x_{2n}))C_t\,\mathrm{sym}_{(n)}[-iD_+(x_i-x_j)\\ &\qquad+tA'(x_i)A'(x_j)],\end{aligned}$$

or

$$\langle(A(x_1)\cdots A(x_{2n-1}))_+\rangle = C_t\,\mathrm{sym}_{(n)}[-iD_+(x_i-x_j)+tA'(x_i)A'(x_j),\quad tA'(x_k)],$$

which is intended to indicate a symmetrant that is obtained from (116) by replacing the elements containing the variables $x_k$, $x_{2n}$ with $tA'(x_k)$.

### The Occupation Number Representation

The diagonal matrix elements referring to the vacuum state are obtained by placing all eigenvalues equal to zero,

$$\begin{aligned}(0|(A(x_1)\cdots A(x_{2n-1}))_+|0) &= 0,\\ (0|(A(x_1)\cdots A(x_{2n}))_+|0) &= (-i)^n\,\mathrm{sym}_{(n)}D_+(x_i-x_j),\end{aligned} \qquad (117)$$

and, in particular,

$$(0|(A(x)A(x'))_+|0) = -iD_+(x-x'). \qquad (118)$$

We introduce the mode functions

$$\begin{aligned}A_{\lambda k}(x)_\mu &= \left(\frac{(d\mathbf{k})}{(2\pi)^3}\frac{1}{2k_0}\right)^{\frac{1}{2}} e_\mu(\lambda k)e^{ikx},\\ \bar{A}_{\lambda k}(x)_\mu &= \left(\frac{(d\mathbf{k})}{(2\pi)^3}\frac{1}{2k_0}\right)^{\frac{1}{2}} e_\mu(\lambda k)e^{-ikx},\end{aligned} \qquad (119)$$

in terms of which the tensor Green's function (III, 23) appears as

$$\begin{aligned}\delta_{\mu\nu}D_+(x-x') &= i\sum_{\lambda k} A_{\lambda k}(x)_\mu \bar{A}_{\lambda k}(x')_\nu, \quad x_0>x_0'\\ &= i\sum_{\lambda k} \bar{A}_{\lambda k}(x)_\mu A_{\lambda k}(x')_\nu, \quad x_0<x_0'. \qquad (120)\end{aligned}$$

Accordingly,

$$A_\nu'(x') = 2\oint d\sigma_\alpha F_{\alpha\mu}'(x)\delta_{\mu\nu}D_+(x-x')$$

becomes (suppressing the vector indices)

$$A'(x) = \sum_{\lambda k}(A_{\lambda k}(x)e^{-ikx_1}A_{\lambda k}^{(+)'}+\bar{A}_{\lambda k}(x)e^{ikx_1}A_{\lambda k}^{(-)'}),$$

where

$$A_{\lambda k}^{(-)'} = 2i\int_{\sigma_1} d\sigma_\mu F_{\mu\nu}^{(-)'}(x)A_{\lambda k}(x)_\nu e^{-ikx_1},$$

and

$$A_{\lambda k}{}^{(+)\prime}=-2i\int_{\sigma_2} d\sigma_\mu F_{\mu\nu}{}^{(+)\prime}(x)\bar{A}_{\lambda k}(x)_\nu e^{ikx_2}.$$

The latter quantities are the negatives of $a_{\lambda k}{}^{(\pm)\prime}$ (III, 27, 28). Since these minus signs would be somewhat unfortunate for our present purposes, we notice that the opposite choice of sign in (III, 27, 28) produces relatively trivial changes in the work of III. Thus, in Eq. (III, 34) and its consequences, the signs of the terms contalnlng $J_{\lambda k}$ and $J_{\lambda k}{}^*$ are to be reversed. We now write the eigenfunctions of the isolated electromagnetic field as

$$(F^{(-)\prime}|n)=\prod_{\lambda k}(n!)^{-\frac{1}{2}}(A_{\lambda k}{}^{(-)\prime})^n,$$

$$(n|F^{(+)\prime})=\prod_{\lambda k}(n!)^{-\frac{1}{2}}(A_{\lambda k}{}^{(+)\prime})^n,$$

while the transformation function is

$$\begin{aligned}(F^{(-)\prime}\sigma_1|F^{(+)\prime}\sigma_2)&=\exp[\textstyle\sum_{\lambda k} A_{\lambda k}{}^{(-)\prime}e^{ikx_1}e^{-ikx_2}A_{\lambda k}{}^{(+)\prime}]\\&=\textstyle\sum_n(F^{(-)\prime}|n)\exp[iP(n)(x_1-x_2)](n|F^{(+)\prime}).\end{aligned}$$

The occupation number matrix of $A$ is derived from (113), written as

$$\begin{aligned}(F^{(-)\prime}\sigma_1|A(x)|F^{(+)\prime}\sigma_2)&=\sum_{n,n'}(F^{(-)\prime}|n)(n\sigma_1|A(x)|n'\sigma_2)(n'|F^{(+)\prime})\\&=A'(x)(F^{(-)\prime}\sigma_1|F^{(+)\prime}\sigma_2).\end{aligned}$$

The substitutions $e^{-ikx_2}A_{\lambda k}{}^{(+)\prime}\rightarrow A_{\lambda k}{}^{(+)\prime}$, $e^{ikx_1}A_{\lambda k}{}^{(-)\prime}\rightarrow A_{\lambda k}{}^{(-)\prime}$ convert this into a generating function for the matrix referring to a standard surface,

$$(n|A(x)|n')=\exp(-iP(n)x_1)\times(n\sigma_1|A(x)|n'\sigma_2)\exp(iP(n')x_2),$$

namely,

$$\begin{aligned}\sum_{n,n'}(F^{(-)\prime}|n)(n|A(x)|n')(n'|F^{(+)\prime})&=\sum_{\lambda k}(A_{\lambda k}(x)A_{\lambda k}{}^{(+)\prime}+\bar{A}_{\lambda k}(x)A_{\lambda k}{}^{(-)\prime})\\&\quad\times\sum_n(F^{(-)\prime}|n)(n|F^{(+)\prime}).\end{aligned}$$

Now

$$A_{\lambda k}{}^{(-)\prime}(F^{(-)\prime}|n-1_{\lambda k})=(F^{(-)\prime}|n)n_{\lambda k}{}^{\frac{1}{2}}$$

and

$$(n-1_{\lambda k}|F^{(+)\prime})A_{\lambda k}{}^{(+)\prime}=n_{\lambda k}{}^{\frac{1}{2}}(n|F^{(+)\prime}),$$

so that the nonvanishing matrix elements of $A(x)$ are

$$(n-1_{\lambda k}|A(x)|n)=n_{\lambda k}{}^{\frac{1}{2}}A_{\lambda k}(x)$$

and

$$(n|A(x)|n-1_{\lambda k})=n_{\lambda k}{}^{\frac{1}{2}}\bar{A}_{\lambda k}(x).$$

The generating function for the matrices of all ordered products, referred to the standard surface, is

$$\begin{aligned}&\sum_{n,n'}(F^{(-)\prime}|n)\times\left(n\left|\left(\exp\left[i\int(dx)J(x)A(x)\right]\right)_+\right|n'\right)(n'|F^{(+)\prime})\\&=(F^{(-)\prime}|F^{(+)\prime})\exp\left[i\tfrac{1}{2}\int(dx)(dx')J(x)\times D_+(x-x')J(x')+i\int(dx)J(x)A'(x)\right]\\&=\exp\left[i\tfrac{1}{2}\int(dx)(dx')JD_+J\right]\\&\quad\times\prod_{\lambda k}\exp\left[A_{\lambda k}{}^{(-)\prime}A_{\lambda k}{}^{(+)\prime}+iA_{\lambda k}{}^{(-)\prime}\times\int(dx)J(x)\bar{A}_{\lambda k}(x)+iA_{\lambda k}{}^{(+)\prime}\times\int(dx)J(x)A_{\lambda k}(x)\right].\end{aligned}\tag{121}$$

The expansion into eigenfunctions is greatly simplified by exploiting the infinitesimal nature of $(d\mathbf{k})$. Thus the second exponential factor of (121) becomes

$$\begin{aligned}\prod_{\lambda k}\sum_{n,n'}\frac{(A^{(-)\prime})^n}{(n!)^{\frac{1}{2}}}&\left[\delta_{n,n'}\left(1-n\int(dx)JA\int(dx)J\bar{A}\right)\right.\\&+\delta_{n,n'+1}n^{\frac{1}{2}}i\int(dx)J\bar{A}+\delta_{n',n+1}n'^{\frac{1}{2}}i\\&\left.\times\int(dx)JA\right]\frac{(A^{(+)\prime})^n}{(n'!)^{\frac{1}{2}}}.\end{aligned}\tag{122}$$

A change in the occupation number of a given mode by two, for example, would lead to a matrix element proportional to $(d\mathbf{k})$, and thus to a transition probability proportional to $(d\mathbf{k})^2$.

We first consider diagonal matrix elements, which are contained in the following terms of (122),

$$\begin{aligned}&\prod_{\lambda k}\sum_n\frac{(A^{(-)\prime})^n}{(n!)^{\frac{1}{2}}}\left[1-n\int(dx)JA\int(dx)J\bar{A}\right]\frac{(A^{(+)\prime})^n}{(n!)^{\frac{1}{2}}}\\&=\sum_n(F^{(-)\prime}|n)\exp\left[-\int(dx)(dx')J(x)\times\left(\sum_{\lambda k}n_{\lambda k}A_{\lambda k}(x)\bar{A}_{\lambda k}(x')\right)J(x')\right](n|F^{(+)\prime}).\end{aligned}$$

Thus

$$\left(n\left|\left(\exp\left[i\int(dx)J(x)A(x)\right]\right)_+\right|n\right)$$
$$=\exp\left[-\tfrac{1}{2}\int(dx)(dx')J(x)(-iD_+(x-x')+\sum_{\lambda k}2n_{\lambda k}A_{\lambda k}(x)\bar{A}_{\lambda k}(x'))J(x')\right],$$

which asserts that

$$(n|(A(x_1)\cdots A(x_{2l-1}))_+|n)=0,$$

and that

$$(n|(A(x_1)\cdots A(x_{2l}))_+|n)=\mathrm{sym}_{(l)}[-iD_+(x_i-x_j)+\textstyle\sum_{\lambda k}n_{\lambda k}\times(A_{\lambda k}(x_i)\bar{A}_{\lambda k}(x_j)+\bar{A}_{\lambda k}(x_i)A_{\lambda k}(x_j))]. \quad (123)$$

The elementary example of the latter result is

$$(n|(A(x)A(x'))_+|n)=-iD_+(x-x')+\textstyle\sum_{\lambda k}n_{\lambda k}(A_{\lambda k}(x)\bar{A}_{\lambda k}(x')+\bar{A}_{\lambda k}(x)A_{\lambda k}(x')). \quad (124)$$

We introduce a classification of modes in the general matrix element: class $a$, those for which $n_{\lambda k}=n_{\lambda k}'-1$; class $b$, modes with $n_{\lambda k}=n_{\lambda k}'$; class $c$, those with $n_{\lambda k}=n_{\lambda k}'+1$. A typical term in the expansion of the product (122) can then be written

$$(F^{(-)\prime}|n)\prod_a\left[n'^{\frac{1}{2}}i\int(dx)JA\right]\times\prod_b\left[1-n\int(dx)JA\int(dx)J\bar{A}\right]\times\prod_c\left[n^{\frac{1}{2}}i\int(dx)J\bar{A}\right](n'|F^{(+)\prime}).$$

Accordingly,

$$\left(n\left|\left(\exp\left[i\int(dx)J(x)A(x)\right]\right)_+\right|n'\right)$$
$$=\exp\left[i\tfrac{1}{2}\int(dx)(dx')J(x)D_+^{(n)}(x,x')J(x')\right]\times\prod_a\left[n'^{\frac{1}{2}}i\int(dx)JA_{\lambda k}\right]\prod_c\left[n^{\frac{1}{2}}i\int(dx)J\bar{A}_{\lambda k}\right], \quad (125)$$

in which we have introduced the symbol

$$D_+^{(n)}(x,x')=D_+(x-x')+i\sum_b n_{\lambda k}\times[A_{\lambda k}(x)\bar{A}_{\lambda k}(x')+\bar{A}_{\lambda k}(x)A_{\lambda k}(x')].$$

We shall use $A_\alpha(x)$ to denote collectively the $a$ mode functions, $A_{\lambda k}(x)$, and the $c$ mode functions $\bar{A}_{\lambda k}(x)$ (a complete unification would be achieved by transposing the matrices with respect to the occupation numbers of the $c$ modes, which would introduce a time-reversed description for emission processes). With this notation, (125) appears as

$$\left(n\left|\left(\exp\left[i\int(dx)J(x)A(x)\right]\right)_+\right|n'\right)$$
$$=\prod_a(n'^{\frac{1}{2}})\prod_c(n^{\frac{1}{2}})\prod_\alpha\left[i\int(dx)J(x)A_\alpha(x)\right]\times\exp\left[i\tfrac{1}{2}\int(dx)(dx')JD_+^{(n)}J\right].$$

The expansion of this generating function leads to the following formula for the general nonvanishing matrix element,

$$(n|(A(x_1)\cdots A(x_m))_+|n')=\prod_a(n'^{\frac{1}{2}})\prod_c(n^{\frac{1}{2}})(-i)^r\,\mathrm{sym}_{(N,r)}\times[A_\alpha(x_i);D_+^{(n)}(x_j,x_k)], \quad (126)$$

where $N$ is the number of $\alpha=a+c$ modes, and $r$ is a nonnegative integer such that

$$m=N+2r.$$

We have introduced an extension of the symmetrant which is constructed from $N$ different functions of a single variable, $A_\alpha(x_i)$, and a symmetrical function of two variables, $D_+^{(n)}(x_j,x_k)$, taken $r$ times,

$$\mathrm{sym}_{(N,r)}[A_\alpha(x_i);D_+^{(n)}(x_j,x_k)]=\sum_{\text{perm}}A_{\alpha 1}(x_{i_1})\cdots A_{\alpha N}(x_{i_N})\times D_+^{(n)}(x_{i_{N+1}},x_{i_{N+2}})\cdots D_+^{(n)}(x_{i_{m-1}},x_{i_m}),$$

where the summation is extended over the $(N+2r)!/2^r r!$ distinct permutations of the indexes $i_1\cdots i_m$, which are some rearrangement of the integers $1\cdots m$.

The diagonal matrix elements correspond to the situation where $N=0$, $m=2r$. At the opposite extreme is $r=0$, $N=m$, where the symmetrant reduces to

$$\mathrm{sym}_{(m,0)}[A_\alpha(x_i)]=\sum_{\text{perm}}A_{\alpha 1}(x_{i_1})\cdots A_{\alpha m}(x_{i_m}).$$

This sum of $m!$ permutations is obtained from the corresponding determinant by omitting the alternating sign factor.

*THE QUANTUM CORRECTION IN THE RADIATION BY ENERGETIC ACCELERATED ELECTRONS*

BY JULIAN SCHWINGER

HARVARD UNIVERSITY

*Communicated December 30, 1953*

The radiation emitted by high energy electrons in a large scale magnetic field is an essentially classical phenomenon, and was so discussed in a paper[1] of the author. It was there remarked that the condition for quantum effects to be unimportant is that the momenta of the radiated quanta be small compared with the electron momentum. At very high electron energies, this requirement reads

$$\hbar\omega << E, \qquad \omega \sim \omega_0(E/mc^2)^3, \tag{1}$$

where

$$\omega_0 = c/R = (eH/mc)(mc^2/E) \tag{2}$$

is the rotational frequency in the orbit determined by the magnetic field **H**. Hence

Reprinted from the *Proceedings of the National Academy of Sciences U.S.A.* **40,** 132–136 (1954).

the magnitude of the quantum correction should be indicated by

$$\frac{\hbar\omega_0}{E}\left(\frac{E}{mc^2}\right)^3 = \frac{\hbar/mc}{R}\left(\frac{E}{mc^2}\right)^2$$

$$= \frac{(e\hbar/mc)H}{mc^2}\frac{E}{mc^2}, \qquad (3)$$

which is extremely small for all existing or contemplated magnetic devices. Recently, there have been several attempts at a quantum mechanical calculation[2] or discussion[3] of this emission process, which have led to varying degrees of doubt concerning the validity of (3) as a measure of the quantum correction. To aid in settling this question, we shall present an explicit calculation of the quantum correction, to the first order in $\hbar$.

At this level of accuracy, the spin degrees of freedom play no role for unpolarized particles. With a suitable canonical transformation, the Dirac Hamiltonian can be replaced by the classical

$$H = \left[c^2\left(\mathbf{p} - \frac{e}{c}\mathbf{A}\right)^2 + (mc^2)^2\right]^{1/2}, \qquad (4)$$

for a field described by a vector potential (this includes the magnetic field and the radiation field). The probability for emitting a photon, during the time interval $T = t_1 - t_2$, into the range $(d\mathbf{k})$ about the propagation vector $\mathbf{k}$ and with polarization vector $\mathbf{e}$, is obtained as the absolute squared matrix element of the operator

$$\left[\frac{(d\mathbf{k})}{(2\pi)^3}\frac{2\pi}{\hbar\omega}\right]^{1/2} e\int_{t_2}^{t_1} dt e^{i\omega t}\,(\mathbf{e}\cdot\mathbf{v}(t), e^{-i\mathbf{k}\cdot\mathbf{r}(t)}), \qquad (5)$$

where $\mathbf{v}(t)$ is the particle velocity, and ( , ) indicates the symmetrized product of the operators. This matrix element refers to specific initial and final particle states. On summing over all final particle states, we obtain the probability for the radiation process as an expectation value computed for the initial particle state,

$$\frac{(d\mathbf{k})}{(2\pi)^3}\frac{2\pi e^2}{\hbar\omega}\langle\int_{t_2}^{t_1} dt\int_{t_2}^{t_1} dt' e^{-i\omega(t-t')}(\mathbf{e}\cdot\mathbf{v}(t), e^{i\mathbf{k}\cdot\mathbf{r}(t)})(\mathbf{e}\cdot\mathbf{v}(t'), e^{-i\mathbf{k}\cdot\mathbf{r}(t')})\rangle \qquad (6)$$

If we are not concerned with the radiation of a specific polarization, one should replace $\mathbf{e}\cdot\mathbf{v}(t)\mathbf{e}\cdot\mathbf{v}(t')$ with $\mathbf{n}\times\mathbf{v}(t)\cdot\mathbf{n}\times\mathbf{v}(t') = \mathbf{v}(t)\cdot\mathbf{v}(t') - \mathbf{n}\cdot\mathbf{v}(t)\mathbf{n}\cdot\mathbf{v}(t')$, where $\mathbf{n}$ is the unit vector directed along $\mathbf{k}$. The relation

$$(i\mathbf{k}\cdot\mathbf{v}(t), e^{i\mathbf{k}\cdot\mathbf{r}(t)}) = (d/dt)e^{i\mathbf{k}\cdot\mathbf{r}(t)} \qquad (7)$$

then enables us to write the probability for radiation, without regard to polarization, as

$$\frac{(d\mathbf{k})}{(2\pi)^3}\frac{2\pi e^2c^2}{\hbar\omega}\langle\int_{t_2}^{t_1} dt\int_{t_2}^{t_1} dt' e^{-i\omega(t-t')}\left[\frac{1}{c^2}(\mathbf{v}(t), e^{i\mathbf{k}\cdot\mathbf{r}(t)})\cdot(\mathbf{v}(t'), e^{-i\mathbf{k}\cdot\mathbf{r}(t')}) - e^{i\mathbf{k}\cdot\mathbf{r}(t)}e^{-i\mathbf{k}\cdot\mathbf{r}(t')}\right]\rangle. \qquad (8)$$

For a sufficiently extended time interval $T$, we can introduce new time variables $t \to t + {}^1/_2\tau$, $t' \to t - {}^1/_2\tau$, and regard the $t$ integrand as the probability per unit time of the event. On multiplication with $\hbar\omega$ we find that the average power radiated into a unit solid angle enclosing $\mathbf{n}$, and contained in a unit frequency interval about $\omega$, is

$$P(\mathbf{n}, \omega, t) = \frac{e^2}{4\pi^2}\frac{\omega^2}{c}\left\langle\int_{-\infty}^{\infty} d\tau e^{-i\omega\tau}\left[\frac{1}{c^2}\left(\mathbf{v}\left(t + \frac{1}{2}\tau\right), e^{i\mathbf{k}\cdot\mathbf{r}(t+{}^1/_2\tau)}\right)\cdot\left(\mathbf{v}\left(t - \frac{1}{2}\tau\right), e^{-i\mathbf{k}\cdot\mathbf{r}(t-{}^1/_2\tau)}\right) - e^{i\mathbf{k}\cdot\mathbf{r}(t+{}^1/_2\tau)}\, e^{-i\mathbf{k}\cdot\mathbf{r}(t-{}^1/_2\tau)}\right]\right\rangle. \quad (9)$$

If the order of operator multiplication is not significant, this result is equivalent to the classical formula (I.38) of reference 1.

We shall now investigate the commutation properties of the operators $\mathbf{r}(t)$ and $\mathbf{v}(t)$, to the first order in $\hbar$. Thus, the equation

$$(\mathbf{v}, H) = \left(\mathbf{p} - \frac{e}{c}\mathbf{A}\right)c^2 \quad (10)$$

is solved, to this accuracy, by

$$\mathbf{v} = \left(H^{-1}, \left(\mathbf{p} - \frac{e}{c}\mathbf{A}\right)c^2\right), \quad (11)$$

from which we derive

$$[r_i, v_j] = i\hbar\frac{\partial v_j}{\partial p_i} = i\hbar\,\frac{c^2}{E}\left[\delta_{ij} - \frac{1}{c^2}v_i v_j\right]. \quad (12)$$

Consistent with the approximation program, we have on the right replaced the operator $H$ by the classical determinate energy $E$, and disregarded the non-commutativity of $v_i$ and $v_j$. The latter is described, to the same approximation, by

$$\frac{1}{c^2}\,\mathbf{v}\times\mathbf{v} = i\hbar\frac{ec}{E^2}\left[(1 - v^2/c^2)\mathbf{H} + \frac{1}{c^2}\mathbf{v}\mathbf{v}\cdot\mathbf{H}\right]. \quad (13)$$

The classical motion occurs in the plane perpendicular to the magnetic field, $\mathbf{v}\cdot\mathbf{H} = 0$, so that (13) implies the following uncertainty principle for the velocity components in this plane,

$$\frac{1}{c^2}\,\Delta v_1\Delta v_2 \geq \frac{1}{2}\,\hbar\,\frac{ecH}{E^2}\,(1 - v^2/c^2), \quad (14)$$

from which we infer

$$\frac{1}{c^2}[(\Delta v_1)^2 + (\Delta v_2)^2] \geq \hbar\,\frac{ecH}{E^2}(1 - v^2/c^2) = \frac{\hbar\omega_0}{E}(1 - v^2/c^2). \quad (15)$$

Since the particle speed occurs significantly only in the combination $1 - v^2/c^2$, the effect of the non-simultaneous specifiability of $v_1$ and $v_2$ is measured by the coefficient $\hbar\omega_0/E = e\hbar cH/E^2$, which is negligible compared to (3), and decreases with increasing energy. Hence, at sufficiently high energies, the components of $\mathbf{v}$ are determinate quantities.

The symmetrized products of $\mathbf{v}$ and $e^{\pm i\mathbf{k}\cdot\mathbf{r}}$ can be replaced by

$$(\mathbf{v}, e^{i\mathbf{k}\cdot\mathbf{r}}) = \left[\mathbf{v} - \frac{\hbar c^2}{2E}\left(\mathbf{k} - \frac{1}{c^2}\,\mathbf{v}\mathbf{k}\cdot\mathbf{v}\right)\right]e^{i\mathbf{k}\cdot\mathbf{r}} \tag{16}$$

and

$$(\mathbf{v}, e^{-i\mathbf{k}\cdot\mathbf{r}}) = e^{-i\mathbf{k}\cdot\mathbf{r}}\left[\mathbf{v} - \frac{\hbar c^2}{2E}\left(k - \frac{1}{c^2}\,\mathbf{v}\mathbf{k}\cdot\mathbf{v}\right)\right]. \tag{17}$$

We then combine the contiguous quantities, $e^{i\mathbf{k}\cdot\mathbf{r}(t+1/2\tau)}$ and $e^{-i\mathbf{k}\cdot\mathbf{r}(t-1/2\tau)}$, with the aid of the theorem

$$e^A e^B = e^{A+B}e^{1/2[A,B]}, \tag{18}$$

which is exact if [A, B] commutes with A and B, and suffices for the purposes of the approximation we are employing. Since the values of $\tau$ that contribute to the integral (9) are such that $\omega\tau(1 - v^2/c^2) \sim \omega_0\tau(1 - v^2/c^2)^{-1/2} \sim 1$, we have $\omega_0\tau << 1$, and one can use the expansion

$$\mathbf{r}\left(t \pm \frac{1}{2}\tau\right) = \mathbf{r}(t) \pm \frac{1}{2}\tau\mathbf{v}(t) + \frac{1}{8}\tau^2\dot{\mathbf{v}}(t) \pm \frac{1}{48}\tau^3\ddot{\mathbf{v}}(t). \tag{19}$$

Thus

$$\begin{aligned}\mathbf{r}\left(t + \frac{1}{2}\tau\right) - \mathbf{r}\left(t - \frac{1}{2}\tau\right) &= \tau\mathbf{v}(t) + \frac{1}{24}\tau^3\ddot{\mathbf{v}}(t)\\ &= \left(\tau - \frac{1}{24}\,\omega_0^2\tau^3\right)\mathbf{v}(t),\end{aligned} \tag{20}$$

and

$$\begin{aligned}\left[\mathbf{k}\cdot\mathbf{r}\left(t + \frac{1}{2}\tau\right), \mathbf{k}\cdot\mathbf{r}\left(t - \frac{1}{2}\tau\right)\right] &= -\tau[\mathbf{k}.\mathbf{r}, \mathbf{k}.\mathbf{v}] - \frac{1}{24}\tau^3[\mathbf{k}.\mathbf{r}, \mathbf{k}.\ddot{\mathbf{v}}]\\ &= -i\hbar\left(\mathbf{k}\cdot\frac{\partial}{\partial\mathbf{p}}\right)\left[\left(\tau - \frac{1}{24}\,\omega_0^2\tau^3\right)\mathbf{k}\cdot\mathbf{v}\right]\\ &= -2i\frac{\hbar\omega}{E}\,\omega\left[\left(1 - \frac{1}{c}\mathbf{n}\cdot\mathbf{v}\right)\tau + \frac{1}{24}\,\omega_0^2\tau^3\right],\end{aligned} \tag{21}$$

in which we have omitted commutators containing only velocity components, and introduced simplifications appropriate to the nature of the angular distribution, $\frac{1}{c}\mathbf{n}\cdot\mathbf{v} \sim 1$. We should also note that the velocity factors multiplying the exponentials in (9) combine into

$$\left(\frac{1}{c^2}\mathbf{v}\left(t + \frac{1}{2}\tau\right)\cdot\mathbf{v}\left(t - \frac{1}{2}\tau\right) - 1\right)\left(1 + \frac{\hbar\omega}{E}\frac{1}{2c}\mathbf{n}\cdot\left(\mathbf{v}\left(t + \frac{1}{2}\tau\right) + \mathbf{v}\left(t - \frac{1}{2}\tau\right)\right)\right) \tag{22}$$

where the second factor can be approximated by $1 + (\hbar\omega/E)$. The resulting form of (9) is

$$P(\mathbf{n}, \omega, t) = \frac{e^2}{4\pi^2}\frac{\omega^2}{c}\left(1 + \frac{\hbar\omega}{E}\right)\int_{-\infty}^{\infty} d\tau\left(\frac{1}{c^2}\mathbf{v}\left(t + \frac{1}{2}\tau\right)\cdot\mathbf{v}\left(t - \frac{1}{2}\tau\right) - 1\right)$$
$$\cdot\exp\left[-i\omega\left(1 + \frac{\hbar\omega}{E}\right)\left(\left(1 - \frac{1}{c}\mathbf{n}\cdot\mathbf{v}\right)\tau + \frac{1}{24}\omega_0^2\tau^3\right)\right]. \quad (23)$$

We have thus obtained the simple result that first order quantum effects are included by making the substitution

$$\omega \rightarrow \omega\left(1 + \frac{\hbar\omega}{E}\right) \quad (24)$$

in the classical formula for $\omega^{-1}P(\mathbf{n}, \omega, t)$. The same statement applies to $\omega^{-1}P(\omega, t)$, since the correction factor does not depend upon the emission direction. On referring to the classical formula (II.16) of reference 1, we see that the power radiated per unit frequency range about $\omega$ is given by

$$P(\omega, t) = \frac{3^{1/2}}{4\pi}\frac{e^2}{R}\left(\frac{E}{mc^2}\right)^4\frac{\omega_0\omega}{\omega_c^2}\int_{\omega/\omega_c(1+(\hbar\omega/E))}^{\infty} d\eta K_{5/3}(\eta), \quad (25)$$

where

$$\omega_c = \frac{3}{2}\omega_0(E/mc^2)^3. \quad (26)$$

Hence, the first order quantum correction is

$$\delta P(\omega, t) = -\frac{3^{1/2}}{4\pi}\frac{e^2}{R}\left(\frac{E}{mc^2}\right)^4\frac{\omega_0}{\omega_c}\left(\frac{\omega}{\omega_c}\right)^3\frac{\hbar\omega_c}{E}K_{5/3}(\omega/\omega_c), \quad (27)$$

and the corresponding correction to the total power emerges as

$$\delta P(t) = -3^{1/2}\frac{55}{36}\omega_0\frac{e^2}{R}\left(\frac{E}{mc^2}\right)^4\frac{\hbar\omega_c}{E}, \quad (28)$$

with the aid of the integral

$$\int_0^{\infty} d\xi\, \xi^3 K_{5/3}(\xi) = \frac{55}{27}\pi. \quad (29)$$

It is thus found that, with regard to the total radiated power, the effect of first order quantum corrections is to multiply the classical value, $\frac{2}{3}\omega_0(e^2/R)(E/mc^2)^4$, by the factor[4]

$$1 - 3^{1/2}\frac{55}{16}\frac{\hbar/mc}{R}\left(\frac{E}{mc^2}\right)^2. \quad (30)$$

[1] Schwinger, J., *Phys. Rev.*, **75**, 1912 (1949).

[2] Parzen, G., *Ibid.*, **84**, 235 (1951) (corrected by Judd, Lepore, Ruderman, and Wolff, *Ibid.*, **86**, 123 (1952), and Olsen, H., and Wergeland, H., *Ibid.*, **86**, 123 (1952)). Neuman, M., *Ibid.*, **90**, 682 (1953).

[3] Schiff, L. I., *Am. J. Phys.*, **20**, 474 (1952).

[4] While writing this paper, the author was delighted to encounter, in the December, 1953, issue of *Physics Abstracts*, a report of a paper by Sokolov, A. A., Klepikov, N. P., and Ternov, I. M., *Doklady Akad. Nauk SSSR*, **89**, 665 (1953), in which the identical result is quoted as obtained by a calculation based on the Dirac equation.

# A Theory of the Fundamental Interactions

JULIAN SCHWINGER

*Harvard University, Cambridge, Massachusetts*

"The axiomatic basis of theoretical physics cannot be extracted from experience but must be freely invented."

A. Einstein.

This note is an account of some developments in an effort to find a description of the present stock of elementary particles within the framework of the theory of quantized fields[1].

The theory of fields suggests that the spin values of ½ for F(ermi)-D(irac) fields, and of 0 and 1 for B(ose)-E(instein) fields are not only exceptional in their simplicity but are likely to be unique in the possibility of constructing a consistent formalism for particles with mass and electric charge. We shall attempt to describe the massive, strongly interacting particles by means of fields with the smallest spin appropriate to the statistics, 0(B.E.) and ½(F.D.). Spin 1 remains a possibility for B.E. fields but will be assumed to refer to a different family of particles, of which the electromagnetic field may be a special example. If the spin values are thus limited, the origin of the diversity of known particles must be sought in internal degrees of freedom. We suppose that the various intrinsic degrees of freedom are dynamically exhibited by specific interactions, each with its characteristic symmetry properties, and that the final effect of interactions with successively lower symmetry[2] is to produce a spectrum of physically distinct particles from initially degenerate states. Thus we attempt to relate the observed masses to the same couplings responsible for the production and interaction of these particles

The general multicomponent Hermitian field $\chi$ separates into the F.D. field $\psi$ and the B.E. field $\phi$. The representation of spin ½ requires 4 components, while spin 0 demands 5 components, decomposable into a scalar and a vector. The existence of internal degrees of freedom is expressed by an additional multiplicity

[1] The initial stages of this work are described in a previous paper (*1*). The further considerations recorded here were first presented in a series of lectures delivered at Harvard and Massachusetts Institute of Technology, Oct.–Dec., 1956.

[2] The concept of a hierarchy of interactions has been discussed before by Pais (*2*), but with reference to particle stability rather than the mass spectrum.



Reprinted from *Annals of Physics* **2, 407–434 (1957).**

or degeneracy of each set of fields, with the corresponding freedom of transformations in the internal symmetry space. In virtue of physical positive-definite requirements, referring to the commutation relations for the $\psi$ field, and to the energy for the $\phi$ field, the metric of the symmetry space is necessarily Euclidean. This is indicated in the structure of the general Lagrange function

$$\mathcal{L} = \tfrac{1}{4}(\chi A^\mu \partial_\mu \chi - \partial_\mu \chi A^\mu \chi) - \mathcal{H}(\chi)$$

by stating that the matrices $A^\mu$ refer only to the space-time properties of the various fields. A classification of internal degrees of freedom follows from the consideration of continuous transformation groups with various numbers of independent infinitesimal operations, as represented by antisymmetrical, imaginary matrices $T_a$. Associated with each symmetry transformation $T$ is a physical quantity $T$, described by a flux vector

$$j_T^\mu = -\frac{i}{2}\chi A^\mu T \chi,$$

which is conserved if the transformation is a dynamical invariance property. Then the total content of property $T$, symbolized by the Hermitian operator

$$\mathbf{T} = \int d\sigma_\mu j_T^\mu,$$

is a constant of the motion. From an evident analogy we speak of an individual property of this type as a charge. The requirement that a number of symmetry operations form a continuous group is expressed by the commutation relations

$$[T_a, T_b] = \textstyle\sum_c t_{abc} T_c,$$

which must also be obeyed by the Hermitian operators $\mathbf{T}_a$, as the generators of an isomorphic group of unitary transformations. The connection between the two groups is provided by the field commutation relations

$$[\chi, \mathbf{T}_a] = T_a \chi.$$

A study of the group requirement shows that the internal symmetry group can be factored into a completely commutative group, and an essentially noncommutative one with a structure characterized by the property that the matrices

$$t_b = (t_{abc})$$

constitute a $T$ matrix representation of dimensionality equal to $n$, the number of transformation parameters.

The first examples of the latter type are encountered for $n = 3$ and $n = 6$. For $n = 3$, the group structure is uniquely that of the three-dimensional Euclidean rotation group. We can conclude that the group with $n = 6$ contains the three-dimensional rotation group as a subgroup, if we make explicit use of the physical requirement that a symmetry group possess the groups of lower order as subgroups in order to effect a systematic reduction of internal symmetry through the addition of interactions with lower symmetry. It then follows that the six parameter group factors into the product of two three-dimensional rotation groups. This group can also be described as the four-dimensional Euclidean rotation group, for which the six independent matrices

$$T_{ab} = -T_{ba} \qquad a, b = 1 \cdots 4$$

obey the commutation relations

$$\frac{1}{i}[T_{ab}, T_{cd}] = \delta_{ac}T_{bd} - \delta_{bc}T_{ad} + \delta_{bd}T_{ac} - \delta_{ad}T_{bc}.$$

The factorization is accomplished by the definitions

$$T_3 = \tfrac{1}{2}(T_{12} + T_{34}), \qquad Z_3 = \tfrac{1}{2}(T_{12} - T_{34}) \tag{1}$$

and their cyclic permutations.

This investigation of the simplest group structures is supplemented by an examination of the symmetries that can be represented by matrices of small dimensionality, $\nu$. The number of independent, antisymmetrical matrices, $\frac{1}{2}\nu(\nu - 1)$, coincides with the number of rotation planes in a $\nu$-dimensional space and we thus encounter elementary representations of the various rotation groups, as provided by the matrices.

$$(T_{ab})_{cd} = \frac{1}{i}(\delta_{ac}\delta_{bd} - \delta_{ad}\delta_{bc}).$$

Each matrix has the eigenvalues 0, $\pm 1$, with the exception of $\nu = 2$ where the 0 eigenvalue does not occur. Hence, beginning with $\nu = 2$, there is one independent antisymmetrical matrix and the symmetry space is two-dimensional; three dimensional symmetries cannot be described. For $\nu = 3$, with three antisymmetrical matrices, we obtain the $T = 1$ representation of the three-dimensional rotation group, appropriate to the rotations of a vector in that space. With $\nu = 4$, and six antisymmetrical matrices, we can describe the rotations of a vector in the four-dimensional space, which combine the $T = 0$ and $T = 1$ representations of the three-dimensional group. Alternatively, we can use (1) to construct the matrices

$$T_k = \tfrac{1}{2}\tau_k, \qquad Z_k = \tfrac{1}{2}\zeta_k, \qquad k = 1 \cdots 3,$$

which give two $T = \frac{1}{2}$ representations of the three-dimensional rotation group. Thus the same four-dimensional matrices can be viewed as referring to the three-dimensional representations $T = 0$, 1, or to a two-fold $T = \frac{1}{2}$ representation.

The distinction between integral and half-integral $T$ would only appear when four-dimensional symmetries are reduced to three-dimensional ones, since we have the choice of selecting the three-dimensional transformations $T_1 = \frac{1}{2}\tau_1, \cdots$ $(T = \frac{1}{2})$ or the transformations $T_1 = T_{23}, \cdots$ $(T = 0, 1)$. It may be noted here that, in the further reduction to two-dimensional symmetry that physically distinguishes $T_{12}$, the relation of this matrix to its three-dimensional origins differs in the two possibilities. Thus, for $T = 0, 1$, we have simply $T_{12} = T_3$ while, for $T = \frac{1}{2}$, there appears $T_{12} = T_3 + Z_3$, where $Z_3 = \frac{1}{2}\zeta_3$. The two forms can be united by writing

$$T_{12} = T_3 + \tfrac{1}{2}Y, \tag{2}$$

in which $Y$, the hypercharge, is defined to be zero for integral $T$ and is represented by the matrix $\zeta_3$, with eigenvalues $\pm 1$, for $T = \frac{1}{2}$.

Since the fundamental $T = \frac{1}{2}$ representation of the three-dimensional rotation group (familiar as isotopic spin) is first encountered within the framework of four-dimensional symmetries, it is natural to suppose that the latter is the underlying symmetry, and that fields exist realizing the two-fold $T = \frac{1}{2}$ representation and the equivalent $T = 0, 1$ representation. The reduction to three-dimensional symmetries should distinguish physically the two $T = \frac{1}{2}$ representations (labelled by $Y = \pm 1$) as well as the $T = 0$ and $T = 1$ representations. Now it is well known that the electromagnetic field is the physical agency that destroys the three-dimensional isotopic space symmetry and leaves two dimensional symmetries. Thus the physical property defined by the two-dimensional rotations is explicitly electrical charge, and (2) automatically provides the identification

$$Q = T_3 + \tfrac{1}{2}Y. \tag{3}$$

This general connection between isotopic spin, electrical charge, and hypercharge enables us to recognize that the simple representations to which we have been led are realized[3] completely for the heavy F.D. particles, where one encounters two charge doublets, $N(Y = +1)$ and $\Xi(Y = -1)$ [although $\Xi^0$ is unknown experimentally], a charge triplet, $\Sigma$, and a charge singlet, $\Lambda$. For the heavy B.E. particles, there exist two charge doublets, $K(Y = +1)$, $\tilde{K}(Y = -1)$, and a charge triplet, $\pi$, but no charge singlet ($\sigma$) is known. Of course, the absence of such a stable or metastable particle does not necessarily mean that a field of the corresponding type does not exist. Thus, if the $\sigma$-field were scalar, as contrasted with the known pseudoscalar character of the $\pi$-field, and the corresponding particles had masses greater than $2\mu_\pi$, these $\sigma$-particles would be highly unstable against rapid disintegration into two $\pi$-mesons. It is also possible that the B.E. field differs from the F.D. field in realizing only $T = \frac{1}{2}$, and not $T = 0, 1$,

[3] Indeed, the first successful classification of the heavy unstable particles (by Gell-Mann and Nishijima) employed an empirical relation having the form of (3).

despite the four-dimensional kinematical equivalence between integral and half-integral $T$. The origin of a $T = 1$ representation can then be found in the matrices defined by the structure of the commutation relations for the four-dimensional rotations, which are themselves representations of the symmetry operations. These matrices are six-dimensional and describe the transformations of an anti-symmetrical tensor. Through the analog of the definitions (1), this tensor can be decomposed into two, each of which is self-dual (to within a sign change). A single self-dual tensor possesses three independent components and provides a basis for the $T = 1$ representation in three dimensions, without an accompanying $T = 0$ representation.

Current field theory describes the dynamics of fields by adding a coupling term in the Lagrange function to those of the uncoupled fields, which are, for spin 0 and spin ½ fields,

$$\mathfrak{L}^{(0)} = -\tfrac{1}{2}[\phi^{\lambda}.\partial_{\lambda}\phi - \phi.\partial_{\lambda}\phi^{\lambda} - \phi^{\lambda}\phi_{\lambda} + \mu_0^2\phi^2]$$

and

$$\mathfrak{L}^{(\frac{1}{2})} = -\tfrac{1}{2}\left[\psi.\beta\gamma^{\mu}\frac{1}{i}\partial_{\mu}\psi + m_0\psi.\beta\psi\right].$$

The dot indicates symmetrical and antisymmetrical multiplication, respectively, as is appropriate to the statistics of these fields. Also $x_k = x^k$, $k = 1 \cdots 3$, $x_0 = -x^0$, and the matrix $\beta = \gamma^0$ is antisymmetrical and imaginary, while the $\gamma^k$ are symmetrical and imaginary. The constants $\mu_0$ and $m_0$ are identified as the masses of the field quanta in the absence of interaction. It should also be noted that, in the natural units we are employing ($\hbar = c = 1$), the B.E. field $\phi$ is an inverse length $[L^{-1}]$ while the F.D. field $\psi$ has the dimensions $[L^{-3/2}]$, since the Lagrange function is in the nature of an inverse four-dimensional volume. In seeking possible forms for the interaction term in $\mathfrak{L}$ we shall be guided by the heuristic principle that the coupling between fields is described by simple algebraic functions of the field operators in which only dimensionless constants appear. This principle expresses the attitude that present theories, which are based on the infinite divisibility of the space-time manifold, contain no intrinsic standard of length. The mass constants of the individual fields are regarded as phenomenological manifestations of the unknown physical agency that produces the failure in the conventional space-time description and establishes the absolute scale of length and of mass. On this view, the coupling terms employed within the present formalism should not embody a unit of length that finds its dynamical origin outside of the domain of physical experience to which the theory of fields is applicable. For interacting spin 0 and spin ½ fields only two types of coupling terms are admitted by this principle, $\phi\psi\psi$ and $\phi\phi\phi\phi$.

We are interested in fields that possess internal degrees of freedom corresponding to the possible $T$ values of 0, ½, and 1. Hence the Yukawa-type coupling term $\phi\psi\psi$, will be composed additively of structures of the type $\phi_{(0,1)}\psi_{(1/2)}\,\psi_{(1/2)}$, $\phi_{(0,1)}\psi_{(0,1)}\psi_{(0,1)}$, and $\phi_{(1/2)}\psi_{(1/2)}\psi_{(0,1)}$, according to the evident requirements stemming from the symmetry of the three-dimensional internal space. We must now examine the possibility of introducing dynamically the four-dimensional symmetry that provides the kinematical foundation for the three-dimensional isotopic space representations. The internal symmetry is superimposed upon the space-time structure which, according to the space reflection properties of the spin 0 B.E. field, can be $\phi\frac{1}{2}\psi\beta\psi$, or $\phi\frac{1}{2}\psi\beta\gamma_5\psi$, $\gamma_5 = \gamma^0\gamma^1\gamma^2\gamma^3$. It is important to remark that, since both matrices $\beta$ and $\beta\gamma_5$ are antisymmetrical and imaginary, which are just the necessary symmetry and reality properties, any additional matrix referring to internal degrees of freedom must be symmetrical and real.

In considering the coupling term $\phi_{(0,1)}\psi_{(1/2)}\psi_{(1/2)}$, let us assume first that the integral $T$ B.E. field is a self-dual antisymmetrical tensor, containing only three independent components $\phi_k$, $k = 1 \cdots 3$, which must be combined with the similar tensor formed from the three matrices $\tau_k$. The latter are antisymmetrical, imaginary matrices, however, and thus unacceptable according to the preceeding comments. Hence, to form an interaction of this type the $T = \frac{1}{2}$ F.D. fields must possess yet another internal degree of freedom in the nature of a charge, which will be identified with nucleonic charge $N$ and which is represented by the antisymmetrical imaginary matrix

$$\nu = \begin{pmatrix} 0 & -i \\ i & 0 \end{pmatrix},$$

defined in an independent two-dimensional nucleonic charge space. The ensuing interaction term appears as

$$g_\pi \phi_{(1)} \frac{1}{2} \psi_{(1/2)}\beta\gamma_5\nu\tau\psi_{(1/2)}$$

where $g_\pi$ occurs as a pure number measuring the strength of the interaction. A definite choice of space-time structure has been made, in accordance with the known pseudoscalar nature of the $\pi$-field, $\phi_{(1)}$.

The field $\psi_{(1/2)}$ now comprises, in its internal multiplicity, $2 \times 4$ Hermitian components which can be acted on by the two-dimensional matrix $\nu = \nu_3$, together with $\nu_1$, $\nu_2$, and the four-dimensional matrices $\tau_k$, $\zeta_k$. But, as we have observed, $T = \frac{1}{2}$ is only one possible interpretation of the transformations in the four-dimensional space to which the latter matrices refer. The matrix representation illustrated by

$$\tau_3 = \begin{pmatrix} 0 & -i & 0 & 0 \\ i & 0 & 0 & 0 \\ 0 & 0 & 0 & -i \\ 0 & 0 & i & 0 \end{pmatrix}$$

can equally be regarded as acting upon a four-dimensional vector, with the first three rows and columns referring to $T = 1$ and the last row and column to $T = 0$. In this way all statements pertaining to the field $\psi_{(1/2)}$ of internal multiplicity $2 \times 4$, can be translated into properties of the $2 \times 4$ fold degenerate field $\psi_{(0,1)}$. We are thus led to the concept of a universal $\pi$-heavy fermion interaction in which both integral and half-integral isotopic spin F.D. fields take part, in a manner that maintains the four-dimensional equivalence between the two distinct three-dimensional interpretations of the common kinematical structure. This universality also implies that nucleonic charge is a general property of the heavy F.D. particle field, without regard to isotopic spin. Thus the $\pi$-field emerges as the dynamical agency that defines nucleonic charge[4], the absolute conservation of which is the quantitative expression of the stability of nuclear matter. The complete $\pi$-coupling term that embodies these considerations is

$$\mathfrak{L}_\pi = g_\pi \phi_{(1)} \frac{1}{2} \Big[ \psi_{(1/2)} \beta \gamma_5 \nu \tau \psi_{(1/2)} + \psi_{(0)} \beta \gamma_5 i \nu \psi_{(1)} - \psi_{(1)} \beta \gamma_5 i \nu \psi_{(0)} + \psi_{(1)} \beta \gamma_5 \nu \frac{1}{i} \times \psi_{(1)} \Big], \tag{4}$$

where the vector operation $(1/i)\times$ can also be expressed by the three-dimensional matrices $t_k$ that provide the $T = 1$ representation. The general concept of the universal $\pi$-interaction is also compatible with the opposite choice of relative sign between the integral $T$ and half-integral $T$ terms.

The internal symmetry properties possessed by this coupling term are described by the following infinitesimal rotations:

$$\begin{aligned}
&N \text{ rotation:} && \psi \to (1 + i\delta\alpha\nu)\psi \\
& && \phi_{(1)} \to \phi_{(1)}; \\
&\tau \text{ rotation:} && \psi_{(1/2)} \to \left(1 + \frac{i}{2}\delta\omega\tau\right)\psi_{(1/2)}, \\
& && \psi_{(0)} \to \psi_{(0)} - \frac{1}{2}\delta\omega\psi_{(1)}, \\
& && \psi_{(1)} \to \left(1 + \frac{i}{2}\delta\omega t\right)\psi_{(1)} + \frac{1}{2}\delta\omega\psi_{(0)}, \\
& && \phi_{(1)} \to (1 + i\delta\omega t)\phi_{(1)};
\end{aligned} \tag{5}$$

[4] This theory can be regarded as the precise realization of some general remarks of Wigner (*3*).

$\zeta$ rotation:

$$\psi_{(1/2)} \rightarrow \left(1 + \frac{i}{2}\,\delta\epsilon\zeta\right)\psi_{(1/2)}\,,$$

$$\psi_{(0)} \rightarrow \psi_{(0)} + \frac{1}{2}\,\delta\epsilon'\psi_{(1)}\,,$$

$$\psi_{(1)} \rightarrow \left(1 + \frac{i}{2}\,\delta\epsilon' t\right)\psi_{(1)} - \frac{1}{2}\,\delta\epsilon'\psi_{(0)}\,, \tag{6}$$

$$\phi_{(1)} \rightarrow \phi_{(1)}\,.$$

The $\zeta$ rotations of the $\psi_{(1/2)}$ and $\psi_{(0,1)}$ fields are quite independent, which enables us to define the three-dimensional $T$ rotation by superimposing the $\zeta$ rotation with $\delta\epsilon' = \delta\omega$ on the four-dimensional $\tau$ rotation:

$T$ rotation:

$$\psi_{(1/2)} \rightarrow \left(1 + \frac{i}{2}\,\delta\omega\tau\right)\psi_{(1/2)}\,,$$

$$\psi_{(0)} \rightarrow \psi_{(0)}\,,$$

$$\psi_{(1)} \rightarrow (1 + i\delta\omega t)\psi_{(1)}\,,$$

$$\phi_{(1)} \rightarrow (1 + i\delta\omega t)\phi_{(1)}\,.$$

Associated with each infinitesimal rotation is a conserved physical property. This is illustrated by the $N$ rotation, which defines the current vector of nucleonic charge

$$j_N^{\mu} = \tfrac{1}{2}\psi\beta\gamma^{\mu}\nu\psi,$$

and thereby the total nucleonic charge,

$$N = \int d\sigma_{\mu}\,\frac{1}{2}\,\psi\beta\gamma^{\mu}\nu\psi\,.$$

The latter operator obeys commutation relations that express the nucleonic charge properties of the various fields,

$$[\psi, N] = \nu\psi, \qquad [\phi, N] = 0.$$

In addition to the symmetry properties described by rotations, there may also exist discrete symmetry operations in the nature of reflections. The introduction of the self-dual tensor $\phi_{(1)}$ excludes any such reflection invariance in the four-dimensional space. But, for the nucleonic charge, as represented by the matrix $\nu_3$ with eigenvalues $\pm 1$, there exists a transformation interchanging these eigenvalues—a nucleonic charge reflection. This is given by

$$\psi \to \nu_1\psi, \qquad \nu_1 = \begin{pmatrix} 1 & 0 \\ 0 & -1 \end{pmatrix},$$

$$\phi_{(1)} \to -\phi_{(1)},$$

the $\pi$-field reversal being required to maintain the structure of $\mathcal{L}_\pi$. Associated with this transformation is a unitary nucleonic charge reflection operator $R_N$, which is such that

$$R_N^{-1}\psi R_N = \nu_1\psi, \qquad R_N^{-1}\phi_{(1)}R_N = -\phi_{(1)}$$
$$R_N^{-1}NR_N = -N.$$

States of zero nucleonic charge can be characterized by the eigenvalues of $R_N$, which are consistently chosen as $\pm 1$, and the $\pi$-field possesses only matrix elements connecting states with opposite values of this nucleonic charge parity.

The kinematical discussion of internal symmetry spaces applies equally to B.E. and F.D. fields. But the manner in which the two types of fields appear dynamically is quite distinct, as emphasized by the structure of the Yukawa coupling, $\phi\psi\psi$. The quadratic dependence upon $\psi$ permits the application of four-dimensional invariance requirements including equivalence between integral and half-integral $T$ representations, whereas the latter have quite different meanings for the field $\phi$. Hence there is no fundamental objection to the four-dimensional representations being realized in the B.E. field only as $\phi_{(1/2)}$. Nevertheless we must ask whether there is some possibility of using the integral $T$ interpretation of the four-dimensional representation and thereby introducing the B.E. field $\phi_{(0,1)}$. If the $\sigma$-field $\phi_{(0)}$ is to be incorporated with the universal $\pi$-coupling, it can only appear as $\phi_{(0)}\frac{1}{2}\psi\beta\psi$ or $\phi_{(0)}\frac{1}{2}\psi\beta\gamma_5\psi$, for, with the exception of the unit matrix there is no matrix of the internal symmetry spaces that is acceptable in its action upon both $\psi_{(1/2)}$ and $\psi_{(0,1)}$. Furthermore, if the complete field $\phi_{(0,1)}$ is pseudoscalar, there is no possibility of exhibiting transformations under which $\phi_{(0,1)}$ behaves as a four-dimensional vector, since $\psi\beta\gamma_5\psi$ is invariant under all transformations referring solely to the internal symmetry space. But such four-vector transformations do exist if the $\sigma$-field is scalar, as distinguished from the pseudoscalar $\pi$-field, and if due account is taken of the dynamical origin of mass. The $\sigma$-field coupling term to be added to $\mathcal{L}$ is then

$$\mathcal{L}_\sigma = g_\pi\phi_{(0)}\tfrac{1}{2}[\psi_{(1/2)}\beta\psi_{(1/2)} + \psi_{(0)}\beta\psi_{(0)} + \psi_{(1)}\beta\psi_{(1)}].$$

First let us note the special features that accompany the assumptions $m_0 = 0$ and $\mu_0 = 0$. The absence of the Dirac field mass term implies invariance of $\mathcal{L}^{(1/2)}$ under the rotation $\psi \to (1 + \delta\varphi\gamma_5)\psi$, while the Lagrange term $\mathcal{L}^{(0)}$ will be invariant to within an added divergence under the spin 0 gauge transformation: $\phi \to \phi + \lambda$, $\phi^\mu \to \phi^\mu$, with $\lambda$ an arbitrary constant, if $\mu_0 = 0$. Now, if both mass constants are zero, the complete Lagrange function is invariant under the following infinitesimal transformation, which represents a partial union of space-time with the Euclidean internal space,

$$\psi_{(1/2)} \to (1 + \tfrac{1}{2}\, \gamma_5 \nu \delta\omega\tau)\, \psi_{(1/2)}\,,$$

$$\psi_{(0)} \to \psi_{(0)} + \frac{i}{2}\, \gamma_5 \nu \delta\omega \psi_{(1)}\,,$$

$$\psi_{(1)} \to \left(1 + \frac{1}{2}\, \gamma_5 \nu \delta\omega t\right) \psi_{(1)} - \frac{i}{2}\, \gamma_5 \nu \delta\omega \psi_{(0)}\,,$$

$$\phi_{(0)} \to \phi_{(0)} + \delta\omega\phi_{(1)}$$

$$\phi_{(1)} \to \phi_{(1)} - \delta\omega\phi_{(0)}\,.$$

If the integral $T$ terms appear with signs opposite from those given in (4), the infinitesimal changes in $\psi_{(0,1)}$ also require the reversed signs. The invariance property we have described for $m_0 = \mu_0 = 0$ persists if only the $\mu_0$ term is included. As to the Dirac mass term, it has the same structure as $\mathfrak{L}_\sigma$ and forms the combination $m_0 - g_\pi \phi_{(0)}$. Hence if $m_0 \neq 0$, it is $\phi_{(0)} - (m_0/g_\pi)$ that transforms with $\phi_{(1)}$ as a four-vector and the addition of the constant to $\phi_{(0)}$ does not upset the invariance of $\mathfrak{L}$ if $\mu_0 = 0$, according to the gauge invariance of $\mathfrak{L}^{(0)}$ appropriate to that circumstance. Of the various possibilities, it is the latter one, $m_0 \neq 0$, $\mu_0 = 0$, that is to be preferred, for, within our dynamical scheme there is no effective source of a Dirac field mass term, whereas the $\phi\phi\phi\phi$ coupling can serve this purpose for the B.E. fields. The former statement is justified by the existence of an invariance transformation for the complete Lagrange function, namely,

$$\psi \to e^{1/2\pi\gamma_5}\psi = \gamma_5\psi, \qquad \phi \to -\phi,$$

which a term of the form $m_0\psi\beta\psi$ would violate. In other words, the Dirac field with $m_0 = 0$ exhibits a space parity symmetry which is not removed by linear coupling with B.E. fields, and thus a mass term must be superimposed to eliminate this nonphysical degeneracy. As to the B.E. fields, a coupling of the form illustrated by

$$\mathfrak{L}_\phi = -g_\phi{}^2 \tfrac{1}{2}(\phi_{(0)}\phi_{(0)} + \phi_{(1)}\phi_{(1)})\tfrac{1}{2}\phi_{(1/2)}\phi_{(1/2)}$$

will produce effective mass terms for each field through the action of the vacuum fluctuations of the other fields. This assertion does not contradict our discussion of the scale-independence of the present formalism, for the theory must be modified artificially (cut off) to convert the infinite quantity $\langle\phi^2\rangle_0$ into a finite inverse square of a length. We conclude that a $\phi_{(0)}$ field can be introduced, in a four-dimensional Euclidean sense, provided it is a scalar field, and that the uncoupled B. E. field is massless. The $\phi^2\,\phi^2$ term that generates masses also disturbs the four-dimensional symmetry and leads to different observed masses for the $\sigma$-meson and the $\pi$-meson. If the former becomes sufficiently heavy ($\mu_\sigma > 2\mu_\pi$), the requirement for practical unobservability of this scalar particle will be satisfied.

The Yukawa coupling referring to the $K$-field, $\phi_{(1/2)}\psi_{(1/2)}\psi_{(0,1)}$, is indicated more precisely by the terms

$$\phi_{(1/2)}\psi_{(0)}\beta(1, \gamma_5)\psi_{(1/2)}, \qquad \phi_{(1/2)}i\tau\psi_{(1)}\beta(1, \gamma_5)\psi_{(1/2)},$$

in which the requirements of three-dimensional isotopic space symmetry are explicitly satisfied. The space reflection characteristics remain to be specified, as does the structure pertaining to other internal degrees of freedom. The physical information needed to determine the form of $\mathfrak{L}_K$ is supplied, in part, by the mass spectrum of the heavy F.D. particles. The coupling with the $K$-field must be the physical agency that differentiates $N$ from $\Xi$, or, effectively decomposes the $\psi_{(1/2)}$ field into two parts labelled by the values of $\zeta_3\nu$, and it must also be the agency that produces different masses for $\Lambda$ and $\Sigma$ or, destroys the four-dimensional symmetry exhibited by $\psi_{(0,1)}$.

In seeking the physical property that distinguishes $N$ from $\Xi$, it is well to notice that, while $\psi_{(0,1)}$ acts as a unit with respect to spatial reflection, a dependence of the reflection properties of $\psi_{(1/2)}$ upon $\zeta_3$ is consistent with the universal $\pi$-coupling. Thus, it is conceivable that the two fields, $\frac{1}{2}(1 + \zeta_3\nu)\psi_{(1/2)}$ and $\frac{1}{2}(1 - \zeta_3\nu)\psi_{(1/2)}$, behave oppositely under spatial reflection, in accordance with the two possibilities available for Hermitian Dirac fields:

$$R_s{:}\,\psi(\mathbf{x}, x^0) \rightarrow \pm i\gamma^0\psi(-\mathbf{x}, x_0).$$

If we accept this tentative interpretation of the physical situation, the choice between the matrices 1 and $\gamma_5$ multiplying $\psi_{(1/2)}$ is resolved by the combination

$$\left(\frac{1 + \zeta_3\nu}{2}\,\gamma_5 + \frac{1 - \zeta_3\nu}{2}\right)\psi_{(1/2)},$$

in which $\gamma_5$ has been (arbitrarily) associated with the nucleon field $\frac{1}{2}(1 + \zeta_3\nu)\psi_{(1/2)}$ and the coefficients have been chosen to permit the invariance operation

$$\psi_{(1/2)} \rightarrow -\zeta_1\nu\gamma_5\psi_{(1/2)}, \qquad \phi_{(1/2)} \rightarrow i\zeta_2\phi_{(1/2)}$$

for $\mathfrak{L}_K$. Of course the mass term in $\mathfrak{L}^{(1/2)}$ is not invariant under this transformation. If the reflection properties of $\psi_{(0,1)}$ are standardized, the complete list of space reflection transformations (omitting the spatial coordinates) is given by

$$\begin{aligned} R_s : \psi_{(1/2)} &\to (\pm)\zeta_3 \nu i\gamma^0 \psi_{(1/2)} , \\ \psi_{(0,1)} &\to i\gamma^0 \psi_{(0,1)} , \\ \phi_{(1/2)} &\to -(\pm)\phi_{(1/2)}, \\ \phi_{(0)} &\to \phi_{(0)} , \qquad \phi_{(1)} \to -\phi_{(1)} , \end{aligned} \tag{7}$$

and the intrinsic space parity of the $K$-field can be specified further only by convention.

The second stage in determining the $K$-coupling term $\mathfrak{L}_K$ is facilitated by considering the behavior of the combinations $\psi_{(0)} \pm i\tau\psi_{(1)}$ under left and right multiplication by the matrices representing infinitesimal transformations. Thus

$$\psi_{(0)} + i\tau\psi_{(1)} \to \left(1 + \frac{i}{2}\,\delta\omega\tau\right)(\psi_{(0)} + i\tau\psi_{(1)})$$

implies

$$\begin{aligned} \psi_{(0)} &\to \psi_{(0)} - \tfrac{1}{2}\delta\omega\psi_{(1)} , \\ \psi_{(1)} &\to \psi_{(1)} - \tfrac{1}{2}\delta\omega \times \psi_{(1)} + \tfrac{1}{2}\delta\omega\psi_{(0)} , \end{aligned} \tag{8}$$

which is the $\tau$ transformation of (5), whereas

$$\psi_{(0)} + i\tau\psi_{(1)} \to (\psi_{(0)} + i\tau\psi_{(1)})\left(1 - \frac{i}{2}\,\delta\epsilon'\tau\right)$$

yields

$$\begin{aligned} \psi_{(0)} &\to \psi_{(0)} + \tfrac{1}{2}\delta\epsilon'\psi_{(1)} , \\ \psi_{(1)} &\to \psi_{(1)} - \tfrac{1}{2}\delta\epsilon' \times \psi_{(1)} - \tfrac{1}{2}\delta\epsilon'\psi_{(0)} , \end{aligned} \tag{9}$$

the $\zeta$ transformation of (6). The similar results for $\psi_{(0)} - i\tau\psi_{(1)}$ can be obtained by transposition. Thus the $\tau$-transformation (8) is generated by

$$\psi_{(0)} - i\tau\psi_{(1)} \to (\psi_{(0)} - i\tau\psi_{(1)})\left(1 - \frac{i}{2}\,\delta\omega\tau\right),$$

while the $\zeta$ transformation (9) emerges from

$$\psi_{(0)} - i\tau\psi_{(1)} \to \left(1 + \frac{i}{2}\,\delta\epsilon'\tau\right)(\psi_{(0)} - i\tau\psi_{(1)}).$$

It may be noted here that the operator structure of the integral $T$ field $\pi$-coupling term is given by

$$\psi_{(1)}\psi_{(0)} - \psi_{(0)}\psi_{(1)} + \psi_{(1)} \times \psi_{(1)} = \tfrac{1}{4} tr[(\psi_{(0)} - i\tau\psi_{(1)}) i\tau(\psi_{(0)} + i\tau\psi_{(1)})],$$

where the trace applies to the four-dimensional matrices. The trace is properly invariant under the $\zeta$ transformation, and the $\tau$-transformation induces the correct three-dimensional rotation.

We now observe that $\phi_{(1/2)}(\psi_{(0)} \pm i\tau\psi_{(1)})\psi_{(1/2)}$ possesses four-dimensional invariance properties with either sign since left or right hand factors multiplying $\psi_{(0)} \pm i\tau\psi_{(1)}$ can be compensated by appropriate transformations of $\phi_{(1/2)}$ or $\psi_{(1/2)}$. But one combination, namely $\psi_{(0)} - i\tau\psi_{(1)}$, is distinguished by leading to the same four-dimensional symmetries as the $\pi$-coupling term. Thus, under a $\tau$-transformation $(\psi_{(0)} - i\tau\psi_{(1)})\psi_{(1/2)}$ is invariant, while a $\zeta$-transformation of the integral $T$ fields implies a corresponding transformation of $\phi_{(1/2)}$ without affecting $\psi_{(1/2)}$. Hence if $\mathfrak{L}_K$ contains only the combination $\psi_{(0)} - i\tau\psi_{(1)}$, the mass degeneracy of $\Lambda$ and $\Sigma$ would persist. We are led thereby, as one simple possibility, to suppose that $\psi_{(0,1)}$ enters $\mathfrak{L}_K$ only in the form $\psi_{(0)} + i\tau\psi_{(1)}$. The mass spectrum of the integral $T$ F.D. particles is then viewed as the result of a clash between the different four-dimensional symmetries possessed by $\mathfrak{L}_\pi$ and $\mathfrak{L}_K$. When the latter is combined with the parity interpretation of the $N - \Xi$ mass splitting, we emerge with the following $K$-field interaction term

$$\mathfrak{L}_K = g_K\phi_{(1/2)}(\psi_{(0)} + i\tau\psi_{(1)})\beta\left(\gamma_5\frac{1+\zeta_3\nu}{2} + \frac{1-\zeta_3\nu}{2}\right)\psi_{(1/2)}. \tag{10}$$

This interaction is qualitatively satisfactory since it produces a complex mass spectrum for the fermions, and also implies a nondegenerate boson mass spectrum in virtue of the difference between the interactions $\mathfrak{L}_K$ and $\mathfrak{L}_\pi$, together with the breakdown of four dimensional symmetry implied by $\mathfrak{L}_\phi$, if the $\phi_{(0,1)}$ description is employed We need hardly emphasize that (10) is but an example of a general class of interaction term, the members of which can be distinguished only by their quantitative implications.

The invariance properties of the complete Lagrange function

$$\mathfrak{L} = \mathfrak{L}^{(0)} + \mathfrak{L}^{(1/2)} + \mathfrak{L}_\pi + \mathfrak{L}_K + \mathfrak{L}_\phi$$

characterize the class of possible interactions. These properties include invariance under the following rotations:

$N$ rotation:
$$\psi \to (1 + i\delta\alpha\nu)\psi, \qquad \phi \to \phi,$$
$$j_N^\mu = \tfrac{1}{2}\psi\beta\gamma^\mu\nu\psi:$$

$T$ rotation:
$$\psi_{(1/2)}, \phi_{(1/2)} \to \left(1 + \frac{i}{2}\delta\omega\tau\right)\psi_{(1/2)}, \phi_{(1/2)},$$
$$\psi_{(1)}, \phi_{(1)} \to (1 + i\delta\omega t)\psi_{(1)}, \phi_{(1)},$$
$$\psi_{(0)}, \phi_{(0)} \to \psi_{(0)}, \phi_{(0)},$$
$$j_T^\mu = \frac{1}{2}\psi_{(1/2)}\beta\gamma^\mu\frac{\tau}{2}\psi_{(1/2)} + \frac{1}{2}\psi_{(1)}\beta\gamma^\mu t\psi_{(1)} + i\phi_{(1/2)}{}^\mu\cdot\frac{\tau}{2}\phi_{(1/2)} + i\phi_{(1)}{}^\mu\cdot t\phi_{(1)};$$

$Y$ rotation:
$$\psi_{(1/2)}\,,\,\phi_{(1/2)} \to (1 + i\delta\eta\zeta_3)\psi_{(1/2)},\,\phi_{(1/2)}$$
$$\psi_{(0,1)}\,,\,\phi_{(0,1)} \to \psi_{(0,1)}\,,\,\phi_{(0,1)}$$
$$j_Y^{\mu} = \tfrac{1}{2}\psi_{(1/2)}\beta\gamma^{\mu}\zeta_3\psi_{(1/2)} + i\phi_{(1/2)}{}^{\mu}\cdot\zeta_3\phi_{(1/2)}\,.$$

The $Y$ rotation, which defines the hypercharge, is equivalent to the $\zeta_3$ rotation of the $T = \frac{1}{2}$ fields, $Z_3 = \frac{1}{2}Y$. There is no invariance under other $\zeta$ rotations, nor with respect to any $\tau$ rotation. Furthermore, in view of the physical coupling between nucleonic charge and hypercharge, nucleonic charge reflection is not an invariance operation, but must be supplemented by the reflection of hypercharge. The latter can be produced by the rotation $e^{\pi i Z_1}$ applied to the $T = \frac{1}{2}$ fields. In this way, we obtain the discrete symmetry operation described by

$$R_T = R_N e^{\pi i Z_1}\,,$$

which we designate as the reflection operator for the three-dimensional isotopic space. Its effect as a unitary operator is indicated by

$$R_T^{-1}\psi_{(1/2)}R_T = i\zeta_1\nu_1\psi_{(1/2)}\,, \qquad R_T^{-1}\phi_{(1/2)}R_T = i\zeta_1\phi_{(1/2)}\,,$$
$$R_T^{-1}\psi_{(0,1)}R_T = \nu_1\psi_{(0,1)}\,, \qquad R_T^{-1}\phi_{(1)}R_T = -\phi_{(1)}\,,$$
$$R_T^{-1}\phi_{(0)}R_T = \phi_{(0)}\,,$$

and

$$R_T^{-1}NR_T = -N, \qquad R_T^{-1}YR_T = -Y.$$

The states of zero nucleonic charge and zero hypercharge can be classified by the eigenvalues of $R_T$. It might be noted here that while

$$R_N^2 = +1,$$

we have

$$R_T^2 = e^{2\pi i Z_1} = e^{2\pi i Z_3}$$
$$= (-1)^Y,$$

which is equivalent to the observation that the unitary operator $R_T^2$ reverses the sign of all fields that carry hypercharge ($T = \frac{1}{2}$), but leaves nonhypercharged fields unaffected. In a similar way, the repetition of the spatial reflection reverses the sign of all F.D. fields ($s = \frac{1}{2}$) and leaves B.E. fields unaltered. Since the former also carry nucleonic charge, this property can be expressed by

$$R_s^2 = (-1)^N. \qquad (11)$$

The electromagnetic field fits easily into this theory of the heavy fermions and bosons. In addition to the Lagrange function of the pure Maxwell field,

$$\mathcal{L}^{(M)} = -\tfrac{1}{2}[\tfrac{1}{2}F^{\mu\nu}\,.\,(\partial_\mu A_\nu - \partial_\nu A_\mu) - A_\nu\,.\,\partial_\mu F^{\mu\nu} - \tfrac{1}{2}F^{\mu\nu}F_{\mu\nu}],$$

we add the classic example of the coupling that contains no dimensional constant,

$$\mathfrak{L}_A = ej_Q{}^\mu \, . \, A^\mu,$$

where the current vector of electrical charge follows from the identification (3),

$$j_Q{}^\mu = j_{T_3}{}^\mu + \frac{1}{2} j_Y{}^\mu = \frac{1}{2}\psi_{(1/2)}\beta\gamma^\mu \frac{\tau_3 + \zeta_3}{2}\psi_{(1/2)} + \frac{1}{2}\psi_{(1)}\beta\gamma^\mu t_3\psi_{(1)}$$

$$+ \, i\phi_{(1/2)}{}^\mu \cdot \frac{\tau_3 + \zeta_3}{2}\phi_{(1/2)} + i\phi_{(1)}{}^\mu \cdot t_3\phi_{(1)} \, .$$

This interaction term makes explicit the dynamical role of the electromagnetic field in reducing the three-dimensional $T$ symmetries to the two-dimensional one described by $T_3$ rotations. At the same time, $R_T$ is no longer a suitable reflection operation since it does not maintain the structure $\tau_3 + \zeta_3$. But if one superimposes the rotation $e^{\pi i T_1}$, the entire operator $j_Q{}^\mu$ reverses sign and an invariance operation is provided by $R_Q$, the electric charge reflection operator, with the following characteristics:

$$\begin{aligned} R_Q^{-1}\psi_{(1/2)}R_Q &= -\,\zeta_1\,\tau_1\,\nu_1\,\psi_{(1/2)}\,, & R_Q^{-1}\phi_{(1/2)}R_Q &= -\,\zeta_1\,\tau_1\,\phi_{(1/2)}\,, \\ R_Q^{-1}\psi_{(1)}R_Q &= (1 - 2t_1^2)\nu_1\psi_{(1)}\,, & R_Q^{-1}\phi_{(1)}R_Q &= (2t_1^2 - 1)\phi_{(1)}\,, \\ R_Q^{-1}\psi_{(0)}R_Q &= \nu_1\psi_{(0)}\,, & R_Q^{-1}\phi_{(0)}R_Q &= \phi_{(0)}\,, \\ R_Q^{-1}A_\mu R_Q &= -A_\mu\,, & R_Q^{-1}F_{\mu\nu}R_Q &= -F_{\mu\nu}\,, \end{aligned}$$

and

$$R_Q^{-1}NR_Q = -N, \qquad R_Q^{-1}YR_Q = -Y, \qquad R_Q^{-1}QR_Q = -Q.$$

We also have

$$R_Q^2 = +1,$$

since

$$R_T^2 e^{2\pi i T_1} = e^{2\pi i (T_3 + 1/2 Y)} = 1.$$

Corresponding to the general characterization of the electric charge operator by rotations in a plane of the four-dimensional symmetry space,

$$Q = T_{12}\,,$$

we can express the electric charge reflection operator as

$$R_Q = R_N e^{\pi i T_{23}}\,, \tag{12}$$

with the interpretation of the reflection of the first axis in that two-dimensional charge space. This description includes the electromagnetic field if the latter is viewed as the [12] = 3 component of an axial vector in the three-dimensional isotopic space. It will be noted that the $\pi$-field acts as a polar vector; the first component reverses sign under the operation $R_Q$ while the other components are unaltered. We observe, in this connection, that states of zero electric charge, zero hypercharge, and zero nucleonic charge can be classified by the eigenvalues of $R_Q$, the charge parity. A familiar example is a single $\pi^0$ meson which, according to the reflection properties of $\phi_{(1)3}$, defines a charge symmetric state. This neutral particle should decay into an even number of photons through the combined action of the $\mathfrak{L}_\pi$ and $\mathfrak{L}_A$ couplings. But, it is characteristic of our theory that the $\pi$-interaction involves the nucleonic charge while the electromagnetic interaction makes no reference to that property. Hence it is specifically the linking of nucleonic charge and hypercharge produced by the K-coupling that enables this transmutation to occur.

The theory thus far devised refers to heavy fermions and heavy bosons, together with the photon, and gives an account of their strong and electromagnetic interactions. Omitted are the light fermions (leptons) and the various physical processes that exhibit a very long time scale. The interactions responsible for these processes are certainly of lower symmetry than those already discussed, the total effect of the latter being described by various two-dimensional rotational symmetries, or charges, and a single charge reflection operation. Since the leptons carry electrical charge, they at least realize a two-dimensional internal symmetry space, which invites an attempt to correlate their properties with the aid of an internal space of higher dimensionality, but one which is presumably of lesser dimensionality than that employed for the heavy particles. Now, in our discussion of matrices with increasing dimensionality that culminated in the four-dimensional space necessary for the characterization of the heavy particles, we encountered the $T = 1$ representation of the three-dimensional rotation group which is thus naturally indicated for the description of the lepton family. According to the physical identification that accompanies the reduction to two-dimensional symmetry, we have

$$Q = T_3,$$

and the three-dimensional matrix $t_3$ has the distinct eigenvalue 1, 0, $-1$. It is tempting to extend to all fermions the property that the sign reversal of the field produced by repetition of a spatial reflection could also be generated by an internal rotation, in the sense of (11). This leads us to assign an analog of nucleonic charge to the leptons, which is called the leptonic charge $L$, and is represented by the matrix

$$\lambda = \begin{pmatrix} 0 & -i \\ i & 0 \end{pmatrix}.$$

Particles labelled by leptonic charge and electrical charge permit a complete identification with the known leptons; $L = +1: \mu^+, \nu^0, e^-$, and $L = -1: e^+, \bar{\nu}^0, \mu^-$, in which the leptonic charge serves to distinguish particles with the same electrical charge. This is analogous to the distinction afforded by the nucleonic charge between $N^+ (N = +1)$ and $\bar{\Xi}^+ (N = -1)$, for example.

The mass spectrum of the leptons is characterized by the zero mass of its electrically neutral members, and the very striking mass asymmetry between the electrically charged particles with opposite values of $t_3\lambda$. Our general viewpoint encourages us to interpret the large mass of $\mu$ mesons ($t_3\lambda = +1$) by means of a comparatively strong interaction that serves to define dynamically the leptonic charge, and to remove the three-dimensional symmetries of the isotopic space. Of course, any proposed interaction must be reconciled with the apparent absence in known $\mu$-meson phenomena of substantial nonelectromagnetic forces. It is at least interesting that an interaction which seems to possess the requisite properties can be exhibited with the aid of the hypothetical $\sigma$-field, $\phi_{(0)}$. To begin with, this scalar field enables us to establish a dynamical relation between the leptons and the strongly interacting particles, without upsetting thereby the higher internal symmetry characteristic of the massive particles. If the coupling of $\phi_{(0)}$ with the lepton field $\psi_l$ is to reduce the lepton internal symmetry, the interaction must contain symmetrical real matrices other than unity, for which the only possibilities are $t_3^2$ and $t_3\lambda$. We choose the interaction as

$$\mathcal{L}_\mu = g_\mu \phi_{(0)} \tfrac{1}{2} \psi_l \beta t_3 \tfrac{1}{2} (t_3 + \lambda) \psi_l ,$$

in which the particular matrix combination serves to select the $\mu$-meson part of the lepton field. The unique properties of the $\sigma$-field can now be called upon again. As a field which is a scalar under all operations in the three-dimensional isotopic space and in space-time, $\phi_{(0)}$ has a nonvanishing expectation value in the vacuum. Although unable to affix the value implied by the strong interactions with heavy fermions, one could at least anticipate that $\langle \phi_{(0)} \rangle$ would have the magnitude of nucleon masses, and thus a suitable $\mu$-meson mass constant might emerge from $g_\mu \langle \phi_{(0)} \rangle$ without requiring a particularly large coupling constant $g_\mu$. It is the latter feature, combined with the supposedly large mass of the $\sigma$-particle, that may enable one to avoid conflict between the dynamical implications of $\mathcal{L}_\mu$ and the present lack of evidence for significant nonelectromagnetic $\mu$ interactions. Naturally, one must eventually find such evidence if our hypothesis is to attain a measure of credibility.

Whatever the interpretation of the $\mu$-meson mass may be, the symmetry properties of the lepton field can be generally stated at the dynamical level that incorporates electromagnetic interactions. These include rotations,

$L$ rotation:
$$\psi_l \to (1 + i\delta\alpha\lambda)\psi_l ,$$
$$j_L^\mu = \tfrac{1}{2} \psi_l \beta \gamma^\mu \lambda \psi_l ,$$

$T_3$ rotation:

$$\psi_l \rightarrow (1 + i\delta\omega t_3)\psi_l\,,$$
$$(j_Q{}^\mu)_l = \tfrac{1}{2}\psi_l\beta\gamma^\mu t_3\psi_l\,,$$

and charge reflection,

$$R_Q{:}\psi_l \rightarrow (1 - 2t_1{}^2)\lambda_1\psi_l\,.$$

The latter transformation is implied by the extension of the general formula (12),

$$R_Q = R_N R_L e^{\pi i T_{23}}\,,$$

where

$$R_L{}^{-1}\psi_l R_L = \lambda_1\psi_l\,,$$

and thus

$$R_Q{}^{-1} L R_Q = -L.$$

There is also the special feature of the zero neutrino mass (although we have yet to produce a mechanism that accounts for the electron mass) which is described by invariance under the infinitesimal transformation

$$\psi_l \rightarrow [1 + i\delta\varphi(1 - t_3{}^2)i\gamma_5]\psi_l\,. \tag{13}$$

But there will be more of this shortly.

The symmetry that exists between the heavy bosons and fermions in their isotopic space properties prompts us to ask: Is there also a family of *bosons* that realizes the $T = 1$ representation of the three-dimensional rotation group? The exceptional position of the electromagnetic field in our scheme, and the formal suggestion that this field is the third component of a three-dimensional isotopic vector, encourage an affirmative answer. We are thus led to the concept of a spin one family of bosons, comprising the massless, neutral, photon and a pair of electrically charged particles that presumably carry mass, in analogy with the leptons (abstracting from the additional complexity that accompanies the fermion property of leptonic charge). These considerations are expressed by the Lagrange function term

$$\mathfrak{L}^{(1)} = -\tfrac{1}{2}[\tfrac{1}{2}Z^{\mu\nu}\,.\,(\partial_\mu Z_\nu - \partial_\nu Z_\mu) - Z_\nu\,.\,\partial_\mu Z^{\mu\nu} - \tfrac{1}{2}Z^{\mu\nu}Z_{\mu\nu}]$$

and the identification

$$Z_3{}^\mu = A^\mu, \qquad Z_3{}^{\mu\nu} = F^{\mu\nu},$$

together with the boson interaction

$$\mathfrak{L}_{Z\phi} = -g_{Z\phi}{}^2\tfrac{1}{2}\phi_{(0)}{}^2\tfrac{1}{2}Z^\mu t_3{}^2 Z_\mu\,,$$

in which we again use the $\sigma$-field to remove three-dimensional internal symmetries and produce masses for charged particles. At the same time, the coupling augments the mass of the $\sigma$-particle. This part of the Lagrange function referring to the $Z$ field, and including the boson interaction that is the electromagnetic coupling with the charged $Z$ particles, possesses the following symmetry properties,

$T_3$ rotation:
$$Z \rightarrow (1 + i\delta\omega t_3)Z,$$
$$(j_Q{}^\mu)_Z = iZ^{\mu\nu} \,.\, t_3 Z_\nu \,,$$

and
$$R_Q : Z \rightarrow (1 - 2t_1{}^2)Z,$$

where the latter contains the known charge reflection property of the electromagnetic field, together with
$$R_Q{}^{-1} Z_1 R_Q = Z_1 \,, \qquad R_Q{}^{-1} Z_2 R_Q = -Z_2 \,.$$

There is also the expression of the null photon mass-invariance of $\mathcal{L}$ (to within an added divergence) under the gauge transformation
$$Z_\mu \rightarrow Z_\mu + \frac{1}{e}(1 - t_3{}^2)\partial_\mu\lambda(x). \qquad Z_{\mu\nu} \rightarrow Z_{\mu\nu} \,,$$

when combined with the general transformation
$$\chi \rightarrow e^{i\lambda(x)T_{12}}\chi$$

of all charged fields, including the $Z$ field.

Now we must face the problem of discovering the specific Yukawa interactions of the massive, charged $Z$ particles. From its role as a partner of the electromagnetic field, we might expect that the charged $Z$ field interacts universally with electric charge, or rather, changes of charge, without particular regard to other internal attributes. If this be so, the coupling with the $Z$ field (henceforth understood to be the charged $Z$ field) will produce further reductions of internal symmetry, which raises the hope that this general mechanism may be the underlying cause of the whole group of physical processes that are characterized by a long time scale. Indeed, that time scale becomes more comprehensible, without invoking inordinately weak interactions, if every observable process requires the virtual creation of a heavy particle. Our general viewpoint regarding the systematic reduction of internal symmetry also impels us to seek some internal symmetry aspect of the $Z$ field that is destroyed by the coupling with various combinations of electrically charged and neutral fields. There is no question, presumably, of a breakdown of invariance under the two-dimensional rotations that define electrical charge, which leaves no choice other than a failure of the charge reflection symmetry property for the $Z$ field interactions. (It must be admitted that, despite its natural place in our scheme, this conclusion required the stimulus of certain recent experiments, and was not drawn in the lectures upon which this article is based.)

To investigate the possibility of charge reflection invariance failure, let us consider the $Z$ field interaction with the lepton field, as illustrated by

$$Z^{\mu}\tfrac{1}{2}\psi_l\beta\gamma_\mu t\psi_l$$
$$(Zt = Z_1t_1 + Z_2t_2),$$

which is invariant under rotations and reflections in the two-dimensional charge space. Another such coupling is obtained on replacing $t_1$ with $t_2 = (1/i)[t_3, t_1]$, and $t_2$ with $-t_1 = (1/i)[t_3, t_2]$, provided the charge reflection properties of $Z$ are reversed to compensate the effect of $t_3$. Although it might appear that reflection invariance would be destroyed if both couplings were operative, that impression is misleading since a suitable relative rotation of the charge axes for the two fields suffices to remove either term. To obtain something fundamentally different, the charge matrix $t$ must multiply $t$ symmetrically, rather than entering in a commutator or alternatively, $t$ could be multiplied by the leptonic charge matrix, although our previous comments would suggest that only electrical charge is relevant. But both of these matrices are symmetrical and are excluded in conjunction with the symmetrical matrices $\beta\gamma_\mu$. A breakdown of charge reflection invariance cannot be produced, therefore, unless one also replaces the symmetrical real matrices $\beta\gamma_\mu$ by the antisymmetrical imaginary matrices $\beta\gamma_\mu i\gamma_5$, which possess the opposite space reflection characteristics. We thus recognize that a failure of invariance under charge reflection must be accompanied by a failure of invariance under space reflection. The common origin of both breakdowns indicates, however, that the combination of the two reflections is still an invariance operation,

$$R = R_s R_Q,$$

which now appears as the ultimate expression of the complete equivalence of oppositely oriented spatial coordinate systems[5]. The eigenvalues of this reflection operator, the union of space parity and charge parity, will be designated as the *parity*.

The vector coupling between the $Z$ field and the lepton field that emerges from these considerations is

$$\mathfrak{L}_{Zl} = g_Z Z^{\mu}\tfrac{1}{2}\psi_l\beta\gamma_\mu(t - i\gamma_5\{t_3, t\})\psi_l, \qquad (14)$$

in which the relative coefficients of the two terms have been chosen to conform

[5] That the indiscernibility of left and right can be reconciled with "nonconservation of parity", in the specific sense of $R_s$, has been emphasized particularly by Landau (4).

with an invariance requirement that expresses the null neutrino mass. The relevant operation is just (13), extended by a suitable transformation for the charged leptons[6],

$$\psi_l \to [1 + i\delta\varphi((1 - t_3^2)i\gamma_5 - t_3)]\psi_l .$$

One could also employ the opposite sign in all $\gamma_5$ terms. This invariance property implies a conservation law for a lepton field quantity that we shall call neutrinic charge, $n$. It is described by the vector

$$j_n^\mu = \tfrac{1}{2}\psi_l\beta\gamma^\mu((1 - t_3^2)i\gamma_5 - t_3)\psi_l .$$

As thus defined, the neutrinic charge of the particles $\mu$ and $e$ is the negative of their electrical charge. The neutrinic charge of the neutrino is represented by the matrix $i\gamma_5$, the eigenvalues of which also have the significance of the spin projection along the direction of motion of the massless particle. Thus a neutrino with $n = +1$ or $-1$ can be designated as a right or left polarized neutrino. In a process involving the creation of a pair of leptons through the intervention of the $Z$ field, the conservation of neutrinic charge states that a lepton of positive electrical charge appears with a right polarized neutrino, and a negatively charged lepton with a left polarized neutrino.

The neutrinic charge conservation law is quite independent of that for leptonic charge[7] which asserts, for example, that a neutrino $\nu(\lambda = +1)$, as distinguished from an antineutrino $\bar{\nu}(\lambda = -1)$, accompanies a positive electron and a negative muon. Thus the detailed charge correlations are

$$Z^+ \leftrightarrow \mu^+ + \bar{\nu}_R , \qquad e^+ + \nu_R ,$$

$$Z^- \leftrightarrow \mu^- + \nu_L , \qquad e^- + \bar{\nu}_L ,$$

where $R$ and $L$ are polarization labels. One consequence of these assignments refers to the self-coupling of the lepton field through the intermediary of the $Z$ field, which implies the physical process $\mu \to e + 2\nu$. The leptonic and neutrinic charge attributes of $\mu$ and $e$ require that the electrically neutral particles possess the same leptonic charge but the opposite neutrinic charge,

$$\mu^+ \to e^+ + \nu_R + \nu_L ,$$

$$\mu^- \to e^- + \bar{\nu}_L + \bar{\nu}_R .$$

[6] A similar invariance requirement has been discussed by Touschek (*5*), but lacks the restriction of the transformation to the lepton field, and the specific form of the transformation properties assigned to the non-neutrino fields.

[7] This shows that the neutrino theory developed here is not to be identified with the so-called two-component theory (*6*). What is called lepton (ic charge) conservation in that formalism is specifically the conservation of neutrinic charge. In this connection, see the paper on the conservation of the lepton charge by Pauli (*7*).

Thus the two neutrinos are oppositely polarized. This is consistent with the finite intensity observed at the high energy end of the electron spectrum, where the neutrinos travel in the same direction. Let us also observe that, when sufficient energy is available, the mass of the electron can be neglected and that particle is produced with a definite polarization. In virtue of the commutativity of $\beta\gamma_\mu$ with $\gamma_5$ and the antisymmetry of the latter matrix, a pair of essentially massless leptons generated by the coupling (14) will be oppositely polarized. The implication for the successive decays $\pi \to \mu + \nu$, $\mu \to e + 2\nu$, arises from the zero spin of the $\pi$ meson according to which $\mu^-$, say, in $\pi^- \to \mu^- + \nu_L$ must have its spin directed oppositely to its direction of motion. At the high energy end of the electron spectrum, it is the electron that carried the angular momentum about the axis of disintegration and, since $e^-$ accompanies $\tilde{\nu}_L$, the electron is right polarized. Hence the electron must be emitted predominantly in the direction of the $\mu$ spin, or oppositely to the initial direction of the $\mu$ meson. This is precisely the asymmetry that was so strikingly revealed in recent experiments (*8*).

The extension of the $Z$ coupling to the heavy F.D. fields must be guided by the concept of universality since there is no analog of the neutrinic charge for these massive particles. The charged $Z$ particles interact equally with all pairs of electrically charged and neutral leptons, thereby destroying charge reflection symmetry. An illustration of the attempt to transfer these properties to the heavy fermions is given by the following coupling term

$$\mathfrak{L}_{ZN} = 2^{-1/2} g_Z Z^\mu \frac{1}{2} \psi\beta\gamma_\mu \left(\tau - i\gamma_5 \left\{\frac{\tau_3 + \zeta_3}{2}, \tau\right\}\right)\psi, \tag{15}$$

where $\psi$ stands for the complete field $\psi_{(1/2)} + \psi_{(0,1)}$. That is, with the physical emphasis placed on electrical charge, the three-dimensional isotopic classification is no longer meaningful and the two sets of fields $\psi_{(1/2)}$ and $\psi_{(0,1)}$ appear on the same footing. The possibility of writing the single term (15) makes very explicit use of the underlying four-dimensional symmetry and the accompanying ability to apply the matrices $\tau$ and $\zeta$ either to $\psi_{(1/2)}$ or to $\psi_{(0,1)}$. The implications of the compact symbolism are made more explicit by observing that we have extended the coupling of $Z^+$ with the fermion pairs $\mu^+\tilde{\nu}$ and $e^+\nu$ to the fermion pairs $N^+\tilde{N}^0$, $\Xi^0\tilde{\Xi}^+$; $\Sigma^+\tilde{\Sigma}^0$, $\Sigma^+\tilde{\Lambda}$, $\Sigma^0\tilde{\Sigma}^+$, $\Lambda\tilde{\Sigma}^+$; $N^+\tilde{\Sigma}^0$, $N^+\tilde{\Lambda}$, $\Sigma^+\tilde{N}^0$, $N^0\tilde{\Sigma}^+$, $\Sigma^0\tilde{\Xi}^+$, $\Lambda\tilde{\Xi}^+$, $\Sigma^+\tilde{\Xi}^0$, $\Xi^0\tilde{\Sigma}^+$.

One formal point still demands attention. The general formula (15) assumes that the complete field $\psi$ behaves as a unit under the reflection operation $R$,

$$R: \qquad \begin{aligned} &\psi \to i\gamma^0\nu_1 && e^{1/2\pi i(\zeta_1+\tau_1)}\psi, \\ &Z_1^0 \to Z_1^0, && Z_2^0 \to -Z_2^0, \\ &Z_1^k \to -Z_1^k, && Z_2^k \to Z_2^k, \end{aligned} \tag{16}$$

(spatial coordinates omitted) and some revision is necessary if we accept the space parity distinction between the $N$ field and the $\Xi$ field. One procedure is to separate the $\psi_{(0,1)}\psi_{(1,2)}$ term from (15) and modify it appropriately. But we can also retain (15) by redefining the $\psi_{(1/2)}$ field.

$$\psi_{(1/2)} \rightarrow e^{\pm\frac{1}{4}\pi i(\zeta_3+\nu)}\psi_{(1/2)},$$

which converts the $K$ coupling term into

$$\mathfrak{L}_K = g_K\phi_{(1/2)}(\psi_{(0)} + i\tau\psi_{(1)})\beta\left(\pm i\gamma_5\nu\,\frac{1+\zeta_3\nu}{2} + \frac{1-\zeta_3\nu}{2}\right)\psi_{(1/2)}\,.$$

The choice of sign is conventional if only strong and electromagnetic interactions are considered but becomes physically significant when the $Z$ field coupling is included. With the new meaning of $\psi_{(1/2)}$, the behavior of this field under the operation $R_T$ reads

$$R_T^{-1}\psi_{(1/2)}R_T = i\zeta_1\nu_1(-\zeta_3\nu_3)\psi_{(1/2)} = i\zeta_2\nu_2\psi_{(1/2)},$$

and the electric charge reflection transformation becomes

$$R_Q^{-1}\psi_{(1/2)}R_Q = -\zeta_1\tau_1\nu_1(-\zeta_3\nu_3)\psi_{(1/2)} = -\zeta_2\tau_1\nu_2\psi_{(1/2)}.$$

On forming the reflection operation $R$, we now find a common behavior for the fields contained in $\psi$ if the lower sign in (7) is chosen. Thus the universal $Z$ coupling determines the *parity* assignments of the heavy particle fields,

$$R:\qquad \begin{aligned} \psi &\rightarrow i\gamma^0\nu_1 e^{1/2\pi i(\zeta_1+\tau_1)}\psi, \\ \phi_{(1/2)} &\rightarrow -\zeta_1\tau_1\phi_{(1/2)}, \\ \phi_{(1)} &\rightarrow (1-2t_1^2)\phi_{(1)}\,, \\ \phi_{(0)} &\rightarrow \phi_{(0)}\,, \end{aligned}$$

to which is added the $Z$ field transformation of (16), and

$$R:\qquad \begin{aligned} A^0 \rightarrow -A^0, &\qquad A^k \rightarrow A^k, \\ \psi_l &\rightarrow i\gamma^0\lambda_1(1-2t_1^2)\psi_l\,. \end{aligned}$$

All these transformations are summarized by

$$R = R_{(s)}R_{(N)}R_{(L)}e^{\pi i T_{23}},$$

where the individual reflection operators are simpler than those previously considered, for $R_{(N)}$ and $R_{(L)}$ induce transformations only on fermion fields, while $R_{(s)}$ generates a standard space reflection transformation that multiplies all fermion fields by $i\gamma^0$ and leaves scalar boson fields unaltered.

Through the intervention of the $Z$ field, physical processes involving heavy particles take place that conserve nucleonic charge and electrical charge only of the list of internal attributes, to which *parity* should be added as a joint internal and space-time property. These processes include known particle decays: $\Sigma$, $\Lambda \to N + \pi$; $\Xi \to \Lambda + \pi$; $K \to 2\pi$, $3\pi$, and the theory must meet various quantitative tests, including its effectiveness in suppressing the decay $\Xi \to N + \pi$. In the latter connection, it is interesting to observe an invariance property that the Lagrange function would possess if the four-dimensional symmetry of $\Lambda$ and $\Sigma$ were not destroyed, which is to say, if only $\psi_{(0)} - i\tau\psi_{(1)}$ occurred in $\mathfrak{L}_K$. The transformation is

$$\psi \to e^{i\varphi\zeta_3}\psi, \qquad \phi_{(1/2)} \to e^{i\varphi(\tau_3+\zeta_3)}\phi_{(1/2)}\,, \tag{17}$$

with all other fields unchanged. The importance of this comment stems from the possibility that such an idealization of processes in which $\Lambda$ and $\Sigma$ do not appear explicitly, and in which the $K$ coupling is not directly responsible for the transition, may be justified by the relatively small $\Lambda - \Sigma$ mass splitting. Accepting this, we conclude that a reaction limited to $T = \frac{1}{2}$ fermions and $\pi$-mesons, with no $K$-particles in evidence, conserves hypercharge. Hence $\Xi \to N + \pi$ is (approximately) forbidden. We can also draw from (17) the conclusion that, in processes where no heavy fermion appears, the electrical charge carried by the $K$-particles is individually conserved. Were we to ignore the direct relevance of the $K$-coupling, we would infer the forbiddenness of charged $K$-particle decay, which may bear on the empirical observation that the shortest $K$-particle lifetime is that of a neutral $K$-particle, with *parity* $+1$.

The $Z$ field also couples the heavy fermions with the light fermions, thereby implying such processes as $\mu + N \to \nu + N$, $\pi \to \mu + \nu$, $K \to \mu + \nu$, $K \to \pi + \mu + \nu$, as well as $N \to N + e + \nu$, $K \to \pi + e + \nu$, and $\pi \to e + \nu$, $K \to e + \nu$. It is the electron phenomena that present an immediate challenge. Thus, the latter decays, $\pi \to e + \nu$, $K \to e + \nu$ have not been observed although their $\mu$-meson counterparts exist. Now it is encouraging that these electron processes would not occur if the electron mass were zero, for then electron and neutrino are oppositely polarized, which produces a net angular momentum about the axis of disintegration and contradicts the zero spin of $\pi$ and $K$. It will be noted that the vector nature of the $Z$ coupling is decisive in this argument. It can be concluded that the theory discriminates against the electron decay of the spinless bosons, but the precise ratio of decay probabilities may depend upon the specific dynamical origins of the electron and muon masses.

We then come to the comparatively well-known $\beta$-decay processes, where there is evidence that the lepton field appears in a tensor form combined with either a vector or a scalar coupling, the latter being currently favored by angular correlation measurements. The empirical tensor interaction fits naturally into the $Z$-particle picture, although two viewpoints are possible. It is familiar that effective tensor interactions are byproducts of fundamental vector couplings and one could imagine this to be the situation, which carries the implication that the weakness of $Z$ couplings is illusory and confined to low-energy phenomena. This interpretation also requires that the electron possess a mass, and the smallness of the latter enhances the quantitative difficulties of the approach. Alternatively, one can invert matters by supposing that with the charged $Z$ field Yukawa interactions we have reached the limits of the principle of scale independence and that a length appears explicitly, in a form analagous to an intrinsic magnetic moment coupling. The combined dynamical effects of the vector and tensor $Z$ couplings then imply a mass difference between the charged and neutral leptons, and the general order of magnitude thus anticipated for the electron mass is not unreasonable[8]. But whether it be a phenomenological description or a fundamental interaction, the form of the tensor coupling of the $Z$ field with leptons is determined by the neutrinic charge invariance property as

$$f\, \tfrac{1}{2} Z^{\mu\nu} \tfrac{1}{2} \psi_l \beta \sigma_{\mu\nu} (t - i\gamma_5 [t_3, t]) \psi_l ,$$

where $f$ has the dimensions of a length. The previous comments referring to the electron mass can now be understood in the following way: If the electron mass is zero, the vector coupling (15) is invariant under the electron field transformation:

$$\psi_l \rightarrow [1 + i\delta\varphi t_3 \lambda (1 - t_3 \lambda)\ (i\gamma_5 + t_3)] \psi_l ,$$

which is equivalent to the possibility of specifying the electron polarization under these circumstances ($e^+$ is left polarized, $e^-$ is right polarized). On the other hand, for a massless electron the tensor coupling term is invariant under the different transformation

$$\psi_l \rightarrow [1 + i\delta\varphi t_3 \lambda (1 - t_3 \lambda)\ (-i\gamma_5 + t_3)] \psi_l ,$$

which also implies the production of electrons with definite polarization ($e_R^+$, $e_L^-$) Hence, the tensor coupling cannot be a dynamical byproduct of the vector coupling if the electron mass is zero, and, if both interactions are fundamental there is no $\gamma_5$ invariance property, which indicates that the interference between vector and tensor couplings generates an electron mass.

The tensor coupling term can be presented in a different way by employing the matrix property

$$\sigma_{\mu\nu}\gamma_5 = \tfrac{1}{2} \epsilon_{\mu\nu\lambda\kappa} \sigma^{\lambda\kappa}, \tag{18}$$

where $\epsilon$ is the alternating symbol specified by $\epsilon^{0123} = +1$. If we also observe

[8] Which is only to say that the value of the cutoff momentum that must be chosen is not exceptionally large.

that electrical charge conservation is expressed by

$$\bar{\psi}[t_3, t]\psi Z = (\bar{\psi}t\psi)t_3 Z,$$

(omitting irrelevant matrices) the following form is obtained,

$$f\tfrac{1}{2}\psi_l\beta\sigma_{\mu\nu}t\psi_l\tfrac{1}{2}[Z^{\mu\nu} - it_3(\epsilon Z)^{\mu\nu}],$$

in which

$$(\epsilon Z)^{\mu\nu} = \tfrac{1}{2}\epsilon^{\mu\nu\lambda\kappa}Z_{\lambda\kappa} \tag{19}$$

is the tensor dual to $Z^{\mu\nu}$. Thus the tensor coupling with leptons involves the $Z$ field in an essentially self-dual combination,

$$\epsilon(Z - it_3\epsilon Z) = it_3(Z - it_3\epsilon Z),$$

where the matrix multiplication is in the sense of (19). The basic property used here,

$$\epsilon^2 = -1,$$

can be understood from the matrix version of (18),

$$\sigma\gamma_5 = \epsilon\sigma,$$

which indicates that $\epsilon$ is a six-dimensional matrix representation of $\gamma_5$.

The tensor combination $(1 - it_3\epsilon)Z$ makes very explicit the role of the $Z$ field in destroying charge reflection and space reflection invariance, while maintaining the reflection property $R$. But the natural extension of the coupling to the heavy fermion field encounters difficulties. A universal coupling of the form $m(1 - it_3\epsilon)Z$ implies an effective self-interaction of the tensor sources of the $Z$ field which, for large $Z$-particle mass, is dominated by

$$\begin{aligned}\tfrac{1}{2}m(1 - it_3\epsilon)(1 + it_3\epsilon)m &= \tfrac{1}{2}m(1 - t_3^2)m \\ &= 0.\end{aligned}$$

Hence we do not obtain direct tensor coupling between heavy and light fermions. While this conclusion is modified by dynamical effects of the $Z$-particles, it is perhaps simpler to say that the tensor coupling is not entirely universal but distinguishes between heavy and light fermions through the appearance of the alternative tensor combinations $(1 \pm it_3\epsilon)Z$. Thus, a possible tensor coupling between the heavy fermions and the charged $Z$-particles is

$$2^{-1/2}f\frac{1}{2}Z^{\mu\nu}\frac{1}{2}\psi\beta\sigma_{\mu\nu}\left(\tau + i\gamma_5\left[\frac{\tau_3 + \zeta_3}{2}, \tau\right]\right)\psi.$$

If both vector and tensor couplings of the heavy F. D. particles are operative, a mass difference between electrically charged and neutral particles is implied beyond that produced by the electromagnetic field. Thus the role initially assigned to the electromagnetic field may be modified by the more complete theory for which it served as a model. We have not discussed linear interactions of the $Z$ field with the heavy bosons. Perhaps it suffices to say that one cannot construct a bilinear vector combination of the spin 0 fields that realized the dynamical function of the $Z$ field to destroy charge and space reflection invariance.

From the general suggestions of a family of bosons that is the isotopic analog of the leptons, and the identification of its neutral member as the photon, we have been led to a dynamics of a charged, unit spin $Z$-particle field that is interpreted as the invisible instrument of the whole class of weak interactions. The direct identification of this hypothetical particle will not be easy. Its linear couplings are neither so strong that it would be produced copiously, nor are they so weak that an appreciable lifetime would be anticipated. And as to the detailed implications of this model for the effective weak interactions that it seeks to comprehend, although the theory is definite enough about the fundamental predominance of vector and tensor coupling, the rest of the structure is hardly unique, and the profound effects of the various strong interactions obscure the actual predictions of the formalism. The definitive results of the group of experiments that exploit the newly discovered lepton polarization properties of the weak interactions will be particularly relevant in judging this hypothesis.

The heavy fermions and bosons, the leptons and the photon-$Z$ particle family, have been viewed as physical realizations of four and three dimensional internal symmetry spaces, respectively. Do particles similarly attached to a two-dimensional symmetry space exist, or does every particle family contain an electrical neutral member? In the latter circumstance, we must descend to a one-dimensional internal space, to a neutral field that presumably possesses no internal properties and responds dynamically only to the space-time attributes of other systems. According to the generalization that introduced leptonic charge, this noninternally degenerate system is a B.E. field. If it follows the example of the electrically neutral fields associated with the other odd-dimensional space, it is massless, and the spin restrictions commented on at the beginning of this discussion need not apply to a single neutral massless field. It appears that in the hierarchy of fields there is a natural place for the gravitational field.

What has been presented here is an attempt to elaborate a complete dynamical theory of the elementary particles from a few general concepts. Such a connected series of speculations can be of value if it provides a convenient frame of reference in seeking a more coherent account of natural phenomena.

RECEIVED: July 31, 1957

REFERENCES

1. J. Schwinger, *Phys. Rev.* **104,** 1164 (1956).
2. A. Pais, *Proc. Nat. Acad. Sci. U. S.* **40,** 484 (1954).
3. E. P. Wigner, *Proc. Nat. Acad. Sci. U. S.* **38,** 449 (1952). See also M. Gell-Mann, *Phys. Rev.* **106,** 1296 (1957).
4. L. Landau, *Nuclear Phys.* **3,** 127 (1957).
5. B. Touschek, *Nuovo cimento* **5,** 1281 (1957).
6. T. Lee and C. Yang, *Phys. Rev.* **105,** 1671 (1957); A. Salam, *Nuovo cimento* **5,** 299 (1957); L. Landau, *Nuclear Phys.* **3,** 127 (1957).
7. W. Pauli, *Nuovo cimento* **6,** 204 (1957).
8. R. Garwin. L. Lederman, and M. Weinrich, *Phys. Rev.* **105,** 1415 (1957).

# *ON THE EUCLIDEAN STRUCTURE OF RELATIVISTIC FIELD THEORY*

BY JULIAN SCHWINGER

HARVARD UNIVERSITY, CAMBRIDGE, MASSACHUSETTS

*Communicated June 2, 1958*

The nature of physical experience is largely conditioned by the topology of space-time, with its indefinite Lorentz metric. It is somewhat remarkable, then, to find that a detailed correspondence can be established between relativistic quantum field theory and a mathematical image based on a four-dimensional Euclidean manifold. The situation can be characterized in the language of group theory, with physical quantities and states appearing as representations of the undeılying Lorentz transformation group. It is well known that some representations of the Lorentz group can be obtained from the attached Euclidean group (the "unitary trick" of Weyl). What is being asserted is that *all* representations of physical interest can be obtained in this way.

The objects that convey this correspondence are the Green's functions of quantum field theory,[1] which contain all possible physical information. We consider a general Hermitian field, $\chi$, which decomposes into a Bose-Einstein field $\phi$, and a Fermi-Dirac field $\psi$. The Green's functions can be defined as vacuum-state expectation values of time-ordered field operator products. There are two types:

$$G_+(x_1 \ldots x_p) = \langle(\chi(x_1) \ldots \chi(x_p))_+ \rangle \epsilon_+(x_1 \ldots x_p)$$

and

$$G_-(x_1 \ldots x_p) = \langle(\chi(x_p) \ldots \chi(x_1))_- \rangle \, \epsilon_-(x_p \ldots x_1),$$

where positive or negative time ordering implies an assignment of multiplication order in accordance with the ascending sense of time, as read from right to left (+) or from left to right (−). The quantities $\epsilon_\pm$ are antisymmetrical functions of the time co-ordinates for the F.D. fields, which assume the value +1 when the time-ordered sense coincides with the written order. The connection between the two Green's functions is simply

Reprinted from the *Proceedings of the National Academy of Sciences U.S.A.* **44**, 956–965 (1958).

$$G_-(x_1 \ldots x_p) = G_+(x_1 \ldots x_p)^*.$$

The definitions as given are actually restricted to those fields for which all components are kinematically independent at a given time. In more general situations, additional terms are necessary,[1] the function of which is to maintain the non-dependence of the Green's functions on the particular timelike direction employed in the time ordering, which is otherwise assured by the commutativity or anticommutativity of fields at points in spacelike relation.

The invariance of the formalism under inhomogeneous Lorentz transformations requires that the Green's functions be translationally invariant functions of the space-time co-ordinates, while the homogeneous, proper, orthochronous Lorentz transformations,

$$\bar{\chi}(lx) = L\chi(x),$$

imply the Green's function invariance property,

$$\left(\prod_{\alpha=1}^{p} L_\alpha\right) G(l^{-1}x) = G(x).$$

The form of the latter for infinitesimal transformations is

$$\left[\sum_{\alpha=1}^{p}\left(x_\mu \frac{1}{i}\,\partial_\nu - x_\nu \frac{1}{i}\,\partial_\mu + S_{\mu\nu}\right)_\alpha\right] G(x) = 0.$$

The theory is also invariant under the proper, antiorthochronous transformation $x^\mu \rightarrow -x^\mu$, provided that a transition to the complex conjugate algebra is included.[2] Now

$$\left[\prod_{\alpha=1}^{p}(R_{st})_\alpha\right] G_+(-x_1 \ldots -x_p)^* = \langle (R_{st}\chi^*(-x_1) \ldots)_+ \rangle\, \epsilon_+(-x_1 \ldots -x_p)$$

$$= G_-(x_1 \ldots x_p),$$

since positive and negative time orderings are interchanged under time reflection, and, with $n$ pairs of F.D. fields,

$$\epsilon_+(-x_1 \ldots -x_p) = (-1)^n \epsilon_-(x_p \ldots x_1),$$

while the sign factor $(-1)^n$ is compensated by the imaginary unit contained in each matrix $R_{st}$ that is associated with a half-integral spin (F.D.) field. Thus, for either type of Green's function.

$$\left[\prod_{\alpha=1}^{p}(R_{st})_\alpha\right] G(-x) = G(x),$$

$$R_{st} = e^{\pi i S_{12}} e^{\pi i S_{34}},$$

which, in its union of two disjoint pieces of the Lorentz group, is a sign of the Euclidean foundation on which the Green's functions rest.

The explicit dependence of the fields on the space-time co-ordinates is governed by the energy-momentum vector, $P^\mu$, according to

$$\chi(x) = e^{-iPx}\chi e^{iPx},$$

while the invariant meaning of the vacuum state is expressed by

$$\langle e^{-iPx} = \langle, \qquad e^{iPx}\rangle = \rangle.$$

Hence, if $x^{(1)} \ldots x^{(p)}$ represents the time-ordered arrangement of $x_1 \ldots x_p$ ($t^{(1)} > \ldots > t^{(p)}$), we have

$$G_+(x) = \langle \chi e^{iP(x^{(1)} - x^{(2)})}\chi \ldots e^{iP(x^{(p-1)} - x^{(p)})}\chi \rangle \epsilon_+(x_1 \ldots x_p)$$

and

$$G_-(x) = \langle \chi e^{iP(x^{(p)} - x^{(p-1)})}\chi \ldots e^{iP(x^{(2)} - x^{(1)})}\chi \rangle \epsilon_-(x_p \ldots x_1),$$

wherein additional indices are needed to distinguish the various types of fields. The time dependence of the Green's function $G_+$ is thus governed by the operators

$$e^{-iP^0(t^{(\alpha)} - t^{(\alpha+1)})},$$

which, in their dependence upon the differences of the consecutively ordered time co-ordinates, contain no negative frequencies ($P^{0'} \geq 0$). The alternative Green's function, $G_-$, analogously constructed from the operators

$$e^{iP^0(t^{(\alpha)} - t^{(\alpha+1)})},$$

contains no positive frequencies.

We shall now use these spectral characteristics of the Green's functions to give a more precise meaning to the assumed existence of the Green's functions, which is in the sense of the summability of Fourier integrals. It is described by the absolute convergence (for distinct $x_1 \ldots x_p$) of the spectral representation obtained on replacing the positive frequency unitary operators in $G_+$ with

$$e^{-iP^0(t^{(\alpha)} - t^{(\alpha+1)})(1 - i\epsilon)}$$

and similarly inserting

$$e^{iP^0(t^{(\alpha)} - t^{(\alpha+1)})(1 + i\epsilon)}$$

in $G_-$, where the limit $\epsilon \to +0$ is to be eventually performed. This modified time dependence is also expressed by the substitutions $t_\alpha \to t_\alpha(1 - i\epsilon)$ in $G_+$ and $t_\alpha \to t_\alpha(1 + i\epsilon)$ in $G_-$. The existence of the Green's functions in this sense is equivalently described by the assumption that the various field operator matrix elements, multiplied by the densities of relevant states, possess no more than an algebraic growth with increasing energy.

The absolute convergence of the spectral representations for the Green's functions $G_+$ and $G_-$ is now assured for the more general time substitution:[3]

$$\left.\begin{array}{ll} G_+: & t_\alpha \to \tau_\alpha e^{-i\vartheta} \\ G_-: & t_\alpha \to \tau_\alpha e^{i\vartheta} \end{array}\right\} \sin\vartheta > 0,$$

in which $\vartheta$ lies in the open interval $0 < \vartheta < \pi$. The new variables $\tau_\alpha$ are real numbers that retain the ordering of the original time variables. We adopt a special notation to accompany the particular choice, $\vartheta = {}^1/_2\pi$, which asserts the existence of the functions $G_+(t \to -ix_4)$ and $G_-(t \to +ix_4)$. In this way there emerges a correspondence between the Green's functions in space-time and functions defined on a four-dimensional Euclidean manifold. To the extent that the two Euclidean functions thus obtained are related, there also appears an analytical continuation that connects the two distinct types of space-time Green's functions, $G_\pm$. Conversely, given one of the Euclidean functions, the substitutions $x_4 \rightarrow e^{i(\pi/2 - \epsilon)}t$ and $x_4 \rightarrow e^{-i(\pi/2 - \epsilon)}t$ will yield functions having the space-time character of $G_+$ and $G_-$, respectively, in the limit as $\epsilon \rightarrow +0$. We must now see how to supply an independent basis for the Euclidean Green's functions, from which has disappeared all reference to the space and time distinctions of the Lorentz metric.

The significance of the latter remark can be appreciated through the form assumed in the Euclidean description by the statement of infinitesimal rotational invariance of the Green's functions. The Hermitian spin matrices, $S_{\mu\nu}$, $\mu, \nu = 1 \ldots 4$ ($x_4 = \pm ix^0$), comprising $S_{kl}$ and $S_{k4} = \pm i\, S_{0k}$, still bear the mark of their Lorentz origin in the symmetry of these matrices; the $S_{kl}$ are antisymmetrical, while the $S_{k4}$ are symmetrical. Hence one must perform a unitary transformation to unite them into six antisymmetrical, imaginary matrices that describe independent infinitesimal orthogonal transformations. The means for distinguishing between the two types of matrices is provided by the space-reflection matrix, $R_s$,

$$R_s^{-1}S_{kl}R_s = S_{kl}, \qquad R_s^{-1}S_{k4}R_s = -S_{k4},$$

or, alternatively, by the time-reflection matrix,

$$R_t = R_{st}R_s.$$

Indeed, for integral spin the necessary transformation is produced by

$$S_{\mu\nu}^{(E)} = e^{\mp \pi i/4R_t}S_{\mu\nu}e^{\pm \pi i/4R_t},$$

where the plus or minus sign applies to the matrices associated with $G_+$ or $G_-$:

$$S_{k4}^{(E)} = e^{\mp \pi i/4R_t}(\pm)iS_{0k}e^{\pm \pi i/4R_t} = R_tS_{0k}.$$

Since the Hermitian matrices $R_s$ and $R_t$ are symmetrical[4] for integer spin and anticommute with $S_{0k}$, the net effect of the transformation is to convert the symmetrical matrix $S_{k4}$ into the desired antisymmetrical $S_{k4}^{(E)}$. For half-integral spin, however, both the skew-Hermitian matrix $R_s$ and the Hermitian matrix $R_t$ are antisymmetrical, and the transformation as stated does not yield the required Euclidean matrices. One must also associate with $R_s$, say, an antisymmetrical matrix that commutes with all $S_{\mu\nu}$ and with $R_s$ (the apparently different possibility of a symmetrical matrix that anticommutes with $R_s$ is simply a change of representation,[4] produced with the aid of $R_{st}$, that effectively replaces $R_s$ with $R_t$). Thus, to permit the complete transformation from the Lorentz to the Euclidean metric, every half-integer spin (F.D.) field must carry a charge. Just such a general fermionic charge property, under the name of nucleonic charge or leptonic charge, is either well established experimentally or has been conjectured on other grounds. The Euclidean formulation may be the proper basis for comprehending this general attribute of F.D. fields. If $l$ is the imaginary, antisymmetrical matrix representing the fermionic charge property, the required transformation for half-integer spin fields is

$$S_{\mu\nu}^{(E)} = e^{\pm \pi/4R_sl}S_{\mu\nu}e^{\mp \pi/4R_sl},$$

and, indeed,

$$S_{k4}^{(E)} = ilR_sS_{0k}$$

has the desired property of antisymmetry.

We should also note the removal of reference to the Lorentz metric from the orthogonal matrices $R_k$, $R_t$ that are associated with the reflections of the individual space-time co-ordinate axes. For a B.E. field all these matrices are commutative, real, symmetrical matrices, and they are unchanged by the transformation to the Euclidean metric. On considering a F.D. field, however, we find that $R_t$ occupies a distinguished position, differing from the anticommuting, real, symmetrical matrices $R_k$ by being imaginary and antisymmetrical. While the matrices $R_k$ are unaltered by the metric change, the Euclidean matrix associated with the reflection of $x_4$ is now

$$R_4^{(E)} = e^{\pm\pi/4R_sl}(\pm)R_te^{\mp\pi/4R_sl} = R_{st}l,$$

which is also a real symmetrical matrix. Thus all the individual Euclidean co-ordinate reflection matrices are real, symmetrical, and orthogonal, with the two classes of fields distinguished by commutativity properties, according to

$$E: \qquad R_\mu R_\nu = e^{\pi iS_{\mu\nu}} = e^{2\pi iS_{\mu\nu}}R_\nu R_\mu$$

To obtain an independent characterization of the Euclidean Green's functions, we can convert to the Euclidean metric the system of differential equations obeyed by the Green's functions. We shall only attempt to outline this process in the following considerations. Let the Lagrange function be written as

$$\mathcal{L} = \frac{1}{4}[\chi A^\mu\partial_\mu\chi - \partial_\mu\chi A^\mu\chi] + \frac{1}{2}\chi B\chi - \mathcal{H}_1(\chi),$$

in which $\mathcal{H}_1$ refers to the interactions between fields. The field equations are

$$A^\mu\partial_\mu\chi + B\chi = \frac{\partial_l\mathcal{H}_1}{\partial\chi},$$

while the commutation properties on a spacelike surface, as expressed in a local co-ordinate system, are given by (for simplicity of description, we adhere to the fiction that all components of $\chi$ are kinematically independent on $\sigma$):

$$[A^0\chi(x), \chi(x')]_\pm = i\delta^0(x - x'),$$

where $\delta^\mu(x - x')$ is defined by

$$\int d\sigma_\mu'\delta^\mu(x - x')f(x') = f(x).$$

Now, uniting the field equations and commutation properties, the differential equations for the Green's functions are obtained:

$$(A^\mu\partial_\mu + B)_1 G_+(x_1 \ldots x_p) + \ldots = i\delta(x_1 - x_2)G_+(x_3 \ldots x_p) \pm i\delta(x_1 - x_3)G_+(x_2 \ldots x_p) + \ldots,$$

in which the omitted terms on the left are the particular Green's function combinations needed to represent the interaction effects in the field equations, while, on the right, the summation is extended over all points that refer to the same field as does $x_1$. The summation is symmetrical or antisymmetrical in these points, according to the statistics of the field. Thus the Green's functions obey an infinite system of equations that are linear and inhomogeneous (since the function with $p = 0$ is simply unity) and which incorporate fully all information concerning the interacting fields.

The analogous differential equations for the $G_-$ can be obtained directly, or by complex conjugation of the equations obeyed by $G_+$, and differ from the latter in the sign of $i$ exhibited by the right-hand B.E. terms. It is worthy of note that, apart from the trivial situation of uncoupled fields, the two sets of differential equations are intrinsically different—the two types of Lorentz Green's functions cannot be characterized in detail as solutions of a common equation that are distinguished by boundary conditions. There is, however, a simple relation between the Green's functions that can be inferred from the differential equations (and from the time-ordered operator definitions), namely,

$$G_-(x) = (-1)^n G_+(-e^{-\pi i}x).$$

The factor $(-1)^n$ is inserted to reverse the sign of the right-hand F.D. terms. The interpretation given to $-e^{-\pi i}$, applied to the space co-ordinates, is unity. But, for the time co-ordinates, it is a combination of two operations: time reflection, which inverts the time order ($t^{(1)} > \ldots t^{(p)}, -t^{(p)} > \ldots -t^{(1)}$), and multiplication by $e^{-\pi i}$, which reverses the sign of all time co-ordinates while retaining the time order. The latter transformation changes the sign of all the delta functions and converts the differential equations for $G_+$ into those describing $G_-$. The negative frequency character of the latter functions is also reproduced. The connection here obtained between $G_+$ and $G_-$ is an analytic continuation stated without reference to intermediate Euclidean functions.

The correspondence between Lorentz and Euclidean Green's functions is now exhibited as

$$G_\pm(x) \longleftrightarrow \Pi_{\text{B.E.}}(e^{\pm \pi i/4 R_t} e^{\mp \pi i/4})\, \Pi_{\text{F.D.}}(e^{\mp \pi/4 R_s l} e^{\mp \pi i/4}) G_\pm(x)^{(E)},$$

in which $x^0$ and $x_4$ are understood as the variables in the appropriate functions. Accompanying the imaginary relations between the variables is the transformation

$$\delta(x) \longleftrightarrow (\pm i)\delta(x)^{(E)}.$$

On performing these substitutions in the differential equations, we encounter the real B.E. matrices,

$$\text{B.E.:} \quad A_\mu{}^{(E)} = -e^{\pm \pi i/4 R_t} e^{\mp \pi i/4} A_\mu e^{\pm \pi i/4 R_t} e^{\mp \pi i/4} = (-A_k R_t,\ A_0),$$
$$B^{(E)} = -e^{\pm \pi i/4 R_t} e^{\mp \pi i/4} B e^{\pm \pi i/4 R_t} e^{\mp \pi i/4} = -BR_t,$$

which are, respectively, antisymmetrical and symmetrical, and the real F.D. matrices,

$$\text{F.D.:} \quad A_\mu{}^{(E)} = \mp e^{\mp \pi/4 R_s l} e^{\mp \pi i/4} A_\mu e^{\mp \pi/4 R_s l} e^{\mp \pi i/4} = (iA_k,\ -A_0 R_s l),$$
$$B^{(E)} l = -e^{\mp \pi/4 R_s l} e^{\mp \pi i/4} B e^{\mp \pi/4 R_s l} e^{\mp \pi i/4} = -BR_s il,$$

which are symmetrical and antisymmetrical, respectively. The resulting form of the Green's function differential equations, as adapted to the Euclidean metric, is

$$[A_\mu \partial_\mu + (1_1 \pm l)B]_1\, G_\pm(x_1 \ldots x_p)^{(E)} + \ldots = \delta(x_1 - x_2) G_\pm(x_3 \ldots x_p)^{(E)} + \ldots,$$

where the choice indicated for the coefficient of $B$ signifies unity for a B.E. field;

$\pm l$ for a F.D. field. If, as we have discussed before,[4] the Lorentz $B$ matrices are constructed as the product of $R_s$ with an invariant, symmetrical matrix that is independent of internal degrees of freedom, the Euclidean $B$ matrices will be completely of the latter type. This remark, together with the observation that the transformation applied to $B^{-1}A_\mu$ is unitary, apart from the F.D. factor $\pm l$, indicates how Lorentz invariance is translated into Euclidean invariance.

The Green's functions of the Lorentz description are intrinsically complex quantities, and, accordingly, there are two linearly independent sets of such functions. It is an indication of the simplification obtained through the introduction of the Euclidean metric that completely real Euclidean Green's functions can be defined—provided that a certain general symmetry restriction is enforced on the field interactions. Conversely, the latter invariance property acquires substantial support through its role in unifying the two classes of Green's functions and eliminating complex numbers from a formulation of the fundamental laws of physics. Let us notice that, aside from the interaction terms, the differential equations for $G_\pm{}^{(E)}$ differ only in the F.D. quantity $\pm l$, referring to the fermionic charge. This sign factor can be removed by introducing the operation of fermionic charge reflection. But, when full account is taken of the variety of field interactions,[5] it appears that all types of charge are dynamically coupled, and the interconversion of the two sets of equations is possible only if the interaction terms differ merely through the effect of general charge reflection. Assuming this property, we conclude that

$$G_-{}^{(E)} = \left[\prod_{\alpha=1}^{p} (R_Q)_\alpha\right] G_+{}^{(E)} = \mathrm{R}_Q G_+{}^{(E)},$$

where the individual charge reflection matrices $R_Q$ are real and orthogonal. The composite matrix $\mathrm{R}_Q$ also describes the reality properties of the Euclidean Green's functions,

$$G_+{}^{(E)*} = \mathrm{R}_Q G_+{}^{(E)},$$

for the mutually complex conjugate relation of $G_\pm$ still applies to the derived functions $G_\pm{}^{(E)}$. In effect, all matrices appearing in the Euclidean formulation of the differential equations are real, with the exception of the imaginary charge matrices, and complex conjugation is equivalent to charge reflection. If we accept the interpretation[2] of the imaginary unit as symbolic of the charge nature of the measurement apparatus (matter—antimatter), the symmetry property we have postulated can be described as the relativistic invariance of the Euclidean formulation with respect to charge reflection, for the application of this transformation to the system under investigation and to the apparatus employed for the purpose produces no discernible change.

Before continuing, we must examine the relation between the matrices, $R_Q{}^{(E)}$, and the charge reflection matrices of the Lorentz description, $R_Q$. The latter, having no reference to space-time properties, are uniformly chosen as real, orthogonal, symmetrical matrices ($R_Q{}^2 = 1$). The distinction between $R_Q$ and $R_Q{}^{(E)}$, which exists only for F.D. fields, arises from the incorporation of the fermionic charge $l$ into $A_4{}^{(E)} = -A_0 R_s l$. To compensate the sign change of $A_4{}^{(E)}$ induced by the reflection of $l$, the F.D. matrix $R_Q{}^{(E)}$ must contain the co-ordinate reflection matrix $R_4{}^{(E)} = R_{st} l$. Thus

$$\text{F.D.:} \qquad R_Q{}^{(E)} = R_Q R_{st} l,$$

which is a real, orthogonal, antisymmetrical matrix ($R_Q{}^{(E)2} = -1$). However, the composite matrix $\mathrm{R}_Q{}^{(E)}$, which is constructed from an even number of F.D. contributions, is a real symmetrical matrix, obeying

$$\mathrm{R}_Q{}^{(E)2} = 1.$$

The hypothesis of Euclidean relativistic charge reflection invariance can now be interpreted as a property of the Lorentz Green's functions and, thereby, of the Lagrange function of the interacting fields. The relation implied between the Green's functions $G_+$ and $G_-$ is

$$\begin{aligned} G_-(ix_4) &= \Pi_{\mathrm{B.E.}}(R_Q R_t)\Pi_{\mathrm{F.D.}}(R_Q i R_t) G_+(-ix_4) \\ &= (-1)^n \mathrm{R}_Q \mathrm{R}_t G_+(-ix_4), \end{aligned}$$

which makes explicit the analytic continuation that connects the two Lorentz Green's functions:

$$G_-(t) = (-1)^n \mathrm{R}_Q \mathrm{R}_t G_+(e^{-\pi i} t),$$

$$G_+(t) = (-1)^n \mathrm{R}_Q \mathrm{R}_t G_-(e^{\pi i} t).$$

When this result is compared with the previously obtained connection,

$$G_\mp(t) = (-1)^n G_\pm(-e^{\mp \pi i} t),$$

we learn that

$$\mathrm{R}_Q \mathrm{R}_t G(-t) = G(t);$$

the Lorentz Green's functions are invariant under charge and time reflection. The same assertion can be made of the combination of charge and space reflection, since space-time reflection is an invariance operation. But in the latter form we are dealing with unitary transformations of Hermitian field operators, and it can be concluded that the invariance of the Lagrange function under space and charge reflection is equivalent to the postulate that Euclidean Green's functions exhibit a relativistic invariance with respect to charge reflection. It is surely significant that we are thus led to a general invariance property which is consistent with all the recent experiments on the so-called parity non-conserving interactions. The existence of an exact invariance transformation involving space reflection is now supplied with a basis that may be considered more substantial than the mere belief in the intrinsic indiscernibility of left and right.

We have not yet exhibited the real Euclidean Green's functions, the existence of which is assured by the presence of a linear transformation equivalent to complex conjugation,

$$G_\pm{}^{(E)*} = \mathrm{R}_Q G_\pm{}^{(E)}.$$

Indeed, functions having the required reality property are given by

$$G^{(E)} = e^{\pi i/4 \mathrm{R}_Q} e^{-\pi i/4} G_+{}^{(E)} = e^{-\pi i/4 \mathrm{R}_Q} e^{\pi i/4} G_-{}^{(E)}.$$

A second such choice is

$$\mathrm{R}_Q G^{(E)} = e^{-\pi i/4 \mathrm{R}_Q} e^{\pi i/4} G_+{}^{(E)} = e^{\pi i/4 \mathrm{R}_Q} e^{-\pi i/4} G_-{}^{(E)},$$

although the latter are not a linearly independent set and can be regarded as pre senting $G^{(E)}$ in a new representation. An essential limitation of the description by real Green's functions must be observed, however. The matrices $e^{\pm \pi i/4R_Q}$ are not composite, and the transformation that introduces $G^{(E)}$ has no simple significance for the differential equations that characterize the Green's functions. Were only B.E. fields involved, the composite transformation formed from the individual $R_Q$ could be employed, but this is not possible for F.D. fields. Nevertheless, it remains true that real Euclidean Green's functions exist from which the physically meaningful Green's functions $G_+$ and $G_-$ can be inferred.

Finally, we shall indicate briefly the possibility of replacing the differential equations, as the characterization of the Euclidean Green's functions, by an explicit, if formal, construction. For this purpose we define fields $\chi(x)$, on the Euclidean manifold, that are completely commutative or anticommutative, as befits the statistics,

$$[\chi(x), \chi(x')]_{\pm} = 0,$$

and a complementary set of fields, $\hat{\chi}(x)$, with the same characteristics, which are such that

$$[\hat{\chi}(x), \chi(x')]_{\pm} = \delta(x - x').$$

The B.E. fields $\phi$ and $i\hat{\phi}$ are Hermitian, while the F.D. fields $\psi$ and $\hat{\psi}$ are mutually Hermitian conjugate. The Euclidean Green's functions are then given by

$$G_{\pm}(x_1 \ldots x_p)^{(E)} = \langle W_{\pm}|\chi(x_1) \ldots \chi(x_p)|0\rangle / \langle W_{\pm}|0\rangle$$

(ordinary operator multiplication!), where $|0\rangle$ is the right eigenvector of the operators $\hat{\chi}(x)$ associated with null eigenvalues,

$$\hat{\chi}(x)\,|0\rangle = 0,$$

and the vector $\langle W_{\pm}|$ is characterized by

$$\langle W_{\pm}|\,[A_\mu \partial_\mu \chi + (1, \pm l)B\chi + \ldots - \hat{\chi}] = 0,$$

in which the omitted terms are the functions of the Euclidean field operators, $\chi(x)$, needed to describe the field interactions. This assertion is verified on observing that

$$\begin{aligned}[A_\mu \partial_\mu + (1, \pm l)B]_1 G_{\pm}(x_1 \ldots x_p)^{(E)} &+ \ldots \\ = \langle W_{\pm}|\,\hat{\chi}(x_1)\chi(x_2) \ldots \chi(x_p)\,|0\rangle &/ \langle W_{\pm}|0\rangle \\ = \delta(x_1 - x_2)G_{\pm}(x_3 \ldots x_p)^{(E)} &+ \ldots,\end{aligned}$$

where commutation relations and the significance of $|0\rangle$ as a $\hat{\chi}$ eigenvector are used to obtain the stated result.

The vector $\langle W_{\pm}|$ can be constructed from the left eigenvector of the $\hat{\chi}$ associated with null eigenvalues,

$$\langle W_{\pm}| = \langle 0|\,e^{-W_{\pm}},$$

where the operator

$$W_{\pm}[\chi] = {}^1/_2\int (dx)\,[\chi A_\mu \partial_\mu \chi + \chi(1, \pm l)B\chi + \ldots]$$

bears an obvious genetic relation to the action operator of the interacting fields. The Euclidean operators $W$ are not Hermitian, but they are related to their Hermitian conjugates by a unitary transformation, which constitutes a self-adjointness property. We now have

$$G_{\pm}(x)^{(E)} = \langle 0|\chi(x_1) \ldots \chi(x_p)e^{-W_{\pm}[x]}|0\rangle / \langle 0|e^{-W_{\pm}}|0\rangle,$$

and, in turn, these Green's functions can be derived from a single generating function, the expansion of which produces the field operator products. In the latter form we make contact with previous developments employing the action principle for quantized fields and the device of external sources[1] (and subsequent work, largely unpublished). A large variety of equivalent forms can now be devised for the Green's functions, based primarily upon the well-established transformation and representation theory[6] for canonical variables of the first and second kind. A discussion of these developments for specific systems will be deferred to another publication, in which the problem of translating quantum electrodynamics into the Euclidean metric is examined.

Although we have emphasized the fundamental implications of the Euclidean representation, it will be evident that the Euclidean Green's functions also have practical advantages. Indeed, the utility of introducing a Euclidean metric has frequently been noticed in connection with various specific problems, but an appreciation of the complete generality of the procedure has been lacking.

[1] J. Schwinger, these PROCEEDINGS, **37**, 452, 1951.

[2] J. Schwinger, these PROCEEDINGS, **44**, 223, 1958; see also R. Jost, *Helv. Phys. Acta*, **30**, 409, 1957.

[3] The full analytic extension of the Green's functions $G_{\pm}$ is produced by $t \rightarrow \xi \mp i\eta$, where $\eta$ retains the initial time order, which is to say that the otherwise arbitrary mapping function $\eta(t)$ is of positive slope.

[4] J. Schwinger, these PROCEEDINGS, **44**, 617–619, 1958.

[5] J. Schwinger, *Ann. Phys.*, **2**, 407, 1957.

[6] An extended discussion is contained in the article "Quantum Theory of Fields," which is being prepared for publication in Vol. V/2 of the *Encyclopedia of Physics* (Springer).

# Euclidean Quantum Electrodynamics

JULIAN SCHWINGER
*Harvard University, Cambridge, Massachusetts**
(Received March 2, 1959)

Quantum electrodynamics is transcribed into a Euclidean metric. A review is presented of the quantum action-principle approach to quantization, with its automatic emphasis on the dynamical variables associated with the physical degrees of freedom. Green's functions of the radiation gauge are defined, and then characterized by differential equations and boundary conditions. These Green's functions are of direct physical significance but involve a distinguished time-like direction. A gauge transformation is then performed to eliminate this dependence, introducing thereby the Green's functions of the Lorentz gauge, which lack immediate physical interpretation. The latter functions are now primarily defined by differential equations and boundary conditions, and form the basis for the analytic extension which is the change from space-time to Euclidean metric. Some properties of anticommuting matrices are discussed in relation to this metric transformation. Real Euclidean Green's functions are defined by correspondence with the Lorentz gauge functions and the appropriate differential equations obtained. Invariance properties of the Euclidean functions are discussed. The individual Euclidean Green's functions are given an operator construction and then combined into a generating Green's functional which is interpreted as the wave function, in a canonical field representation, of a state characterized by the Euclidean action operator. Differential operator realizations and some other benefits of a canonical variable description are exhibited.

## INTRODUCTION

In a recent note[1] the author has remarked on the possibility of establishing a correspondence between the quantum theory of fields in space-time, and a mathematical structure that employs a four-dimensional Euclidean coordinate manifold. In that note the simplifying fiction was adopted, for the purposes of exposition, that all field components are kinematically independent in the standard form of the action principle that produces first order differential field equations. While this is true of the only known kind of F.D. (Fermi-Dirac) field, with spin $\frac{1}{2}$, there is no B.E. (Bose-Einstein) field of this type. Accordingly, we must supply some assurance that the discussion applies to real systems, and the Maxwell field is naturally indicated as an example of more than routine interest. Thus the content of this paper is the detailed transcription of quantum electrodynamics into the Euclidean formulation.

## THE ACTION PRINCIPLE

We shall need to review some aspects of the development of quantum electrodynamics from the action principle. (Perhaps one should record at this point the author's opinion that the currently popular indefinite-metric quantization of the electromagnetic field is unphysical and unnecessary.) The Lagrange function for the system of Maxwell and Dirac fields is

$$\mathcal{L}=-\tfrac{1}{2}F^{\mu\nu}(\partial_\mu A_\nu-\partial_\nu A_\mu)+\tfrac{1}{4}F^{\mu\nu}F_{\mu\nu}+\tfrac{1}{2}\psi\beta\gamma^\mu\partial_\mu\psi$$
$$+\tfrac{1}{2}im\psi\beta\psi-\tfrac{1}{2}ie\psi\beta\gamma^\mu q\psi A_\mu,$$

where symmetrized or antisymmetrized multiplication is to be understood for B.E. and F.D. terms, respectively. The electric current operator formed from the Hermitian field $\psi$,

$$j^\mu=-\tfrac{1}{2}ie\psi\beta\gamma^\mu q\psi,$$

is a B.E. quantity in the latter context. The Dirac matrices $\gamma^\mu$ have the algebraic property

$$\{\gamma^\mu,\gamma^\nu\}=-2g^{\mu\nu},$$
$$g^{00}=-1,\quad g^{kl}=\delta^{kl},$$

while

$$\beta=i\gamma^0$$

is a real, antisymmetrical matrix and the matrices $\beta\gamma^\mu$ are symmetrical and imaginary. The imaginary, antisymmetrical charge matrix $q$ possesses integer eigenvalues. This way of defining the charge characteristics of the $\psi$ field is designed to emphasize that the coupling constant $e$ is primarily a property of the electromagnetic field. Indeed, by a suitable scale change of the fields, $e$ is removed from the coupling term to reappear only in the field strength term:

$$\tfrac{1}{4}e^2F^{\mu\nu}F_{\mu\nu}.$$

With the latter field definition, the gauge invariance of the theory refers to the purely kinematical transformation

$$\psi\to e^{iq\lambda}\psi,$$
$$A_\mu\to A_\mu+\partial_\mu\lambda.$$

The application of the stationary action principle extracts from the Lagrange function the field equations

$$F_{\mu\nu}=\partial_\mu A_\nu-\partial_\nu A_\mu,\quad \partial_\nu F^{\mu\nu}=j^\mu,$$
$$[\gamma^\mu(-i\partial_\mu-eqA_\mu)+m]\psi=0,$$

in which the product $A\psi$ is to be symmetrized, and the infinitesimal generator

$$G=\int_\sigma d\sigma_\mu[-F^{\mu\nu}\delta A_\nu+\tfrac{1}{2}\psi\beta\gamma^\mu\delta\psi]$$
$$=\int d\sigma[-F^{0k}\delta A_k+\tfrac{1}{2}i\psi\delta\psi],$$

* This paper was largely written during the summer of 1958 at the University of Wisconsin, Madison. The hospitality of the Department of Physics is gratefully acknowledged.

[1] J. Schwinger, Proc. Nat. Acad. Sci. U. S. 44, 956 (1958); and *1958 Annual International Conference on High-Energy Physics at CERN*, edited by B. Ferretti (CERN, Geneva, 1958).



Reprinted from *The Physical Review* **115**, 721–731 (1959).

where the second form refers to the special coordinate system that identifies $n_\mu$, the unit time-like vector normal to the surface $\sigma$, with the time axis. To facilitate the interpretation of this generator it should be noted that, while all components of $\psi$ obey explicit equations of motion, the only Maxwell field equations of that type are

$$\partial_0 A_k = \partial_k A_0 + F_{0k}, \quad \partial_0 F^{k0} = j^k - \partial_l F^{kl},$$

the remaining equations,

$$F_{kl} = \partial_k A_l - \partial_l A_k, \quad \partial_k F^{0k} = j^0,$$

containing no time derivatives. Thus the magnetic field (in this coordinate system) is not an independent dynamical quantity, and the longitudinal part of the electric field is an explicit function of the Dirac field through the charge density, as given by

$${}^L F^{0k}(x) = -\partial^k \int (dx')\mathfrak{D}(x-x')j^0(x'),$$

where

$$-[\partial^2 + (n\partial)^2]\mathfrak{D}(x-x') = -\partial^k\partial_k\mathfrak{D}(x-x') = \delta(x-x').$$

From the longitudinal component of the equation of motion for $A_k$, we learn that

$${}^L A_k(x) = \partial_k \Lambda(x),$$

$$A_0(x) = \int (dx')\mathfrak{D}(x-x')j_0(x') + \partial_0\Lambda(x),$$

in which $\Lambda(x)$ must remain arbitrary to express the freedom of gauge transformations. Hence it is only the gauge-invariant transverse vector potential ${}^T A_k$, together with the complementary variables $-{}^T F^{0k}$, that qualify as the independent dynamical variables of the electromagnetic field.

Since a variation of the longitudinal part of the vector potential is a gauge transformation

$$\delta_\Lambda {}^L A_k = \partial_k \delta\Lambda,$$

we remove from $\delta\psi$ the corresponding infinitesimal gauge transformation

$$\delta_\Lambda \psi = ieq\delta\Lambda\psi,$$

and observe that these contributions to the infinitesimal generator are

$$\int d\sigma[-F^{0k}\partial_k\delta\Lambda - \tfrac{1}{2}e\psi q\psi\delta\Lambda] = \int d\sigma[\partial_k F^{0k} - j^0]\delta\Lambda = 0.$$

Thus the generator of independent variations of the dynamical variables at a given time is

$$G = \int d\sigma[-{}^T F^{0k}\delta^T A_k + \tfrac{1}{2}i\psi\delta\psi]$$

where the $\delta^T A_k$ and $\delta\psi$ possess the properties of commutativity or anticommutativity characteristic of the field statistics. Only one set of complementary variables for the Maxwell field is changed in this form and the interpretation of the generator[2] appropriate to that circumstance yields for the nonvanishing commutators

$$\delta(x^0 - x^{0\prime})\frac{1}{i}[{}^T A_k(x), -{}^T F^{0l}(x')]$$
$$= {}^T(\delta_k{}^l\delta(x-x')) = \delta_k{}^l\delta(x-x') - \partial_k\partial^{l\prime}\mathfrak{D}(x-x'),$$

while the treatment of all components of $\psi$ on the same footing implies that

$$\delta(x^0 - x^{0\prime})\{\psi(x), \psi(x')\} = \delta(x-x').$$

Various additional commutation properties which can be derived from these fundamental ones will be stated as they are needed.

### GREEN'S FUNCTIONS

We now proceed to the Green's functions of the Maxwell-Dirac system. Although the definitions and relevant properties of these functions are most naturally and compactly obtained from the device of external sources used in conjunction with the action principle, that procedure may not provide, for some, the conviction that accompanies the following explicit considerations. The Green's functions are symmetrical functions of B.E. coordinates, $\xi_1 \cdots \xi_\nu$, and antisymmetrical functions of F.D. coordinates, $x_1 \cdots x_{2n}$, defined as vacuum expectation values of time ordered products:

$$G_\pm(x_1 \cdots x_{2n}, \xi_1 \cdots \xi_\nu)$$
$$= \langle(\psi(x_1) \cdots \psi(x_{2n})\bar{A}(\xi_1) \cdots \bar{A}(\xi_\nu))_\pm\rangle\epsilon_\pm(x).$$

The vacuum referred to here is the lowest-energy state of the fully interacting system[3]—no use is made of the device of adiabatic decoupling. Positive or negative time ordering is the assignment of multiplication order in conformity with the sequence of projections on a time-like vector $n_\mu$, the positive sense of multiplication being from right to left. The quantities

$$\epsilon_+(x_1 \cdots x_{2n}) = \epsilon_-(x_{2n} \cdots x_1),$$

are antisymmetrical functions of the F.D. coordinates that assume the value $+1$ when the time-ordered arrangement coincides with the written order. The symbols

$$\bar{A}_\mu(\xi) = A_\mu(\xi) \pm in_\mu n_\nu \int (d\xi')\mathfrak{D}(\xi-\xi')\delta/\delta A_\nu(\xi')$$

are used to give a compact expression to a sum of terms.

[2] J. Schwinger, Phil. Mag. 44, 1171 (1953).

[3] Thus there are no apparent difficulties of physical interpretation, such as P.A.M. Dirac [Quantum Mechanics (Clarendon Press, Oxford, 1958), fourth edition,] believes exist, which arise from an unwarranted physical identification of states that do not include the effects of interaction.

The simplest example is

$$G_\pm(\xi_1\xi_2)_{\mu_1\mu_2}=\langle(A_{\mu_1}(\xi_1)A_{\mu_2}(\xi_2))_\pm\rangle\pm in_\mu n_\nu\mathfrak{D}(\xi_1-\xi_2).$$

The operators $A_\mu(\xi)$ are defined, in the special coordinate system when the unit time-like vector $n_\mu$ coincides with the time axis, by assigning the value zero to the arbitrary gauge function $\Lambda(\xi)$. Thus $A^0$ is the instantaneous Coulomb potential of the charges, and $A_k$ is entirely transverse, which properties characterize the *radiation gauge*. The property of transversality is described more generally by

$$\boldsymbol{\partial}_\mu A^\mu=\boldsymbol{\partial}A=0,$$
$$\boldsymbol{\partial}_\mu=\partial_\mu+n_\mu n^\nu\partial_\nu,$$

which also applies to the symbolic quantities $\bar{A}$, since

$$n\boldsymbol{\partial}=0.$$

Hence

$$(\boldsymbol{\partial})_{\xi_\alpha}G_\pm(x,\xi)=0,\quad \alpha=1\cdots\nu$$

distinguishes the Green's functions of the radiation gauge, $G^{(R)}$.

The two classes of Green's functions are complex conjugate,

$$G_-=G_+^*,$$

and, in the simplest case of two points, the real functions formed by addition of $G_+$ and $G_-$ become definite functions on removing the alternating sign factor,

$$G_+^{(R)}(\xi_1\xi_2)+G_-^{(R)}(\xi_1\xi_2)=\langle\{A(\xi_1),A(\xi_2)\}\rangle\geq0,$$
$$\epsilon_+(x_1x_2)[G_+^{(R)}(x_1x_2)+G_-^{(R)}(x_1x_2)]=\langle\{\psi(x_1),\psi(x_2)\}\rangle\geq0.$$

The differential equations obeyed by the Green's functions combine the field equations and the commutation relations, the latter appearing to characterize the discontinuities encountered at equal times. In applying the Maxwell field equations the following commutators are required:

$$\begin{aligned}\delta(x^0-x^{0\prime})i[F^{\mu0}(x),A_\nu(x')]&=(\delta_\nu{}^\mu+n^\mu n_\nu)\delta(x-x')+\boldsymbol{\partial}^\mu\boldsymbol{\partial}_\nu\mathfrak{D}(x-x')\\&=[\delta_\nu{}^\mu+n^\mu n_\nu-\boldsymbol{\partial}^\mu\boldsymbol{\partial}_\nu(\boldsymbol{\partial}^2)^{-1}]\delta(x-x'),\end{aligned}$$

and

$$\delta(x^0-x^{0\prime})[F^{\mu0}(x),\psi(x')]=-\boldsymbol{\partial}^\mu\mathfrak{D}(x-x')eq\psi(x'),$$

which express the operator properties of the transverse and longitudinal electric field, respectively. The latter result has been derived from, and in turn implies, the charge density commutator

$$\delta(x^0-x^{0\prime})[j^0(x),\psi(x')]=-\delta(x-x')eq\psi(x').$$

The Maxwell differential equations for the Green's functions $G_+^{(R)}$ emerge as

$$\begin{aligned}&(\partial\partial-\partial^2)_{\xi_1}G_+^{(R)}(x,\xi)\\&=(1-\boldsymbol{\partial}\partial(\boldsymbol{\partial}^2)^{-1})_{\xi_1}(1/i)\delta(\xi_1-\xi_2)G_+^{(R)}(x,\xi_3\cdots)+\cdots\\&\quad+(\boldsymbol{\partial}(\boldsymbol{\partial}^2)^{-1})_{\xi_1}\sum_{a=1}^{2n}\delta(\xi_1-x_a)eq_aG_+^{(R)}(x,\xi_2\cdots)\\&\quad+\operatorname{tr}\tfrac{1}{2}i\beta\gamma eqG_+^{(R)}(x_1\cdots x_{2n}\xi_1\xi_1,\xi_2\cdots),\end{aligned}$$

where the dots following the first right-hand term signify the $\nu-2$ similar ones containing $\delta(\xi_1-\xi_\alpha)$, $\alpha=3\cdots\nu$. These terms are produced partly by the transverse field commutation properties, and partly by the differential operator contained in $\bar{A}$, with the aid of the relation

$$(\partial\partial-\partial^2)nn(\boldsymbol{\partial}^2)^{-1}=-nn+\boldsymbol{\partial}(\boldsymbol{\partial}-\partial)(\boldsymbol{\partial}^2)^{-1}.$$

The last term of this equation is the Green's function expression for

$$\langle(\psi(x_1)\cdots\psi(x_{2n})j(\xi_1)\bar{A}(\xi_2)\cdots)_+\rangle\epsilon_+(x),$$

which is obtained on writing

$$\begin{aligned}j(x)&=-\tfrac{1}{2}ie\psi(x)\beta\gamma q\psi(x)=\operatorname{tr}\tfrac{1}{2}i\beta\gamma eq\psi(x)\psi(x)\\&=\lim_{x'\to x}\operatorname{tr}\tfrac{1}{2}i\beta\gamma eq(\psi(x)\psi(x'))_+\epsilon_+(xx').\end{aligned}$$

The limiting approach of $x'$ to $x$ can be performed symmetrically from the past and the future.

In consequence of the conservation, and of the operator properties, of the current vector we have

$$\begin{aligned}&(\partial_\mu)_{\xi_1}\operatorname{tr}\tfrac{1}{2}i\beta\gamma^\mu eqG_+^{(R)}(\cdots x_{2n}\xi_1\xi_1,\xi_2\cdots)\\&\qquad=-\sum_{a=1}^{2n}\delta(\xi_1-x_a)eq_aG_+^{(R)}(x,\xi_2\cdots),\end{aligned}$$

which enables the Maxwell differential equations for the Green's functions to be presented as

$$\begin{aligned}&(\partial\partial-\partial^2)_{\xi_1}G_+^{(R)}(x,\xi)\\&=(1-\boldsymbol{\partial}\partial(\boldsymbol{\partial}^2)^{-1})_{\xi_1}[(1/i)\delta(\xi_1-\xi_2)G_+^{(R)}(x,\xi_3\cdots)\\&\qquad+\cdots+\operatorname{tr}\tfrac{1}{2}i\beta\gamma eqG_+^{(R)}(\cdots\xi_1\xi_1,\xi_2\cdots)].\end{aligned}$$

These equations still need to be supplemented by the condition of transversality characteristic of the radiation gauge. This requirement is explicitly satisfied in the following form of the Maxwell equations,

$$\begin{aligned}&(-\partial^2)_{\xi_1}G_+^{(R)}(x,\xi)\\&=[(1-\partial\boldsymbol{\partial}(\boldsymbol{\partial}^2)^{-1})(1-\boldsymbol{\partial}\partial(\boldsymbol{\partial}^2)^{-1})]_{\xi_1}\\&\qquad\times[(1/i)\delta(\xi_1-\xi_2)G_+^{(R)}(x,\xi_3\cdots)\\&\qquad+\cdots+\operatorname{tr}\tfrac{1}{2}i\beta\gamma eqG_+^{(R)}(\cdots\xi_1\xi_1,\xi_2\cdots)].\end{aligned}$$

Complex conjugation produces the analogous differential equation for $G_-^{(R)}$,

$$\begin{aligned}&(-\partial^2)_{\xi_1}G_-^{(R)}(x,\xi)\\&=[(1-\partial\boldsymbol{\partial}(\boldsymbol{\partial}^2)^{-1})(1-\boldsymbol{\partial}\partial(\boldsymbol{\partial}^2)^{-1})]_{\xi_1}\\&\qquad\times[i\delta(\xi_1-\xi_2)G_-^{(R)}(x,\xi_3\cdots)\\&\qquad+\cdots-\operatorname{tr}\tfrac{1}{2}i\beta\gamma eqG_-^{(R)}(\cdots\xi_1\xi_1,\xi_2\cdots)].\end{aligned}$$

The Dirac equations for the Green's functions are

$$\begin{aligned}&(\beta\gamma(1/i)\partial+m\beta)_{x_1}G_\pm^{(R)}(x,\xi)-(\beta\gamma eq)_1G_\pm^{(R)}(x,x_1\xi_1\cdots)\\&\qquad=\delta(x_1-x_2)G_\pm^{(R)}(x_3\cdots,\xi)-\cdots,\end{aligned}$$

where the dots indicate the $2n-2$ similar terms containing $(-1)^a\delta(x_1-x_a)$, $a=3\cdots2n$. Involved here are the Dirac field commutation properties and the relation

$$\delta(x^0-x^{0\prime})[\psi(x),A_\mu(x')]=n_\mu n_\nu\mathfrak{D}(x-x')ieq\beta\gamma^\nu\psi(x).$$

The latter is the source of the additional terms necessary to convert

$$\langle(\tfrac{1}{2}\{A(x_1),\psi(x_1)\}\psi(x_2)\cdots\bar{A}(\xi_1)\cdots)_+\rangle\epsilon_+(x)$$

into the Green's function $G_+^{(R)}(x_1\cdots x_{2n},x_1\xi_1\cdots\xi_\nu)$.

The Green's functions are also characterized by their spectral properties. When all time coordinates are distinct, the Green's functions assume the general form[1]

$$\langle(\chi(z_1)\chi(z_2)\cdots)_\pm\rangle\epsilon_\pm,$$

where $\chi(z)$ refers either to the B.E. field $A(\xi)$ or to the F.D. field $\psi(x)$. Hence $G_+^{(R)}$ contains no negative frequencies in its dependence upon the differences of the consecutive time coordinates, while $G_-^{(R)}$ contains no positive frequencies. The connection between the spectral characteristics of the two types of Green's functions is expressed by complex conjugation, and also by the analytic continuation[1]

$$G_-^{(R)}(z)=(-1)^nG_+^{(R)}(-e^{-\pi i}z).$$

The verification of the latter involves more than the comparison of the two time-ordered forms at distinct times since the Green's functions also contain delta functions of the time differences. The differential equations take account of these terms and we observe that the operation $-e^{-\pi i}$, which reverses the sign of all delta functions, together with the factor $(-1)^n$, produces the $G_-$ differential equations from those for $G_+$.

## THE LORENTZ GAUGE

We now begin the task of subjecting the radiation-gauge Green's functions to a gauge transformation that is designed to remove the explicit dependence upon the unit time-like vector $n_\mu$ and thereby introduce the Green's functions of the Lorentz gauge. The preliminary transformation,

$$G^{(R)}(x,\xi)=\prod_{\alpha=1}^{\nu}(1-\partial\boldsymbol{\partial}(\boldsymbol{\partial}^2)^{-1})_{\xi_\alpha}G(x,\xi),$$

exhibits radiation-gauge functions in terms of new functions which, in their dependence upon each variable $\xi_\alpha$, remain arbitrary to the extent of added gradients. If the latter functions are restricted by the differential equations,

$$\begin{aligned}(-\partial^2)_{\xi_1}G_+(x,\xi)&\\ =-i\delta(\xi_1-\xi_2)&G_+(x,\xi_3\cdots)+\cdots\\ &+(1-\boldsymbol{\partial}\partial(\boldsymbol{\partial}^2)^{-1})_{\xi_1}\operatorname{tr}\tfrac{1}{2}i\beta\gamma eqG_+(\cdots\xi_1\xi_1,\xi_2\cdots),\end{aligned}$$

the radiation-gauge differential equations will be reproduced. To present the analogous Dirac field equations most conveniently, we introduce the symbol $\mathfrak{A}(\xi)$, which is defined by

$$\mathfrak{A}(\xi)G(x,\xi_1\cdots\xi_\nu)=G(x,\xi\xi_1\cdots\xi_\nu).$$

According to the resulting product

$$\mathfrak{A}(\xi)\mathfrak{A}(\xi')G(x,\xi_1\cdots\xi_\nu)=G(x,\xi\xi'\xi_1\cdots\xi_\nu),$$

and the symmetry of the Green's functions in B.E. coordinates, these symbols are commutative (we are now imitating the external source procedure). The Dirac differential equations for the functions then appear as

$$\begin{aligned}(\beta\gamma(1/i)\partial+\beta m)_{x_1}G_+(x,\xi)-(\beta\gamma eq)_1G_+(x,x_1\xi_1\cdots)&\\ +(\beta\gamma eq\partial\boldsymbol{\partial}(\boldsymbol{\partial}^2)^{-1}\mathfrak{A})_1G_+(x,\xi)&\\ =\delta(x_1-x_2)G_+(x_3\cdots,\xi)-\cdots&,\end{aligned}$$

which indicates the Dirac field gauge transformation implied by that of the Maxwell field,

$$G_+(x,\xi)=\prod_{a=1}^{2n}\exp[-ieq_a\boldsymbol{\partial}(\boldsymbol{\partial}^2)^{-1}\mathfrak{A}(x_a)]G_+^{(L)}(x,\xi).$$

The utility of the $\mathfrak{A}$ symbols is clearly shown in this result which constructs the $G$ functions by means of an infinite series of $G^{(L)}$ functions, with increasing numbers of B.E. coordinates. The new Green's functions obey the Dirac equation

$$\begin{aligned}(\beta\gamma(1/i)\partial+m\beta)_{x_1}G_+^{(L)}(x,\xi)-(\beta\gamma eq)_1G_+^{(L)}(x,x_1\xi_1\cdots)&\\ =\delta(x_1-x_2)G_+^{(L)}(x_3\cdots,\xi)-\cdots&,\end{aligned}$$

and the same form applies to the complex conjugate functions $G_-^{(L)}(x,\xi)$.

Before obtaining the Maxwell field differential equations obeyed by the $G^{(L)}$, which are the desired Lorentz-gauge functions, we must notice another aspect of the symbols $\mathfrak{A}(\xi)$. If we compare the differential equation characterizing $G_+(x,\xi_1\cdots\xi_\nu)$ with the one for $G_+(x,\xi_1\cdots\xi_\nu\xi)=\mathfrak{A}(\xi)G_+(x,\xi_1\cdots\xi_\nu)$, we recognize that

$$[(-\partial^2)_{\xi_1},\ \mathfrak{A}(\xi)]\mathfrak{A}(\xi_1)=-i\delta(\xi-\xi_1).$$

The required form of this property is

$$\left\{\exp\left[-\int(d\xi)\lambda(\xi)\mathfrak{A}(\xi)\right](-\partial^2)_{\xi_1}\times\exp\left[\int(d\xi)\lambda(\xi)\mathfrak{A}(\xi)\right]-(-\partial^2)_\xi\right\}\mathfrak{A}(\xi_1)=-i\lambda(\xi_1),$$

where we choose

$$-i\lambda(\xi)=\textstyle\sum_a eq_a\boldsymbol{\partial}(\boldsymbol{\partial}^2)^{-1}\delta(\xi-x_a).$$

Accordingly, the insertion of

$$G_+=\exp\left[\int(d\xi)\lambda(\xi)\mathfrak{A}(\xi)\right]G_+^{(L)}$$

into the Maxwell differential equations yields a similar

set for the $G_+^{(L)}$ that contains in the left-hand member the additional term $-i\lambda(\xi_1)\times G_+^{(L)}(x,\xi_2\cdots)$. But in view of the relation that follows directly from the Dirac Green's function equation (in contrast with the deduction of the identical radiation gauge result from operator properties),

$$(\partial_\mu)_{\xi_1}\operatorname{tr}\tfrac{1}{2}i\beta\gamma^\mu eqG_+^{(L)}(\cdots\xi_1\xi_1,\xi_2\cdots)=-\sum_{a=1}^{2n}\delta(\xi_1-x_a)eq_aG_+^{(L)}(x,\xi_2\cdots),$$

this additional term finds an exact counterpart already present, and

$$(-\partial^2)_{\xi_1}G_\pm^{(L)}(x,\xi)=\mp i\delta(\xi_1-\xi_2)G_\pm^{(L)}(x,\xi_3\cdots)+\cdots \pm\operatorname{tr}\tfrac{1}{2}i\beta\gamma eqG_\pm^{(L)}(x\xi_1\xi_1,\xi_2\cdots)$$

is the desired set of Maxwell differential equations for the Green's functions of the Lorentz gauge. To these equations must be added boundary conditions that will reproduce the spectral characteristics of the radiation-gauge functions. We specify the Lorentz gauge completely by requiring that the $G_+^{(L)}$ contain no negative frequencies and the $G_-^{(L)}$ no positive frequencies in their dependence upon the differences of consecutive time coordinates. The two classes of Lorentz-gauge functions are then connected by complex conjugation and by the analytic continuation

$$G_-^{(L)}(z)=(-1)^nG_+^{(L)}(-e^{-\pi i}z).$$

We have now shown[4] that the radiation-gauge Green's functions, which have a direct operator definition and corresponding physical interpretation, but involve a distinguished time-like direction, can be constructed from Lorentz-gauge Green's functions,

$$G_\pm^{(R)}(x,\xi)=\prod_{\alpha=1}^{\nu}(1-\partial\boldsymbol{\partial}(\partial^2)^{-1})_{\xi_\alpha}\times\prod_{a=1}^{2n}\exp[-ieq\boldsymbol{\partial}(\partial^2)^{-1}\mathfrak{A}]_{x_a}G_\pm^{(L)}(x,\xi),$$

where the latter functions do not depend upon the time-like vector $n_\mu$ but have no immediate physical significance. The simplest examples of this construction are

$$G_\pm^{(R)}(\xi_1\xi_2)=\int(d\xi_1')(d\xi_2')[\delta(\xi_1-\xi_1')+\partial\boldsymbol{\partial}\mathfrak{D}(\xi_1-\xi_1')]\times[\delta(\xi_2-\xi_2')+\partial\boldsymbol{\partial}\mathfrak{D}(\xi_2-\xi_2')]G_\pm^{(L)}(\xi_1'\xi_2'),$$

and

$$G_\pm^{(R)}(x_1x_2)=\exp\left\{ie\int(d\xi)[q_1\boldsymbol{\partial}\mathfrak{D}(x_1-\xi)+q_2\boldsymbol{\partial}\mathfrak{D}(x_2-\xi)]\mathfrak{A}(\xi)\right\}G_\pm^{(L)}(x_1x_2)$$
$$=G_\pm^{(L)}(x_1x_2)+ie\int(d\xi)[q_1\boldsymbol{\partial}\mathfrak{D}(x_1-\xi)+q_2\boldsymbol{\partial}\mathfrak{D}(x_2-\xi)]G_\pm^{(L)}(x_1x_2,\xi)$$
$$-\tfrac{1}{2}e^2\int(d\xi)(d\xi')[q_1\boldsymbol{\partial}\mathfrak{D}(x_1-\xi)+q_2\boldsymbol{\partial}\mathfrak{D}(x_2-\xi)]\times[q_1\boldsymbol{\partial}\mathfrak{D}(x_1-\xi')+q_2\boldsymbol{\partial}\mathfrak{D}(x_2-\xi')]G_\pm^{(L)}(x_1x_2,\xi\xi')+\cdots.$$

The same connections between radiation-gauge and Lorentz-gauge functions apply to the linear combinations $G_+(\xi_1\xi_2)+G_-(\xi_1\xi_2)$, and $\epsilon_+(x_1x_2)[G_+(x_1x_2)+G_-(x_1x_2)]$, the latter relations also containing the infinite sequence of functions

$$\epsilon_+(x_1x_2)[G_+^{(L)}(x_1x_2,\xi_1\cdots\xi_\nu)+G_-^{(L)}(x_1x_2,\xi_1\cdots\xi_\nu)],\quad \nu=1,2\cdots.$$

The combinations of radiation-gauge functions are non-negative, but there is no assurance that such properties extend to the Lorentz-gauge functions. Indeed, the general loss of the positiveness conditions that accompany physical realizability is made evident by the attempt to supply a time-ordered operator construction for the Green's functions of the Lorentz gauge.

The differential equations characterizing the $G^{(L)}$ are satisfied by the following structure:

$$G_\pm^{(L)}(x,\xi)=\langle(\psi(x_1)\cdots\psi(x_{2n})A(\xi_1)\cdots A(\xi_2))_\pm\rangle\epsilon_\pm(x)$$

where the operators $\psi(x)$ and $A_\mu(\xi)$ obey

$$[\gamma((1/i)\partial-eqA)+m]\psi=0,$$
$$-\partial^2A=-\tfrac{1}{2}ie\psi\beta\gamma q\psi,$$

and (among others)

$$\delta(\xi^0-\xi^{0\prime})i[\partial_0A_\mu(\xi),A_\nu(\xi')]=g_{\mu\nu}\delta(\xi-\xi'),$$
$$\delta(x^0-x^{0\prime})\{\psi(x),\psi(x')\}=\delta(x-x'),$$
$$\delta(\xi^0-x^0)[A(\xi),\psi(x)]=0.$$

The symbol $\langle\ \rangle$ signifies a linear mapping of operators onto numbers, including the correspondence $\langle 1\rangle=1$, which must possess the property

$$\langle\chi_1\cdots\chi_k\rangle^*=\langle\chi_k\cdots\chi_1\rangle$$

in order to reproduce the complex conjugate relationship of $G_+^{(L)}$ and $G_-^{(L)}$. The spectral requirements on the $G^{(L)}$ indicate that the space-time variation of the fields is represented by

$$\chi(x)=e^{-iPx}\chi e^{iPx},$$
$$(1/i)\partial_\mu\chi(x)=[\chi(x),P_\mu],$$

[4] This result was first obtained some years ago by K. Johnson and the author (unpublished), using the method of external currents.

where the operator $P^0$ has a non-negative eigenvalue spectrum, and the more specific interpretation is attached to $\langle$ and $\rangle$ of the left and right eigenvector of $P^0$ associated with the eigenvalue zero. The field operators are self-adjoint with respect to the operation that interchanges the left and right eigenvectors. We now remark on the following consequence of the commutation relations,

$$\delta(\xi^0-\xi^{0\prime})\langle A_\mu(\xi)P^0A_\nu(\xi')+A_\nu(\xi')P^0A_\mu(\xi)\rangle=g_{\mu\nu}\delta(\xi-\xi'),$$

which is to be compared with the implications of the hypothesis that the adjoint operation is Hermitian conjugation. The evident contradiction between the non-negative nature of the left-hand side and the value $g_{00}=-1$ demands a more general interpretation of the adjoint, corresponding to the introduction of an indefinite metric in the vector space. We shall not continue this approach, with its inevitable requirement that the consistency of the various assumptions concerning operator properties be established for the physical situation of interacting fields. It is our view that the physical operator basis used in the definition of the radiation gauge Green's functions is entirely adequate, the introduction of the Lorentz-gauge functions being an application of the freedom of gauge transformations, and not an occasion for a somewhat dubious reconstruction of the mathematical foundation of the theory.

## EUCLIDEAN GREEN'S FUNCTIONS

The Lorentz-gauge Green's functions referring to $2n+\nu=p$ space-time points involve $p-1$ linearly independent coordinate differences, and these appear in $p!$ distinct functional forms corresponding to the various time orderings. Each continuous function associated with a particular time order, $t^{(1)}>\cdots>t^{(p)}$, is formed from harmonic functions of the time differences, $t^{(\alpha)}-t^{(\alpha+1)}$, which contain only non-negative frequencies ($G_+$), or only nonpositive frequencies ($G_-$). These functions are also defined outside the special time domain where they reproduce the Green's function, and can be identified as boundary values, on the real axis, of complex variable functions which are regular in various half-planes. But the Green's function is more than the union of its several parts. In particular, it possesses possibilities of analytic extension that are not available to the functions[5] associated with a particular time order. Thus, for distinct space-time points, but with no restriction on the $p$ time variables, the Green's function $G_+(t_1\cdots t_p)$ emerges as the boundary value on the positive real axis of a function $G_+(\zeta t_1\cdots\zeta t_p)$ which is regular in the lower-half $\zeta$ plane (it is sufficient to let $\zeta\to+1$). If $\zeta$ approaches the limit $-1$ from the lower half-plane we obtain the function $(-1)^nG_-(-t_1\cdots-t_p)$. Similarly there exists a function of $\zeta$, regular in the upper half-plane, that yields $G_-(t_1\cdots t_p)$ as $\zeta\to1$ and $(-1)^nG_+(-t_1\cdots-t_p)$ as $\zeta\to-1$. These analyticity properties imply, in particular, that the values obtained when $\zeta$ occupies the appropriate imaginary axis ($\zeta=\pm i$ suffices) completely determine the Green's function, and conversely.

A very simple illustration of these remarks may be helpful. The functions

$$g_\pm(t_1t_2)=\frac{1}{2\omega}e^{\mp i\omega|t_1-t_2|},\quad \omega>0,$$

obey

$$(\partial^2/\partial t_1^2+\omega^2)g_\pm(t_1t_2)=\mp i\delta(t_1-t_2),$$

and have the frequency characteristics appropriate to their designation. As a function of the single time difference $t=t_1-t_2$, $2\omega g_+$ for $t>0$, and $2\omega g_-$ for $t<0$, coincide with the function $e^{-i\omega t}$, which is defined for all real $t$ and is the value on the real axis of a function that is regular and bounded in the lower-half $t$ plane. The quantities $2\omega g_+(t<0)$ and $2\omega g_-(t>0)$ are similarly related to the function $e^{i\omega t}$ which possesses a bounded analytic extension into the upper half plane. The functions $g_\pm$ are represented for all $t$ by

$$g_\pm(t_1t_2)=\mp\frac{i}{2\pi}\int_{-\infty}^{\infty}d\nu\frac{e^{-i\nu(t_1-t_2)}}{\omega^2-\nu^2},$$

where the integration contour for $g_+$ passes below $-\omega$ and above $+\omega$; it is to be reflected in the real axis for $g_-$. An analytic extension of $g_+$ is now obtained by making the substitutions $t\to\zeta t$, $\zeta=\rho e^{i\vartheta}$, $-\pi<\vartheta<0$, together with $\nu\to\zeta^{-1}\nu$, which gives the contour a positive rotation. The resulting integral extended along the real axis,

$$g_+(\zeta t_1\zeta t_2)=\frac{1}{2\pi i}\int_{-\infty}^{\infty}d\nu\frac{\zeta}{\omega^2\zeta^2-\nu^2}e^{-i\nu(t_1-t_2)},$$

defines for all $t$ an analytic function of $\zeta[=(2\omega)^{-1}e^{-i\zeta\omega|t_1-t_2|}]$ which is regular and bounded in the lower half plane and reproduces $g_+$ as $\zeta\to+1$, or $g_-$ as $\zeta\to-1$. Similarly, the substitutions $t\to\zeta t$, $\nu\to\zeta^{-1}\nu$, $\zeta=\rho e^{i\theta}$, $\pi>\theta>0$, performed in the integral representation of $g_-$ yields

$$g_-(\zeta t_1\zeta t_2)=-\frac{1}{2\pi i}\int_{-\infty}^{\infty}d\nu\frac{\zeta}{\omega^2\zeta^2-\nu^2}e^{-i\nu(t_1-t_2)},$$

and this is a regular, bounded function of $\zeta$, in the upper half-plane, for all $t[=(2\omega)^{-1}e^{i\zeta\omega|t_1-t_2|}]$. As $\zeta$ approaches $+1$ or $-1$, we obtain $g_-$ or $g_+$. The evident relation between these analytic extensions of $g_+$ and $g_-$ is such that

$$g_+(-it_1-it_2)=g_-(it_1it_2)=\frac{1}{2\pi}\int_{-\infty}^{\infty}d\nu\frac{e^{-i\nu(t_1-t_2)}}{\nu^2+\omega^2}$$

$$=\frac{1}{2\omega}e^{-\omega|t_1-t_2|}.$$

[5] When the Green's function is directly connected with a vacuum expectation value of time-ordered field operator products, these functions are the unordered product expectation values that have been discussed particularly by A. Wightman.

which is a bounded quantity for all $t$. In contrast, the real functions that are analogously produced from $e^{\pm i\omega t}$ are not bounded for all values of $t$. It should also be noted that

$$(-\partial^2/\partial t_1^2+\omega^2)g_\pm(\mp it_1\mp it_2)=\delta(t_1-t_2),$$

from which we infer the formal transformation

$$\mp i\delta(\mp it)=\delta(t).$$

To facilitate the conversion of the Green's function differential equations to the Euclidean metric, some remarks about the structure of Dirac matrices are needed. The matrices $\gamma^\mu$ have simple algebraic properties, as determined by the metric tensor $g^{\mu\nu}$, while it is the matrices $\beta\gamma^\mu$ that possess the property of symmetry. The matrix $\beta$ appears here as representative of the indefinite Lorentz (or Minkowski) metric. Accordingly, the replacement of the latter by the Euclidean metric will introduce matrices $\alpha_\mu$ that are symmetrical and have the simple algebraic property

$$\{\alpha_\mu,\alpha_\nu\}=2\delta_{\mu\nu}.$$

We shall inquire generally about the possibility of constructing such matrices. It is well known that $2n$ anticommuting matrices of unit square generate an algebra with the dimensionality

$$1+2n+2n(2n-1)/2+\cdots+2n+1=2^{2n},$$

corresponding to the enumeration of the operator basis formed by the independent products. The last element of this collection is

$$\alpha_{2n+1}=i^{-n}\alpha_1\cdots\alpha_{2n},$$

which extends by one the set of anticommuting matrices. We now want to recognize that when these $2n+1$ matrices are irreducible and possess a particular symmetry, $n+1$ of them are symmetrical, and the remaining $n$ are antisymmetrical.

An elementary proof employs the construction of the algebra as the product of $n$ independent algebras of dimensionality $2^2$, represented by the $2\times2$ Pauli matrices, $1$, $\sigma_1$, $\sigma_2$, $\sigma_3$. The ascent from $2n+1$ anticommuting matrices $\alpha_\kappa$ to $2n+3$ such matrices is produced, for example, by

$$\sigma_1,\ \sigma_2,\ \sigma_3\alpha_\kappa,\quad \kappa=1\cdots2n+1.$$

If $\sigma_3$ is symmetrical the set of $2n+1$ matrices has the same symmetry distribution as the $\alpha_\kappa$, $\kappa=1\cdots2n+1$. The two additional matrices constitute one symmetrical and one antisymmetrical matrix. Hence the number of each type grows by one when $n$ is increased by unity. The stated result now follows from the remark that we can begin with unity, for $n=0$, and it is symmetrical. The particular construction method employed does not influence the dimensionality of the two symmetry categories.

We first consider $n=2$, which produces the familiar set of five anticommuting $4\times4$ matrices. Since there are only three symmetrical matrices, this set is adapted to the $3+1$ Lorentz space but cannot be applied to the four-dimensional Euclidean space. To have four symmetrical anticommuting matrices, we must choose $n\geq3$, corresponding to $8\times8$ matrices, at least. Thus the requirement of a Euclidean formulation excludes the simplest field in space-time, the four-component Hermitian spin-$\frac{1}{2}$ field (Majorana). In this context, a trivial observation may be worth repeating—a four-component Hermitian field is fully equivalent to a two-component non-Hermitian field.

For $n=3$, the three symmetrical matrices $i\gamma^1$, $i\gamma^2$, $i\gamma^3$ and the antisymmetrical matrix $\gamma^0$ are supplemented by $i\gamma_5l_3$, $i\gamma_5l_1$, $i\gamma_5l_2$, of which the first is symmetrical. The matrices $l_1l_2l_3$ are $2\times2$ Pauli matrices, with $l_3$ antisymmetrical, and we have defined

$$\gamma_5=\gamma^0\gamma^1\gamma^2\gamma^3.$$

There are a variety of unitary transformations that will effectively change the order of these seven anticommuting matrices and supply four symmetrical real matrices followed by three antisymmetrical imaginary matrices. Thus, the unitary transformation by the matrix $\exp[\frac{1}{2}\pi i\gamma^0\gamma_5l_3]$ produces the sequence: $i\gamma^1$, $i\gamma^2$, $i\gamma^3$, $i\gamma_5l_3$; $-\gamma^0$, $i\gamma_5l_1$, $i\gamma_5l_2$. Subsequent unitary-orthogonal transformations will not alter this symmetry partitioning. The transformation matrix $\exp[\frac{1}{2}\pi i\gamma^0]$, for example, supplies the list: $\gamma^0\gamma^1$, $\gamma^0\gamma^2$, $\gamma^0\gamma^3$, $\gamma^0\gamma_5l_3$; $-\gamma^0$, $\gamma^0\gamma_5l_1$, $\gamma^0\gamma_5l_2$. With $n>3$, we gain the possibility of describing additional internal symmetry properties, which appear rather differently in Lorentz or Euclidean form, but we shall not discuss it here.

The analytic extension of the time variables to the appropriate imaginary axis introduces the Euclidean metric,

$$G_\pm:\ x^0\leftrightarrow\mp ix_4.$$

The Green's functions also undergo additional matrix transformations. For the vector indices related to the Maxwell field the transformation is just that formally associated with the redefinition of the coordinates, and we shall not indicate it explicitly. With this understanding, a suitable correspondence is

$$\prod_{a=1}^{2n}(\exp(\pm\tfrac{1}{4}\pi i\gamma^0\gamma_5l_3\mp\tfrac{1}{4}\pi i))_aG_\pm{}^{(L)}(x)\leftrightarrow G^{(E)}(x),$$

where, as the right side suggests, the same real Euclidean Green's function is obtained from the two complex Lorentz Green's functions. The effect of this substitution on the various Dirac matrices is indicated by

$$\begin{aligned}\exp(\mp\tfrac{1}{4}\pi i\gamma^0\gamma_5l_3)(1/i)\beta(\gamma^k,\pm i\gamma^0,1)\exp(\mp\tfrac{1}{4}\pi i\gamma^0\gamma_5l_3)\\ =\gamma^0\exp(\pm\tfrac{1}{4}\pi i\gamma^0\gamma_5l_3)(\gamma^k,\pm i\gamma^0,1)\exp(\mp\tfrac{1}{4}\pi i\gamma^0\gamma_5l_3)\\ =\gamma^0\gamma^k,\gamma^0\gamma_5l_3,\gamma^0,\end{aligned}$$

and the resulting set can be labelled consistently as $\alpha_\mu$,

$\mu=1\cdots 5$. The real differential equations thus obtained are

$$(\alpha\partial+mi\alpha_5)_{x1}G^{(E)}(x,\xi)-e(\alpha iq)_1G^{(E)}(x,x_1\xi)$$
$$=\delta(x_1-x_2)G^{(E)}(x_3\cdots,\xi)-\cdots,$$

and

$$(-\partial^2)_{\xi 1}G^{(E)}(x,\xi)+e\,\mathrm{tr}\tfrac{1}{2}\alpha iqG^{(E)}(x\xi_1\xi_1,\xi_2\cdots)$$
$$=\delta(\xi_1-\xi_2)G^{(E)}(x,\xi_3\cdots)+\cdots.$$

The accompanying regularity conditions demand the boundedness of each $G^{(E)}(x_1\cdots x_{2n},\xi_1\cdots\xi_\nu)$ when neighborhoods of coincidence of the $2n+\nu$ points are excluded.

The Euclidean Green's functions are invariant under a variety of transformations, in the sense of

$$\prod_{\alpha=1}^{p}(R)_\alpha G(r^{-1}z)=G(z).$$

Four-dimensional rotational invariance is described in the evident manner with the Dirac field spin matrices given by $\frac{1}{2}\sigma_{\mu\nu}$, where each

$$\sigma_{\mu\nu}=(1/2i)[\alpha_\mu,\alpha_\nu],\quad \mu,\,\nu=1\cdots 4$$

is antisymmetrical and imaginary. If the matrix $q$ is invariant, the reflection of any coordinate axis implies the corresponding vector transformation for the Maxwell field, while the Dirac field reflection matrices can be chosen as

$$R_\mu=i\alpha_\mu\alpha_6,\quad \mu=1\cdots 4$$

which are real, symmetrical, anticommuting matrices. We also observe that the geometrical connection between reflection and rotation is correctly described,

$$R_\mu R_\nu=e^{\frac{1}{2}\pi i\sigma_{\mu\nu}}.$$

Invariance under the coordinate-independent transformation of the Dirac field that is generated by the imaginary antisymmetrical rotation matrix

$$(1/i)\alpha_6\alpha_7=l(=l_3),$$

implies the conservation of the fermionic charge represented by $l$. There is a similar transformation associated with the electric charge matrix $q$. The two charges may be identical, but we need not insist on this. Indeed, the coordinate reflection transformation that we have described is an invariance operation only when $l$ and $q$ are independent. If these charges are the same (or are coupled together) the reflection transformation must be accompanied by an additional sign reversal of the Maxwell field, for each $R_\mu$ induces a reflection of the fermionic charge. This is a combined coordinate and charge reflection transformation.

The matrix $\alpha_6$, or $\alpha_7$, induces a coordinate-independent invariance transformation that also implies fermionic charge reflection. Hence one could combine $R_\mu$ with either $\alpha_6$ or $\alpha_7$, yielding $i\alpha_\mu$ or $l\alpha_\mu$, which describe coordinate reflections without charge reflection. It is interesting that the matrices now under discussion are imaginary, and yet the reality of the Green's functions is not disturbed since a transformation involves an even number of such Dirac matrix factors. Let us also note here that the algebraic sign of $m$ is without physical significance since the Green's function transformation described by the imaginary Dirac matrix $\alpha_5$, or $-il\alpha_5=\alpha_1\alpha_2\alpha_3\alpha_4$, has no other effect than to reverse this sign.

The full equivalence of all directions in the Euclidean space makes it unnatural to relate the Green's functions to operators ordered by coordinate projection on some line, and one might seek to introduce an invariant ordering parameter (proper time). We shall follow a different course, however, which also has its counterpart in space-time where it is the formulation to which the source techniques lead. To avoid emulation of the latter procedure, we set down directly the following operator construction of the Euclidean Green's functions

$$G(x,\xi)=\langle 0|\psi(x_1)\cdots\psi(x_{2n})A(\xi_1)\cdots A(\xi_\nu)|W\rangle/\langle 0|W\rangle,$$

where

$$[A_\mu(\xi),A_\nu(\xi')]=[A_\mu(\xi),\psi(x)]=\{\psi(x),\psi(x')\}=0,$$

throughout the four-dimensional space. The symmetry properties of the Green's functions are thereby reproduced. In addition to these commutative or anticommutative fields, there is a complementary set of fields, $B_\mu(\xi)$, $\phi(x)$, which are also everywhere commutative or anticommutative, and obey

$$i[B_\mu(\xi),A_\nu(\xi')]=\delta_{\mu\nu}\delta(\xi-\xi'),$$
$$\{\phi(x),\psi(x')\}=\delta(x-x').$$

All other commutators vanish. The state $\langle 0|$ is characterized by zero eigenvalues of the second operator set,

$$\langle 0|B(\xi)=0,\quad \langle 0|\phi(\xi)=0,$$

while $|W\rangle$ is described by

$$[(\alpha\partial+mi\alpha_5)\psi(x)-eiq\alpha A(x)\psi(x)+\phi(x)]|W\rangle=0,$$

and

$$[-\partial^2A(\xi)-\tfrac{1}{2}e\psi(\xi)iq\alpha\psi(\xi)+iB(\xi)]|W\rangle=0.$$

One verifies immediately that the differential equations are reproduced by these definitions. In virtue of the complementary field commutation properties, the operator definitions of the vector $|W\rangle$ are satisfied by the construction

$$|W\rangle=e^{-W}|0\rangle,$$

where

$$W=\tfrac{1}{2}\int(dx)[(\partial_\mu A_\nu)^2+\psi(\alpha_\mu\partial_\mu+mi\alpha_5)\psi-eA_\mu\psi iq\alpha_\mu\psi],$$

and $|0\rangle$ is the right eigenvector of $B$ and $\phi$ associated with zero eigenvalues. If these states are to exist it is necessary that the $B$ and $A$ operators be Hermitian. That property, together with the choice of $\psi$ and $\phi$ as mutually Hermitian conjugate operators,

$$\phi(x)=\psi(x)^{\dagger},$$

assures the reality of the Green's functions. We verify the latter statement by remarking that

$$G(x,\xi)=\frac{\langle\phi'=B'=0|\psi(x_1)\cdots\psi(x_{2n})A(\xi_1)\cdots A(\xi_\nu)e^{-W[\psi,A]}|0\rangle}{\langle 0|e^{-W[\psi,A]}|0\rangle},$$

implies

$$G(x,\xi)^*=\frac{\langle\psi'=B'=0|i\phi(x_1)\cdots i\phi(x_{2n})A(\xi_1)\cdots A(\xi_\nu)e^{-W[i\phi,A]}|0\rangle}{\langle 0|e^{-W[i\phi,A]}|0\rangle}$$

where the factor of $i$ associated with each F.D. field refers to the sign reversal induced in every pair of such field products by the adjoint operation. Since the conversion of $i\phi$ to $\psi$ and the interchange of the null-eigenvalue states is a canonical transformation, the equivalence of the two forms, and the reality of the Green's functions, follows.

The latter discussion shows that while the Euclidean action operator $W$ is not Hermitian, it is connected with its Hermitian adjoint by a unitary transformation. In contrast with this expression of reality, the Green's function invariance properties to which we have referred are associated with unitary transformations that leave the action operator invariant. An example that has not been mentioned is provided by the unitary operator

$$U=\exp(i\epsilon_\mu P_\mu),$$

where $P_\mu$, the Euclidean total linear momentum vector, is the Hermitian operator

$$P_\mu=\int(dx)\left[-B_\nu\partial_\mu A_\nu+\phi\frac{1}{i}\partial_\mu\psi\right]$$

$$=\int(dx)\left[A_\nu\partial_\mu B_\nu+\psi\frac{1}{i}\partial_\mu\phi\right].$$

Evidently,

$$[A,P_\mu]=(1/i)\partial_\mu A,\quad [\psi,P_\mu]=(1/i)\partial_\mu\psi,$$

and generally

$$U^{-1}\chi(z)U=\chi(z+\epsilon).$$

The action operator is invariant under this transformation, and since

$$\langle 0|P_\mu=0,\quad P_\mu|0\rangle=0,$$

we verify that the Green's functions are translationally invariant. We give this property another form on replacing the coordinate variables with the complementary momentum variables. This is expressed by the transformation

$$G(x_1\cdots x_{2n},\xi_1\cdots\xi_\nu)=\int\frac{(dp_1)}{(2\pi)^2}\cdots\frac{(dk_\nu)}{(2\pi)^2}\exp[i(p_1x_1+\cdots+k_\nu\xi_\nu)]\times G(p_1\cdots p_{2n},k_1\cdots k_\nu).$$

Under the related field transformation, the action operator acquires the form

$$W[\psi,A]=\tfrac{1}{2}\int(dk)A(-k)k^2A(k)$$
$$+\tfrac{1}{2}i\int(dp)\psi(-p)(\alpha p+m\alpha_5)\psi(p)$$
$$-\tfrac{1}{2}ie\int(dp)(dp')(dk)\times\frac{1}{(2\pi)^2}\delta(p+p'+k)A(k)\psi(p)q\alpha\psi(p').$$

The translational invariance of $G(x,\xi)$ requires that $G(p,k)$ contain the factor $\delta(p_1+\cdots p_{2n}+k_1+\cdots+k_\nu)$, and this follows directly from the invariance of the operator expression for $G(p,k)$ under the transformation

$$U^{-1}A(k)U=e^{i\epsilon k}A(k),$$
$$U^{-1}\psi(p)U=e^{i\epsilon p}\psi(p).$$

A compact expression of the totality of Green's functions is obtained by defining the generating function(al)

$$G[\phi'B']=1+\sum_{n\nu}\int\frac{(dx_1)\cdots(dx_{2n})}{(2n)!}\frac{(d\xi_1)\cdots(d\xi_\nu)}{\nu!}\times\frac{1}{i}B(\xi_1)'\cdots\frac{1}{i}B(\xi_\nu)'\phi(x_{2n})'\cdots\phi(x_1)'\times G(x_1\cdots x_{2n},\xi_1\cdots\xi_\nu).$$

Here the $B_\mu(\xi)'$ are a continuous set of arbitrary real numbers, while the $\phi(x)'$ are completely anticommuting symbols formed from an algebra external to that of the F.D. fields and thus are commutative with the latter. We shall also use the notation $\phi(x)'$ for reference to symbols that are anticommutative with the operator fields $\psi(x)$, $\phi(x)$. The connection between the two sets is produced by the operator that generates the canonical transformation of sign reversal of all F.D. fields, under which the action operator is invariant, namely

$$\rho=\exp\left[\pi i\int(dx)\phi(x)\psi(x)\right]=\exp\left[-\pi i\int(dx)\psi(x)\phi(x)\right].$$

The equivalence of these two forms depends explicitly on the even number of components possessed by the

F.D. field. (The integrals are given a meaning by an equivalent summation over arbitrarily small four-dimensional cells.) When acting upon the zero-eigenvalue states of the operators $\phi$, or $\psi$, the field reflection operator $\rho$ exhibits the eigenvalue $+1$,

$$\langle 0|\rho=\langle 0|, \quad \rho|0\rangle=|0\rangle.$$

Accordingly, we write

$$\phi(x)'\langle 0| = \langle 0|\phi(x)',$$

and

$$|0\rangle\phi(x)'=\phi(x)'|0\rangle,$$

where $\phi(x)'$ in the position of an operator contains $\rho$, and thus is anticommutative with all $\psi$ and $\phi$ operators. The summation defining the generating function, or Green's functional, can now be performed and we get

$$G[\phi'B']=\left\langle 0\left|\exp\left[-i\int(d\xi)B_\mu(\xi)'A_\mu(\xi)\right]\right.\right.$$
$$\left.\left.\times\exp\left[\int(dx)\phi(x)'\psi(x)\right]\right|W\right\rangle\Big/\langle 0|W\rangle,$$

or

$$G[\phi'B']=\langle\phi'B'|W\rangle/\langle 0|W\rangle,$$

in which we have recognized that the exponentials are operators of the special canonical group,[6] which translate eigenvalues of canonical variables. Indeed,

$$\langle\phi'B'|B_\mu(\xi)=\langle\phi'0|e^{-iB'A}B_\mu(\xi)$$
$$=\langle\phi'B'|B_\mu(\xi)',$$

and

$$\langle\phi'B'|\phi(x)=\langle 0B'|e^{\phi'\psi}\phi(x)$$
$$=\langle\phi'B'|\phi(x)'.$$

Thus the generating function is exhibited as the wave function representing the state $|W\rangle$ in the $\phi'B'$ description, apart from a constant that normalizes $G[00]$ to unity.

The operators $A$ and $\psi$ can be given differential operator realizations in the $\phi'B'$ description. Infinitesimal eigenvalue changes induce the variation

$$\delta\langle\phi'B'|=\langle\phi'B'|\left[-i\int(d\xi)\delta B_\mu(\xi)'A_\mu(\xi)\right.$$
$$\left.+\int(dx)\delta\phi(x)'\psi(x)\right],$$

which we express as

$$\langle\phi'B'|A_\mu(\xi)=i\frac{\delta}{\delta B_\mu(\xi)'}\langle\phi'B'|$$

and

$$\langle\phi'B'|\psi(x)=\frac{\delta_l}{\delta\phi(x)'}\langle\phi'B'|.$$

The latter form also involves the conversion of the eigenvalue variations into elements of the external algebra,

$$\langle\phi'B'|\delta\phi(x)'=\delta\phi(x)'\langle\phi'B'|.$$

The abstract operator equations defining the state $|W\rangle$ acquire thereby a differential operator realization that characterizes the Green's functional,

$$\left[(\alpha\partial+mi\alpha_5)\frac{\delta_l}{\delta\phi(x)'}+eq\alpha\frac{\delta}{\delta B(x)'}\frac{\delta_l}{\delta\phi(x)'}+\phi(x)'\right]$$
$$\times G[\phi'B']=0,$$

$$\left[-\partial^2\frac{\delta}{\delta B(\xi)'}-\tfrac{1}{2}e\frac{\delta_l}{\delta\phi(\xi)'}q\alpha\frac{\delta_l}{\delta\phi(\xi)'}+B(\xi)'\right]G[\phi'B']=0,$$

and from which the original Green's function equations are recovered with the aid of the correspondence

$$G(x_1\cdots x_{2n},\xi_1\cdots\xi_\nu)$$
$$=\frac{\delta_l}{\delta\phi(x_1)'}\cdots\frac{\delta_l}{\delta\phi(x_{2n})'}$$
$$\times i\frac{\delta}{\delta B(\xi_1)'}\cdots i\frac{\delta}{\delta B(\xi_\nu)'}G[\phi'B']\Big|_{\phi'=B'=0}.$$

A formal solution of the functional differential equations appears on applying the differential operator realization to the explicit operator construction of the state $|W\rangle$,

$$\langle 0|W\rangle G[\phi'B']=e^{-W[\delta_l/\delta\phi',\,i\delta/\delta B']}\delta[\phi']\delta[B'].$$

We have used the notation

$$\delta[\phi']\delta[B']=\langle\phi'B'|0(\phi'B')\rangle$$
$$=\langle 0|e^{-iB'A+\phi'\psi}|0\rangle,$$

which is in conformity with the fundamental property

$$B_\mu(\xi)'\delta[\phi']\delta[B']=\langle\phi'B'|B_\mu(\xi)|0\rangle=0,$$
$$\phi(x)'\delta[\phi']\delta[B']=\langle\phi'B'|\phi(x)|0\rangle=0.$$

An alternative expression of the relation between $G$ and $W$ is obtained by writing

$$\langle 0|W\rangle\langle\phi'B'|G[\phi,B]|0(\psi'A')\rangle=\langle\phi'B'|e^{-W[\psi,A]}|0(\phi'B')\rangle,$$

which utilizes the unit value ascribed to the transformation function

$$\langle\phi'B'|\psi'A'\rangle=e^{\phi'\psi'-iB'A'},$$

when either eigenvalue set is placed equal to zero. As a vector equation,

$$\langle 0|W\rangle G[\phi,B]|0(\psi'A')\rangle=e^{-W[\psi,A]}|0(\phi'B')\rangle$$

can also be displayed in the $\psi'A'$ representation, where

[6] A fuller discussion of this group appears in an article being prepared for publication in the *Encyclopedia of Physics* [Springer-Verlag (to be published)] Vol. 5, Part 2.

it takes the form

$$\langle 0|W\rangle G[\delta_l/\delta\psi',(1/i)\delta/\delta A']\delta[\psi']\delta[A']=e^{-W[\psi'A']},$$

with the aid of the following consequences of the canonical transformation $B\rightarrow A$, $A\rightarrow -B$; $\phi\leftrightarrow\psi$:

$$\langle\psi'A'|\phi'B'\rangle=e^{-\phi'\psi'+iB'A'}$$
$$\langle\psi'A'|\psi''A''\rangle=\delta[\psi'-\psi'']\delta[A'-A''].$$

It should also be mentioned that an integration concept can be devised for both types of field variable, such that

$$\delta[\phi']\delta[B']=\langle 0|e^{\phi'\psi-iB'A}|0\rangle$$
$$=\int d[\psi']d[A']e^{\phi'\psi'-iB'A'},$$

and

$$\int d[\phi']d[B']\delta[\phi']\delta[B']=1.$$

The generating function, for example, thereby acquires the integral representation

$$G[\phi'B']=\frac{\int d[\psi']d[A']e^{-W[\psi'A']+\phi'\psi'-iB'A'}}{\int d[\psi']d[A']e^{-W[\psi'A']}}.$$

The discussion of Euclidean Green's functions will be continued in another publication.

*Note added in proof.*—It has not been sufficiently emphasized in this paper that the term "Lorentz gauge," as descriptive of a gauge in which there is no distinguished time-like vector, refers to a class rather than just the special gauge used in the paper. One can also introduce, for example, the transverse Lorentz gauge (here "transverse" has a four-dimensional space-time significance), characterized by

$$(\partial)_{\xi_\alpha}G_{\pm}{}^{(L)}(x,\xi)=0,\quad \alpha=1\ .\ .\ .\ \nu$$

together with appropriately modified Maxwell differential equations. The radiation gauge functions constructed from Lorentz gauge Green's functions are clearly independent of the specific Lorentz gauge employed. This subject will be discussed further in a later paper.

# *THE ALGEBRA OF MICROSCOPIC MEASUREMENT*

BY JULIAN SCHWINGER

HARVARD UNIVERSITY

*Communicated August 28, 1959*

This note initiates a brief account of the fundamental mathematical structure of quantum mechanics, not as an independent mathematical discipline with physical applications, but evolved naturally as the symbolic expression of the physical laws that govern the microscopic realm.[1]

The classical theory of measurement is implicitly based upon the concept of an interaction between the system of interest and the measurement apparatus that can be made arbitrarily small, or at least precisely compensated, so that one can speak meaningfully of an idealized measurement that disturbs no property of the system. The classical representation of physical quantities by numbers is the identification of all properties with the results of such nondisturbing measurements. It is characteristic of atomic phenomena, however, that the interaction between system and instrument cannot be indefinitely weakened. Nor can the disturbance produced by the interaction be compensated since it is only statistically predictable. Accordingly, a measurement of one property can produce uncontrollable changes in the value previously assigned to another property, and it is without meaning to ascribe numerical values to all the attributes of a microscopic system. The mathematical language that is appropriate to the atomic domain is found in the symbolic transcription of the laws of microscopic measurement.

The basic concepts are developed most simply in the context of idealized physical systems which are such that any physical quantity $A$ assumes only a finite number of distinct values, $a', \ldots a^n$. In the most elementary type of measurement, an

Reprinted from the *Proceedings of the National Academy of Sciences U.S.A.* **45**, 1542–1553 (1959).

ensemble of independent similar systems is sorted by the apparatus into subensembles, distinguished by definite values of the physical quantity being measured. Let $M(a')$ symbolize the selective measurement that accepts systems possessing the value $a'$ of property $A$ and rejects all others. We define the addition of such symbols to signify less specific selective measurements that produce a subensemble associated with any of the values in the summation, none of these being distinguished by the measurement. The multiplication of the measurement symbols represents the successive performance of measurements (read from right to left). It follows from the physical meaning of these operations that addition is commutative and associative, while multiplication is associative. With 1 and 0 symbolizing the measurements that, respectively, accept and reject all systems, the properties of the elementary selective measurements are expressed by

$$M(a')M(a') = M(a') \tag{1}$$

$$M(a')M(a'') = 0, \qquad a' \neq a'' \tag{2}$$

$$\sum_{a'} M(a') = 1. \tag{3}$$

Indeed, the measurement symbolized by $M(a')$ accepts every system produced by $M(a')$ and rejects every system produced by $M(a'')$, $a'' \neq a'$, while a selective measurement that does not distinguish any of the possible values of $a'$ is the measurement that accepts all systems.

According to the significance of the measurements denoted as 1 and 0, these symbols have the algebraic properties

$$\begin{gathered} 1\,1 = 1, \qquad 0\,0 = 0 \\ 1\,0 = 0\,1 = 0 \\ 1 + 0 = 1, \end{gathered}$$

and

$$\begin{gathered} 1M(a') = M(a')1 = M(a'), \; 0M(a') = M(a')0 = 0 \\ M(a') + 0 = M(a'), \end{gathered}$$

which justifies the notation. The various properties of 0, $M(a')$, and 1 are consistent, provided multiplication is distributive. Thus,

$$\sum_{a''} M(a')M(a'') = M(a') = M(a')1 = M(a')\sum_{a''} M(a'').$$

The introduction of the numbers 1 and 0 as multipliers, with evident definitions, permits the multiplication laws of measurement symbols to be combined in the single statement

$$M(a')M(a'') = \delta(a',a'')M(a'),$$

where

$$\delta(a',a'') = \begin{cases} 1, & a' = a'' \\ 0, & a' \neq a''. \end{cases}$$

Two physical quantities $A_1$ and $A_2$ are said to be compatible when the measurement of one does not destroy the knowledge gained by prior measurement of the other. The selective measurements $M(a_1')$ and $M(a_2')$, performed in either order, produce an ensemble of systems for which one can simultaneously assign the values $a_1'$ to $A_1$ and $a_2'$ to $A_2$. The symbol of this compound measurement is

$$M(a_1'a_2') = M(a_1')M(a_2') = M(a_2')M(a_1').$$

By a complete set of compatible physical quantities, $A_1, \ldots A_k$, we mean that every pair of these quantities is compatible and that no other quantities exist, apart from functions of the set $A$, that are compatible with every member of this set. The measurement symbol

$$M(a') = \prod_{i=1}^{k} M(a_i')$$

then describes a complete measurement, which is such that the systems chosen possess definite values for the maximum number of attributes; any attempt to determine the value of still another independent physical quantity will produce uncontrollable changes in one or more of the previously assigned values. Thus the optimum state of knowledge concerning a given system is realized by subjecting it to a complete selective measurement. The systems admitted by the complete measurement $M(a')$ are said to be in the state $a'$. The symbolic properties of complete measurements are also given by equations (1), (2), and (3).

A more general type of measurement incorporates a disturbance that produces a change of state. (It is here that we go beyond previous developments along these lines.) The symbol $M(a',a'')$ indicates a selective measurement in which systems are accepted only in the state $a''$ and emerge in the state $a'$. The measurement process $M(a')$ is the special case for which no change of state occurs,

$$M(a') = M(a',a').$$

The properties of successive measurements of the type $M(a',a'')$ are symbolized by

$$M(a',a'')M(a''',a^{\text{iv}}) = \delta(a'',a''')M(a',a^{\text{iv}}), \tag{4}$$

for, if $a'' \neq a'''$, the second stage of the compound apparatus accepts none of the systems that emerge from the first stage, while if $a'' = a'''$, all such systems enter the second stage and the compound measurement serves to select systems in the state $a^{\text{iv}}$ and produce them in the state $a'$. Note that if the two stages are reversed, we have

$$M(a''',a^{\text{iv}})M(a',a'') = \delta(a',a^{\text{iv}})M(a''',a''),$$

which differs in general from equation (4). Hence the multiplication of measurement symbols is noncommutative.

The physical quantities contained in one complete set $A$ do not comprise the totality of physical attributes of the system. One can form other complete sets, $B, C, \ldots$, which are mutually incompatible, and for each choice of noninterfering physical characteristics there is a set of selective measurements referring to systems in the appropriate states, $M(b',b''), M(c',c'') \ldots$. The most general selective measurement involves two incompatible sets of properties. We symbolize by $M(a',b')$ the measurement process that rejects all impinging systems except those in the state $b'$, and permits only systems in the state $a'$ to emerge from the apparatus. The compound measurement $M(a',b')M(c',d')$ serves to select systems in the state $d'$ and produce them in the state $a'$, which is a selective measurement of the type $M(a',d')$. But, in addition, the first stage supplies systems in the state $c'$ while the second stage accepts only systems in the state $b'$. The examples of compound measurements that we have already considered involve the passage of all systems or no systems between the two stages, as represented by the multiplicative numbers 1 or 0. More generally, measurements of properties $B$, performed on a system in a state $c'$ that refers to properties incompatible with $B$, will yield a statistical distribution of the possible values. Hence, only a determinate fraction of the systems emerging from the first stage will be accepted by the second stage. We express this by the general multiplication law

$$M(a',b')M(c',d') = \langle b'|c'\rangle M(a',d'), \tag{5}$$

where $\langle b'|c'\rangle$ is a number characterizing the statistical relation between the states $b'$ and $c'$. In particular,

$$\langle a'|a''\rangle = \delta(a',a'').$$

Special examples of (5) are

$$M(a')M(b',c') = \langle a'|b'\rangle M(a',c')$$

and

$$M(a',b')M(c') = \langle b'|c'\rangle M(a',c').$$

We infer from the fundamental measurement symbol property (3) that

$$\sum_{a'}\langle a'|b'\rangle M(a',c') = \sum_{a'} M(a')M(b',c') = M(b',c')$$

and similarly

$$\sum_{c'}\langle b'|c'\rangle M(a',c') = M(a',b'),$$

which shows that measurement symbols of one type can be expressed as a linear combination of the measurement symbols of another type. The general relation is

$$M(c',d') = \sum_{a'b'} M(a')M(c',d')M(b') = \sum_{a'b'}\langle a'|c'\rangle\langle d'|b'\rangle M(a',b'). \tag{6}$$

From its role in effecting such connections, the totality of numbers $\langle a'|b'\rangle$ is called the transformation function relating the $a$- and $b$-descriptions, where the phrase "$a$-description" signifies the description of a system in terms of the states produced by selective measurements of the complete set of compatible physical quantities $A$.

A fundamental composition property of transformation functions is obtained on comparing

$$\sum_{b'} M(a')M(b')M(c') = \sum_{b'}\langle a'|b'\rangle\langle b'|c'\rangle M(a',c')$$

with

$$M(a')(\sum_{b'} M(b'))M(c') = M(a')M(c')$$
$$= \langle a'|c'\rangle M(a',c'),$$

namely

$$\sum_{b'}\langle a'|b'\rangle\langle b'|c'\rangle = \langle a'|c'\rangle.$$

On identifying the $a$- and $c$-descriptions this becomes

$$\sum_{b'}\langle a'|b'\rangle\langle b'|a''\rangle = \delta(a',a'')$$

and similarly

$$\sum_{a'}\langle b'|a'\rangle\langle a'|b''\rangle = \delta(b',b'').$$

As a consequence, we observe that

$$\sum_{a'}\sum_{b'}\langle a'|b'\rangle\langle b'|a'\rangle = \sum_{a'} 1$$
$$= \sum_{b'}\sum_{a'}\langle b'|a'\rangle\langle a'|b'\rangle = \sum_{b'} 1,$$

which means that $N$, the total number of states obtained in a complete measurement, is independent of the particular choice of compatible physical quantities that are measured. Hence the total number of measurement symbols of any specified type is $N^2$. Arbitrary numerical multiples of measurement symbols in additive combination thus form the elements of a linear algebra of dimensionality $N^2$—the algebra of measurement. The elements of the measurement algebra are called operators.

The number $\langle a'|b'\rangle$ can be regarded as a linear numerical function of the operator $M(b',a')$. We call this linear correspondence between operators and numbers the trace,

$$\langle a'|b'\rangle = trM(b',a'), \tag{7}$$

and observe from the general linear relation (6) that

$$trM(c',d') = \sum_{a'b'}\langle a'|c'\rangle\langle d'|b'\rangle trM(a',b')$$
$$= \sum_{a'b'}\langle d'|b'\rangle\langle b'|a'\rangle\langle a'|c'\rangle$$
$$= \langle d'|c'\rangle,$$

which verifies the consistency of the definition (7). In particular,

$$trM(a',a'') = \delta(a',a'')$$
$$trM(a') = 1.$$

The trace of a measurement symbol product is

$$trM(a',b')M(c',d') = \langle b'|c'\rangle trM(a',d')$$
$$= \langle b'|c'\rangle\langle d'|a'\rangle, \tag{8}$$

which can be compared with

$$trM(c',d')M(a',b') = \langle d'|a'\rangle trM(c',b') = \langle d'|a'\rangle\langle b'|c'\rangle.$$

Hence, despite the noncommutativity of multiplication, the trace of a product of two factors is independent of the multiplication order. This applies to any two elements $X$, $Y$, of the measurement algebra,

$$trXY = trYX.$$

A special example of (8) is

$$trM(a')M(b') = \langle a'|b'\rangle\langle b'|a'\rangle \tag{9}$$

It should be observed that the general multiplication law and the definition of the trace are preserved if we make the substitutions

$$M(a',b') \rightarrow \lambda(a')^{-1}M(a',b')\lambda(b')$$
$$\langle a'|b'\rangle \rightarrow \lambda(a')\langle a'|b'\rangle\lambda(b')^{-1}, \tag{10}$$

where the numbers $\lambda(a')$ and $\lambda(b')$ can be given arbitrary nonzero values. The elementary measurement symbols $M(a')$ and the transformation function $\langle a'|a''\rangle$ are left unaltered. In view of this arbitrariness, a transformation function $\langle a'|b'\rangle$ cannot, of itself, possess a direct physical interpretation but must enter in some combination that remains invariant under the substitution (10).

The appropriate basis for the statistical interpretation of the transformation function can be inferred by a consideration of the sequence of selective measurements $M(b')M(a')M(b')$, which differs from $M(b')$ in virtue of the disturbance attendant upon the intermediate $A$-measurement. Only a fraction of the systems selected in the initial $B$-measurement is transmitted through the complete apparatus. Correspondingly, we have the symbolic equation

$$M(b')M(a')M(b') = p(a',b')M(b'),$$

where the number

$$p(a',b') = \langle a'|b'\rangle\langle b'|a'\rangle \tag{11}$$

is invariant under the transformation (10). If we perform an $A$-measurement that does not distinguish between two (or more) states, there is a related additivity of the numbers $p(a',b')$,

$$M(b')(M(a') + M(a''))M(b') = (p(a',b') + p(a'',b'))M(b)',$$

and, for the $A$-measurement that does not distinguish among any of the states, there appears

$$M(b')(\sum_{a'} M(a'))M(b') = M(b'),$$

whence

$$\sum_{a'} p(a',b') = 1.$$

These properties qualify $p(a',b')$ for the role of the probability that one observes the state $a'$ in a measurement performed on a system known to be in the state $b'$.

But a probability is a real, nonnegative number. Hence we shall impose an admissible restriction on the numbers appearing in the measurement algebra, by requiring that $\langle a'|b'\rangle$ and $\langle b'|a'\rangle$ form a pair of complex conjugate numbers.[2]

$$\langle b'|a'\rangle = \langle a'|b'\rangle^*, \tag{12}$$

for then

$$p(a',b') = |\langle a'|b'\rangle|^2 \geq 0.$$

To maintain the complex conjugate relation (12), the numbers $\lambda(a')$ of (10) must obey

$$\lambda(a')^* = \lambda(a')^{-1},$$

and therefore have the form

$$\lambda(a') = e^{i\varphi(a')}$$

in which the phases $\varphi(a')$ can assume arbitrary real values.

Another satisfactory aspect of the probability formula (11) is the symmetry property

$$p(a',b') = p(b',a').$$

Let us recall the arbitrary convention that accompanies the interpretation of the measurement symbols and their products—the order of events is read from right to left (sinistrally). But any measurement symbol equation is equally valid if interpreted in the opposite sense (dextrally), and no physical result should depend upon which convention is employed. On introducing the dextral interpretation, $\langle a'|b'\rangle$ acquires the meaning possessed by $\langle b'|a'\rangle$ with the sinistral convention. We conclude that the probability connecting states $a'$ and $b'$ in given sequence must be constructed symmetrically from $\langle a'|b'\rangle$ and $\langle b'|a'\rangle$. The introduction of the opposite convention for measurement symbols will be termed the adjoint operation, and is indicated by†. Thus,

$$M(a',b')^\dagger = M(b',a')$$

and

$$M(a',a'')^\dagger = M(a'',a').$$

In particular,

$$M(a')^\dagger = M(a'),$$

which characterizes $M(a')$ as a self-adjoint or Hermitian operator. For measurement symbol products we have

$$\begin{aligned}(M(a',b')M(c',d'))^\dagger &= M(d',c')M(b',a')\\ &= M(c',d')^\dagger M(a',b')^\dagger,\end{aligned}$$

or equivalently,

$$\begin{aligned}(\langle b'|c'\rangle M(a',d'))^\dagger &= \langle c'|b'\rangle M(d',a')\\ &= \langle b'|c'\rangle^* M(a',d')^\dagger.\end{aligned}$$

The significance of addition is uninfluenced by the adjoint procedure, which permits us to extend these properties to all elements of the measurement algebra:

$$(X + Y)\dagger = X\dagger + Y\dagger,\ (XY)\dagger = Y\dagger X\dagger,\ (\lambda X)\dagger = \lambda^* X\dagger,$$

in which $\lambda$ is an arbitrary number.

The use of complex numbers in the measurement algebra implies the existence of a dual algebra in which all numbers are replaced by the complex conjugate numbers. No physical result can depend upon which algebra is employed. If the operators of the dual algebra are written $X^*$, the correspondence between the two algebras is governed by the laws

$$(X + Y)^* = X^* + Y^*,\ (XY)^* = X^*Y^*,\ (\lambda X)^* = \lambda^* X^*.$$

The formation of the adjoint within the complex conjugate algebra is called transposition,

$$X^T = X^*\dagger = X\dagger^*.$$

It has the algebraic properties

$$(X + Y)^T = X^T + Y^T,\ (XY)^T = Y^T X^T,\ (\lambda X)^T = \lambda X^T.$$

The measurement symbols of a given description provide a basis for the representation of an arbitrary operator by $N^2$ numbers, and the abstract properties of operators are realized by the combinatorial laws of these arrays of numbers, which are those of matrices. Thus

$$X = \sum_{a'a''} \langle a'|X|a''\rangle M(a',a'')$$

defines the matrix of $X$ in the $a$-description or $a$-representation, and the product

$$\begin{aligned} XY &= \Sigma\langle a'|X|a''\rangle M(a',a'')\Sigma\langle a^{\text{iv}}|Y|a'''\rangle M(a^{\text{iv}},a''') \\ &= \Sigma\langle a'|X|a''\rangle \delta(a'',a^{\text{iv}})\langle a^{\text{iv}}|Y|a'''\rangle M(a',a''') \end{aligned}$$

shows that

$$\langle a'|XY|a'''\rangle = \sum_{a''} \langle a'|X|a''\rangle\langle a''|Y|a'''\rangle.$$

The elements of the matrix that represents $X$ can be expressed as

$$\langle a'|X|a''\rangle = trXM(a'',a'),$$

and in particular

$$\langle a'|X|a'\rangle = trXM(a').$$

The sum of the diagonal elements of the matrix is the trace of the operator. The corresponding basis in the dual algebra is $M(a',a'')^*$, and the matrices that represent $X^*$ and $X^T$ are the complex conjugate and transpose, respectively, of the matrix representing $X$. The operator $X\dagger = X^{T*}$, an element of the same algebra as $X$, is represented by the transposed, complex conjugate, or adjoint matrix.

The matrix of $X$ is the mixed $ab$-representation is defined by

$$X = \sum_{a'b'} \langle a'|X|b'\rangle M(a',b')$$

where

$$\langle a'|X|b'\rangle = trXM(b',a').$$

The rule of multiplication for matrices in mixed representations is

$$\langle a'|XY|c'\rangle = \sum_{b'}\langle a'|X|b'\rangle\langle b'|Y|c'\rangle.$$

On placing $X = Y = 1$ we encounter the composition property of transformation functions, since

$$\begin{aligned}\langle a'|1|b'\rangle &= trM(b',a')\\ &= \langle a'|b'\rangle.\end{aligned}$$

If we set $X$ or $Y$ equal to 1, we obtain examples of the connection between the matrices of a given operator in various representations. The general result can be derived from the linear relations among measurement symbols. Thus,

$$\begin{aligned}\langle a'|X|d'\rangle &= trXM(d',a')\\ &= trX\sum_{b'c'}\langle c'|d'\rangle\langle a'|b'\rangle M(c',b')\\ &= \sum_{b'c'}\langle a'|b'\rangle\langle b'|X|c'\rangle\langle c'|d'\rangle.\end{aligned}$$

The adjoint of an operator $X$, displayed in the mixed *ab*-basis, appears in the *ba*-basis with the matrix

$$\langle b'|X\dagger|a'\rangle = \langle a'|X|b'\rangle^*.$$

As an application of mixed representations, we present an operator equivalent of the fundamental properties of transformation functions:

$$\begin{aligned}\sum_{b'}\langle a'|b'\rangle\langle b'|c'\rangle &= \langle a'|c'\rangle\\ \langle a'|b'\rangle^* &= \langle b'|a'\rangle,\end{aligned}$$

which is achieved by a differential characterization of the transformation functions. If $\delta\langle a'|b'\rangle$ and $\delta\langle b'|c'\rangle$ are any conceivable infinitesimal alteration of the corresponding transformation functions, the implied variation of $\langle a'|c'\rangle$ is

$$\delta\langle a'|c'\rangle = \sum_{b'}[\delta\langle a'|b'\rangle\langle b'|c'\rangle + \langle a'|b'\rangle\delta\langle b'|c'\rangle], \qquad (13)$$

and also

$$\delta\langle a'|b'\rangle^* = \delta\langle b'|a'\rangle.$$

One can regard the array of numbers $\delta\langle a'|b'\rangle$ as the matrix of an operator in the *ab*-representation. We therefore write

$$\delta\langle a'|b'\rangle = i\langle a'|\delta W_{ab}|b'\rangle,$$

which is the definition of an infinitesimal operator $\delta W_{ab}$. If infinitesimal operators $\delta W_{bc}$ and $\delta W_{ac}$ are defined similarly, the differential property (13) becomes the matrix equation

$$\langle a'|\delta W_{ac}|c'\rangle = \sum_{b'}[\langle a'|\delta W_{ab}|b'\rangle\langle b'|c'\rangle + \langle a'|b'\rangle\langle b'|\delta W_{bc}|c'\rangle],$$

from which we infer the operator equation

$$\delta W_{ac} = \delta W_{ab} + \delta W_{bc}. \tag{14}$$

Thus the multiplicative composition law of transformation functions is expressed by an additive composition law for the infinitesimal operators $\delta W$.

On identifying the $a$- and $b$-descriptions in (14), we learn that

$$\delta W_{aa} = 0$$

or

$$\delta\langle a' | a'' \rangle = 0,$$

which expresses the fixed numerical values of the transformation function

$$\langle a' | a'' \rangle = \delta(a',a'').$$

Indeed, the latter is not an independent condition on transformation functions but is implied by the composition property and the requirement that transformation functions, as matrices, be nonsingular. If we identify the $a$- and $c$-descriptions we are informed that

$$\delta W_{ba} = -\delta W_{ab}.$$

Now

$$\begin{aligned} \delta\langle a' | b' \rangle^* &= -i\langle a' | \delta W_{ab} | b' \rangle^* \\ &= -i\langle b' | \delta W_{ab}\dagger | a' \rangle, \end{aligned}$$

which must equal

$$\delta\langle b' | a' \rangle = i\langle b' | \delta W_{ba} | a' \rangle,$$

and therefore

$$\delta W_{ab}\dagger = -\delta W_{ba} = \delta W_{ab}.$$

The complex conjugate property of transformation functions is thus expressed by the statement that the infinitesimal operators $\delta W$ are Hermitian.

The expectation value of property $A$ for systems in the state $b'$ is the average of the possible values of $A$, weighted by the probabilities of occurrence that are characteristic of state $b'$. On using (9) to write the probability formula as

$$p(a',b') = trM(a')M(b'),$$

the expectation value becomes

$$\begin{aligned} \langle A \rangle_{b'} &= \sum_{a'} a' p(a',b') = trAM(b') \\ &= \langle b' | A | b' \rangle, \end{aligned}$$

where the operator $A$ is

$$A = \sum_{a'} a' M(a').$$

The correspondence thus obtained between operators and physical quantities is such that a function $f(A)$ of the property $A$ is assigned the operator $f(A)$, and the operators associated with a complete set of compatible physical quantities form a complete set of commuting Hermitian operators. In particular, the function of $A$ that exhibits the value unity in the state $a'$, and zero otherwise, is characterized by the operator $M(a')$.

The physical operation symbolized by $M(a')$ involves the functioning of an apparatus capable of separating an ensemble into subensembles that are distinguished by the various values of $a'$, together with the act of selecting one subensemble and rejecting the others. The measurement process prior to the stage of selection, which we call a nonselective measurement, will now be considered for the purpose of finding its symbolic counterpart. It is useful to recognize a general quantitative interpretation attached to the measurement symbols. Let a system in the state $c'$ be subjected to the selective $M(b')$ measurement and then to an $A$-measurement. The probability that the system will exhibit the value $b'$ and then $a'$, for the respective properties, is given by

$$\begin{aligned} p(a',b',c') &= p(a',b')p(b',c') = |\langle a'|b'\rangle\langle b'|c'\rangle|^2 \\ &= |\langle a'|M(b')|c'\rangle|^2. \end{aligned}$$

If, in contrast, the intermediate $B$-measurement accepts all systems without discrimination, which is equivalent to performing no $B$-measurement, the relevant probability is

$$\begin{aligned} p(a',1,c') &= |\langle a'|c'\rangle|^2 \\ &= |\langle a'|\sum_{b'} M(b')|c'\rangle|^2. \end{aligned}$$

There are examples of the relation between the symbol of any selective measurement and a corresponding probability,

$$p(a',\ ,c') = |\langle a'|M|c'\rangle|^2.$$

Now let the intervening measurement be nonselective, which is to say that the apparatus functions but no selection of systems is performed. Accordingly,

$$\begin{aligned} p(a',b,c') &= \sum_{b'} p(a',b')p(b',c') \\ &= \sum_{b'} |\langle a'|M(b')|c'\rangle|^2 \end{aligned}$$

which differs from

$$p(a',1,c') = |\sum_{b'}\langle a'|M(b')|c'\rangle|^2$$

by the absence of interference terms between different $b'$ states. This indicates that the symbol to be associated with the nonselective $B$-measurement is

$$M_b = \sum_{b'} e^{i\varphi_{b'}} M(b')$$

where the real phases $\varphi_{b'}$ are independent, randomly distributed quantities. The uncontrollable nature of the disturbance produced by a measurement thus finds its mathematical expression in these random phase factors. Since a nonselective measurement does not discard systems we must have

$$\sum_{a'} p(a',b,c') = 1$$

which corresponds to the unitary property of the $M_b$ operators,

$$M_b\dagger M_b = M_b M_b\dagger = 1.$$

It should also be noted that, within this probability context, the symbols of the elementary selective measurements are derived from the nonselective symbol by replacing all but one of the phases by positive infinite imaginary numbers, which is an absorptive description of the process of rejecting subensembles.

The general probability statement for successive measurements is

$$p(a',b', \ldots s',t') = |\langle a'|M(b') \ldots M(s')|t'\rangle|^2$$

which is applicable to any type of observation by inserting the appropriate measurement symbol. Other versions are

$$p(a', \ldots t') = \langle t'|(M(a') \ldots M(s'))\dagger(M(a') \ldots M(s'))|t'\rangle$$

and

$$p(a', \ldots t') = tr(M(a') \ldots M(t'))\dagger(M(a') \ldots M(t')),$$

each of which can also be extended to all types of selective measurements, and to nonselective measurements (the adjoint form is essential here). The expectation value construction shows that a quantity which equals unity if the properties $A$, $B, \ldots S$ successively exhibit, in the sinistral sense, the values $a'$, $b'$, $\ldots s'$, and is zero otherwise, is represented by the Hermitian[3] operator $(M(a') \ldots M(s'))\dagger$-$(M(a') \ldots M(s'))$.

Measurement is a dynamical process, and yet the only time concept that has been used is the primitive relationship of order. A detailed formulation of quantum dynamics must satisfy the consistency requirement that its description of the interactions that constitute measurement reproduces the symbolic characterizations that have emerged at this elementary stage. Such considerations make explicit reference to the fact that all measurement of atomic phenomena ultimately involves the amplification of microscopic effects to the level of macroscopic observation.

Further analysis of the measurement algebra leads to a geometry associated with the states of systems.

[1] This development has been presented in numerous lecture series since 1951, but is heretofore unpublished.

[2] Here we bypass the question of the utility of the real number field. According to a comment in THESE PROCEEDINGS, **44,** 223 (1958), the appearance of complex numbers, or their real equivalents, may be an aspect of the fundamental matter-antimatter duality, which can hardly be discussed at this stage.

[3] Compare P. A. M. Dirac, *Rev. Mod. Phys,* **17,** 195 (1945), where non-Hermitian operators and complex "probabilities" are introduced.

# *THE SPECIAL CANONICAL GROUP***

BY JULIAN SCHWINGER

HARVARD UNIVERSITY

*Communicated August 30, 1960*

This note is concerned with the further development and application of an operator group described in a previous paper.[1] It is associated with the quantum degree of freedom labeled $\nu = \infty$ which is characterized by the complementary pair of operators $q$, $p$ with continuous spectra.[2] The properties of such a degree of freedom are obtained as the limit of one with a finite number of states, specifically given by a prime integer $\nu$. We recall that unitary operators $U$ and $V$ obeying

$$U^\nu = V^\nu = 1, \; VU = e^{2\pi i/\nu}UV$$

define two orthonormal coordinate systems $\langle u^k|$ and $\langle v^k|$, where

$$u^k = v^k = e^{2\pi ik/\nu}$$

and

$$\langle u^k|v^l\rangle = \nu^{-1/2}e^{2\pi ikl/\nu}.$$

For any prime $\nu > 2$ we can choose the integers $k$ and $l$ to range from $-{}^1/_2(\nu - 1)$ to ${}^1/_2(\nu - 1)$, rather than from 0 to $\nu - 1$. An arbitrary state $\Psi$ can be represented alternatively by the wave functions

$$\psi(u^k) = \langle u^k|\Psi, \; \psi(v^k) = \langle v^k|\Psi$$

where

$$\Psi^\dagger\Psi = \sum_k |\psi(u^k)|^2 = \sum_k |\psi(v^k)|^2,$$

and the two wave functions are reciprocally related by

$$\psi(u^k) = \sum_l \nu^{-1/2}\, e^{2\pi ikl/\nu}\, \psi(v^l)$$

$$\psi(v^l) = \sum_k \nu^{-1/2}\, e^{-2\pi ikl/\nu}\, \psi(u^k).$$

We now shift our attention to the Hermitian operators $q$, $p$ defined by

$$U = e^{i\epsilon q}, \quad V = e^{i\epsilon p}, \quad \epsilon = (2\pi/\nu)^{1/2}$$

and the spectra

$$q', p' = \epsilon[-1/2(\nu - 1) \;..\; 0 \;..\; 1/2(\nu - 1)].$$

Furthermore, we redefine the wave functions so that

$$\psi(u^k) = \epsilon^{1/2}\psi(q'), \; \psi(v^k) = \epsilon^{1/2}\psi(p')$$

where $\epsilon = \Delta q' = \Delta p'$, the interval between adjacent eigenvalues. Then we have

$$\Psi^\dagger\Psi = \sum_{q'} \Delta q'|\psi(q')|^2 = \sum_{p'} \Delta p'|\psi(p')|^2$$



Reprinted from the *Proceedings of the National Academy of Sciences U.S.A.* **46**, 1401–1415 (1960).

and

$$\psi(q') = \sum_{p'} \Delta p' (2\pi)^{-1/2} e^{iq'p'} \psi(p')$$
$$\psi(p') = \sum_{q'} \Delta q' (2\pi)^{-1/2} e^{-iq'p'} \psi(q').$$

As $\nu$ increases without limit the spectra of $q$ and $p$ become arbitrarily dense and the eigenvalues of largest magnitude increase indefinitely. Accordingly we must restrict all further considerations to that physical class of states, or physical subspace of vectors, for which the wave functions $\psi(q')$ and $\psi(p')$ are sufficiently well behaved with regard to continuity and the approach of the variable to infinity that a uniform transition to the limit $\nu = \infty$ can be performed, with the result

$$\Psi^\dagger\Psi = \int_{-\infty}^{\infty} dq' |\psi(q')|^2 = \int_{-\infty}^{\infty} dp' |\psi(p')|^2$$

and

$$\psi(q') = \int_{-\infty}^{\infty} dp' \, (2\pi)^{-1/2} e^{iq'p'} \psi(p')$$
$$\psi(p') = \int_{-\infty}^{\infty} dq' \, (2\pi)^{-1/2} e^{-iq'p'} \psi(q').$$

We shall not attempt here to delimit more precisely the physical class of states. Note however that the reciprocal relation between wave functions can be combined into

$$\psi(q') = \int_{-\infty}^{\infty} \frac{dp'}{2\pi} e^{iq'p'} \int_{-\infty}^{\infty} dq'' \, e^{-iq''p'} \psi(q'')$$

which must be an identity for wave functions of the physical class when the operations are performed as indicated. There will also be a class of functions $K(p', \epsilon)$, such that

$$K(p', \epsilon) \rightarrow \begin{cases} 1, & \epsilon \rightarrow 0 \\ 0, & |p'| \rightarrow \infty \end{cases}$$

and

$$\psi(q') = \lim_{\epsilon \to 0} \int_{-\infty}^{\infty} \frac{dp'}{2\pi} e^{iq'p'} K(p', \epsilon) \int_{-\infty}^{\infty} dq'' \, e^{-iq''p'} \psi(q'')$$
$$= \lim_{\epsilon \to 0} \int_{-\infty}^{\infty} dq'' \left[ \int_{-\infty}^{\infty} \frac{dp'}{2\pi} K(p', \epsilon) \, e^{ip'(q'-q'')} \right] \psi(q'').$$

This is what is implied by the symbolic notation (Dirac)

$$\psi(q') = \int_{-\infty}^{\infty} dq'' \delta(q' - q'') \psi(q'')$$
$$\delta(q' - q'') = \int_{-\infty}^{\infty} \frac{dp'}{2\pi} e^{ip'(q'-q'')}$$

We shall also use the notation that is modeled on the discrete situation, with integrals replacing summations, as in

$$\langle q' | q'' \rangle = \int_{-\infty}^{\infty} \langle q' | p' \rangle dp' \langle p' | q'' \rangle = \delta(q' - q''),$$

with

$$\langle q' | p' \rangle = \langle p' | q' \rangle^* = (2\pi)^{-1/2} e^{iq'p'}.$$

There are other applications of the limit $\nu \rightarrow \infty$. The reciprocal property of the operators $U$ and $V$ is expressed by

$$\langle u^k | V = \langle u^{k+1}|, \ \langle v^k | U^{-1} = \langle v^{k+1}|$$

or

$$\langle q' | e^{i\epsilon p} = \langle q' + \epsilon|, \ \langle p' | e^{-i\epsilon q} = \langle p' + \epsilon|,$$

with an exception when $q'$ or $p'$ is the greatest eigenvalue, for then $q' + \epsilon$ or $p' + \epsilon$ is identified with the least eigenvalue, $-q'$ or $-p'$. We write these relations as wave function statements, of the form

$$\frac{1}{\epsilon}[\langle q' + \epsilon | \Psi - \langle q' | \Psi] = \langle q' | \frac{1}{\epsilon}(e^{i\epsilon p} - 1)\Psi$$

and

$$\frac{1}{\epsilon}[\langle p' + \epsilon | \Psi - \langle p' | \Psi] = \langle p' | \frac{1}{\epsilon}(e^{-i\epsilon q} - 1)\Psi.$$

In the transition to the limit $\nu = \infty$, the subspace of physical vectors $\Psi$ is distinguished by such properties of continuity and behavior at infinity of the wave functions that the left-hand limits $\epsilon \rightarrow 0$, exist as derivatives of the corresponding wave function. We conclude, for the physical class of states, that

$$\frac{1}{i}\frac{\partial}{\partial q'}\langle q' | \Psi = \langle q' | p\Psi$$

and

$$i\frac{\partial}{\partial p'}\langle p' | \Psi = \langle p' | q\Psi.$$

It will also be evident from this application of the limiting process to the unitary operators $\exp(i\epsilon q)$, $\exp(i\epsilon p)$ that a restriction to a physical subspace is needed for the validity of the commutation relation

$$[q, p] = i.$$

The elements of an orthonormal operator basis are given by

$$\nu^{-1/2} e^{\pi imn/\nu} U^m V^n = \nu^{-1/2} e^{-(\pi imn/\nu)} V^n U^m$$

or $\nu^{-1/2}U(q'p')$, with

$$U(q'p') = e^{1/2ip'q'} e^{ipq'} e^{-ip'q} = e^{-1/2ip'q'} e^{-ip'q} e^{ipq'}$$

$$= (\nu \rightarrow \infty) \ e^{i(pq' - p'q)}.$$

Thus, since $\nu^{-1} = \Delta q' \Delta p'/2\pi$, we have, for an arbitrary function $f(q'p')$ of the discrete variables $q'$, $p'$,

$$\sum_{q''p''} \frac{\Delta q'' \Delta p''}{2\pi} \, tr \, U(q'p')^\dagger U(q''p'') f(q''p'') = f(q'p').$$

In the limit $\nu = \infty$ there is a class of functions $f(q'p')$ such that

$$tr\ U(q'p')^{\dagger} \int_{-\infty}^{\infty} \frac{dq''dp''}{2\pi}\ U(q''p'')f(q''p'') = f(q'p'),$$

which we express symbolically by

$$tr\ U(q'p')^{\dagger}U(q''p'') = 2\pi\delta(q' - q'')\delta(p' - p'').$$

In particular,

$$tr\ e^{i(pq'-p'q)} = 2\pi\delta(q')\delta(p')$$
$$= \int_{-\infty}^{\infty} \frac{dqdp}{2\pi}\ e^{i(pq'-p'q)}$$

where $q$ and $p$ on the right-hand side are numerical integration variables.

The completeness of the operator basis $U(q'p')$ is expressed by

$$\int_{-\infty}^{\infty} \frac{dq'dp'}{2\pi}\ U(q'p')\ X\ U(q'p')^{\dagger} = 1\ tr\ X.$$

The properties of the $U(q'p')$ basis are also described, with respect to an arbitrary discrete operator basis $X(\alpha)$, by

$$\int_{-\infty}^{\infty} \frac{dq'dp'}{2\pi}\ \langle\alpha| q'p'\rangle \langle q'p'| \alpha'\rangle = \delta(\alpha, \alpha')$$

and

$$\sum_{\alpha} \langle q'p'| \alpha\rangle \langle\alpha| q''p''\rangle = 2\pi\delta(q' - q'')\delta(p' - p'').$$

When $X$ is given as $F(q, p)$, the operations of the special canonical group can be utilized to bring the completeness expression into the form

$$\int_{-\infty}^{\infty} \frac{dq'dp'}{2\pi}\ F(q + q', p + p') = 1\ tr\ F(q, p).$$

This operator relation implies a numerical one if it is possible to order $F(q, p)$, so that all $q$ operators stand to the left, for example, of the operator $p$: $F(q; p)$. Then the evaluation of the $\langle q' = 0| \quad |p' = 0\rangle$ matrix element gives

$$tr\ F(q, p) = \int_{-\infty}^{\infty} \frac{dqdp}{2\pi}\ F(q; p)$$

which result also applies to a system with $n$ continuous degrees of freedom if it is understood that

$$\frac{dqdp}{2\pi} = \prod_{k=1}^{n} \frac{dq_k dp_k}{2\pi}.$$

As an example of this ordering process other than the one already given by $tr\ U(q'p')$, we remark that

$$e^{-1/2(q^2+p^2)\beta} = (\cosh\beta)^{-\frac{1}{2}}\ e^{-\frac{1}{2}q^2 \tanh\beta}\ e^{iq;\ p(\operatorname{sech}\beta-1)}\ e^{-1/2p^2 \tanh\beta}$$

where

$$e^{a;\ b} = \sum_{0}^{\infty} \frac{1}{n!}\ a^n b^n,$$

and therefore

$$tr\ e^{-1/2(q^2+p^2)\beta} = \int_{-\infty}^{\infty} \frac{dqdp}{2\pi} (\cosh \beta)^{-1/2}\ e^{-1/2(q^2+p^2)\tanh \beta}\ e^{iqp\ (\operatorname{sech} \beta - 1)}$$

$$= \frac{1}{2 \sinh {}^1/_2\beta} = \sum_{n=0}^{\infty} e^{-(n+1/2)\beta},$$

which reproduces the well-known non-degenerate spectrum of the operator $^1/_2(q^2 + p^2)$.

Now we shall consider the construction of finite special canonical transformations from a succession of infinitesimal ones, as represented by the variation of a parameter $\tau$. Let the generator of the transformation associated with $\tau \to \tau + d\tau$ be

$$d\tau G_s = d\tau(qP - pQ)$$

where $Q(\tau)$ and $P(\tau)$ are arbitrary numerical functions of $\tau$. That is, the infinitesimal transformation is

$$q(\tau + d\tau) - q(\tau) = -\frac{1}{i}[q, d\tau G_s] = d\tau Q(\tau)$$

$$p(\tau + d\tau) - p(\tau) = -\frac{1}{i}[p, d\tau G_s] = d\tau P(\tau),$$

which implies the finite transformation

$$q(\tau_1) - q(\tau_2) = \int_{\tau_2}^{\tau_1} d\tau Q(\tau)$$

$$p(\tau_1) - p(\tau_2) = \int_{\tau_2}^{\tau_1} d\tau P(\tau).$$

Some associated transformation functions are easily constructed. We have

$$\langle \tau + d\tau| = \langle \tau| [1 + id\tau G(x(\tau), \tau)]$$

or

$$\frac{1}{i}\frac{\partial}{\partial \tau}\langle \tau| = \langle \tau| [q(\tau)P(\tau) - p(\tau)Q(\tau)],$$

and therefore

$$\frac{1}{i}\frac{\partial}{\partial \tau}\langle q'\tau|p'\tau_2\rangle$$

$$= \langle q'\tau| [q'P(\tau) - (p' + \textstyle\int_{\tau_2}^{\tau} d\tau' P(\tau'))Q(\tau)]|p'\tau_2\rangle$$

which, in conjunction with the initial condition

$$\tau = \tau_2:\ \langle q'\tau|p'\tau_2\rangle = \langle q'|p'\rangle = (2\pi)^{-1/2}e^{iq'p'}$$

gives

$$\langle q'\tau_1|p'\tau_2\rangle^{QP} = (2\pi)^{-1/2} \exp [i(q'p' + q' \textstyle\int_{\tau_2}^{\tau_1} d\tau P(\tau) - p' \int_{\tau_2}^{\tau_1} d\tau Q(\tau) - \int_{\tau_2}^{\tau_1} d\tau d\tau' Q(\tau)\eta_+(\tau - \tau')P(\tau'))]$$

with

$$\eta_+(\tau - \tau') = \begin{cases} 1, & \tau > \tau' \\ 0, & \tau < \tau' \end{cases}.$$

From this result we derive

$$\langle q'\tau_1 | q''\tau_2\rangle^{QP} = \int_{-\infty}^{\infty} \langle q'\tau_1 | p'\tau_2\rangle^{QP} dp' \langle p' | q''\rangle = \delta(q' - q'' - \int_{\tau_2}^{\tau_1} d\tau Q(\tau))\, e^{iq'\int d\tau P}\, e^{-i\int d\tau d\tau' Q\eta_+ P},$$

and

$$\langle p'\tau_1 | p''\tau_2\rangle^{QP} = \int_{-\infty}^{\infty} \langle p' | q'\rangle dq' \langle q'\tau_1 | p''\tau_2\rangle^{QP} = \delta(p' - p'' - \int_{\tau_2}^{\tau_1} d\tau P(\tau))\, e^{-ip''\int d\tau Q}\, e^{-i\int d\tau d\tau' Q\eta_+ P}$$

or, alternatively,

$$\delta(p' - p'' - \int d\tau P)\, e^{-ip'\int d\tau Q}\, e^{i\int d\tau d\tau' P\eta_+ Q},$$

on using the fact that

$$\eta_-(\tau - \tau') = 1 - \eta_+(\tau - \tau') = \eta_+(\tau' - \tau).$$

These transformation functions can also be viewed as matrix elements of the unitary operator, an element of the special canonical group, that produces the complete transformation. That operator, incidentally, is

$$\exp[i(q \int_{\tau_2}^{\tau_1} d\tau P(\tau) - p \int_{\tau_2}^{\tau_1} d\tau Q(\tau) - {}^1/_2 \int_{\tau_2}^{\tau_1} d\tau d\tau' Q(\tau)\epsilon(\tau - \tau')P(\tau'))]$$

where $\epsilon(\tau - \tau')$ is the odd step function

$$\epsilon = \eta_+ - \eta_-.$$

We can compute the trace of a transformation function, regarded as a matrix, and this will equal the trace of the associated unitary operator provided the otherwise arbitrary representation is not an explicit function of $\tau$. Thus

$$tr\langle \tau_1 | \tau_2\rangle^{QP} = \int_{-\infty}^{\infty} dq' \langle q'\tau_1 | q'\tau_2\rangle^{QP} = \int_{-\infty}^{\infty} dp' \langle p'\tau_1 | p'\tau_2\rangle^{QP} = 2\pi\delta\,(\int_{\tau_2}^{\tau_1} d\tau Q(\tau))\delta(\int_{\tau_2}^{\tau_1} d\tau P(\tau))\, e^{-i\int d\tau d\tau' Q\eta_+ P}$$

where, in view of the delta function factors, $\eta_+(\tau - \tau')$ can be replaced by other equivalent functions, such as $\eta_+ - {}^1/_2 = {}^1/_2\epsilon$, or ${}^1/_2\epsilon(\tau - \tau') - (\tau - \tau')/T$, with $T = \tau_1 - \tau_2$. The latter choice has the property of giving a zero value to the double integral whenever $Q(\tau)$ or $P(\tau)$ is a constant. As an operator statement, the trace formula is the known result

$$tr\, e^{i(pq' - p'q)} = 2\pi\delta(q')\delta(p').$$

It is important to recognize that the trace, which is much more symmetrical than any individual transformation function, also implies specific transformation functions. Thus, let us make the substitutions

$$\left.\begin{aligned} Q(\tau) &\rightarrow Q(\tau) - (q' - q'')\delta(\tau - \tau_1 + \epsilon T) \\ P(\tau) &\rightarrow P(\tau) + p'\delta(\tau - \tau_2 - \epsilon T) \end{aligned}\right\}, \quad \epsilon \rightarrow +0$$

and then indicate the effect of the additional localized transformations by the equivalent unitary operators, which gives

$$tr\langle \tau_1 |\; e^{i(q' - q'')p(\tau_1)}\, e^{ip'q(\tau_2)} | \tau_2\rangle^{QP} = \int_{-\infty}^{\infty} dq \langle q + q' - q'' \quad \tau_1 | q\tau_2\rangle^{QP}\, e^{ip'q}.$$

Accordingly, if we also multiply by $\exp(-ip'q'')$ and integrate with respect to $dp'/2\pi$, what emerges is $\langle q'\tau_1 | q''\tau_2\rangle^{QP}$, as we can verify directly.

We shall find it useful to give an altogether different derivation of the trace formula. First note that

$$tr\langle\tau_1|(q(\tau_1) - q(\tau_2))|\tau_2\rangle = \int_{-\infty}^{\infty} dq'\langle q'\tau_1|(q' - q')|q'\tau_2\rangle = 0$$

and similarly

$$tr\langle\tau_1|(p(\tau_1) - p(\tau_2))|\tau_2\rangle = 0$$

which is a property of periodicity over the interval $T = \tau_1 - \tau_2$. Let us, therefore, represent the operators $q(\tau)$, $p(\tau)$ by the Fourier series

$$q(\tau) = q_0 + \sum_1^{\infty} (\pi n)^{-1/2}\left[q_n \cos\frac{2\pi n}{T}(\tau - \tau_2) + q_{-n}\sin\frac{2\pi n}{T}(\tau - \tau_2)\right]$$

$$p(\tau) = p_0 + \sum_1^{\infty} (\pi n)^{-1/2}\left[-p_n \sin\frac{2\pi n}{T}(\tau - \tau_2) + p_{-n}\cos\frac{2\pi n}{T}(\tau - \tau_2)\right],$$

where the coefficients are so chosen that the action operator for an arbitrary special canonical transformation

$$W_{12} = \int_{\tau_2}^{\tau_1} [{}^1/{}_2(pdq - dpq) + d\tau(qP - pQ)]$$

acquires the form

$$W_{12} = \sum_{-\infty}^{\infty}{}' p_n q_n + \sum_{-\infty}^{\infty} (q_n P_n - p_n Q_n).$$

Here the dash indicates that the term $n = 0$ is omitted, and

$$Q_0 = \int_{\tau_2}^{\tau_1} d\tau Q(\tau), \quad P_0 = \int_{\tau_2}^{\tau_1} d\tau P(\tau)$$

$$Q_{n,-n} = (\pi n)^{-1/2}\int_{\tau_2}^{\tau_1} d\tau Q(\tau)(-\sin, \cos)\left(\frac{2\pi n}{T}(\tau - \tau_2)\right)$$

$$P_{n,-n} = (\pi n)^{-1/2}\int_{\tau_2}^{\tau_1} d\tau P(\tau)(\cos, \sin)\left(\frac{2\pi n}{T}(\tau - \tau_2)\right).$$

The action principle for the trace is

$$\delta\, tr\,\langle\tau_1|\tau_2\rangle = i\, tr\,\langle\tau_1|\delta[W_{12}]|\tau_2\rangle$$

and the principle of stationary action asserts that

$$Q_0\, tr\,\langle\tau_1|\tau_2\rangle^{QP} = P_0\, tr\,\langle\tau_1|\tau_2\rangle^{QP} = 0,$$

together with

$$tr\,\langle\tau_1|(q_n - Q_n)|\tau_2\rangle^{QP} = tr\,\langle\tau_1|(p_n + P_n)|\tau_2\rangle^{QP} = 0.$$

The first of these results implies that the trace contains the factors $\delta(Q_0)$ and $\delta(P_0)$. The dependence upon $Q_n$, $P_n$, $n \neq 0$, is then given by the action principle as

$$\frac{\partial}{\partial Q_n}(tr) = i\, tr\,\langle\tau_1|(-p_n)|\tau_2\rangle = i\, P_n\,(tr)$$

$$\frac{\partial}{\partial P_n}(tr) = i\, tr\,\langle\tau_1|q_n|\tau_2\rangle = iQ_n(tr).$$

Therefore

$$tr\,\langle\tau_1|\,\tau_2\rangle^{QP} = 2\pi\delta(Q_0)\delta(P_0)\,e^{i\sum_{-\infty}^{\infty}{}' Q_nP_n}$$

where the factor of $2\pi$ is supplied by reference to the elementary situation with constant $Q(\tau)$ and $P(\tau)$. We note, for comparison with previous results, that

$$Q_nP_n + Q_{-n}P_{-n} = -\frac{1}{\pi n}\int_{\tau_2}^{\tau_1} d\tau d\tau' Q(\tau)\sin\left(\frac{2\pi n}{T}(\tau-\tau')\right)P(\tau')$$

and

$$\sum_{1}^{\infty}\frac{1}{\pi n}\sin\frac{2\pi n}{T}(\tau-\tau') = {}^1/_2\,\epsilon(\tau-\tau') - \frac{\tau-\tau'}{T}.$$

The new formula for the trace can be given a uniform integral expression by using the representation

$$e^{iQP} = \int_{-\infty}^{\infty}\frac{dqdp}{2\pi}\,e^{i(pq+qP-pQ)}$$

for now we can write

$$tr\,\langle\tau_1|\,\tau_2\rangle^{QP} = \int d[q,p]\,e^{iW[q,p]},$$

where

$$d[q,p] = \prod_{-\infty}^{\infty}\frac{dq_ndp_n}{2\pi}$$

and $W[q, p]$ is the numerical function formed in the same way as the action operator,

$$W[q,p] = \sum_{-\infty}^{\infty}{}' p_nq_n + \sum_{-\infty}^{\infty}(q_nP_n - p_nQ_n).$$

Alternatively, we can use the Fourier series to define the numerical functions $q(\tau)$, $p(\tau)$. Then

$$W[q,p] = \int_{\tau_2}^{\tau_1}[{}^1/_2(pdq - dpq) + d\tau(qP - pQ)]$$

and $d[q, p]$ appears as a measure in the quantum phase space of the functions $q(\tau)$, $p(\tau)$.

It is the great advantage of the special canonical group that these considerations can be fully utilized in discussing arbitrary additional unitary transformations, as described by the action operator

$$W_{12} = \int_{\tau_2}^{\tau_1}[{}^1/_2(pdq - dpq) + d\tau(qP - pQ) + d\tau G(x(\tau),\tau)].$$

First, let us observe how an associated transformation function $\langle\tau_1|\,\tau_2\rangle_G{}^{QP}$ depends upon the arbitrary functions $Q(\tau)$, $P(\tau)$. The action principle asserts that

$$\delta QP\,\langle\tau_1|\,\tau_2\rangle = i\langle\tau_1|\int_{\tau_2}^{\tau_1}d\tau(q\delta P - p\delta Q)|\,\tau_2\rangle = i\int_{\tau_2}^{\tau_1}d\tau[\delta P(\tau)\langle\tau_1|\,q(\tau)|\,\tau_2\rangle - \delta Q(\tau)\langle\tau_1|\,p(\tau)|\,\tau_2\rangle]$$

which we express by the notation

$$-i\frac{\delta}{\delta P(\tau)}\langle\tau_1|\,\tau_2\rangle = \langle\tau_1|\,q(\tau)\,|\,\tau_2\rangle$$

$$i\frac{\delta}{\delta Q(\tau)}\langle\tau_1|\,\tau_2\rangle = \langle\tau_1|\,p(\tau)\,|\,\tau_2\rangle.$$

More generally, if $F(\tau')$ is an operator function of $x(\tau')$ but not of $Q$, $P$ we have

$$\langle\tau_1|\,F(\tau')\,|\,\tau_2\rangle_G{}^{QP} = \langle\tau_1|\,\tau'\rangle_G{}^{QP}\times\langle\tau'|\,F(\tau')\,|\,\tau'\rangle\times\langle\tau'|\,\tau_2\rangle_G{}^{QP},$$

where $|\,\tau'\rangle\times\langle\tau'|$ symbolizes the summation over a complete set of states, and therefore

$$-i\frac{\delta}{\delta P(\tau)}\langle\tau_1|\,F(\tau')\,|\,\tau_2\rangle = \langle\tau_1|\,(q(\tau)F(\tau'))_o\,|\,\tau_2\rangle$$

$$i\frac{\delta}{\delta Q(\tau)}\langle\tau_1|\,F(\tau')\,|\,\tau_2\rangle = \langle\tau_1|\,(p(\tau)F(\tau'))_o\,|\,\tau_2\rangle.$$

Here $(\quad)_o$ is an ordered product that corresponds to the sense of progression from $\tau_2$ to $\tau_1$. If $\tau$ follows $\tau'$ in sequence, the operator function of $\tau$ stands to the left, while if $\tau$ precedes $\tau'$ the associated operator appears on the right. This description covers the two algebraic situations: $\tau_1 > \tau_2$, where we call the ordering positive, $(\quad)_+$, and $\tau_1 < \tau_2$, which produces negative ordering, $(\quad)_-$. For the moment take $\tau_1 > \tau_2$ and compare

$$i\frac{\delta}{\delta Q(\tau)}\langle\tau_1|\,q(\tau+0)\,|\,\tau_2\rangle = \langle\tau_1|\,q(\tau)p(\tau)\,|\,\tau_2\rangle$$

with

$$i\frac{\delta}{\delta Q(\tau)}\langle\tau_1|\,q(\tau-0)\,|\,\tau_2\rangle = \langle\tau_1|\,p(\tau)q(\tau)\,|\,\tau_2\rangle.$$

The difference of these expressions,

$$i\frac{\delta}{\delta Q(\tau)}\langle\tau_1|\,[q(\tau+0)-q(\tau-0)]\,|\,\tau_2\rangle = \langle\tau_1|\,[q(\tau),\,p(\tau)]\,|\,\tau_2\rangle,$$

refers on one side to the noncommutativity of the complementary variables $q$ and $p$, and on the other to the "equation of motion" of the operator $q(\tau)$. According to the action principle

$$\frac{dq}{d\tau}+\frac{\partial G}{\partial p} = Q$$

and therefore

$$q(\tau+0)-q(\tau-0) = \lim_{\epsilon\to0}\int_{\tau-\epsilon}^{\tau+\epsilon}d\tau'\left[-\frac{\partial G}{\partial p}+Q(\tau')\right]$$

which yields the expected result,

$$i\frac{\delta}{\delta Q(\tau)}\langle\tau_1|\,[q(\tau+0)-q(\tau-0)]\,|\,\tau_2\rangle = i\langle\tau_1|\,\tau_2\rangle.$$

Thus, through the application of the special canonical group, we obtain functional differential operator representations for all the dynamical variables. The general statement is

$$\langle\tau_1|\,F(q,\,p)_o|\,\tau_2\rangle = F\left(-i\frac{\delta}{\delta P},\,i\frac{\delta}{\delta Q}\right)\langle\tau_1|\,\tau_2\rangle$$

where $F(q,\,p)_o$ is an ordered function of the $q(\tau)$, $p(\tau)$ throughout the interval between $\tau_2$ and $\tau_1$, and, as the simple example of $q(\tau)p(\tau)$ and $p(\tau)q(\tau)$ indicates, the particular order of multiplication for operators with a common value of $\tau$ must be reproduced by a suitable limiting process from different $\tau$ values.

The connection with the previous considerations emerges on supplying $G$ with a variable factor $\lambda$. For states at $\tau_1$ and $\tau_2$ that do not depend explicitly upon $\lambda$ we use the action principle to evaluate

$$\frac{\partial}{\partial\lambda}\langle\tau_1|\,\tau_2\rangle_{\lambda G}{}^{QP} = i\langle\tau_1|\int_{\tau_2}^{\tau_1}d\tau G(qp\tau)|\,\tau_2\rangle_{\lambda G}{}^{QP} = i\int_{\tau_2}^{\tau_1}d\tau G\left(-i\frac{\delta}{\delta P},\,i\frac{\delta}{\delta Q},\,\tau\right)\langle\tau_1|\,\tau_2\rangle_{\lambda G}{}^{QP}.$$

The formal expression that gives the result of integrating this differential equation from $\lambda = 0$ to $\lambda = 1$ is

$$\langle\tau_1|\,\tau_2\rangle_G{}^{QP} = \exp\left[i\int_{\tau_2}^{\tau_1}d\tau G\left(-i\frac{\delta}{\delta P},\,i\frac{\delta}{\delta Q},\,\tau\right)\right]\langle\tau_1|\,\tau_2\rangle^{QP}$$

where the latter transformation function is that for $\lambda = 0$ and therefore refers only to the special canonical group. An intermediate formula, corresponding to $G = G_1 + G_2$, contains the functional differential operator constructed from $G_1$ acting on the transformation function associated with $G_2$. The same structure applies to the traces of the transformation functions. If we use the integral representation for the trace of the special canonical transformation function, and perform the differentiations under the integration sign, we obtain the general integral formula[3]

$$tr\,\langle\tau_1|\,\tau_2\rangle_G{}^{QP} = \int d[q,\,p]e^{iW[q,\,p]}$$

Here the action functional $W[q,\,p]$ is

$$W[q,\,p] = \int_{\tau_2}^{\tau_1}[{}^1/{}_2(pdq - dpq) + d\tau(q(\tau)P(\tau) - p(\tau)Q(\tau)) + d\tau G(q(\tau)p(\tau)\tau)],$$

which is formed in essentially the same way as the action operator $W_{12}$, the multiplication order of noncommutative operator factors in $G$ being replaced by suitable infinitesimal displacements of the parameter $\tau$. The Hermitian operator $G$ can always be constructed from symmetrized products of Hermitian functions of $q$ and of $p$, and the corresponding numerical function is real. Thus the operator ${}^1/{}_2\{f_1(q(\tau),\,f_2(p(\tau)\}$ is represented by ${}^1/{}_2(f_1(q(\tau + \epsilon)) + f_1(q(\tau - \epsilon)))f_2(p(\tau))$, for example. One will expect to find that this averaged limit, $\epsilon \rightarrow 0$, implies no more than the direct use of $f_1(q(\tau)f_2(p(\tau))$, although the same statement is certainly not true of either term containing $\epsilon$. Incidentally, it is quite sufficient to construct the action from $pdq$, for example, rather than the more symmetrical version, in virtue of the periodicity.

As a specialization of this trace formula, we place $Q = P = 0$ and consider the class of operators $G$ that do not depend explicitly upon $\tau$. Now we are computing

$$tr\, e^{iTG} = \int d[q, p] e^{iW[q, p]}$$

$$W[q, p] = \int_0^T (pdq + d\tau G),$$

in which we have used the possibility of setting $\tau_2 = 0$. A simple example is provided by $G = {}^1/_2(p^2 + q^2)$, where

$$W = {}^1/_2 T(p_0^2 + q_0^2) + \sum_{-\infty}^{\infty}{}' \left[ p_n q_n + \frac{T}{4\pi|n|} (p_n^2 + q_n^2) \right]$$

and

$$tr\, e^{iT1/2(p^2+q^2)} = \int_{-\infty}^{\infty} \frac{dqdp}{2\pi} e^{iT1/2(p^2+q^2)} \times \prod_1^{\infty} \left[ \int_{-\infty}^{\infty} \frac{dqdp}{2\pi} e^{1/2i(2qp + T/2\pi n(p^2+q^2))} \right]^2 =$$

$$\frac{i}{T} \prod_1^{\infty} \frac{1}{1 - \left(\dfrac{T}{2\pi n}\right)^2} = \frac{i}{2 \sin {}^1/_2 T}.$$

This example also illustrates the class of Hermitian operators with spectra that are bounded below and for which the trace of exp $(iTG)$ continues to exist on giving $T$ a positive imaginary component, including the substitution $T \rightarrow i\beta$, $\beta > 0$. The trace formula can be restated for the latter situation on remarking that the Fourier series depend only upon the variable $(\tau - \tau_2)/T = \lambda$, which varies from 0 to 1, and therefore

$$tr\, e^{-\beta G} = \int d[q, p] e^{-w[q, p]}$$

$$w[q, p] = \int_0^1 d\lambda \left[ -ip \frac{dq}{d\lambda} + \beta G(q(\lambda)p(\lambda)) \right].$$

Another property of the example is that the contributions to the trace of all Fourier coefficients except $n = 0$ tend to unity for sufficiently small $T$ or $\beta$. This is also true for a class of operators of the form $G = {}^1/_2 p^2 + f(q)$. By a suitable translation of the Fourier coefficients for $p(\lambda)$ we can write $w[q, p]$ as

$$w = \int_0^1 d\lambda \left[ {}^1/_2 \beta p(\lambda)^2 + \frac{1}{2\beta} \left(\frac{dq}{d\lambda}\right)^2 + \beta f(q(\lambda)) \right].$$

For sufficiently small $\beta$, the term involving $dq/d\lambda$, to which all the Fourier coefficients of $q(\lambda)$ contribute except $q_0$, will effectively suppress these Fourier coefficients provided appropriate restrictions are imposed concerning singular points in the neighborhood of which $f(q)$ acquires arge negative values. Then $f(q(\lambda) \sim f(q_0))$ and we can reduce the integrations to just the contribution of $q_0$ and $p_0$, as expressed by

$$tr\, e^{-\beta G(q, p)} \sim \int_{-\infty}^{\infty} \frac{dqdp}{2\pi} e^{-\beta G(q, p)}.$$

Comparison with the previously obtained trace formula involving the ordering of operators shows that the noncommutativity of $q$ and $p$ is not significant in this limit. Thus we have entered the classical domain, where the incompatibility of physical properties at the microscopic level is no longer detectible. Incidentally, a first correction to the classical trace evaluation, stated explicitly for one degree of freedom, is

$$tr\, e^{-\beta[1/2p^2+f(q)]} \sim \int_{-\infty}^{\infty} \frac{dq}{(2\pi\beta)^{1/2}}\, e^{-\beta f(q)}\, \frac{{}^1/_2\beta(f''(q))^{1/2}}{\sinh {}^1/_2\beta(f''(q))^{1/2}},$$

which gives the exact value when $f(q)$ is a positive multiple of $q^2$.

Another treatment of the general problem can be given on remarking that the equations of motion implied by the stationary action principle,

$$\frac{dq}{d\tau} + \frac{\partial G}{\partial p} = Q,\quad \frac{dp}{d\tau} - \frac{\partial G}{\partial q} = P,$$

can be represented by functional differential equations[4]

$$\left[\frac{d}{d\tau}(-i)\frac{\delta}{\delta P(\tau)} + \frac{\partial G}{\partial p}\left(-i\frac{\delta}{\delta P}, i\frac{\delta}{\delta Q}\right) - Q(\tau)\right]\langle\tau_1|\,\tau_2\rangle_G^{QP} = 0$$

$$\left[\frac{d}{d\tau}\, i\frac{\delta}{\delta Q(\tau)} - \frac{\partial G}{\partial q}\left(-i\frac{\delta}{\delta P}, i\frac{\delta}{\delta Q}\right) - P(\tau)\right]\langle\tau_1|\,\tau_2\rangle_G{}^{QP} = 0.$$

These equations are valid for any such transformation function. The trace is specifically distinguished by the property of periodicity,

$$\left(\frac{\delta}{\delta Q(\tau_1)} - \frac{\delta}{\delta Q(\tau_2)}\right) tr\,\langle\tau_1|\,\tau_2\rangle_G{}^{QP} = \left(\frac{\delta}{\delta P(\tau_1)} - \frac{\delta}{\delta P(\tau_2)}\right)(tr) = 0$$

which asserts that the trace depends upon $Q(\tau)$, $P(\tau)$ only through the Fourier coefficients $Q_n$, $P_n$, and that the functional derivatives can be interpreted by means of ordinary derivatives:

$$\frac{\delta}{\delta Q(\tau)} = \frac{\partial}{\partial Q_0} + \sum_1^{\infty}(\pi n)^{-1/2}\left[-\sin\frac{2\pi n}{T}(\tau-\tau_2)\frac{\partial}{\partial Q_n} + \cos\frac{2\pi n}{T}(\tau-\tau_2)\frac{\partial}{\partial Q_{-n}}\right]$$

$$\frac{\delta}{\delta P(\tau)} = \frac{\partial}{\partial P_0} + \sum_1^{\infty}(\pi n)^{-1/2}\left[\cos\frac{2\pi n}{T}(\tau-\tau_2)\frac{\partial}{\partial P_n} + \sin\frac{2\pi n}{T}(\tau-\tau_2)\frac{\partial}{\partial P_{-n}}\right].$$

Now introduce the functional differential operator that is derived from the numerical action function $W_G[q, p]$ which refers only to the transformation generated by $G$, namely

$$W_G\left[-i\frac{\delta}{\delta P}, i\frac{\delta}{\delta Q}\right] = \int_{\tau_2}^{\tau_1}\left[\frac{\delta}{\delta Q}\, d\,\frac{\delta}{\delta P} + d\tau G\left(-i\frac{\delta}{\delta P}, i\frac{\delta}{\delta Q}, \tau\right)\right],$$

and observe that the differential equations are given by

$$(i[W_G, Q(\tau)] + Q(\tau))\, tr\,\langle\tau_1|\,\tau_2\rangle_G{}^{QP} = 0$$

$$(i[W_G, P(\tau)] + P(\tau))\, tr\,\langle\tau_1|\,\tau_2\rangle_G{}^{QP} = 0,$$

or by

$$e^{iW_G}Q(\tau)e^{-iW_G}(tr) = e^{iW_G}P(\tau)e^{-iW_G}(tr) = 0.$$

The latter form follows from the general expansion

$$e^{A}Be^{-A} = B + [A, B] + {}^1/_2![A, [A, B]] + \ldots$$

on noting that $[W_G, Q(\tau)]$, for example, is constructed entirely from differential operators and is commutative with the differential operator $W_G$. Accordingly,

$$Q(\tau)e^{-iW_G}(tr) = P(\tau)e^{-iW_G}(tr) = 0,$$

which asserts that $\exp(-iW_G)(tr)$ vanishes when multiplied by any of the Fourier coefficients $Q_n$, $P_n$ and therefore contains a delta function factor for each of these variables. We conclude that

$$tr\,\langle\tau_1|\,\tau_2\rangle_G{}^{QP} = e^{iW_G[-i\,\delta/\delta P,\; i\,\delta/\delta Q]}\,\delta[Q, P]$$

where, anticipating the proper normalization constants,

$$\delta[Q, P] = \prod_{-\infty}^{\infty} 2\pi\delta(Q_n)\delta(P_n).$$

A verification of these factors can be given by placing $G = 0$, which returns us to the consideration of the special canonical transformations. In this procedure we encounter the typical term

$$e^{i\,\partial/\partial Q\,\partial/\partial P}\,2\pi\delta(Q)\delta(P) = e^{iQP}$$

the proof of which follows from the remarks that

$$\left(Q + i\,\frac{\partial}{\partial P}\right)e^{i\,\partial/\partial Q\,\partial/\partial P} = e^{i\,\partial/\partial Q\,\partial/\partial P}\,Q$$

and

$$\int_{-\infty}^{\infty}\frac{dP}{2\pi}\,e^{i\,\partial/\partial Q\,\partial/\partial P}\,2\pi\delta(Q)\delta(P) = \delta(Q).$$

The result is just the known form of the transformation function trace

$$tr\,\langle\tau_1|\,\tau_2\rangle^{QP} = 2\pi\delta(Q_0)\delta(P_0)e^{\,i-\sum_{\infty}^{\infty}{}'\,Q_nP_n}$$

When integral representations are inserted for each of the delta function factors in $\delta[Q, P]$, we obtain

$$\delta[Q, P] = \prod_{-\infty}^{\infty}\int_{-\infty}^{\infty}\frac{dq_ndp_n}{2\pi}\,e^{i(q_nP_n - p_nQ_n)} = \int d[q, p]\,e^{i\int_{\tau_2}^{\tau_1} d\tau(q(\tau)P(\tau) - p(\tau)Q(\tau))}$$

and the consequence of performing the differentiations in $W_G$ under the integration signs is[5]

$$tr\,\langle\tau_1|\,\tau_2\rangle_G{}^{QP} = \int d[q, p]\,e^{iW[q,\,p]}$$

where the action function $W[q, p]$ now includes the special canonical transformation described by $Q$ and $P$.

We shall also write this general integral formula as

$$tr \langle \tau_1 | \tau_2 \rangle_G^{QP} = \int d[q, p]\, e^{i\int_{\tau_2}^{\tau_1} d\tau (qP - pQ)}\, e^{iW_G[q, p]}$$

in order to emphasize the reciprocity between the trace, as a function of $Q_n$, $P_n$ or functional of $Q(\tau)$, $P(\tau)$ and $\exp(iW_G[q, p])$ as a function of $q_n$, $p_n$ or functional of $q(\tau)$, $p(\tau)$. Indeed,

$$e^{iW_G[q, p]} = \int d[Q, P] e^{-i\int_{\tau_2}^{\tau_1} d\tau (qP - pQ)}\, tr \langle \tau_1 | \tau_2 \rangle_G^{QP},$$

where

$$d[Q, P] = \prod_{-\infty}^{\infty} \frac{dQ_n dP_n}{2\pi}$$

is such that

$$\int d[Q, P] \delta[Q, P] = 1$$

A verification of the reciprocal formula follows from the latter property on inserting for the trace the formal differential operator construction involving $\delta[Q, P]$. The reality of $W_G[q, p]$ now implies that

$$\int d[Q, P] d[Q', P'] e^{i\int d\tau (q(P-P') - p(Q-Q'))}\, (tr)_{QP}^* \, (tr)_{Q'P'} = 1$$

or, equivalently,

$$\int d[Q, P] (tr)_{X+X_1}^* (tr)_{X+X_2} = \delta[Q_1 - Q_2, P_1 - P_2]$$

where $X$ combines $Q$ and $P$.

The trace possesses the composition property

$$tr \langle \tau_0 | \tau_1 \rangle_G^{QP} \times tr \langle \tau_1 | \tau_2 \rangle_G^{QP} = tr \langle \tau_0 | \tau_2 \rangle_G^{QP}.$$

The operation involved is the replacement of $Q(\tau)$, $P(\tau)$ in the respective factors by $Q(\tau) \mp q'\delta(\tau - \tau_1)$, $P(\tau) \mp p'\delta(\tau - \tau_1)$ followed by integration with respect to $dq'dp'/2\pi$. The explicit form of the left side is therefore

$$\int_{-\infty}^{\infty} \frac{dq'dp'}{2\pi}\, tr \langle \tau_0 | e^{i[p(\tau_1)q' - p'q(\tau_1)]} | \tau_1 \rangle_G^{QP}\, tr \langle \tau_1 | e^{-i[p(\tau_1)q' - p'q(\tau_1)]} | \tau_2 \rangle_G^{QP}$$

or

$$\sum_{a'a''} \int_{-\infty}^{\infty} \frac{dq'dp'}{2\pi} \langle a'\tau_0 | U(q'p') | a'\tau_1 \rangle \langle a''\tau_1 | U(q'p')^\dagger | a''\tau_2 \rangle = \sum_{a'a''} \langle a'\tau_0 | \delta(a', a'') | a''\tau_2 \rangle = tr \langle \tau_0 | \tau_2 \rangle_G^{QP},$$

in view of the completeness of the operator basis formed from $q(\tau_1)$, $p(\tau_1)$.

No special relation has been assumed among $\tau_0$, $\tau_1$ and $\tau_2$. If $\tau_0$ and $\tau_2$ are equated, one transformation function in the product is the complex conjugate of the other. We must do more than this, however, to get a useful result. The most general procedure would be to choose the special canonical transformation in $tr \langle \tau_2 | \tau_1 \rangle_G^{QP} = tr \langle \tau_1 | \tau_2 \rangle_G^{QP*}$ arbitrarily different from that in $tr \langle \tau_1 | \tau_2 \rangle_G^{QP}$. The calculational advantages that appear in this way will be explored elsewhere.

Here we shall be content to make the special canonical transformations differ only at $\tau_2$. The corresponding theorem is

$$tr\ \langle\tau_2|\ \tau_1\rangle^{X+X'} \times tr\ \langle\tau_1|\ \tau_2\rangle^{X+X''} = 2\pi\delta(q' - q'')\delta(p' - p'')$$

where

$$Q'(\tau) = -q'\delta(\tau - \tau_2),\ P'(\tau) = -p'\delta(\tau - \tau_2)$$

and similarly for $Q''(\tau)$, $P''(\tau)$. This statement follows immediately from the orthonormality of the $U(q'p')$ operator basis on evaluating the left-hand side as

$$tr\ e^{-i[p(\tau_2)q' - p'q(\tau_2)]}\ e^{i[p(\tau_2)q'' - p''q(\tau_2)]} = 2\pi\delta(q' - q'')\delta(p' - p'').$$

* Supported by the Air Force Office of Scientific Research (ARDC).

[1] These PROCEEDINGS, **46,** 883 (1960).

[2] These PROCEEDINGS, **46,** 570 (1960).

[3] This formulation is closely related to the algarithms of Feynman, *Phys. Rev.*, **84,** 108 (1951), *Rev. Mod. Phys.*, **20,** 36 (1948). It differs from the latter in the absence of ambiguity associated with noncommutative factors, but primarily in the measure that is used. See Footnote 5.

[4] These are directly useful as differential equations only when $G(qp)$ is a sufficiently simple algebraic function of $q$ and $p$. The kinematical, group foundation for the representation of equations of motion by functional differential equations is to be contrasted with the dynamical language used in these PROCEEDINGS, **37,** 452 (1951).

[5] In this procedure, $q(\tau)$ and $p(\tau)$ are continuous functions of the parameter $\tau$ and the Fourier coefficients that represent them are a denumerably infinite set of integration variables. An alternative approach is the replacement of the continuous parameter $\tau$ by a discrete index while interpreting the derivative with respect to $\tau$ as a finite difference and constructing $\delta[Q, P]$ as a product of delta functions for each discrete $\tau$ value. With the latter, essentially the Feynman-Wiener formulation, the measure $d[q, p]$ is the product of $dq(\tau)dp(\tau)/2\pi$ for each value of $\tau$, periodicity is explicitly imposed at the boundaries, and the limit is eventually taken of an infinitely fine partitioning of the interval $T = \tau_1 - \tau_2$. The second method is doubtless more intuitive, since it is also the result of directly compounding successive infinitesimal transformations but it is more awkward as a mathematical technique.

## *ON THE BOUND STATES OF A GIVEN POTENTIAL**

JULIAN SCHWINGER

HARVARD UNIVERSITY

*Communicated November 29, 1960*

It is a fundamental property of any spherically symmetrical potential V(r) for which $\int_0^\infty dr\, r|V(r)|$ exists that there are only a finite number of bound states. This has been expressed by V. Bargmann[1] in the form of the inequality[2]

$$(2l + 1)n_l < \int_0^\infty dr\, r|V(r)|$$

where $n_l$ is the number of bound states for given $l$ (and magnetic quantum number $m_l$). The method of derivation of the latter result is sufficiently specialized, however, that it is not easily applied to nonspherically-symmetrical potentials, for

Reprinted from the *Proceedings of the National Academy of Sciences U.S.A.* **47**, 122–129 (1961).

example, or to the tensor forces that appear in the two-nucleon problem. Accordingly, we propose to give another derivation of this inequality, together with some of its extensions.

Any discussion of this subject has as its foundation the elementary fact that a decrease of the potential in some region must lower the energies of the bound states and therefore cannot lessen their number. Thus, we conclude that

$$n_l(V) \leq n_l(-|V|)$$

since the substitution of $-|V(r)|$ for $V(r)$ can either leave the potential unchanged or decrease it. Next, let us replace $-|V(r)|$ by $-\lambda|V(r)|$ with $0 < \lambda \leq 1$. An increase of $\lambda$ lowers the energies of the bound states and cannot lessen their number. As $\lambda$ increases from 0, we reach a critical value, $\lambda_1$, at which a bound state first appears at $E = -0$. With further growth of $\lambda$, the energy of this state decreases until we reach a second critical value, $\lambda_2$, at which a second bound state appears, and so on. When $\lambda$ has attained the value unity and

$$\lambda_n \leq 1 < \lambda_{n+1},$$

there are $n$ bound states.

The eigenvalue problem for $\lambda$, associated with $E = 0$ and orbital angular momentum $l$, is given by

$$\left(-\frac{d^2}{dr^2} + \frac{l(l+1)}{r^2}\right) u_l(r) = \lambda|V(r)|u_l(r)$$

or, on incorporating the boundary conditions,

$$u_l(r) = \lambda\int_0^\infty dr'\, g_l\,(r, r')|V(r')|u_l(r').$$

The Green's function, $g_l(r, r')$, obeys

$$\left(-\frac{d^2}{dr^2} + \frac{l(l+1)}{r^2}\right) g_l(r, r') = \delta(r - r')$$

and is explicitly given by

$$g_l(r, r') = \frac{1}{2l+1}\, r_<^{\,l+1} r_>^{\,-l}.$$

The integral equation can also be written with a real, symmetrical kernel,

$$\lambda^{-1}\phi_l(r) = \int_0^\infty dr' K_l(r, r')\phi_l(r'),$$

where

$$K_l(r, r') = |V(r)|^{1/2} g_l(r, r')|V(r')|^{1/2}$$

and

$$\phi_l(r) = |V(r)|^{1/2}\, u_l(r).$$

Thus, the eigenvalues of the Hermitian, positive kernel $K_l$ are the reciprocals of the critical numbers $\lambda_1, \lambda_2, \ldots$ just described.

The trace of $K_l$ is the sum of the eigenvalues,

$$\sum_1^\infty \frac{1}{\lambda_\alpha} = \int_0^\infty drK_l(r, r) = \frac{1}{2l+1}\int_0^\infty dr\, r|V(r)|.$$

But, if there are $n$ bound states, $\lambda_1 < \lambda_2 \ldots < \lambda_n < 1$, while $0 < \lambda_\alpha \leq \infty$, which supplies the inequality

$$\sum_1^\infty \frac{1}{\lambda_\alpha} \geq \sum_1^n \frac{1}{\lambda_\alpha} > n$$

and the theorem

$$n_l(V) \leq n_l(-|V|) < \frac{1}{2l+1}\int_0^\infty dr\, r\,|V(r)|.$$

When the potential is positive over an appreciable range of $r$, the upper limit given by this theorem may not be very realistic. Then, one would do better to compare $V(r)$ with a potential that equals $V(r)$ where $V(r) < 0$ but is zero wherever $V(r) > 0$. The same considerations show that

$$n_l(V) < \frac{1}{2l+1}\int dr\, r|V(r)|]_{V<0},$$

and now the integration is extended only over the regions of negative $V(r)$. This form of the inequality requires no restriction on the behavior of $V(r)$ in the domain of positive values.

A potential that realizes the upper limit to $n_l$ as closely as one wishes, for any particular $l$, is given by

$$V(r) = -\sum_{\nu=1}^n V_\nu\, \delta(r - r_\nu), \qquad V_\nu > 0.$$

The eigenvalue problem is equivalent to the $n$-dimensional determinantal equation

$$\det[\lambda^{-1}\delta_{\mu\nu} - V_\mu^{1/2} g_l(r_\mu, r_\nu) V_\nu^{1/2}] = 0.$$

and there are just $n$ eigenvalues. We can now choose each of the ratios $r_{\mu+1}/r_\mu$ to be sufficiently large that the nondiagonal elements of the determinant are as small as desired. Furthermore, we make all the diagonal elements equal by requiring of every $V_\nu$ that

$$V_\nu g_l(r_\nu, r_\nu) = \frac{1}{2l+1} V_\nu r_\nu \rightarrow 1$$

and in this way attain, with arbitrary precision, a single $n$-fold degenerate eigenvalue at unity. The resulting situation is one with $n$ bound states of essentially zero binding energy and with $\sum \lambda_\alpha^{-1} \cong n$.

The number of bound states is the number of states that lie at or below zero energy, and a similar problem can be posed for any negative energy, $-\kappa^2$. All the previous arguments continue to apply, provided one replaces the Green's function with the one defined by

$$\left(-\frac{d^2}{dr^2} + \kappa^2 + \frac{l(l+1)}{r^2}\right) g_l(r, r', \kappa) = \delta(r - r').$$

The required function is

$$g_l(r, r', \kappa) = \frac{1}{\kappa} S_l(\kappa r_<)E_l(\kappa r_>)$$

where

$$S_l(x) = i^{-l}\, x\, j_l(ix)$$

$$E_l(x) = -i^l\, x\, h_l(ix)$$

in the notation of spherical Bessel functions. Thus,

$$n_l(E \leq -\kappa^2) < \int_0^\infty dr g_l(rr\kappa)|V(r)| = \int_0^\infty dr\ r|V(r)|(1/\kappa r)S_l(\kappa r)E_l(\kappa r),$$

and

$$\frac{1}{x} S_0(x)\, E_0(x) = \frac{1}{2x}\, [1 - e^{-2x}]$$

$$\frac{1}{x} S_1(x)\, E_1(x) = \frac{1}{2x}\left[1 - \frac{1}{x^2} + e^{-2x}\left(1 + \frac{1}{x}\right)^2\right]$$

for example. Here again, and in the following, integrations only over the regions of negative $V$ can be used.

The function $g_l(rr\kappa)$, and thereby the integral $\int_0^\infty dr g_l(rr\kappa)|V(r)|$, has two characteristics that are physically necessary as properties of the number of states with any specified $l$ below the energy $-\kappa^2$. It is a monotonically decreasing function of $\kappa$ and of $l$. On regarding $l$ together with $\kappa$ as continuous parameters, we deduce from the Green's function differential equation that for any infinitesimal increase of $\kappa$ or $l$,

$$\delta g_l(rr\kappa) = -\int_0^\infty dr'[2\kappa\delta\kappa + r^{-2}(2l + 1)\delta l](g_l(rr'\kappa))^2 < 0.$$

We shall apply the monotonic dependence on $\kappa$ to a situation with $n$ bound states for a given $l$, so that

$$\int_0^\infty dr g_l(rro)|V(r)| > n \geq 1.$$

Then, there is a unique solution of the equation

$$\int_0^\infty dr g_l(rr\kappa_1)|V(r)| = 1,$$

and

$$n_l(E \leq -\kappa_1^2) < 1,$$

which is to say that the lowest bound state, the ground state for the given $l$, lies above the energy $-\kappa_1^2$, or

$$E_1 > -\kappa_1^2.$$

The deepest of these ground states is the one for $l = 0$, and a lower limit to the ground state energy is obtained by solving the equation

$$\int_0^\infty dr(1 - e^{-2\kappa_1 r})|V(r)| = 2\kappa_1.$$

For the class of potentials defined by $\int_0^\infty dr|V(r)| < \infty$, we have the crude estimate

$$\kappa_1 < \frac{1}{2}\int_0^\infty dr|V(r)|$$

and
$$E_1 > -\frac{1}{4}\left(\int_0^\infty dr|V(r)|\right)^2.$$

The potential
$$V(r) = -V_1\delta(r - r_1), \qquad V_1r_1 > 1$$

shows that even this limit can be approached as closely as desired, by choosing $V_1r_1$ to be sufficiently large. In a similar way, the equation
$$\int_0^\infty drg_l(rr\kappa_m)|V(r)| = m = 2, 3, \ldots \leq n$$

has a unique solution, and
$$n_l(E \leq -\kappa_m^2) < m$$

or
$$E_m > -\kappa_m^2.$$

Accordingly, we have obtained lower limits to the energies of all the bound states of a given potential.[3]

The transference of these ideas to an arbitrary nonspin-dependent three-dimensional potential $V(\mathbf{r})$ requires only one major modification. The Hermitian, positive kernel of the integral equation
$$\lambda^{-1}\phi(\mathbf{r}) = \int (d\mathbf{r}')K(\mathbf{r}\mathbf{r}'\kappa)\phi(\mathbf{r}'),$$

is
$$K(\mathbf{r}\mathbf{r}'\kappa) = |V(\mathbf{r})|^{1/2}G(\mathbf{r}\mathbf{r}'\kappa)|V(\mathbf{r}')|^{1/2}$$

and the Green's function, which is defined by
$$(-\nabla^2 + \kappa^2)G(\mathbf{r}\mathbf{r}'\kappa) = \delta(\mathbf{r} - \mathbf{r}')$$

and explicitly presented as
$$G(\mathbf{r}\mathbf{r}'\kappa) = \frac{e^{-\kappa|\mathbf{r}-\mathbf{r}'|}}{4\pi|\mathbf{r} - \mathbf{r}'|},$$

implies that the kernel is singular on the diagonal. We therefore shift our attention to the iterated kernel, the trace of which is
$$\int (d\mathbf{r})(d\mathbf{r}')(K(\mathbf{r}\mathbf{r}'\kappa))^2 = \sum \frac{1}{\lambda_\alpha^2},$$

and deduce for a suitable class of potentials that
$$N(E \leq -\kappa^2) < \frac{1}{(4\pi)^2}\int (d\mathbf{r})(d\mathbf{r}')|V(\mathbf{r})| \frac{e^{-2\kappa|\mathbf{r}-\mathbf{r}'|}}{|\mathbf{r} - \mathbf{r}'|^2} V(\mathbf{r}')|,$$

which includes an upper limit to the total number of bound states,
$$N < \frac{1}{(4\pi)^2}\int (d\mathbf{r})(d\mathbf{r}') \frac{|V(\mathbf{r})||V(\mathbf{r}')|}{|\mathbf{r} - \mathbf{r}'|^2}.$$

The latter quantity exists for potentials that decrease more rapidly than $|\mathbf{r}|^{-2}$ as $|\mathbf{r}| \to \infty$ and that in the neighborhoods of a finite number of points $\mathbf{r}_0$ are less singular than $|\mathbf{r} - \mathbf{r}_0|^{-2}$. These statements are to be interpreted by such inequalities as

$$|\mathbf{r}| > R, \; |V(\mathbf{r})| < C|\mathbf{r}|^{-a}, \; a > 2.$$

A potential of this class that is spherically symmetrical about its only singularity, at the origin, satisfies the condition $\int_0^\infty dr\, r|V(r)| < \infty$. It should be noted that the upper limit to $N$, deduced from the individual $n_l$ limits and from the fact that no bound state can occur for $2l + 1 > \int_0^\infty dr\, r|V|$, is

$$N < \frac{1}{2}\int_0^\infty dr\, r|V| \left[\int_0^\infty dr\, r|V| + 1\right].$$

The energy of the ground state is bounded below by

$$E_1 > -\kappa_1^2$$

where $\kappa_1$ is the unique solution of the equation

$$\frac{1}{(4\pi)^2}\int (d\mathbf{r})(d\mathbf{r}')|V(\mathbf{r})| \frac{e^{-2\kappa_1|\mathbf{r}-\mathbf{r}'|}}{|\mathbf{r} - \mathbf{r}'|^2} |V(\mathbf{r}')| = 1,$$

given that
$$\frac{1}{(4\pi)^2}\int (d\mathbf{r})(d\mathbf{r}')| \frac{|V(\mathbf{r})||V(\mathbf{r}')|}{|\mathbf{r} - \mathbf{r}'|^2} > 1.$$

More generally, when the last integral exceeds the integer $N$, the solution of the equation

$$\frac{1}{(4\pi)^2}\int (d\mathbf{r})(d\mathbf{r}')V(\mathbf{r}) \frac{e^{-2\kappa_m|\mathbf{r}-\mathbf{r}'|}}{|\mathbf{r} - \mathbf{r}'|^2} |V(\mathbf{r}')| = m = 1, 2, \ldots N$$

supplies a lower limit to the energy of the $m^{\text{th}}$ of the $N$ discrete states,

$$E_m > -\kappa_m^2.$$

There is a simple variant of these procedures that is worth mentioning. A comparison potential to $V(\mathbf{r})$ can be defined that equals $V(\mathbf{r})$ wherever $V(\mathbf{r}) < -\kappa^2$ and equals $-\kappa^2$ in the regions for which $V(\mathbf{r}) \geq -\kappa^2$. The energy values associated with the latter potential are depressed by the amount $\kappa^2$ relative to those of the potential that equals $V(\mathbf{r}) + \kappa^2$ wherever this quantity is negative but is zero otherwise. Thus, an upper bound to the total number of states associated with the last potential serves to limit the number of states that lie at or below the energy $-\kappa^2$ for the potential $V(\mathbf{r})$. The explicit statement is

$$N(E \leq \kappa^2) < \frac{1}{(4\pi)^2}\int (d\mathbf{r})(d\mathbf{r}') \left.\frac{|V(\mathbf{r}) + \kappa^2||V(\mathbf{r}') + \kappa^2|}{|\mathbf{r} - \mathbf{r}'|^2}\right]_{V + \kappa^2 < 0},$$

and, for the states of given $l$ in a spherically symmetrical potential,

$$n_l(E \leq -\kappa^2) < \frac{1}{2l + 1}\int dr\, r|V(r) + \kappa^2|\Big]_{V + \kappa^2 < 0},$$

from which we obtain lower limits to the energies of the various states. Note that no matter how slowly the potential approaches zero at great distances, there is a finite upper limit to the number of states that lie at or below any energy $-\kappa^2 < 0$. Of course, as $\kappa \to 0$, this limit will approach infinity if the conditions for a finite total number of states are not satisfied.

As an important example of spin-dependent potentials we consider

$$V(\mathbf{r}) = V_a(r) + V_b(r)S_{12}$$

$$S_{12} = 3\,\frac{\sigma_1\cdot\mathbf{r}\sigma_2\cdot\mathbf{r}}{r^2} - \sigma_1\cdot\sigma_2,$$

which refers to a pair of particles with spin angular momenta $^1/_2\sigma$. In the spin singlet state, $S_{12} = 0$ and we need consider only the triplet state, for which $\sigma_1\cdot\sigma_2 = 1$. We then note the algebraic property $(S_{12} + 1)^2 = 9$, so that the triplet eigenvalues of $S_{12}$ are 2 and $-4$. Just for simplicity, we shall assume that $V_b(r)$ is everywhere negative, but $V_a(r)$ is not restricted in this way. The notation $A \geq B$ will be used for spin matrices to mean that $A - B$ is a positive matrix, one that can never realize a negative expectation value. Thus,

$$V(\mathbf{r}) \geq V_a(r) + 2\,V_b(r),$$

and the spin-independent spherically symmetrical potential that equals $V_a + 2V_b$ where this quantity is negative and that equals zero otherwise provides a comparison potential to which the preceding three-dimensional considerations can be applied, since no classification of states is involved.

More detailed results can be obtained by considering specific states, such as the even parity, $J = 1$ states, $^3S_1 + {}^3D_1$. The wave function for energy $-\kappa^2$ is described by the pair of radial functions $u_0(r)$, $u_2(r)$ that obey the coupled integral equations

$$u(r) = \int_0^\infty dr' g(rr'\kappa)(-V(r')u(r')).$$

Here,
$$u(r) = \begin{pmatrix} u_0(r) \\ u_2(r) \end{pmatrix}, \qquad g = \begin{pmatrix} g_0 & 0 \\ 0 & g_2 \end{pmatrix},$$

and
$$V = \begin{pmatrix} V_a & 2^{3/2}V_b \\ 2^{3/2}\,V_b & V_a - 2V_b \end{pmatrix}.$$

If we exhibit a matrix, $V_c(r)$, such that

$$V(r) \geq V_c(r)$$

and
$$-V_c(r) \geq 0,$$

the evident matrix generalization of the previous arguments supplies the limit

$$n(E) \leq -\kappa^2 < \int dr\; tr\;[g(rr\kappa) - V_c(r)]$$

where the trace refers to the two-dimensional matrices.

A suitable choice is the comparison potential just described, written as a multiple of the unit matrix, which gives

$$n(E \leq -\kappa^2) < \int_0^\infty dr(g_0(rr\kappa) + g_2(rr\kappa))|V_a(r) + 2V_b(r)]|_{V_a + 2V_b < 0}$$

and, in particular,

$$n < \frac{6}{5}\int dr\; r|V_a(r) + 2V_b(r)|\Big]_{V_a + 2V_b < 0}.$$

The latter also implies the alternative bound

$$n(E < -\kappa^2) < \frac{6}{5} \int dr\, r|V_a(r) + 2V_b(r) + \kappa^2| \Big]_{V_a + 2V_b + \kappa^2 < 0}$$

A somewhat more elaborate treatment follows from the remark that the matrix $V(r)$ defines three regions. In region I, $V_a < -4|V_b|$, and $-V$ is positive definite. Region II is characterized by $2|V_b| > V_a > -4|V_b|$, and here the matrix $V$ is indefinite, while in region III, $V_a > 2|V_b|$ and $V$ is positive definite. For a comparison potential, we use $V$ itself in region I, the multiple $V_a + 2V_b$ of the unit matrix in region II, and zero in region III. There results the upper bound

$$n(E \leq -\kappa^2) < \int_{\mathrm{I}} dr[g_0(rr\kappa)|V_a(r)| + g_2(rr\kappa)|V_a(r) - 2V_b(r)|] + \int_{\mathrm{II}} dr(g_0(rr\kappa) + g_2(rr\kappa))|V_a(r) + 2V_b(r)|$$

and

$$n < \int_{\mathrm{I}} dr\, r\left[|V_a(r)| + \frac{1}{5}|V_a(r) - 2V_b(r)|\right] + \frac{6}{5}\int_{\mathrm{II}} dr\, r|V_a(r) + 2V_b(r)|.$$

Again, an alternative limit is obtained for $n(E \leq -\kappa^2)$ on replacing $V_a(r)$ with $V_a(r) + \kappa^2$ in the latter formula, with a corresponding redefinition of regions I and II.

In an application to a physical system, such as the deuteron, for which the distribution of energy values is known, these inequalities provide simple bounds on the potential used to represent the data.

* Supported in part by the Air Force Office of Scientific Research (ARDC).

[1] These PROCEEDINGS, **38**, 961 (1952).

[2] The physical constant $2m/\hbar^2$ is absorbed into the definitions of potential and energy.

[3] Some remarks in a very recent paper, L. Rosenberg and L. Spruch, *Phys. Rev.*, **120**, 474 (1960), footnote 21, indicate that these authors have considered similar questions.

# Brownian Motion of a Quantum Oscillator

JULIAN SCHWINGER*
*Harvard University, Cambridge, Massachusetts*
(Received November 28, 1960)

An action principle technique for the direct computation of expectation values is described and illustrated in detail by a special physical example, the effect on an oscillator of another physical system. This simple problem has the advantage of combining immediate physical applicability (e.g., resistive damping or maser amplification of a single electromagnetic cavity mode) with a significant idealization of the complex problems encountered in many-particle and relativistic field theory. Successive sections contain discussions of the oscillator subjected to external forces, the oscillator loosely coupled to the external system, an improved treatment of this problem and, finally, there is a brief account of a general formulation.

## INTRODUCTION

THE title of this paper refers to an elementary physical example that we shall use to illustrate, at some length, a solution of the following methodological problem. The quantum action principle[1] is a differential characterization of transformation functions, $\langle a't_1|b't_2\rangle$, and thus is ideally suited to the practical computation of transition probabilities (which includes the determination of stationary states). Many physical questions do not pertain to individual transition probabilities, however, but rather to expectation values of a physical property for a specified initial state,

$$\langle X(t_1)\rangle_{b't_2}=\sum_{a'a''}\langle b't_2|a't_1\rangle\langle a't_1|X(t_1)|a''t_1\rangle\langle a''t_1|b't_2\rangle,$$

or, more generally, a mixture of states. Can one devise an action principle technique that is adapted to the direct computation of such expectation values, without requiring knowledge of the individual transformation functions?

The action principle asserts that ($\hbar=1$),

$$\delta\langle a't_1|b't_2\rangle=i\left\langle a't_1\left|\delta\left[\int_{t_2}^{t_1}dtL\right]\right|b't_2\right\rangle,$$

and

$$\delta\langle b't_2|a't_1\rangle=-i\left\langle b't_2\left|\delta\left[\int_{t_2}^{t_1}dtL\right]\right|a't_1\right\rangle,$$

in which we shall take $t_1>t_2$. These mutually complex-conjugate forms correspond to the two viewpoints whereby states at different times can be compared, either by progressing forward from the earlier time, or backward from the later time. The relation between the pair of transformation functions is such that

$$\delta\left[\sum_{a'}\langle b't_2|a't_1\rangle\langle a't_1|b''t_2\rangle\right]=0,$$

which expresses the fixed numerical value of

$$\langle b't_2|b''t_2\rangle=\delta(b',b'').$$

But now, imagine that the positive and negative senses of time development are governed by different dynamics. Then the transformation function for the closed circuit will be described by the action principle

$$\delta\langle t_2|t_2\rangle=\delta[\langle t_2|t_1\rangle\times\langle t_1|t_2\rangle]$$
$$=i\left\langle t_2\left|\delta\left[\int_{t_2}^{t_1}dtL_+-\int_{t_2}^{t_1}dtL_-\right]\right|t_2\right\rangle,$$

in which abbreviated notation the multiplication sign symbolizes the composition of transformation functions by summation over a complete set of states. If, in particular, the Lagrangian operators $L_\pm$ contain the dynamical term $\lambda_\pm(t)X(t)$, we have

$$\delta_\lambda\langle t_2|t_2\rangle=i\left\langle t_2\left|\int_{t_2}^{t_1}dt(\delta\lambda_+-\delta\lambda_-)X(t)\right|t_2\right\rangle,$$

and, therefore,

$$-i\frac{\delta}{\delta\lambda_+(t_1)}\langle t_2|t_2\rangle=i\frac{\delta}{\delta\lambda_-(t_1)}\langle t_2|t_2\rangle$$
$$=\langle t_2|X(t_1)|t_2\rangle,$$

where $\lambda_\pm$ can now be identified. Accordingly, if a system is suitably perturbed[2] in a manner that depends upon the time sense, a knowledge of the transformation function referring to a closed time path determines the expectation value of any desired physical quantity for a specified initial state or state mixture.

## OSCILLATOR

To illustrate this remark we first consider an oscillator subjected to an arbitrary external force, as described by the Lagrangian operator

$$L=iy^\dagger(dy/dt)-\omega y^\dagger y-y^\dagger K(t)-yK^*(t),$$

* Supported by the Air Force Office of Scientific Research (ARDC).

[1] Some references are: Julian Schwinger, Phys. Rev. **82**, 914 (1951); **91**, 713 (1953); Phil. Mag. **44**, 1171 (1953). The first two papers also appear in *Selected Papers on Quantum Electrodynamics* (Dover Publications, New York, 1958). A recent discussion is contained in Julian Schwinger, Proc. Natl. Acad. Sci. U. S. **46**, 883 (1960).

[2] Despite this dynamical language, a change in the Hamiltonian operator of a system can be kinematical in character, arising from the consideration of another transformation along with the dynamical one generated by the Hamiltonian. See the last paper quoted in footnote 1, and Julian Schwinger, Proc. Natl. Acad. Sci. U. S. **46**, 1401 (1960).



Reprinted from the *Journal of Mathematical Physics* **2**, 407–432 (1961).

in which the complementary pair of non-Hermitian operators $y$, $iy^\dagger$, are constructed from Hermitian operators $q$, $p$ by

$$y=2^{-\frac{1}{2}}(q+ip)$$
$$iy^\dagger=2^{-\frac{1}{2}}(p+iq).$$

The equations of motion implied by the action principle are

$$i(dy/dt)-\omega y=K$$
$$-i(dy^\dagger/dt)-\omega y^\dagger=K^*,$$

and solutions are given by

$$y(t)=e^{-i\omega(t-t_2)}y(t_2)-i\int_{t_2}^{t}dt'e^{-i\omega(t-t')}K(t'),$$

together with the adjoint equation. Since we now distinguish between the forces encountered in the positive time sense, $K_+(t)$, $K_+{}^*(t)$, and in the reverse time direction, $K_-(t)$, $K_-{}^*(t)$, the integral must be taken along the appropriate path. Thus, when $t$ is reached first in the time evolution from $t_2$, we have

$$y_+(t)=e^{-i\omega(t-t_2)}y_+(t_2)-i\int_{t_2}^{t}dt'e^{-i\omega(t-t')}K_+(t'),$$

while on the subsequent return to time $t$,

$$y_-(t)=e^{-i\omega(t-t_2)}y_+(t_2)-i\int_{t_2}^{t_1}dt'e^{-i\omega(t-t')}K_+(t')$$
$$+i\int_{t}^{t_1}dt'e^{-i\omega(t-t')}K_-(t').$$

Note that

$$y_-(t_1)-y_+(t_1)=0,$$
$$y_-(t_2)-y_+(t_2)=i\int_{t_2}^{t_1}dte^{i\omega(t-t_2)}(K_--K_+)(t).$$

We shall begin by constructing the transformation function referring to the lowest energy state of the unperturbed oscillator, $\langle 0t_2|0t_2\rangle^{K\pm}$. This state can be characterized by

$$\langle 0t_2|y^\dagger y(t_2)|0t_2\rangle=0$$

or, equivalently, by the eigenvector equations

$$y(t_2)|0t_2\rangle=0,\quad \langle 0t_2|y^\dagger(t_2)=0.$$

Since the transformation function simply equals unity if $K_+=K_-$ and $K_+{}^*=K_-{}^*$, we must examine the effect of independent changes in $K_+$ and $K_-$, and of $K_+{}^*$ and $K_-{}^*$, as described by the action principle

$$\delta_K\langle 0t_2|0t_2\rangle^{K\pm}=-i\left\langle 0t_2\left|\left[\int_{t_2}^{t_1}dt(\delta K_+{}^*y_+-\delta K_-{}^*y_-)\right.\right.\right.$$
$$\left.\left.\left.+\int_{t_2}^{t_1}dt(y_+{}^\dagger\delta K_+-y_-{}^\dagger\delta K_-)\right]\right|0t_2\right\rangle^{K\pm}.$$

The choice of initial state implies effective boundary conditions that supplement the equations of motion,

$$y_+(t_2)\to 0,\quad y_-{}^\dagger(t_2)\to 0.$$

Hence, in effect we have

$$y_+(t)=-i\int_{t_2}^{t_1}dt'e^{-i\omega(t-t')}\eta_+(t-t')K_+(t')$$

and

$$y_-(t)=-i\int_{t_2}^{t_1}dt'e^{-i\omega(t-t')}K_+(t')$$
$$+i\int_{t_2}^{t_1}dt'e^{-i\omega(t-t')}\eta_-(t-t')K_-(t'),$$

together with the similar adjoint equations obtained by interchanging the $\pm$ labels. For convenience, step functions have been introduced:

$$\eta_+(t-t')=\begin{cases}1, & t-t'>0\\ 0, & t-t'<0\end{cases},$$

$$\eta_-(t-t')=\begin{cases}1, & t-t'<0\\ 0, & t-t'>0\end{cases},$$

$$\eta_+(t-t')+\eta_-(t-t')=1,\quad \eta_+(0)=\eta_-(0)=\tfrac{1}{2}.$$

We shall also have occasion to use the odd function

$$\epsilon(t-t')=\eta_+(t-t')-\eta_-(t-t').$$

The solution of the resulting integrable differential expression for $\log\langle 0t_2|0t_2\rangle^{K\pm}$ is given by

$$\langle 0t_2|0t_2\rangle^{K\pm}=\exp\left[-i\int_{t_2}^{t_1}dtdt'K^*(t)G_0(t-t')K(t')\right],$$

in a matrix notation, with

$$K(t)=\begin{pmatrix}K_+(t)\\ K_-(t)\end{pmatrix}$$

and

$$iG_0(t-t')=e^{-i\omega(t-t')}\begin{pmatrix}\eta_+(t-t') & 0\\ -1 & \eta_-(t-t')\end{pmatrix}.$$

The requirement that the transformation function reduce to unity on identifying $K_+$ with $K_-$, $K_+{}^*$ with $K_-{}^*$, is satisfied by the null sum of all elements of $G_0$, as assured by the property $\eta_++\eta_-=1$.

An operator interpretation of $G_0$ is given by the second variation

$$-\delta_{K^*}\delta_K\langle 0t_2|0t_2\rangle^{K\pm}|_{K=K^*=0}$$
$$=i\int dtdt'\delta K^*(t)G_0(t-t')\delta K(t').$$

Generally, on performing two distinct variations in the structure of $L$ that refer to parameters upon which the dynamical variables at a given time are not explicitly dependent, we have

$$-\delta_1\delta_2\langle t_2|t_2\rangle=\left\langle t_2\left|\int_{t_2}^{t_1}dtdt'\{(\delta_1L_+(t)\delta_2L_+(t'))_+ -\delta_1L_-(t)\delta_2L_+(t')-\delta_2L_-(t)\delta_1L_+(t') +(\delta_1L_-(t)\delta_2L_-(t'))_-\}\right|t_2\right\rangle,$$

in which the multiplication order follows the sense of time development. Accordingly,

$$iG_0(t-t')=\begin{pmatrix}\langle(y(t)y^\dagger(t'))_+\rangle_0 & -\langle y^\dagger(t')y(t)\rangle_0\\ -\langle y(t)y^\dagger(t')\rangle_0 & \langle(y(t)y^\dagger(t'))_-\rangle_0\end{pmatrix},$$

where the expectation values and operators refer to the lowest state and the dynamical variables of the unperturbed oscillator. The property of $G_0$ that the sum of rows and columns vanishes is here a consequence of the algebraic property

$$(y(t)y^\dagger(t'))_+ + (y(t)y^\dagger(t'))_- = \{y(t),y^\dagger(t')\}.$$

The choice of oscillator ground state is no essential restriction since we can now derive the analogous results for any initial oscillator state. Consider, for this purpose, the impulse forces

$$K_+(t)=iy''\delta(t-t_2),$$
$$K_-^*(t)=-iy^{\dagger\prime}\delta(t-t_2),$$

the effects of which are described by

$$y_+(t_2+0)-y_+(t_2)=y'',$$
$$y_-^\dagger(t_2+0)-y_-^\dagger(t_2)=y^{\dagger\prime}.$$

Thus, under the influence of these forces, the states $|0t_2\rangle$ and $\langle 0t_2|$ become, at the time $t_2+0$, the states $|y''t_2\rangle$ and $\langle y^{\dagger\prime}t_2|$, which are right and left eigenvectors, respectively, of the operators $y(t_2)$ and $y^\dagger(t_2)$. On taking into account arbitrary additional forces, the transformation function for the closed time path can be expressed as

$$\langle y^{\dagger\prime}t_1|y''t_2\rangle^{K\pm} =\exp\left[y^{\dagger\prime}y''-y^{\dagger\prime}\left(\int_{t_2}^{t_1}dtG_0(t_2-t)K(t)\right)_- +\left(\int_{t_2}^{t_1}dtK^*(t)G_0(t-t_2)\right)_+y'' -i\int_{t_2}^{t_1}dtdt'K^*(t)G_0(t-t')K(t')\right],$$

in which

$$\left(\int_{t_2}^{t_1}dtG_0(t_2-t)K(t)\right)_-=-i\int_{t_2}^{t_1}dte^{i\omega(t-t_2)}(K_--K_+)(t)$$

and

$$\left(\int dtK^*(t)G_0(t-t_2)\right)_+ =-i\int_{t_2}^{t_1}dte^{-i\omega(t-t_2)}(K_+^*-K_-^*)(t).$$

The eigenvectors of the non-Hermitian canonical variables are complete and have an intrinsic physical interpretation in terms of $q$ and $p$ measurements of optimum compatibility.[3] For our immediate purposes, however, we are more interested in the unperturbed oscillator energy states. The connection between the two descriptions can be obtained by considering the unperturbed oscillator transformation function

$$\langle y^{\dagger\prime}t_1|y''t_2\rangle=\langle y^{\dagger\prime}|\exp[-i(t_1-t_2)\omega y^\dagger y]|y''\rangle.$$

Now

$$i(\partial/\partial t_1)\langle y^{\dagger\prime}t_1|y''t_2\rangle=\langle y^{\dagger\prime}t_1|\omega y^\dagger(t_1)y(t_1)|y''t_2\rangle =\omega y^{\dagger\prime}e^{-i\omega(t_1-t_2)}y''\langle y^{\dagger\prime}t_1|y''t_2\rangle,$$

since

$$y(t_1)=e^{-i\omega(t_1-t_2)}y(t_2),$$

which yields

$$\langle y^{\dagger\prime}t_1|y''t_2\rangle=\exp[y^{\dagger\prime}e^{-i\omega(t_1-t_2)}y''] =\sum_{n=0}^{\infty}\frac{(y^{\dagger\prime})^n}{(n!)^{\frac12}}e^{-in\omega(t_1-t_2)}\frac{(y'')^n}{(n!)^{\frac12}}.$$

We infer the nonnegative integer spectrum of $y^\dagger y$, and the corresponding wave functions

$$\langle y^{\dagger\prime}|n\rangle=(y^{\dagger\prime})^n/(n!)^{\frac12},\quad \langle n|y'\rangle=(y')^n/(n!)^{\frac12}.$$

Accordingly, a non-Hermitian canonical variable transformation function can serve as a generator for the transformation function referring to unperturbed oscillator energy states,

$$\langle y^{\dagger\prime}t_2|y''t_2\rangle^{K\pm}=\sum_{n,n'=0}^{\infty}\frac{(y^{\dagger\prime})^n}{(n!)^{\frac12}}\langle nt_2|n't_2\rangle^{K\pm}\frac{(y'')^{n'}}{(n'!)^{\frac12}}.$$

If we are specifically interested in $\langle nt_2|nt_2\rangle^{K\pm}$, which supplies all expectation values referring to the initial state $n$, we must extract the coefficient of $(y^{\dagger\prime}y'')^n/n!$ from an exponential of the form

$$\exp[y^{\dagger\prime}y''+y^{\dagger\prime}\alpha+\beta y''+\gamma] =\sum_{kl}\frac{(y^{\dagger\prime})^k}{k!}\frac{(y'')^l}{l!}\alpha^k\beta^l\exp[y^{\dagger\prime}y''+\gamma].$$

All the terms that contribute to the required coefficient

[3] A discussion of non-Hermitian representations is given in *Lectures on Quantum Mechanics* (Les Houches, 1955), unpublished.

are contained in

$$\sum_{k=0}^{\infty} \frac{(y^{\dagger\prime}y'')^k}{(k!)^2}(\alpha\beta)^k \exp[y^{\dagger\prime}y''+\gamma]$$
$$=\frac{1}{2\pi i}\oint\frac{d\lambda}{\lambda}e^{\lambda}\exp[y^{\dagger\prime}y''(1+\lambda^{-1}\alpha\beta)+\gamma],$$

where the latter version is obtained from

$$\frac{1}{2\pi i}\oint\frac{d\lambda}{\lambda^{k+1}}e^{\lambda}=\frac{1}{k!},$$

and

$$\langle nt_2|nt_2\rangle^{K\pm}=\exp\left[-i\int dtdt'K^*(t)G_0(t-t')K(t')\right]$$
$$\times L_n\left[\left(\int dtK^*(t)G_0(t-t_2)\right)_+ \times\left(\int dtG_0(t_2-t)K(t)\right)_-\right],$$

in which the $n$th Laguerre polynomial has been introduced on observing that

$$\frac{1}{2\pi i}\oint\frac{d\lambda}{\lambda}e^{\lambda}(1-\lambda^{-1}x)^n=\frac{1}{n!}e^x\left(\frac{d}{dx}\right)^n x^n e^{-x}=L_n(x).$$

One obtains a much neater form, however, from which these results can be recovered, on considering an initial mixture of oscillator energy states for which the $n$th state is assigned the probability

$$(1-e^{-\beta\omega})e^{-n\beta\omega},$$

and

$$\beta^{-1}=\vartheta$$

can be interpreted as a temperature. Then, since

$$(1-e^{-\beta\omega})\sum_{n=0}^{\infty}e^{-n\beta\omega}L_n(x)=(1-e^{-\beta\omega})\frac{1}{2\pi i}\oint\frac{d\lambda}{\lambda}$$
$$\times e^{\lambda}[1-e^{-\beta\omega}+\lambda^{-1}e^{-\beta\omega}x]^{-1}=\exp\left[-\frac{x}{e^{\beta\omega}-1}\right],$$

we obtain

$$\langle t_2|t_2\rangle_\vartheta^{K\pm}=\exp\left[-i\int_{t_2}^{t_1}dtdt'K^*(t)G_\vartheta(t-t')K(t')\right],$$

with

$$iG_\vartheta(t-t')=iG_0(t-t')+(e^{\beta\omega}-1)^{-1}G_0(t-t_2)_+\,{}_-G_0(t_2-t'),$$

and in which

$$iG_0(t-t_2)_+=e^{-i\omega(t-t_2)}\begin{pmatrix}1\\-1\end{pmatrix},$$

$$i_-G_0(t_2-t)=e^{i\omega(t-t_2)}\begin{pmatrix}-1 & 1\end{pmatrix}.$$

Thus,

$$iG_\vartheta(t-t')=e^{-i\omega(t-t')}\begin{pmatrix}\eta_+(t-t')+\langle n\rangle_\vartheta, & -\langle n\rangle_\vartheta\\ -1-\langle n\rangle_\vartheta, & \eta_-(t-t')+\langle n\rangle_\vartheta\end{pmatrix},$$

where we have written

$$\langle n\rangle_\vartheta=(e^{\beta\omega}-1)^{-1},$$

and since the elements of $G_\vartheta$ are also given by unperturbed oscillator thermal expectation values

$$iG_\vartheta(t-t')=\begin{pmatrix}\langle(y(t)y^\dagger(t'))_+\rangle_\vartheta & -\langle y^\dagger(t')y(t)\rangle_\vartheta\\ -\langle y(t)y^\dagger(t')\rangle_\vartheta & \langle(y(t)y^\dagger(t'))_-\rangle_\vartheta\end{pmatrix},$$

the designation $\langle n\rangle_\vartheta$ is consistent with its identification as $\langle y^\dagger y\rangle_\vartheta$.

The thermal forms can also be derived directly by solving the equations of motion, in the manner used to find $\langle 0t_2|0t_2\rangle^{K\pm}$. On replacing the single diagonal element

$$\langle 0t_2|0t_2\rangle^{K\pm}=\langle 0t_2|U|0t_2\rangle$$

by the statistical average

$$(1-e^{-\beta\omega})\sum_0^\infty e^{-n\beta\omega}\langle nt_2|nt_2\rangle^{K\pm}$$
$$=(1-e^{-\beta\omega})\ \mathrm{tr}[\exp(-\beta\omega y^\dagger y)U],$$

we find the following relation,

$$y_-(t_2)=e^{\beta\omega}y_+(t_2),$$

instead of the effective initial condition $y_+(t_2)=0$. This is obtained by combining

$$\exp(-\beta\omega y^\dagger y)y\exp(\beta\omega y^\dagger y)=\exp(\beta\omega)y$$

with the property of the trace

$$\mathrm{tr}[\exp(-\beta\omega y^\dagger y)yU]=\mathrm{tr}[\exp(\beta\omega)y\exp(-\beta\omega y^\dagger y)U]$$
$$=\mathrm{tr}[\exp(-\beta\omega y^\dagger y)U\exp(\beta\omega)y].$$

We also have

$$y_-(t_2)-y_+(t_2)=-i\int_{t_2}^{t_1}dte^{i\omega(t-t_2)}(K_+-K_-)(t),$$

and therefore, effectively,

$$y_+(t_2)=-i\frac{1}{e^{\beta\omega}-1}\int_{t_2}^{t_1}dte^{i\omega(t-t_2)}(K_+-K_-)(t).$$

Hence, to the previously determined $y_\pm(t)$ is to be added the term

$$-i\langle n\rangle_\vartheta\int_{t_2}^{t_1}dt'e^{-i\omega(t-t')}(K_+-K_-)(t'),$$

and correspondingly

$$\langle t_2|t_2\rangle_\vartheta^{K\pm}=\langle t_2|t_2\rangle_0^{K\pm}\exp\left[-\langle n\rangle_\vartheta\int_{t_2}^{t_1}dtdt' \times(K_+^*-K_-^*)(t)e^{-i\omega(t-t')}(K_+-K_-)(t')\right],$$

which reproduces the earlier result.

As an elementary application let us evaluate the expectation value of the oscillator energy at time $t_1$ for a system that was in thermal equilibrium at time $t_2$ and is subsequently disturbed by an arbitrarily time-varying force. This can be computed as

$$\langle t_2|\omega y^\dagger y(t_1)|t_2\rangle_\vartheta{}^K = \omega\frac{\delta}{\delta K_-(t_1)}\frac{\delta}{\delta K_+{}^*(t_1)}\langle t_2|t_2\rangle_\vartheta{}^{K\pm}|_{K_+=K_-,\, K_+{}^*=K_-{}^*}.$$

The derivative $\delta/\delta K_+{}^*(t_1)$ supplies the factor

$$-i\left(\int_{t_2}^{t_1} dt G_\vartheta(t_1-t')K(t')\right)_+,$$

the subsequent variation with respect to $K_-(t_1)$ gives

$$-iG_\vartheta(0)_{+-}+\left(\int dt K^*(t)G_\vartheta(t-t_1)\right)_- \times\left(\int dt' G_\vartheta(t_1-t')K(t')\right)_+,$$

and the required energy expectation value equals

$$\omega\langle n\rangle_\vartheta+\omega\left|\int_{t_2}^{t_1} dt e^{i\omega t}K(t)\right|^2.$$

More generally, the expectation values of all functions of $y(t_1)$ and $y^\dagger(t_1)$ are known from that of

$$\exp\{-i[\lambda y^\dagger(t_1)+\mu y(t_1)]\},$$

and this quantity is obtained on supplementing $K_+$ and $K_+{}^*$ by the impulsive forces (note that in this use of the formalism a literal complex-conjugate relationship is not required)

$$K_+(t)=\lambda\delta(t-t_1),$$
$$K_+{}^*(t)=\mu\delta(t-t_1).$$

Then,

$$\langle t_2|\exp\{-i[\lambda y^\dagger(t_1)+\mu y(t_1)]\}|t_2\rangle_\vartheta{}^K = \exp\left[-\lambda\mu(\langle n\rangle_\vartheta+\tfrac{1}{2})+\lambda\int_{t_2}^{t_1} dt e^{i\omega(t_1-t)}K^*(t) -\mu\int_{t_2}^{t_1} dt e^{-i\omega(t_1-t)}K(t)\right],$$

which involves the special step-function value

$$\eta_+(0)=\tfrac{1}{2}.$$

Alternatively, if we choose

$$K_+(t)=\lambda\delta(t-t_1),$$
$$K_+{}^*(t)=\mu\delta(t-t_1+0),$$

there appears

$$\langle t_2|\exp[-i\lambda y^\dagger(t_1)]\exp[-i\mu y(t_1)]|t_2\rangle_\vartheta{}^K = \exp\left[-\lambda\mu\langle n\rangle_\vartheta+\lambda\int_{t_2}^{t_1} dt e^{i\omega(t_1-t)}K^*(t) -\mu\int_{t_2}^{t_1} dt e^{-i\omega(t_1-t)}K(t)\right].$$

It may be worth remarking, in connection with these results, that the attention to expectation values does not deprive us of the ability to compute individual probabilities. Indeed, if probabilities for specific oscillator energy states are of interest, we have only to exhibit, as functions of $y$ and $y^\dagger$, the projection operators for these states, the expectation values of which are the required probabilities. Now

$$P_n=|n\rangle\langle n|$$

is represented by the matrix

$$\langle y^{\dagger\prime}|P_n|y''\rangle=(y^{\dagger\prime}y'')^n/n! =[(y^{\dagger\prime}y'')^n/n!]\exp(-y^{\dagger\prime}y'')\langle y^{\dagger\prime}|y''\rangle,$$

and, therefore,

$$P_n=\frac{1}{n!}(y^\dagger)^n\left[\sum_{k=0}^{\infty}\frac{(-1)^k}{k!}(y^\dagger)^k y^k\right]y^n =\frac{1}{n!}(y^\dagger)^n\exp(-y^\dagger;y)y^n,$$

in which we have introduced a notation to indicate this ordered multiplication of operators. A convenient generating function for these projection operators is

$$\sum_{n=0}^{\infty}\alpha^n P_n=\exp[-(1-\alpha)y^\dagger;y],$$

and we observe that

$$\sum_0^\infty \alpha^n P_n=\exp\left[(1-\alpha)\frac{\partial}{\partial\lambda}\frac{\partial}{\partial\mu}\right] \times\exp(-i\lambda y^\dagger)\exp(-i\mu y)|_{\lambda=\mu=0}.$$

Accordingly,

$$\sum_0^\infty \alpha^n p(n,\vartheta,K)=\exp\left[(1-\alpha)\frac{\partial}{\partial\lambda}\frac{\partial}{\partial\mu}\right]\exp\left[-\lambda\mu\langle n\rangle_\vartheta +\lambda e^{i\omega t_1}\int dt e^{-i\omega t}K^*(t) -\mu e^{-i\omega t_1}\int dt e^{i\omega t}K(t)\right]\Bigg|_{\lambda=\mu=0}$$

gives the probability of finding the oscillator in the $n$th energy state after an arbitrary time-varying force has acted, if it was initially in a thermal mixture of states.

To evaluate

$$X=\exp\left[(1-\alpha)\frac{\partial}{\partial\lambda}\frac{\partial}{\partial\mu}\right]\exp[-\lambda\mu\langle n\rangle+\lambda\gamma^*-\mu\gamma]|_{\lambda=\mu=0},$$

we first remark that

$$\frac{\partial}{\partial\gamma^*}X=\exp\left[(1-\alpha)\frac{\partial}{\partial\lambda}\frac{\partial}{\partial\mu}\right]\lambda\exp[\ ]|_{\lambda=\mu=0}$$

$$=(1-\alpha)\exp\left[(1-\alpha)\frac{\partial}{\partial\lambda}\frac{\partial}{\partial\mu}\right]\frac{\partial}{\partial\mu}\exp[\ ]|$$

$$=(1-\alpha)\exp\left[(1-\alpha)\frac{\partial}{\partial\lambda}\frac{\partial}{\partial\mu}\right](-\lambda\langle n\rangle-\gamma)\exp[\ ]|,$$

from which follows

$$\frac{\partial}{\partial\gamma^*}X=-\frac{\gamma(1-\alpha)}{1+\langle n\rangle(1-\alpha)}X$$

or

$$X=X_0\exp\left[-|\gamma|^2\frac{1-\alpha}{1+\langle n\rangle(1-\alpha)}\right].$$

Here

$$X_0=\exp\left[(1-\alpha)\frac{\partial}{\partial\lambda}\frac{\partial}{\partial\mu}\right]\exp[-\lambda\mu\langle n\rangle]|_{\lambda=\mu=0}$$

$$=[1+\langle n\rangle(1-\alpha)]^{-1},$$

as one shows with a similar procedure, or by direct series expansion. Therefore,

$$\sum_0^\infty \alpha^n p(n,\vartheta,K)=\frac{1-e^{-\beta\omega}}{1-\alpha e^{-\beta\omega}}\exp\left[-|\gamma|^2\frac{1-e^{-\beta\omega}}{1-\alpha e^{-\beta\omega}}(1-\alpha)\right],$$

where

$$|\gamma|^2=\left|\int dt e^{i\omega t}K(t)\right|^2,$$

and on referring to the previously used Laguerre polynomial sum formula, we obtain

$$p(n,\vartheta,K)=(1-e^{-\beta\omega})e^{-n\beta\omega}\exp[-|\gamma|^2(1-e^{-\beta\omega})]\times L_n[-4|\gamma|^2\sinh^2(\beta\omega/2)].$$

In addition to describing the physical situation of initial thermal equilibrium, this result provides a generating function for the individual transition probabilities between oscillator energy states,

$$\sum_{n'=0}^\infty p(n,n',K)e^{-(n'-n)\beta\omega}$$

$$=\exp[-|\gamma|^2(1-e^{-\beta\omega})]L_n[-(1-e^{-\beta\omega})(e^{\beta\omega}-1)|\gamma|^2].$$

This form, and the implied transition probabilities, have already been derived in another connection,[4] and we shall only state the general result here:

$$p(n,n',K)=\frac{n_<!}{n_>!}(|\gamma|^2)^{n_>-n_<}[L_{n_<}^{(n_>-n_<)}(|\gamma|^2)]^2\times\exp(-|\gamma|^2),$$

in which $n_>$ and $n_<$ represent the larger and smaller of the two integers $n$ and $n'$.

Another kind of probability is also easily identified, that referring to the continuous spectrum of the Hermitian operator

$$q=2^{-\frac{1}{2}}(y+y^\dagger)$$

[or $p=-2^{-\frac{1}{2}}i(y-y^\dagger)$]. For this purpose, we place $\lambda=\mu=-2^{-\frac{1}{2}}p'$ and obtain

$$\langle t_2|e^{ip'q(t_1)}|t_2\rangle_\vartheta{}^K=\exp[-\tfrac{1}{2}p'^2(\langle n\rangle_\vartheta+\tfrac{1}{2})+ip'\langle q(t_1)\rangle^K],$$

with

$$\langle q(t_1)\rangle^K=2^{-\frac{1}{2}}i\left[e^{i\omega t_1}\int_{t_2}^{t_1}dt e^{-i\omega t}K^*(t)-e^{-i\omega t_1}\int_{t_2}^{t_1}dt e^{i\omega t}K(t)\right].$$

If we multiply this result by $\exp(-ip'q')$ and integrate with respect to $p'/2\pi$ from $-\infty$ to $\infty$, we obtain the expectation value of $\delta[q(t_1)-q']$ which is the probability of realizing a value of $q(t_1)$ in a unit interval about $q'$:

$$p(q't_1,\beta,K)=(\pi^{-1}\tanh\tfrac{1}{2}\beta\omega)^{\frac{1}{2}}\times\exp[-(\tanh\tfrac{1}{2}\beta\omega)(q'-\langle q(t_1)\rangle^K)^2].$$

Still another derivation of the formula giving thermal expectation values merits attention. Now we let the return path terminate at a different time $t_2'=t_2-T$, and on regarding the resulting transformation function as a matrix, compute the trace, or rather the trace ratio

$$\mathrm{tr}\langle t_2'|t_2\rangle^{K\pm}/\mathrm{tr}\langle t_2'|t_2\rangle,$$

which reduces to unity in the absence of external forces. The action principle again describes the dependence upon $K_\pm^*(t)$, $K_\pm(t)$ through the operators $y_\pm(t)$, $y_\pm^\dagger(t)$ which are related to the forces by the solutions of the equations of motion, and, in particular,

$$y_-(t_2')=e^{-i\omega(t_2'-t_2)}y_+(t_2)-i\int_{t_2}^{t_1}dt e^{i\omega(t-t_2')}K_+(t)+i\int_{t_2'}^{t_1}dt e^{i\omega(t-t_2')}K_-(t).$$

Next we recognize that the structure of the trace implies the effective boundary condition

$$y_-(t_2')=y_+(t_2).$$

[4] Julian Schwinger, Phys. Rev. **91**, 728 (1953).

Let us consider

$$\mathrm{tr}\langle t_2'|y_-(t_2')|t_2\rangle=\sum_{a'}\langle a't_2'|y_-(t_2')|a't_2\rangle,$$

where we require of the $a$ representation only that it have no explicit time dependence. Then

$$\langle a't_2'|y_-(t_2')=\sum_{a''}\langle a'|y|a''\rangle\langle a''t_2'|$$

and

$$\begin{aligned}\mathrm{tr}\langle t_2'|y_-(t_2')|t_2\rangle&=\sum_{a'a''}\langle a''t_2'|a't_2\rangle\langle a'|y|a''\rangle\\&=\mathrm{tr}\langle t_2'|y_+(t_2)|t_2\rangle,\end{aligned}$$

which is the stated result.

The effective initial condition now appears as

$$y_+(t_2)=-\frac{1}{e^{-i\omega T}-1}i\left[\int_{t_2}^{t_1}dt\,e^{i\omega(t-t_2)}K_+(t)-\int_{t_2'}^{t_1}dt\,e^{i\omega(t-t_2)}K_-(t)\right],$$

and the action principle supplies the following evaluation of the trace ratio:

$$\exp\left[-i\int dtdt'K^*(t)G_0(t-t')K(t')\right]\times\exp\left[-(e^{-i\omega T}-1)^{-1}\left|\int dt\,e^{i\omega t}(K_+-K_-)(t)\right|^2\right],$$

where the time variable in $K_+$ and $K_-$ ranges from $t_2$ to $t_1$ and from $t_2'$ to $t_1$, respectively. To solve the given physical problem we require that $K_-(t)$ vanish in the interval between $t_2'$ and $t_2$ so that all time integrations are extended between $t_2$ and $t_1$. Then, since

$$\langle t_2'|=\langle t_2|e^{-i\omega(t_2'-t_2)n},\quad n=y^\dagger y(t_2),$$

what has been evaluated equals

$$\mathrm{tr}\langle t_2|e^{i\omega Tn}|t_2\rangle^{K\pm}/\mathrm{tr}\langle t_2|e^{i\omega Tn}|t_2\rangle,$$

and by adding the remark that this ratio continues to exist on making the complex substitution

$$-iT\to\beta>0,$$

the desired formula emerges as

$$\mathrm{tr}\langle t_2|e^{-\beta\omega n}|t_2\rangle^{K\pm}/\mathrm{tr}\langle t_2|e^{-\beta\omega n}|t_2\rangle=\exp\left[-i\int dtdt'K^*(t)G_\vartheta(t-t')K(t')\right].$$

### EXTERNAL SYSTEM

This concludes our preliminary survey of the oscillator and we turn to the specific physical problem of interest: An oscillator subjected to prescribed external forces and loosely coupled to an essentially macroscopic external system. All oscillator interactions are linear in the oscillator variables, as described by the Lagrangian operator

$$L=iy^\dagger(dy/dt)-\omega_0y^\dagger y-y^\dagger K(t)-yK^*(t)-2^{\frac{1}{2}}qQ+L_{\mathrm{ext}},$$

in which $L_{\mathrm{ext}}$ characterizes the external system and $Q(t)$ is a Hermitian operator of that system.

We begin our treatment with a discussion of the transformation function $\langle t_2|t_2\rangle_{\vartheta_0\vartheta}{}^{K\pm}$ that refers initially to a thermal mixture at temperature $\vartheta$ for the external system, and to an independent thermal mixture at temperature $\vartheta_0$ for the oscillator. The latter temperature can be interpreted literally, or as a convenient parametric device for obtaining expectation values referring to oscillator energy states. To study the effect of the coupling between the oscillator and the external system we supply the coupling term with a variable parameter $\lambda$, and compute

$$\frac{\partial}{\partial\lambda}\langle t_1|t_2\rangle^{K\pm}=-i\left\langle t_2\left|\int_{t_2}^{t_1}dt[2^{\frac{1}{2}}q_+(t)Q_+(t)-2^{\frac{1}{2}}q_-(t)Q_-(t)]\right|t_2\right\rangle^{K\pm},$$

where the distinction between the forward and return paths arises only from the application of different external forces $K_\pm(t)$ on the two segments of the closed time contour. The characterization of the external system as essentially macroscopic now enters through the assumption that this large system is only slightly affected by the coupling to the oscillator. In a corresponding first approximation, we would replace the operators $Q_\pm(t)$ by the effective numerical quantity $\langle Q(t)\rangle_\vartheta$. The phenomena that appear in this order of accuracy are comparatively trivial, however, and we shall suppose that

$$\langle Q(t)\rangle_\vartheta=0,$$

which forces us to proceed to the next approximation.

A second differentiation with respect to $\lambda$ gives

$$-\frac{1}{2}\frac{\partial^2}{\partial\lambda^2}\langle t_2|t_2\rangle^{K\pm}=\left\langle t_2\left|\int_{t_2}^{t_1}dtdt'[(qQ(t)qQ(t'))_+-2q_-Q_-(t)q_+Q_+(t')+(qQ(t)qQ(t'))_-]\right|t_2\right\rangle^{K\pm}.$$

The introduction of an approximation based upon the slight disturbance of the macroscopic system converts this into

$$\begin{aligned}-\frac{1}{2}\frac{\partial^2}{\partial\lambda^2}\langle t_2|t_2\rangle^{K\pm}=\Big\langle t_2\Big|\int_{t_2}^{t_1}dtdt'[&(y(t')y^\dagger(t))_+A_{++}(t-t')\\&-y_-(t')y_+^\dagger(t)A_{+-}(t-t')-y_-^\dagger(t)y_+(t')A_{-+}(t-t')\\&+(y(t')y^\dagger(t))_-A_{--}(t-t')]\Big|t_2\Big\rangle^{K\pm},\end{aligned}$$

where

$$A(t-t')=\begin{pmatrix}\langle(Q(t)Q(t'))_+\rangle_\vartheta & \langle Q(t')Q(t)\rangle_\vartheta\\ \langle Q(t)Q(t')\rangle_\vartheta & \langle(Q(t)Q(t'))_-\rangle_\vartheta\end{pmatrix},$$

and we have also discarded all terms containing $y(t)y(t')$ and $y^\dagger(t)y^\dagger(t')$. The latter approximation refers to the assumed weakness of the coupling of the oscillator to the external system, for, during the many periods that are needed for the effect of the coupling to accumulate, quantities with the time dependence $e^{\pm i\omega_0(t+t')}$ will become suppressed in comparison with those varying as $e^{\pm i\omega_0(t-t')}$. At this point we ask what effective term in an action operator that refers to the closed time path of the oscillator would reproduce this approximate value of $(\partial/\partial\lambda)^2\langle t_2|t_2\rangle$ at $\lambda=0$. The complete action that satisfies this requirement, with $\lambda^2$ set equal to unity, is given by

$$W=\int_{t_2}^{t_1}dt\left[iy^\dagger\frac{dy}{dt}-\omega_0y^\dagger y-y^\dagger K-yK^*\Big|_+-\Big|_-\right]$$
$$+i\int_{t_2}^{t_1}dtdt'[(y^\dagger(t)y(t'))_+A_{++}(t-t')$$
$$-y_-^\dagger(t)y_+(t')A_{-+}(t-t')-y_-(t')y_+^\dagger(t)A_{+-}(t-t')$$
$$+(y^\dagger(t)y(t'))_-A_{--}(t-t')].$$

The application of the principle of stationary action to this action operator yields equations of motion that are nonlocal in time, namely,

$$i\frac{dy_+}{dt}-\omega_0y_++i\int_{t_2}^{t_1}dt'[A_{++}(t-t')y_+(t')-A_{+-}(t-t')y_-(t')]=K_+(t)$$
$$i\frac{dy_-}{dt}-\omega_0y_--i\int_{t_2}^{t_1}dt'[A_{--}(t-t')y_-(t')-A_{-+}(t-t')y_+(t')]=K_-(t),$$

together with

$$-i\frac{dy_+^\dagger}{dt}-\omega_0y_+^\dagger+i\int_{t_2}^{t_1}dt'[y_+^\dagger(t')A_{++}(t'-t)-y_-^\dagger(t')A_{-+}(t'-t)]=K_+^*(t)$$
$$-i\frac{dy_-^\dagger}{dt}-\omega_0y_-^\dagger-i\int_{t_2}^{t_1}dt'[y_-^\dagger(t')A_{--}(t'-t)-y_+^\dagger(t')A_{+-}(t'-t)]=K_-^*(t).$$

The latter set is also obtained by combining the formal adjoint operation with the interchange of the + and − labels attached to the operators and $K(t)$. Another significant form is conveyed by the pair of equations

$$\left(i\frac{d}{dt}-\omega_0\right)(y_--y_+)-i\int_{t_2}^{t_1}dt'(A_{--}-A_{+-})(t-t')\times(y_--y_+)(t')=K_--K_+$$

and

$$\left(i\frac{d}{dt}-\omega_0\right)(y_++y_-)+i\int_{t_2}^{t_1}dt'(A_{++}-A_{+-})(t-t')\times(y_++y_-)(t')-i\int_{t_2}^{t_1}dt'(A_{+-}+A_{-+})(t-t')\times(y_--y_+)(t')=K_++K_-,$$

where

$$(A_{--}-A_{+-})(t-t')=-(A_{++}-A_{-+})(t-t')=\langle[Q(t),Q(t')]\rangle_\vartheta\eta_-(t-t'),$$
$$(A_{++}-A_{+-})(t-t')=-(A_{--}-A_{-+})(t-t')=\langle[Q(t),Q(t')]\rangle_\vartheta\eta_+(t-t'),$$

and

$$(A_{+-}+A_{-+})(t-t')=\langle\{Q(t),Q(t')\}\rangle_\vartheta.$$

The nonlocal character of these equations is not very marked if, for example, the correlation between $Q(t)$ and $Q(t')$ in the macroscopic system disappears when $|t-t'|$ is still small compared with the period of the oscillator. Then, since the behavior of $y(t)$ over a short time interval is given approximately by $e^{-i\omega t}$, the matrix $A(t-t')$ is effectively replaced by

$$\int_{-\infty}^{\infty}d(t-t')e^{i\omega(t-t')}A(t-t')=A(\omega),$$

and the equations of motion read

$$[i(d/dt)-\omega_-](y_--y_+)=K_--K_+,$$
$$[i(d/dt)-\omega_+](y_++y_-)-ia(y_--y_+)=K_++K_-.$$

Here we have defined

$$\omega_-=\omega_0+i(A_{--}-A_{+-})(\omega)=\omega+\tfrac{1}{2}i\gamma,$$
$$\omega_+=\omega_0-i(A_{++}-A_{+-})(\omega)=\omega-\tfrac{1}{2}i\gamma,$$

and

$$a(\omega)=(A_{+-}+A_{-+})(\omega).$$

It should be noted that $A_{+-}(\omega)$ and $A_{-+}(\omega)$ are real positive quantities since

$$A_{-+}(\omega)=\lim_{T\to\infty}\frac{1}{T}\left\langle\left(\int_{-\frac{1}{2}T}^{\frac{1}{2}T}dte^{-i\omega t}Q(t)\right)^\dagger\times\left(\int_{-\frac{1}{2}T}^{\frac{1}{2}T}dte^{-i\omega t}Q(t)\right)\right\rangle$$

and

$$A_{+-}(\omega)=A_{-+}(-\omega).$$

One consequence is

$$a(\omega)=a(-\omega)\geq 0.$$

It also follows from

$$\omega_--\omega_+=i(A_{--}+A_{++}-2A_{+-})(\omega)=i(A_{-+}-A_{+-})(\omega)$$

that

$$\gamma(\omega)=A_{-+}(\omega)-A_{+-}(\omega)=-\gamma(-\omega)$$

is real. Furthermore

$$\omega=\omega_0-\tfrac{1}{2}i(A_{++}-A_{--})(\omega),$$

where

$$(A_{++}-A_{--})(t-t')=\langle[Q(t),Q(t')]\rangle_\vartheta\epsilon(t-t')=(A_{-+}-A_{+-})(t-t')\epsilon(t-t'),$$

so that

$$-i(A_{++}-A_{--})(\omega)=\frac{1}{\pi}P\int_{-\infty}^{\infty}\frac{d\omega'}{\omega-\omega'}\gamma(\omega'),$$

and $\omega$ emerges as the real quantity

$$\omega=\omega_0-\frac{1}{\pi}P\int_0^{\infty}\frac{\omega' d\omega'}{\omega'^2-\omega^2}\gamma(\omega').$$

We have not yet made direct reference to the nature of the expectation value for the macroscopic system, which is now taken as the thermal average:

$$\langle X\rangle_\vartheta=C\ \mathrm{tr}e^{-\beta H}X$$
$$C^{-1}=\mathrm{tr}e^{-\beta H},$$

where $H$ is the energy operator of the external system. The implication for the structure of the expectation values is contained in

$$\langle Q(t)Q(t')\rangle_\vartheta=C\ \mathrm{tr}e^{-\beta H}Q(t)Q(t')$$
$$=\langle Q(t')Q(t+i\beta)\rangle_\vartheta,$$

which employs the formal property

$$e^{-\beta H}Q(t)e^{\beta H}=Q(t+i\beta).$$

On introducing the time Fourier transforms, however, this becomes the explicit relation

$$A_{-+}(\omega)=e^{\beta\omega}A_{+-}(\omega),$$

and we conclude that

$$e^{-\frac{1}{2}\beta\omega}A_{-+}(\omega)=e^{\frac{1}{2}\beta\omega}A_{+-}(\omega)$$
$$=a(\omega)/2\cosh\tfrac{1}{2}\beta\omega,$$

which is a positive even function of $\omega$. As a consequence, we have

$$\gamma(\omega)=a(\omega)\tanh\tfrac{1}{2}\beta\omega,$$
$$\geq 0,\quad \beta\omega>0,$$

which can also be written as

$$a(\omega)=2\gamma(\omega)[(e^{\beta\omega}-1)^{-1}+\tfrac{1}{2}].$$

The net result of this part of the discussion is to remove all explicit reference to the external system as a dynamical entity. We are given effective equations of motion for $y_+$ and $y_-$ that contain the prescribed external forces and three parameters, the angular frequency $\omega(\simeq\omega_0)$, $\gamma$, and $a$, the latter pair being related by the temperature of the macroscopic system. The accompanying boundary conditions are

$$(y_- - y_+)(t_1)=0$$

and, for the choice of an initial thermal mixture,

$$y_-(t_2)=e^{\beta_0\omega}y_+(t_2).$$

We now find that

$$(y_- - y_+)(t)=i\int_{t_2}^{t_1}dt' e^{i\omega_-(t'-t)}\eta_-(t-t')(K_- - K_+)(t'),$$

which supplies the initial condition for the second equation of motion,

$$(y_+ + y_-)(t_2)=\coth(\tfrac{1}{2}\beta_0\omega)i\int_{t_2}^{t_1}dt e^{i\omega_-(t-t_2)}(K_- - K_+)(t),$$

and the required solution is given by

$$i(y_+ + y_-)(t)$$
$$=\int_{t_2}^{t_1}dt' e^{-i\omega_+(t-t')}\eta_+(t-t')(K_+ + K_-)(t')$$
$$-\coth(\tfrac{1}{2}\beta\omega)\int_{t_2}^{t_1}dt'[e^{-i\omega_+(t-t')}\eta_+(t-t')$$
$$+e^{i\omega_-(t'-t)}\eta_-(t-t')](K_- - K_+)(t')$$
$$+(\coth\tfrac{1}{2}\beta\omega-\coth\tfrac{1}{2}\beta_0\omega)\int_{t_2}^{t_1}dt' e^{-i\omega_+(t-t_2)}$$
$$\times e^{i\omega_-(t'-t_2)}(K_- - K_+)(t').$$

The corresponding solutions for $y_\pm^\dagger(t)$ are obtained by interchanging the $\pm$ labels in the formal adjoint equation.

The differential dependence of the transformation function $\langle t_2|t_2\rangle_{\vartheta_0\vartheta}{}^{K\pm}$ upon the external forces is described by these results, and the explicit formula obtained on integration is

$$\langle t_2|t_2\rangle_{\vartheta_0\vartheta}{}^{K\pm}$$
$$=\exp\left[-i\int dt dt' K^*(t)G_{\vartheta_0\vartheta}(t-t_2, t'-t_2)K(t')\right],$$

where $[n_0=\langle n\rangle_{\vartheta_0},\ n=\langle n\rangle_\vartheta]$

$$iG_{\vartheta_0\vartheta}(t-t_2, t'-t_2)$$
$$=e^{-i\omega_+(t-t')}\eta_+(t-t')\begin{pmatrix} n+1, & -n \\ -n-1, & n\end{pmatrix}$$
$$+e^{-i\omega_-(t-t')}\eta_-(t-t')\begin{pmatrix} n, & -n \\ -n-1, & n+1\end{pmatrix}$$
$$+e^{-i\omega_+(t-t_2)}e^{i\omega_-(t'-t_2)}(n_0-n)\begin{pmatrix} 1 & -1 \\ -1 & 1\end{pmatrix}.$$

Another way of presenting this result is

$$iG_{\vartheta_0\vartheta}(t-t_2, t'-t_2)$$
$$=e^{-i\omega(t-t')}e^{-\frac{1}{2}\gamma|t-t'|}\begin{pmatrix} \eta_+(t-t')+n, & -n \\ -n-1, & \eta_-(t-t')+n\end{pmatrix}$$
$$+e^{-i\omega(t-t')}e^{-\gamma[\frac{1}{2}(t+t')-t_2]}(n_0-n)\begin{pmatrix} 1 & -1 \\ -1 & 1\end{pmatrix},$$

although the simplest description of $G$ is supplied by

the differential equation

$$\left[\left(i\frac{d}{dt}-\omega_+\right)G-\delta(t-t')\begin{pmatrix}1 & 0\\ 0 & -1\end{pmatrix}\right]\left(-i\frac{d^T}{dt'}-\omega_-\right)$$
$$=-i\delta(t-t')\gamma\begin{pmatrix}n, & -n\\ -n-1, & n+1\end{pmatrix},$$

(where $d^T$ indicates differentiation to the left) in conjunction with the initial value

$$iG_{\vartheta_0\vartheta}(0,0)=\begin{pmatrix}n_0+\frac{1}{2}, & -n_0\\ -n_0-1, & n_0+\frac{1}{2}\end{pmatrix}$$

and the boundary conditions

$$[i(d/dt)-\omega_+]G=0,\quad t>t'$$
$$[i(d/dt)-\omega_-]G=0,\quad t<t'.$$

A more symmetrical version of this differential equation is given by

$$\left(i\frac{d}{dt}-\omega_+\right)\left(-i\frac{d}{dt'}-\omega_-\right)G$$
$$-\begin{pmatrix}1 & 0\\ 0 & -1\end{pmatrix}\left(i\frac{d}{dt}-\omega\right)\delta(t-t')$$
$$=-i\delta(t-t')\gamma\begin{pmatrix}n+\frac{1}{2}, & -n\\ -n-1, & n+\frac{1}{2}\end{pmatrix}.$$

We note the vanishing sum of all $G$ elements, and that the role of complex conjugation in exchanging the two segments of the closed time path is expressed by

$$-\begin{pmatrix}0 & 1\\ 1 & 0\end{pmatrix}G(t',t)^{T*}\begin{pmatrix}0 & 1\\ 1 & 0\end{pmatrix}=G(t,t'),$$

which is to say that

$$-G(t',t)_{+-}{}^*=G(t,t')_{+-},\quad -G(t',t)_{-+}{}^*=G(t,t')_{-+}$$
$$-G(t',t)_{--}{}^*=G(t,t')_{++}.$$

It will be observed that only when

$$\langle n\rangle_{\vartheta_0}=\langle n\rangle_\vartheta$$

is $G_{\vartheta_0\vartheta}(t-t_2,t'-t_2)$ independent of $t_2$ and a function of $t-t'$. This clearly refers to the initial physical situation of thermal equilibrium between the oscillator and the external system at the common temperature $\vartheta_0=\vartheta>0$, which equilibrium persists in the absence of external forces. If the initial circumstances do not constitute thermal equilibrium, that will be established in the course of time at the macroscopic temperature $\vartheta>0$. Thus, all reference to the initial oscillator temperature disappears from $G_{\vartheta_0\vartheta}(t-t_2,t'-t_2)$ when, for fixed $t-t'$,

$$\gamma[\tfrac{1}{2}(t+t')-t_2]\gg 1.$$

The thermal relaxation of the oscillator energy is derived from

$$\langle t_2|y^\dagger y(t_1)|t_2\rangle_{\vartheta_0\vartheta}=\frac{\delta}{\delta K_-(t_1)}\frac{\delta}{\delta K_+{}^*(t_1)}\langle t_2|t_2\rangle_{\vartheta_0\vartheta}{}^{K\pm}\Big|_{K_\pm=0}$$
$$=-iG_{\vartheta_0\vartheta}(t_1-t_2,t_1-t_2)_{+-},$$

and is expressed by

$$\langle n(t_1)\rangle=\langle n\rangle_\vartheta+(\langle n\rangle_{\vartheta_0}-\langle n\rangle_\vartheta)e^{-\gamma(t_1-t_2)}.$$

The previously employed technique of impulsive forces applied at the time $t_1$ gives the more general result

$$\langle t_2|\exp[-i(\lambda y^\dagger(t_1)+\mu y(t_1))]|t_2\rangle_{\vartheta_0\vartheta}{}^K$$
$$=\exp\left[-\lambda\mu(\langle n(t_1)\rangle+\tfrac{1}{2})+\lambda\int_{t_2}^{t_1}dt e^{i\omega_-(t_1-t)}K^*(t)\right.$$
$$\left.-\mu\int_{t_2}^{t_1}dt e^{-i\omega_+(t_1-t)}K(t)\right],$$

from which a variety of probability distributions and expectation values can be obtained.

The latter calculation illustrates a general characteristic of the matrix $G(t,t')$, which is implied by the lack of dependence on the time $t_1$. Indeed, such a terminal time need not appear explicitly in the structure of the transformation function $\langle t_2|t_2\rangle^{K\pm}$ and all time integrations can range from $t_2$ to $+\infty$. Then $t_1$ is implicit as the time beyond which $K_+$ and $K_-$ are identified, and the structure of $G$ must be such as to remove any reference to a time greater than $t_1$. In the present situation, the use of an impulsive force at $t_1$ produces, for example, the term

$$\int_{t_2}^{\infty}dtG(t_1-t_2,t-t_2)K(t),$$

in which $K_+$ and $K_-$ are set equal. Hence it is necessary that

$$G(t,t')\begin{pmatrix}1\\ 1\end{pmatrix}=0,\quad t<t'$$

and similarly that

$$\overbrace{1\quad 1}\,G(t,t')=0,\quad t>t',$$

which says that adding the columns of $G(t,t')$ gives retarded functions of $t-t'$, while the sum of rows supplies a vector that is an advanced function of $t-t'$. In each instance, the two components must have a zero sum. These statements are immediately verified for the explicitly calculated $G_{\vartheta_0\vartheta}(t-t_2,t'-t_2)$ and follow more generally from the operator construction

$$iG=\begin{pmatrix}\langle(y(t)y^\dagger(t'))_+\rangle, & -\langle y^\dagger(t')y(t)\rangle\\ -\langle y(t)y^\dagger(t')\rangle, & \langle(y(t)y^\dagger(t'))_-\rangle\end{pmatrix},$$

for, as we have already noted in connection with $Q$ products,

$$(y(t)y^\dagger(t'))_+-y^\dagger(t')y(t)=y(t)y^\dagger(t')-(y(t)y^\dagger(t'))_-$$
$$=\eta_+(t-t')[y(t),y^\dagger(t')]$$

and

$$(y(t)y^\dagger(t'))_+ - y(t)y^\dagger(t') = y^\dagger(t')y(t) - (y(t)y^\dagger(t'))_- = -\eta_-(t-t')[y(t),y^\dagger(t')].$$

Our results show, incidentally, that

$$\langle [y(t),y^\dagger(t')]\rangle_{\vartheta_0\vartheta} = e^{-i\omega_+(t-t')}\eta_+(t-t') + e^{-i\omega_-(t-t')}\eta_-(t-t') = e^{-i\omega(t-t')}e^{-\frac{1}{2}\gamma|t-t'|}.$$

Another general property can be illustrated by our calculation, the positiveness of $-iG(t,t')_{+-}$,

$$-\int dtdt' K(t) iG(t-t_2,\, t'-t_2)_{+-} K^*(t') = \left\langle t_2 \left| \left(\int dt K(t) y(t)\right)^\dagger \left(\int dt K(t) y(t)\right) \right| t_2 \right\rangle > 0.$$

We have found that

$$-iG_{\vartheta_0\vartheta}(t-t_2,\, t'-t_2)_{+-} = \exp\{-i\omega(t-t') - \gamma[\tfrac{1}{2}(t+t') - t_2]\}\langle n\rangle_\vartheta + e^{-i\omega(t-t')}[e^{-\frac{1}{2}\gamma|t-t'|} - e^{-\gamma[\frac{1}{2}(t+t')-t_2]}]\langle n\rangle_{\vartheta_0},$$

and it is clearly necessary that each term obey separately the positiveness requirement. The first term is trivial,

$$\int dtdt' K(t) \exp\{-i\omega(t-t') - \gamma[\tfrac{1}{2}(t+t') - t_2]\} K^*(t') = \left|\int dt e^{-i\omega_+(t-t_2)} K(t)\right|^2 > 0,$$

and the required property of the second term follows from the formula

$$e^{-\frac{1}{2}\gamma|t-t'|} - e^{-\gamma[\frac{1}{2}(t+t')-t_2]} = \frac{2\gamma}{\pi}\int^\infty d\omega' \frac{\sin\omega'(t-t_2)\,\sin\omega'(t'-t_2)}{\omega'^2 + (\frac{1}{2}\gamma)^2}.$$

All the information that has been obtained about the oscillator is displayed on considering the forces

$$K_\pm(t) = \lambda_\pm(t) + K(t), \quad K_\pm^*(t) = \mu_\pm(t) + K^*(t),$$

and making explicit the effects of $\lambda_\pm(t)$, $\mu_\pm(t)$ by equivalent time-ordered operators:

$$\left\langle t_2 \left| \left(\exp\left[i\int_{t_2}^\infty dt(\lambda_- y^\dagger + \mu_- y)\right]\right)_- \times \left(\exp\left[-i\int_{t_2}^\infty dt(\lambda_+ y^\dagger + \mu_+ y)\right]\right)_+ \right| t_2 \right\rangle_{\vartheta_0\vartheta}^K$$

$$= \exp\Big[-i\int dtdt' \mu(t) G(t-t_2,\, t'-t_2)\lambda(t') + \int dtdt' K^*(t) e^{-i\omega_-(t-t')}\eta_-(t-t')(\lambda_+ - \lambda_-)(t') - \int dtdt' (\mu_+ - \mu_-)(t) e^{-i\omega_+(t-t')}\eta_+(t-t') K(t')\Big].$$

This is a formula for the direct computation of expectation values of general functions of $y(t)$ and $y^\dagger(t)$. A less explicit but simpler result can also be given by means of expectation values for functions of the operators

$$[i(d/dt) - \omega_+]y(t) - K(t) = K_f(t),$$
$$[-i(d/dt) - \omega_-]y^\dagger(t) - K^*(t) = K_f^\dagger(t).$$

Let us recognize at once that

$$\langle K_f(t)\rangle = 0, \quad \langle K_f^\dagger(t)\rangle = 0,$$

and therefore that the fluctuations of $y(t)$, $y^\dagger(t)$ can be ascribed to the effect of the forces $K_f$, $K_f^\dagger$, which appear as the quantum analogs of the random forces in the classical Langevin approach to the theory of the Brownian motion. The change in viewpoint is accomplished by introducing

$$\lambda_\pm(t) = [i(d/dt) - \omega_-]u_\pm(t)$$
$$\mu_\pm(t) = [-i(d/dt) - \omega_+]v_\pm(t),$$

where we assume, just for simplicity, that the functions $u(t)$, $v(t)$ vanish at the time boundaries. Then, partial time integrations will replace the operators $y$, $y^\dagger$ with $K_f$, $K_f^\dagger$.

To carry this out, however, we need the following lemma on time-ordered products:

$$\left(\exp\left[\int dt(A(t) + (d/dt)B(t))\right]\right)_+ = \left(\exp\left[\int dt A(t)\right]\right)_+ \exp\left(\int dt[A + \tfrac{1}{2}(dB/dt),\, B]\right),$$

which involves the unessential assumption that $B(t)$ vanishes at the time terminals, and the hypothesis that $[A(t), B(t)]$ and $[dB(t)/dt, B(t)]$ are commutative with all the other operators. The proof is obtained by replacing $B(t)$ with $\lambda B(t)$ and differentiating with respect to $\lambda$,

$$\frac{\partial}{\partial\lambda}\left(\exp\left[\int_{t_2}^{t_1} dt\left(A + \lambda\frac{d}{dt}B\right)\right]\right)_+ = \int_{t_2}^{t_1} dt \left(\exp\left[\int_t^{t_1}\right]\right)_+ \frac{d}{dt}B(t)\left(\exp\left[\int_{t_2}^{t}\right]\right)_+.$$

Then, a partial integration yields

$$\int_{t_2}^{t_1} dt \left(\exp\left[\int_t^{t_1}\right]\right)_+ \left[A(t) + \lambda\frac{dB(t)}{dt},\, B(t)\right] \times \left(\exp\left[\int_{t_2}^{t}\right]\right)_+ = \left(\exp\left[\int_{t_2}^{t_1}\right]\right)_+ \int_{t_2}^{t_1} dt\left[A + \lambda\frac{dB}{dt},\, B\right],$$

according to the hypothesis, and the stated result follows on integrating this differential equation.

The structure of the lemma is given by the rearrangement

$$-i(\lambda y^\dagger+\mu y)=-i[u(K^*+K_f^\dagger)+v(K+K_f)] +(d/dt)(uy^\dagger-vy),$$

and we immediately find a commutator that is a multiple of the unit operator,

$$[A+(d/dt)B,\,B]=-i[\lambda y^\dagger+\mu y,\,uy^\dagger-vy] =-i(\mu u+\lambda v)\to -2iv(i(d/dt)-\omega)u.$$

The last form involves discarding a total time derivative that will not contribute to the final result. To evaluate $[A,B]$ we must refer to the meaning of $K_f$ and $K_f^\dagger$ that is supplied by the actual equations of motion,

$$K_f(t)=Q(t)+(\omega_0-\omega_+)y(t)$$
$$K_f^\dagger(t)=Q(t)+(\omega_0-\omega_-)y^\dagger(t),$$

for then

$$[A(t),B(t)]=-i[u(\omega_0-\omega_-)y^\dagger+v(\omega_0-\omega_+)y,\,uy^\dagger-vy] =2ivu(\omega-\omega_0),$$

which is also proportional to the unit operator. Accordingly,

$$\left(\exp\left[-i\int dt(\lambda y^\dagger+\mu y)\right]\right)_+ =\left(\exp\left[-i\int dt[u(K^*+K_f^\dagger)+v(K+K_f)]\right]\right)_+ \times\exp\left[i(\omega-\omega_0)\int dtvu-i\int dtv\left(i\frac{d}{dt}-\omega\right)u\right],$$

and complex conjugation yields the analogous result for negatively time-ordered products.

With the aid of the differential equation obeyed by $G$, we now get

$$\left\langle t_2\left|\left(\exp\left[i\int dt(uK_f^\dagger+vK_f)\right]\right)_-\times\left(\exp\left[-i\int dt(uK_f^\dagger+vK_f)\right]\right)_+\right|t_2\right\rangle_{\vartheta_0\vartheta}^K =\exp\left[-i\int_{t_2}^{t_1}dtv(t)\kappa u(t)\right],$$

where

$$\kappa=\gamma\begin{pmatrix} n+\frac{1}{2}, & -n\\ -n-1, & n+\frac{1}{2}\end{pmatrix}+i(\omega-\omega_0)\begin{pmatrix}1 & 0\\ 0 & -1\end{pmatrix}.$$

The elements of this matrix are also expressed by

$$\kappa\delta(t-t')=\begin{pmatrix}\langle(K_f(t)K_f^\dagger(t'))_+\rangle, & -\langle K_f^\dagger(t')K_f(t)\rangle\\ -\langle K_f(t)K_f^\dagger(t')\rangle, & \langle(K_f(t)K_f^\dagger(t'))_-\rangle\end{pmatrix}.$$

Such expectation values are to be understood as effective evaluations that serve to describe the properties of the oscillator under the circumstances that validate the various approximations that have been used.

It will be observed that when $n$ is sufficiently large to permit the neglect of all other terms,

$$\kappa\simeq\tfrac{1}{2}a\begin{pmatrix}1 & -1\\ -1 & 1\end{pmatrix}\quad[\tfrac{1}{2}a=\gamma(n+\tfrac{1}{2})],$$

and the sense of operator multiplication is no longer significant. This is the classical limit, for which

$$\left\langle\exp\left[-i\int dt(uK_f^\dagger+vK_f)\right]\right\rangle_\vartheta =\exp\left[-\int dt\tfrac{1}{2}av(t)u(t)\right],$$

where we have placed $u_+-u_-=u$, $v_+-v_-=v$. On introducing real components of the random force

$$K_f=2^{-\frac{1}{2}}(K_1+iK_2),\quad K_f^\dagger=2^{-\frac{1}{2}}(K_1-iK_2),$$

the classical limiting result reads

$$\left\langle\exp\left[-i\int dt(u_1K_1+u_2K_2)\right]\right\rangle_\vartheta =\exp\left[-\int dt\tfrac{1}{4}a(u_1^2+u_2^2)\right].$$

The fluctuations at different times are independent. If we consider time-averaged forces,

$$\bar{K}=\frac{1}{\Delta t}\int_t^{t+\Delta t}dt'K(t'),$$

we find by Fourier transformation that

$$\langle\delta(\bar{K}_1-K_1')\delta(\bar{K}_2-K_2')\rangle_\vartheta=\frac{\Delta t}{\pi a}\exp\left[-\frac{\Delta t}{a}(K_1'^2+K_2'^2)\right],$$

which is the Gaussian distribution giving the probability that the force averaged over a time interval $\Delta t$ will have a value within a small neighborhood of the point $K'$. In this classical limit the fluctuation constant $a$ is related to the damping or dissipation constant $\gamma$ and the macroscopic temperature $\vartheta$ by

$$a=(2\gamma/\omega)\vartheta.$$

Our simplified equations can also be applied to situations in which the external system is not at thermal equilibrium. To see this possibility let us return to the real positive functions $A_{-+}(\omega)$, $A_{+-}(\omega)$ that describe the external system and remark that, generally,

$$\frac{A_{-+}(-\omega)}{A_{+-}(-\omega)}=\left[\frac{A_{-+}(\omega)}{A_{+-}(\omega)}\right]^{-1}\geq 0.$$

These properties can be expressed by writing

$$A_{-+}(\omega)/A_{+-}(\omega)=e^{\omega\beta(\omega)},$$

where $\beta(\omega)$ is a real even function that can range from $-\infty$ to $+\infty$. When only one value of $\omega$ is of interest, all conceivable situations for the external system can be described by the single parameter $\beta$, the reciprocal of which appears as an effective temperature of the macroscopic system. A new physical domain that appears in this way is characterized by negative temperature, $\beta<0$. Since $a$ is an intrinsically positive constant, it is $\gamma$ that will reverse sign

$$-\gamma=a(1-e^{-|\beta|\omega})/(1+e^{-|\beta|\omega})>0,$$

and the effect of the external system on the oscillator changes from damping to amplification.

We shall discuss the following physical sequence. At time $t_2$ the oscillator, in a thermal mixture of states at temperature $\vartheta_0$, is acted on by external forces which are present for a time, short in comparison with $1/|\gamma|$. After a sufficiently extended interval $\sim(t_1-t_2)$ such that the amplification factor or gain is very large,

$$k=e^{\frac{1}{2}|\gamma|(t_1-t_2)}\gg 1,$$

measurements are made in the neighborhood of time $t_1$. A prediction of all such measurements is contained in the general expectation value formula. Approximations that convey the physical situation under consideration are given by

$$\int dtdt'(\mu_+-\mu_-)(t)e^{-i\omega_+(t-t')}\eta_+(t-t')K(t')$$
$$\simeq k\int dt(\mu_+-\mu_-)(t)e^{-i\omega t}\int dt'e^{i\omega t'}K(t'),$$

$$\int dtdt'K^*(t)e^{-i\omega_-(t-t')}\eta_-(t-t')(\lambda_+-\lambda_-)(t')$$
$$\simeq k\int dtK^*(t)e^{-i\omega t}\int dt'e^{i\omega t'}(\lambda_+-\lambda_-)(t'),$$

and

$$i\int dtdt'\mu(t)G(t-t_2,t'-t_2)\lambda(t')$$
$$\simeq k^2(\langle n\rangle_{\vartheta_0}+(1-e^{-|\beta|\omega})^{-1})\int dt(\mu_+-\mu_-)(t)e^{-i\omega t}$$
$$\times\int dt'(\lambda_+-\lambda_-)(t')e^{i\omega t'}.$$

From the appearance of the combinations $\mu_+-\mu_-=\mu$, $\lambda_+-\lambda_-=\lambda$ only, we recognize that noncommutativity of operator multiplication is no longer significant, and thus the motion of the oscillator has been amplified to the classical level. To express the consequences most simply, we write

$$y(t)=ke^{-i\omega t}(y_s+y_n)$$
$$y^\dagger(t)=ke^{i\omega t}(y_s^*+y_n^*),$$

with

$$y_s=-i\int_{t_2}^{\infty}dt'e^{i\omega t'}K(t'),$$

and, on defining

$$u=k\int dte^{i\omega t}\lambda(t),\quad v=k\int dte^{-i\omega t}\mu(t),$$

we obtain the time-independent result

$$\langle\exp[-i(uy_n^*+vy_n)]\rangle$$
$$=\exp[-(\langle n\rangle_{\vartheta_0}+(1-e^{-|\beta|\omega})^{-1})vu],$$

which implies that

$$\langle y_n\rangle=\langle y_n^*\rangle=0$$
$$\langle|y_n|^2\rangle=\langle n\rangle_{\vartheta_0}+(1-e^{-|\beta|\omega})^{-1}\geq\langle n\rangle_{\vartheta_0}+1.$$

Thus, the oscillator coordinate $y(t)$ is the amplified superposition of two harmonic terms, one of definite amplitude and phase (signal), the other with random amplitude and phase (noise), governed by a two-dimensional Gaussian probability distribution.

These considerations with regard to amplification can be viewed as a primitive model of a maser device,[5] with the oscillator corresponding to a single mode of a resonant electromagnetic cavity, and the external system to an atomic ensemble wherein, for a selected pair of levels, the thermal population inequality is reversed by some means such as physical separation or electromagnetic pumping.

## AN IMPROVED TREATMENT

In this section we seek to remove some of the limitations of the preceding discussion. To aid in dealing successfully with the nonlocal time behavior of the oscillator, it is convenient to replace the non-Hermitian operator description with one employing Hermitian operators. Accordingly, we begin the development again, now using the Lagrangian operator

$$L=p(dq/dt)-\tfrac{1}{2}(p^2+\omega_0^2q^2)+qF(t)+qQ+L_{\text{ext}},$$

where $Q$ has altered its meaning by a constant factor. One could also include an external prescribed force that is coupled to $p$. We repeat the previous approximate construction of the transformation function $\langle t_2|t_2\rangle_{\vartheta_0\vartheta}{}^{F\pm}$ which proceeds by the introduction of an effective action operator that retains only the simplest correlation aspects of the external system, as comprised in

$$A(t-t')=\begin{pmatrix}\langle(Q(t)Q(t'))_+\rangle_\vartheta & \langle Q(t')Q(t)\rangle_\vartheta\\ \langle Q(t)Q(t')\rangle_\vartheta & \langle(Q(t)Q(t'))_-\rangle_\vartheta\end{pmatrix}.$$

[5] A similar model has been discussed recently by R. Serber and C. H. Townes, *Symposium on Quantum Electronics* (Columbia University Press, New York, 1960).

The action operator, with no other approximations, is

$$W=\int_{t_2}^{t_1} dt\left[p\frac{dq}{dt}-\tfrac{1}{2}(p^2+\omega_0^2q^2)+qF(t)|_+-|_-\right]$$

$$+\tfrac{1}{2}i\int_{t_2}^{t_1} dtdt'[(q(t)q(t'))_+A_{++}(t-t')$$

$$-2q_-(t)q_+(t')A_{-+}(t-t')+(q(t)q(t'))_-A_{--}(t-t')],$$

and the implied equations of motion, presented as second-order differential equations after eliminating

$$p=dq/dt,$$

are

$$\left(\frac{d^2}{dt^2}+\omega_0^2\right)q_+(t)$$

$$-i\int_{t_2}^{t_1} dt'[A_{++}(t-t')q_+(t')-A_{+-}(t-t')q_-(t')]=F_+(t)$$

and

$$\left(\frac{d^2}{dt'}+\omega_0^2\right)q_-(t)$$

$$+i\int_{t_2}^{t_1} dt[A_{--}(t-t')q_-(t')-A_{-+}(t-t')q_+(t')]=F_-(t).$$

It will be seen that the adjoint operation is equivalent to the interchange of the $\pm$labels.

We define

$$-iA_r(t-t')=\langle[Q(t),Q(t')]\rangle_\vartheta\eta_+(t-t')$$
$$=A_{++}-A_{+-}=A_{-+}-A_{--}$$

and

$$-iA_a(t-t')=-\langle[Q(t),Q(t')]\rangle_\vartheta\eta_-(t-t')$$
$$=A_{++}-A_{-+}=A_{+-}-A_{--},$$

together with

$$a(t-t')=\langle\{Q(t),Q(t')\}\rangle_\vartheta$$
$$=A_{+-}+A_{-+},$$

which enables us to present the integro-differential equations as

$$\left(\frac{d^2}{dt^2}+\omega_0^2\right)(q_--q_+)(t)-\int_{t_2}^{t_1} dt'A_a(t-t')(q_--q_+)(t')$$
$$=(F_--F_+)(t)$$

and

$$\left(\frac{d^2}{dt^2}+\omega_0^2\right)(q_++q_-)(t)-\int_{t_2}^{t_1} dt'A_r(t-t')(q_++q_-)(t')$$

$$+i\int_{t_2}^{t_1} dt'a(t-t')(q_--q_+)(t')=(F_++F_-)(t).$$

The accompanying boundary conditions are

$$(q_--q_+)(t_1)=0,\quad (d/dt)(q_--q_+)(t_1)=0$$

and

$$q_-(t_2)=q_+(t_2)\cosh\beta_0\omega_0+\frac{i}{\omega_0}\frac{d}{dt}q_+(t_2)\sinh\beta_0\omega_0$$

$$\frac{d}{dt}q_-(t_2)=-i\omega_0q_+(t_2)\sinh\beta_0\omega_0+\frac{d}{dt}q_+(t_2)\cosh\beta_0\omega_0,$$

or, more conveniently expressed,

$$(q_++q_-)(t_2)=\frac{i}{\omega_0}\coth(\tfrac{1}{2}\beta_0\omega_0)\frac{d}{dt}(q_--q_+)(t_2)$$

$$\frac{d}{dt}(q_++q_-)(t_2)=-i\omega_0\coth(\tfrac{1}{2}\beta_0\omega_0)(q_--q_+)(t_2),$$

which replace the non-Hermitian relations

$$y_-(t_2)=e^{\beta_0\omega_0}y_+(t_2),\quad y_-^\dagger(t_2)=e^{-\beta_0\omega_0}y_+^\dagger(t_2).$$

Note that it is the intrinsic oscillator frequency $\omega_0$ that appears here since the initial condition refers to a thermal mixture of unperturbed oscillator states.

The required solution of the equation for $q_--q_+$ can be written as

$$(q_--q_+)(t)=\int_{-\infty}^{\infty} dt'G_a(t-t')(F_--F_+)(t'),$$

where $G_a(t-t')$ is the real Green's function defined by

$$\left(\frac{d^2}{dt^2}+\omega_0^2\right)G_a(t-t')-\int_{-\infty}^{\infty} d\tau A_a(t-\tau)G_a(\tau-t')=\delta(t-t')$$

and

$$G_a(t-t')=0,\quad t>t'.$$

Implicit is the time $t_1$ as one beyond which $F_--F_+$ equals zero. The initial conditions for the second equation, which this solution supplies, are

$$(q_++q_-)(t_2)=\frac{i}{\omega_0}\coth(\tfrac{1}{2}\beta_0\omega_0)\int_{t_2}^{\infty} dt'\frac{\partial}{\partial t_2}$$
$$\times G_a(t_2-t')(F_--F_+)(t')$$

and

$$\frac{d}{dt}(q_++q_-)(t_2)$$

$$=-i\omega_0\coth(\tfrac{1}{2}\beta_0\omega_0)\int_{t_2}^{\infty} dt'G_a(t_2-t')(F_--F_+)(t').$$

The Green's function that is appropriate for the equation obeyed by $q_++q_-$ is defined by

$$\left(\frac{d^2}{dt^2}+\omega_0^2\right)G_r(t-t')-\int_{-\infty}^{\infty} d\tau A_r(t-\tau)G_r(\tau-t')=\delta(t-t'),$$

$$G_r(t-t')=0,\quad t<t',$$

and the two real functions are related by

$$G_a(t-t')=G_r(t'-t).$$

The desired solution of the second differential equation is

$$(q_++q_-)(t)=\int_{t_2}^{\infty}dt'G_r(t-t')(F_++F_-)(t')$$
$$-i\int_{t_2}^{\infty}dt'w(t-t_2,\,t'-t_2)(F_--F_+)(t'),$$

where

$$w(t-t_2,\,t'-t_2)$$
$$=\int_{t_2}^{\infty}d\tau d\tau' G_r(t-\tau)a(\tau-\tau')G_a(\tau'-t')$$
$$+\frac{1}{\omega_0}\coth(\tfrac{1}{2}\beta_0\omega_0)\left[\frac{\partial}{\partial t_2}G_r(t-t_2)\frac{\partial}{\partial t_2}G_a(t_2-t')+\omega_0^2G_r(t-t_2)G_a(t_2-t')\right]$$

is a real symmetrical function of its two arguments.

The differential description of the transformation function that these solutions imply is indicated by

$$\delta_{F\pm}\langle t_2|t_2\rangle^{F\pm}=i\left\langle t_2\left|\int dt(\delta F_+q_+-\delta F_-q_-)\right|t_2\right\rangle$$
$$=-\tfrac{1}{2}i\left\langle t_2\left|\int dt[\delta(F_--F_+)(q_++q_-)+\delta(F_++F_-)(q_--q_+)]\right|t_2\right\rangle,$$

and the result of integration is

$$\langle t_2|t_2\rangle_{\vartheta_0\vartheta}{}^{F\pm}$$
$$=\exp\left\{-\tfrac{1}{2}i\int dtdt'(F_--F_+)(t)G_r(t-t')(F_++F_-)(t')\right.$$
$$\left.-\tfrac{1}{4}\int dtdt'(F_--F_+)(t)w(t-t_2,\,t'-t_2)(F_--F_+)(t')\right\}.$$

This can also be displayed in the matrix form

$$\langle t_2|t_2\rangle_{\vartheta_0\vartheta}{}^{F\pm}=\exp\left\{\tfrac{1}{2}i\int dtdt'F(t)G_{\vartheta_0\vartheta}(t-t_2,\,t'-t_2)F(t')\right\},$$

with

$$G_{\vartheta_0\vartheta}(t-t_2,\,t'-t_2)$$
$$=\tfrac{1}{2}G_r(t-t')\begin{pmatrix}1 & 1\\ -1 & -1\end{pmatrix}+\tfrac{1}{2}G_a(t-t')\begin{pmatrix}1 & -1\\ 1 & -1\end{pmatrix}$$
$$+\tfrac{1}{2}iw(t-t_2,\,t'-t_2)\begin{pmatrix}1 & -1\\ -1 & 1\end{pmatrix}.$$

The latter obeys

$$G(t',t)^T=G(t,t')$$
$$-\begin{pmatrix}0 & 1\\ 1 & 0\end{pmatrix}G(t,t')^*\begin{pmatrix}0 & 1\\ 1 & 0\end{pmatrix}=G(t,t'),$$

and its elements are given by

$$G=i\begin{pmatrix}\langle(q(t)q(t'))_+\rangle_{\vartheta_0\vartheta}, & -\langle q(t')q(t)\rangle_{\vartheta_0\vartheta}\\ -\langle q(t)q(t')\rangle_{\vartheta_0\vartheta}, & \langle(q(t)q(t'))_-\rangle_{\vartheta_0\vartheta}\end{pmatrix}.$$

We note the identifications

$$G_r(t-t')=i\langle[q(t),q(t')]\rangle\eta_+(t-t')$$
$$G_a(t-t')=-i\langle[q(t),q(t')]\rangle\eta_-(t-t')$$
$$w(t-t_2,\,t'-t_2)=\langle\{q(t),q(t')\}\rangle.$$

It is also seen that the sum of the columns of $G$ is proportional to $G_r(t-t')$, while the sum of the rows contains only $G_a(t-t')$.

We shall suppose that $G_r(t-t')$ can have no more than exponential growth, $\sim e^{\alpha(t-t')}$, as $t-t'\to\infty$. Then the complex Fourier transform

$$G(\zeta)=\int_{-\infty}^{\infty}d(t-t')e^{i\zeta(t-t')}G_r(t-t')$$

exists in the upper half-plane

$$\mathrm{Im}\zeta>\alpha$$

and is given explicitly by

$$G(\zeta)=[\omega_0^2-\zeta^2-A(\zeta)]^{-1}.$$

Here

$$A(\zeta)=\int_{-\infty}^{\infty}d(t-t')e^{i\zeta(t-t')}A_r(t-t')$$
$$=i\int_0^{\infty}d\tau e^{i\zeta\tau}\int_{-\infty}^{\infty}\frac{d\omega}{2\pi}e^{-i\omega\tau}(A_{-+}-A_{+-})(\omega)$$
$$=\int_{-\infty}^{\infty}\frac{d\omega}{2\pi}\frac{(A_{-+}-A_{+-})(\omega)}{\omega-\zeta}$$

or, since $(A_{-+}-A_{+-})(\omega)$ is an odd function of $\omega$,

$$A(\zeta)=\int_0^{\infty}\frac{d\omega}{\pi}\frac{\omega(A_{-+}-A_{+-})(\omega)}{\omega^2-\zeta^2}.$$

We have already remarked on the generality of the representation

$$A_{-+}(\omega)/A_{+-}(\omega)=e^{\omega\beta(\omega)},\quad \beta(-\omega)=\beta(\omega),$$

and thus we shall write

$$(A_{-+}-A_{+-})(\omega)=a(\omega)\tanh[\tfrac{1}{2}\omega\beta(\omega)]$$
$$(A_{-+}+A_{+-})(\omega)=a(\omega)=a(-\omega)\geq 0,$$

which gives

$$G(\zeta)^{-1}=\omega_0^2-\zeta^2-\int_0^\infty \frac{d\omega}{\pi}\frac{\omega a(\omega)\tanh[\frac{1}{2}\omega\beta(\omega)]}{\omega^2-\zeta^2}.$$

Since this is an even function of $\zeta$, it also represents the Fourier transform of $G_a$ in the lower half-plane $\text{Im}\zeta<-\alpha$.

If the effective temperature is positive and finite at all frequencies, $\beta(\omega)>0$, $G(\zeta)$ can have no complex poles as a function of the variable $\zeta^2$. A complex pole at $\zeta^2=x+iy$, $y\neq 0$, is a zero of $G(\zeta)^{-1}$ and requires that

$$y\left[1+\int_0^\infty \frac{d\omega}{\pi}\frac{\omega a(\omega)\tanh[\frac{1}{2}\omega\beta(\omega)]}{(\omega^2-x)^2+y^2}\right]=0,$$

which is impossible since the quantity in brackets exceeds unity. On letting $y$ approach zero, we see that a pole of $G(\zeta)$ can occur at a point $x=\omega'^2>0$ only if $a(\omega')=0$. If the external system responds through the oscillator coupling to any impressed frequency, $a(\omega)>0$ for all $\omega$ and no pole can appear on the positive real axis of $\zeta^2$. As to the negative real axis, $G(\zeta)^{-1}$ is a monotonically decreasing function of $\zeta^2=x$ that begins at $+\infty$ for $x=-\infty$ and will therefore have no zero on the negative real axis if it is still positive at $x=0$. The corresponding condition is

$$\omega_0^2>\int_0^\infty \frac{d\omega}{\pi}a(\omega)\frac{\tanh[\frac{1}{2}\omega\beta(\omega)]}{\omega}.$$

Under these circumstances $\alpha=0$, for $G(\zeta)$, *qua* function of $\zeta^2$, has no singularity other than the branch line on the positive real axis, and the $\zeta$ singularities are therefore confined entirely to the real axis. This is indicated by

$$G(\zeta)=\int_0^\infty d\omega^2\frac{B(\omega^2)}{\omega^2-\zeta^2}$$

$$=\int_{-\infty}^\infty d\omega\,\epsilon(\omega)\frac{B(\omega^2)}{\omega-\zeta},$$

and $B(\omega^2)$ is the positive quantity

$$B(\omega^2)=\frac{(2\pi)^{-1}a(\omega)\tanh[\frac{1}{2}|\omega|\beta(\omega)]}{\left[\omega_0^2-\omega^2-P\int_0^\infty \frac{d\omega'^2}{2\pi}\frac{\tanh(\frac{1}{2}\omega'\beta(\omega'))}{\omega'^2-\omega^2}a(\omega')\right]^2+[\frac{1}{2}a(\omega)\tanh\frac{1}{2}\omega\beta(\omega)]^2}.$$

Some integral relations are easily obtained by comparison of asymptotic forms. Thus

$$\int_0^\infty d\omega^2 B(\omega^2)=1,$$

$$\int_0^\infty d\omega^2\omega^2 B(\omega^2)=\omega_0^2,$$

and

$$\int_0^\infty d\omega^2\omega^4 B(\omega^2)=\omega_0^4+\int_0^\infty \frac{d\omega^2}{2\pi}a(\omega)\tanh[\tfrac{1}{2}\omega\beta(\omega)]$$

$$=\omega_0^4+\langle[i\dot{Q},Q]\rangle_\vartheta,$$

while setting $\zeta=0$ yields

$$\int_0^\infty d\omega^2\frac{B(\omega^2)}{\omega^2}=\left[\omega_0^2-\int_0^\infty \frac{d\omega}{\pi}a(\omega)\frac{\tanh\frac{1}{2}\omega\beta(\omega)}{\omega}\right]^{-1}.$$

The Green's functions are recovered on using the inverse Fourier transformation

$$G(t-t')=\int_{-\infty}^\infty \frac{d\zeta}{2\pi}e^{-i\zeta(t-t')}G(\zeta),$$

where the path of integration is drawn in the half-plane of regularity. Accordingly,

$$G_r(t-t')=\int_0^\infty d\omega^2 B(\omega^2)\frac{\sin\omega(t-t')}{\omega}\eta_+(t-t')$$

and

$$G_a(t-t')=-\int_0^\infty d\omega^2 B(\omega^2)\frac{\sin\omega(t-t')}{\omega}\eta_-(t-t').$$

The integral relations mentioned previously can be expressed in terms of these Green's functions. Thus,

$$\int_0^\infty d\tau G_r(\tau)=\left[\omega_0^2-\int_0^\infty \frac{d\omega}{\pi}a\frac{\tanh(\frac{1}{2}\omega\beta)}{\omega}\right]^{-1},$$

while, in the limit of small positive $\tau$,

$$G_r(\tau)-(1/\omega_0)\sin\omega_0\tau\sim(\tau^5/5!)\langle[i\dot{Q},Q]\rangle_\vartheta,$$

which indicates the initial effect of the coupling to the external system.

The function $B(\omega^2)$ is bounded, and the Green's functions must therefore approach zero as $|t-t'|\to\infty$. Accordingly, all reference to the initial oscillator condition and to the time $t_2$ must eventually disappear. For sufficiently large $t-t_2$, $t'-t_2$, the function $w(t-t_2$,

$t'-t_2$) reduces to

$$w(t-t')=\int_{-\infty}^{\infty} d\tau d\tau' G_r(t-\tau)a(\tau-\tau')G_a(\tau'-t')$$

$$=\int_{-\infty}^{\infty}\frac{d\omega}{2\pi}e^{-i\omega(t-t')}G(\omega+i\epsilon)a(\omega)G(\omega-i\epsilon)|_{\epsilon\to 0}.$$

But

$$(1/2\pi)a(\omega)|G(\omega+i\epsilon)|^2=B(\omega^2)\coth[\tfrac{1}{2}|\omega|\beta(\omega)],$$

and, therefore,

$$w(t-t')=\int_0^{\infty} d\omega^2 B(\omega^2)\coth[\tfrac{1}{2}\omega\beta(\omega)]\frac{1}{\omega}\cos\omega(t-t').$$

The corresponding asymptotic form of the matrix $G(t-t_2, t'-t_2)$ is given by

$$G(t-t')=i\int_{-\infty}^{\infty} d\omega B(\omega^2)e^{-i\omega(t-t')}\begin{pmatrix}\eta_+(\omega)\eta_+(t-t')+\eta_-(\omega)\eta_-(t-t')+n, & -\eta_-(\omega)-n \\ -\eta_+(\omega)-n, & \eta_+(\omega)\eta_-(t-t')+\eta_-(\omega)\eta_+(t-t')+n\end{pmatrix},$$

with

$$n(\omega)=(e^{|\omega|\beta(\omega)}-1)^{-1},$$

which describes the oscillator in equilibrium at each frequency with the external system. When the temperature is frequency independent, this is thermal equilibrium. Note also that at zero temperature $n(\omega)=0$, and $G(t-t')_{++}$ is characterized by the temporal outgoing wave boundary condition—positive (negative) frequencies for positive (negative) time difference. The situation is similar for $G(t-t')_{--}$ as a function of $t'-t$.

It can no longer be maintained that placing $\beta_0=\beta$ removes all reference to the initial time. An interval must elapse before thermal equilibrium is established at the common temperature. This can be seen by evaluating the $t_2$ derivative of $w(t-t_2, t'-t_2)$:

$$\frac{\partial}{\partial t_2}w=-G_r(t-t_2)\int_{t_2}^{\infty} d\tau' a(t_2-\tau')G_a(\tau'-t')$$

$$-\int_{t_2}^{\infty} d\tau G_r(t-\tau)a(\tau-t_2)G_a(t_2-t')$$

$$+\frac{1}{\omega_0}\coth(\tfrac{1}{2}\omega_0\beta_0)$$

$$\times\left\{\frac{\partial}{\partial t_2}G_r(t-t_2)\int_{-\infty}^{\infty} d\tau' A_a(t_2-\tau')G_a(\tau'-t')\right.$$

$$\left.+\int_{-\infty}^{\infty} d\tau A_r(t-\tau)G_r(\tau-t_2)\frac{\partial}{\partial t_2}G_a(t_2-t')\right\},$$

for if this is to vanish, the integrals involving $G_r$, say, must be expressible as linear combinations of $G_r(t-t_2)$ and its time derivative, which returns us to the approximate treatment of the preceding section, including the approximate identification of $\omega_0$ with the effective oscillator frequency. Hence $\vartheta_0=\vartheta$ does not represent the initial condition of thermal equilibrium between oscillator and external system. While it is perfectly clear that the latter situation is described by the matrix $G_\vartheta(t-t')$, a derivation that employs thermal equilibrium as an initial condition would be desirable.

The required derivation is produced by the device of computing the trace of the transformation function $\langle t_2'|t_2\rangle^{F\pm}$, in which the return path terminates at the different time $t_2'=t_2-T$, and the external force $F_-(t)$ is zero in the interval between $t_2$ and $t_2'$. The particular significance of the trace appears on varying the parameter $\lambda$ that measures the coupling between oscillator and external system:

$$\frac{\partial}{\partial\lambda}\langle t_2'|t_2\rangle^{F\pm}=i\left\langle t_2'\left|\left[\int_{t_2}^{t_1} dtq_+Q_+(t)-\int_{t_2'}^{t_1} dtq_-Q_-(t)+G_\lambda(t_2')-G_\lambda(t_2)\right]\right|t_2\right\rangle^{F\pm}.$$

The operators $G_\lambda$ are needed to generate infinitesimal transformations of the individual states at the corresponding times, if these states are defined by physical quantities that depend upon $\lambda$, such as the total energy. There is no analogous contribution to the trace, however, for the trace is independent of the representation, which is understood to be defined similarly at $t_2$ and $t_2'$, and one could use a complete set that does not refer to $\lambda$. More generally, we observe that $G_\lambda(t_2')$ bears the same relation to the $\langle t_2'|$ states as does $G_\lambda(t_2)$ to the states at time $t_2$, and therefore

$$\mathrm{tr}\langle t_2'|G_\lambda(t_2')|t_2\rangle-\mathrm{tr}\langle t_2'|G_\lambda(t_2)|t_2\rangle=0.$$

Accordingly, the construction of an effective action operator can proceed as before, with appropriately modified ranges of time integration, and, for the external system, with

$$\langle Q(t)Q(t')\rangle=\frac{\mathrm{tr}\langle t_2'|Q(t)Q(t')|t_2\rangle}{\mathrm{tr}\langle t_2'|t_2\rangle}.$$

This trace structure implies that

$$\langle Q(t)Q(t_2)\rangle=\langle Q(t_2')Q(t)\rangle$$

or, since these correlation functions depend only on

time differences, that

$$A_{-+}(t-t_2)=A_{+-}(t-t_2'),$$

which is also expressed by

$$A_{-+}(\omega)=e^{-i\omega T}A_{+-}(\omega).$$

The equations of motion for $t>t_2$ are given by

$$\left(\frac{d^2}{dt^2}+\omega_0{}^2\right)(q_--q_+)(t)-\int_{-\infty}^{\infty}dt'A_a(t-t')(q_--q_+)(t')$$
$$=(F_--F_+)(t)$$

and

$$\left(\frac{d^2}{dt^2}+\omega_0{}^2\right)(q_++q_-)(t)-\int_{t_2}^{\infty}dt'A_r(t-t')(q_++q_-)(t')$$
$$+i\int_{t_2}^{\infty}dt'a(t-t')(q_--q_+)(t')$$
$$=(F_++F_-)(t)-2i\int_{t_2'}^{t_2}dt'A_{+-}(t-t')q_-(t').$$

These are supplemented by the equation for $q_-(t)$ in the interval from $t_2'$ to $t_2$:

$$\left(\frac{d^2}{dt^2}+\omega_0{}^2\right)q_-(t)+i\int_{t_2'}^{t_2}dt'A_{--}(t-t')q_-(t')$$
$$=-i\int_{t_2}^{\infty}dt'A_{-+}(t-t')(q_--q_+)(t'),$$

and the effective boundary condition

$$q_-(t_2')=q_+(t_2).$$

The equation for $q_--q_+$ is solved as before,

$$(q_--q_+)(t)=\int_{-\infty}^{\infty}dt'G_a(t-t')(F_--F_+)(t'),$$

whereas

$$(q_++q_-)(t)$$
$$=\int_{-\infty}^{\infty}dt'G_r(t-t')(F_++F_-)(t')$$
$$-i\int_{t_2}^{\infty}d\tau G_r(t-\tau)\int_{t_2}^{\infty}dt'a(\tau-t')(q_--q_+)(t')$$
$$-2i\int_{t_2}^{\infty}d\tau G_r(t-\tau)\int_{t_2-T}^{t_2}dt'A_{+-}(\tau-t')q_-(t')$$
$$+G_r(t-t_2)\frac{\partial}{\partial t_2}(q_++q_-)(t_2)$$
$$-\frac{\partial}{\partial t_2}G_r(t-t_2)(q_++q_-)(t_2),$$

which has been written for external forces that are zero until the moment $t_2$ has passed.

Perhaps the simplest procedure at this point is to ask for the dependence of the latter solution upon $t_2$, for fixed $T$. We find that

$$\frac{\partial}{\partial t_2}(q_++q_-)(t)=-\int_{t_2}^{\infty}dt'G_r(t-t')A_r(t'-t_2)(q_++q_-)(t_2)$$
$$+i\int_{t_2}^{\infty}dt'G_r(t-t')a(t'-t_2)(q_--q_+)(t_2)$$
$$-2i\int_{t_2}^{\infty}dt'G_r(t-t')[A_{+-}(t'-t_2)q_-(t_2)$$
$$-A_{-+}(t'-t_2)q_+(t_2)],$$

on using the relations

$$\int_{-\infty}^{\infty}d\tau A_r(t-\tau)G_r(\tau-t')=\int_{-\infty}^{\infty}d\tau G_r(t-\tau)A_r(\tau-t'),$$
$$A_{+-}(t-t_2')q_-(t_2')=A_{-+}(t-t_2)q_+(t_2).$$

Therefore,

$$(\partial/\partial t_2)(q_++q_-)(t)=0,$$

since, with positive time argument,

$$a-iA_r=2A_{-+}$$
$$a+iA_r=2A_{+-}.$$

The utility of this result depends upon the approach of the Green's functions to zero with increasing magnitude of the time argument, which is assured, after making the substitution $T\to i\beta$, under the circumstances we have indicated. Then we can let $t_2\to-\infty$ and obtain

$$(q_++q_-)(t)=\int_{-\infty}^{\infty}dt'G_r(t-t')(F_++F_-)(t')$$
$$-i\int_{-\infty}^{\infty}dt'w(t-t')(F_--F_+)(t')$$

with

$$w(t-t')=\int_{-\infty}^{\infty}d\tau d\tau'G_r(t-\tau)a(\tau-\tau')G_a(\tau'-t'),$$

as anticipated.

Our results determine the trace ratio

$$\frac{\mathrm{tr}\langle t_2'|t_2\rangle^{F\pm}}{\mathrm{tr}\langle t_2'|t_2\rangle}=\frac{\mathrm{tr}\langle t_2|e^{iTH}|t_2\rangle^{F\pm}}{\mathrm{tr}e^{iTH}},$$

where $H$ is the Hamiltonian operator of the complete system, and the substitution $T\to i\beta$ yields the transformation function

$$\langle t_2|t_2\rangle_\vartheta{}^{F\pm}=\exp\left[\tfrac{1}{2}i\int dtdt'F(t)G_\vartheta(t-t')F(t')\right]$$

with

$$G_\vartheta(t-t')=\tfrac{1}{2}G_r(t-t')\begin{pmatrix}1 & 1\\ -1 & -1\end{pmatrix}+\tfrac{1}{2}G_a(t-t')\begin{pmatrix}1 & -1\\ 1 & -1\end{pmatrix}$$
$$+\tfrac{1}{2}iw(t-t')\begin{pmatrix}1 & -1\\ -1 & 1\end{pmatrix}$$

and

$$w(t-t')=\int_{-\infty}^{\infty}d\omega B(\omega^2)\coth(\tfrac{1}{2}|\omega|\beta)e^{-i\omega(t-t')}.$$

We can also write

$$w(t-t')=\int_{-\infty}^{\infty}d\tau C(t-\tau)(G_a-G_r)(\tau-t'),$$

where

$$C(t-t')=\frac{i}{2\pi}P\int_{-\infty}^{\infty}d\omega\coth(\tfrac{1}{2}\omega\beta)e^{-i\omega(t-t')}$$
$$=\frac{1}{\beta}\coth\left[\frac{\pi}{\beta}(t-t')\right].$$

What is asserted here about expectation values in the presence of an external field $F(t)$ becomes explicit on writing

$$F_\pm(t)=f_\pm(t)+F(t)$$

and indicating the effect of $f_\pm(t)$ by equivalent time-ordered operators,

$$\left\langle\left(\exp\left[-i\int dtf_-(t)q(t)\right]\right)_-\times\left(\exp\left[i\int dtf_+(t)q(t)\right]\right)_+\right\rangle_\vartheta^F$$
$$=\exp\left\{\tfrac{1}{2}i\int dtdt'f(t)G_\vartheta(t-t')f(t')+i\int dtdt'(f_+-f_-)(t)G_r(t-t')F(t')\right\}.$$

Thus

$$\langle q(t)\rangle_\vartheta{}^F=\int_{-\infty}^{\infty}dt'G_r(t-t')F(t')$$

and the properties of $q-\langle q\rangle_\vartheta{}^F$, which are independent of $F$, are given by setting $F=0$ in the general result. In particular, we recover the matrix identity

$$G_\vartheta(t-t')=i\begin{pmatrix}\langle(q(t)q(t'))_+\rangle_\vartheta, & -\langle q(t')q(t)\rangle_\vartheta\\ \langle q(t)q(t')\rangle_\vartheta, & \langle(q(t)q(t'))_-\rangle_\vartheta\end{pmatrix}.$$

The relation between $w$ and $G_a-G_r$ can then be displayed as a connection between symmetrical product and commutator expectation values

$$\langle\{q(t),q(t')\}\rangle_\vartheta=\int_{-\infty}^{\infty}d\tau C(t-\tau)\left\langle\frac{1}{i}[q(\tau),q(t')]\right\rangle_\vartheta.$$

In addition to the trace ratio, which determines the thermal average transformation function $\langle t_2|t_2\rangle_\vartheta{}^{F\pm}$ with its attendant physical information, it is possible to compute the trace

$$\mathrm{tr}\langle t_2'|t_2\rangle=\mathrm{tr}e^{iTH}\to\mathrm{tr}e^{-\beta H}$$

which describes the complete energy spectrum and thereby the thermostatic properties of the oscillator in equilibrium with the external system. For this purpose we set $F_\pm=0$ for $t>t_2$ and apply an arbitrary external force $F_-(t)$ in the interval from $t_2'$ to $t_2$. Moreover, the coupling term between oscillator and external system in the effective action operator is supplied with the variable factor $\lambda$ (formerly $\lambda^2$). Then we have

$$\frac{\partial}{\partial\lambda}\mathrm{tr}\langle t_2'|t_2\rangle^{F-}$$
$$=-\tfrac{1}{2}\,\mathrm{tr}\left\langle t_2'\left|\int_{t_2'}^{t_2}dtdt'A_{--}(t-t')(q(t)q(t'))_-\right|t_2\right\rangle^{F-}$$
$$=-\tfrac{1}{2}i\int_{t_2'}^{t_2}dtdt'A_{--}(t-t')\frac{\delta}{\delta F_-(t')}\mathrm{tr}\langle t_2'|q_-(t)|t_2\rangle^{F-},$$

where $q_-(t)$ obeys the equation of motion

$$\left(\frac{d^2}{dt^2}+\omega_0^2\right)q_-(t)+i\lambda\int_{t_2'}^{t_2}dt'A_{--}(t-t')q_-(t')=F_-(t)$$

with the accompanying boundary condition

$$q_-(t_2')=q_+(t_2)=q_-(t_2),$$

which is a statement of periodicity for the interval $T=t_2'-t_2$. The solution of this equation is

$$q_-(t)=\int_{t_2'}^{t_2}dt'G(t-t')F_-(t'),$$

where the Green's function obeys

$$\left(\frac{d^2}{dt^2}+\omega_0^2\right)G(t-t')+i\lambda\int_{t_2'}^{t_2}d\tau A_{--}(t-\tau)G(\tau-t')=\delta(t-t')$$

and the requirement of periodicity. We can now place $F_-=0$ in the differential equation for the trace, and obtain

$$\frac{\partial}{\partial\lambda}\log\mathrm{tr}\langle t_2'|t_2\rangle=-\tfrac{1}{2}i\int_{t_2'}^{t_2}dtdt'A_{--}(t-t')G(t-t').$$

The periodic Green's function is given by the Fourier series

$$G(t-t')=\frac{1}{T}\sum_{n=-\infty}^{\infty}\exp\left[-\frac{2\pi in}{T}(t-t')\right]G(n)$$

with

$$G(n)=\left[\omega_0^2-\left(\frac{2\pi n}{T}\right)^2-\lambda A(n)\right]^{-1}=G(-n)$$

and

$$A(n)=-\int_{t_2'}^{t_2}dt\exp\left[\frac{2\pi in}{T}(t-t')\right]iA_{--}(t-t')$$

$$=\int_0^\infty\frac{\omega d\omega}{\pi}\frac{(A_{-+}-A_{+-})(\omega)}{\omega^2-(2\pi n/T)^2},$$

where, it is to be recalled,

$$A_{-+}(\omega)=e^{-i\omega T}A_{+-}(\omega),$$

so that the integrand has no singularities at $\omega T=2\pi|n|$. Now we have

$$\frac{\partial}{\partial\lambda}\log\mathrm{tr}=\tfrac{1}{2}\sum_{-\infty}^{\infty}A(n)G(n)$$

$$=-\frac{1}{2}\frac{\partial}{\partial\lambda}\sum_{-\infty}^{\infty}\log\left[\omega_0^2-\left(\frac{2\pi n}{T}\right)^2-\lambda A(n)\right]$$

which, together with the initial condition

$$\lambda=0:\quad \mathrm{tr}e^{iTH}=(\mathrm{tr}_e e^{iTH_{\mathrm{ext}}})\sum_{n=0}^{\infty}e^{i(n+\frac{1}{2})\omega_0 T}$$
$$=(\mathrm{tr}_e)(i/2\sin\tfrac{1}{2}\omega_0 T),$$

yields

$$\mathrm{tr}e^{iTH}=(\mathrm{tr}_e)(i/2\sin\tfrac{1}{2}\omega_0 T)$$
$$\times\exp\left\{-\tfrac{1}{2}\sum_{-\infty}^{\infty}\log\left[\frac{\omega_0^2-(2\pi n/T)^2-A(n)}{\omega_0^2-(2\pi n/T)^2}\right]\right\}.$$

We have already introduced the function

$$G^{-1}(\zeta)=\omega_0^2-\zeta^2-\int_0^\infty\frac{\omega d\omega}{\pi}\frac{(A_{-+}-A_{+-})(\omega)}{\omega^2-\zeta^2}$$

and examined some of its properties for real and positive $A_{-+}(\omega)$, $A_{+-}(\omega)$. This situation is recovered on making the substitution $T\rightarrow i\beta$, and thus

$$Z=\mathrm{tr}e^{-\beta H}=Z_e(1/2\sinh\tfrac{1}{2}\beta\omega_0)$$
$$\times\exp\left\{-\tfrac{1}{2}\sum_{-\infty}^{\infty}\log\left[\frac{G^{-1}(i2\pi n/\beta)}{\omega_0^2+(2\pi n/\beta)^2}\right]\right\},$$

the existence of which for all $\beta>0$ requires that $G^{-1}(\zeta)$ remain positive at every value comprised in $\zeta^2=-(2\pi n/\beta)^2$, which is to say the entire negative $\zeta^2$ axis including the origin. The condition

$$G^{-1}(0)>0$$

is thereby identified as a stability criterion. To evaluate the summation over $n$ most conveniently we shall give an alternative construction for the function $\log(G^{-1}(\zeta)/-\zeta^2)$, which, as a function of $\zeta^2$, has all its singularities located on the branch line extending from 0 to $\infty$ and vanishes at infinity in this cut plane. Hence

$$\log(G^{-1}(\zeta)/-\zeta^2)=\frac{1}{\pi}\int_0^\infty d\omega^2\frac{\varphi(\omega)}{\omega^2-\zeta^2},$$

where the value

$$\varphi(0)=\pi$$

reproduces the pole of $G^{-1}(\zeta)/(-\zeta^2)$ at $\zeta^2=0$. We also recognize, on relating the two forms,

$$G^{-1}(\zeta)=(-\zeta^2)\exp\left[\frac{1}{\pi}\int_0^\infty d\omega^2\frac{\varphi(\omega)}{\omega^2-\zeta^2}\right]$$

$$=\omega_0^2-\zeta^2-\int_0^\infty\frac{d\omega^2}{2\pi}\frac{a(\omega)\tanh(\frac{1}{2}\omega\beta)}{\omega^2-\zeta^2},$$

that

$$-\tfrac{1}{2}a(\omega)\tanh(\tfrac{1}{2}\omega\beta)\cot\varphi(\omega)$$
$$=\omega_0^2-\omega^2-P\int_0^\infty\frac{d\omega'}{\pi}\frac{\omega' a(\omega')\tanh(\frac{1}{2}\omega'\beta)}{\omega'^2-\omega^2}.$$

The positive value of the right-hand side as $\omega\rightarrow 0$ shows that $\varphi(\omega)$ approaches the zero frequency limiting value of $\pi$ from below, and the assumption that $a(\omega)>0$ for all $\omega$ implies

$$\pi\geq\varphi(\omega)>0,$$

where the lower limit is approached as $\omega\rightarrow\infty$.

A comparison of asymptotic forms for $G^{-1}(\zeta)$ shows that

$$\omega_0^2=\frac{1}{\pi}\int_0^\infty d\omega^2\varphi(\omega)=\int_0^\infty d\omega\left(-\frac{1}{\pi}\frac{d\varphi(\omega)}{d\omega}\right)\omega^2,$$

while

$$\int_0^\infty d\omega\left(-\frac{1}{\pi}\frac{d\varphi(\omega)}{d\omega}\right)\omega^4=\omega_0^4+2\langle[i\dot{Q},Q]\rangle_\vartheta.$$

The introduction of the phase derivative can also be performed directly in the structure of $G^{-1}(\zeta)$,

$$G^{-1}(\zeta)=\exp\left[\int_0^\infty d\omega\left(-\frac{1}{\pi}\frac{d\varphi(\omega)}{d\omega}\right)\log(\omega^2-\zeta^2)\right],$$

and equating the two values for $G^{-1}(0)$ gives

$$\int_0^\infty d\omega\left(-\frac{1}{\pi}\frac{d\varphi(\omega)}{d\omega}\right)\log\omega^2$$
$$=\log\left[\omega_0^2-\int_0^\infty\frac{d\omega}{\pi}a(\omega)\frac{\tanh\frac{1}{2}\omega\beta}{\omega}\right].$$

We now have the representation

$$\log\left[\frac{G^{-1}(i2\pi n/\beta)}{\omega_0^2+(2\pi n/\beta)^2}\right]$$
$$=\int_0^\infty d\omega\left(-\frac{1}{\pi}\frac{d\varphi(\omega)}{d\omega}\right)\log\frac{\omega^2+(2\pi n/\beta)^2}{\omega_0^2+(2\pi n/\beta)^2},$$

and the summation formula derived from the product form of the hyperbolic sine function,

$$\tfrac{1}{2}\sum_{-\infty}^{\infty}\log\frac{\omega^2+(2\pi n/\beta)^2}{\omega_0^2+(2\pi n/\beta)^2}=\log\left[\frac{\sinh\frac{1}{2}\omega\beta}{\sinh\frac{1}{2}\omega_0\beta}\right],$$

gives us the desired result

$$Z=Z_e\exp\left[-\int_0^\infty d\omega\left(-\frac{1}{\pi}\frac{d\varphi(\omega)}{d\omega}\right)\log 2\sinh\tfrac{1}{2}\omega\beta\right].$$

The second factor can be ascribed to the oscillator, with its properties modified by interaction with the external system. The average energy of the oscillator at temperature $\vartheta=\beta^{-1}$ is therefore given by

$$E=\frac{\partial}{\partial\beta}\int_0^\infty d\omega\left(-\frac{1}{\pi}\frac{d\varphi}{d\omega}\right)\log 2\sinh\tfrac{1}{2}\omega\beta$$

in which the temperature dependence of the phase $\varphi(\omega)$ is not to be overlooked. In an extreme high-temperature limit, such that $\omega\beta\ll 1$ for all significant frequencies, we have

$$E\simeq\frac{\partial}{\partial\beta}[\log\beta+\tfrac{1}{2}\log(\omega_0^2-\beta\langle Q^2\rangle_\vartheta)],$$

and the simple classical result $E=\vartheta$ appears when $\langle Q^2\rangle_\vartheta$ is proportional to $\vartheta$. The oscillator energy at zero temperature is given by

$$E_0=\tfrac{1}{2}\int_0^\infty d\omega\left(-\frac{1}{\pi}\frac{d\varphi}{d\omega}\right)_{\vartheta=0}\omega,$$

and the oscillator contribution to the specific heat vanishes.

The following physical situation has consequences that resemble the simple model of the previous section. For values of $\omega\lesssim\omega_0$, $a(\omega)\tanh(\frac{1}{2}\omega\beta)\ll\omega_0^2$, and $a(\omega)$ differs significantly from zero until one attains frequencies that are large in comparison with $\omega_0$. The magnitudes that $a(\omega)$ can assume at frequencies greater than $\omega_0$ is limited only by the assumed absence of rapid variations and by the requirement of stability. The latter is generally assured if

$$\frac{1}{\pi}\int_{\sim\omega_0}^\infty\frac{d\omega}{\omega}a(\omega)<\omega_0^2.$$

We shall suppose that the stability requirement is comfortably satisfied, so that the right-hand side of the equation for $\cot\varphi(\omega)$ is an appreciable fraction of $\omega_0^2$ at sufficiently low frequencies. Then $\tan\varphi$ is very small at such frequencies, or $\varphi(\omega)\sim\pi$, and this persists until we reach the immediate neighborhood of the frequency $\omega_1<\omega_0$ such that

$$\omega_0^2-\omega_1^2-P\int_0^\infty\frac{d\omega}{\pi}\frac{\omega a(\omega)\tanh(\frac{1}{2}\omega\beta)}{\omega^2-\omega_1^2}=0.$$

That the function in question, $\mathrm{Re}G^{-1}(\omega+i0)$, has a zero, follows from its positive value at $\omega=0$ and its asymptotic approach to $-\infty$ with indefinitely increasing frequency. Under the conditions we have described, with the major contribution to the integral coming from high frequencies, the zero point is given approximately as

$$\omega_1^2\simeq\omega_0^2-\int_0^\infty\frac{d\omega}{\pi}a(\omega)\frac{\tanh(\frac{1}{2}\omega\beta)}{\omega},$$

and somewhat more accurately by

$$\omega_1^2=B\left[\omega_0^2-\int_0^\infty\frac{d\omega}{\pi}a(\omega)\frac{\tanh(\frac{1}{2}\omega\beta)}{\omega}\right],$$

where

$$B^{-1}=1+P\int_0^\infty\frac{d\omega}{2\pi}\frac{1}{\omega^2-\omega_1^2}\frac{d}{d\omega}[a(\omega)\tanh(\tfrac{1}{2}\omega\beta)].$$

As we shall see, $B$ is less than unity, but only slightly so under the circumstances assumed.

In the neighborhood of the frequency $\omega_1$, the equation that determines $\varphi(\omega)$ can be approximated by

$$-\tfrac{1}{2}a(\omega_1)\tanh(\tfrac{1}{2}\omega_1\beta)\cot\varphi(\omega)=B^{-1}(\omega_1^2-\omega^2)$$

or

$$\cot\varphi(\omega)=(\omega^2-\omega_1^2)/\gamma\omega_1\simeq(\omega-\omega_1)/\tfrac{1}{2}\gamma,$$

with the definition

$$\gamma=\tfrac{1}{2}Ba(\omega_1)[\tanh(\tfrac{1}{2}\omega_1\beta)/\omega_1]\ll\omega_1.$$

Hence, as $\omega$ rises through the frequency $\omega_1$, $\varphi$ decreases abruptly from a value close to $\pi$ to one near zero. The subsequent variations of the phase are comparatively gradual, and $\varphi$ eventually approaches zero as $\omega\to\infty$. A simple evaluation of the average oscillator energy can be given when the frequency range $\omega>\omega_1$ over which $a(\omega)$ is appreciable in magnitude is such that $\beta\omega\gg 1$. There will be no significant temperature variation in the latter domain and in particular $\omega_1$ should be essentially temperature independent. Then, since $-(1/\pi)(d\varphi/d\omega)$ in the neighborhood of $\omega_1$ closely resembles $\delta(\omega-\omega_1)$, we have approximately

$$E\doteq\frac{\partial}{\partial\beta}\left[\log(2\sinh\tfrac{1}{2}\omega_1\beta)+\beta\int_{>\omega_1}^\infty d\omega\left(-\frac{1}{\pi}\frac{d\varphi}{d\omega}\right)\tfrac{1}{2}\omega\right]$$

$$=\omega_1\left(\frac{1}{e^{\beta\omega_1}-1}+\tfrac{1}{2}\right)+\int_{>\omega_1}^\infty d\omega\left(-\frac{1}{\pi}\frac{d\varphi}{d\omega}\right)\tfrac{1}{2}\omega,$$

which describes a simple oscillator of frequency $\omega_1$, with a displaced origin of energy.

Note that with $\varphi(\omega)$ very small at a frequency slightly greater than $\omega_1$ and zero at infinite frequency, we have

$$\int_{>\omega_1}^\infty d\omega\left(-\frac{1}{\pi}\frac{d\varphi}{d\omega}\right)\tfrac{1}{2}\omega\simeq\frac{1}{2\pi}\int_{>\omega_1}^\infty d\omega\,\varphi(\omega)>0.$$

Related integrals are

$$\omega_0^2-\omega_1^2 \simeq \frac{2}{\pi}\int_{>\omega_1}^{\infty} d\omega\omega\varphi(\omega)>0$$

and

$$\log B^{-1} \simeq \frac{2}{\pi}\int_{>\omega_1}^{\infty} \frac{d\omega}{\omega}\varphi(\omega)>0.$$

The latter result confirms that $B<1$. A somewhat more accurate formula for $B$ is

$$B=\exp\left[-\int_{>\omega_1}^{\infty} d\omega\left(-\frac{1}{\pi}\frac{d\varphi(\omega)}{d\omega}\right)\log(\omega^2-\omega_1^2)\right].$$

If the major contributions to all these integrals come from the general vicinity of a frequency $\bar{\omega}\gg\omega_0$, we can make the crude estimates

$$\frac{1}{2\pi}\int_{>\omega_1}^{\infty} d\omega\varphi(\omega)\sim\frac{\omega_0^2}{\bar{\omega}}\ll\omega_1,\ \log B^{-1}\sim\left(\frac{\omega_0}{\bar{\omega}}\right)^2\ll 1.$$

Then neither the energy shift nor the deviation of the factor $B$ from unity are particularly significant effects.

The approximation of $\mathrm{Re}G(\omega+i0)$ as $B^{-1}(\omega_1^2-\omega^2)$ evidently holds from zero frequency up to a frequency considerably in excess of $\omega_1$. Throughout this frequency range we have

$$-\tfrac{1}{2}a(\omega)\tanh(\tfrac{1}{2}\omega\beta)\cot\varphi(\omega)=B^{-1}(\omega_1^2-\omega^2)$$

or

$$\cot\varphi(\omega)=(\omega_1^2-\omega^2)/\gamma\omega$$

with

$$\gamma(\omega)=\tfrac{1}{2}Ba(\omega)\tanh(\tfrac{1}{2}\omega\beta)/\omega.$$

If in particular $\beta\omega_1\ll 1$, the frequencies under consideration are in the classical domain and $\gamma$ is the frequency independent constant

$$\gamma=\tfrac{1}{4}Ba(0)\beta.$$

To regard $\gamma$ as constant for a quantum oscillator requires a suitable frequency restriction to the vicinity of $\omega_1$. The function $B(\omega^2)$ can be computed from

$$B(\omega^2)=\frac{2}{\pi}\frac{\sin^2\varphi(\omega)}{a(\omega)\tanh(\frac{1}{2}\omega\beta)}=\frac{1}{\pi}\frac{B}{\gamma\omega}\frac{1}{\cot^2\varphi+1},$$

and accordingly is given by

$$B(\omega^2)=B\frac{1}{\pi}\frac{\gamma\omega}{(\omega^2-\omega_1^2)^2+(\gamma\omega)^2}=B\frac{\gamma\omega}{\pi}\left|\frac{1}{\omega^2+i\gamma\omega-\omega_1^2}\right|^2.$$

The further concentration on the immediate vicinity of $\omega_1$, $|\omega-\omega_1|\sim\gamma$, gives

$$B(\omega^2)=\frac{B}{2\omega_1}\frac{1}{\pi}\frac{\frac{1}{2}\gamma}{(\omega-\omega_1)^2+(\frac{1}{2}\gamma)^2}$$

which clearly identifies $B<1$ with the contribution to the integral $\int d\omega^2 B(\omega^2)$ that comes from the vicinity of this resonance of width $\gamma$ at frequency $\omega_1$, although the same result is obtained without the last approximation. The remainder of the integral, $1-B$, arises from frequencies considerably higher than $\omega_1$ according to our assumptions.

There is a similar decomposition of the expressions for the Green's functions. Thus, with $t>t'$,

$$G_r(t-t')\simeq B\int_{-\infty}^{\infty}\frac{d\omega}{2\pi}e^{-i\omega(t-t')}\frac{1}{-\omega^2-i\gamma\omega+\omega_1^2}+\int_{\gg\omega_1^2}^{\infty} d\omega^2 B(\omega^2)\frac{\sin\omega(t-t')}{\omega}.$$

The second high-frequency term will decrease very quickly on the time scale set by $1/\omega_1$. Accordingly, in using this Green's function, say in the evaluation of

$$\langle q(t)\rangle_\beta{}^F=\int_{-\infty}^{\infty} dt' G_r(t-t')F(t')$$

for an external force that does not vary rapidly in relation to $\omega_1$, the contribution of the high-frequency term is essentially given by

$$F(t)\int_0^{\infty} d(t-t')\int_{\gg\omega_1^2}^{\infty} d\omega^2 B(\omega^2)\frac{\sin\omega(t-t')}{\omega}=F(t)\int_{\gg\omega_1^2}^{\infty} d\omega^2\frac{B(\omega^2)}{\omega^2}.$$

But

$$\int_{\gg\omega_1^2}^{\infty} d\omega^2\frac{B(\omega^2)}{\omega^2}\simeq\left[\omega_0^2-\int_0^{\infty}\frac{d\omega}{\pi}a(\omega)\frac{\tanh\frac{1}{2}\omega\beta}{\omega}\right]^{-1}-\frac{B}{\omega_1^2}\simeq 0,$$

and the response to such an external force is adequately described by the low-frequency part of the Green's function. We can represent this situation by an equivalent differential equation

$$\left(\frac{d^2}{dt^2}+\gamma\frac{d}{dt}+\omega_1^2\right)\langle q(t)\rangle_\beta{}^F=BF(t)$$

which needs no further qualification when the oscillations are classical but implies a restriction to a frequency interval within which $\gamma$ is constant, for quantum oscillations. We note the reduction in the effectiveness of the external force by the factor $B$. Under the circumstances outlined this effect is not important and we shall place $B$ equal to unity.

One can make a general replacement of the Green's functions by their low-frequency parts:

$$G_r(t-t') \to e^{-\frac{1}{2}\gamma(t-t')}\frac{1}{\omega_1}\sin(\omega_1(t-t'))\eta_+(t-t')$$

$$G_a(t-t') \to -e^{-\frac{1}{2}\gamma|t-t'|}\frac{1}{\omega_1}\sin(\omega_1(t-t'))\eta_-(t-t'),$$

if one limits the time localizability of measurements so that only time averages of $q(t)$ are of physical interest. This is represented in the expectation value formula by considering only functions $f_\pm(t)$ that do not vary too quickly. The corresponding replacement for $w(t-t')$ is

$$w(t-t') \to \coth(\tfrac{1}{2}\omega_1\beta)e^{-\frac{1}{2}\gamma|t-t'|}\frac{1}{\omega_1}\cos\omega_1(t-t'),$$

and the entire matrix $G_\vartheta(t-t')$ obtained in this way obeys the differential equation

$$\left(\frac{d^2}{dt^2}+\gamma\frac{d}{dt}+\omega_1^2\right)\left(\frac{d^2}{dt'^2}+\gamma\frac{d}{dt'}+\omega_1^2\right)G_\vartheta(t-t')$$

$$=\begin{pmatrix}1 & 0\\ 0 & -1\end{pmatrix}\left(\frac{d^2}{dt^2}+\omega_1^2\right)\delta(t-t')$$

$$+\begin{pmatrix}0 & -1\\ 1 & 0\end{pmatrix}\gamma\frac{d}{dt}\delta(t-t')+\begin{pmatrix}1 & -1\\ -1 & 1\end{pmatrix}\tfrac{1}{2}ia\delta(t-t'),$$

where $a=a(\omega_1)$.

The simplest presentation of results is again to be found in the Langevin viewpoint, which directs the emphasis from the coordinate operator $q(t)$ to the fluctuating force defined by

$$\left(\frac{d^2}{dt^2}+\gamma\frac{d}{dt}+\omega_1^2\right)q(t)=F(t)+F_f(t),$$

which is to say

$$F_f(t)=Q(t)+\gamma\frac{d}{dt}q(t)+(\omega_1^2-\omega_0^2)q(t).$$

This change is introduced by the substitution

$$f_\pm(t)=\left(\frac{d^2}{dt^2}-\gamma\frac{d}{dt}+\omega_1^2\right)k_\pm(t),$$

and the necessary partial integrations involve the previously established lemma on time-ordered operators, which here asserts that

$$\left(\exp\left[i\int dt fq\right]\right)_+=\left(\exp\left[i\int dtk(F+F_f)\right]\right)_+$$

$$\times\exp\left\{\tfrac{1}{2}i\int dt[(\omega_1^2-\omega_0^2)k^2+\omega_1^2k^2-(dk/dt)^2]\right\}.$$

We now find

$$\left\langle\left(\exp\left[-i\int dtk_-F_f\right]\right)_-\left(\exp\left[i\int dtk_+F_f\right]\right)_+\right\rangle_\vartheta$$

$$=\exp\left[-\tfrac{1}{2}\int dtdt'k(t)\zeta(t-t')k(t')\right]$$

with

$$\zeta(t-t')=\tfrac{1}{2}a\begin{pmatrix}1 & -1\\ -1 & 1\end{pmatrix}\delta(t-t')$$

$$-i(\omega_0^2-\omega_1^2)\begin{pmatrix}1 & 0\\ 0 & -1\end{pmatrix}\delta(t-t')$$

$$+i\gamma\begin{pmatrix}0 & 1\\ -1 & 0\end{pmatrix}\frac{d}{dt}\delta(t-t').$$

The latter matrix can also be identified as

$$\zeta(t-t')=\begin{pmatrix}\langle(F_f(t)F_f(t'))_+\rangle_\vartheta, & -\langle F_f(t')F_f(t)\rangle_\vartheta\\ -\langle F_f(t)F_f(t')\rangle_\vartheta, & \langle(F_f(t)F_f(t'))_-\rangle_\vartheta\end{pmatrix}.$$

In the classical limit

$$\left\langle\exp\left[i\int dtkF_f\right]\right\rangle_\vartheta=\exp\left[-\tfrac{1}{4}a\int dtk^2\right]$$

and

$$\tfrac{1}{4}a=\gamma\vartheta.$$

If a comparison is made with the similar results of the previous section it can be appreciated that the frequency range has been extended and the restriction $\omega_1\simeq\omega_0$ removed.

We return from these extended considerations on thermal equilibrium and consider one extreme example of negative temperature for the external system. This is described by

$$a(\omega)=a\delta(\omega-\omega_1),\quad \omega>0$$

and

$$-\beta(\omega_1)=|\beta|>0.$$

With the definition

$$(1/\pi)\omega_1 a\tanh(\tfrac{1}{2}\omega_1|\beta|)=(\omega_1\mu)^2,$$

we have

$$G(\zeta)=\left[\omega_0^2-\zeta^2+\frac{(\omega_1\mu)^2}{\omega_1^2-\zeta^2}\right]^{-1}$$

$$=\frac{\omega_1^2-\zeta^2}{[\zeta^2-\frac{1}{2}(\omega_0^2+\omega_1^2)]^2+(\omega_1\mu)^2-[\frac{1}{2}(\omega_0^2-\omega_1^2)]^2}.$$

As a function of $\zeta^2$, $G(\zeta)$ now has complex poles if

$$\tfrac{1}{2}|\omega_0^2-\omega_1^2|<\omega_1\mu.$$

We shall suppose, for simplicity, that $\omega_1=\omega_0$ and $\mu\ll\omega_0$. Then the poles of

$$G(\zeta)=\tfrac{1}{2}[(\omega_0^2+i\omega_0\mu-\zeta^2)^{-1}+(\omega_0^2-i\omega_0\mu-\zeta^2)^{-1}]$$

are located at $\zeta=\pm(\omega_0+\frac{1}{2}i\mu)$ and $\zeta=\pm(\omega-\frac{1}{2}i\mu)$. Accordingly, $G(\zeta)$ is regular outside a strip of width $2\alpha=\mu$. The associated Green's functions are given by

$$G_r(t-t')=\cosh(\tfrac{1}{2}\mu(t-t'))(\omega_0)^{-1}\sin(\omega_0(t-t'))\eta_+(t-t'),$$
$$G_a(t-t')=-\cosh(\tfrac{1}{2}\mu(t-t'))(\omega_0)^{-1}\sin(\omega_0(t-t'))\eta_-(t-t'),$$

and the function $w(t-t_2, t'-t_2)$, computed for $\omega_0(t-t_2)$, $\omega_0(t'-t_2)\gg1$, is

$$\begin{aligned}w(t-t_2, t'-t_2)&\\ \simeq(\omega_0)^{-1}\cos(\omega_0(t-t'))&[\coth(\tfrac{1}{2}\omega_0|\beta|)\sinh(\tfrac{1}{2}\mu(t-t_2))\\ &\times\sinh(\tfrac{1}{2}\mu(t'-t_2))+\coth(\tfrac{1}{2}\omega_0\beta_0)\\ &\times\cosh(\tfrac{1}{2}\mu(t-t_2))\cosh(\tfrac{1}{2}\mu(t'-t_2))].\end{aligned}$$

After the larger time intervals $\mu(t-t_2)$, $\mu(t'-t_2)\gg1$, we have

$$w(t-t_2, t'-t_2)\sim(2\omega_0)^{-1}e^{+\frac{1}{2}\mu(t-t_2)}e^{+\frac{1}{2}\mu(t'-t_2)}\cos\omega_0(t-t')\times[n_0+(1-e^{-\omega_0|\beta|})^{-1}],$$

with

$$n_0=(e^{\omega_0\beta_0}-1)^{-1}.$$

When $t$ is in the vicinity of a time $t_1$, such that the amplification factor

$$k\simeq\tfrac{1}{2}e^{\frac{1}{2}\mu(t_1-t_2)}\gg1,$$

the oscillator is described by the classical coordinate

$$q(t)=k[q_s(t)+q_n(t)].$$

Here

$$q_s(t)=\int_{t_2}^{\infty}dt'\frac{1}{\omega_0}\sin(\omega_0(t-t'))F(t')e^{-\frac{1}{2}\mu(t'-t_2)}$$

and

$$q_n(t)=q_1\cos\omega t+q_2\sin\omega t,$$

where $q_1$ and $q_2$ are characterized by the expectation value formula

$$\langle e^{i(q_1f_1+q_2f_2)}\rangle=\exp[-(\nu/\omega_0)\tfrac{1}{2}(f_1^2+f_2^2)],$$

in which

$$\nu=n_0+(1-e^{-\omega_0|\beta|})^{-1}.$$

Accordingly, the probability of observing $q_1$ and $q_2$ within the range $dq_1$, $dq_2$ is

$$p(q_1q_2)dq_1dq_2=\frac{1}{2\pi}\frac{\omega_0}{\nu}\exp\left[-\frac{\omega_0}{\nu}\frac{1}{2}(q_1^2+q_2^2)\right]dq_1dq_2$$
$$=\frac{\omega_0}{\nu}\exp\left(-\frac{1}{2}\frac{\omega_0}{\nu}q_n^2\right)q_ndq_n\frac{1}{2\pi}d\varphi,$$

where $q_n$ and $\varphi$ are the amplitude and phase of $q_n(t)$. Despite rather different assumptions about the external system, these are the same conclusions as before, apart from a factor of $\frac{1}{2}$ in the formula for the gain.

## GENERAL THEORY

The whole of the preceding discussion assumes an external system that is only slightly influenced by the presence of the oscillator. Now we must attempt to place this simplification within the framework of a general formulation. A more thorough treatment is also a practical necessity in situations such as those producing amplification of the oscillator motion, for a sizeable reaction in the external system must eventually appear, unless a counter mechanism is provided.

It is useful to supplement the previous Lagrangian operator with the term $q'(t)Q$, in which $q'(t)$ is an arbitrary numerical function of time, and also, to imagine the coupling term $qQ$ supplied with a variable factor $\lambda$. Then

$$\begin{aligned}\frac{\partial}{\partial\lambda}\langle t_2|t_2\rangle^{F\pm q\pm'}&=i\left\langle\left|\int dt(q_+Q_+-q_-Q_-)\right|\right\rangle,\\ &=-i\int_{t_2}^{t_1}dt\left(\frac{\delta}{\delta F_+(t)}\frac{\delta}{\delta q_+'(t)}-\frac{\delta}{\delta F_-(t)}\frac{\delta}{\delta q_-'(t)}\right)\\ &\qquad\times\langle t_2|t_2\rangle^{F\pm q\pm'},\end{aligned}$$

provided that the states to which the transformation function refers do not depend upon the coupling between the systems, or that the trace of the transformation function is being evaluated. A similar statement would apply to a transformation function with different terminal times. This differential equation implies an integrated form, in which the transformation function for the fully coupled system ($\lambda=1$) is expressed in terms of the transformation function for the uncoupled system ($\lambda=0$). The latter is the product of transformation functions for the independent oscillator and external system. The relation is

$$\langle t_2|t_2\rangle^{F\pm}=\exp\left[-i\int_{t_2}^{t_1}dt\left(\frac{\delta}{\delta F_+}\frac{\delta}{\delta q_+'}-\frac{\delta}{\delta F_-}\frac{\delta}{\delta q_-'}\right)\right]\times\langle t_2|t_2\rangle_{\text{osc}}^{F\pm}\langle t_2|t_2\rangle_{\text{ext}}^{q\pm'}|_{q\pm'=0},$$

and we have indicated that $q_\pm'$ is finally set equal to zero if we are concerned only with measurements on the oscillator.

Let us consider for the moment just the external system with the perturbation $q'Q$, the effect of which is indicated by[6]

$$\langle t_2|t_2\rangle^{q\pm'}=\left\langle t_2\left|\left(\exp\left[-i\int dtq_-'Q\right]\right)_-\right.\right.$$
$$\times\left.\left.\left(\exp\left[i\int dtq_+'Q\right]\right)_+\right|t_2\right\rangle.$$

We shall define

$$Q_+(t,q_\pm')=\frac{\langle t_2|Q_+(t)|t_2\rangle^{q\pm'}}{\langle t_2|t_2\rangle^{q\pm'}}$$
$$=\frac{1}{i}\frac{\delta}{\delta q_+'(t)}\log\langle t_2|t_2\rangle^{q\pm'}$$

and similarly

$$Q_-(t,q_\pm')=\frac{\langle t_2|Q_-(t)|t_2\rangle^{q\pm'}}{\langle t_2|t_2\rangle^{q\pm'}}$$
$$=-\frac{1}{i}\frac{\delta}{\delta q_-'(t)}\log\langle t_2|t_2\rangle^{q\pm'}.$$

When $q_\pm'(t)=q'(t)$, we have

$$Q_+(t,q')=Q_-(t,q')=\langle t_2|Q(t)|t_2\rangle^{q'},$$

which is the expectation value of $Q(t)$ in the presence of the perturbation described by $q'(t)$. This is assumed to be zero for $q'(t)=0$ and depends generally upon the history of $q'(t)$ between $t_2$ and the given time.

The operators $q_\pm(t)$ are produced within the transformation function by the functional differential operators $(\pm 1/i)\ \delta/\delta F_\pm(t)$, and since the equation of motion for the uncoupled oscillator is

$$\left(\frac{d^2}{dt^2}+\omega_0^2\right)q(t)=F(t),$$

we have

$$\left(\frac{\partial^2}{\partial t^2}+\omega_0^2\right)\left(\pm\frac{1}{i}\right)\frac{\delta}{\delta F_\pm(t)}\langle t_2|t_2\rangle^{F\pm}$$
$$=\exp\left[-i\int dt\left(\frac{\delta}{\delta F_+}\frac{\delta}{\delta q_+'}-\frac{\delta}{\delta F_-}\frac{\delta}{\delta q_-'}\right)\right]$$
$$\times F_\pm(t)\langle t_2|t_2\rangle_{\text{osc}}{}^{F\pm}\langle t_2|t_2\rangle_{\text{ext}}{}^{q\pm'}|_{q_\pm'=0}.$$

On moving $F_\pm(t)$ to the left of the exponential, this becomes

$$F_\pm(t)\langle t_2|t_2\rangle^{F\pm}+\exp[\ ]\langle t_2|t_2\rangle_{\text{osc}}{}^{F\pm}$$
$$\times\left(\pm\frac{1}{i}\right)\frac{\delta}{\delta q_\pm'(t)}\langle t_2|t_2\rangle_{\text{ext}}{}^{q\pm'}|_{q_\pm'=0}.$$

But

$$\left(\pm\frac{1}{i}\right)\frac{\delta}{\delta q_\pm'(t)}\langle t_2|t_2\rangle_{\text{ext}}{}^{q\pm'}=Q_\pm(t,q_\pm')\langle t_2|t_2\rangle_{\text{ext}}{}^{q\pm'},$$

and furthermore,

$$\exp[\ ]Q(t,q_\pm')\langle t_2|t_2\rangle_{\text{osc}}{}^{F\pm}\langle t_2|t_2\rangle_{\text{ext}}{}^{q\pm'}|_{q_\pm'=0}$$
$$=Q\left(t,\pm\frac{1}{i}\frac{\delta}{\delta F_\pm}\right)\langle t_2|t_2\rangle^{F\pm},$$

which leads us to the following functional differential equation for the transformation function $\langle t_2|t_2\rangle^{F\pm}$, in which a knowledge is assumed of the external system's reaction to the perturbation $q_\pm'(t)$:

$$\left[\left(\frac{\partial^2}{\partial t^2}+\omega_0^2\right)\left(\pm\frac{1}{i}\right)\frac{\delta}{\delta F_\pm(t)}-Q_\pm\left(t,\pm\frac{1}{i}\frac{\delta}{\delta F_\pm}\right)-F_\pm(t)\right]$$
$$\times\langle t_2|t_2\rangle^{F\pm}=0.$$

Throughout this discussion one must distinguish between the $\pm$ signs attached to particular components and those involved in the listing of complete sets of variables.

The differential equations for time development are supplemented by boundary conditions which assert, at a time $t_1$ beyond which $F_+(t)=F_-(t)$, that

$$\left(\frac{\delta}{\delta F_+(t_1)}+\frac{\delta}{\delta F_-(t_1)}\right)\langle t_2|t_2\rangle^{F\pm}=0$$

while, for the example of the transformation function $\langle t_2|t_2\rangle_{\beta_0}{}^{F\pm}$, we have the initial conditions

$$\left[\left(\frac{\delta}{\delta F_+}-\frac{\delta}{\delta F_-}\right)(t_2)\right.$$
$$\left.+\frac{i}{\omega_0}\coth(\tfrac{1}{2}\omega_0\beta_0)\frac{\partial}{\partial t_2}\left(\frac{\delta}{\delta F_+}+\frac{\delta}{\delta F_-}\right)(t_2)\right]\langle t_2|t_2\rangle^{F\pm}=0$$

and

$$\left[\frac{\partial}{\partial t_2}\left(\frac{\delta}{\delta F_+}-\frac{\delta}{\delta F_-}\right)(t_2)\right.$$
$$\left.-i\omega_0\coth(\tfrac{1}{2}\omega_0\beta_0)\left(\frac{\delta}{\delta F_+}+\frac{\delta}{\delta F_-}\right)(t_2)\right]\langle t_2|t_2\rangle^{F\pm}=0.$$

The previous treatment can now be identified as the

[6] Such positive and negative time-ordered products occur in a recent paper by K. Symanzik [J. Math. Phys. 1, 249 (1960)], which appeared after this paper had been written and its contents used as a basis for lectures delivered at the Brandeis Summer School, July, 1960.

approximation of the $Q_\pm(t,q_\pm')$ by linear functions of $q_\pm'$,

$$Q_+(t,q_\pm')=i\int dt'[A_{++}(t-t')q_+'(t')-A_{+-}(t-t')q_-'(t')]$$

$$Q_-(t,q_\pm')=i\int dt'[A_{-+}(t-t')q_+'(t')-A_{--}(t-t')q_-'(t')],$$

wherein the linear equations for the operators $q_\pm(t)$ and their meaning in terms of variations of the $F_\pm$ have been united in one pair of functional differential equations. This relation becomes clearer if one writes

$$\langle t_2|t_2\rangle^{F_\pm}=\exp[iW(F_\pm)]$$

and, with the definition

$$q_\pm(t,F_\pm)=(\pm)\frac{\delta}{\delta F_\pm(t)}W(F_\pm)$$

$$=\left(\pm\frac{1}{i}\right)\frac{\delta}{\delta F_\pm(t)}\log\langle t_2|t_2\rangle^{F_\pm},$$

converts the functional differential equations into

$$\left(\frac{\partial^2}{\partial t^2}+\omega_0^2\right)q_\pm(t,F_\pm)-Q_\pm\left(t,q_\pm\pm\frac{1}{i}\frac{\delta}{\delta F_\pm}\right)-F_\pm(t)=0.$$

The boundary conditions now appear as

$$(q_+-q_-)(t_1,F_\pm)=0$$

and

$$(q_++q_-)(t_2,F_\pm)+\frac{i}{\omega_0}\coth(\tfrac{1}{2}\omega_0\beta_0)\frac{\partial}{\partial t_2}(q_+-q_-)(t_2,F_\pm)=0$$

$$\frac{\partial}{\partial t_2}(q_++q_-)(t_2,F_\pm)-i\omega_0\coth(\tfrac{1}{2}\omega_0\beta_0)(q_+-q_-)(t_2,F_\pm)=0.$$

When the $Q_\pm$ are linear functions of $q_\pm$, the functional differential operators disappear[7] and we regain the linear equations for $q_\pm(t)$, which in turn imply the quadratic form of $W(F_\pm)$ that characterizes the preceding discussion.

[7] The degeneration of the functional equations into ordinary differential equations also occurs when the motion of the oscillator is classical and free of fluctuation.

# Gauge Invariance and Mass

JULIAN SCHWINGER
*Harvard University, Cambridge, Massachusetts, and University of California, Los Angeles, California*
(Received July 20, 1961)

It is argued that the gauge invariance of a vector field does not necessarily imply zero mass for an associated particle if the current vector coupling is sufficiently strong. This situation may permit a deeper understanding of nucleonic charge conservation as a manifestation of a gauge invariance, without the obvious conflict with experience that a massless particle entails.

DOES the requirement of gauge invariance for a vector field coupled to a dynamical current imply the existence of a corresponding particle with zero mass? Although the answer to this question is invariably given in the affirmative,[1] the author has become convinced that there is no such necessary implication, once the assumption of weak coupling is removed. Thus the path to an understanding of nucleonic (baryonic) charge conservation as an aspect of a gauge invariance, in strict analogy with electric charge,[2] may be open for the first time.

One potential source of error should be recognized at the outset. A gauge-invariant system is not the continuous limit of one that fails to admit such an arbitrary function transformation group. The discontinuous change of invariance properties produces a corresponding discontinuity of the dynamical degrees of freedom and of the operator commutation relations. No reliable conclusions about the mass spectrum of a gauge-invariant system can be drawn from the properties of an apparently neighboring system, with a smaller invariance group. Indeed, if one considers a vector field coupled to a divergenceless current, where gauge invariance is destroyed by a so-called mass term with parameter $m_0$, it is easily shown[3] that the mass spectrum must extend below $m_0$. The lowest mass value will therefore become arbitrarily small as $m_0$ approaches zero. Nevertheless, if $m_0$ is exactly zero the commutation relations, or equivalent properties, upon which this conclusion is based become entirely different and the argument fails.

If invariance under arbitrary gauge transformations is asserted, one should distinguish sharply between numerical gauge functions and operator gauge functions, for the various operator gauges are not on the same quantum footing. In each coordinate frame there is a unique operator gauge, characterized by three-dimensional transversality (radiation gauge), for which one has the standard operator construction in a vector space of positive norm, with a physical probability interpretation. When the theory is formulated with the aid of vacuum expectation values of time-ordered operator products, the Green's functions, the freedom of formal gauge transformation can be restored.[4] The Green's functions of other gauges have more complicated operator realizations, however, and will generally lack the positiveness properties of the radiation gauge.

Let us consider the simplest Green's function associated with the field $A_\mu(x)$, which can be derived from the unordered product

$$\langle A_\mu(x)A_\nu(x')\rangle = \int \frac{(dp)}{(2\pi)^3} e^{ip(x-x')} dm^2\, \eta_+(p)\delta(p^2+m^2)A_{\mu\nu}(p),$$

where the factor $\eta_+(p)\delta(p^2+m^2)$ enforces the spectral restriction to states with mass $m \geq 0$ and positive energy. The requirement of non-negativeness for the matrix $A_{\mu\nu}(p)$ is satisfied by the structure associated with the radiation gauge, in virtue of the gauge-dependent asymmetry between space and time (the time axis is specified by the unit vector $n_\mu$):

$$A_{\mu\nu}{}^R(p) = B(m^2)\left[g_{\mu\nu} - \frac{(p_\mu n_\nu + p_\nu n_\mu)(np) + p_\mu p_\nu}{p^2+(np)^2}\right].$$

Here $B(m^2)$ is a real non-negative number. It obeys the sum rule

$$1 = \int_0^\infty dm^2\, B(m^2),$$

which is a full expression of all the fundamental equal-time commutation relations.

The field equations supply the analogous construction for the vacuum expectation value of current products $\langle j_\mu(x)j_\nu(x')\rangle$, in terms of the non-negative matrix

$$j_{\mu\nu}(p) = m^2 B(m^2)(p_\mu p_\nu - g_{\mu\nu}p^2).$$

The factor $m^2$ has the decisive consequence that $m=0$ is not contained in the current vector's spectrum of vacuum fluctuations. The latter determines $B(m^2)$ for $m>0$, but leaves unspecified a possible delta function contribution at $m=0$,

$$B(m^2) = B_0\delta(m^2) + B_1(m^2).$$

The non-negative constant $B_0$ is then fixed by the sum rule,

$$1 = B_0 + \int_0^\infty dm^2\, B_1(m^2).$$

[1] For example, J. Schwinger, Phys. Rev. **75**, 651 (1949).
[2] T. D. Lee and C. N. Yang, Phys. Rev. **98**, 1501 (1955).
[3] K. Johnson, Nuclear Phys. **25**, 435 (1961).
[4] J. Schwinger, Phys. Rev. **115**, 721 (1959).



Reprinted from *The Physical Review* **125**, 397–398 (1962).

We have now recognized that the vacuum fluctuations of the vector $A_\mu$ are composed of two parts. One, with $m>0$, is directly related to corresponding current fluctuations, while the other part, with $m=0$, can be associated with a pure radiation field, which is transverse in both three- and four-dimensional senses and has no accompanying current. Imagine that the current vector contains a variable numerical factor. If this is set equal to zero, we have $B_1(m^2)=0$ and $B_0=1$ or, just the radiation field. For a sufficiently small nonzero value of the parameter, $B_0$ will be slightly less than unity, which may be the situation for the electromagnetic field. Or it may be that the electrodynamic coupling is quite considerable and gives rise to a small value of $B_0$, which has the appearance of a fairly weak coupling. Can we increase further the magnitude of the variable parameter until $\int dm^2\, B_1(m^2)$ attains its limiting value of unity, at which point $B_0=0$, and $m=0$ disappears from the spectrum of $A_\mu$? The general requirement of gauge invariance no longer seems to dispose of this essentially dynamical question.

Would the absence of a massless particle imply the existence of a stable, unit spin particle of nonzero mass? Not necessarily, since the vacuum fluctuation spectrum of $A_\mu$ becomes identical with that of $j_\mu$, which is governed by all of the dynamical properties of the fields that contribute to this current. For the particularly interesting situation of a vector field that is coupled to the current of nucleonic charge, the relevant spectrum, in the approximate strong-interaction framework, is that of the states with $N=Y=T=0$, $R_T=-1$, $J=1$, and odd parity. This is a continuum, beginning at three pion masses.[5] It is entirely possible, of course, that $B(m^2)$ shows a more or less pronounced maximum which could be characterized approximately as an unstable particle.[6] But the essential point is embodied in the view that the observed physical world is the outcome of the dynamical play among underlying primary fields, and the relationship between these fundamental fields and the phenomenological particles can be comparatively remote, in contrast to the immediate correlation that is commonly assumed.

---

[5] The very short range of the resulting nuclear interaction together with the qualitative inference that like nucleonic charges are thereby repelled suggests that the vector field which defines nucleonic charge is also the ultimate instrument of nuclear stability.

[6] *Note added in proof*. Experimental evidence for an unstable particle of this type has recently been announced by B. C. Maglić, L. W. Alvarez, A. H. Rosenfeld, and M. L. Stevenson, in Phys. Rev. Letters 7, 178 (1961).

# Non-Abelian Gauge Fields. Commutation Relations*

JULIAN SCHWINGER†
*Harvard University, Cambridge, Massachusetts, and Institute of Theoretical Physics, Department of Physics, Stanford University, Stanford, California*

(Received August 25, 1961)

The question is raised for non-Abelian vector gauge fields whether gauge invariance necessarily implies a massless physical particle. As a preliminary to studying this problem, the action principle is used to discover the independent dynamical variables of such gauge fields and construct their commutation relations.

## INTRODUCTION

It is well known that gauge invariance intimately ties the electromagnetic field $A_\mu(x)$, $F_{\mu\nu}(x)$ to the set of all fields $\chi(x)$ that bear electrical charge. This internal property is described by a finite imaginary Hermitian matrix $q$ with integer eigenvalues. A gauge transformation involves an arbitrary numerical function $\lambda(x)$. It is a linear homogeneous transformation for the charged fields $\chi(x)$, but an inhomogeneous one for the gauge field $A_\mu(x)$,

$$\chi(x)\to e^{iq\lambda(x)}\chi(x),\quad A_\mu(x)\to A_\mu(x)+\partial_\mu\lambda(x),$$
$$F_{\mu\nu}(x)\to F_{\mu\nu}(x).$$

Such transformations form an Abelian group, in which the gauge function,

$$\lambda(x)=\lambda^{(1)}(x)+\lambda^{(2)}(x),$$

describes the superposition of two individual transformations. The integer spectrum of charge is related to the compact structure of this group, which has the topology of the circle. Gauge invariance implies that local conservation of charge is not just a consequence of the equations of motion of the charge bearing fields but appears as an identity characteristic of the gauge field differential equations.

In this familiar situation the gauge field does not carry the internal property to which it is coupled. A different example is furnished by the gravitational field, for this couples with energy and momentum, to which all physical systems must contribute. In other respects, however, the requirement of general coordinate invariance is quite analogous to that of gauge invariance. There is an intermediate possibility in which the gauge field is coupled to, and also carries, internal rather than space-time properties. Then the gauge field retains the space-time transformation properties of the electromagnetic field. This is indicated by the tensor notation $\phi_{\mu a}$, $G_{\mu\nu a}$, where the index $a=1\cdots n$ refers to the internal space. For the gravitational field the latter is also a coordinate index, which requires fields of more complicated space-time transformation properties.

The gauge transformations of a field $\chi(x)$ that supports a number of internal properties, as represented by finite linearly independent matrices $T_a$, $a=1\cdots n$, can generally be stated explicitly only for infinitesimal transformations,

$$\chi(x)\to\left[1+i\sum_{a=1}^{n}T_a\delta\lambda_a(x)\right]\chi(x).$$

If these are to generate a transformation group, two

* Supported in part by the Air Force Office of Scientific Research (Air Research and Development Command).

† Visiting professor, Stanford University, Stanford, California, summer, 1961.

Reprinted from *The Physical Review* **125**, 1043–1048 (1962).

successive infinitesimal transformations, performed in opposite order, must be connected by another such transformation. This implies the commutation relations,

$$[T_b,T_c]=\sum_{a=1}^{n} T_a t_{abc},$$

and the constants,

$$t_{abc}=-t_{acb},$$

characterize the structure of the group.

The statement that the gauge field also carries these internal properties is conveyed by the infinitesimal gauge transformation,

$$\phi_\mu(x)\rightarrow[1+i\sum_{a=1}^{n} t_a\delta\lambda_a(x)]\phi_\mu(x)+\partial_\mu\delta\lambda(x),$$

$$G_{\mu\nu}(x)\rightarrow[1+i\sum t_a\delta\lambda_a(x)]G_{\mu\nu}(x),$$

which uses a matrix notation for the $n$-dimensional internal space. The homogeneous transformation of $G_{\mu\nu}$ implies that the matrices $t_a$ obey the group commutation relations,

$$[t_b,t_c]=\sum_{a=1}^{n} t_a t_{abc}.$$

But the inhomogeneous transformations of $\phi_\mu$ must also represent the group structure. The corresponding condition is

$$\sum_{bc}[(t_b)_{ac}\delta\lambda_b^{(1)}\partial_\mu\delta\lambda_c^{(2)}-(t_b)_{ac}\delta\lambda_b^{(2)}\partial_\mu\delta\lambda_c^{(1)}]=\partial_\mu[\sum_{bc} t_{abc}\delta\lambda_b^{(1)}\delta\lambda_c^{(2)}],$$

which asserts that

$$(t_b)_{ac}=t_{abc}.$$

Thus the matrices $t_a$ are derived from the structure constants of the group. To verify that these matrices do obey the group commutation relations, we write the latter in the matrix form

$$[T_b,T]=Tt_b.$$

Then

$$[T_b,[T_c,T]]-[T_c,[T_b,T]]=T[t_b,t_c],$$

which also equals

$$[[T_b,T_c],T]=T\sum_a t_a t_{abc},$$

and the desired result follows from the linear independence of the $T$ matrices.

A general intuition about the space of internal properties can be formulated as the requirement that its symmetry group be compact, in contrast to the open Lorentz group. It is then possible to make all matrix representations be unitary, so that the matrices $T_a$ are Hermitian. This includes the matrices $t_a$, which generate an $n$-dimensional representation. But we should also note that the structure constants associated with the Hermitian $T$ matrices are imaginary, and thus the imaginary Hermitian $t$ matrices must be antisymmetrical, or

$$(t_b)_{ac}=t_{abc}=-t_{cba}.$$

This property, in conjunction with

$$t_{abc}=-t_{acb},$$

expresses the total antisymmetry of the set of $n^3$ numbers $t_{abc}$. In order to construct nonzero $t$ matrices it is necessary that $n\geq 3$, and for $n=3$ the structure of a non-Abelian group is uniquely that of the three-dimensional Euclidean rotation group.[1]

The concept of an internal symmetry group has long been considered a possible basis for describing the non-space-time properties of physical particles. To relate such a group to gauge transformations of vector fields is an attractive idea, but one which seems to run into difficulty immediately if it is accepted that a gauge field implies a corresponding massless particle. Only the photon is known as an example of this class of physical particle. It is hard to agree that the objection is overcome by destroying completely[2] the gauge invariance which is the entire motivation of the gauge fields. But there may be an escape from this dilemma. The author has remarked that gauge-invariant systems of the electromagnetic or, more generally expressed, Abelian type need not have an accompanying massless particle if the coupling is sufficiently strong.[3] The question is whether a similar possibility exists for systems with non-Abelian gauge groups. To discuss this problem requires at least a full knowledge of the operator properties of the gauge field, treated as a physical quantum-mechanical system without reference to weak coupling approximations. These commutation relations are not known. And it is not a trivial query whether a consistent quantum field theory is possible at all for a system that admits a non-Abelian gauge group. But the latter can hardly be answered until a set of commutation relations has been displayed, for, without these, the nature of the operator description, with its necessary attribute of completeness, remains unknown. It is the purpose of this paper to produce such commutation relations, but we shall leave untouched the more difficult question of consistency.

### THE ACTION PRINCIPLE

In order to construct an invariant Lagrange function in the standard first order differential form we must combine the antisymmetrical tensor of Hermitian operators $G_{\mu\nu}$ with a similarly transforming differential construct of the Hermitian operators $\phi_\mu$. Unlike the electromagnetic situation, the antisymmetrical gradient or curl will not suffice, since its infinitesimal gauge

[1] It is this context that non-Abelian gauge groups were first discussed, C. N. Yang and R. Mills, Phys. Rev. **96**, 191 (1954).
[2] See, for example, J. Sakurai, Ann. Phys. **11**, 1 (1960).
[3] J. Schwinger, Phys. Rev. **125**, 397 (1962).

transformation is

$$\partial_\mu\phi_\nu-\partial_\nu\phi_\mu \to [1+i\sum_a t_a\delta\lambda_a](\partial_\mu\phi_\nu-\partial_\nu\phi_\mu) + i\sum_a t_a(\phi_\nu\partial_\mu\delta\lambda_a-\phi_\mu\partial_\nu\delta\lambda_a),$$

and the last term has no counterpart in the $G_{\mu\nu}$ transformation law. We are thus led to consider the compensating gauge transformation of the expression

$$i(\phi_\mu\,.\,t\phi_\nu),$$

which employs a notation for a vector in the $n$-dimensional internal space. The components are

$$i(\phi_\mu\,.\,t_b\phi_\nu)=\sum_{ac}\phi_{\mu a}\,.\,it_{abc}\phi_{\nu c}.$$

In addition, the dot symbolizes symmetrized multiplication of the operators,

$$\phi_{\mu a}\,.\,\phi_{\nu c}=\tfrac{1}{2}\{\phi_{\mu a},\phi_{\nu c}\},$$

so that the whole structure is a Hermitian operator. In view of the complete antisymmetry of $t_{abc}$, the same vector can be written alternatively as

$$i(\phi_\mu\,.\,t\phi_\nu)=-i`t\phi_\mu'\,.\,\phi_\nu=-i\phi_\mu\,.\,`t\phi_\nu'=-i(\phi_\nu\,.\,t\phi_\mu).$$

Here we have introduced a notation for a matrix,

$$`t\phi'=\sum_a t_a\phi_a,$$

which has the elements,

$$(t\phi)_{bc}=\sum_a t_{bac}\phi_a.$$

On transcribing the gauge transformation of the curl into this notation, it reads

$$\partial_\mu\phi_\nu-\partial_\nu\phi_\mu \to (1+i`t\delta\lambda')(\partial_\mu\phi_\nu-\partial_\nu\phi_\mu) + i(\phi_\nu t\partial_\mu\delta\lambda)-i(\phi_\mu t\partial_\nu\delta\lambda),$$

which is to be compared with

$$i(\phi_\mu\,.\,t\phi_\nu) \to i(\phi_\mu\,.\,t\phi_\nu)-(\phi_\mu\,.\,[t,`t\delta\lambda']\phi_\nu) - i(\phi_\nu t\partial_\mu\delta\lambda)+i(\phi_\mu t\partial_\nu\delta\lambda).$$

The commutation properties of the $t$-matrices are expressed by

$$(\phi_\mu\,.\,[t,`t\delta\lambda']\phi_\nu)=`t\delta\lambda'(\phi_\mu\,.\,t\phi_\nu),$$

and we have the desired result

$$\partial_\mu\phi_\nu-\partial_\nu\phi_\mu+i(\phi_\mu\,.\,t\phi_\nu) \to (1+i`t\delta\lambda')[\partial_\mu\phi_\nu-\partial_\nu\phi_\mu+i(\phi_\mu\,.\,t\phi_\nu)].$$

A possible Lagrange function is given by

$$\mathcal{L}=-\tfrac{1}{2}G^{\mu\nu}\,.\,[\partial_\mu\phi_\nu-\partial_\nu\phi_\mu+i(\phi_\mu\,.\,t\phi_\nu)] + \tfrac{1}{4}f^2G^{\mu\nu}G_{\mu\nu}+k^\mu\,.\,\phi_\mu+\mathcal{L}(\chi),$$

where scalar products of vectors in the internal space are understood. The contributory Lagrange function,

$$\mathcal{L}(\chi)=\tfrac{1}{4}(\chi A^\mu\partial_\mu\chi-\partial_\mu\chi A^\mu\chi)-\mathcal{H}(\chi),$$

is that of the systems carrying the properties represented by the matrices $T_a$. The latter, incidentally, are imaginary and antisymmetrical if all fields are chosen to be Hermitian. The flux of these properties is described by the Hermitian current vector

$$k_a{}^\mu(x)=-\tfrac{1}{2}i\chi(x)A^\mu T_a\chi(x),$$

the structure of which follows from the requirement of gauge invariance. If $\mathcal{H}(\chi)$ is a gauge scalar, the response of $\mathcal{L}(\chi)$ to an infinitesimal gauge transformation is

$$\mathcal{L}(\chi)\to\mathcal{L}(\chi)-k^\mu\partial_\mu\delta\lambda,$$

under the assumption that the kinematical matrices $A^\mu$ operate entirely in space-time, or that

$$[A^\mu,T_a]=0.$$

A compensating term is produced by $k^\mu\,.\,\phi_\mu$, on taking into account the homogeneous gauge transformation of the current,

$$k^\mu\to(1+i`t\delta\lambda')k^\mu,$$

which follows from the commutation properties of the $T$ matrices, displayed as

$$[T,`T\delta\lambda']=`t\delta\lambda'T.$$

The dimensionless number $f^2$ appears as an arbitrary coupling constant. We shall see that it must be a positive quantity.

Until the commutation properties of the fields are known, a Lagrange function $\mathcal{L}$ or action operator

$$W_{12}=\int_{\sigma_2}^{\sigma_1}(dx)\mathcal{L},$$

constructed from formal invariance arguments, has only a heuristic significance, leading through the principle of stationary action,

$$\delta W_{12}=G_1-G_2,$$

to the tentative statement of covariant field equations and infinitesimal transformation generators. Let us consider first the infinitesimal gauge transformations of the $\chi$ field alone,

$$\delta W_{12}=\int(dx)[-k^\mu\partial_\mu\delta\lambda+\phi_\mu\,.\,i`t\delta\lambda'k^\mu],$$

which imply the extended conservation law

$$\partial_\mu k^\mu-i`t\phi_\mu'\,.\,k^\mu=0,$$

and the infinitesimal generator

$$G_\lambda=\int d\sigma_\mu(-k^\mu\delta\lambda)=\int(d\mathbf{x})(-k^0\delta\lambda).$$

The significance of the latter is expressed by

$$(1/i)[\chi,G_\lambda]=\delta\chi = i`T\delta\lambda'\chi,$$

which gives the commutation rule

$$x^0=x^{0\prime}:\quad [\chi(x),k_a{}^0(x')]=\delta(\mathbf{x}-\mathbf{x}')T_a\chi(x).$$

The corresponding integral form is

$$[\chi(x),K_a]=T_a\chi(x),$$

where

$$K_a=\int d\sigma_\mu\, k_a{}^\mu=\int (d\mathbf{x})k_a{}^0.$$

These Hermitian operators obey the group commutation relations,

$$[K_b,K_c]=\textstyle\sum_a K_a t_{abc}.$$

They are not constants of the motion, however, since $k^\mu$ is not governed by a true conservation equation.

The differential equations of the gauge field implied by the action principle are

$$\partial_\mu\phi_\nu-\partial_\nu\phi_\mu+i(\phi_\mu.t\phi_\nu)=f^2G_{\mu\nu},$$

and

$$\partial_\nu G^{\mu\nu}-i^{\prime}t\phi_\nu{}^{\prime}.G^{\mu\nu}=k^\mu.$$

These equations are also conveyed by the following matrix statements:

$$[\partial_\mu-i^{\prime}t\phi_\mu{}^{\prime},\,\partial_\nu-i^{\prime}t\phi_\nu{}^{\prime}]=-if^{2\prime}tG_{\mu\nu}{}^{\prime},$$
$$[\partial_\nu-i^{\prime}t\phi_\nu{}^{\prime},\,{}^{\prime}tG^{\mu\nu\prime}]={}^{\prime}tk^{\mu\prime},$$

in which it must be clearly understood that the commutators refer only to the coordinate and matrix indices; the operators, on the contrary, are to be symmetrically multiplied. In this manner of displaying the operators, an infinitesimal gauge transformation is generated by orthogonal transformation with the matrix $1+i^{\prime}t\delta\lambda^{\prime}$.

According to the antisymmetry of $G^{\mu\nu}$, the vector

$$j^\mu=k^\mu-i(\phi_\nu.tG^{\mu\nu})$$

is divergenceless,

$$\partial_\mu j^\mu=0,$$

and thus it is the Hermitian operators,

$$\mathbf{T}_a=\int d\sigma_\mu\, j_a{}^\mu$$
$$=K_a-i\int(d\mathbf{x})\,\phi_k.t_aG^{0k},$$

that are constants of the motion. It is natural to expect that these operators also obey the group commutation relations, but the verification must await the specification of the gauge field's operator properties.

The field equations can be decomposed into apparent equations of motion,

$$\partial_0\phi_k=\partial_k\phi_0-i^{\prime}t\phi_k{}^{\prime}.\phi_0+f^2G_{0k},$$
$$-\partial_0G^{0k}=-\partial_lG^{kl}+i^{\prime}t\phi_l{}^{\prime}.G^{kl}-i^{\prime}t\phi_0{}^{\prime}.G^{0k}+k^k,$$

and equations of constraint,

$$f^2G_{kl}=\partial_k\phi_l-\partial_l\phi_k+i(\phi_k.t\phi_l),$$
$$\partial_kG^{0k}=i^{\prime}t\phi_k{}^{\prime}.G^{0k}+k^0.$$

The latter show that neither $G_{kl}$ nor the longitudinal part of the three-dimensional vector $G^{0k}$ are independent dynamical variables. It will be useful, then, to write

$$G^{0k}=G^{0kT}+G^{0kL},$$

where

$$\partial_kG^{0kT}=0,$$

and

$$G^{0kL}=-\partial^k\psi.$$

In an ordinary three-dimensional notation, the equation to determine the Hermitian operator $\psi(x)$ is

$$-\nabla^2\psi+i^{\prime}t\boldsymbol{\phi}^{\prime}\cdot\nabla.\psi=i^{\prime}t\boldsymbol{\phi}^{\prime}:\mathbf{G}^T+k^0.$$

This information can be utilized in the equation of motion for $\phi^k$ by taking the divergence,

$$\partial_0\nabla\cdot\boldsymbol{\phi}-i^{\prime}t\phi_0{}^{\prime}.\nabla\cdot\boldsymbol{\phi}=\nabla^2\phi_0-i^{\prime}t\boldsymbol{\phi}^{\prime}\cdot\nabla.\phi_0-f^2\nabla\cdot\mathbf{G}.$$

But we must still reckon with the freedom of gauge transformation, which shows the impossibility of a complete specification of $\phi^k$ and $\phi^0$ by the field equations. In order to obtain a definite set of operators we adopt a specific gauge, and the naturally indicated choice is the three-dimensional transverse or radiation gauge,

$$\partial_k\phi^k=\nabla\cdot\boldsymbol{\phi}=0,$$

for this extracts from the apparent equations of motion another equation of constraint,

$$-\nabla^2\phi^0+i^{\prime}t\boldsymbol{\phi}^{\prime}\cdot\nabla.\phi^0=-f^2\nabla^2\psi,$$

which must serve to determine $\phi^0$.

It has now become clear that the independent dynamical variables of the gauge field are the three-dimensional transverse vectors $\phi_a{}^k$ and $G_a{}^{0kT}$, in close analogy with electrodynamics. If the variations of $\phi^k$ are performed within the radiation gauge,

$$\partial_k\delta\phi^k=0,$$

only the transverse part of $G^{0k}$ appears in the generator,

$$G_\phi=-\int(d\mathbf{x})\,G^{0kT}\delta\phi_k,$$

the interpretation of which is given by

$$(1/i)[\phi_k(x),G_\phi]=\delta\phi_k(x),\quad (1/i)[G^{0kT}(x),G_\phi]=0.$$

When these statements are combined with the similar properties of the alternative generator

$$G_{G^T}=\int(d\mathbf{x})\,\phi_k\delta G^{0kT},$$

we obtain the full set of commutation relations for the

fundamental dynamical variables of the gauge field:

$$x^0=x^{0\prime}:\quad [\phi_{ka}(x),\phi_{lb}(x')]=0,$$
$$[G_a^{0k}(x)^T,G_b^{0l}(x')^T]=0,$$
$$i[\phi_{ka}(x),G_b^{0l}(x')^T]=\delta_{ab}(\delta_k{}^l\delta(\mathbf{x}-\mathbf{x}'))^T.$$

Apart from the multiplicity of the internal space, these are identical with the electromagnetic field commutation relations.

The ability to display the fundamental commutation relations is of the greatest importance, for it provides the assurance that the operators designated as the basic dynamical variables do constitute the generators of a complete operator basis, which is their primary role as operators. Otherwise stated, it makes explicit the infinitesimal transformations of the quantum transformation group which, together with the coordinate and other invariance groups, largely characterize the physical system.

We can now confirm a previous expectation. The contribution of the gauge field to the operators $\mathbf{T}_a$ involves only the transverse components of $G^{0k}$, since $\phi^k$ is divergenceless. The canonical structure of the commutation relations implies that

$$\left[-i\int(d\mathbf{x})\ \boldsymbol{\phi}\cdot t_a\mathbf{G}^T,\ -i\int(d\mathbf{x})\ \boldsymbol{\phi}\cdot t_b\mathbf{G}^T\right]$$
$$=-i\int(d\mathbf{x})\ \boldsymbol{\phi}\cdot[t_a,t_b]\mathbf{G}^T,$$

which, together with statements of kinematical independence between the gauge field and the other systems,

$$x^0=x^{0\prime}:\quad [\phi_a{}^k(x),k_b{}^0(x')]=[G_a{}^{0k}(x)^T,k_b{}^0(x')]=0$$

supplies the verification that the group commutation relations are obeyed by the conserved $\mathbf{T}$-operators,

$$[\mathbf{T}_b,\mathbf{T}_c]=\textstyle\sum_a \mathbf{T}_a t_{abc}.$$

To complete this part of our study we must give the explicit operator construction of the longitudinal part of $G^{0k}$ and of $\phi^0$. Consider then $(x^0=x^{0\prime})$

$$(-\nabla^2+i^{\prime}t\boldsymbol{\phi}(x)^{\prime}\cdot\nabla).[\psi(x),\boldsymbol{\phi}(x')]$$
$$=i^{\prime}t\boldsymbol{\phi}(x)^{\prime}\cdot[\mathbf{G}^T(x),\boldsymbol{\phi}(x')],$$

and note that the solution of this equation requires no reference to operator properties since it is concerned only with the completely commutative components of $\boldsymbol{\phi}(x)$ at a common time. The relevant Green's function is defined by the matrix differential equation

$$[-\nabla^2+i^{\prime}t\boldsymbol{\phi}(x)^{\prime}\cdot\nabla]\mathfrak{D}_\phi(\mathbf{x},\mathbf{x}')=\delta(\mathbf{x}-\mathbf{x}'),$$

in which the indicated functional dependence on $\boldsymbol{\phi}$ also produces a time dependence. The self-adjointness of the defining equation implies the symmetry of this real function,

$$\mathfrak{D}_\phi(\mathbf{x},\mathbf{x}')_{ab}=\mathfrak{D}_\phi(\mathbf{x}',\mathbf{x})_{ba}.$$

Now return to the $\psi$ equation and remove the symmetrization of $\boldsymbol{\phi}$ with $\psi$, which gives

$$[-\nabla^2+i^{\prime}t\boldsymbol{\phi}^{\prime}\cdot\nabla]\psi=i^{\prime}t\boldsymbol{\phi}^{\prime}:\mathbf{G}^T+k^0+ih(\boldsymbol{\phi}),$$

where $h(\boldsymbol{\phi})$ is a Hermitian function of $\boldsymbol{\phi}$. On solving this equation the last term will contribute a skew-Hermitian function of $\boldsymbol{\phi}$, which disappears on combining $\psi$ with its identical Hermitian conjugate operator. The result is the symmetrized form

$$\psi(x)=\int(d\mathbf{x}')\ \mathfrak{D}_\phi(\mathbf{x},\mathbf{x}').[i^{\prime}t\phi_k(x')^{\prime}.G^{0k}(x')^T+k^0(x')],$$

or, more symbolically,

$$\psi=\mathfrak{D}_\phi.[i^{\prime}t\boldsymbol{\phi}^{\prime}:\mathbf{G}^T+k^0].$$

It is immaterial whether the symmetrization of $\boldsymbol{\phi}$ with $\mathbf{G}^T$ is independently performed, as indicated, or accompanies the symmetrization of the $\mathfrak{D}_\phi$ product. That is,

$$A.(B.C)-(A.B).C=\tfrac{1}{4}[[A,C],B],$$

which vanishes when $A$, $B$, $C$ are $\mathfrak{D}_\phi$, $\boldsymbol{\phi}$, and $\mathbf{G}^T$, respectively.

The form of the complete operator $G^{0k}$, as it has now been obtained, is given symbolically by

$$\mathbf{G}=(1-\nabla\mathfrak{D}_\phi i^{\prime}t\boldsymbol{\phi}^{\prime}):\mathbf{G}^T-\nabla\mathfrak{D}_\phi k^0,$$

or, equally well, by the versions

$$\mathbf{G}=[1+\nabla\mathfrak{D}_\phi(\nabla-i^{\prime}t\boldsymbol{\phi}^{\prime})]:\mathbf{G}^T-\nabla\mathfrak{D}_\phi k^0$$
$$=\mathbf{G}^T:[1+(\nabla-i^{\prime}t\boldsymbol{\phi}^{\prime})\mathfrak{D}_\phi\nabla]-\nabla\mathfrak{D}_\phi k^0,$$

which are related through the symmetry of $\mathfrak{D}_\phi$. The significance of the bracketed structures is indicated by

$$\nabla\cdot[1+(\nabla-i^{\prime}t\boldsymbol{\phi}^{\prime})\mathfrak{D}_\phi\nabla]=0,$$

and

$$(\nabla-i^{\prime}t\boldsymbol{\phi}^{\prime})\cdot[1+\nabla\mathfrak{D}_\phi(\nabla-i^{\prime}t\boldsymbol{\phi}^{\prime})]=0.$$

The equal-time commutator of $\boldsymbol{\phi}$ with $\mathbf{G}$ is immediately evaluated as

$$i[\phi_{ka}(x),G_b{}^{0l}(x')]$$
$$=[\delta_k{}^l\delta(\mathbf{x}-\mathbf{x}')-(\partial_k-i^{\prime}t\phi_k(x)^{\prime})\mathfrak{D}_\phi(\mathbf{x},\mathbf{x}')\partial^{\prime l}]_{ab},$$

where the last gradient acts on the function to its left.

The construction of $\phi_a{}^0(x)$ proceeds from the equation

$$(-\nabla^2+i^{\prime}t\boldsymbol{\phi}^{\prime}\cdot\nabla).\boldsymbol{\phi}^0=f^2\nabla\cdot\mathbf{G},$$

and the solution is

$$\boldsymbol{\phi}^0=f^2\mathfrak{D}_\phi.\nabla\cdot\mathbf{G},$$

or alternatively

$$\boldsymbol{\phi}^0=f^2\mathfrak{D}_\phi i^{\prime}t\boldsymbol{\phi}^{\prime}\cdot[1+\nabla\mathfrak{D}_\phi(\nabla-i^{\prime}t\boldsymbol{\phi}^{\prime})]:\mathbf{G}^T$$
$$+f^2(\mathfrak{D}_\phi-\mathfrak{D}_\phi i^{\prime}t\boldsymbol{\phi}^{\prime}\cdot\nabla\mathfrak{D}_\phi)k^0.$$

It follows directly that

$$(1/i)[\boldsymbol{\phi}^0,\boldsymbol{\phi}]=f^2\mathfrak{D}_\phi i^{\prime}t\boldsymbol{\phi}^{\prime}\cdot[1+\nabla\mathfrak{D}_\phi(\nabla-i^{\prime}t\boldsymbol{\phi}^{\prime})],$$

which symbolic equation has the following explicit meaning as an equal-time commutator:

$$(1/i)[\phi_a{}^0(x),\phi_b{}^k(x')]=f^2\Big[\mathfrak{D}_\phi(\mathbf{x},\mathbf{x}')i^ct\phi^k(x')'$$
$$+\int(d\mathbf{x}'')\,\mathfrak{D}_\phi(\mathbf{x},\mathbf{x}'')i^ct\phi(x'')'$$
$$\cdot\nabla''\mathfrak{D}_\phi(\mathbf{x}'',\mathbf{x}')(-\partial'^k-i^ct\phi^k(x')')\Big]_{ab}.$$

If the expression for $\phi^0$ is substituted in the equation for $\partial_0\boldsymbol{\phi}$, one obtains a fundamental transverse equation of motion,

$$-\partial_0\boldsymbol{\phi}=f^2[1+(\nabla-i^ct\boldsymbol{\phi}')\mathfrak{D}_\phi\nabla]:\mathbf{G},$$

or, equivalently,

$$-\partial_0\boldsymbol{\phi}=f^2\{[1+(\nabla-i^ct\boldsymbol{\phi}')\mathfrak{D}_\phi\nabla]$$
$$\cdot[1+\nabla\mathfrak{D}_\phi(\nabla-i^ct\boldsymbol{\phi}')]\}:\mathbf{G}^T$$
$$-f^2[1+(\nabla-i^ct\boldsymbol{\phi}')\mathfrak{D}_\phi\nabla]\cdot i^ct\boldsymbol{\phi}'\mathfrak{D}_\phi k^0.$$

One implication of this equation of motion is the equal-time commutator

$$[i\partial_0\boldsymbol{\phi},\boldsymbol{\phi}]=f^2[1+(\nabla-i^ct\boldsymbol{\phi}')\mathfrak{D}_\phi\nabla]\cdot[1+\nabla\mathfrak{D}_\phi(\nabla-i^ct\boldsymbol{\phi}')].$$

If we introduce an arbitrary real numerical transverse vector function $\mathbf{a}_a(\mathbf{x})$ and define the Hermitian operator

$$A(x^0)=\int(d\mathbf{x})\,\mathbf{a}\cdot\boldsymbol{\phi},$$

this commutator becomes

$$[i\partial_0A,A]=f^2\int(d\mathbf{x})\,\mathbf{b}\cdot\mathbf{b},$$

where the Hermitian vector $\mathbf{b}$ is

$$\mathbf{b}=[1+\nabla\mathfrak{D}_\phi(\nabla-i^ct\boldsymbol{\phi}')]\cdot\mathbf{a}.$$

But the vacuum expectation value of such a commutator can never be negative,

$$\langle[[A,P^0],A]\rangle=2\langle AP^0A\rangle>0,$$

and therefore

$$f^2>0.$$

Finally, we state the equal-time commutator

$$i[G_a{}^{0k}(x),G_b{}^{0l}(x')]=[\partial^k\mathfrak{D}_\phi(\mathbf{x},\mathbf{x}')\cdot i^ctG^{0l}(x')'$$
$$+i^ctG^{0k}(x)'\cdot\mathfrak{D}_\phi(\mathbf{x},\mathbf{x}')\partial'^l]_{ab},$$

and leave the proof to the reader.

# EXTERIOR ALGEBRA AND THE ACTION PRINCIPLE, I*

By Julian Schwinger

HARVARD UNIVERSITY

*Communicated February 27, 1962*

The quantum action principle[1] that has been devised for quantum variables of type $\nu = 2$ lacks one decisive feature that would enable it to function as an instrument of calculation. To overcome this difficulty we shall enlarge the number system by adjoining an exterior or Grassmann algebra.[2]

An exterior algebra is generated by $N$ elements $\epsilon_\kappa$, $\kappa = 1 \cdot\cdot N$, that obey

$$\{\epsilon_\kappa, \epsilon_\lambda\} = 0,$$

which includes $\epsilon_\kappa^2 = 0$.

A basis for the exterior algebra is supplied by the unit element and the homogeneous products of degree $d$

$$\epsilon_{\kappa_1}\epsilon_{\kappa_2}\cdot\cdot\epsilon_{\kappa_d}, \quad \kappa_1 < \kappa_2 \cdot\cdot < \kappa_d$$

for $d = 1 \cdot\cdot N$. The total number of linearly independent elements is counted as

$$\sum_{d=0}^{N} \frac{N!}{d!(N-d)!} = 2^N.$$

The algebraic properties of the generators are unaltered by arbitrary nonsingular linear transformations.

To suggest what can be achieved in this way, we consider a new class of special

Reprinted from the *Proceedings of the National Academy of Sciences U.S.A.* **48**, 603–611 (1962).

variations, constructed as the product of $\xi_{2n+1}$ with an arbitrary real infinitesimal linear combination of the exterior algebra generators,

$$^1/_2\delta\xi_k = -\xi_{2n+1}\sum_{\kappa=1}^{N}\delta\omega_{k\kappa}\epsilon_\kappa.$$

The members of this class anticommute with the $\xi_k$ variables, owing to the factor $\xi_{2n+1}$, and among themselves,

$$\{\delta^{(1)}\xi_k, \quad \delta^{(2)}\xi_l\} = 0,$$

since they are linear combinations of the exterior algebra generators. Accordingly, the generator of a special variation,

$$G_\xi = {}^1/_2 i\sum_k \xi_k\delta\xi_k = -{}^1/_2 i\sum \delta\xi_k\xi_k$$

$$= \sum_{k\kappa}\delta\omega_{k\kappa}\epsilon_\kappa(1/i)\xi_k\xi_{2n+1},$$

commutes with any such variation,

$$[\delta^{(1)}\xi_k, G_\xi^{(2)}] = 0.$$

If we consider the commutator of two generators we get

$$(1/i)[G_\xi^{(1)}, G_\xi^{(2)}] = -{}^1/_2 i\sum\delta^{(1)}\xi_k(1/i)[\xi_k, G_\xi^{(2)}] = -{}^1/_4 i\sum\delta^{(1)}\xi_k\delta^{(2)}\xi_k,$$

where the right-hand side is proportional to the unit operator, and to a bilinear function of the exterior algebra generating elements,

$$\sum\delta^{(1)}\xi_k\delta^{(2)}\xi_k = \sum(\delta^{(1)}\omega_{k\lambda}\delta^{(2)}\omega_{k\mu} - \delta^{(1)}\omega_{k\mu}\delta^{(2)}\omega_{k\lambda})\epsilon_\lambda\epsilon_\mu.$$

The latter structure commutes with all operators and with all exterior algebra elements. The commutator therefore commutes with any generator $G_\xi$, and the totality of the new special variations has a group structure which is isomorphic to that of the special canonical group for the degrees of freedom of type $\nu = \infty$. This, rather than the rotation groups previously discussed, is the special canonical group for the $\nu = 2$ variables.

We are thus led to reconsider the action principle, now using the infinitesimal variations of the special canonical group. The class of transformation generators $G$ that obey

$$[\delta\xi_k, G] = 0$$

includes not only all even operator functions of the $2n$ variables $\xi_k$, but also even functions of the $\xi_k$ that are multiplied by even functions of the $N$ exterior algebra generators $\epsilon_\kappa$, and odd functions of the $\xi_k$ multiplied by odd functions of the $\epsilon_\kappa$. The generators of the special canonical transformations are included in the last category.

The concept of a Hermitian operator requires generalization to accommodate the noncommutative numerical elements. The reversal in sense of multiplication that is associated with the adjoint operation now implies that

$$(\lambda A)^\dagger = A^\dagger\lambda^*,$$

where $\lambda$ is a number, and we have continued to use the same notation and language despite the enlargement of the number system. Complex conjugation thus has the algebraic property

$$(\lambda_1\lambda_2)^* = \lambda_2^*\lambda_1^*.$$

The anticommutativity of the generating elements is maintained by this operation, and we therefore regard complex conjugation in the exterior algebra as a linear mapping of the $N$-dimensional subspace of generators,

$$\epsilon_\kappa^* = \sum_\lambda R_{\kappa\lambda}\epsilon_\lambda.$$

The matrix $R$ obeys

$$R^*R = 1,$$

which is not a statement of unitarity. Nevertheless, a Cayley parametrization exists,

$$R = \frac{1 + ir}{1 - ir}, \quad r^* = r,$$

provided $\det(1 + R) \neq 0$. Then,

$$\bar{\epsilon}_\kappa = \epsilon_\kappa + i \sum r_{\kappa\lambda}\epsilon_\lambda$$

obeys

$$\bar{\epsilon}_\kappa^* = \bar{\epsilon}_\kappa,$$

which asserts that a basis can always be chosen with real generators. There still remains the freedom of real nonsingular linear transformations.

This conclusion is not altered when $R$ has the eigenvalue $-1$. In that circumstance, we can construct $p(R)$, a polynomial in $R$ that has the properties

$$(1 + R)p = 0, \quad p(1 - p) = 0, \quad p^* = p, \quad \text{and} \quad p(-1) = 1.$$

The matrix

$$R' = R(1 - 2p) = R(1 - p) + p$$

also obeys

$$R'^*R' = 1,$$

while

$$\det(1 + R') \neq 0,$$

since the contrary would imply the existence of a nontrivial vector $v$ such that

$$(1 + R')v = ((1 + R)(1 - p) + 2p)v = 0$$

or, equivalently,

$$pv = 0, \quad (1 + R)v = 0,$$

which is impossible since $p(-1) = 1$. Now,

$$1 - 2p = \frac{1 - p - ip}{1 - p + ip},$$

and if we use the Cayley construction for $R'$ in terms of a real matrix $r'$, which is a function of $R$ with the property $r'(-1) = 0$, we get

$$R = \frac{1 - p + ip}{1 - p - ip}\frac{1 + ir'}{1 - ir'} = \frac{1 - p + i(p + (1 - p)r')}{1 - p - i(p + (1 - p)r')}.$$

This establishes the generality of the representation

$$R = \rho/\rho^*,$$

where the nonsingular matrices $\rho$ and $\rho^*$ are commutative, and thereby proves the reality of the generator set

$$\bar{\epsilon}_\kappa = \sum \rho_{\kappa\lambda}\epsilon_\lambda.$$

When real generators are chosen, the other elements of a real basis are supplied by the nonvanishing products

$$i^{d(d-1)/2}\, \epsilon_{\kappa 1} \cdots \epsilon_{\kappa d},$$

for

$$(\epsilon_{\kappa 1} \cdots \epsilon_{\kappa d})^* = \epsilon_{\kappa d} \cdots \epsilon_{\kappa 1} = (-1)^{d(d-1)/2}\, \epsilon_{\kappa 1} \cdots \epsilon_{\kappa d}.$$

A Hermitian operator, in the extended sense, is produced by linear combinations of conventional Hermitian operators multiplied by real elements of the exterior algebra. The generators $G_\xi$ are Hermitian, as are the commutators $i[G_\xi^{(1)}, G_\xi^{(2)}]$.

The transformation function $\langle \tau_1 | \tau_2 \rangle$ associated with a generalized unitary transformation is an element of the exterior algebra. It possesses the properties

$$\langle \tau_1 | \tau_2 \rangle^* = \langle \tau_2 | \tau_1 \rangle$$

and

$$\langle \tau_1 | \tau_3 \rangle = \langle \tau_1 | \tau_2 \rangle \times \langle \tau_2 | \tau_3 \rangle,$$

where $\times$ symbolizes the summation over a complete set of states which, for the moment at least, are to be understood in the conventional sense. These attributes are consistent with the nature of complex conjugation, since

$$(\langle \tau_1 | \tau_2 \rangle \times \langle \tau_2 | \tau_3 \rangle)^* = \langle \tau_3 | \tau_2 \rangle \times \langle \tau_2 | \tau_1 \rangle = \langle \tau_3 | \tau_1 \rangle.$$

For an infinitesimal transformation, we have

$$\langle \tau + d\tau | \tau \rangle = \langle | 1 + id\tau G(\xi, \tau) | \rangle,$$

where $G$ is Hermitian, which implies that its matrix array of exterior algebra elements obeys

$$\langle a' | G | a'' \rangle^* = \langle a'' | G | a' \rangle.$$

The subsequent discussion of the action principle requires no explicit reference to the structure of the special variations, and the action principle thus acquires a dual significance, depending upon the nature of the number system.

We shall need some properties of differentiation in an exterior algebra. Since $\epsilon_\lambda^2 = 0$, any given function of the generators can be displayed uniquely in the alternative forms

$$f(\epsilon) = f_0 + \epsilon_\lambda f_l = f_0 + f_r \epsilon_\lambda,$$

where $f_0$, $f_l$, $f_r$ do not contain $\epsilon_\lambda$. By definition, $f_r$ and $f_l$ are the right and left derivatives, respectively, of $f(\epsilon)$ with respect to $\epsilon_\lambda$,

$$\frac{\partial_l f(\epsilon)}{\partial \epsilon_\lambda} = f_l, \quad \frac{\partial_r f(\epsilon)}{\partial \epsilon_\lambda} = f_r.$$

If $f(\epsilon)$ is homogeneous of degree $d$, the derivatives are homogeneous of degree $d - 1$, and left and right derivatives are equal for odd $d$, but of opposite sign when $d$ is even. For the particular odd function $\epsilon_\mu$, we have

$$\epsilon_\mu = \epsilon_\mu(1 - \delta_{\lambda\mu}) + \epsilon_\lambda\delta_{\lambda\mu},$$

so that

$$\partial\epsilon_\mu/\partial\epsilon_\lambda = \delta_{\lambda\mu}.$$

Since a derivative is independent of the element involved, repetition of the operation annihilates any function,

$$(\partial/\partial\epsilon_\lambda)^2 f(\epsilon) = 0.$$

To define more general second derivatives we write, for $\lambda \neq \mu$,

$$f(\epsilon) = f_0 + \epsilon_\lambda f_\lambda + \epsilon_\mu f_\mu + \epsilon_\lambda\epsilon_\mu f_{\lambda\mu},$$

where the coefficients are independent of $\epsilon_\lambda$ and $\epsilon_\mu$. The last term has the alternative forms

$$\epsilon_\lambda\epsilon_\mu f_{\lambda\mu} = \epsilon_\mu\epsilon_\lambda f_{\mu\lambda}$$

with

$$f_{\mu\lambda} = -f_{\lambda\mu}.$$

Then,

$$\frac{\partial_l f}{\partial\epsilon_\lambda} = f_\lambda + \epsilon_\mu f_{\lambda\mu}, \quad \frac{\partial_l f}{\partial\epsilon_\mu} = f_\mu + \epsilon_\lambda f_{\mu\lambda}$$

and

$$\frac{\partial_l}{\partial\epsilon_\mu}\frac{\partial_l}{\partial\epsilon_\lambda} f = f_{\lambda\mu}, \quad \frac{\partial_l}{\partial\epsilon_\lambda}\frac{\partial_l}{\partial\epsilon_\mu} f = f_{\mu\lambda},$$

which shows that different derivatives are anticommutative,

$$\left\{\frac{\partial_l}{\partial\epsilon_\mu}, \frac{\partial_l}{\partial\epsilon_\lambda}\right\} f(\epsilon) = 0.$$

A similar statement applies to right derivatives.

The definition of the derivative has been given in purely algebraic terms. We now consider $f(\epsilon + \delta\epsilon)$, where $\delta\epsilon_\lambda$ signifies a linear combination of the exterior algebra elements with arbitrary infinitesimal numerical coefficients, and conclude that

$$f(\epsilon + \delta\epsilon) - f(\epsilon) = \sum \delta\epsilon_\lambda \frac{\partial_l f}{\partial\epsilon_\lambda} = \sum \frac{\partial_r f}{\partial\epsilon_\lambda}\delta\epsilon_\lambda,$$

to the first order in the infinitesimal numerical parameters. If this differential property is used to identify derivatives, it must be supplemented by the requirement that the derivative be of lower degree than the function, for any numerical multiple of $\epsilon_1\epsilon_2\cdots\epsilon_N$ can be added to a derivative without changing the differential form. Let us also note the possibility of using arbitrary nonsingular linear combinations of the $\epsilon_\lambda$ in differentiation, as expressed by the matrix formula

$$(\partial/\partial a\epsilon)f = (a^{-1})^T(\partial/\partial\epsilon)f,$$

which follows directly from the differential expression.

We shall now apply the extended action principle to the superposition of two transformations, one produced by a conventional Hermitian operator $G$, an even function of the $\xi_k$, while the other is a special canonical transformation performed arbitrarily, but continuously in $\tau$. The effective generator is

$$G(\xi(\tau), \tau) - i\sum_{k=1}^{2n} \xi_k(\tau)X_k(\tau),$$

where $-X_k(\tau)d\tau$ is the special variation induced in $\xi_k(\tau)$ during the interval $d\tau$. The objects $X_k(\tau)$ are constructed by multiplying $\xi_{2n+1}$ with a linear combination of the $\epsilon_\lambda$ containing real numerical coefficients that are arbitrary continuous functions of $\tau$. To use the stationary action principle, we observe that each $X_k(\tau)$, as a special variation, is commutative with a generator of special variations and thus

$$\delta' X_k(\tau) = 0.$$

The action principle then asserts that

$$\sum_k \delta\xi_k \left[ i\frac{d\xi_k}{d\tau} + \frac{\partial_l G}{\partial \xi_k} - iX_k \right] = 0,$$

where the $\delta\xi_k$ are variations constructed from the exterior algebra elements. We cannot entirely conclude that

$$\frac{d\xi_k}{d\tau} = i\frac{\partial_l G}{\partial\xi_k} + X_k,$$

since there remains an arbitrariness associated with multiples of $\epsilon_1 \cdots \epsilon_N$, as in the identification of derivatives from a differential form. No such term appears, however, on evaluating

$$\frac{d\xi_k}{d\tau} = -\frac{1}{i}\left[\xi_k, G - i\sum \xi_l X_l\right] = i\frac{\partial_l G}{\partial \xi_k} + X_k.$$

This apparent incompleteness of the action principle is removed on stating the obvious requirement that the transformation function $\langle \tau_1 | \tau_2 \rangle$ be an ordinary number when all $X_k(\tau)$ vanish. Thus, the elements of the exterior algebra enter only through the products of the $X_k(\tau)$ with $\xi_k(\tau)$, as the latter are obtained by integrating the equations of motion, and terms in these equations containing $\epsilon_1 \cdots \epsilon_N$ are completely without effect.

Let us examine how the transformation function $\langle \tau_1 | \tau_2 \rangle_G{}^X$ depends upon the $X_k(\tau)$. It is well to keep in mind the two distinct factors that compose $X_k(\tau)$,

$$X_k(\tau) = \rho X_k'(\tau).$$

Here,
$$\rho = 2^{1/2}\xi_{2n+1}, \quad \rho^2 = 1,$$

while $X_k'(\tau)$ is entirely an element of the exterior algebra. Thus, it is really the $X'(\tau)$ upon which the transformation function depends. An infinitesimal change of the latter induces

$$\delta\langle\tau_1|\tau_2\rangle_G{}^X = \langle\tau_1|\int_{\tau_2}^{\tau_1} d\tau\xi(\tau)\delta X(\tau)|\tau_2\rangle_G{}^X = \int_{\tau_2}^{\tau_1} d\tau\langle\tau_1|\xi\rho\delta X'|\tau_2\rangle_G{}^X,$$

where the summation index $k$ has been suppressed. The repetition of such variations gives the ordered products

$$\delta^{2m}\langle\tau_1|\tau_2\rangle = \int_{\tau_2}^{\tau_1} d\tau^1\cdot\cdot d\tau^{2m}\langle\tau_1|(\xi\delta X(\tau^1)\cdot\cdot\xi\delta X(\tau^{2m}))_0|\tau_2\rangle = \int d\tau^1\cdot\cdot d\tau^{2m}(\delta X'(\tau^1)\cdot\cdot\delta X'(\tau^{2m}))_0{}^*\langle\tau_1|(\xi(\tau^1)\cdot\cdot\xi(\tau^{2m}))_0|\tau_2\rangle,$$

which form is specific to an even number of variations. Complex conjugation is applied to the exterior algebra elements in order to reverse the sense of multiplication. In arriving at the latter form, we have exploited the fact that special canonical variations are not implicit functions of $\tau$ and therefore anticommute with the $\xi_k$ without regard to the $\tau$ values. Thus, one can bring together the $2m$ quantities $\delta X(\tau^1), \cdot\cdot\delta X(\tau^{2m})$, and this product is a multiple of the unit operator since $\rho^2 = 1$. The multiple is the corresponding product of the exterior algebra elements $\delta X'(\tau^1), \cdot\cdot\delta X'(\tau^{2m})$, which, as an even function, is completely commutative with all elements of the exterior algebra and therefore can be withdrawn from the matrix element. The reversal in multiplication sense of the exterior algebra elements gives a complete account of the sign factors associated with anticommutativity.

The notation of functional differentiation can be used to express the result. With left derivatives, we have

$$\left(\frac{\delta_l}{\delta X'(\tau^1)}\cdot\cdot\frac{\delta_l}{\delta X'(\tau^{2m})}\right)_0\langle\tau_1|\tau_2\rangle^X = \langle\tau_1|(\xi(\tau^1)\cdot\cdot\xi(\tau^{2m}))_0|\tau_2\rangle^X,$$

where the anticommutativity of exterior algebra derivatives implies that

$$\left(\frac{\delta}{\delta X'(\tau^1)}\cdot\cdot\frac{\delta}{\delta X'(\tau^{2m})}\right)_0 = \epsilon_0(\tau^1\cdot\cdot\tau^{2m})\frac{\delta}{\delta X'(\tau^1)}\cdot\cdot\frac{\delta}{\delta X'(\tau^{2m})}.$$

Here,

$$\epsilon_0(\tau^1\cdot\cdot\tau^{2m}) = \pm 1$$

according to whether an even or odd permutation is required to bring $\tau^1, \cdot\cdot\tau^{2m}$ into the ordered sequence. This notation ignores one vital point, however. In an exterior algebra with $N$ generating elements, no derivative higher than the $N$th exists. If we wish to evaluate unlimited numbers of derivatives, in order to construct correspondingly general functions of the dynamical variables, we must choose $N = \infty$; the exterior algebra is of infinite dimensionality. Then we can assert of arbitrary even ordered functions of the $\xi_k(\tau)$ that

$$\langle\tau_1|(F(\xi))_0|\tau_2\rangle^X = F(\delta_l/\delta X')_0\langle\tau_1|\tau_2\rangle^X = \langle\tau_1|\tau_2\rangle^X F(\delta_r{}^T/\delta X')_0.$$

In the alternative right derivative form, $\delta^T$ signifies that successive differentiations are performed from left to right rather than in the conventional sense. We also note that the particular order of multiplication for operator products at a common time is to be produced by limiting processes from unequal $\tau$ values. As an application to the transformation function $\langle\tau_1|\tau_2\rangle_G{}^X$, we supply the even operator function $G$ with a variable factor $\lambda$ and compute

$$\frac{\partial}{\partial\lambda}\langle\tau_1|\tau_2\rangle_{\lambda G}^X = i\langle\tau_1|\int_{\tau_2}^{\tau_1} d\tau G(\xi,\tau)|\tau_2\rangle = i\int_{\tau_2}^{\tau_1} d\tau\, G\left(\frac{\delta_l}{\delta X'},\tau\right)\langle\tau_1|\tau_2\rangle,$$

which gives the formal construction

$$\langle\tau_1|\tau_2\rangle_{G_\ell}^X = \exp[i\int_{\tau_2}^{\tau_1} d\tau G(\delta_l/\delta X',\tau)]\langle\tau_1|\tau_2\rangle^X,$$

where the latter transformation function refers entirely to the special canonical group.

The corresponding theorems for odd ordered functions of the $\xi_k(\tau)$ are

$$F(\delta_l/\delta X')_0\langle\tau_1|\tau_2\rangle^X = -\langle\tau_1|\rho(\tau_1)(F(\xi))_0|\tau_2\rangle^X$$

and

$$\langle\tau_1|\tau_2\rangle^X F(\delta_r{}^T/\delta X')_0 = \langle\tau_1|(F(\xi))_0\rho(\tau_2)|\tau_2\rangle^X,$$

provided the states are conventional ones. Since the variations $\delta X(\tau)$ have no implicit $\tau$ dependence, the factors $\rho\epsilon$ can be referred to any $\tau$ value. If this is chosen as $\tau_1$ or $\tau_2$ there will remain one $\rho$ operator while the odd product of exterior algebra elements can be withdrawn from the matrix element if, as we assume, the product commutes with the states $\langle\tau_1|$ and $|\tau_2\rangle$. The sign difference between the left and right derivative forms stems directly from the property $\xi\delta X = -\delta X\xi$.

As an example, let $F(\xi)$ be the odd function that appears on the left-hand side in the equation of motion

$$\frac{d\xi}{d\tau} - i\frac{\partial_l G}{\partial\xi} = X.$$

The corresponding functional derivatives are evaluated as

$$\langle\tau_1|\rho(\tau_1)X(\tau)|\tau_2\rangle = X'(\tau)\langle\tau_1|\tau_2\rangle$$

or

$$\langle\tau_1|X(\tau)\rho(\tau_2)|\tau_2\rangle = \langle\tau_1|\tau_2\rangle X'(\tau),$$

since the $X(\tau)$ are special canonical displacements, and this yields the functional differential equations obeyed by a transformation function $\langle\tau_1|\tau_2\rangle_G{}^X$, namely,

$$\left[-\frac{d}{d\tau}\frac{\delta_l}{\delta X'(\tau)} + i\frac{\partial_l G}{\partial\xi}\left(\frac{\delta_l}{\delta X'}\right) - X'(\tau)\right]\langle\tau_1|\tau_2\rangle_G{}^X = 0$$

and

$$\langle\tau_1|\tau_2\rangle_G{}^X\left[\frac{d}{d\tau}\frac{\delta_r{}^T}{\delta X'(\tau)} + i\frac{\partial_r G}{\partial\xi}\left(\frac{\delta_r{}^T}{\delta X'}\right) - X'(\tau)\right] = 0.$$

Some properties that distinguish the trace of the transformation function can also be derived from the statement about odd functions. Thus,

$$tr\langle\tau_1|\rho(\tau_1)F|\tau_2\rangle = tr\langle\tau_1|F\rho(\tau_2)|\tau_2\rangle,$$

since either side is evaluated as

$$\sum_{a'a''}\langle a'|\rho|a''\rangle\langle a''\tau_1|F|a'\tau_2\rangle.$$

Accordingly,

$$F(\delta_l/\delta X')_0\, tr\,\langle\tau_1|\,\tau_2\rangle^X = -tr\langle\tau_1|\,\tau_2\rangle^X F(\delta_r{}^T/\delta X')_0$$

and, in particular,

$$\left(\frac{\delta_l}{\delta X'(\tau)} + \frac{\delta_r}{\delta X'(\tau)}\right) tr\langle\tau_1|\,\tau_2\rangle_G{}^X = 0,$$

which shows that the trace is an even function of the $X'(\tau)$. The nature of the trace is involved again in the statement

$$tr\langle\tau_1|\,\rho(\tau_1)(\xi(\tau_1) + \xi(\tau_2))\,|\,\tau_2\rangle^X = tr\langle\tau_1|\,\{\rho(\tau_1),\ \xi(\tau_1)\}\,|\,\tau_2\rangle = 0,$$

which is an assertion of effective antiperiodicity for the operators $\xi(\tau)$ over the interval $T = \tau_1 - \tau_2$. The equivalent restriction on the trace of the transformation function is

$$\left(\frac{\delta_l}{\delta X'(\tau_1)} + \frac{\delta_l}{\delta X'(\tau_2)}\right) tr\,\langle\tau_1|\,\tau_2\rangle_G{}^X = 0,$$

or

$$\left(\frac{\delta_l}{\delta X'(\tau_1)} - \frac{\delta_r}{\delta X'(\tau_2)}\right) tr\,\langle\tau_1|\,\tau_2\rangle_G{}^X = 0.$$

* Publication assisted by the Air Force Office of Scientific Research.

[1] These PROCEEDINGS, **47**, 1075 (1961).

[2] A brief mathematical description can be found in the publication *The Construction and Study of Certain Important Algebras*, by C. Chevalley (1955, the Mathematical Society of Japan). Although such an extension of the number system has long been employed in quantum field theory (see, for example, these PROCEE :NGS, **37**, 452 (1951)), there has been an obvious need for an exposition of the general algebraic and group theoretical basis of the device.

# Non-Abelian Gauge Fields. Relativistic Invariance

JULIAN SCHWINGER*
*Harvard University, Cambridge, Massachusetts*
(Received January 19, 1962)

A simple criterion for Lorentz invariance in quantum field theory is stated as a commutator condition relating the energy density to the momentum density. With its aid a relativistically invariant radiation-gauge formulation is devised for a non-Abelian vector-gauge field coupled to a spin-½ Fermi field.

## INTRODUCTION

It has been the historical role of gauge-variant field systems to pose the greatest challenge to relativistic quantum-field theory. From the first beginnings of a general quantum electrodynamics in the hands of Heisenberg and Pauli, difficulties were encountered, owing to the absence in the Lagrange function of the time derivative of some of the field variables, which frustrated the application of the simplest canonical quantization scheme. Two general responses to this situation can be distinguished. In the first of these, the physical system is accepted for what it is; the gauge variance of the field is interpreted to mean that not all the field components at a given time are fundamental dynamical variables, and the latter are identified. We shall describe this view point as the method of the radiation gauge.[1] It can be characterized by the desire to clarify the quantum nature of the system, be it at the expense of manifest Lorentz invariance. The second approach reverses this order of priority. Although there are several versions, all share the feature that the physical system is modified in order to restrict the group of gauge transformations and thereby extend the status of fundamental dynamical variable to all field components. The states of physical interest must then be related to the states of this larger system. These devices will be described collectively as the method of the Lorentz gauge.[2]

It is usual to assert that the two viewpoints are equivalent, and that a Lorentz gauge method has the advantage of calculational simplicity. But, against the validity of the Lorentz gauge methods as the basis of a general theory, must be arrayed the body of experience which indicates that the nature of the quantum vector space for a system with an infinite number of degrees of freedom is intimately associated with the dynamics and that no operator transformation connecting the states of different dynamical systems can be guaranteed to exist.[3] For this reason, combined with the conviction that an intrinsic method is superior in economy of concept to the artifice of imbedding the physical system in another kinematical and dynamical framework, we reject all Lorentz gauge formulations as unsuited to the role of providing the fundamental operator foundation for a gauge variant field system.

The radiation gauge formulation is three dimensional in structure, and Lorentz invariance must be verified by explicit calculation. In the electromagnetic or Abelian gauge-field situation one can easily exhibit the operator gauge transformation that is induced by a Lorentz transformation. The covariance of all aspects of the theory can then be checked directly. Such a program is vastly more complicated for non-Abelian gauge fields, particularly since the Lagrange function is ambiguous, in a manner that influences Lorentz transformation properties. Thus, we are in grave need of a simple criterion for Lorentz invariance if we are to select a satisfactory theory from a class of theories that are acceptable in three dimensions. This is what we propose to supply in the present paper.

## THE COMMUTATOR CONDITION

Consider a field system for which the fundamental dynamical variables obey equal time commutator or anticommutator relations of the form

$$x^0=x^{0\prime}:\quad [\chi(x),\chi(x')]_\pm=c(\mathbf{x}-\mathbf{x}'),$$

where $c(\mathbf{x}-\mathbf{x}')$ is a numerical matrix function. Let the system be characterized by Hermitian momentum density operators $T^0{}_k(x)$, $k=1, 2, 3$, such that the operators of linear momentum,

$$P_k=\int(d\mathbf{x})T^0{}_k(x),$$

and angular momentum,

$$J_{kl}=\int(d\mathbf{x})[x_kT^0{}_l-x_lT^0{}_k],$$

obey the commutation relations appropriate to the three-dimensional translation-rotation group,

$$[P_k,P_l]=0,$$
$$-i[P_k,J_{lm}]=\delta_{km}P_l-\delta_{kl}P_m,$$
$$-i[J_{kl},J_{mn}]=\delta_{km}J_{ln}-\delta_{lm}J_{kn}-\delta_{kn}J_{lm}+\delta_{ln}J_{km}.$$

* Supported in part by the Air Force Office of Scientific Research (Air Research and Development Command), under contract number A.F. 49(638)-589.

[1] For a recent discussion of the gravitational field, in this spirit, see R. Arnowitt, S. Deser, and C. W. Misner, Phys. Rev. **117**, 1595 (1960).

[2] See, for example, J. M. Jauch and F. Rohrlich, *Theory of Photons and Electrons* (Addison-Wesley Publishing Company, Inc., Reading, Massachusetts, 1955).

[3] Some remarks of this nature are made by R. Haag, Kgl. Danske Videnskab. Selskab, Mat.-fys. Medd. **29**, No. 12 (1955).



Reprinted from *The Physical Review* **127**, 324–330 (1962).

The significance of these operators as generators of translations and rotations should also be expressed by

$$[\chi(x),P_k]=(1/i)\partial_k\chi(x),$$
$$[\chi(x),J_{kl}]=[x_k(1/i)\partial_l-x_l(1/i)\partial_k+S_{kl}]\chi(x),$$

where the finite-dimensional Hermitian spin matrices $S_{kl}$ obey the angular-momentum commutation relations.

We now assert, as a sufficient condition for invariance under the group of proper, orthochronous Lorentz transformations, that the Hermitian energy density operator $T^{00}(x)$ obeys the equal-time commutator condition[4]:

$$-i[T^{00}(x),T^{00}(x')]=-(T^0{}_k(x)+T^0{}_k(x'))\partial^k\delta(\mathbf{x}-\mathbf{x}'),$$

at least for the systems now under consideration, in which only spin values of $\frac{1}{2}$ and 1 occur. It is also required, of course, that $T^{00}(x)$ be a three-dimensional scalar function of the field operators at the given time, with no explicit coordinate dependence.

The latter property implies that the energy operator,

$$P^0=\int(d\mathbf{x})T^{00}(x),$$

obeys

$$[P^0,P_k]=[P^0,J_{kl}]=0.$$

Furthermore, the three infinitesimal generators of Lorentz transformations,

$$J^0{}_k=\int(d\mathbf{x})[x^0T^0{}_k-x_kT^{00}]=x^0P_k-\int(d\mathbf{x})x_kT^{00},$$

evidently constitute a three-dimensional vector and thus

$$-i[J^0{}_k,J_{lm}]=\delta_{km}J^0{}_l-\delta_{kl}J^0{}_m.$$

In addition,

$$-i[J^0{}_k,P_l]=\int(d\mathbf{x})x_k\partial_lT^{00}=-\delta_{kl}P^0.$$

Lacking from the list of commutation relations obeyed by the ten infinitesimal generators of the inhomogeneous Lorentz group,

$$[P_\mu,P_\nu]=0,$$
$$-i[P_\mu,J_{\nu\lambda}]=g_{\mu\lambda}P_\nu-g_{\mu\nu}P_\lambda,$$
$$-i[J_{\mu\nu},J_{\lambda\kappa}]=g_{\mu\lambda}J_{\nu\kappa}-g_{\nu\lambda}J_{\mu\kappa}-g_{\mu\kappa}J_{\nu\lambda}+g_{\nu\kappa}J_{\mu\lambda};$$

are the commutators $[P^0,J^0{}_k]$, $[J^0{}_k,J^0{}_l]$; and it is just these that are supplied by the commutator condition on the energy density.

On integrating over the three-dimensional domain of the variable $x'$, the commutator condition becomes

$$-i[T^{00}(x),P^0]=-\partial_kT^{0k}(x),$$

which is the assertion of local energy conservation,

$$\partial_0T^{00}+\partial_kT^{0k}=0.$$

A subsequent $x$ integration, including the factor $x_k$, gives

$$-i[x^0P_k-J^0{}_k,\ P^0]=-\int(d\mathbf{x})x_k\partial_lT^{0l},$$

or

$$-i[P^0,J^0{}_k]=P_k.$$

If the additional factor of $x_l'$ is included in the $x'$ integration, the result is

$$-i[T^{00}(x),x^0P_l-J^0{}_l]=-T^0{}_l(x)-\partial_k[x_lT^{0k}(x)],$$

or equivalently

$$[T^{00}(x),J^0{}_l]=[x^0(1/i)\partial_l-x_l(1/i)\partial^0]T^{00}(x)+2iT^0{}_l(x).$$

This is an infinitesimal transformation statement about the tensor character of $T^{\mu\nu}(x)$. An $x$ integration, with the factor $x_k$, now yields

$$-i[J^0{}_k,J^0{}_l]=-\int(d\mathbf{x})(x_kT^0{}_l-x_lT^0{}_k)=-J_{kl},$$

which completes the set of commutation relations obeyed by the infinitesimal Lorentz generators.

These unitary group properties, combined with the invariance of the fundamental field commutation relations under unitary transformations, comprise the content of the requirement of Lorentz invariance. It will be noted that the energy density commutator equation could contain additional terms, which do not contribute to the various three-dimensional integrals. No such terms will appear, however, for the system we now examine,[5] a non-Abelian vector gauge field coupled with a spin-$\frac{1}{2}$ field.

### THE ENERGY DENSITY OPERATOR

The fundamental dynamical variables of our system are a spin-$\frac{1}{2}$ Hermitian Fermi field $\psi(x)$, and a transverse vector Hermitian Bose field $\phi_k(x)$, $G^{0kT}(x)$, $k=1, 2, 3$. The former obeys the equal-time anti-commutation relation,

$$x^0=x^{0\prime}:\quad \{\psi(x),\psi(x')\}=\delta(\mathbf{x}-\mathbf{x}'),$$

as a matrix equation in the four-component spinor indices. This field also has an additional internal multiplicity in order to realize the properties that are represented by $n$ imaginary antisymmetrical matrices $T_a$, which obey the group commutation properties,

$$[T_b,T_c]=\sum_{a=1}^{n}T_at_{abc}.$$

[4] I am unaward of any similar statement in the literature. Although the work of P. A. M. Dirac, Phys. Rev. **73**, 1092 (1948), is certainly directly related, the possibility of its application to the tensor $T_{\mu\nu}$ is there specifically rejected.

[5] This discussion is a continuation of a previous paper, J. Schwinger, Phys. Rev. **125**, 1043 (1962).

The $n$-dimensional matrix,

$$t_b = (t_{abc}),$$

also obeys this group commutation law.

At a given time, the transverse vector fields commute with $\psi$, and obey the commutation relations

$$[\phi_k(x),\phi_l(x')] = [G^{0kT}(x),G^{0lT}(x')] = 0,$$
$$i[\phi_k(x),G^{0lT}(x')] = [\delta^l{}_k\delta(\mathbf{x}-\mathbf{x}')]^T.$$

These are matrix equations in the $n$-dimensional internal space to which the matrices $t_b$ refer. We also employ the fields defined by

$$f^2G_{kl} = \partial_k\phi_l - \partial_l\phi_k + i(\phi_k t\phi_l)$$

and

$$\partial_k G^{0k} - i' t\phi_k' . G^{0k} = k^0 = \tfrac{1}{2}\psi T\psi$$

where

$$G^{0k} = G^{0kT} - \partial^k\Psi.$$

The explicit form of $G^{0k}$ is given symbolically by

$$\mathbf{G} = [1 + \nabla\mathfrak{D}_\phi(\nabla - i' t\boldsymbol{\phi}')]:\mathbf{G}^T - \nabla\mathfrak{D}_\phi k^0,$$

in which

$$(-\nabla^2 + i' t\boldsymbol{\phi}(x)'\cdot\nabla)\mathfrak{D}_\phi(\mathbf{x},\mathbf{x}') = \delta(\mathbf{x}-\mathbf{x}').$$

We propose the following Hermitian operators as candidates for the momentum and energy densities of this system:

$$T^0{}_k(x) = \tfrac{1}{2}\psi(x).[(1/i)\partial_k - `T\phi_k(x)']\psi(x) + \tfrac{1}{2}\partial^l[\tfrac{1}{4}\psi(x)\sigma_{kl}\psi(x)] + f^2G^{0m}(x).G_{km}(x) = T^0{}_k(x)^F + T^0{}_k(x)^B,$$

and

$$T^{00}(x) = \tfrac{1}{2}\psi(x).\alpha^k[(1/i)\partial_k - `T\phi_k(x)']\psi(x) - \tfrac{1}{2}im\psi(x)\beta\psi(x) + \tfrac{1}{2}f^2[(G^{0k}(x))^2 + \tfrac{1}{2}(G^{kl}(x))^2] + t_\phi(\mathbf{x}) = T^{00}(x)^F + T^{00}(x)^B.$$

Note that symmetrized and antisymmetrized multiplications are called for in Bose and Fermi terms, respectively. The scalar function $t_\phi(\mathbf{x})$ will not be specified here. Its determination is the essential contribution of this paper. The Dirac matrices $\alpha^k$ and $\beta$ are real, and are respectively symmetrical and antisymmetrical. The spin matrices,

$$\sigma_{kl} = (1/2i)[\alpha_k,\alpha_l],$$

are imaginary and antisymmetrical.

In order to verify that $T^0{}_k$ produces correct three-dimensional transformation properties we remark that

$$f^2G^{0m}(x).G_{km}(x) = G^{0m}(x).\partial_k\phi_m(x) + k^0(x)\phi_k(x) - \partial_m[G^{0m}(x).\phi_k(x)],$$

in which a reordering of symmetrized products is involved. Accordingly,

$$T^0{}_k(x) = \tfrac{1}{2}\psi(x).(1/i)\partial_k\psi(x) + \tfrac{1}{2}\partial^l[\tfrac{1}{4}\psi(x)\sigma_{kl}\psi(x)] + G^{0m}(x).\partial_k\phi_m(x) - \partial_m[G^{0m}(x).\phi_k(x)],$$

and therefore

$$P_k = \int(d\mathbf{x})T^0{}_k = \int(d\mathbf{x})\left[\frac{1}{2}\psi.\frac{1}{i}\partial_k\psi + G^{0mT}.\partial_k\phi_m\right],$$

which exhibits the momentum operator in terms of the canonical variables. The immediate result is

$$[\chi(x),P_k] = (1/i)\partial_k\chi(x),$$

in which $\chi$ may be $\psi$, $\phi_l$, or $G^{0lT}$. Similarly,

$$J_{kl} = \int(d\mathbf{x})\{\tfrac{1}{2}\psi.[x_k(1/i)\partial_l - x_l(1/i)\partial_k + \tfrac{1}{2}\sigma_{kl}]\psi + G^{0mT}.(x_k\partial_l - x_l\partial_k)\phi_m + G^0{}_k{}^T.\phi_l - G^0{}_l{}^T.\phi_k\}$$

from which we derive

$$[\psi(x),J_{kl}] = [x_k(1/i)\partial_l - x_l(1/i)\partial_k + \tfrac{1}{2}\sigma_{kl}]\psi(x),$$
$$[\phi^m(x),J_{kl}] = [x_k(1/i)\partial_l - x_l(1/i)\partial_k]\phi^m(x) - i\delta_k{}^m\phi_l(x) + i\delta_l{}^m\phi_k(x),$$

and a similar equation for $G^{0mT}(x)$. These properties ensure the appropriate transformation behavior for $P_k$ and $J_{kl}$, and thereby the validity of the commutation relations for the infinitesimal generators of the three-dimensional translation-rotation group.

It is worth noting here that a momentum density must obey equal-time commutation relations of the form

$$-i[T^0{}_k(x),T^0{}_l(x')] = -T^0{}_k(x')\partial_l\delta(\mathbf{x}-\mathbf{x}') - T^0{}_l(x)\partial_k\delta(\mathbf{x}-\mathbf{x}') + \tau_{kl}(x,x'),$$

where

$$\tau_{kl}(x,x') = -\tau_{lk}(x',x)$$

and

$$\int(d\mathbf{x})\tau_{kl}(x,x') = \int(d\mathbf{x})[x_k\tau_{lm}(x,x') - x_l\tau_{km}(x,x')] = 0.$$

The latter are the conditions demanded by the infinitesimal transformation equations,

$$[T^0{}_k(x),P_l] = (1/i)\partial_l T^0{}_k(x),$$
$$[T^0{}_m(x),J_{kl}] = [x_k(1/i)\partial_l - x_l(1/i)\partial_k]T^0{}_m(x) - i\delta_{km}T^0{}_l(x) + i\delta_{lm}T^0{}_k(x),$$

which also imply the commutation properties of $P_k$ and $J_{kl}$. The Hermitian operators $\tau_{kl}(x,x')$ should vanish at finite $|\mathbf{x}-\mathbf{x}'|$ but generally are not identically zero for field systems with spin. Concerning the system now under discussion we shall only remark that $\tau_{kl}(x,x')$ has terms containing no higher than second derivatives of $\delta(\mathbf{x}-\mathbf{x}')$.

The Fermi field contribution to the energy density obeys the commutator condition,

$$-i[T^{00}(x)^F,T^{00}(x')^F] = -(T^0{}_k(x)^F + T^0{}_k(x')^F)\partial^k\delta(\mathbf{x}-\mathbf{x}'),$$

as one can easily confirm. In order to evaluate $[T^{00}(x)^F, T^{00}(x')^B]$, for example, we first compute $[T^{00}(x)^F, G^{0l}(x')]$ which differs from zero in virtue of the commutator

$$i[\phi_k(x), G^{0l}(x')] = \delta^l{}_k \delta(\mathbf{x}-\mathbf{x}') - [\partial_k - i\text{`}t\phi_k(x)\text{'}]\mathfrak{D}_\phi(\mathbf{x},\mathbf{x}')\partial'^l,$$

and of the density operators $k^0{}_a$ contained in $G^{0l}(x')$. The result obtained by combining the two contributions is simply

$$-i[T^{00}(x)^F, G^{0l}(x')] = \delta(\mathbf{x}-\mathbf{x}')k^l(x),$$

where

$$k_a{}^l(x) = \tfrac{1}{2}\psi(x)\alpha^l T_a \psi(x).$$

It follows that

$$-i[T^{00}(x)^F, \tfrac{1}{2}(G^{0l}(x'))^2] = \delta(\mathbf{x}-\mathbf{x}')k_l(x).G^{0l}(x),$$

which is a symmetrical function of $x$ and $x'$, and accordingly

$$[T^{00}(x)^F, T^{00}(x')^B] + [T^{00}(x)^B, T^{00}(x')^F] = 0.$$

Turning to the evaluation of $[T^{00}(x)^B, T^{00}(x')^B]$, we observe that

$$-i[f^2 G_{kl}(x), G^{0m}(x')] = -[\partial_k - i\text{`}t\phi_k(x)\text{'}]\delta_l{}^m\delta(\mathbf{x}-\mathbf{x}') + [\partial_l - i\text{`}t\phi_l(x)\text{'}]\delta_k{}^m\delta(\mathbf{x}-\mathbf{x}') - i\text{`}tG_{kl}(x)\text{'}\mathfrak{D}_\phi(\mathbf{x},\mathbf{x}')\partial'^m,$$

and therefore

$$-i[\tfrac{1}{4}f^2(G_{kl}(x))^2, G^{0m}(x')] = -G^{km}(x)[\partial_k - i\text{`}t\phi_k(x)\text{'}]\delta(\mathbf{x}-\mathbf{x}')$$

since

$$(G_{kl}(x)tG_{kl}(x)) = 0.$$

As a result,

$$(1/i)[\tfrac{1}{4}f^2(G_{kl}(x))^2, \tfrac{1}{2}f^2(G^{0m}(x'))^2] + (1/i)[\tfrac{1}{2}f^2(G^{0m}(x))^2, \tfrac{1}{4}f^2(G_{kl}(x'))^2] = -[T^0{}_k(x)^B + T^0{}_k(x')^B]\partial^k\delta(\mathbf{x}-\mathbf{x}'),$$

and the verification of the commutator condition would be completed, were it known that

$$[\tfrac{1}{2}f^2(G^{0k}(x))^2 + t_\phi(\mathbf{x}), \tfrac{1}{2}f^2(G^{0l}(x'))^2 + t_\phi(\mathbf{x}')] = 0.$$

Let us recall the commutator

$$i[G^{0k}(x), G^{0l}(x')] = \partial^k\mathfrak{D}_\phi(\mathbf{x},\mathbf{x}').i\text{`}tG^{0l}(x')\text{'} + i\text{`}tG^{0k}(x)\text{'}.\mathfrak{D}_\phi(\mathbf{x},\mathbf{x}')\partial'^l,$$

of which one important consequence is that

$$(G^{0k}(x)tG^{0k}(x)),$$

unlike the similar construction with $G_{kl}$, does not vanish. Indeed,

$$(G^{0k}(x)tG^{0k}(x)) = -\textstyle\sum_a t_a\partial_k\mathfrak{D}_\phi(\mathbf{x},\mathbf{x})t_a.G^{0k}(x),$$

where only the first variable of $\mathfrak{D}_\phi$ is to be differentiated. We shall now simply record the outcome of computing $[(G^{0k}(x))^2, (G^{0l}(x'))^2]$ and invite the reader to duplicate this calculation:

$$-i[\tfrac{1}{2}(G^{0k}(x))^2, \tfrac{1}{2}(G^{0l}(x'))^2] = -\gamma_\phi(\mathbf{x},\mathbf{x}')_k.G^{0k}(x') + G^{0k}(x).\gamma_\phi(\mathbf{x}',\mathbf{x})_k,$$

in which

$$f^2\gamma_\phi(\mathbf{x},\mathbf{x}')^k = (1/i)[t_\phi(\mathbf{x}), G^{0k}(x')]$$

and

$$t_\phi(\mathbf{x}) = \tfrac{1}{8}f^2\textstyle\sum_a \mathrm{Tr}[t_a\partial^k\mathfrak{D}_\phi(\mathbf{x},\mathbf{x})t_a\partial_k\mathfrak{D}_\phi(\mathbf{x},\mathbf{x})].$$

In this result we recognize the statement that a function $t_\phi(\mathbf{x})$ exists such that the set of operators, $\tfrac{1}{2}f^2(G^{0k}(x))^2 + t_\phi(\mathbf{x})$ for all $\mathbf{x}$, are commutative.

The intimate relationship that must exist between the two terms $\tfrac{1}{2}f^2(G^{0k}(x))^2$ and $t_\phi(\mathbf{x})$ is emphasized by writing the sum as the positive Hermitian operator,

$$\tfrac{1}{2}f^2\textstyle\sum_a [G_a{}^{0k}(x) - \tfrac{1}{2}\mathrm{Tr}t_a\partial^k\mathfrak{D}_\phi(\mathbf{x},\mathbf{x})][G_a{}^{0k}(x) + \tfrac{1}{2}\mathrm{Tr}t_a\partial^k\mathfrak{D}_\phi(\mathbf{x},\mathbf{x})] = \tfrac{1}{2}f^2(G^{0k}(x))^2 + t_\phi(\mathbf{x}),$$

where, it will be remembered, the elements of the matrices $t_a$ are imaginary numbers. The possibility of this rearrangement is equivalent to a statement of the identity

$$\textstyle\sum_a [\mathrm{Tr}t_a\partial^k\mathfrak{D}_\phi(\mathbf{x},\mathbf{x}), G_a{}^{0k}(x)] = -\tfrac{1}{2}\sum_a \mathrm{Tr}[t_a\partial^k\mathfrak{D}_\phi(\mathbf{x},\mathbf{x})t_a\partial^k\mathfrak{D}_\phi(\mathbf{x},\mathbf{x})] - \tfrac{1}{2}\sum_a [\mathrm{Tr}t_a\partial^k\mathfrak{D}_\phi(\mathbf{x},\mathbf{x})]^2.$$

Its verification will require the following theorem:

$$\textstyle\sum_a \mathrm{Tr}[t_a\partial^k\mathfrak{D}_\phi(\mathbf{x},\mathbf{x})t_a(\partial^k\mathfrak{D}_\phi(\mathbf{x},\mathbf{x}) - \mathfrak{D}_\phi(\mathbf{x},\mathbf{x})\partial^k)] = \sum_a [\mathrm{Tr}t_a\partial^k\mathfrak{D}_\phi(\mathbf{x},\mathbf{x})]^2.$$

## FIELD TRANSFORMATION PROPERTIES

Now that we are in possession of the explicit operators[6] for $P^0$ and $J^0{}_k$, the equations of motion and the Lorentz transformation behavior of the various field quantities can be derived. Let us begin with the Fermi field $\psi(x)$ and remark that $(x^0 = x^{0'})$

$$[\psi(x), k^0(x')] = \delta(\mathbf{x}-\mathbf{x}')T\psi(x),$$

which has the consequence

$$[\psi(x), G^{0l}(x')] = -\partial'^l\mathfrak{D}_\phi(\mathbf{x}',\mathbf{x})T\psi(x).$$

Then we get

$$i\partial_0\psi(x) = [\psi(x), P^0] = \text{`}T\phi^0(x)\text{'}.\psi(x) + \alpha^k[(1/i)\partial_k - \text{`}T\phi_k(x)\text{'}]\psi(x) - im\beta\psi(x),$$

where

$$\phi^0(x) = f^2\int (d\mathbf{x}')\mathfrak{D}_\phi(\mathbf{x},\mathbf{x}').\partial_l'G^{0l}(x').$$

[6] Despite the criticism of Lorentz gauge methods, it seems reasonable to suppose that no difficulty of a formal nature will appear in a Lorentz gauge treatment analogous to that given by Fermi for the electromagnetic field, provided one avoids all reference to state vector norms, for these cannot be finite. Accordingly, it behooves one to show that the elimination of longitudinal modes from the Lorentz gauge formulation reproduces the energy operator of the radiation gauge method. This I have succeeded in doing.

This equation of motion is also expressed by

$$\{\alpha^\mu[\partial_\mu - i\,{}^{\backprime}T\phi_\mu(x)'] + \beta m\}\,.\,\psi(x) = 0,$$

with

$$\alpha^0 = 1.$$

In evaluating the commutator of any field operator $F(x)$ with $J_{0l}$, it is convenient to write the latter as

$$J_{0l} = x_0 P_l - x_l P_0 + \int (d\mathbf{x}')(x_l' - x_l) T^{00}(x'),$$

so that

$$[F(x), J_{0l}] = [x_0(1/i)\partial_l - x_l(1/i)\partial_0] F(x) + \int (d\mathbf{x}')(x_l' - x_l)[F(x), T^{00}(x')].$$

The result of applying this procedure to $\psi(x)$ is

$$[\psi(x), J_{0l}] = [x_0(1/i)\partial_l - x_l(1/i)\partial_0 + \tfrac{1}{2}\sigma_{0l}]\psi(x) + {}^{\backprime}T\Lambda_l(x)'\,.\,\psi(x),$$

in which

$$\Lambda_l(x) = f^2 \int (d\mathbf{x}')\mathfrak{D}_\phi(\mathbf{x},\mathbf{x}')\,.\,\partial_k'[(x_l' - x_l)G^{0k}(x')]$$

appears as an operator gauge function. We have also defined

$$\sigma_{0l} = -i\alpha_l.$$

A similar calculation can be performed for the densities $k^0(x)$ with the aid of the equal-time commutation relation

$$[k_b{}^0(x), k_c{}^0(x')] = \delta(\mathbf{x} - \mathbf{x}')\textstyle\sum_a k_a{}^0(x) t_{abc}.$$

We find

$$i\partial_0 k^0(x) = {}^{\backprime}t\phi^0(x)'\,.\,k^0(x) + [(1/i)\partial_l - {}^{\backprime}t\phi_l(x)']k^l(x),$$

or

$$[\partial_\mu - i\,{}^{\backprime}t\phi_\mu(x)']\,.\,k^\mu(x) = 0,$$

and the Lorentz transformation law

$$[k^0(x), J_{0l}] = [x_0(1/i)\partial_l - x_l(1/i)\partial_0]k^0(x) - ik_l(x) + {}^{\backprime}t\Lambda_l(x)'\,.\,k^0(x).$$

A related result is

$$[k_m(x), J_{0l}] = [x_0(1/i)\partial_l - x_l(1/i)\partial_0]k_m(x) - i\delta_{ml}k^0(x) + {}^{\backprime}t\Lambda_l(x)'\,.\,k_m(x).$$

Turning to the transverse vector field $\phi_k(x)$, we first note that

$$i[\phi_k(x), \tfrac{1}{2}(G^{0m}(x'))^2] = \{\delta_{km}\delta(\mathbf{x} - \mathbf{x}') - [\partial_k - i\,{}^{\backprime}t\phi_k(x)']\mathfrak{D}_\phi(\mathbf{x}\,\mathbf{x}')\partial_m'\}\,.\,G^{0m}(x')$$

in which $\partial_m'$ is still understood to act on the function to its left. The immediate consequences are

$$\partial_0\phi_k(x) = f^2 G_{0k}(x) + [\partial_k - i\,{}^{\backprime}t\phi_k(x)']\,.\,\phi_0(x)$$

or equivalently,

$$f^2 G_{0k} = \partial_0\phi_k - \partial_k\phi_0 + (\phi_0 it\phi_k),$$

and

$$[\phi_k(x), J_{0l}] = [x_0(1/i)\partial_l - x_l(1/i)\partial_0]\phi_k(x) - i\delta_{kl}\phi^0(x) + [(1/i)\partial_k - {}^{\backprime}t\phi_k(x)']\,.\,\Lambda_l(x).$$

Since both sides of the latter equation must be divergenceless we learn, incidentally, that

$$0 = \partial_l\phi^0 - \partial_0\phi_l + (\nabla^2 - i\,{}^{\backprime}t\boldsymbol{\phi}'\cdot\nabla)\Lambda_l,$$

which supplies an alternative construction for the operator gauge function,

$$\Lambda_l(x) = \int (d\mathbf{x}')\mathfrak{D}_\phi(\mathbf{x},\mathbf{x}')\,.\,[\partial_l'\phi^0(x') - \partial_0\phi_l(x')],$$

as one can verify directly. This version of the gauge function can be used to rewrite the Lorentz transformation property of $\phi_k$ in the symbolic form

$$[\phi_k, J_{0l}] = [1 + (\nabla - i\,{}^{\backprime}t\boldsymbol{\phi}')\mathfrak{D}_\phi\nabla]_{km}\,.\,\{[x_0(1/i)\partial_l - x_l(1/i)\partial_0]\phi^m - i\delta_l{}^m\phi^0\}$$

which makes explicit the origin of the operator gauge transformation in the radiation gauge requirement of transversality. A closely related transformation law is that of $G_{km}(x)$, since

$$f^2\delta G_{km} = (\partial_k - i\,{}^{\backprime}t\phi_k')\delta\phi_m - (\partial_m - i\,{}^{\backprime}t\phi_m')\delta\phi_k$$

and the result of this calculation is

$$[G_{km}(x), J_{0l}] = [x_0(1/i)\partial_l - x_l(1/i)\partial_0]G_{km}(x) + i\delta_{kl}G_{0m}(x) - i\delta_{ml}G_{0k}(x) + {}^{\backprime}t\Lambda_l(x)'\,.\,G_{km}(x).$$

Thus far the Lorentz transformation properties present a comparatively simple picture. In addition to the anticipated geometrical transformations the various fields are subjected to an operator gauge transformation. Indeed, this would be a completely valid assertion for an Abelian gauge field, but it is not true for the time components of a non-Abelian gauge field. If we consider the field $G^{0k}(x)$, commutation properties which have already been employed show that

$$-i[G^{0k}(x), T^{00}(x')] = G^{mk}(x')[\partial_m' - i\,{}^{\backprime}t\phi_m(x')']\delta(\mathbf{x} - \mathbf{x}') - \delta(\mathbf{x} - \mathbf{x}')k^k(x) - f^2 i\,{}^{\backprime}tG^{0k}(x)'\,.\,[\mathfrak{D}_\phi(\mathbf{x},\mathbf{x}')\partial'^l\,.\,G^{0l}(x')] + r_\phi{}^k(\mathbf{x},\mathbf{x}'),$$

where the last two terms appear as an evaluation of

$$-i[G^{0k}(x), \tfrac{1}{2}f^2(G^{0l}(x'))^2 + t_\phi(\mathbf{x}')].$$

The notation anticipates the structure of the function $r_\phi{}^k(\mathbf{x},\mathbf{x}')$. It is again understood that $\partial'^l$ acts on the function to its left. The result of an $x'$ integration is the equation of motion

$$\partial_0 G^{0k}(x) = i\,{}^{\backprime}t\phi_0(x)'\,.\,G^{0k}(x) - [\partial_l - i\,{}^{\backprime}t\phi_l(x)']G^{lk}(x) - k^k(x) + r_\phi{}^k(\mathbf{x}),$$

while the additional factor $x_l' - x_l$ produces the Lorentz transformation formula

$$[G^{0k}(x),J_{0l}]=[x_0(1/i)\partial_l-x_l(1/i)\partial_0]G^{0k}(x)+iG^k{}_l(x)+{}^{\backprime}t\Lambda_l(x)^{\prime}.G^{0k}(x)+ir_\phi{}^k{}_l(\mathbf{x}).$$

Here we have defined

$$r_\phi{}^k(\mathbf{x})=\int(d\mathbf{x}')r_\phi{}^k(\mathbf{x},\mathbf{x}')$$

and

$$r_\phi{}^k{}_l(\mathbf{x})=\int(d\mathbf{x}')(x_l'-x_l)r_\phi{}^k(\mathbf{x},\mathbf{x}').$$

The novel transformation aspects of the operators $G^{0k}(x)$ are thus associated with the appearance of the function $r_\phi{}^k(\mathbf{x},\mathbf{x}')$.

This function can be evaluated by remarking that

$$-f^2G^{0k}(x).[i^{\backprime}tG^{0k}(x)^{\prime}.(\mathfrak{D}_\phi(\mathbf{x},\mathbf{x}')\partial'^l.G^{0l}(x'))]+G^{0k}(x).r_\phi{}^k(\mathbf{x},\mathbf{x}')=-i[\tfrac{1}{2}(G^{0k}(x))^2,\tfrac{1}{2}f^2(G^{0l}(x'))^2+t_\phi(\mathbf{x}')]=i[t_\phi(\mathbf{x}),G^{0l}(x')].G^{0l}(x'),$$

so that it suffices to identify the coefficients of $G^{0k}(x)$, after a reduction of the first term. A fairly explicit statement of the result is given by

$$r_\phi{}^k(\mathbf{x},\mathbf{x}')_a=-\tfrac{1}{8}f^2\sum_{bc}[it_a\partial^k\mathfrak{D}_\phi(\mathbf{x},\mathbf{x}')]_{bc}\times\mathrm{Tr}[t_b\mathfrak{D}_\phi(\mathbf{x},\mathbf{x}')\partial'^lt_c\partial'^l\mathfrak{D}_\phi(\mathbf{x}',\mathbf{x})]+\tfrac{1}{4}f^2\sum_b\mathrm{Tr}\{t_at_b\mathfrak{D}_\phi(\mathbf{x},\mathbf{x}')\partial'^li[G_b{}^{0k}(x),\partial'^l\mathfrak{D}_\phi(\mathbf{x}',\mathbf{x})]\}.$$

In this and in previous manipulations the following relation has been useful:

$$i[\partial^k\mathfrak{D}_\phi(\mathbf{x},\mathbf{x}')_{ab},G_c{}^{0l}(x'')]-i[\partial''^l\mathfrak{D}_\phi(\mathbf{x}'',\mathbf{x}')_{cb},G_a{}^{0k}(x)]=-[\partial^k\mathfrak{D}_\phi(\mathbf{x},\mathbf{x}'')it_c\partial''^l\mathfrak{D}_\phi(\mathbf{x}'',\mathbf{x}')]_{ab}-[\partial^k\mathfrak{D}_\phi(\mathbf{x},\mathbf{x}')it_b\mathfrak{D}_\phi(\mathbf{x}',\mathbf{x}'')\partial''^l]_{ac}-[\mathfrak{D}_\phi(\mathbf{x}',\mathbf{x})\partial^kit_a\mathfrak{D}_\phi(\mathbf{x},\mathbf{x}'')\partial''^l]_{bc}.$$

Another aspect of the function $r_\phi{}^k(\mathbf{x},\mathbf{x}')$ should be noted. It is a consequence of the definition that

$$[\partial_k-i^{\backprime}t\phi_k(x)^{\prime}]r_\phi{}^k(\mathbf{x},\mathbf{x}')=\sum_{ab}t_bt_a\tfrac{1}{4}[\phi_a{}^k(x),[G_b{}^{0k}(x),f^2\mathfrak{D}_\phi(\mathbf{x},\mathbf{x}')\partial'^l.G^{0l}(x')]]-\sum_{ab}t_at_b\tfrac{1}{4}[G_b{}^{0k}(x),[\phi_a{}^k(x),f^2\mathfrak{D}_\phi(\mathbf{x},\mathbf{x}')\partial'^l.G^{0l}(x')]].$$

Beyond remarking that an evaluation of these double commutators would indeed yield a function of the field $\boldsymbol{\phi}$, we shall not record the explicit result, for the important thing is the commutator structure, which is such that the constraint equation,

$$[\partial_k-i^{\backprime}t\phi_k(x)^{\prime}].G^{0k}(x)=k^0(x),$$

is maintained in time and under Lorentz transformations. Thus, a detailed term-by-term evaluation of

$$[(\partial_k-i^{\backprime}t\phi_k(x)^{\prime}).G^{0k}(x)-k^0(x),\ J_{0l}]$$

will finally yield zero only in virtue of the implied property,

$$[\partial_k-i^{\backprime}t\phi_k(x)^{\prime}]r_\phi{}^k{}_l(\mathbf{x})+r_{\phi l}(\mathbf{x})=-\sum_{ab}t_bt_a\tfrac{1}{4}[\phi_{ka}(x),[G_b{}^{0k}(x),\Lambda_l(x)]]+\sum_{ab}t_at_b\tfrac{1}{4}[G_b{}^{0l}(x),[\phi_{ka}(x),\Lambda_l(x)]],$$

which expression, incidentally, also equals

$$\sum_{ab}t_bt_a\tfrac{1}{4}[\phi_{ka}(x),[\Lambda_{lb}(x),G^{0k}(x)]]-\sum_{ab}t_at_b\tfrac{1}{4}[\Lambda_{lb}(x),[\phi_{ka}(x),G^{0k}(x)]].$$

The complete density of the internal property represented by the matrices $T_a$ and $t_a$ is given by

$$j_a{}^0(x)=\partial_kG_a{}^{0k}(x)=k_a{}^0(x)-i(\phi_k(x).t_aG^{0k}(x)),$$

and the total content of this property is described by the constant Hermitian operator,

$$\mathbf{T}_a=\int(d\mathbf{x})j_a{}^0(x).$$

Whenever the contributions to this integral are effectively confined to a spatially bounded region the scalar function $\Psi(x)$, which specifies the longitudinal component of $G^{0k}(x)$,

$$G^{0k}=G^{0kT}-\partial^k\Psi,$$

must have the asymptotic behavior

$$|\mathbf{x}|\to\infty:\quad \Psi_a(x)\sim\frac{1}{4\pi|\mathbf{x}|}\mathbf{T}_a.$$

No such slowly decreasing term will appear in the time derivative of $G^{0k}$, however, and we can identify the current $j^k$, which obeys the conservation law

$$\partial_0j^0+\partial_kj^k=0,$$

as

$$j^k(x)=-\partial_0G^{0k}(x)-\partial_lG^{lk}(x)=k^k(x)+i(\phi_0(x).tG^{0k}(x))+i(\phi_l(x)tG^{lk}(x))-r_\phi{}^k(\mathbf{x}).$$

We see that the function $r_\phi{}^k(\mathbf{x})$ also intervenes here.

Although the current $j^\mu(x)$ obeys a local conservation law, in contrast with $k^\mu(x)$, the density $j^0(x)$ does not have a localized character with respect to equal-time commutation relations, as does $k^0(x)$. Thus,

$$[j^0(x),j^0(x')]=-{}^{\backprime}tj^0(x)^{\prime}\delta(\mathbf{x}-\mathbf{x}')+\partial_k\partial_l'i[{}^{\backprime}t\phi^k(x)^{\prime}\mathfrak{D}_\phi(\mathbf{x},\mathbf{x}').{}^{\backprime}tG^{0l}(x')^{\prime}-{}^{\backprime}tG^{0k}(x)^{\prime}.\mathfrak{D}_\phi(\mathbf{x},\mathbf{x}')^{\backprime}t\phi^l(x')^{\prime}].$$

The additional divergence term does not contribute to integrated quantities, and we reaffirm the commutation properties of the **T** operators,

$$[\mathbf{T}_b,\mathbf{T}_c]=\sum_a \mathbf{T}_a t_{abc},$$

while obtaining

$$[j^0(x),\mathbf{T}_a]=t_a j^0(x).$$

The Lorentz transformation properties of $j^\mu$ are more complicated than those of $k^\mu$. For example,

$$[j^0(x),J_{0l}]=[x_0(1/i)\partial_l-x_l(1/i)\partial_0]j^0(x)-ij_l(x)$$
$$+{}'t\Lambda_l(x)'.j^0(x)+{}'t\partial_k\Lambda_l(x)'.G^{0k}(x)+i\partial_k r_\phi{}^k{}_l(\mathbf{x}),$$

and by writing the latter as

$$[j^0(x),J_{0l}]=(1/i)\partial_k[\delta_l{}^k x_0 j^0(x)+x_l j^k(x)$$
$$+i'tA_l(x)'.G^{0k}(x)-r_\phi{}^k{}_l(\mathbf{x})]$$

one can verify the Lorentz invariance of the conserved **T** operators,

$$[\mathbf{T}_a,J_{0l}]=0.$$

We consider, finally, the Lorentz transformation behavior of the field $\phi^0(x)$. One might begin with the explicit construction,

$$\phi^0(x)=f^2\int(d\mathbf{x}')\mathfrak{D}_\phi(\mathbf{x},\mathbf{x}').\partial_k'G^{0k}(x'),$$

which, incidentally, has a counterpart in

$$\phi^k(x)=f^2\int(d\mathbf{x}')\mathfrak{D}_\phi(\mathbf{x},\mathbf{x}')\partial_l'G^{kl}(x'),$$

but it is simpler to return to the differential equation

$$f^2G_{0k}=\partial_0\phi_k-(\partial_k-i't\phi_k').\phi_0.$$

From the known characteristics of $\phi_k$ and $G_{0k}$ we now find that

$$[\phi^0(x),J_{0l}]=[x_0(1/i)\partial_l-x_l(1/i)\partial_0]\phi^0(x)-i\phi_l(x)$$
$$+i[\partial_0-i't\phi_0(x)'].\Lambda_l(x)+i\rho_{\phi l}(\mathbf{x}),$$

where

$$[\partial_k-i't\phi_k(x)']\rho_{\phi l}(\mathbf{x})=-f^2r_{\phi kl}(\mathbf{x})$$
$$+\sum_{ab} t_b t_a \tfrac{1}{4}[\phi_{ka}(x),[\Lambda_{lb}(x),\phi^0(x)]]$$
$$-\sum_{ab} t_a t_b \tfrac{1}{4}[\Lambda_{lb}(x),[\phi_{ka}(x),\phi^0(x)]],$$

and the divergence of this equation will supply the additional function of the field $\phi$ that transcends the elementary geometrical and gauge transformations. The asymptotic behavior of $\phi^0(x)$ is given by

$$|\mathbf{x}|\to\infty:\quad \phi^0(x)\sim\frac{1}{4\pi|\mathbf{x}|}f^2\mathbf{T},$$

according to the differential equation

$$-\nabla^2\phi^0=f^2j^0-\nabla\cdot(i''\boldsymbol{\phi}'\phi^0).$$

*Note added in proof.* The energy density commutator equation can be given a general dynamical basis by examining the modification in the $T^{\mu\nu}$ conservation equation produced by an external gravitational field. Also involved is a minimum assumption of time locality. For an analogous derivation of the null charge density equal-time commutator from the current conservation equation, one must consider systems such that the current is not an explicit function of the time derivative of an external vector potential.

# Gauge Invariance and Mass. II*

JULIAN SCHWINGER
*Harvard University, Cambridge, Massachusetts*
(Received July 2, 1962)

The possibility that a vector gauge field can imply a nonzero mass particle is illustrated by the exact solution of a one-dimensional model.

IT has been remarked[1] that the gauge invariance of a vector field does not necessarily require the existence of a massless physical particle. In this note we shall add a few related comments and give a specific model for which an exact solution affirms this logical possibility. The model is the physical, if unworldly situation of electrodynamics in one spatial dimension, where the charge-bearing Dirac field has no associated mass constant. This example is rather unique since it is a simple model for which there is an exact divergence-free solution.[2]

## GENERAL DISCUSSION

The Green's function of an Abelian vector gauge field has the structure

$$\mathcal{G}_{\mu\nu}(x,x')=\pi_{\mu\nu}(-i\partial)\mathcal{G}(-i\partial)\delta(x-x'),$$

where $\pi_{\mu\nu}(p)$ is a gauge-dependent projection matrix and

$$\mathcal{G}(p)=\int_0^\infty dm^2\,\frac{B(m^2)}{p^2+m^2-i\epsilon},$$

which is subject to the sum rule

$$1=\int_0^\infty dm^2\,B(m^2).$$

An alternative form of $\mathcal{G}(p)$ is

$$\mathcal{G}(p)=\left[p^2+\lambda^2-i\epsilon+(p^2-i\epsilon)\int_0^\infty dm^2\,\frac{s(m^2)}{p^2+m^2-i\epsilon}\right]^{-1},$$

where the function $s(m^2)$ and the constant $\lambda^2$ are non-negative. The latter has been derived[3] with the understanding that the pole at $z=0$ of the expression

$$-\frac{\lambda^2}{z}+\int_0^\infty dm^2\,\frac{s(m^2)}{m^2-z}=\int_0^\infty \frac{dm^2}{m^2-z}[s(m^2)+\lambda^2\delta(m^2)]$$

is completely described by the parameter $\lambda$. Accordingly,

$$\mathcal{G}(0)=\frac{1}{\lambda^2}=\int_0^\infty dm^2\,\frac{B(m^2)}{m^2},$$

and $\lambda^2>0$ unless $m=0$ is contained in the spectrum. Thus, it is necessary that $\lambda$ vanish if $m=0$ is to appear as an isolated mass value in the physical spectum. But it is also necessary that

$$s(0)=0,$$

such that

$$\int_{\to 0}^\infty \frac{dm^2}{m^2}s(m^2)<\infty,$$

for only then do we have a pole at $p^2=0$,

$$p^2\sim 0:\quad \mathcal{G}(p)\sim B_0/(p^2-i\epsilon),\quad 0<B_0<1.$$

Under these conditions,

$$B(m^2)=B_0\delta(m^2)+B_1(m^2),$$

where

$$B_0=\left(1+\int_0^\infty \frac{dm^2}{m^2}s(m^2)\right)^{-1}$$

and

$$B_1(m^2)=[s(m^2)/m^2]\Bigg/\left[1+P\int_0^\infty dm'^2\frac{s(m'^2)}{m'^2-m^2}\right]+[\pi s(m^2)]^2.$$

The physical interpretation of $s(m^2)$ derives from the relation of the Green's function to the vacuum transformation function in the presence of sources. For sufficiently weak external currents $J_\mu(x)$,

$$\langle 0|0\rangle^J=\exp\left[\tfrac{1}{2}i\int(dx)(dx')J^\mu(x)\mathcal{G}_{\mu\nu}(x,x')J^\nu(x')\right]$$

$$=\exp\left[\tfrac{1}{2}i\int(dp)J^\mu(p)^*\mathcal{G}(p)J_\mu(p)\right],$$

which involves the reduction of the projection matrix $\pi_{\mu\nu}(p)$ to $g_{\mu\nu}$ for a conserved current, or equivalently

$$p_\mu J^\mu(p)=0.$$

We shall present this transformation function as a measure of the response to the external vector potential

$$A_\mu(p)=\mathcal{G}(p)J_\mu(p),$$

namely,

$$\langle 0|0\rangle^J=\exp\left[\tfrac{1}{2}i\int(dp)A^\mu(p)^*\mathcal{G}(p)^{-1}A_\mu(p)\right].$$

---

* Supported in part by the Air Force Office of Scientific Research (Air Research and Development Command), under contract number AF-49(638)-589.

[1] J. Schwinger, Phys. Rev. **125**, 397 (1962).

[2] There is a divergence in the so-called Thirring model [W. E. Thirring, Ann. Phys. (New York) **3**, 91 (1958)], which uses local current interactions rather than a Bose field.

[3] J. Schwinger, Ann. Phys. (New York) **9**, 169 (1960).



Reprinted from *The Physical Review* **128**, 2425–2429 (1962).

The probability that the vacuum state shall persist despite the disturbance is

$$|\langle 0|0\rangle^J|^2 = \exp\left[-\int (dp) A_\mu(p)^* A_\mu(p) \operatorname{Im} \mathcal{G}(p)^{*-1}\right]$$
$$= \exp\left[-\pi \int (dp) dm^2 \delta(p^2+m^2) \times s(m^2)(-\tfrac{1}{2}) F^{\mu\nu}(p)^* F_{\mu\nu}(p)\right],$$

which exhibits $s(m^2)$ as a measure of the probability that an external field $F_{\mu\nu}$ will produce a vacuum excitation involving an energy-momentum transfer measured by the mass $m$.

The vanishing of $s(m^2)$ at $m=0$ is normal threshold behavior for an excitation function. If a zero-mass particle is not to exist, $m=0$ must be an abnormal threshold. Two possibilities can be distinguished. In the first of these, $s(m^2)$ is finite or possibly singular at $m=0$, but in such a way that

$$\lim_{z\to 0} z \int_0^\infty dm^2 \frac{s(m^2)}{m^2-z} = 0.$$

Then the physical mass spectrum begins at $m=0$ but there is no recognizable zero-mass particle. For the second situation, $s(m^2)$ has a delta-function singularity at $m^2=0$,

$$s(m^2) = \lambda^2 \delta(m^2) + s_1(m^2),$$

and

$$s_1(m^2) = 0, \quad m^2 < m_0^2.$$

If the threshold mass $m_0$ is zero, the restriction of the previous situation applies to the function $s_1(m^2)$. Now, $m=0$ is not contained in the spectrum at all. This statement is true even if $m_0=0$ for, according to the structure of $B_1(m^2)=B(m^2)$,

$$B(m^2) = \frac{m^2 s_1(m^2)}{[R(m^2)]^2 + [\pi m^2 s_1(m^2)]^2},$$

in which

$$R(m^2) = m^2 - \lambda^2 + m^2 P \int_{m_0^2}^\infty dm'^2 \frac{s_1(m^2)}{m'^2 - m^2},$$

we have

$$\lim_{m^2\to 0} B(m^2) = \lim_{m^2\to 0} \frac{m^2 s_1(m^2)}{\lambda^4} = 0.$$

Let us suppose that $m_0$ is the threshold of a continuous spectrum. A stable particle of mass $m < m_0$ will exist if $R(m_0^2) > 0$. Should both $R(m_0^2)$ and $s_1(m_0^2)$ be zero there would be a stable particle of mass $m_0$. No stable particle exists if $R(m_0^2) < 0$. But there is always an unstable particle, in a certain sense. By this we mean that $R(m^2)$ vanishes at some mass value $m_1 > m_0$, under the general restrictions required for the continuity of the function $R(m^2)$, as a consequence of this function's asymptotic approach to $+\infty$ with increasing $m^2$. The mass $m_1$ will be physically recognizable as the mass of an unstable particle if the mass width

$$\gamma = \frac{\pi m_1 s_1(m_1^2)}{[dR(m_1^2)/dm_1^2]}$$

is sufficiently small. [We take the derivative of $R(m_1^2)$ to be positive, which is appropriate for the simplifying assumption that only one zero occurs.] The contribution of such a fairly sharp resonance to the sum rule for $B(m^2)$ is given by

$$\int_{m\sim m_1} dm^2 B(m^2) = [dR(m_1^2)/dm_1^2]^{-1} < 1.$$

## SIMPLE MODELS

Some of these possibilities can be illustrated in very simple physical contexts. We consider the linear approximation to the problem of electromagnetic vacuum polarization for spaces of dimensionality $n=2$ and 1. A modification of a technique[4] previously applied to three-dimensional space yields for $m > m_0$:

$$s(m^2) = \int_0^{(1-m_0^2/m^2)^{1/2}} dv (1-v^2)(e^2/8\pi^2) \quad \text{for} \quad n=3$$
$$= \int_0^{(1-m_0^2/m^2)^{1/2}} dv (1-v^2)(e^2/4\pi^2) \times [m^2(1-v^2)-m_0^2]^{-1/2} \quad \text{for} \quad n=2$$
$$= \int_0^{(1-m_0^2/m^2)^{1/2}} dv (1-v^2)(e^2/\pi) \delta[m^2(1-v^2)-m_0^2] \quad \text{for} \quad n=1;$$

for $m < m_0$:

$$s(m^2) = 0,$$

where the known result for $n=3$ has been included for comparison. The threshold mass $m_0$ is that for single pair creation. It should be noted that the coupling constant $e^2$ of electrodynamics in $n$-dimensional space has the dimensions of a mass raised to the power $3-n$. For $n<3$ this single pair approximation does not lead to difficulties concerning the existence of such integrals as

$$B_0^{-1} - 1 = \int_0^{\to\infty} \frac{dm^2}{m^2} B(m^2),$$

since, for $m \gg m_0$:

$$s(m^2) \sim (e^2/12\pi^2) \quad \text{for} \quad n=3,$$
$$\sim (e^2/16\pi)(1/m) \quad \text{for} \quad n=2,$$
$$\sim (e^2/2\pi)(m_0^2/m^4) \quad \text{for} \quad n=1.$$

The particular situation in which we are interested appears at the limit $m_0 \to 0$. Then we have

$$s(m^2) = (e^2/16\pi)(1/m) \quad \text{for} \quad n=2,$$
$$= (e^2/\pi)\delta(m^2) \quad \text{for} \quad n=1.$$

[4] *Selected Papers on Quantum Electrodynamics* (Dover Publications, New York, 1958), p. 209.

Two-dimensional electrodynamics illustrates the first of the two possibilities for an anomalous threshold at $m=0$. The spectral function $B(m^2)$ describes a purely continuous spectrum,

$$dm^2\, B(m^2)=\frac{2}{\pi}\frac{e^2}{16}\frac{dm}{m^2+(e^2/16)^2},$$

and an $m$ integration from 0 to $\infty$ satisfies the sum rule. In one-dimensional electrodynamics we meet a special case of the second possibility, with

$$\lambda^2=e^2/\pi,\quad s_1(m^2)=0.$$

Accordingly,

$$B(m^2)=\delta(m^2-(e^2/\pi))$$

and the mass spectrum is localized at one point, describing a stable particle of mass $e/\pi^{1/2}$.

The basis indicated for the latter conclusion will not be very convincing, but it is an exact result. To prove this we first compute for one spatial dimension the electric current induced by an arbitrary external potential in the vacuum state of a massless charged Dirac field. The appropriate gauge-invariant expression for the current[5] is

$$j_\mu(x)=-\tfrac{1}{2}e\,\mathrm{tr}q\alpha_\mu G(x,x')\exp\left[-ieq\int_{x'}^{x}d\xi^\mu A_\mu(\xi)\right]\Bigg|_{x'\to x},$$

in which the approach of $x'$ to $x$ is performed from a spatial direction in order to maintain time locality. The Green's function is defined by the differential equation

$$\alpha^\mu[\partial_\mu-ieqA_\mu(x)]G(x,x')=\delta(x-x'),$$

together with the outgoing wave boundary condition, in the absence of the potential. Only two Dirac matrices appear here, $\alpha^0=-\alpha_0=1$ and $\alpha^1=\alpha_1$, which has the eigenvalues $\pm1$. Those are also the eigenvalues of the independent charge matrix $q$. The Green's function equation can be satisfied by writing

$$G(x,x')=G^0(x,x')\exp\{ieq[\phi(x)-\phi(x')]\},$$

where

$$\alpha^\mu\partial_\mu\phi=\alpha^\mu A_\mu(x)$$

and

$$\alpha^\mu\partial_\mu G^0(x,x')=\delta(x-x').$$

The latter defines the free Green's function, which is given explicitly by

$$G^0(x,x')=\int_0^\infty\frac{dp}{2\pi}\exp[ip\alpha^\mu(x_\mu-x_\mu')]\quad\text{for}\quad x^0>x^{0\prime},$$

$$=-\int_{-\infty}^0\frac{dp}{2\pi}\exp[ip\alpha^\mu(x_\mu-x_\mu')]\quad\text{for}\quad x^0<x^{0\prime}.$$

At equal times, and for sufficiently small $x_1-x_1'$, we have

$$G(x,x')\exp\left[-ieq\int_{x'}^{x}d\xi^\mu A_\mu(\xi)\right]$$

$$\cong\frac{i}{2\pi}\frac{\alpha_1}{x_1-x_1'}-\frac{eq}{2\pi}\alpha_1[\partial_1\phi(x)-A_1(x)].$$

The first term does not contribute to the vacuum current when the limit $x_1'\to x_1$ is performed symmetrically. On utilizing the relation

$$\alpha_1(\partial_1\phi-A_1)=-(\partial_0\phi-A_0),$$

we find that

$$j_\mu(x)=-\frac{e^2}{\pi}A_\mu(x)+\partial_\mu\left[\frac{e^2}{4\pi}\,\mathrm{tr}\phi(x)\right].$$

This expression for the induced current is Lorentz covariant, gauge invariant, and obeys the equation of conservation. It is also a linear function of the external field. To verify these statements we construct a differential equation for $\mathrm{tr}\phi(x)$ by multiplying the $\phi$ equation with $\partial_0-\alpha_1\partial_1$ and evaluating the trace. The result is

$$\partial^2\tfrac{1}{4}\,\mathrm{tr}\phi(x)=\partial_\mu A^\mu(x),$$

and therefore

$$\tfrac{1}{4}\,\mathrm{tr}\phi(x)=-\int(dx')D(x,x')\partial_\mu' A^\mu(x'),$$

in which $D$ is the outgoing-wave Green's function defined by

$$-\partial^2 D(x,x')=\delta(x-x').$$

By using a symbolic matrix notation for coordinates and vector indices, we can write

$$j=-(e^2/\pi)(1+\partial D\partial)A,$$

which exhibits the symmetrical projection matrix

$$\pi=1+\partial D\partial,$$

$$\partial\pi=\pi\partial=0,$$

that guarantees gauge invariance and current conservation.

We shall insert this result in the functional differential equation obeyed by the Green's functional $G[J]$, the vacuum transformation function in the presence of external currents. It is convenient to use the particular system of equations that refer to the Lorentz gauge,

$$\left\{(\partial\partial-\partial^2)\frac{1}{i}\frac{\delta}{\delta J}-(1+\partial D\partial)\left[J+j\left(\frac{1}{i}\frac{\delta}{\delta J}\right)\right]\right\}G[J]=0,$$

$$\partial\frac{\delta}{\delta J}G[J]=0,$$

[5] The necessity for the line integral factor has been noted before [J. Schwinger, Phys. Rev. Letters 3, 296 (1959)].

which also utilize a symbolic notation for vectorial coordinate functions. We have written $j(-i\delta/\delta J)$ to indicate the conversion of $j(A)$ into a functional differential operator by the substitution $A\to -i\delta/\delta J$. The functional differential equation implied by the known structure of this operator is

$$\pi\left[\left(-\partial^2+\frac{e^2}{\pi}\right)\frac{1}{i}\frac{\delta}{\delta J}-J\right]G[J]=0,$$

or, on uniting the two defining properties of the functional,

$$\left(\frac{1}{i}\frac{\delta}{\delta J}-\pi\mathcal{G}J\right)G[J]=0,$$

in which

$$[-\partial^2+(e^2/\pi)]\mathcal{G}(x,x')=\delta(x-x').$$

The Green's functional $G[J]$ is therefore given exactly by

$$G[J]=\exp\left[\tfrac{1}{2}i\int(dx)(dx')J^\mu(x)\mathcal{G}_{\mu\nu}(x,x')J^\nu(x')\right],$$

with

$$\mathcal{G}_{\mu\nu}(x,x')=\pi_{\mu\nu}(-i\partial)\mathcal{G}(-i\partial)\delta(x-x')$$

and

$$\mathcal{G}(p)=\frac{1}{p^2+(e^2/\pi)-i\epsilon}.$$

Thus, all states that can be excited by vector currents are fully described as noninteracting ensembles of Bose particles with the mass $e/\pi^{1/2}$.

Concerning the complete Green's functional including Fermi sources, $G[\eta J]$, we shall only remark that

$$G[\eta J]=\exp\left[-\frac{1}{2}\int(dx)(dx')\eta(x)\times G\left(x,x',\frac{1}{i}\frac{\delta}{\delta J}\right)\eta(x')\right]G[J],$$

in which the Green's function can be presented as

$$G(x,x',A)=G^0(x,x')\exp\left[i\int(d\xi)j^\mu(\xi,x,x')A_\mu(\xi)\right]$$

with

$$j^\mu(\xi,x,x')=eq\alpha^\mu\left(\alpha^1\frac{\partial}{\partial\xi^1}-\frac{\partial}{\partial\xi^0}\right)[D(\xi,x)-D(\xi,x')].$$

On expanding the Green's functional in even powers of the Fermi source, we encounter functional differential operators that are contained in one or more factors of the type

$$\exp\left[\int(d\xi)j^\mu(\xi,x,x')\delta/\delta J^\mu(\xi)\right],$$

the effect of which is simply to produce the translation $J\to J+j$ in $G[J]$. The first Fermi Green's function is

$$G(x,x')=G(x,x',-i\delta/\delta J)G[J]|_{J=0}$$
$$=G^0(x,x')\exp\left[\tfrac{1}{2}i\int(d\xi)(d\xi')\times j^\mu(\xi,x,x')\mathcal{G}_{\mu\nu}(\xi,\xi')j^\nu(\xi',x,x')\right].$$

The latter exponential factor is given by

$$\exp\left[-\frac{i}{4\pi}\int(dp)\left(\frac{1}{p^2-i\epsilon}-\frac{1}{p^2+(e^2/\pi)-i\epsilon}\right)\times(1-e^{ip(x-x')})\right].$$

We shall be content to note that this integral and the similar integrals encountered in more general Green's functions are completely convergent. The detailed physical interpretation of the Green's functions is rather special and apart from our main purpose.

These simple examples are quite uninformative in one important respect. They do not exhibit a critical dependence upon the coupling constant. As we have discussed previously, one can view the electromagnetic field as undercoupled and the hypothetical vector field that relates to nucleonic charge as overcoupled, in the sense of a critical value at which the massless Bose particle ceases to exist. The corresponding appearance of an anomalous zero-mass threshold must be attributed to a dynamical mechanism. We can supply an artificial mathematical model that illustrates the situation. Let the following be a contributory term in $s(m^2)$:

$$s_0(m^2)=\frac{\lambda^2}{\pi}\frac{m\gamma}{(m^2-m_0^2\kappa)^2+(m\gamma)^2},$$

in which $m_0$ is a characteristic physical fermion mass, and $\lambda/m_0$, $\gamma/m_0$, and $\kappa$ are positive functions of the (dimensionless) coupling constant. In electrodynamics the near-resonant contributions of such a term can be identified with the creation of a unit angular momentum positronium state, while the values far below resonance refer to the creation of three-photon states (the model falsifies the latter, which should vary as $m^8$ for $m\ll m_0$). It is reasonable to suppose that $\kappa$ decreases with increasing strength of the coupling, and we can imagine that a critical value exists for which both $\kappa$ and $\gamma$ reach zero, with finite $\lambda$. In that circumstance,

$$s_0(m^2)=\lambda^2\delta(m^2),$$

and the null-mass particle disappears from the spectrum. Since this argument requires that one type of excitation move down to zero mass at the critical coupling strength, it is plausible that some other types of excitation will then be located at fairly small fractions of $m_0$. Thus, one could anticipate that the known spin-0 bosons, for example, are secondary dynamical manifestations of strongly coupled primary fermion fields and vector gauge fields. This line of thought emphasizes that the question "Which particles are fundamental?" is incorrectly formulated. One should ask "What are the fundamental fields?"

**ACKNOWLEDGMENTS**

I have had the benefit of conversations on this and related topics with Kenneth Johnson and Charles Sommerfield.

# Quantum Variables and Group Parameters.

J. SCHWINGER (*)

*Institut des Hautes Etudes Scientifiques - Bures-sur-Yvette (S.-et-O.)*

(ricevuto il 22 Aprile 1963)

**Summary.** — Mathematical and physical difficulties in the usual treatments of field theories that admit infinite-dimensional transformation groups are overcome by combining the physical operator variables with the mathematical group parameters. There is also a discussion, for non-Abelian gauge theories, of the limited, asymptotic, physical significance of total internal properties, and a derivation of the energy-density commutator condition.

## 1. – Introduction.

Quantum field theories that admit infinite-dimensional transformation groups in the nature of gauge or co-ordinate transformations have a characteristic feature—not all spatial components of certain tensor fields are fundamental quantum variables. As a consequence of making the fundamental variables explicit, the manipulations needed to show the consistency of the formalism become somewhat involved. This complexity increases as one progresses through the physical examples provided by Abelian gauge fields (electrodynamics), non-Abelian gauge fields, and the gravitational field. In an alternative approach, for which Fermi's treatment of the electromagnetic field is the model, the quantum vector space is enlarged in order to place all spatial components of fields on the same footing, while physical consideration is limited to a certain class of states. Although questions relating to consistency can be studied more simply in this formulation, the nonphysical enlargement of the quantum vector space poses new problems, both conceptual and practical.

(*) Permanent address: Harvard University.

Reprinted from *Il Nuovo Cimento* **30**, 278–291 (1963).

Is there a method that retains the physical immediacy of the first viewpoint while exploiting the mathematical simplicity of the latter? It is our purpose to show that such can be done by combining appropriately the physical operator variables with the mathematical numerical parameters of the transformation group. The technique will be developed in the various physical contexts.

## 2. – Abelian gauge fields.

In the three-dimensional notation of electrodynamics, a generator of vector field variations appears as

$$G_A = \int (\mathrm{d}\boldsymbol{x})(-\boldsymbol{E})\,\delta\boldsymbol{A}\,,$$

which must be interpreted within the framework of the charge-density constraint equation

$$\boldsymbol{\nabla}\cdot\boldsymbol{E} = j^0$$

and the freedom of gauge transformations. This is facilitated by the transverse-longitudinal decomposition

$$\boldsymbol{A}(x) = \boldsymbol{A}^T(x) + \boldsymbol{\nabla}\lambda(x)\,,$$

$$\boldsymbol{E}(x) = \boldsymbol{E}^T(x) - \boldsymbol{\nabla}\int(\mathrm{d}\boldsymbol{x}')\mathscr{D}(\boldsymbol{x}-\boldsymbol{x}')\,j^0(x')\,, \qquad -\boldsymbol{\nabla}^2\mathscr{D}(\boldsymbol{x}-\boldsymbol{x}') = \delta(\boldsymbol{x}-\boldsymbol{x}')\,,$$

since

$$G_A = \int(\mathrm{d}\boldsymbol{x})(-\boldsymbol{E}^T)\,\delta\boldsymbol{A}^T + \int(\mathrm{d}\boldsymbol{x})\,j^0\,\delta\lambda = G_{A^T} + G_\lambda\,,$$

makes explicit the separation into variations of the quantum variables and infinitesimal alterations of the numerical gauge parameter $\lambda(x)$. The interpretation of the generator $G_{A^T}$ is conveyed by the equal-time commutation relation

$$i[A_k^T(x),\, E_l^T(x')] = \big(\delta_{kl}\,\delta(\boldsymbol{x}-\boldsymbol{x}')\big)^T = \delta_{kl}\,\delta(\boldsymbol{x}-\boldsymbol{x}') - \partial_k\,\partial'_l\,\mathscr{D}(\boldsymbol{x}-\boldsymbol{x}')\,,$$

while that of $G_\lambda$ can be expressed by the state vector differential equation

$$\delta_\lambda\langle\lambda x^0| = i\langle\lambda x^0|\int(\mathrm{d}\boldsymbol{x})\,j^0\,\delta\lambda\,,$$

or

$$-i\big(\delta/\delta\lambda(x)\big)\langle\lambda x^0| = \langle\lambda x^0|\,j^0(x)\,.$$

In order to deal with numerical group parameters and operators on a similar footing we shall adopt a representation of the quantum variables by means of numbers and functional differential operators, which are commutative or anticommutative in accordance with the statistics. This is symbolized by

$$\langle | F = F \langle | ,$$

in which position serves to distinguish the operator from its representative. Thus, we now write

$$[-i(\delta/\delta\lambda(x)) - j^0(x)]\langle \lambda x^0 | = 0 .$$

Commutation relations for $j^0$ are implied by the integrability conditions for these differential equations,

$$[-i(\delta/\delta\lambda(x)) - j^0(x), -i(\delta/\delta\lambda(x')) - j^0(x')] =$$
$$= i\,\delta j^0(x')/\delta\lambda(x) - i\,\delta j^0(x)/\delta\lambda(x') + [j^0(x), j^0(x')] = 0 .$$

If, as we shall assume, $j^0$ is not an explicit function of $\lambda$, we obtain the equal-time commutator

$$[j^0(x), j^0(x')] = 0 .$$

The operator and group transformation aspects of the generator $G_A$ are combined in the equation

$$(\mathscr{E}(x) - \boldsymbol{E}(x))\langle \lambda x^0 | = 0 ,$$

where

$$\mathscr{E}(x) = \boldsymbol{E}^T(x) + \boldsymbol{\nabla}\int (\mathrm{d}\boldsymbol{x}')\mathscr{D}(\boldsymbol{x}-\boldsymbol{x}')\,i(\delta/\delta\lambda(x')) .$$

This construct, which we term an extended operator, satisfies the commutation relation

$$i[A_k(x), \mathscr{E}_l(x')] = \delta_{kl}\delta(\boldsymbol{x}-\boldsymbol{x}') .$$

The same form would apply if all three spatial components of the vector fields were independent dynamical variables. The two infinitesimally similar situations differ in the integral aspects expressed by the completeness relation. The eigenvalues of $\boldsymbol{A}^T$ constitute a complete set for the electromagnetic field. With the understanding that these are supplemented by eigenvalues referring to the other fields, the completeness relation reads

$$1 = \int [\mathrm{d}\boldsymbol{A}^{T'}] | \boldsymbol{A}^{T'} \lambda x^0 \rangle \langle \boldsymbol{A}^{T'} \lambda x^0 | ,$$

for each value of $x^0$ and for each gauge function $\lambda(x)$. Were all components of $\boldsymbol{A}$ independent variables, functional integration with respect to $\lambda(x)$ would also be required. This distinction is fundamental, for the basic objection to the Fermi procedure concerns the impossibility of normalizing states that obey the constraint equation

$$(\nabla \cdot \mathscr{E}(x) - j^0(x))\Psi = 0 \ .$$

With our interpretation, this is the functional differential equation

$$[-i(\delta/\delta\lambda(x)) - j^0(x)]\langle\lambda x^0|\Psi = 0 \ ,$$

in which $\lambda$, as a functional transformation parameter, does not enter the integrations required for the normalization of the state. Then no contradiction with a finite norm is implied.

The explicit verification of relativistic invariance requires the exhibition of an energy density $T^{00}(x)$ and a momentum density $T^{0k}(x)$ that obey the equal-time commutation relation (see the Appendix)

$$-i[T^{00}(x),\ T^{00}(x')] = -\left(T^{0k}(x) + T^{0k}(x')\right)\partial_k\delta(\boldsymbol{x} - \boldsymbol{x}') \ .$$

These operators must be constructed from the dynamical variables $\boldsymbol{A}^T$, $\boldsymbol{E}^T$, together with field variables of the charged fields. The exceptional difficulty of this procedure stems from the spatially nonlocal nature of the commutation relation between $\boldsymbol{A}^T$ and $\boldsymbol{E}^T$, and between $\boldsymbol{E}$ and the charge-bearing fields. We now propose to devise extended operators $\Theta^{00}$ and $\Theta^{0k}$ by the use of $\boldsymbol{A}$ and $\mathscr{E}$, with their local commutation properties, in such a way that

$$-i[\Theta^{00}(x),\ \Theta^{00}(x')] = -\left(\Theta^{0k}(x) + \Theta^{0k}(x')\right)\partial_k\delta(\boldsymbol{x} - \boldsymbol{x}')$$

and

$$[i(\delta/\delta\lambda(x)) + j^0(x),\ \Theta^{00}(x')] = 0 \ .$$

Then the operators $T^{00}$ and $T^{0k}$ defined by

$$\left(\Theta^{00}(x) - T^{00}(x)\right)\langle\lambda x^0| = 0 \ ,$$

$$\left(\Theta^{0k}(x) - T^{0k}(x)\right)\langle\lambda x^0| = 0 \ ,$$

will obey the necessary commutation relations, since

$$\Theta^{00}(x)\ \Theta^{00}(x')\langle\lambda x^0| = T^{00}(x)\ \Theta^{00}(x')\langle\lambda x^0|$$

$$= T^{00}(x)\,T^{00}(x')\langle\lambda x^0|\,.$$

It is essential to this reduction that

$$[i(\delta/\delta\lambda(x)) + j^0(x)]\, \Theta^{00}(x')\langle\lambda x^0| = 0 .$$

This is assured by the second commutator condition for $\Theta^{00}$, which is effectively the requirement of gauge-invariance.

The latter acquires a more specific physical form if we take into account the local dependence of $\Theta^{00}$ upon $\nabla\lambda$, as expressed by

$$\delta\Theta^{00}(x')/\delta\lambda(x) = \boldsymbol{j}(x')\cdot\boldsymbol{\nabla}\delta(\boldsymbol{x} - \boldsymbol{x}) .$$

The resulting commutator

$$-\, i[j^0(x),\, \Theta^{00}(x')] = -\boldsymbol{j}(x')\cdot\boldsymbol{\nabla}\delta(\boldsymbol{x} - \boldsymbol{x}')$$

also maintains its form on reducing extended operators,

$$-\, i[j^0(x),\, T^{00}(x')] = -\boldsymbol{j}(x')\cdot\boldsymbol{\nabla}\delta(\boldsymbol{x} - \boldsymbol{x}') ,$$

since both $j^0$ and $\Theta^{00}$ are gauge-invariant objects. The physical interpretation of $\boldsymbol{j}$ as the spatial current is confirmed by the following consequences of this commutation relation:

$$\partial_0 j^0(x) + \boldsymbol{\nabla}\cdot\boldsymbol{j}(x) = 0 ,$$

$$-\, i[j^0(x),\, J_{0k}] = -\,(x_0\,\partial_k - x_k\partial_0)\, j^0(x) - j_k(x) .$$

We may also note the integrated commutator

$$-\, i[T^{00}(x), Q] = -\,\boldsymbol{j}(x)\cdot\boldsymbol{\nabla}\int_V (\mathrm{d}\boldsymbol{x}')\,\delta(\boldsymbol{x} - \boldsymbol{x}') , \qquad Q = \int_V (\mathrm{d}\boldsymbol{x})\, j^0(x) ,$$

which equals zero whenever $x$ lies within (or outside) the spatial volume $V$. Thus the total charge associated with a region is unaltered by any dynamical action in its interior.

An operator such as $T^{00}$, which has been derived from $\Theta^{00}$ by eliminating the functional derivatives, $\delta/\delta\lambda$, will still be an explicit function of $\lambda$: $T^{00}(x, \lambda)$. The latter dependence disappears on performing the unitary transformation that expresses the explicit gauge variation of the state $\langle\lambda x^0|$. Thus,

$$\langle\lambda x^0| = \exp\left[i\int(\mathrm{d}\boldsymbol{x})\, j^0(x)\,\lambda(x)\right]\langle x^0|$$

and

$$T^{00}(x,\lambda)\langle\lambda x^0| = \exp\left[i\int(\mathrm{d}\boldsymbol{x})\,j^0\lambda\right]T^{00}(x)\langle x^0|\,,$$

so that the final physical operator $T^{00}(x)$ is obtained just by setting $\lambda = 0$.

This procedure is illustrated by the electrodynamics of the charged Dirac field, for which

$$\Theta^{00} = \tfrac{1}{2}(\mathscr{E}^2 + \boldsymbol{H}^2) + \tfrac{1}{2}\Psi\boldsymbol{\alpha}\cdot(-i\boldsymbol{\nabla} - eq\boldsymbol{A})\Psi - \tfrac{1}{2}im\Psi\beta\Psi\,,$$

$$\Theta^0_k = (\mathscr{E}\times\boldsymbol{H})_k + \tfrac{1}{2}\Psi(-i\boldsymbol{\nabla} - eq\boldsymbol{A})_k\Psi + \tfrac{1}{4}(\boldsymbol{\nabla}\times\tfrac{1}{2}\Psi\boldsymbol{\sigma}\Psi)_k\,.$$

The verification of the $\Theta^{00}$ commutation condition is elementary since it reduces additively to that of the individual fields. It is essential that $\mathscr{E}$ commutes with $\Psi$, as contrasted with the nonlocal commutator,

$$[\boldsymbol{E}(x),\,\Psi(x')] = \boldsymbol{\nabla}\mathscr{D}(\boldsymbol{x}-\boldsymbol{x}')\,eq\Psi(x')\,.$$

One also checks immediately the commutator that relates

$$j^0(x) = e\tfrac{1}{2}\Psi(x)q\Psi(x)$$

with

$$\boldsymbol{j}(x) = e\tfrac{1}{2}\Psi(x)q\boldsymbol{\alpha}\Psi(x)\,.$$

It is the simplicity of the Abelian gauge situation that the reduction of $\Theta^{00}$ and $\Theta^{0k}$ requires no more than the substitution of $\boldsymbol{E}$ for $\mathscr{E}$, since $\mathscr{E}$ is gauge-invariant. This, together with the replacement of $\boldsymbol{A}$ by $\boldsymbol{A}^T$ produced by setting $\lambda = 0$, gives $T^{00}$ and $T^{0k}$ in the radiation gauge.

The appearance of the radiation gauge is not accidental; it is demanded by the manner in which the gauge parameter was originally introduced. There are other possibilities, however. Thus, let us write

$$A_3(x) = \partial_3\lambda(x)\,,$$

and present the generator of field variations as

$$G_A = \int(\mathrm{d}x)(-\boldsymbol{E}^\perp)\,\delta\boldsymbol{A}^\perp + \int(\mathrm{d}\boldsymbol{x})(j^0 - \boldsymbol{\nabla}\cdot\boldsymbol{E}^\perp)\,\delta\lambda\,.$$

Here, $\boldsymbol{E}^\perp$ denotes the projection of $\boldsymbol{E}$ perpendicular to the axial direction designated as 3. There is again a separation into a generator of operator variations and a generator of changes in a gauge parameter. The latter produces, by an operator transformation, the gauge variations that must accompany the change of $A_3$, given as an explicit function of $\lambda$. We should expect the commutation relations and infinitesimal gauge transformations to be described by

$$i[A_k^\perp(x),\, E_l^\perp(x')] = \delta_{kl}\delta(\boldsymbol{x}-\boldsymbol{x}')\,, \qquad k, l = 1, 2$$

and

$$[-i(\delta/\delta\lambda(x)) - j^0(x) + \nabla\cdot\boldsymbol{E}^\perp(x)]\langle\lambda x^0| = 0\,.$$

Now we can write

$$(\boldsymbol{\mathscr{E}}(x) - \boldsymbol{E}(x))\langle\lambda x^0| = 0\,,$$

where the extended operator $\boldsymbol{\mathscr{E}}(x)$ is defined by

$$\boldsymbol{\mathscr{E}}^\perp(x) = \boldsymbol{E}^\perp(x)\,,$$

$$\mathscr{E}_3(x) = \int_{-\infty}^{\infty} \mathrm{d}x_3'\,\eta(x_3 - x_3')(-i)\,\delta/\delta\lambda(x_1x_2x_3')\,,$$

together with

$$(\partial/\partial x_3)\,\eta(x_3 - x_3') = \delta(x_3 - x_3')\,.$$

This extended operator also obeys

$$i[A_k(x),\, \mathscr{E}_l(x')] = \delta_{kl}\,\delta(\boldsymbol{x}-\boldsymbol{x}')\,,$$

and the subsequent discussion is identical with the previous one. Thus the extended operators $\Theta^{00}$ and $\Theta^{0k}$ are common to all operator gauges, and the gauge functional differential equation can always be identified with

$$(\nabla\cdot\boldsymbol{\mathscr{E}}(x) - j^0(x))\langle\lambda x^0| = 0\,.$$

The various gauges are distinguished by the manner of separating the three components of $\boldsymbol{A}$ into two operators and one explicit function of the gauge parameter together with the complementary decomposition of $\boldsymbol{\mathscr{E}}$ into two operators and a functional differential operator. With the second way of introducing the gauge parameter, the reduction of the extended operators yields $T^{00}$ and $T^{0k}$ in an axial gauge.

There is an annoying difficulty, however, which we have deferred mentioning in order to arrive most directly at the unified picture provided by the extended operator formulation. This is indicated by the construction

$$E_3(x) = \int_{-\infty}^{\infty} \mathrm{d}x_3' \, \eta(x_3 - x_3')(j^0 - \boldsymbol{\nabla}\cdot\boldsymbol{E}^{\perp})(x_1 x_2 x_3') \,,$$

and the implied asymptotic behaviour in the axial direction

$$E_3(x_1 x_2 + \infty) - E_3(x_1 x_2 - \infty) = \int_{-\infty}^{\infty} \mathrm{d}x_3 (j^0 - \boldsymbol{\nabla}\cdot\boldsymbol{E}^{\perp})(x_1 x_2 x_3) \,.$$

States in which the latter operator exhibits nonzero values are physically unacceptable. (The equation of motion for $\boldsymbol{A}^{\perp}$ would be divergent, for example.) The resulting two-dimensional constraint modifies the statement of the axial gauge with respect to field components that are independent of $x_3$. There should be no alteration in the abstract extended operator description, however, and further consideration is best reserved for a detailed discussion of the axial gauge.

## 3. – Non-Abelian gauge fields.

The field variation generator and constraint equation are (1)

$$G_{\boldsymbol{\Phi}} = \int (\mathrm{d}\boldsymbol{x})(-\boldsymbol{G})\,\delta\boldsymbol{\Phi} \,,$$

and

$$(\boldsymbol{\nabla} - i\,\text{`}t\boldsymbol{\Phi}\text{'}) \cdot \boldsymbol{G} = k^0 \,.$$

If we regard the longitudinal component of $\boldsymbol{G}$ as determined by the constraint equation, the decomposition

$$\boldsymbol{\Phi}(x) = \boldsymbol{\Phi}^T(x) + \nabla\lambda(x) \,,$$

$$\boldsymbol{G}(x) = \boldsymbol{G}^T(x) - \boldsymbol{\nabla}\int (\mathrm{d}\boldsymbol{x}')\mathscr{D}(\boldsymbol{x} - \boldsymbol{x}')\boldsymbol{\nabla}'\cdot\boldsymbol{G}(x') \,,$$

converts the generator into

$$G_{\boldsymbol{\Phi}} = \int (\mathrm{d}\boldsymbol{x})(-\boldsymbol{G}^T)\,\delta\boldsymbol{\Phi}^T + \int (\mathrm{d}\boldsymbol{x})\,\boldsymbol{\nabla}\cdot\boldsymbol{G}\,\delta\lambda \,.$$

(1) The notation is that introduced in J. Schwinger: *Phys. Rev.*, **125**, 1043 (1962).

The interpretation of this structure is given by

$$i[\Phi^T_{ka}(x),\ G^T_{lb}(x')] = (\delta_{ab}\,\delta_{kl}\ \delta(\boldsymbol{x}-\boldsymbol{x}'))^T$$

and

$$[-i(\delta/\delta\lambda(x)) - \boldsymbol{\nabla}\cdot\boldsymbol{G}(x)]\langle\lambda x^0| = 0\ .$$

The alternative axial decomposition is described by

$$\Phi_3(x) = \partial_3\lambda(x)\ ,$$

$$G_3(x) = \int_{-\infty}^{\infty} dx_3'\eta(x_3 - x_3')\ \partial_3' G(x_1 x_2 x_3')\ ,$$

and

$$G_\Phi = \int(d\boldsymbol{x})(-\boldsymbol{G}^\perp)\,\delta\boldsymbol{\Phi}^\perp + \int(d\boldsymbol{x})\ \partial_3 G_3\,\delta\lambda\ ,$$

which implies, essentially,

$$i[\Phi^\perp_{ka}(x),\ G^\perp_{lb}(x')] = \delta_{ab}\,\delta_{kl}\,\delta(\boldsymbol{x}-\boldsymbol{x}') \qquad k, l = 1, 2$$

and

$$[-i(\delta/\delta\lambda(x)) - \partial_3\, G_3(x)]\langle\lambda x^0| = 0\ .$$

We can define extended operators

$$\mathscr{G} = \boldsymbol{G}^T(x) + \boldsymbol{\nabla}\int(d\boldsymbol{x}')\mathscr{D}(\boldsymbol{x}-\boldsymbol{x}')\ i\delta/\delta\lambda(x')\ ,$$

or

$$\mathscr{G}^\perp(x) = \boldsymbol{G}^\perp(x)\ ,$$

$$\mathscr{G}_3(x) = \int_{-\infty}^{\infty} dx_3'\eta(x_3 - x_3')(-i)\ \delta/\delta\lambda(x_1 x_2 x_3')\ ,$$

such that

$$i[\Phi_{ka}(x), \mathscr{G}_{lb}(x')] = \delta_{ab}\,\delta_{kl}\,\delta(\boldsymbol{x}-\boldsymbol{x}')$$

and

$$(\mathscr{G}(x) - \boldsymbol{G}(x))\,\langle\lambda x^0| = 0\ .$$

In order to convert the constraint equation into an extended operator statement, we must remove the operator symmetrization. On using the longitudinal decomposition, we have

$$(\boldsymbol{\nabla} - i\text{`}t\boldsymbol{\Phi}\text{'})\dot{:}\boldsymbol{G} = (\boldsymbol{\nabla} - i\text{`}t\boldsymbol{\Phi}\text{'})\cdot(\boldsymbol{G} + \tfrac{1}{2}\operatorname{tr} t\,\boldsymbol{\nabla}\mathscr{D}_\Phi)\ ,$$

in which

$$-(\nabla - i\text{`}t\boldsymbol{\Phi}\text{'})\cdot\nabla\mathscr{D}_{\Phi}(\boldsymbol{x},\boldsymbol{x}') = \delta(\boldsymbol{x}-\boldsymbol{x}')$$

and

$$\operatorname{tr} t\,\nabla\mathscr{D}_{\Phi}(\boldsymbol{x},\boldsymbol{x}) = \lim_{\boldsymbol{x}'\to\boldsymbol{x}} \operatorname{tr} t\,\nabla\mathscr{D}_{\Phi}(\boldsymbol{x},\boldsymbol{x}') .$$

The required equation is, therefore,

$$[(\nabla - i\text{`}t\boldsymbol{\Phi}\text{'})\cdot(\mathscr{G} + \tfrac{1}{2}\operatorname{tr} t\,\nabla\mathscr{D}_{\Phi}) - k^0]\langle\lambda x^0| = 0$$

and the additive function of $\boldsymbol{\Phi}$ can be removed by a transformation,

$$\langle\lambda x^0| = \exp[-\tfrac{1}{2}v(\boldsymbol{\Phi})]\langle\lambda x^0\| ,$$

where

$$\delta v/\delta\boldsymbol{\Phi} = -\operatorname{tr} it\,\nabla\mathscr{D}_{\Phi}$$

is an integrable system of functional differential equations (²). Thus,

$$[(\nabla - i\text{`}t\boldsymbol{\Phi}\text{'})\cdot\mathscr{G} - k^0]\langle\lambda x^0\| = 0 ,$$

which is also applicable to the axial gauge decomposition, without distinction between $\langle|$ and $\langle\|$.

The commutation properties of $k^0$ are implied by the consistency conditions for these equations, regarded as a closed system of functional differential equations. We note that

$$[{}_a(-\nabla + i\text{`}t\boldsymbol{\Phi}\text{'})\cdot\mathscr{G}(x),\ {}_b(-\nabla + i\text{`}t\boldsymbol{\Phi}\text{'})\cdot\mathscr{G}(x')] =$$
$$= \delta(\boldsymbol{x}-\boldsymbol{x}')\sum_c t_{abc}\,{}_c(-\nabla + i\text{`}t\boldsymbol{\Phi}\text{'})\cdot\mathscr{G}(x) .$$

Accordingly, we must have

$$[k_a^0(x),\ k_b^0(x')] = \delta(\boldsymbol{x}-\boldsymbol{x}')\sum t_{abc}k_c^0(x) ,$$

under the assumption of commutativity between $k^0$ and the vector fields, if the integrability conditions are to possess the group structure that asserts the closure of the system.

We require extended operators $\Theta^{00}(x)$ and $\Theta^{0k}(x)$ that satisfy

$$-i[\Theta^{00}(x),\ \Theta^{00}(x')] = -\left(\Theta^{0k}(x) + \Theta^{0k}(x')\right)\partial_k\delta(\boldsymbol{x}-\boldsymbol{x}')$$

---

(²) J. Schwinger: *Phys. Rev.*, **130**, 402 (1963).

and

$$[(\nabla - i\,`t\boldsymbol{\Phi}'(x))\mathcal{G}^{(x)} - k^0(x),\ \Theta^{00}(x')] = 0\ ,$$

which is a statement of gauge invariance for $\Theta^{00}$.

From these we obtain operators $T^{00}$ and $T^{0k}$,

$$(\Theta^{00}(x) - T^{00}(x))\,\langle\lambda x^0\| = 0\ , \qquad (\Theta^{0k}(x) - T^{0k}(x))\,\langle\lambda x^0\| = 0\ ,$$

that obey the fundamental energy-density commutation relation. After performing the transformation that restores the states $\langle\lambda x^0|$, the operators will be Hermitian. The further transformation that exhibits the gauge variation of the states,

$$\langle\lambda x^0| = V(\lambda)\,\langle x^0|\ ,$$

will yield finally the physical energy and momentum density operators, expressed in the particular gauge that corresponds to the manner of exhibiting the numerical gauge parameters in the structure of $\Phi$.

For the example of a spin $\frac{1}{2}$ field,

$$\Theta^{00} = \tfrac{1}{2}f^2(\mathcal{G}^2 + B^2) + \tfrac{1}{2}\Psi\boldsymbol{\alpha}\cdot(-i\nabla - `T\boldsymbol{\Phi}')\Psi - \tfrac{1}{2}im\,\Psi\beta\Psi\ ,$$

$$\Theta^0_k = f^2(\mathcal{G}\times\boldsymbol{B})_k + \tfrac{1}{2}\Psi(-i\,\nabla - `T\boldsymbol{\Phi}')_k\Psi + \tfrac{1}{4}(\nabla\times\tfrac{1}{2}\Psi\boldsymbol{\sigma}\Psi)_k\ ,$$

where

$$f^2B = \nabla\times\boldsymbol{\Phi} + i(\boldsymbol{\Phi}\times t\boldsymbol{\Phi})\ .$$

Despite the change of interpretation, the detailed reduction to the operators $T^{00}$ and $T^0_k$ is performed much as before ([2]), and need not be repeated here.

Unlike the Abelian gauge situation, neither of the local quantities $k^0$ or the extended operator

$$\nabla\cdot\mathcal{G} = J^0$$

is gauge invariant. But the spatial integral of $J^0$, being a surface quantity, is invariant with respect to arbitrary gauge transformations in the interior of the region. Indeed,

$$\left[(\nabla - i`t\boldsymbol{\Phi}(x)')\cdot\mathcal{G}(x) - k^0(x),\ \int_V(\mathrm{d}\boldsymbol{x}')J^0(x')\right] = `t\,\mathcal{G}(x)'\cdot\nabla\int_V(\mathrm{d}\boldsymbol{x}')\,\delta(\boldsymbol{x}-\boldsymbol{x}')$$

vanishes for all points $x$ in the interior of $V$. From the commutator

$$-i[\Theta^{00}(x),\ J^0(x')] = -\boldsymbol{j}(x)\cdot\nabla\delta(\boldsymbol{x}-\boldsymbol{x}')\ ,$$

where

$$\boldsymbol{j} = \boldsymbol{k} + i\,`t\boldsymbol{\Phi}' \times \boldsymbol{B}$$

and

$$\boldsymbol{k} = \tfrac{1}{2}\Psi\boldsymbol{\alpha}T\Psi$$

for a spin $\frac{1}{2}$ field, we see that

$$-i\left[\Theta^{00}(x), \int_V (\mathrm{d}\boldsymbol{x}')J^0(x')\right] = -\boldsymbol{j}(x)\cdot\boldsymbol{\nabla}\int_V (\mathrm{d}\boldsymbol{x}')\,\delta(\boldsymbol{x}-\boldsymbol{x}')$$

is also zero when $x$ is in the interior of $V$. We should like to infer from these observations a corresponding operator statement about interior points of $V$,

$$[T^{00}(x), \boldsymbol{T}] = 0\,, \qquad \boldsymbol{T}_a = \int_V (\mathrm{d}\boldsymbol{x})\, j^0_a(x)\,,$$

where

$$j^0 = k^0 + i\,`t\boldsymbol{\Phi}'\cdot\boldsymbol{G} = \boldsymbol{\nabla}\cdot\boldsymbol{G}$$

is obtained by the reduction of $J^0$. Such a conclusion can only be maintained in an asymptotic sense, however, with sufficiently large volumes and for points that are sufficiently far from the surface, since the reduction of extended operators is a spatially nonlocal process. In the longitudinal gauge description, for example, we have

$$\mathscr{G}(x) - \boldsymbol{G}(x) = -\boldsymbol{\nabla}\int(\mathrm{d}\boldsymbol{x}')\mathscr{D}_{\Phi}(\boldsymbol{x}-\boldsymbol{x}')[(\boldsymbol{\nabla} - i\,`t\boldsymbol{\Phi}')\cdot\mathscr{G} - k^0](x')\,.$$

The same conclusion, that the $\boldsymbol{T}_a$ are physically meaningful only when defined over a suitably large domain, can be drawn from their commutation properties. Those of $k^0 + i`t\boldsymbol{\Phi}'\cdot\mathscr{G}$ can be expressed as

$$[j^0(x) + i\,`t\boldsymbol{\Phi}'\cdot(\mathscr{G}-\boldsymbol{G})(x),\; j^0(x') + i\,`t\boldsymbol{\Phi}'\cdot(\mathscr{G}-\boldsymbol{G})(x')] =$$
$$= -\,\delta(\boldsymbol{x}-\boldsymbol{x}')\,`t(j^0 + i\,`t\boldsymbol{\Phi}'\cdot(\mathscr{G}-\boldsymbol{G}))'\,,$$

or

$$\{[j^0(x), j^0(x')] + \delta(\boldsymbol{x}-\boldsymbol{x}')\,`tj^0(x)' + i\,`t\boldsymbol{\Phi}(x)'\cdot[\mathscr{G}(x) - \boldsymbol{G}(x), J^0(x')] -$$
$$-\, i\,`t\boldsymbol{\Phi}(x')'\cdot[\mathscr{G}(x') - \boldsymbol{G}(x'), J^0(x)]\}\langle\lambda x^0\| = 0\,.$$

The latter is a matrix equation in the internal variables apart from the last term, which requires transposition. The commutators involving $J^0$ that appear there do not vanish in general, and depend upon the particular operator gauge that is employed in their reduction. But

$$\left[\mathscr{G}(x)-\boldsymbol{G}(x),\int_V(\mathrm{d}\boldsymbol{x}')\,J^0(x')\right]\langle\lambda x^0\|\to 0$$

asymptotically, as the boundary surface of $V$ recedes. To the extent that this remains valid despite an additional integration over $x$, owing to the factor $\boldsymbol{\Phi}(x)$, we shall obtain for the total quantities $\boldsymbol{T}$ the group commutation relations

$$[\boldsymbol{T},\boldsymbol{T}]=-\text{‘}t\boldsymbol{T}\text{’}\,.$$

#### 4. – Gravitational field.

The method discussed here has been developed as a compact way of expressing known results. But when one turns to the problem of the gravitational field, it becomes the indispensible instrument for the discovery of a consistent formulation. We shall present those considerations separately.

## Appendix

The charge-density commutation relations arise from the integrability conditions for gauge transformations. In the same way, the energy-density commutation relations express integrability conditions for arbitrary co-ordinate transformations. The general theory is not needed, however, to derive the commutation relations of special relativistic field theory. It suffices to consider the generator of infinitesimal displacements for a spacelike surface,

$$G_x=\int_\sigma \mathrm{d}\sigma_\mu\,T^{\mu\nu}\,\delta x_\nu\,,$$

and then only for infinitesimal deviations from plane surfaces.

Let

$$x^0=\sigma(\boldsymbol{x})\,,$$

be the equation of a spacelike surface, in which $\sigma(\boldsymbol{x})$ only differs infinitesimally from a constant. The directed element of area normal to this surface is

$$\mathrm{d}\sigma_\mu=n_\mu(\mathrm{d}\boldsymbol{x})\,,\qquad n_\mu=\partial_\mu\big(x^0-\sigma(\boldsymbol{x})\big)\,,$$

or

$$d\sigma_0 = (d\boldsymbol{x}) , \qquad d\sigma_k = -(d\boldsymbol{x})\,\partial_k\sigma(\boldsymbol{x}) .$$

The generator of a transformation in which only time co-ordinates are displaced is then given by

$$G_x = \int (d\boldsymbol{x})(-\delta\sigma(\boldsymbol{x}))\big(T^{(0)(0)}(x) + T^{0k}(x)\,\partial_k\sigma(\boldsymbol{x})\big) ,$$

where we have introduced the normal component of $T^{\mu\nu}$,

$$T^{(0)(0)} = n_\mu T^{\mu\nu} n_\nu .$$

The corresponding differential equation for a state,

$$\delta\langle\sigma| = i\langle\sigma|G_x = iG_x\langle\sigma| ,$$

can be presented as the functional differential equation

$$[i(\delta/\delta\sigma(x)) - \big(T^{(0)(0)}(x) + T^{0k}(x)\,\partial_k\sigma(\boldsymbol{x})\big)]\langle\sigma| = 0 .$$

The analogue of the class of physical systems for which the charge density does not depend explicitly upon the gauge parameter is the distinguished set for which $T^{(0)(0)}$ is not explicitly dependent upon the spacelike surface. The integrability conditions of such systems, evaluated for $\boldsymbol{\nabla}\sigma = 0$, yield the anticipated result

$$-i[T^{00}(x), T^{00}(x')] = i[i\delta/\delta\sigma(x) - T^{0k}(x)\,\partial_k\sigma(\boldsymbol{x}),\ i\delta/\delta\sigma(x') - T^{0k}(x')\,\partial'_k\sigma(\boldsymbol{x}')] =$$
$$= -\big(T^{0k}(x) + T^{0k}(x')\big)\,\partial_k\delta(\boldsymbol{x}-\boldsymbol{x}') .$$

According to an earlier discussion ([3]), this class contains only fields of spin 0, $\frac{1}{2}$ or 1.

---

([3]) J. SCHWINGER: *Phys. Rev.*, **130**, 800 (1963).

---

## RIASSUNTO (*)

Combinando le variabili dell'operatore fisico con i parametri matematici di gruppo, si superano le difficoltà fisiche e matematiche dell'usuale trattazione delle teorie di campo che ammettono gruppi di trasformazione ad infinite dimensioni. Si discute anche, per le teorie di gauge non abeliane, il significato fisico limitato ed asintotico delle proprietà interne totali, ed una deduzione delle condizioni del commutatore della densità di energia.

---

(*) *Traduzione a cura della Redazione.*

# Commutation Relations and Conservation Laws

JULIAN SCHWINGER*
*Harvard University, Cambridge, Massachusetts*
(Received 16 October 1962)

The response of a physical system to external electromagnetic and gravitational fields, as embodied in the electric current and stress tensor conservation laws, is used to derive the equal-time commutation relations for charge density and energy density.

## INTRODUCTION

AMONG the more important physical properties in relativistic quantum field theory are the conserved local quantities, such as the electric charge flux vector $j^\mu(x)$ and the stress tensor $T^{\mu\nu}(x)$. In order to answer questions about the simultaneous measurability of these quantities one needs the commutation relations of the operators on a space-like surface or, more specifically, at a common time. The physical independence of different points on a space-like surface guarantees the compatability of any associated localized physical properties. That is a general assertion of commutability under such circumstances. A complete treatment of equal-time commutators has been lacking, however. Thus, although it has long been remarked that the electric charge density at all spatial points obeys

$$x^0=x^{0\prime}: \quad [j^0(x),j^0(x')]=0,$$

a corresponding statement about the energy density had not been recorded until it was observed,[1] for a particular system, that

$$x^0=x^{0\prime}: \quad -i[T^{00}(x),T^{00}(x')] = -(T^{0k}(x)+T^{0k}(x'))\partial_k\delta(\mathbf{x}-\mathbf{x}').$$

It is our intention to supply a general basis for this and other equal-time commutators.

The measurement theory of the electric current vector and the stress tensor is founded upon the specific dynamical nature of these properties as the sources of the electromagnetic and gravitational fields, respectively. More precisely, we exploit the reciprocal dynamical aspect of $j^\mu$ and $T^{\mu\nu}$ whereby they determine the response of a system to external electromagnetic and gravitational fields. What is characteristic of these dynamical agencies, and equivalent to the existence of the local conservation laws for the properties of interest, is the freedom in description associated with gauge and coordinate transformations.

## ELECTRIC CURRENT

The electric current provides the simpler illustration of the method. Let $W$ be the action operator of all charge-bearing fields $\chi(x)$, excluding the purely electromagnetic action term. The vector potential $A_\mu(x)$ appears as an external quantity in this action operator,

$$W=\int (dx)\,\mathfrak{L}_{\text{ch}}(\chi,A_\mu),$$

and the infinitesimal numerical variation

$$\delta_A W=\int (dx)\,j^\mu(x)\delta A_\mu(x)$$

defines the electric current vector. The requirement of gauge invariance, applied to the infinitesimal gauge transformation

$$\delta A_\mu(x)=-\partial_\mu\delta\lambda(x),$$

yields the charge conservation equation

$$\partial_\mu j^\mu(x)=0.$$

The gravitational potential $g_{\mu\nu}$ replaces $A_\mu$ in the analogous discussion of $T^{\mu\nu}$. For that circumstance the use of an external field is quite justified by the weak dynamical influence of the gravitational field in a special relativistic context. This argument does not apply to the electromagnetic field, of course, and we must remove the implication that a weak-coupling treatment of the electromagnetic field is necessarily involved. To do that we have only to rephrase our procedure by replacing $W$ with the total action operator

$$W=\int (dx)[\mathfrak{L}_{\text{ch}}(\chi,\,A_\mu+A_\mu')+\mathfrak{L}_{\text{em}}(A_\mu,F_{\mu\nu})]$$

in which $A_\mu'(x)$ is an arbitrary numerical external potential. Infinitesimal variations of the latter can now be used to define the electric current vector while incorporating the full dynamical effect of the electromagnetic field.

The charge conservation equation

$$\partial_0 j^0(x)=-\partial_k j^k(x)$$

is an example of a relationship between operators, of the type

$$\partial_0 A(x)=B(x),$$

that is maintained for arbitrary values of certain parameters—the external potentials, in this example. Now, the quantum action principle,

$$\delta\langle\sigma_1|\sigma_2\rangle=i\langle\sigma_1|\delta W_{12}|\sigma_2\rangle,$$

* Supported by the Air Force Office of Scientific Research (ARDC) under contract A. F. 49(638)-589.
[1] J. Schwinger, Phys. Rev. 127, 324 (1962).



Reprinted from *The Physical Review* **130**, 406–409 (1963).

applies, in particular, to infinitesimal alterations in the structure of the Lagrange function, as realized by variations of numerical parameters. It is a corollary of the action principle that

$$\delta\langle\sigma_1|F(x)|\sigma_2\rangle=\langle\sigma_1|\delta' F(x)+i\int_{\sigma_2}^{\sigma_1}(dx')(F(x)\delta\mathfrak{L}(x'))_+|\sigma_2\rangle,$$

where $\delta' F(x)$ refers to an explicit dependence of the operator $F(x)$ upon these parameters. To maintain the relationship between $A(x)$ and $B(x)$ then requires that

$$\partial_0\left[\delta' A(x)+i\int(dx')(A(x)\delta\mathfrak{L}(x'))_+\right]=\delta' B(x)+i\int(dx')(B(x)\delta\mathfrak{L}(x'))_+.$$

But, the time derivative of the ordered product is given by

$$\partial_0(A(x)\delta\mathfrak{L}(x'))_+=(\partial_0 A(x)\delta\mathfrak{L}(x'))_+ +\delta(x^0-x^{0\prime})[A(x),\delta\mathfrak{L}(x')],$$

and therefore,

$$-i\int(d\mathbf{x}')[A(x),\delta\mathfrak{L}(x')]|_{x^0=x^{0\prime}}=\partial_0\delta' A(x)-\delta' B(x).$$

This statement supplies a general foundation for equal-time commutation relations. Note, incidentally, that $A(x)$ cannot depend explicitly upon the parameters unless $B(x)$ correspondingly involves the time derivative of these parameters. In the absence of such a dependence, the right-hand side of the above equation is just $-\delta' B(x)$.

When a number of parameters are involved, the explicit dependence upon the parameters is subject to certain integrability conditions or reciprocity relations. We illustrate this with the continuum of parameters constituted by the external potential $A_\mu(x)$. The calculation of the second variation for a transformation function proceeds from the action principle as

$$\delta^2\langle\sigma_1|\sigma_2\rangle=\delta\left[i\int(dx)\delta A_\mu(x)\langle\sigma_1|j^\mu(x)|\sigma_2\rangle\right]$$
$$=\int(dx)(dx')\delta A_\mu(x)\delta A_\nu(x')[-\langle\sigma_1|(j^\mu(x)j^\nu(x'))_+|\sigma_2\rangle+i\langle\sigma_1|\delta' j^\mu(x)/\delta A_\nu(x')|\sigma_2\rangle],$$

and the necessary symmetry of this result supplies the reciprocity relation

$$\delta' j^\mu(x)/\delta A_\nu(x')=\delta' j^\nu(x')/\delta A_\mu(x).$$

In order to obtain explicit equal-time commutation relations for components of the electric current vector, we must be somewhat more specific about the dependence of the current upon the external potential. The major consideration here is locality. The current usually does not involve the potential at relatively space-like points, but we only insist here that $j_\mu(x)$ does not refer to the potential at neighboring times, which is to say that it does not contain the time derivative of the potential. That restriction defines a certain class of electric charge-bearing physical systems (which may well be without exception). The immediate implication from the conservation equation is that $j^0(x)$ cannot be an explicit function of the external potential. The reciprocity relation then asserts that

$$\delta' j^k(x)/\delta A_0(x')=\delta' j^0(x')/\delta A_k(x)=0.$$

The equal-time commutation relation supplied by the conservation equation for charge now reads

$$-i\int(d\mathbf{x}')[j^0(x),j^\mu(x')]\delta A_\mu(x')=\partial_k\int(d\mathbf{x}')[\delta_3' j^k(x)/\delta A_l(x')]\delta A_l(x'),$$

and therefore, $(x^0=x^{0\prime})$

$$[j^0(x),j^0(x')]=0,$$
$$-i[j^0(x),j^l(x')]=\partial_k[\delta_3' j^k(x)/\delta A_l(x')]=\partial_k[\delta_3' j^l(x')/\delta A_k(x)].$$

The variational derivatives that appear here are the three-dimensional ones defined by

$$\delta' j^k(x)/\delta A_l(x')=\delta(x^0-x^{0\prime})\delta_3' j^k(x)/\delta A_l(x').$$

Despite the use of an external potential, these commutators are assertions about an isolated physical system, if the potential is set equal to zero after differentiation.

One should recognize that an explicit dependence of the current upon an external potential occurs for all physical systems. Let us use the conservation equation again, and convert the second commutator into

$$[j^0(x),\,-i\partial_0 j^0(x')]=-\partial_k\partial_l'[\delta_3' j^l(x')/\delta A_k(x)],$$

which is symmetrical in the two points $x$ and $x'$. A contradiction to the hypothetical vanishing of the right-hand member of this equation arises from the positiveness exhibited by the vacuum expectation value of the left-hand member. Thus, if $\varphi(x)$ is an arbitrary real function, with which one forms the Hermitian operator

$$J(x^0)=\int(d\mathbf{x})\varphi(x)j^0(x),$$

the equation of motion

$$i\partial_0 j^0(x)=[j^0(x),P^0],$$

combined with the null energy of the vacuum, yields

$$\int (d\mathbf{x})(d\mathbf{x}')\varphi(x)\langle[j^0(x), -i\partial_0 j^0(x')]\rangle\varphi(x') = 2\langle JP^0J\rangle \geq 0.$$

Now it is essential to call upon the relativistic principle that any sufficiently localized act must excite the vacuum, which implies that functions $\varphi(x)$ surely exist for which the states $\langle J$ have an energy expectation value greater than zero. It can be shown that the explicit dependence of the current on the potential is completely local,

$$-\delta_3' j^k(x)/\delta A_l(x') = \delta(\mathbf{x}-\mathbf{x}')j^{kl}(x).$$

The expectation value of the symmetrical tensor $j^{kl}(x)$ in the invariant vacuum state is of the form

$$\langle j^{kl}(x)\rangle = \delta^{kl}C.$$

Accordingly,

$$2\langle JP^0J\rangle = C\int (d\mathbf{x})[\nabla\varphi(x)]^2$$

and the constant $C$ must be a positive number,

$$C>0,$$

which shows, incidentally, that a positive energy expectation value is realized for every nonconstant function $\varphi(x)$.

## STRESS TENSOR

Through the agency of an external gravitational field, the stress-energy-momentum tensor $T^{\mu\nu}(x)$ is defined by the variational equation

$$\delta_g W = \int (dx)(-g)^{1/2}\tfrac{1}{2}T^{\mu\nu}\delta g_{\mu\nu},$$

in which

$$g = \det g_{\mu\nu}.$$

The role formerly played by gauge invariance is now taken over by the requirement of general coordinate invariance. The infinitesimal coordinate transformation

$$\bar{x}^\mu = x^\mu + \delta x^\mu(x)$$

induces

$$\delta g_{\mu\nu} = \delta x^\lambda \partial_\lambda g_{\mu\nu} + g_{\lambda\nu}\partial_\mu \delta x^\lambda + g_{\mu\lambda}\partial_\nu \delta x^\lambda,$$

from which we infer the extended conservation equations

$$\partial_\mu[(-g)^{1/2}g_{\lambda\nu}T^{\mu\nu}] = \tfrac{1}{2}(-g)^{1/2}T^{\mu\nu}\partial_\lambda g_{\mu\nu}.$$

Alternative forms are

$$\partial_\mu(g_{\lambda\nu}T^{\mu\nu}) = \tfrac{1}{2}T^{\mu\nu}(\partial_\lambda g_{\mu\nu} - g_{\lambda\nu}g^{\alpha\beta}\partial_\mu g_{\alpha\beta})$$

and

$$\partial_\mu[(-g)^{1/2}T^{\mu\nu}] = -(-g)^{1/2}T^{\alpha\beta}\Gamma_{\alpha\beta}{}^\nu$$

where, of course,

$$\Gamma_{\alpha\beta}{}^\nu = \tfrac{1}{2}g^{\nu\lambda}[\partial_\alpha g_{\beta\lambda} + \partial_\beta g_{\alpha\lambda} - \partial_\lambda g_{\alpha\beta}].$$

As a first application, consider an infinitesimal deviation, $\delta g_{\mu\nu}(x)$, from the Minkowski metric. The extended conservation equations can then be presented as

$$\partial_\lambda \mathbf{T}^{\lambda\nu} = -\tfrac{1}{2}\delta g_{\lambda\kappa}\partial^\nu T^{\lambda\kappa},$$

where

$$\mathbf{T}^{\mu\nu} = (-g)^{1/2}T^{\mu\nu} - g^{\mu\nu}\tfrac{1}{2}\delta g_{\lambda\kappa}T^{\lambda\kappa} + T^{\mu\lambda}\delta g_{\lambda\kappa}g^{\nu\kappa}.$$

Let us also observe that

$$\partial_\lambda(x^\mu \mathbf{T}^{\lambda\nu} - x^\nu \mathbf{T}^{\lambda\mu}) = -\tfrac{1}{2}\delta g_{\lambda\kappa}(x^\mu\partial^\nu - x^\nu\partial^\mu)T^{\lambda\kappa} + \mathbf{T}^{\mu\nu} - \mathbf{T}^{\nu\mu},$$

in which

$$\mathbf{T}^{\mu\nu} - \mathbf{T}^{\nu\mu} = \tfrac{1}{2}\delta g_{\lambda\kappa}(T^{\mu\lambda}g^{\nu\kappa} - T^{\nu\lambda}g^{\mu\kappa} + T^{\mu\kappa}g^{\nu\lambda} - T^{\nu\kappa}g^{\mu\lambda}).$$

An integration over all three-dimensional space removes the space derivative terms and yields

$$\partial_0\int (d\mathbf{x})\mathbf{T}^0{}_\nu = -\int (d\mathbf{x})\tfrac{1}{2}\delta g_{\lambda\kappa}\partial_\nu T^{\lambda\kappa},$$

$$\partial_0\int (d\mathbf{x})(x_\mu \mathbf{T}^0{}_\nu - x_\nu \mathbf{T}^0{}_\mu) = -\int (d\mathbf{x})[\tfrac{1}{2}\delta g_{\lambda\kappa}(x_\mu\partial_\nu - x_\nu\partial_\mu)T^{\lambda\kappa} - \mathbf{T}_{\mu\nu} + \mathbf{T}_{\nu\mu}].$$

These forms lead immediately to commutation relations between the components of the stress tensor and the generators of the special relativistic infinitesimal coordinate transformations,

$$P_\nu = \int (d\mathbf{x})T^0{}_\nu, \quad J_{\mu\nu} = \int (d\mathbf{x})(x_\mu T^0{}_\nu - x_\nu T^0{}_\mu),$$

namely,

$$[T^{\lambda\kappa}(x), P_\nu] = -i\partial_\nu T^{\lambda\kappa}(x)$$

and

$$[T^{\lambda\kappa}, J_{\mu\nu}] = -i(x_\mu\partial_\nu - x_\nu\partial_\mu)T^{\lambda\kappa} + i(\delta_\nu{}^\lambda T_\mu{}^\kappa - \delta_\mu{}^\lambda T^\kappa{}_\nu + \delta_\nu{}^\kappa T_\mu{}^\lambda - \delta_\mu{}^\kappa T^\lambda{}_\nu).$$

These commutators, representing the transformation properties of the stress tensor, produce, through integration, the commutation relations of the ten infinitesimal generators of the inhomogeneous Lorentz group. In this way special relativistic kinematics emerges from gravitational dynamics.

To obtain more detailed information, let us choose the special gravitational field

$$g_{kl} = \delta_{kl}, \quad g_{0k} = 0, \quad -g_{00}(x) \neq 1,$$

so that properties of the energy density can be inferred by variation of $g_{00}(x)$. The extended conservation equations are

$$\partial_0[(-g_{00})T^{00}]=-\partial_k[(-g_{00})T^{0k}]+\tfrac{1}{2}T^{0k}\partial_k g_{00}$$

and

$$\partial_0[(-g_{00})^{1/2}T^{0k}]=-\partial_l[(-g_{00})^{1/2}T^{kl}]+\tfrac{1}{2}(-g_{00})^{1/2}T^{00}\partial_k g_{00},$$

where each form is chosen to avoid the explicit appearance of $\partial_0 g_{00}$. We confine our attention to the class of physical systems which are such that $T^{kl}$ does not contain explicitly the time derivative of $g_{00}$, although it may be an explicit function of $g_{00}$ at the same time.[2] It can be concluded that neither $(-g_{00})T^{00}$ nor $(-g_{00})^{1/2}T^{0k}$ are explicit functions of $g_{00}$ for this distinguished class of material system, which is to say that these local quantities are the same functions of the fundamental dynamical variables as in the absence of an external gravitational field. The equation of motion for $(-g_{00})T^{00}$ now implies the equal-time commutator

$$-i\left[(-g_{00}T^{00})(x),\int(d\mathbf{x}')(-g_{00}T^{00})(x')\delta(-g_{00}(x'))^{1/2}\right]$$
$$=-\partial_k[(-g_{00})^{1/2}T^{0k}(x)\delta(-g_{00}(x))^{1/2}]-(-g_{00})^{1/2}T^{0k}(x)\partial_k\delta(-g_{00}(x))^{1/2},$$

where, it is noted, there is no explicit dependence upon $g_{00}(x)$, which indicates the consistency of the physical restriction. On setting $-g_{00}=1$, we obtain

$$-i[T^{00}(x),T^{00}(x')]=-[T^{0k}(x)+T^{0k}(x')]\partial_k\delta(\mathbf{x}-\mathbf{x}').$$

This derivation of the energy density commutator condition, for a class of physical systems, supplies a simple and general basis for what may well be considered the most fundamental equation of relativistic quantum field theory.

[2] In fact, $T^{kl}$ must be an explicit local function of the second spatial derivatives of $g_{00}$.

# Energy and Momentum Density in Field Theory*

JULIAN SCHWINGER
*Harvard University, Cambridge, Massachusetts*
(Received 27 November 1962)

It is shown that the energy density commutator condition in its simplest form is valid for interacting spin 0, ½, 1 field systems, but not for higher spin fields. The action principle is extended, for this purpose, to arbitrary coordinate frames. There is a discussion of four categories of fields and some explicit consideration of spin $\frac{3}{2}$ as the simplest example that gives additional terms in the energy density commutator. As the fundamental equation of relativistic quantum field theory, the commutator condition makes explicit the greater physical complexity of higher spin fields.

## INTRODUCTION

AFTER it had been noticed[1] that the energy and momentum density of a particular field system obeyed the equal time commutation relation

$$-i[T^{00}(x),T^{00}(x')]=-(T^{0k}(x)+T^{0k}(x'))\partial_k\delta(\mathbf{x}-\mathbf{x}'),$$

a general proof was constructed[2] by considering the response to an external gravitational field. The commutator condition applies to all systems for which $(-g^{00})T^{00}$ and $(-g_{00})^{1/2}T^{0k}$ are independent of the gravitational field, when it is of the special type

$$g_{kl}=\delta_{kl},\quad g_{0k}=0,\quad -g_{00}(x)\neq 1.$$

How extensive is this distinguished class of physical systems? We shall find that fields with spin 0, ½, 1 are included, but not fields of larger spin. Thus, higher spin fields can now be characterized, not merely as mathematically more complicated structures, but by their greater physical complexity.

The technical problem encountered here is the extension of the action principle to arbitrary coordinate frames, subject to the requirement of coordinate invariance. Even more is involved for, as Weyl[3] was the first to recognize, the description of spin entails the introduction, at each point, of an independent Lorentz coordinate frame, combined with the demand of invariance under local Lorentz transformations. This is the ultimate expression of the local field concept.

The relation between the local and the global coordinate systems is conveyed by a family of vector fields $e_a{}^\mu(x)$, which respond to general coordinate transformations and local Lorentz transformations as

$$\bar{e}_a{}^\mu(\bar{x})=(\partial\bar{x}^\mu/\partial x^\nu)e_a{}^\nu(x),$$

and

$$\bar{e}_a{}^\mu(x)=l_a{}^b(x)e_b{}^\mu(x),$$

respectively. In the latter, the matrix $l$ obeys the Lorentz invariance condition

$$l^Tgl=g,$$

where $g^{ab}$ is the Minkowski metric tensor, which we take to have the value $-1$ for its temporal component. The inverse vector set $e_\mu{}^a(x)$,

$$e_\mu{}^a e_b{}^\mu=\delta_b{}^a,\quad e_\mu{}^a e_a{}^\nu=\delta_\mu{}^\nu,$$

* Supported in part by the Air Force Office of Scientific Research under contract number A.F. 49(638)-589.

[1] J. Schwinger, Phys. Rev. 127, 324 (1962).

[2] J. Schwinger, Phys. Rev. 130, 406 (1963).

[3] H. Weyl, Z. Physik. 56, 330 (1929). Our approach derives most directly from this source rather than the later developments of Schrödinger, Bargmann, Belinfante, and others.

Reprinted from *The Physical Review* **130**, 800–805 (1963).

has analogous transformation properties,

$$\bar e_\mu{}^a(\bar x) = (\partial x^\nu/\partial \bar x^\mu)\, e_\nu{}^a(x),$$
$$\bar e_\mu{}^a(x) = l^a{}_b(x)\, e_\mu{}^b(x).$$

In particular,

$$(dx)\,\det e_\mu{}^a(x), \quad (dx)=dx^0dx^1dx^2dx^3,$$

is an invariant for both types of transformations (with det $l=+1$).

The change of the invariant action operator $W$ for an infinitesimal alteration of the $e_\mu{}^a(x)$ can be written as

$$\delta_e W = \int (dx)(\det e_\nu{}^b(x))\,\delta e_\mu{}^a(x)\, T^\mu{}_a(x),$$

which defines the set of Hermitian operators $T^\mu{}_a$. If $\delta e_\mu{}^a$ refers to an infinitesimal local Lorentz transformation,

$$\delta e_\mu{}^a(x) = \delta\omega^a{}_b(x)\, e_\mu{}^b(x),$$
$$\delta\omega_{ab} = -\delta\omega_{ba},$$

local Lorentz invariance requires that

$$T^{ab}(x)=T^{ba}(x),$$

where

$$T^{ab}=e_\mu{}^a T^{\mu b}.$$

The same symmetry property applies to the tensor

$$T^{\mu\nu} = e_a{}^\mu e_b{}^\nu T^{ab} = T^{\mu b} e_b{}^\nu = T^{\nu\mu}.$$

That symmetry can be exploited by writing

$$\delta e_\mu{}^a T^\mu{}_a = \delta e_\mu{}^a e_{\nu a} T^{\mu\nu} = \tfrac12 \delta(e_\mu{}^a e_{\nu a})\, T^{\mu\nu},$$

which introduces the symmetrical tensor

$$g_{\mu\nu} = e_\mu{}^a g_{ab} e_\nu{}^b = e_\mu{}^a e_{\nu a}.$$

Since

$$g = \det g_{\mu\nu} = -(\det e_\mu{}^a)^2,$$

we can conclude that

$$\delta_e W = \delta_g W = \int (dx)(-g)^{1/2}\tfrac12 \delta g_{\mu\nu} T^{\mu\nu},$$

which makes contact with the more familiar definition of the stress tensor.

### ACTION PRINCIPLE

We take the action operator to have the following standard form in a Minkowski space:

$$W=\int (dx)[\tfrac14(\chi A^\mu\partial_\mu\chi - \partial_\mu\chi A^\mu\chi) - \mathcal{H}(\chi)],$$

where the $A^\mu$ are four constant numerical matrices and the $\chi(x)$ are a set of Hermitian fields. To generalize this to arbitrary coordinate frames, we write

$$W=\int (dx)\,\det e_\mu{}^a(x)\,\mathcal{L}(x),$$

where $\mathcal{L}$ is tentatively constructed as

$$\mathcal{L}(x) = \tfrac14 e_a{}^\mu(x)[\chi(x)A^a\partial_\mu\chi(x) - \partial_\mu\chi(x)A^a\chi(x)] - \mathcal{H}(\chi(x)).$$

Here the $A^a$ are the same set of four constant numerical matrices now defined relative to the local coordinate system. The requirement of invariance under general coordinate transformations is trivially satisfied if the $\chi(x)$ behave as scalars,

$$\bar\chi(\bar x)=\chi(x).$$

But what of invariance under local Lorentz transformations?

For such a transformation,

$$\bar e_a{}^\mu(x) = l_a{}^b(x)\, e_b{}^\mu(x),$$

we have

$$\bar\chi(x) = L(x)\chi(x),$$

where

$$\mathcal{H}(\bar\chi)=\mathcal{H}(\chi)$$

and

$$L^T A^a L = l^a{}_b A^b$$

ensure invariance if the Lorentz transformation does not depend upon position. But our tentative Lagrange function $\mathcal{L}$ is not invariant under arbitrary local Lorentz transformations:

$$\bar{\mathcal{L}} - \mathcal{L} = \tfrac14 e_a{}^\mu[\chi A^a(L^{-1}\partial_\mu L)\chi - \chi(L^{-1}\partial_\mu L)^T A^a\chi].$$

To identify the structure of $L^{-1}\partial_\mu L$, we remark that two infinitesimally neighboring Lorentz transformations are connected by an infinitesimal transformation,

$$l_{ab}+\delta l_{ab} = l_{ac}(\delta_b{}^c + \delta\omega^c{}_b),$$

where

$$\delta\omega_{ab} = l^c{}_a \delta l_{cb} = -\delta\omega_{ba}.$$

The corresponding group composition property of $L$, expressed in terms of the spin matrices, is

$$L+\delta L = L(1+\tfrac12 i\delta\omega_{ab}S^{ab}),$$

and therefore

$$L^{-1}\partial_\mu L = \tfrac12 i S^{ab} l^c{}_a \partial_\mu l_{cb}.$$

This result indicates that the structure of the Lagrange function must be modified into

$$\mathcal{L} = \tfrac14 e_a{}^\mu[\chi A^a(\partial_\mu - \tfrac12 i\omega_{\mu bc}S^{bc})\chi - (\partial_\mu - \tfrac12 i\omega_{\mu bc}S^{bc})\chi A^a\chi] - \mathcal{H}(\chi),$$

where the vector fields $\omega_{\mu ab}=-\omega_{\mu ba}$ are given such local Lorentz transformation properties that

$$L^{-1}(\partial_\mu - \tfrac12 i\bar\omega_{\mu ab}S^{ab})L = \partial_\mu - \tfrac12 i\omega_{\mu ab}S^{ab}.$$

The required transformation law is

$$\bar{\omega}_{\mu ab} = l_a{}^{a'} l_b{}^{b'} (\omega_{\mu a'b'} + l_{ca'} \partial_\mu l^c{}_{b'})$$
$$= l_a{}^{a'} l_b{}^{b'} \omega_{\mu a'b'} + l_b{}^{b'} \partial_\mu l_{ab'}.$$

An alternative statement, in terms of the functions

$$\omega_{abc}(x) = e_a{}^\mu(x) \omega_{\mu bc}(x),$$

is

$$\bar{\omega}_{abc} = l_a{}^{a'} l_b{}^{b'} l_c{}^{c'} \omega_{a'b'c'} + l_a{}^{a'} l_c{}^{c'} e_{a'}{}^\mu \partial_\mu l_{bc'}.$$

Now let us observe the following local Lorentz transformation property:

$$(\bar{e}_a{}^\mu \partial_\mu \bar{e}_b{}^\nu) \bar{e}_{\nu c} = l_a{}^{a'} l_b{}^{b'} l_c{}^{c'} (e_{a'}{}^\mu \partial_\mu e_{b'}{}^\nu) e_{\nu c'} + l_a{}^{a'} l_c{}^{c'} e_{a'}{}^\mu \partial_\mu l_{bc'}.$$

Accordingly,

$$\omega_{abc} - (e_a{}^\mu \partial_\mu e_b{}^\nu) e_{\nu c} = \Gamma_{abc}$$

has the simple behavior

$$\bar{\Gamma}_{abc} = l_a{}^{a'} l_b{}^{b'} l_c{}^{c'} \Gamma_{a'b'c'}.$$

One should not be mislead by the notation to conclude that $\Gamma_{abc}(x)$, like $\omega_{abc}(x)$, is a scalar under general coordinate transformations. Rather

$$\bar{\Gamma}_{abc}(\bar{x}) - \Gamma_{abc}(x) = -e_a{}^\mu e_b{}^\lambda (\partial_\mu \partial_\lambda \bar{x}^\nu) e_{\kappa c} (\partial x^\kappa / \partial \bar{x}^\nu),$$

which also shows that the combination $\Gamma_{abc} - \Gamma_{bac}$ is a scalar. The antisymmetry of $\omega_{abc}$ in $b$ and $c$ implies that

$$\Gamma_{abc} + \Gamma_{acb} = -e_a{}^\mu [(\partial_\mu e_b{}^\nu) e_{\nu c} - (\partial_\mu e_{\nu b}) e_c{}^\nu]$$
$$= e_a{}^\mu e_b{}^\lambda e_c{}^\nu \partial_\mu g_{\nu\lambda}$$

or, with an evident definition,

$$\Gamma_{\mu\nu\lambda} + \Gamma_{\mu\lambda\nu} = \partial_\mu g_{\nu\lambda}.$$

If we adopt the invariant symmetry restriction

$$\Gamma_{\mu\nu\lambda} - \Gamma_{\nu\mu\lambda} = 0,$$

we can construct $\Gamma_{\mu\nu\lambda}$ explicitly. It is the Christoffel symbol

$$\Gamma_{\mu\nu\lambda} = \tfrac{1}{2}(\partial_\mu g_{\nu\lambda} + \partial_\nu g_{\mu\lambda} - \partial_\lambda g_{\mu\nu}).$$

The same symmetry restriction in the form

$$\Gamma_{abc} - \Gamma_{bac} = 0$$

gives

$$\omega_{abc} - \omega_{bac} = \Omega_{cab},$$

where

$$\Omega_{cab} = e_{\nu c} [e_a{}^\mu \partial_\mu e_b{}^\nu - e_b{}^\mu \partial_\mu e_a{}^\nu]$$
$$= -\Omega_{cba}.$$

From this we derive

$$\omega_{abc} = \tfrac{1}{2}[\Omega_{bca} + \Omega_{cab} - \Omega_{abc}].$$

## ENERGY AND MOMENTUM DENSITY

It suffices, for present purposes, to choose the field $e_a{}^\mu(x)$ as

$$e_{(0)}{}^0(x) \neq 1, \quad e_{(k)}{}^0 = 0,$$
$$e_{(0)}{}^k(x) \simeq 0, \quad e_{(l)}{}^k = \delta_l{}^k,$$

where (0) and $(k)$ are local coordinate indices and $\simeq 0$ implies that $e_{(0)}{}^k$ differs only infinitesimally from zero. The inverse vector system is

$$e_0{}^{(0)} = (e_{(0)}{}^0)^{-1}, \quad e_k{}^{(0)} = 0,$$
$$e_0{}^{(k)} = -e_0{}^{(0)} e_{(0)}{}^k \simeq 0, \quad e_l{}^{(k)} = \delta_l{}^k,$$

and

$$-g_{00} = (e_0{}^{(0)})^2, \quad g_0{}^k = e_0{}^{(k)}, \quad g_{kl} = \delta_{kl}.$$

The dependence of the action operator upon the variable functions is given by

$$\delta_e W = \int (dx) [-T^{(0)(0)} \delta e_0{}^{(0)} + T^{(0)}{}_{(k)} \delta e_0{}^{(k)}],$$

where, with $g_{0k} = 0$,

$$T^{(0)(0)} = (-g_{00}) T^{00}, \quad T^{(0)}{}_{(k)} = (-g_{00})^{1/2} T^0{}_k$$

are just the quantities of interest.

The elements of $\Omega_{abc}$ that differ from zero are

$$\Omega_{(0)(k)(0)} = -\Omega_{(0)(0)(k)} = e_{(0)}{}^0 \partial_k e_0{}^{(0)},$$

and

$$\Omega_{(l)(k)(0)} = -\Omega_{(l)(0)(k)} = -e_{(0)}{}^0 \partial_k e_0{}^{(l)}.$$

The action operator is now given by

$$W = \int (dx) [\tfrac{1}{4}(\chi A^{(0)} \partial_0 \chi - \partial_0 \chi A^{(0)} \chi) - e_0{}^{(0)} T^{(0)(0)} + e_0{}^{(k)} T^{(0)}{}_{(k)}],$$

where

$$T^{(0)(0)} = \mathcal{H} - \tfrac{1}{4}(\chi A^{(k)} \partial_k \chi - \partial_k \chi A^{(k)} \chi) - \tfrac{1}{4} i \partial_k [\chi (A^{(0)} S^{(0)(k)} - S^{(0)(k)T} A^{(0)}) \chi],$$

and

$$T^{(0)}{}_{(k)} = -\tfrac{1}{4}(\chi A^{(0)} \partial_k \chi - \partial_k \chi A^{(0)} \chi) - \tfrac{1}{8} i \partial_l [\chi (A^{(0)} S^{(k)(l)} - S^{(k)(l)T} A^{(0)}) \chi + \chi (A^{(k)} S^{(0)(l)} + A^{(l)} S^{(0)(k)} - S^{(0)(l)T} A^{(k)} - S^{(0)(k)T} A^{(l)}) \chi]$$

are the desired densities. It would be incorrect to conclude from these results that the local energy and momentum densities are independent of $g_{00}$ for any system, since not all the components of $\chi$ are fundamental dynamical variables in general, and the construction of the dependent components may involve $g_{00}$ explicitly.

## FUNDAMENTAL VARIABLES

The identification of the fundamental dynamical variables depends upon the structure of the generator of field variations

$$G_\chi = \int (d\mathbf{x}) \tfrac{1}{2} (\chi A^{(0)} \delta \chi - \delta \chi A^{(0)} \chi),$$

and of the field equations. According to the action

principle, these are ($g_{0k}=0$)

$$A^{(0)}\partial_0\chi=e_0^{(0)}[(\partial_l\mathcal{H}/\partial\chi)-A^{(k)}\partial_k\chi]$$
$$+\tfrac{1}{2}\partial_k e_0^{(0)}[-A^{(k)}\chi+i(A^{(0)}S^{(0)(k)}-S^{(0)(k)T}A^{(0)})\chi].$$

It is also useful to note the relations between the matrices $A^{(0)}$ and $A^{(k)}$ that are implied by invariance under infinitesimal Lorentz transformations,

$$-i(A^{(0)}S^{(0)(k)}+S^{(0)(k)T}A^{(0)})=A^{(k)},$$
$$-i(A_{(l)}S^{(0)(k)}+S^{(0)(k)T}A_{(l)})=\delta_l^{\,k}A^{(0)}.$$

Thus, the coefficient of $\partial_k e_0^{(0)}$ in the field equations is alternatively written

$$iA^{(0)}S^{(0)(k)}\chi.$$

We must now discuss several categories of fields.[4] In the first of these $A^{(0)}$ is a nonsingular matrix—all field equations are equations of motion for the fundamental variables—and the local energy and momentum densities are independent of $g_{00}$, validating the commutator condition. When $A^{(0)}$, or a completely reduced portion of it, refers to a Fermi-Dirac (F.D.) field $\psi$, the real symmetrical matrix $\alpha^{(0)}=-iA^{(0)}$ must be positive-definite according to the equal-time anticommutator implied by the action principle,

$$\{\psi(x),\psi(x')\}=(\alpha^{(0)})^{-1}\delta(\mathbf{x}-\mathbf{x}').$$

This is impossible for other than spin $\frac{1}{2}$ fields.

On introducing the real symmetrical Euclidean spin matrix

$$S_{k4}=iS_{0k}$$

(we omit the designation of local components), the transformation properties of the $A$ matrices read

$$\{A^0,S_{k4}\}=A_k,\quad \{A_l,S_{k4}\}=\delta_{kl}A^0,$$

and therefore (no $k$ summation)

$$A^0=\{S_{k4},\{S_{k4},A^0\}\}.$$

A corresponding matrix statement is

$$[(S_{k4}'+S_{k4}'')^2-1]\langle S_{k4}'|A^0|S_{k4}''\rangle=0,$$

and all such matrix elements of $A^0$ vanish, unless

$$S_{k4}'+S_{k4}''=\pm 1.$$

When the spin is $\frac{1}{2}$, the eigenvalues of $S_{k4}$ are $\pm\frac{1}{2}$ and the diagonal elements of $A^0$, $S_{k4}'=S_{k4}''=\pm\frac{1}{2}$, differ from zero (in fact, $\alpha^0=1$). But with higher spin values all other diagonal matrix elements must vanish and $\alpha^0$ cannot be positive-definite.

Can $a^0$, the reduced submatrix of $A^0$ associated with Bose-Einstein (B.E.) fields, be nonsingular? An argument denying this possibility is based upon the physical existence of the vacuum state. The equal-time commutation relations for the B.E. field $\phi(x)$ would be

$$[\phi(x),\phi(x')]=i(a^0)^{-1}\delta(\mathbf{x}-\mathbf{x}'),$$

where $a^0$ is real and antisymmetrical. We combine these with the equations of motion ($-g_{00}=1$)

$$a^0\partial_0\phi=-a^k\partial_k\phi+(\partial\mathcal{H}/\partial\phi)$$

to get

$$i[a^0\partial_0\phi(x),a^0\phi(x')]$$
$$=-a^k\partial_k\delta(\mathbf{x}-\mathbf{x}')+(\partial^2\mathcal{H}/\partial\phi\partial\phi)\delta(\mathbf{x}-\mathbf{x}'),$$

and thereby the vacuum expectation value

$$\langle a^0\phi(x)P^0a^0\phi(x')+a^0\phi(x')P^0a^0\phi(x)\rangle$$
$$=-a^k\partial_k\delta(\mathbf{x}-\mathbf{x}')+\langle\partial^2\mathcal{H}/\partial\phi\partial\phi\rangle\delta(\mathbf{x}-\mathbf{x}').$$

The left-hand member of the latter equation must be positive definite. But the term $-a^k\partial_k\delta(\mathbf{x}-\mathbf{x}')$ does not have this character and can lead to arbitrarily large negative values of a corresponding quadratic form (provided $\langle\partial^2\mathcal{H}/\partial\phi\partial\phi\rangle$ is finite, which is assured at least for the usual dynamical applications of B.E. fields with linear couplings). The following example is cited to indicate that purely structural considerations will not suffice, in general:

$$\mathcal{L}=-\tfrac{1}{2}F^{\mu\nu}(\partial_\mu A_\nu-\partial_\nu A_\mu)-\tfrac{1}{2}F_d^{\,\mu\nu}(\partial_\mu B_\nu-\partial_\nu B_\mu)$$
$$-F_A\partial_\mu A^\mu-F_B\partial_\mu B^\mu-\mathcal{H},$$

in which $F_d^{\,\mu\nu}$ is the tensor dual to $F_{\mu\nu}$,

$$F_d^{\,01}=F_{23},\cdots.$$

The matrix $a^0$ is nonsingular, since every field variable appears in the time derivative term.

The matrix $A^0$ must be singular, then, apart from spin $\frac{1}{2}$ fields. This matrix can be reduced completely into a non-singular submatrix and a null submatrix, corresponding to the field decomposition

$$\chi=\chi_n+\chi_s,$$

where $\chi_s$ refers to the singular subspace. The latter imply field equations that are not equations of motion but equations of constraint,

$$_s[-A^k\partial_k\chi+\partial_l\mathcal{H}/\partial\chi]=0,$$

and these are independent of $g_{00}$. The second category consists of those physical systems for which the constraint equations suffice to determine the $\chi_s$ explicitly in terms of the $\chi_n$. Since that connection does not refer to $g_{00}$, the commutator condition for $T^{00}$ will be valid.

There are no F.D. fields in this category. The action principle gives the anticommutation relations for the $\psi_n$ in terms of $(\alpha^0)_n$, the nonsingular submatrix, which must be positive-definite. But, as a consequence of the Lorentz transformation equations,

$$0={}_s(S_{k4})_n(\alpha^0)_{n\;n}(S_{k4})_s,$$

[4] It may be of interest to compare these remarks with some earlier ones, which were limited to uncoupled fields. They are summarized by E. M. Corson, *Introduction to Tensors, Spinors, and Relativistic Wave Equations* (Blackie and Son, Limited, London, 1953).

which is impossible unless

$$_s(S_{k4})_n = {}_n(S_{k4})_s = 0.$$

Thus, the separation into two subspaces has a Lorentz invariant meaning and

$$\alpha^\mu = (\alpha^\mu)_n,$$

which is a complete reduction to the subspace of positive-definite $\alpha^0$, and spin $\frac{1}{2}$.

To discuss B.E. fields we must be more explicit about the structure of $\mathcal{H}(\chi)$. Let

$$\mathcal{H} = \tfrac{1}{2}\phi B\phi + \mathcal{H}_1,$$

where $\mathcal{H}_1$ is no more than linear in the components of any particular type of B.E. field. The real symmetrical matrix $B$ must satisfy conditions expressing the invariance of the corresponding term under infinitesimal Lorentz transformations,

$$BS_{\mu\nu} + S_{\mu\nu}{}^T B = 0.$$

With the permissible choice of imaginary antisymmetrical $S_{kl}$ and imaginary symmetrical $S_{0k}$, these conditions are satisfied by

$$B = R_{sp} I,$$

$$[I, S_{\mu\nu}] = 0,$$

where $R_{sp}$ is the geometrical space reflection matrix. It obeys

$$R_{sp}{}^T = R_{sp}{}^* = R_{sp}{}^{-1} = R_{sp}.$$

We shall adopt the general principle that the question of parity conservation is a purely dynamical one, referring only to the coupling among various fields, so that

$$[R_{sp}, I] = 0.$$

Then the invariant matrix $I$, like $R_{sp}$, is real and symmetrical and both matrices can be brought into diagonal form. Such a representation of $I$ has a Lorentz invariant meaning, while the diagonalization of $R_{sp}$ is automatic in the tensor description of $\phi$.

The equations of motion $(-g_{00}=1)$ and constraint equations are

$$(a^0)_n \partial_0 \phi_n = -(a^k)_n \partial_k \phi_n - {}_n(a^k)_s \partial_k \phi_s + B\phi_n + \partial\mathcal{H}_1/\partial\phi_n,$$

and

$$0 = -{}_s(a^k)_n \partial_k \phi_n + B_s \phi_s + \partial\mathcal{H}_1/\partial\phi_s,$$

respectively. The $\phi_s$ are explicitly determined by the latter equations provided $B_s$ is nonsingular. In that situation the commutation relations of the fundamental variables are

$$[\phi_n(x), \phi_n(x')] = i(a^0)_n{}^{-1}\delta(\mathbf{x}-\mathbf{x}').$$

We can now derive the commutator

$$i[(a^0)_n \partial_0 \phi_n(x), (a^0)_n \phi_n(x')]$$
$$= -(a^k)_n \partial_k \delta(\mathbf{x}-\mathbf{x}') + B_n \delta(\mathbf{x}-\mathbf{x}')$$
$$+ {}_n(a^k)_s B_s{}^{-1} {}_s(a^l)_n \partial_k \partial_l{}' \delta(\mathbf{x}-\mathbf{x}'),$$

with the assumption that

$$[\phi_n(x), \partial\mathcal{H}_1/\partial\phi(x')] = 0.$$

This result also gives the vacuum expectation value of the commutator. The corresponding positive-definite quadratic form can be dominated by either of the last two terms if one considers sufficiently slowly or rapidly varying functions. Accordingly, it is necessary that

$$B_n > 0, \quad -B_s > 0.$$

But $B = R_{sp} I$, where $I$ has a common value for all components of a given type of tensor field. Every component of $\phi_n$ selected from a particular tensor must, therefore, have the same value of the space parity while the $\phi_s$ components possess the opposite value. If we now observe that

$$\{a^\mu, R_{st}\} = 0$$

$$R_{st} = e^{\pi i S_{12}} e^{\pi i S_{34}},$$

which refers to the invariance property of B.E. fields with respect to the proper transformation $x^\mu \to -x^\mu$, it follows that the matrices $a^\mu$ only connect different tensors, characteristized by opposite values of $R_{st}$. And, since

$$[a^0, R_{sp}] = 0$$

we can conclude, for an irreducible field, that all components of $\phi_n$ have a common parity. In that circumstance,

$$(a^k)_n = 0.$$

For which B.E. fields can $a^0$ and $R_{sp}a^0$ be identical, or differ merely by a sign? If we compare

$$a^0 = \{S_{k4}, \{S_{k4}, a^0\}\}$$

with

$$R_{sp} a^0 = [S_{k4}, [S_{k4}, R_{sp} a^0]],$$

it is seen that nonvanishing matrix elements $\langle S_{k4}' | a^0 \times | S_{k4}'' \rangle$ can only occur when both of the following conditions are satisfied:

$$(S_{k4}' + S_{k4}'')^2 = 1,$$

$$(S_{k4}' - S_{k4}'')^2 = 1.$$

Thus, one of the quantum numbers must be zero and the other of unit magnitude. Since no larger value can appear, the spins of such fields are limited to zero and unity.

There is a third category, of spin one fields, for which $B_s$ is singular corresponding to the intrinsic arbitrariness of vector fields that admit gauge transformations. We shall not consider these systems here since the commutator condition has already been verified (indeed, discovered) in connection with the general non-Abelian vector gauge field coupled to a spin $\frac{1}{2}$ field.

The fourth category contains F.D. fields of spin $\geq\frac{3}{2}$ and B.E. fields with spins $\geq 2$. The matrices $(\alpha^0)_n$ of F.D. fields and $B_n$ for B.E. fields will be indefinite and not all of the $\chi_n$ can be fundamental variables. Hence, the constraint equations must act to diminish the space of independent variables by introducing relations among the components of $\chi_n$. Since all of these obey equations of motion, the introduction of the restrictions on the $\chi_n$ will yield further constraint equations. But the latter, being consequences of the equations of motion, involve $g_{00}$ and the needed reassurance that $(-g_{00})T^{00}$ and $(-g_{00})^{1/2}T^{0k}$ are independent of $g_{00}$ can no longer be given.

The $\partial_k e_0{}^{(0)}$ term in the field equations is essential to this argument for otherwise the existence of linear relations among the $\chi_n$ would imply constraint equations that were still independent of $g_{00}$. The process of deriving new constraint conditions from the equations of motion may require several repetitions depending upon the spin of the field, and the explicit construction of the dependent variables will involve corresponding numbers of space and time derivatives of $g_{00}$.

### SPIN $\frac{3}{2}$

We shall briefly examine the field in the fourth category with the lowest spin value, in order to obtain the simplest example of the additional terms in the commutator condition. It will suffice to consider an uncoupled field for, unlike the situation with lower spin fields, our task is no longer to show that the commutator condition is valid despite the effect of field interactions. The preceding discussion indicates that $\psi_s$ becomes an explicit function of $g_{00}$, but that time derivatives of $g_{00}$ do not yet appear, for spin $\frac{3}{2}$. Accordingly, $T^{(0)(0)}$ must remain independent of $g_{00}$. The explicit dependence of $T^{(0)(k)}$ is given by

$$\delta_3 T^{(0)(k)}(x)/\delta e_0{}^{(0)}(x') = -\tfrac{1}{2}i\partial_l[\psi(x)(A^kS^{0l}+A^lS^{0k})_s\delta_3\psi_s(x)/\delta e_0{}^{(0)}(x')],$$

where

$$-i(A^kS^{0l})_s = i(S^{0l}A^k)_s = (S^{0l}A^0S^{0k})_s.$$

We choose

$$\mathcal{H} = \tfrac{1}{2}\psi B\psi,$$

where $B$ is an antisymmetrical imaginary matrix, and the constraint equation becomes

$${}_s(A^k\partial_k - B)_n\psi_n = 0.$$

The further constraint equation derived from the equations of motion is

$$0 = {}_s(A^k\partial_k - B)_n(A_n{}^0)^{-1}{}_n[e_0{}^{(0)}(-A^l\partial_l + B)\psi + (\partial_l e_0{}^{(0)})iA^0S^{0l}\psi].$$

The local nature of the theory requires that this be an explicit expression for $\psi_s$, rather than a differential equation. It can also be anticipated that the only significant terms are those containing the second spatial derivatives of $e_0{}^{(0)}$. These considerations give, for $-g_{00}=1$,

$$(BA_n{}^0B)_s\frac{\delta_3\psi_s(x)}{\delta e_0{}^{(0)}(x')} = {}_s(S^{0m}A^0S^{0n}\psi(x))\partial_m\partial_n\delta(\mathbf{x}-\mathbf{x}'),$$

and

$$\frac{\delta_3 T^{(0)(k)}(x)}{\delta e_0{}^{(0)}(x')} = \partial_l\partial_m{}'\partial_n{}'[\delta(\mathbf{x}-\mathbf{x}')f^{kl,mn}(x)].$$

Here

$$f^{kl,mn}(x) = \psi(x)S^{0(k}A^0S^{0l)}(BA_n{}^{0-1}B)_s{}^{-1}S^{0(m}A^0S^{0n)}\psi(x),$$

which uses the notation

$$S^{0(k}A^0S^{0l)} = \tfrac{1}{2}(S^{0k}A^0S^{0l} + S^{0l}A^0S^{0k}).$$

We notice that $f^{kl,mn}$ has the symmetries

$$f^{kl,mn} = f^{lk,mn} = f^{kl,nm} = -f^{mn,kl},$$

where the last statement is a consequence of the F.D. anticommutation relations. One can verify with a specific example that $f^{kl,mn}$ is not identically zero.

The resulting equal-time energy density commutator equation is

$$-i[T^{00}(x),T^{00}(x')] = -(T^{0k}(x)+T^{0k}(x'))\partial_k\delta(\mathbf{x}-\mathbf{x}') - \partial_k\partial_l\partial_m{}'\partial_n{}'\{\delta(\mathbf{x}-\mathbf{x}')f^{kl,mn}(x)\}.$$

The additional term has the anticipated properties. It vanishes for finite $|\mathbf{x}-\mathbf{x}'|$, it is an antisymmetrical function of $x$ and $x'$, and it does not contribute to commutators containing the integrated quantities $P^0$ or $J_{0k}$.

The considerations of this paper apply to all fields save one—the gravitational field.

# Quantized Gravitational Field. II

JULIAN SCHWINGER*
*Institut des Hautes Etudes Scientifiques, Bures-sur-Yvette, Seine et Oise, France*

(Received 17 June 1963)

A consistent formulation is given for the quantized gravitational field in interaction with integer spin fields. Lorentz transformation equivalence within a class of physically distinguished coordinate systems is verified.

## INTRODUCTION

THE quantization problem posed by the gravitational field is not that of exhibiting canonical variables but rather consists in verifying that the generators of coordinate transformations, which are only known implicitly, satisfy the necessary commutation properties. A technique appropriate to this problem has been devised,[1] in which canonical operator variables are combined with mathematical parameters of a functional-transformation group. We shall apply this method to construct a consistent formulation for the quantized gravitational field coupled to matter fields of spin 0 and 1.

The following is a summary of results obtained by a heuristic application of the quantum action principle to the gravitational field and a spin 0 matter field.[2] The operator reduces to

$$W=\int (dx)[\Pi_{kl}\partial_0 q^{kl}+\phi^0\partial_0\phi]$$

action subject to the constraints

$$\tau_k+T_k=0,\quad \tau^0+T^0=0,$$

where

$$\tau_k=-\Pi_{lm}\partial_k q^{lm}+\partial_k(2\Pi_{lm}q^{lm})-\partial_l(2\Pi_{km}q^{lm})$$

and

$$q^s\tau^0=1/(2\kappa)q^s(\partial_k\partial_l q^{kl}+Q)-2\kappa\Pi_{kl}q^s(q^{kl}q^{mn}-q^{kn}q^{lm})\Pi_{mn},$$

in which

$$Q=-\tfrac{1}{4}q^{mn}\partial_m q^{kl}\partial_n q_{kl}-\tfrac{1}{2}\partial_m q^{kl}q_{ln}\partial_k q^{mn}$$
$$-\tfrac{1}{2}q^{kl}\partial_k \ln q^{1/2}\partial_l \ln q^{1/2}.$$

We have also included in the definition of $\tau^0$ an arbitrary power of the quantity

$$q=\det q^{kl},$$

in order to suggest, in a potentially constructive way, the ambiguity thus far implicit in the discussion. The corresponding operators of the spinless-matter field are

$$T_k=-\phi^0\partial_k\phi$$
$$T^0=\tfrac{1}{2}[(\phi^0)^2+\partial_k\phi q^{kl}\partial_l\phi+q^{1/2}m^2\phi^2].$$

They obey the equal-time commutation relations

$$-i[T^0(x),T^0(x')]$$
$$=-(q^{kl}(x)T_l(x)+q^{kl}(x')T_l(x'))\partial_k\delta(\mathbf{x}-\mathbf{x}'),$$
$$-i[T_k(x),T_l(x')]$$
$$=-T_l(x)\partial_k\delta(\mathbf{x}-\mathbf{x}')-T_k(x')\partial_l\delta(\mathbf{x}-\mathbf{x}').$$

The generality of these relations can be inferred from the alternative example of a unit spin matter field.

## SPIN-1 MATTER FIELD

We consider only an Abelian-gauge field. The action operator in a prescribed metric field $g_{\mu\nu}$ is

$$W=\int (dx)[-\tfrac{1}{2}F^{\mu\nu}H_{\mu\nu}+\tfrac{1}{4}F^{\mu\nu}(-g)^{-1/2}g_{\mu\lambda}g_{\nu\kappa}F^{\lambda\kappa}],$$

where

$$H_{\mu\nu}=\partial_\mu A_\nu-\partial_\nu A_\mu$$

and $F_{\mu\nu}$ is a tensor density. The constraint equations obtained by variation of $A_0$ and $F^{kl}$ are, respectively,

$$\partial_k F^{0k}=0$$

and

$$H_{kl}=(-g)^{-1/2}g_{k\lambda}g_{l\kappa}F^{\lambda\kappa}.$$

The latter appears in the time gauge as[3]

$$e_{(m)}{}^k e_{(n)}{}^l H_{kl}=e_{(0)}{}^0 g^{-1/2}e_{\lambda(m)}e_{\kappa(n)}F^{\lambda\kappa},$$

and two algebraic consequences are given by

$$H_{kl}q^{-1}q^{km}q^{ln}H_{mn}=(-g)^{-1}F^{\mu\nu}g_{\mu\lambda}g_{\nu\kappa}F^{\lambda\kappa}+2F^{0k}q_{kl}F^{0l}$$
$$=e_{(0)}{}^0 g^{-1/2}H_{kl}(F^{kl}-2e_0{}^{(0)}e_{(0)}{}^k F^{0l}).$$

The resulting canonical variable form of the action operator is

$$W=\int (dx)[-F^{0k}\partial_0 A_k-e_0{}^{(0)}e_{(0)}{}^k T_k-e_0{}^{(0)}g^{-1/2}T^0],$$

where

$$T_k=H_{kl}F^{0l},$$
$$T^0=\tfrac{1}{2}[F^{0k}q^{1/2}q_{kl}F^{0l}+\tfrac{1}{2}H_{kl}q^{-1/2}q^{km}q^{ln}H_{mn}].$$

With the aid of the canonical commutation relations, effectively given by

$$-i[H_{kl}(x),F^{0m}(x')]=(\delta_k{}^m\partial_l-\delta_l{}^m\partial_k)\delta(\mathbf{x}-\mathbf{x}'),$$

* Permanent address: Harvard University, Cambridge, Massachusetts.
[1] J. Schwinger, Nuovo Cimento (to be published).
[2] J. Schwinger, Phys. Rev. 130, 1253 (1963).
[3] Notation: $(-g)=-\det g_{\mu\nu}$, $g=\det g_{kl}$, and $(-g)^{-1/2}=e_{(0)}{}^0 g^{1/2}$.



Reprinted from *The Physical Review* **132**, 1317–1321 (1963).

one verifies that $T_k$ and $T^0$ obey the previously stated commutation properties.

## EXTENDED OPERATORS

The significance of $q^{kl}$ and $\Pi_{kl}$ is obtained by writing

$$q^{kl}=q^{klT}+\tfrac{1}{2}(\partial_k q_l+\partial_l q_k)-\delta_{kl}\partial_m q_m+\partial_k\partial_l q$$
$$\partial_k q^{klT}=q^{kkT}=0,$$

and, similarly,

$$\Pi_{kl}=\Pi_{kl}{}^T+\tfrac{1}{2}(\partial_k\Pi_l+\partial_l\Pi_k)-\delta_{kl}\partial_m\Pi_m+\partial_k\partial_l\Pi .$$

One also recasts the constraint equations in the forms

$$\partial_l\Pi_{kl}-\partial_k\Pi_{ll}=\tfrac{3}{2}\partial_k\partial_l\Pi_l+\tfrac{1}{2}\nabla^2\Pi_k=\tfrac{1}{2}\theta_k,$$
$$\partial_k\partial_l q^{kl}=(\nabla^2)^2 q=-2\kappa\theta^0,$$

with the aid of the definitions

$$\tau_k=t_k-2(\partial_l\Pi_{kl}-\partial_k\Pi_{ll}),$$
$$\tau^0=t^0+1/(2\kappa)\partial_k\partial_l q^{kl},$$

and

$$\theta_k=t_k+T_k,\quad \theta^0=t^0+T^0.$$

As a result we have, to within an additive total differential,

$$\int(d\mathbf{x})\Pi_{kl}dq^{kl}=\int(d\mathbf{x})[\Pi_{kl}{}^T dq^{klT}+\theta_k d(-\tfrac{1}{2}q_k)-\theta^0 d(-2\kappa\Pi)],$$

which is the sum of a generator of operator variations and the generator of an infinitesimal transformation parameterized by

$$-\tfrac{1}{2}q_k=\xi_k,\quad -2\kappa\Pi=\xi^0.$$

This description is conveyed by the operator commutation relation

$$-i[q^{klT}(x),\Pi_{mn}{}^T(x')]=(\delta_{mn}{}^{kl}\delta(\mathbf{x}-\mathbf{x}'))^T$$

and the differential-state-vector equation

$$\delta_\xi\langle\xi|=i\langle\xi|\int(d\mathbf{x})(\theta_k\delta\xi_k-\theta^0\delta\xi^0).$$

Equivalent versions of the latter are

$$-i(\delta/\delta\xi_k(x))\langle\xi|=\langle\xi|\theta_k(x)$$
$$i(\delta/\delta\xi^0(x))\langle\xi|=\langle\xi|\theta^0(x),$$

and

$$[-i(\delta/\delta\xi_k(x))-\theta_k(x)]\langle\xi|=0,$$
$$[i(\delta/\delta\xi^0(x))-\theta^0(x)]\langle\xi|=0.$$

In the last form a representation of the field operators by means of eigenvalues and functional differential operators is understood.

We can now interpret $q^{kl}$ and $\Pi_{kl}$ as extended operators by introducing the functional differential operator constructions of $\Pi_k$ and $q$,

$$\Pi_k=[\delta_{kl}\mathfrak{D}_1+\tfrac{3}{4}\partial_k\partial_l\mathfrak{D}_2]i\delta/\delta\xi_l,$$
$$q=(-2\kappa)\mathfrak{D}_2 i\delta/\delta\xi^0-\tfrac{3}{2}\mathbf{x}^2.$$

These are written in a symbolic notation with the aid of the functions defined (apart from boundary conditions) by

$$-\nabla^2\mathfrak{D}_1(\mathbf{x}-\mathbf{x}')=\delta(\mathbf{x}-\mathbf{x}'),$$
$$-\nabla^2\mathfrak{D}_2(\mathbf{x}-\mathbf{x}')=\mathfrak{D}_1(\mathbf{x}-\mathbf{x}').$$

Thus,

$$q^{kl}=q^{klT}-(\partial_k\xi_l+\partial_l\xi_k)+2\delta_{kl}\partial_m\xi_m-3\delta_{kl}-2\kappa\partial_k\partial_l\mathfrak{D}_2 i\delta/\delta\xi_0$$

and

$$\Pi_{kl}=\Pi_{kl}{}^T+\tfrac{1}{2}i\mathfrak{D}_1(\partial_k\delta/\delta\xi_l+\partial_l\delta/\delta\xi_k-\tfrac{1}{2}\delta_{kl}\partial_m\delta/\delta\xi_m)+\tfrac{3}{4}i\partial_k\partial_l\mathfrak{D}_2\partial_m\delta/\delta\xi_m-1/(2\kappa)\partial_k\partial_l\xi^0.$$

As one can verify, with the aid of the explicit construction,

$$\begin{aligned}(\delta_{mn}{}^{kl}\delta(\mathbf{x}-\mathbf{x}'))^T &=\delta_{mn}{}^{kl}\delta(\mathbf{x}-\mathbf{x}')-\tfrac{1}{2}\delta^{kl}\delta_{mn}\delta(\mathbf{x}-\mathbf{x}')\\ &\quad+\tfrac{1}{2}(\delta_n{}^l\partial_k\partial_m+\delta_n{}^k\partial_l\partial_m+\delta_m{}^l\partial_k\partial_n+\delta_m{}^k\partial_l\partial_n)\mathfrak{D}_1(\mathbf{x}-\mathbf{x}')\\ &\quad-\tfrac{1}{2}(\delta^{kl}\partial_m\partial_n+\delta_{mn}\partial_k\partial_l)\mathfrak{D}_1(\mathbf{x}-\mathbf{x}')\\ &\quad+\tfrac{1}{2}\partial_k\partial_l\partial_m\partial_n\mathfrak{D}_2(\mathbf{x}-\mathbf{x}'),\end{aligned}$$

these extended operators obey the simple canonical commutation relation

$$-i[q^{kl}(x),\Pi_{mn}(x')]=\delta_{mn}{}^{kl}\delta(\mathbf{x}-\mathbf{x}').$$

## CONSISTENCY

The fundamental problem in formulating, the theory has now resolved itself into verifying, or imposing consistency on the four functional differential equations that govern the states $\langle\xi|$,

$$(\tau_k(x)+T_k(x))\langle\xi|=0$$
$$(\tau^0(x)+T^0(x))\langle\xi|=0.$$

Let us consider first the extended operator

$$G_x=\int(d\mathbf{x})(\tau_k+T^k)\delta x^k$$

and observe that it generates the transformation accompanying the arbitrary infinitesimal spatial coordinate transformation $\delta x^k$. Thus

$$-i[q^{kl},G_x]=-\delta x^m\partial_m q^{kl}+q^{ml}\partial_m\delta x^k+q^{km}\partial_m\delta x^l-2q^{kl}\partial_m\delta x^m,$$

$$-i[\Pi_{kl},G_x]=-\delta x^m\partial_m\Pi_{kl}-\Pi_{ml}\partial_k\delta x^m-\Pi_{km}\partial_k\delta x^m+\Pi_{kl}\partial_m\delta x^m,$$

and, for the example of the spin-0 matter field,

$$-i[\phi,G_x]=-\delta x^m\partial_m\phi$$
$$-i[\phi^0,G_x]=-\delta x^m\partial_m\phi^0-\phi^0\partial_m\delta x^m.$$

These are infinitesimal transformation laws of the various three-dimensional tensor densities. (We speak of a tensor density of degree $\delta$ if the object is obtained from the corresponding tensor by multiplication with $(g^{1/2})^\delta$). Indeed, $q^{kl}$ and $\Pi_{kl}$ are tensor densities of degree $+2$ and $-1$, respectively, while $\phi$ and $\phi^0$ are scalar densities of degree 0 and $+1$, respectively. The commutation properties of the set of operators $G_x$ corresponds to the composition law of successive infinitesimal transformations for the group of general coordinate transformations. Two successive infinitesimal coorinate transformations, performed in alternative order, are connected by another infinitesimal transformation,

$$\delta^{[12]}x^k = \delta^{(2)}x^l\partial_l\delta^{(1)}x^k - \delta^{(1)}x^l\partial_l\delta^{(2)}x^k,$$

and correspondingly

$$-i[G_x^{(1)},G_x^{(2)}] = G_x^{[12]}.$$

The implied commutation relations are

$$-i[(\tau_k+T_k)(x),\ (\tau_l+T_l)(x')] = -(\tau_l+T_l)(x)\partial_k\delta(\mathbf{x}-\mathbf{x}') - (\tau_k+T_k)(x')\partial_l\delta(\mathbf{x}-\mathbf{x}'),$$

which can also be derived from the transformation properties of $\tau_k+T_k$, a vector density of degree $+1$,

$$-i[\tau_k+T_k,\ G_x] = -\delta x^m\partial_m(\tau_k+T_k) - (\tau_m+T_m)\partial_k\delta x^m - (\tau_k+T_k)\partial_m\delta x^m.$$

It will be noted that the commutation relations are obeyed separately by $\tau_k$ and $T_k$. The group structure of these commutators confirms the consistency of the three functional differential equations,

$$(\tau_k+T_k)\langle\xi| = 0.$$

The various contributions to $\tau^0+T^0$ are, individually, scalar densities of degree $+2$. (It should be recalled that $\partial_k\partial_l q^{kl}+Q = g_{(3)}R$.) The corresponding commutation relation,

$$-i[(\tau^0+T^0)(x),\ (\tau_k+T_k)(x')] = -((\tau^0+T^0)(x)+(\tau^0+T^0)(x'))\partial_k\delta(\mathbf{x}-\mathbf{x}'),$$

shows the consistency between the functional differential equation

$$(\tau^0+T^0)\langle\xi| = 0$$

and the set of three referring to spatial coordinate transformations.

All this reflects the automatic way in which three-dimensional covariance is assured by the formalism. The essential problem is contained in the commutation properties of the operator set $(\tau^0+T^0)(x)$. Let us note first that $T^0(x)$, for both examples of integer spin fields, involves $q^{kl}(x)$ without spatial derivatives. The contributions to $[\tau^0(x),T^0(x')]$ will then come entirely from the terms in $\tau^0(x)$ involving $\Pi_{kl}(x)$, and thus are proportional to $\delta(\mathbf{x}-\mathbf{x}')$. Such a result is symmetrical between $x$ and $x'$, and

$$[\tau^0(x),T^0(x')]+[T^0(x),\tau^0(x')] = 0.$$

Hence,

$$[(\tau^0+T^0)(x),\ (\tau^0+T^0)(x')] = [\tau^0(x),\tau^0(x')]+[T^0(x),T^0(x')],$$

and the necessity of a resulting group structure demands that the $\tau^0$ commutators have the same form as those of $T^0$ in relation to $\tau_k$ and $T_k$, respectively.

It is more convenient to consider $q^s\tau^0$. We first note that

$$[q^s\tau^0(x),q^s\tau^0(x')] = [q^s(\partial_k\partial_l q^{kl}+Q)(x),\ \Pi_{kl}q^s(q^{kn}q^{lm}-q^{kl}q^{mn})\Pi_{mn}(x')] - (x\leftrightarrow x'),$$

where the last term indicates the interchange of $x$ and $x'$ in the preceding commutator. The result is a linear function of the $\Pi_{kl}$ symmetrically multiplying a function of the $q^{kl}$, and it is not difficult to verify that

$$-i[q^s\tau^0(x),q^s\tau^0(x')] = -(q^{2s}q^{kl}(x)\ .\ \tau_l(x)+q^{2s}q^{kl}(x')\ .\ \tau_l(x'))\partial_k\delta(\mathbf{x}-\mathbf{x}')$$

in which the dot appears to indicate the symmetrization of multiplication. Symmetrization is also applied to the extended operator expression for $\tau_l$, but this is not significant if it is agreed that

$$-i[\partial_p q^{kl}(x),\Pi_{mn}(x)] = \lim_{\mathbf{x}'\to\mathbf{x}} \delta_{mn}{}^{kl}\partial_p\delta(\mathbf{x}-\mathbf{x}') = 0,$$

as will be the result of any symmetrical approach to the limit. It must also be remarked that there are various equivalent ways of writing the coefficient of $\nabla\delta(\mathbf{x}-\mathbf{x}')$, since

$$(f(x)g(x')+f(x')g(x))\nabla\delta(\mathbf{x}-\mathbf{x}') = (f(x)g(x)+f(x')g(x'))\nabla\delta(\mathbf{x}-\mathbf{x}').$$

Thus,

$$-i[q^s\tau^0(x),q^s\tau^0(x')] = -(q^{2s}q^{kl}(x)\ .\ \tau_l(x')+q^{2s}q^{kl}(x')\ .\ \tau_l(x))\partial_k\delta(\mathbf{x}-\mathbf{x}').$$

The $q^s$ factors can also be included in the $T^0$ commutation relation and the result will indeed have the analogous form. There is one basic difference, however. Although symmetrization with $q^{2s}q^{kl}$ is trivial for $T_l$, it is not for $\tau_l$ since the latter does not generally commute with its factor. But the verification of consistency for the equation

$$q^s(\tau^0+T^0)(x)\langle\xi| = 0,$$

which is equivalent to

$$(\tau^0+T^0)(x)\langle\xi| = 0,$$

demands that the commutator of two such extended operators yield the combination $\tau_l+T_l$ on the right-hand side only, in position to annihilate the state $\langle\xi|$. Thus, all depends on the commutation relation between $q^{2s}q^{kl}$ and $\tau_l$.

The product $q^{2s}q^{kl}$ is a tensor density of degree $8s+2$,

$$-i[q^{2s}q^{kl},G_x]=-\delta x^m\partial_m(q^{2s}q^{kl})+q^{2s}q^{ml}\partial_m\delta x^k+q^{2s}q^{km}\partial_m\delta x^l-(8s+2)q^{2s}q^{kl}\partial_m\delta x^m,$$

which asserts that

$$-i[q^{2s}q^{kl}(x),\tau_m(x')]=-q^{2s}q^{kl}(x')\partial_m\delta(\mathbf{x}-\mathbf{x}')+q^{2s}q^{nl}(x)\delta_m{}^k\partial_n\delta(\mathbf{x}-\mathbf{x}')+q^{2s}q^{kn}(x)\delta_m{}^l\partial_n\delta(\mathbf{x}-\mathbf{x}')-(8s+1)q^{2s}q^{kl}(x)\partial_m\delta(\mathbf{x}-\mathbf{x}').$$

The commutator of interest is

$$-i[q^{2s}q^{kl}(x),\tau_l(x')]=-q^{2s}q^{kl}(x')\partial_l\delta(\mathbf{x}-\mathbf{x}')-(8s-3)q^{2s}q^{kl}(x)\partial_l\delta(\mathbf{x}-\mathbf{x}').$$

It can now be seen that there is a unique value of $s$ for which the right-hand side is an antisymmetrical function of $x$ and $x'$, and

$$[q^{2s}q^{kl}(x),\tau_l(x')]+[q^{2s}q^{kl}(x'),\tau_l(x)]=0,$$

namely,

$$s=\tfrac{1}{2}.$$

With this choice, we have

$$-i[q^{1/2}(\tau^0+T^0)(x),\,q^{1/2}(\tau^0+T^0)(x')]=-(qq^{kl}(\tau_l+T_l)(x)+qq^{kl}(\tau_l+T_l)(x'))\partial_k\delta(\mathbf{x}-\mathbf{x}')$$

and all consistency tests are satisfied.

The addition of an arbitrary numerical multiple of $q^{1/2}=g$ to

$$\tau^0=1/(2\kappa)(\partial_k\partial_lq^{kl}+Q)-2\kappa q^{-1/2}\Pi_{kl}q^{1/2}(q^{kl}q^{mn}-q^{km}q^{ln})\Pi_{mn}$$

will not alter this conclusion. This is also true of the additive term $q^{1/4}q^{kl}\Pi_{kl}$, in any multiplication order. But if one uses the particular combination

$$\tfrac{1}{2}(q^{1/4}q^{kl}\Pi_{kl}+q^{-1/2}\Pi_{kl}q^{3/4}q^{kl})$$

that term can be removed completely from $\tau^0$, without affecting $\tau^k$, by the canonical transformation

$$\Pi_{kl}(x)\to\exp\left[-i\lambda\int(d\mathbf{x})g^{1/2}\right]\Pi_{kl}(x)\exp\left[i\lambda\int(d\mathbf{x})g^{1/2}\right]=\Pi_{kl}(x)+\tfrac{1}{4}\lambda q^{1/4}q_{kl}(x).$$

## LORENTZ INVARIANCE

The coordinate conditions $\xi^\mu=x^\mu(\xi^k=\xi_k,\,\xi^0=-\xi_0)$ define a physically distinguished class of Lorentz transformation equivalent coordinate systems. The explicit verification of Lorentz invariance, in its four-dimensional aspects, concerns volume integrated properties of the energy density equal-time commutator. The energy and momentum density operators $\vartheta^\mu(x)$ are to be obtained through the reduction of the extended operators $\theta^\mu(x)$ by means of the four functional differential equations

$$(\Theta^\mu(x)-\theta^\mu(z))\langle\xi|=0,$$

in which we have written

$$-1/(2\kappa)\partial_k\partial_lq^{kl}=\Theta^0=-\Theta_0$$
$$\partial_l2(\Pi_{kl}-\delta_{kl}\Pi_{mm})=\Theta^k=\Theta_k.$$

We first note the equal-time commutator equation

$$[(\Theta^0-\theta^0)(x),\,(\Theta^0-\theta^0)(x')]\langle\xi|=0,$$

where

$$-i[\Theta^0(x),\theta^0(x')]=-2\pi^{kl}(x')\partial_k\partial_l\delta(\mathbf{x}-\mathbf{x}')$$

and

$$2\pi^{kl}=(q^{km}q^{ln}-q^{kl}q^{mn})\Pi_{mn}+q^{-1/2}\Pi_{mn}q^{1/2}(q^{km}q^{ln}-q^{kl}q^{mn}).$$

Accordingly, we have

$$\{-i[\theta^0(x),\theta^0(x')]+2(\partial_l\pi^{kl}(x)+\partial_l'\pi^{kl}(x'))\partial_k\delta(\mathbf{x}-\mathbf{x}')\}\langle\xi|=0.$$

It should also be observed that

$$2\partial_l\pi^{kl}=\Theta^k+\partial_lf^{kl},$$

where

$$f^{kl}=f^{lk}=(q^{km}q^{ln}-q^{kl}q^{mn})\Pi_{mn}+q^{-1/2}\Pi_{mn}q^{1/2}(q^{km}q^{ln}-q^{kl}q^{mn})-2(\delta_{km}\delta_{ln}-\delta_{kl}\delta_{mn})\Pi_{mn},$$

and a further rearrangement of the commutator equation yields

$$\{-i[\vartheta^0(x),\vartheta^0(x')]+(\vartheta^k(x)+\vartheta^k(x'))\partial_k\delta(\mathbf{x}-\mathbf{x}')-i[(\theta^0-\vartheta^0)(x),\,\Theta^0(x')]+i[(\theta^0-\vartheta^0)(x'),\,\Theta^0(x)]+(\partial_lf^{kl}(x)+\partial_l'f^{kl}(x'))\partial_k\delta(\mathbf{x}-\mathbf{x}')\}\langle\xi|=0.$$

Extended operators have spatially-localized commutation properties. But the reduction of extended operators is a nonlocal process, and, consequently, the individual commutators in the preceding equation will not vanish for finite $|\mathbf{x}-\mathbf{x}'|$. This effectively denies physical significance to the detailed specification of energy distributions by means of the operator $\vartheta^0(x)$. The situation differs with regard to integral aspects, however, since

$$\int_V(d\mathbf{x})\Theta^0=-1/(2\kappa)\int_S d\sigma_k\partial_lq^{kl}$$

and

$$\int_V(d\mathbf{x})x^k\Theta^0=-1/(2\kappa)\int_S d\sigma_l(x^k\partial_mq^{lm}-q^{kl})$$

refer to extended operators localized on the boundary surface. It is reasonable to presume that the nonlocal commutators connecting surface and internal points of a region tend to zero asymptotically, with increasing volume. The resulting integral commutators will involve the combinations

$$\int_V (d\mathbf{x})(\vartheta^k+\partial_l f^{kl}) = P^k + \int_S d\sigma_l f^{kl}$$

and

$$\int_V (d\mathbf{x})[x^k(\vartheta^l+\partial_m f^{ml}) - x^l(\vartheta^k+\partial_m f^{mk})]$$

$$= J^{kl} + \int_S d\sigma_m (x^k f^{ml} - x^l f^{mk}).$$

The asymptotic vanishing of these surface integrals is in the nature of a boundary condition characterizing a physically closed system. This property can be verified, if one retains only the slowly decreasing terms in the asymptotic behavior of the fields,

$$|\mathbf{x}| \to \infty: \quad q^{kl} \sim \delta_{kl} + (\kappa/4\pi)P^0\partial_k\partial_l|\mathbf{x}|,$$

$$\Pi_{kl} \sim -1/(8\pi)P_m[\delta_{lm}\partial_k|\mathbf{x}|^{-1} + \delta_{km}\partial_l|\mathbf{x}|^{-1} - \tfrac{1}{2}\delta_{kl}\partial_m|\mathbf{x}|^{-1} - \tfrac{3}{4}\partial_k\partial_l\partial_m|\mathbf{x}|].$$

The outcome of these considerations is the commutation properties

$$-i[P^0, J^{0k}] = P^k,$$

$$-i[J^{0k}, J^{0l}] = -J^{kl},$$

which completes the formal verification of Lorentz invariance. But a much more careful examination will be required to test whether the loosely stated physical boundary conditions can be maintained as assertions about operators in relation to a class of physical states.

# Coulomb Green's Function

Julian Schwinger
*Harvard University, Cambridge, Massachusetts*
(Received 19 June 1964)

A one-parameter integral representation is given for the momentum space Green's function of the nonrelativistic Coulomb problem.

It has long been known that the degeneracy of the bound states in the nonrelativistic Coulomb problem can be described by a four-dimensional Euclidean rotation group, and that the momentum representation is most convenient for realizing the connection. It seems not to have been recognized, however, that the same approach can be used to obtain an explicit construction for the Green's function of this problem. The derivation[1] is given here.

The momentum representation equation for the Green's function is ($\hbar = 1$)

$$\left(E - \frac{p^2}{2m}\right)G(\mathbf{p}, \mathbf{p}') + \frac{Ze^2}{2\pi^2} \int (d\mathbf{p}'') \frac{1}{(\mathbf{p} - \mathbf{p}'')^2} \times G(\mathbf{p}'', \mathbf{p}') = \delta(\mathbf{p} - \mathbf{p}').$$

We shall solve this equation by assuming, at first, that

$$E = -(p_0^2/2m)$$

is real and negative. The general result is inferred by analytic continuation.

The parameters

$$\xi = -\frac{2p_0\mathbf{p}}{p_0^2 + p^2}, \qquad \xi_0 = \frac{p_0^2 - p^2}{p_0^2 + p^2}$$

define the surface of a unit four-dimensional Eu-

[1] It was worked out to present at a Harvard quantum mechanics course given in the late 1940's. I have been stimulated to rescue it from the quiet death of lecture notes by recent publications in this Journal, which give alternative forms of the Green's function: E. H. Wichmann and C. H. Woo, J. Math. Phys. **2**, 178 (1961); L. Hostler, *ibid.* **5**, 591 (1964).

Reprinted from the *Journal of Mathematical Physics* 5, 1606–1608 (1964).

clidean sphere,

$$\xi_0^2 + \boldsymbol{\xi}^2 = 1,$$

the points of which are in one to one correspondence with the momentum space. The element of area on the sphere is

$$d\Omega = \frac{(d\boldsymbol{\xi})}{|\xi_0|} = \left(\frac{2p_0}{p_0^2 + p^2}\right)^3 (d\mathbf{p}),$$

if one keeps in mind that $p \gtrless p_0$ corresponds to the two semispheres $\xi_0 = \mp(1 - \boldsymbol{\xi}^2)^{\frac{1}{2}}$. As another form of this relation, we write the delta function connecting two points on the unit sphere as

$$\delta(\Omega - \Omega') = \left(\frac{p_0^2 + p^2}{2p_0}\right)^3 \delta(\mathbf{p} - \mathbf{p}').$$

Next, observe that

$$(\xi - \xi')^2 = (\xi_0 - \xi_0')^2 + (\boldsymbol{\xi} - \boldsymbol{\xi}')^2 = \frac{4p_0^2}{(p_0^2 + p^2)(p_0^2 + p'^2)} (\mathbf{p} - \mathbf{p}')^2.$$

Then, if we define

$$\Gamma(\Omega, \Omega') = -\frac{1}{16mp_0^3} (p_0^2 + p^2)^2 G(\mathbf{p}, \mathbf{p}')(p_0^2 + p'^2)^2,$$

that function obeys a four-dimensional Euclidean surface integral equation,

$$\Gamma(\Omega, \Omega') - 2\nu \int d\Omega''\, D(\xi - \xi'')\Gamma(\Omega'', \Omega') = \delta(\Omega - \Omega'),$$

where

$$D(\xi - \xi') = \frac{1}{4\pi^2} \frac{1}{(\xi - \xi')^2}$$

and

$$\nu = Ze^2 m/p_0.$$

The function $D$ that is defined similarly throughout the Euclidean space is the Green's function of the four-dimensional Poisson equation,

$$-\partial^2 D(\xi - \xi') = \delta(\xi - \xi').$$

It can be constructed in terms of a complete set of four-dimensional solid harmonics. In the spherical coordinates indicated by $\rho$, $\Omega$, these are

$$(\rho^{n-1}, \rho^{-n-1}) Y_{nlm}(\Omega), \qquad n = 1, 2, \cdots,$$

where the quantum numbers $l$, $m$ provide a three-dimensional harmonic classification of the four-dimensional harmonics. The largest value of $l$ contained in the homogeneous polynomial $\rho^{n-1}Y_{nlm}(\Omega)$ is the degree of the polynomial, $n - 1$. Thus,

$$-l \le m \le l, \qquad 0 \le l \le n - 1$$

label the $n^2$ distinct harmonics that have a common value of $n$.

The Green's function $D$ is exhibited as

$$D(\xi - \xi') = \sum_{n=1}^{\infty} \frac{\rho_<^{n-1}}{\rho_>^{n+1}} \frac{1}{2n} \sum_{lm} Y_{nlm}(\Omega) Y_{nlm}(\Omega')^*,$$

where

$$\delta(\Omega - \Omega') = \sum_{nlm} Y_{nlm}(\Omega) Y_{nlm}(\Omega')^*$$

conveys the normalization and completeness of the surface harmonics. One can verify that $D$ has the radial discontinuity implied by the delta function inhomogeneity of the differential equation,

$$-\rho^3 \frac{\partial}{\partial \rho} D(\xi - \xi')\Big]_{\rho'-0}^{\rho'+0} = \delta(\Omega - \Omega').$$

The function $D$ is used in the integral equation for $\Gamma$ with $\rho = \rho' = 1$. The equation is solved by

$$\Gamma(\Omega, \Omega') = \sum_{nlm} \frac{Y_{nlm}(\Omega) Y_{nlm}(\Omega')^*}{1 - (\nu/n)}.$$

The singularities of this function at $\nu = n = 1, 2, \cdots$ give the expected negative energy eigenvalues. The residues of $G$ at the corresponding poles in the $E$ plane provide the normalized wavefunctions, which are

$$\Psi_{nlm}(p) = \frac{4p_0^{5/2}}{(p_0^2 + p^2)^2} Y_{nlm}(\Omega),$$

$$p_0 = Ze^2 m/n.$$

One can exhibit $\Gamma(\Omega, \Omega')$ in essentially closed form with the end of the expansion for $D$. We use the following version of this expansion:

$$\frac{1}{2\pi^2} \frac{1}{(1 - \rho)^2 + \rho(\xi - \xi')^2} = \sum_{n=1}^{\infty} \rho^{n-1} \frac{1}{n} \sum_{lm} Y_{nlm}(\Omega) Y_{nlm}(\Omega')^*,$$

where $\xi$ and $\xi'$ are of unit length and $0 < \rho < 1$. Note, incidentally, that if we set $\xi = \xi'$ and integrate over the unit sphere, of area $2\pi^2$, we get

$$\frac{1}{(1 - \rho)^2} = \sum_{n=1}^{\infty} \rho^{n-1} \frac{1}{n} m_n,$$

where $m_n$ is the multiplicity of the quantum number $n$. This confirms that $m_n = n^2$.

The identity

$$\frac{1}{1 - (\nu/n)} = 1 + \frac{\nu}{n} + \nu^2 \frac{1}{n(n - \nu)},$$

together with the integral representation

$$\frac{1}{n - \nu} = \int_0^1 d\rho \rho^{-\nu} \rho^{n-1},$$

valid for $\nu < 1$, gives

$$\Gamma(\Omega,\Omega') = \delta(\Omega-\Omega') + \frac{\nu}{2\pi^2}\frac{1}{(\xi-\xi')^2} + \frac{\nu^2}{2\pi^2}\int_0^1 d\rho\rho^{-\nu}\frac{1}{(1-\rho)^2+\rho(\xi-\xi')^2}. \qquad (1)$$

Equivalent forms, produced by partial integrations, are

$$\Gamma(\Omega,\Omega') = \delta(\Omega-\Omega') + \frac{\nu}{2\pi^2}\int_0^1 d\rho\rho^{-\nu}\frac{d}{d\rho}\frac{\rho}{(1-\rho)^2+\rho(\xi-\xi')^2}, \qquad (2)$$

and

$$\Gamma(\Omega,\Omega') = \frac{1}{2\pi^2}\int_0^1 d\rho\rho^{-\nu}\frac{d}{d\rho}\frac{\rho(1-\rho^2)}{[(1-\rho)^2+\rho(\xi-\xi')^2]^2}, \qquad (3)$$

which uses the limiting relation

$$\delta(\Omega-\Omega') = \lim_{\rho\to1}\frac{1}{2\pi^2}\frac{1-\rho^2}{[(1-\rho)^2+\rho(\xi-\xi')^2]^2}.$$

Note that $\Gamma$ is a function of a single variable, $(\xi-\xi')^2$.

The restriction $\nu<1$ can be removed by replacing the real integrals with contour integrals,

$$\int_0^1 d\rho\rho^{-\nu}(\ \ ) \to \frac{i}{2\sin\pi\nu}e^{\pi i\nu}\int_C d\rho\rho^{-\nu}(\ \ ).$$

The path $C$ begins at $\rho = 1+0i$, where the phase of $\rho$ is zero and terminates at $\rho = 1-0i$, after encircling the origin within the unit circle.

The Green's function expressions implied by (1), (2), and (3) are

$$G(\mathbf{p},\mathbf{p}') = \frac{\delta(\mathbf{p}-\mathbf{p}')}{E-T} - \frac{Ze^2}{2\pi^2}\frac{1}{E-T}\frac{1}{(\mathbf{p}-\mathbf{p}')^2}\frac{1}{E-T'} - \frac{Ze^2}{2\pi^2}\frac{1}{E-T}\left[i\eta\int_0^1 d\rho\rho^{-i\eta} \times \frac{1}{(\mathbf{p}-\mathbf{p}')^2\rho-(m/2E)(E-T)(E-T')(1-\rho)^2}\right]\frac{1}{E-T'}, \qquad (1')$$

where

$$T = p^2/2m, \qquad \eta = -i\nu = Ze^2m/k;$$

$$G(\mathbf{p},\mathbf{p}') = \frac{\delta(\mathbf{p}-\mathbf{p}')}{E-T} - \frac{Ze^2}{2\pi^2}\frac{1}{E-T}\left[\int_0^1 d\rho\rho^{-i\eta}\frac{d}{d\rho} \times \frac{\rho}{(\mathbf{p}-\mathbf{p}')^2\rho-(m/2E)(E-T)(E-T')(1-\rho)^2}\right]\frac{1}{E-T'}; \qquad (2')$$

and

$$G(\mathbf{p},\mathbf{p}') = -\frac{i}{4\pi^2}\frac{k}{E}\int_0^1 d\rho\rho^{-i\eta}\frac{d}{d\rho} \times \frac{\rho(1-\rho^2)}{[(\mathbf{p}-\mathbf{p}')^2\rho-(m/2E)(E-T)(E-T')(1-\rho)^2]^2}. \qquad (3')$$

The Green's function is regular everywhere in the complex $E$ plane with the exception of the physical energy spectrum. This consists of the negative-energy eigenvalues already identified and the positive-energy continuum. The integral representations (1′), (2′), and (3′) are not completely general since it is required that

$$\operatorname{Re} i\eta = -\operatorname{Im}\eta < 1.$$

As we have indicated, this restriction can be removed. It is not necessary to do so, however, if one is interested in the limit of real $k$. These representations can therefore be applied directly to the physical scattering problem.

The asymptotic conditions that characterize finite angle deflections are

$$E-T\sim0, \qquad E-T'\sim0, \qquad (\mathbf{p}-\mathbf{p}')^2>0.$$

The second of the three forms given for $G$ is most convenient here. The asymptotic behavior is dominated by small $\rho$ values, and one immediately obtains

$$G(\mathbf{p},\mathbf{p}') \sim G^0(\mathbf{p})(-1/4\pi^2m)f(\mathbf{p},\mathbf{p}')G^0(\mathbf{p}'),$$

where

$$G^0(\mathbf{p}) = \frac{1}{E-T}\exp\left[-i\eta\log\frac{E-T}{4E}\right]\left(\frac{2\pi\eta}{e^{2\pi\eta}-1}\right)^{\frac{1}{2}}$$

and

$$f(\mathbf{p},\mathbf{p}') = \frac{2mZe^2}{(\mathbf{p}-\mathbf{p}')^2}\exp\left[-i\eta\log\frac{4k^2}{(\mathbf{p}-\mathbf{p}')^2}\right],$$

$$p^2 = p'^2 = k^2.$$

One would have found the same asymptotic form for any potential that decreases more rapidly than the Coulomb potential at large distances, but with $G^0(\mathbf{p}) = (E-T)^{-1}$. The factors $G^0(\mathbf{p}')$ and $G^0(\mathbf{p})$ describe the propagation of the particle before and after the collision, respectively, and $f$ is identified as the scattering amplitude. The same interpretation is applicable here since the modified $G^0$ just incorporates the long-range effect of the Coulomb potential. This is most evident from the asymptotic behavior of the corresponding spatial function, which is a distorted spherical wave,

$$\int\frac{(d\mathbf{p})}{(2\pi)^3}e^{i\mathbf{p}\cdot\mathbf{r}}G(\mathbf{p}) \sim (-m/2\pi r)\exp\left[i(kr+\eta\log 2kr+\zeta)\right],$$

$$\zeta = \arg\Gamma(1-i\eta).$$

The scattering amplitude obtained in this way coincides with the known result,

$$f(\vartheta) = (Ze^2/4E)\csc^2\tfrac{1}{2}\vartheta\exp\left[-i\eta\log\csc^2\tfrac{1}{2}\vartheta\right].$$

# Non-Abelian Vector Gauge Fields and the Electromagnetic Field

JULIAN SCHWINGER*
*Harvard University, Cambridge, Massachusetts*

## INTRODUCTION

Hypothetical vector gauge fields have been introduced in order to give a deeper dynamical foundation for such internal properties as isotopic spin.[1] An essential aspect of isotopic spin is electrical charge, and there is no doubt about the dynamical relation of this property to the electromagnetic field. Do these different types of vector fields simply coexist, or can they be combined to form a more unified theory of vector gauge fields? An integrated formulation can indeed be given, and it is not a trivial one since there are definite dynamical implications with regard to electromagnetic properties and the structure of the non-Abelian transformation group. The unification can encompass all fields that partake in both strong and electromagnetic interactions.[2] This success poses a physical problem, however. As one member of a set of gauge fields, the electromagnetic field is not physically distinguished and fails to perform its physical role of destroying the conservation of isotopic spin. Perhaps it is in this apparent dilemma that we find the clue to the existence in nature of other sets of fields which possess electromagnetic interactions, but no strong interactions. Is it the presence of charged leptonic fields that denies the higher symmetry transformations, relating the electromagnetic field with the non-Abelian fields, and gives to the electromagnetic field its characteristic physical influence?

The inclusion of electromagnetic lepton interactions produces a new difficulty, one of consistency. The gauge invariance of all terms in the Lagrange function save one contradicts the principle of stationary action. Another term that violates gauge invariance must be included. The simplest choice is a mass term in which the mass constant is presumably small, on the scale of strongly interacting particle masses, if a domain of approximate gauge invariance is to exist. And this modification raises again the physical mass problem of gauge fields: Are unit spin particles of small mass implied by the theory?

## UNIFIED THEORY

The Lagrange function of a non-Abelian vector gauge field coupled with a spin $\frac{1}{2}$ field is[3]

$$\mathcal{L} = -\tfrac{1}{2}\, G^{\mu\nu}[\partial_\mu \phi_\nu - \partial_\nu \phi_\mu + (\phi_\mu i t' \phi_\nu)] + \tfrac{1}{4} G^{\mu\nu} G_{\mu\nu} + \tfrac{1}{2}\, i\psi \alpha^\mu (\partial_\mu - i\,``T'\phi_\mu")\,\psi + \tfrac{1}{2}\, i m \psi \beta \psi\,,$$

where the matrices $t'$ and $T'$ include coupling con-

* Supported in part by the Air Force Office of Scientific Research under contract number AF49(638)-589.

[1] C. N. Yang and R. Mills, Phys. Rev. 96, 191 (1954).

[2] The problem of compatability has been given a more restricted discussion by R. Arnowitt and S. Deser, to be published.

[3] The notation follows J. Schwinger, Phys. Rev. 125, 1043 (1962).



Reprinted from *Reviews of Modern Physics* **36**, 609–613 (1964).

stants. We now interpret the internal space to be $(n+1)$ dimensional and use the notation

$$G_0^{\mu\nu} = F^{\mu\nu}, \quad \phi_0^\mu = A^\mu,$$

together with

$$t_0' = eq, \quad T_0' = eQ,$$
$$t_a' = ft_a, \quad T_a' = fT_a, \quad a = 1, \cdots n.$$

The total antisymmetry of $t'_{abc}$, $a, b, c = 0, 1 \cdots n$, should not be overlooked, for

$$(t_a')_{0b} = (t_b')_{a0} = -t'_{a0b} = -(eq)_{ab}.$$

The result obtained by writing the Lagrange function in terms of the $n$-dimensional internal space is given by

$$\begin{aligned}\mathcal{L} = & -\tfrac{1}{2}F^{\mu\nu}[\partial_\mu A_\nu - \partial_\nu A_\mu + (\phi_\mu ieq\phi_\nu)] + \tfrac{1}{4}F^{\mu\nu}F_{\mu\nu} \\ & - \tfrac{1}{2}G^{\mu\nu}[(\partial_\mu - ieqA_\mu)\phi_\nu - (\partial_\nu - ieqA_\nu)\phi_\mu \\ & \qquad + (\phi_\mu ift\phi_\nu)] \\ & + \tfrac{1}{4}G^{\mu\nu}G_{\mu\nu} + \tfrac{1}{2}i\psi\alpha^\mu(\partial_\mu - ieQA_\mu - if\,\text{“}T\,\phi_\mu\text{”})\psi \\ & \qquad + \tfrac{1}{2}im\psi\beta\psi,\end{aligned}$$

which exhibits the required electromagnetic gauge structure, and also implies an intrinsic magnetic moment for the vector field.[4]

The infinitesimal gauge variations of the unified theory,

$$\begin{aligned}\delta\phi_\mu &= (\partial_\mu - i\,\text{“}t'\,\phi_\mu\text{”})\,\delta\lambda \\ \delta G_{\mu\nu} &= i\,\text{“}t'\,\delta\lambda\text{”}\,G_{\mu\nu} \\ \delta\psi &= i\,\text{“}T'\,\delta\lambda\text{”}\,\psi,\end{aligned}$$

are written out as

$$\begin{aligned}\delta A_\mu &= \partial_\mu\,\delta\lambda_0 + (\phi_\mu ieq\,\delta\lambda), \\ \delta\phi_\mu &= ieq\,\delta\lambda_0\,\phi_\mu + (\partial_\mu - ieq\,A_\mu - if\,\text{“}t\phi_\mu\text{”})\,\delta\lambda,\end{aligned}$$

and

$$\begin{aligned}\delta F_{\mu\nu} &= (G_{\mu\nu}ieq\,\delta\lambda), \\ \delta G_{\mu\nu} &= ieq\,\delta\lambda_0\,G_{\mu\nu} - ieq\,\delta\lambda F_{\mu\nu} + if\,\text{“}t\,\delta\lambda\text{”}\,G_{\mu\nu},\end{aligned}$$

together with

$$\delta\psi = (ieQ\,\delta\lambda_0 + if\,\text{“}T\,\delta\lambda\text{”})\,\psi.$$

The single functional subgroup of electromagnetic gauge transformations is evident.

The transformations generated by $\delta\lambda_a$, $a = 1 \cdots n$ do not form a subgroup, unless $e = 0$. Yet, if the neglect of electromagnetic effects is a reasonable approximation to the complete theory, an $n$-parameter transformation group should be implicit in the latter. We interpret this somewhat vague physical requirement to have the following meaning. The group commutation properties of the $(n+1)$-dimensional matrices $t_a'$,

$$[t_a', t_b'] = \sum_{c=0}^{n} t'_{abc} t_c',$$

and of the $n$-dimensional matrices $t_a$,

$$[t_a, t_b] = \sum_{c=1}^{n} t_{abc} t_c,$$

are both valid, with $e \neq 0$. More symmetrically expressed, we require the compatability of

$$\sum_{g=0}^{n} (t'_{dag}t'_{gbc} + t'_{dbg}t'_{gca} + t'_{dcg}t'_{gab}) = 0 \qquad a, b, c, d = 0, 1, \cdots n$$

with

$$\sum_{g=1}^{n} (t_{dag}t_{gbc} + t_{dbg}t_{gca} + t_{dcg}t_{gab}) = 0 \qquad a, b, c, d = 1, \cdots n.$$

If $a, b, c, d \neq 0$ in the $(n+1)$-dimensional set of structure constant equations, the term in the summation with $g = 0$ must vanish separately, or

$$q_{da}q_{bc} + q_{db}q_{ca} + q_{dc}q_{ab} = 0.$$

On setting one of these four variables equal to 0, we learn that

$$[t_a, q] = \sum_{b=1}^{n} q_{ab} t_b,$$

and this exhausts the relations implied by the unified theory.

The restrictions imposed on the matrix $q$ can best be appreciated by multiplying the quadratic equation with $q_{gd}$ to form

$$(q^2)_{ga}q_{bc} + (q^2)_{gb}q_{ca} + (q^2)_{gc}q_{ab} = 0,$$

and then summing over $g = c$, which yields

$$q^3 = \lambda^2 q, \quad \lambda^2 = \tfrac{1}{2}\,\mathrm{tr}\,q^2.$$

With the permissible normalization $\lambda^2 = 1$, the eigenvalues of $q$, the electric charge matrix of the vector field, can only be $0, \pm 1$. Furthermore,

$$1 = \tfrac{1}{2}[m(+1) + m(-1)]$$

where $m(\pm 1)$ are the multiplicities of the respective $q$ eigenvalues. On invoking symmetry between positive and negative charge we learn that

$$m(+1) = m(-1) = 1.$$

[4] It is interesting to encounter here just the special electromagnetic interaction that I first considered in collaboration with H. C. Corben [Phys. Rev. **58**, 953 (1940)], while we were research fellows at Berkeley. That work was inspired by the cosmic ray investigations of J. R. Oppenheimer, R. Serber, and H. Snyder [Phys. Rev. **57**, 75 (1940)].

Thus the vector gauge field contains one pair of charged components, and all other components are electrically neutral. With an appropriate labeling of field components, the charge matrix $q$ is

$$q = \begin{pmatrix} 0 & -i & \vdots & 0 \\ i & 0 & \vdots & \\ \hline & 0 & \vdots & 0 \end{pmatrix}$$

and it can be verified that the quadratic conditions on $q_{ab}$ are satisfied.

The structure of the charge matrix asserts that

$$[q,t_1] = it_2\,, \quad [q,t_2] = -it_1\,, \quad [q,t_a] = 0\,, \quad a = 3,\cdots n\,.$$

According to these properties,

$$[t_a,t_b] - \sum_{c=3}^{n} t_{abc}t_c = t_{ab1}t_1 + t_{ab2}t_2 = 0 \qquad a, b = 3,\cdots n\,,$$

since the left-hand member commutes with $q$. The implied vanishing of the structure constants $t_{ab1}$, $t_{ab2}$, $a$, $b = 3,\cdots n$, also asserts that

$$[t_1,t_a] = t_{1a2}t_2\,, \quad [t_2,t_a] = t_{2a1}t_1\,, \quad a = 3,\cdots n.$$

By performing a linear transformation of the $t_a$ it can be arranged that

$$t_{12a} = i\delta_{a3}\,.$$

Then

$$[t_3,t_1] = it_2\,, \quad [t_2,t_3] = it_1\,, \quad [t_1,t_2] = it_3\,,$$

while

$$[t_{1,2,3},t_a] = 0 \qquad a = 4,\cdots n$$

and

$$[t_a,t_b] = \sum_{c=4}^{n} t_{abc}t_c \qquad a, b = 4,\cdots n\,.$$

Thus the group structure is completely factored into the three-dimensional isotopic spin group and a group with $(n - 3)$ parameters. According to the identification

$$q = t_3\,,$$

only zero and unit isotopic spin representations are contained in the vector field.

It may appear that these structural results refer just to a particular group representation rather than the group, for the general commutation relations of the unified theory

$$[T'_a,T'_b] = \sum_{c=0}^{n} t'_{abc}T'_c$$

imply that

$$[T_a,T_b] = \sum_{c=1}^{n} t_{abc}T_c - (e^2/f^2)q_{ab}Q$$

together with

$$[T_a,Q] = \sum_{b=1}^{n} q_{ab}T_b\,.$$

In more detail, these are

$$-i[T_1,T_2] = T_3 + (e^2/f^2)Q\,,$$
$$-i[T_3,T_1] = T_2, \qquad -i[T_2,T_3] = T_1\,,$$
$$[T_1,Q] = -iT_2\,, \qquad [T_2,Q] = iT_1\,,$$
$$[Q,T_3] = 0\,,$$

and

$$[T_{1,2,3},T_a] = 0\,, \quad [T_a,T_b] = \sum_{c=4}^{n} t_{abc}T_c \qquad a, b = 4,\cdots n\,.$$

Thus, the general commutation relations differ by the presence of the term $(e^2/f^2)Q$, added to $T_3$. But, it suffices to define

$$^*T_3 = \gamma^{-2}[T_3 + (e^2/f^2)Q]\,, \quad ^*T_{1,2} = \gamma^{-1}T_{1,2}\,,$$

where

$$\gamma = [1 + (e^2/f^2)]^{\frac{1}{2}}\,,$$

in order to regain the commutator structure of the three-dimensional isotopic spin group for $^*T_{1,2,3}$. It is essential that $Q$ and $T_3$ have the same commutation properties with $T_{1,2}$. We place

$$Q - T_3 = \gamma^2\,\tfrac{1}{2}\,Y\,, \quad [Y,{}^*T_{1,2,3}] = 0\,,$$

which yields

$$Q = {}^*T_3 + \tfrac{1}{2}\,Y$$

and identifies $Y$ with the hypercharge. Then

$$T_{1,2} = \gamma{}^*T_{1,2}\,, \quad T_3 = {}^*T_3 - (\gamma^2 - 1)\,\tfrac{1}{2}\,Y\,.$$

The group commutation relations govern the conserved integral quantities of the theory. In the unified formulation the latter are given by

$$\mathbf{T}'_a = \int (d\mathbf{x})[G^{0k}it'_a\phi_k + \tfrac{1}{2}\,\psi T'_a\psi]\,,$$

which includes

$$\mathbf{Q} = \int (d\mathbf{x})[G^{0k}iq\phi_k + \tfrac{1}{2}\,\psi Q\psi]$$

and

$$\mathbf{T}_a = \int (d\mathbf{x})[G^{0k}it_a\phi_k + (e/f)(iqG^{0k})_a A_k - (e/f)F^{0k}(iq\phi_k)_a + \tfrac{1}{2}\,\psi T_a\psi]\,.$$

The operators $*\mathbf{T}_{1,2,3}$ that obey isotopic spin commutation relations then appear as

$$*\mathbf{T}_3 = \int (d\mathbf{x})[G^{0k} it_3\phi_k + \tfrac{1}{2}\psi *T_3\psi]$$

and

$$*\mathbf{T}_{1,2} = \int (d\mathbf{x})[\gamma^{-1}G^{0k}it_{1,2}\phi_k \pm (e/f)\gamma^{-1}(G^{0k}_{2,1}A_k - F^{0k}\phi_{k2,1}) + \tfrac{1}{2}\psi *T_{1,2}\psi] .$$

In the last formula, the lower sign applies to $*\mathbf{T}_2$.

One can hardly fail now to recognize the variables that make explicit the three-dimensional isotopic spin invariance of the theory. The fields

$$*\phi_{\mu 3} = \gamma^{-1}[\phi_{\mu 3} + (e/f)A_\mu] ,$$
$$*G_3^{\mu\nu} = \gamma^{-1}[G_3^{\mu\nu} + (e/f)F^{\mu\nu}] ,$$
$$*A_\mu = \gamma^{-1}[A_\mu - (e/f)\phi_{\mu 3}] ,$$
$$*F^{\mu\nu} = \gamma^{-1}[F^{\mu\nu} - (e/f)G_3^{\mu\nu}] ,$$

together with

$$*\phi_{\mu 1,2} = \phi_{\mu 1,2} , \quad *G^{\mu\nu}_{1,2} = G^{\mu\nu}_{1,2}$$

are such that

$$*\mathbf{T}_a = \int (d\mathbf{x})[*G^{0k}it_a *\phi_k + \tfrac{1}{2}\psi *T_a\psi] ,$$

$$\mathbf{Q} = \int (d\mathbf{x})[*G^{0k}iq *\phi_k + \tfrac{1}{2}\psi Q\psi] .$$

For simplicity of presentation, the transformed Lagrange function is written without the fields $\phi_a^\mu$, $a = 4\cdots n$, which can easily be reinstated, and omitting the star designation on the fields. The result is

$$\mathfrak{L} = -\tfrac{1}{2}F^{\mu\nu}(\partial_\mu A_\nu - \partial_\nu A_\mu) + \tfrac{1}{4}F^{\mu\nu}F_{\mu\nu} - \tfrac{1}{2}G^{\mu\nu}[\partial_\mu\phi_\nu - \partial_\nu\phi_\mu + (\phi_\mu if\gamma t\phi_\nu)] + \tfrac{1}{4}G^{\mu\nu}G_{\mu\nu} + \tfrac{1}{2}i\psi\alpha^\mu(\partial_\mu - ie\gamma\tfrac{1}{2}YA_\mu - if\gamma\text{“}*T\phi_\mu\text{”})\psi + \tfrac{1}{2}im\psi\beta\psi .$$

All electromagnetic effects for the gauge field have disappeared, and the field $*A_\mu$, $*F_{\mu\nu}$ is identified as a hypercharge field!

It is tempting to assert that $*A_\mu$, $*F_{\mu\nu}$ is *the* hypercharge gauge field. The electromagnetic field would then be constructed as a linear combination of the hypercharge gauge field and a component of the isotopic spin gauge field. The alternative possibility is to include in the set $\phi_a^\mu$, $a = 4,\cdots n$, an Abelian gauge field that is coupled to hypercharge, with coupling constant $g$. Then two linear combinations of the hypercharge fields exist, one of which has the coupling constant $[g^2 + (\tfrac{1}{2}e\gamma)^2]^{\frac{1}{2}}$, while the other is completely uncoupled.

### LEPTON INTERACTIONS

The existence of three-dimensional isotopic spin invariance is conveyed by the unusual transformations in which $A_\mu$, $F_{\mu\nu}$ is combined with $\phi_\mu$, $G_{\mu\nu}$. A purely electromagnetic interaction with another system destroys this symmetry and gives the electromagnetic field its specific dynamical significance. Let the Lagrange function contain additional terms that are invariant under electromagnetic gauge transformations

$$\mathfrak{L}_e + j_e^\mu A_\mu ,$$

such as those representing charged lepton fields. The complete Lagrange function responds to an $(n + 1)$ parameter infinitesimal gauge transformation as

$$\delta\mathfrak{L} = j_e^\mu(\phi_\mu ieq\,\delta\lambda) .$$

Alternatively, we can write

$$A_\mu = \gamma^{-1}[*A_\mu + (e/f)*\phi_{\mu 3}] ,$$

which shows how a preferential direction is introduced in the isotopic space. The response to the three-dimensional isotopic spin gauge transformation

$$\delta *\phi_\mu = (\partial_\mu - if\gamma\,\text{“}t *\phi_\mu\text{”})\,\delta *\lambda$$

is again

$$\delta\mathfrak{L} = j_e^\mu(\phi_\mu ieq\,\delta\lambda)_3 ,$$

since $q = t_3$ and $*\phi_{\mu 1,2} = \phi_{\mu 1,2}$.

As we have remarked in the Introduction, this result contradicts the principle of stationary action, which demands that $\delta\mathfrak{L}$ be zero, to within a divergence term. Another nongauge invariant expression must be added, such as

$$-\tfrac{1}{2}\phi^\mu m_0^2\phi_\mu = -\tfrac{1}{2}m_0^2\sum_{a=1}^{3}\phi_a^\mu\phi_{\mu a} .$$

Then

$$\delta\mathfrak{L} = j_e^\mu(\phi_\mu ieq\,\delta\lambda) - \phi^\mu m_0^2(\partial_\mu - ieq\,A_\mu)\,\delta\lambda$$

and the stationary action principle asserts that

$$m_0^2(\partial_\mu - ieq\,A_\mu)\phi^\mu = ieq\,\phi_\mu j_e^\mu .$$

The field equation that is modified by the $m_0$ term is

$$(\partial_\nu - ieqA_\nu - if\,\text{“}t\phi_\nu\text{”})G^{\mu\nu} = -ieqF^{\mu\nu}\phi_\nu + k^\mu - m_0^2\phi^\mu ,$$
$$k^\mu = \tfrac{1}{2}\psi\alpha^\mu fT\psi .$$

This can be written as

$$\partial_\nu G^{\mu\nu} + m_0^2\phi^\mu = j^\mu ,$$

where

$$j^\mu = (G^{\mu\nu}ift\phi_\nu) + ieqA_\nu G^{\mu\nu} - ieqF^{\mu\nu}\phi_\nu + k^\mu ,$$

from which we derive

$$\partial_\mu j^\mu = m_0^2 \partial_\mu \phi^\mu = ieq(j_e^\mu + m_0^2 A_\mu)\phi^\mu .$$

It will be noted how the presence of the $m_0$ term is required to obtain the physically necessary equations of nonconservation. The corresponding integral relations are

$$\partial_0 f\mathbf{T} = \int (d\mathbf{x}) ieq(j_e^\mu + m_0^2 A_\mu)\phi^\mu ,$$

which asserts that

$$\partial_0 {}^*\mathbf{T}_{1,2} = \pm (e/f)\gamma^{-1} \int (d\mathbf{x}) (j_e^\mu + m_0^2 A_\mu)\phi_{2,1}^\mu$$
$$= -i[{}^*\mathbf{T}_{1,2}, P^0] .$$

We do not try to discuss here whether this mechanism implies reasonable magnitudes for the $T_3$ dependence of mass multiplets.

We make only one comment about the physical mass problem. For an Abelian gauge field coupled to a conserved current, there are sum rules[5] that require the existence of unit spin excitations with mass $m$ less than $m_0$. The sum rules are different, however, if the current is not conserved,[6] or if the field is non-Abelian. Then, no simple conclusions about the spectrum in relation to $m_0$ can be drawn.

[5] K. Johnson, Nucl. Phys. **25**, 435 (1961).

[6] D. G. Boulware and W. Gilbert, Phys. Rev. **126**, 1563 (1962).

# Field Theory of Matter

JULIAN SCHWINGER*
*Harvard University, Cambridge, Massachusetts*
(Received 23 March 1964)

A speculative field theory of matter is developed. Simple computational methods are used in a preliminary survey of its consequences. The theory exploits the known properties of leptons by means of a principle of symmetry between electrical and nucleonic charge. There are fundamental fields with spins 0, $\frac{1}{2}$, 1. The spinless field is neutral. Spin $\frac{1}{2}$ and 1 fields can carry both electrical and nucleonic charge. The multiplicity of any nonzero charge is 3. Explicit dynamical mechanisms for the breakdown of unitary symmetry and for the muon-electron mass difference are given. A more general view of lepton properties is proposed. Mass relations for baryon and meson multiplets are derived, together with approximate couplings among the multiplets. The weakness of $\phi$ production in $\pi-N$ collisions and the suppression of the $\phi\rightarrow\rho+\pi$ decay is explained.

## THE THEORY

TWO exact conservation laws of nature, those of electrical and nucleonic charge, have received similar dynamical interpretations through the structure of vector gauge fields.[1] An obvious physical difference between these charges is also accounted for dynamically by the observation that a massless physical particle will exist only for sufficiently weak interactions.[2] It is worth asking how far this qualified analogy can be pursued. The significance of the attempt lies in the basic concept of field theory[3] that a dynamical link must be established between the fundamental field variables and the observed particles. That relationship may be quite immediate for particles having predominantly electromagnetic interactions, with corresponding implications about the primitive fields, while it may be rather remote for the strongly interacting entities. The postulated analogy thus becomes a tool for analyzing the structure of the otherwise inaccessible substratum of fundamental fields and interactions.

The analogy between nucleonic charge $N$ and electrical charge $Q$ will be used first to explore the structure of the fundamental spin $\frac{1}{2}$ Fermi field $\psi$. Based upon the known leptons, $\mu^{\pm}$, $e^{\pm}$, $\nu$, $\bar{\nu}$, it can be assumed that such a field carries $Q=1$, 0, $-1$, and therefore also carries $N=1$, 0, $-1$. Furthermore, each value of $Q$ occurs twice among the leptons, which have $N=0$. A table of multiplicities based upon this information and the postulated symmetry between $N$ and $Q$ would then appear as

| $N \backslash Q$ | 1 | 0 | $-1$ |
|---|---|---|---|
| 1 | $n_+$ | 2 | $n_-$ |
| 0 | 2 | 2 | 2 |
| $-1$ | $n_-$ | 2 | $n_+$ |

which also incorporates a requirement of symmetry for the reflection of both charges. The total number of charged fields, of either type, is $4+2n$, $n=n_++n_-$. The simplest assumption is that $n=1$. Then twelve fields can be divided, for each sense of charge, into six charged fields and six neutral fields. The choice between the two possibilities $n_+=1$, $n_-=0$ and $n_-=1$, $n_+=0$ is a physical one since it must imply the correlation between $N$ and $Q$ that is embodied in the distinction between the proton ($N=Q$) and the charged cascade particle ($N=-Q$). We shall make the provisional assumption that $n_-=1$, $n_+=0$. Then the $N=\pm1$ parts of the fundamental Fermi field have three components each, one with $Q=\mp1$ and two with $Q=0$.

It has been proposed that the vector electromagnetic field $A_\mu$, the dynamical instrument of electrical charge, is the neutral partner of a unit charged vector field $Z_\mu$, which is coupled universally to unit changes of charge.[4] That idea, which also anticipated the existence of two neutrinos, implies the phenomenological current—current couplings of the weak interactions if a sufficiently massive particle is associated with the $Z$ field. Searches for more direct manifestations of this particle are in progress.

The qualitative analogy between electric and nucleonic charge suggests that $B_\mu$, the vector gauge field that is strongly coupled to $N$, is the neutral partner of a multicomponent vector field $V_\mu$, which is coupled to changes of nucleonic charge. The Bose field $V_\mu$ must carry compensating amounts of nucleonic charge. Now, there appears to be an approximate meaning to a world without leptons and the electromagnetic field, in the sense of a restricted time scale, which implies that rapid exchanges of nucleonic charge are limited to the

* Supported in part by the U. S. Air Force Office of Scientific Research under contract number A. F. 49(638)-589. This work was reported at the 30–31 January 1964 meeting at the University of Miami on Symmetry Principles at High Energies, and in Phys. Rev. Letters 12, 237 (1964). See also *ibid.* 12, 630 (1964).

[1] T. D. Lee and C. N. Yang, Phys. Rev. 98, 1501 (1955).

[2] J. Schwinger, Phys. Rev. 125, 397 (1962); 128, 2425 (1962).

[3] The term field theory must be qualified, since there are diverse opinions on the subject. I contend that the fundamental dynamical variables are field operators, while particles are identified as the stable or quasistable excitations of the coupled field system. There is no *a priori* relation between the primary dynamical fields and the secondary phenomenological fields that can be associated with the observed particles. For some further discussion see the *Miramare, Trieste, Lectures* (International Atomic Energy Agency, Vienna, 1963).

[4] J. Schwinger, Ann. Phys. 2, 407 (1957).



Reprinted from *The Physical Review* 135, B816–B830 (1964).

nonleptonic world, and therefore only connect the $N=\pm 1$ components of the fundamental Fermi field. It is for this reason that we deviate from the strict analogy and postulate that $V_\mu$ carries *two* units of nucleonic charge. One might speculate on the role that a unit nucleonic charged boson field could play for weak interactions, but we shall assume that it is not relevant to strong interactions.

What values of electrical charge can we reasonably suppose the $V_\mu$ field to carry? Here we shall only indicate the most plausible hypothesis, in the light of its physical consequences. It is that $V_\mu(N=\pm 2)$ has the same charge structure as $\psi(N=\pm 1)$; there are three components, one with $Q=\mp 1$ and two with $Q=0$. This is a threefold way.[5]

A simplified picture of the nucleonic world is obtained by considering the Fermi field $\psi(N=\pm 1; Q=\mp 1, 0, 0)$ and the Bose field $V_\mu(N=\pm 2; Q=\mp 1, 0, 0)$, each coupled to the neutral gauge field $B_\mu$, but without the direct interaction between $\psi$ and $V_\mu$ that exchanges nucleonic charge. The nucleonic charge interaction term in the Lagrange function is

$$fB_\mu[\tfrac{1}{2}\psi\alpha^\mu\nu_3\psi + iV^{\mu\nu}2\nu_3V_\nu] = fB_\mu[\bar\psi\gamma^\mu\psi + i(\bar V^{\mu\nu}2V_\nu - \bar V_\nu 2V^{\mu\nu})], \quad (1)$$

where the two equivalent forms correspond to the use of Hermitian fields that are eigenvectors of charge reflection matrices, or non-Hermitian fields that are charge eigenvectors. The matrix $\nu_3$ is the antisymmetrical nucleonic charge matrix with eigenvalues $\pm 1$. Symmetrization or antisymmetrization of operator multiplication in accordance with statistics is understood. Also understood is a summation over the three components contained in each non-Hermitian field, as illustrated by

$$\bar\psi\gamma^\mu\psi = \sum_{a=1}^{3} \bar\psi_a\gamma^\mu\psi_a .$$

This structure is invariant under three-dimensional unitary transformations of the $\psi_a$.

With appropriate restrictions on the kinematical part of the Lagrange function, the group of three-dimensional unitary transformations of the $\psi_a$ with coordinate-independent coefficients, the group $U_3$, is an exact invariance group in the simplified description. That property is expressed by the local conservation laws

$$\partial_\mu {}^{(\psi)}j_{ab}^{\mu}(x) = 0 ,$$

where

$${}^{(\psi)}j_{ab}^{\mu} = {}^{(\psi)}j_{ba}^{\mu\dagger} = -\bar\psi_b\gamma^\mu\psi_a = \psi_a(\bar\psi_b\gamma^\mu).$$

The associated integral quantities

$${}^{(\psi)}T_{ab} = {}^{(\psi)}T_{ba}^{\dagger} = \int (d\mathbf{x})(-)\bar\psi_b\gamma^0\psi_a$$

obey the unitary group-generator commutation relations

$$[T_{ab}, T_{cd}] = \delta_{bc}T_{ad} - \delta_{ad}T_{cb}.$$

According to our hypothesis about the structure of the Bose field $V_\mu$, a similar but independent $U_3$ invariance group can be assumed to exist for that field. The corresponding current operators are given by

$${}^{(V)}j_{ab}^{\mu} = -i(\bar V_b^{\mu\nu}V_{\nu a} - \bar V_{\nu b}V_a^{\mu\nu}).$$

The nucleonic charge currents for the two fields are

$${}^{(\psi)}j_N^{\mu} = -\sum_{a=1}^{3} {}^{(\psi)}j_{aa}^{\mu},$$

$${}^{(V)}j_N^{\mu} = -2\sum_{a=1}^{3} {}^{(V)}j_{aa}^{\mu}.$$

They are conserved independently in this idealization. The additional invariance transformations that do not refer to nucleonic charge belong to the subgroup of $U_3$, the unimodular or special unitary group, $SU_3$. The infinitesimal generators of this group, $'T_{ab}$, obey the commutation relations of the $T_{ab}$, together with the trace condition

$$\sum_{a=1}^{3} {}'T_{aa} = 0.$$

A few properties of these operators will be noted here. Eight independent Hermitian operators are introduced by the notation[6]

$$'T_{12} = T_1 + iT_2, \quad 'T_{21} = T_1 - iT_2, \quad 'T_{11} - 'T_{22} = 2T_3,$$

$$'T_{23} = U_1 + iU_2, \quad 'T_{32} = U_1 - iU_2, \quad 'T_{22} - 'T_{33} = 2U_3,$$

$$'T_{31} = V_1 + iV_2, \quad 'T_{13} = V_1 - iV_2, \quad 'T_{33} - 'T_{11} = 2V_3,$$

where

$$T_3 + U_3 + V_3 = 0.$$

We also define

$$'T_{11} = Q, \quad 'T_{11} + 'T_{22} = -'T_{33} = Y$$

so that

$$'T_{22} = Y - Q$$

and

$$T_3 = Q - \tfrac{1}{2}Y, \quad U_3 = Y - \tfrac{1}{2}Q, \quad V_3 = -\tfrac{1}{2}(Q + Y).$$

[5] That a threefold internal multiplicity could be basic was suggested quite long ago [S. Sakata, Progr. Theoret. Phys. (Kyoto) **16**, 686 (1956)], as was the symmetry represented by the group $SU_3$ [S. Ogawa, Progr. Theoret. Phys. (Kyoto) **21**, 209 (1959)]. An inability to describe the known baryons transferred attention to the eight-dimensional representation of this group [Y. Ne'eman, Nucl. Phys. **26**, 222 (1961); M. Gell-Mann, Phys. Rev. **125**, 1067 (1962)]. The threefold multiplicity is introduced here at a deeper dynamical level than the observed particles. An independent attempt in this direction has been made by M. Gell-Mann, Phys. Letters (to be published). He introduces particles of fractional charge which can be detected, presumably, only by their "palpitant piping, chirrup, croak, and quark." [Hartley Burr Alexander, quoted in *Webster's New International Dictionary of the English Language* (G. & C. Merriam Company, Publishers, Springfield, Massachusetts, 1944), 2nd. ed., p. 2033].

[6] Compare S. P. Rosen, Phys. Rev. Letters **11**, 100 (1963) and references therein to contributions of C. A. Levinson, H. J. Lipkin and S. Meshkov.

The various ordered sets, $T_{1,2,3}$; $U_{1,2,3}$; $V_{1,2,3}$; and $2T_2$, $2U_2$, $2V_2$, each obey the commutation relations of three-dimensional angular momentum. The $T$ operators commute with $Y$, the $U$ operators commute with $Q$, and the $V$ operators commute with $Y-Q$. An operator that commutes with all eight generators is

$$\frac{1}{2}\sum_{a,b=1}^{3} {}'T_{ab}\,{}'T_{ba}$$

$$=\sum_{a=1}^{3}(T_a^2+U_a^2+V_a^2) -\tfrac{1}{9}[(T_3-U_3)^2+(U_3-V_3)^2+(V_3-T_3)^2].$$

The relation between gauge invariance and a zero-mass, unit-spin particle is such that the latter ceases to exist when the coupling becomes sufficiently strong that an excitation produced by the field descends to zero mass. Under these conditions of strong binding between oppositely charged fields, we can anticipate the existence of low-lying boson excitations with spin and parity $0^\pm$, $1^\pm$, $\cdots$. Odd-parity, spin-zero, and spin-one excitations, for example, will be produced by the effect on the vacuum state of operators such as

$$\bar\psi_a\gamma_5\psi_b-\tfrac{1}{3}\delta_{ab}\bar\psi\gamma_5\psi$$

and

$$\bar\psi_a\gamma^\mu\psi_b-\tfrac{1}{3}\delta_{ab}\bar\psi\gamma^\mu\psi,$$

or

$$\bar V_a{}^{\mu\nu}\epsilon_{\mu\nu\lambda\kappa}V_b{}^{\lambda\kappa}-\tfrac{1}{3}\delta_{ab}\bar V^{\mu\nu}\epsilon_{\mu\nu\lambda\kappa}V^{\lambda\kappa},$$

and

$$\bar V_a{}^{\mu\nu}V_{\nu b}-\bar V_{\nu a}V_b{}^{\mu\nu}-\tfrac{1}{3}\delta_{ab}(\bar V^{\mu\nu}V_\nu-\bar V_\nu V^{\mu\nu}),$$

which form independent unitary octuplets. It is not possible to decide in any general way which of these octuplets is least massive. The spin-one, odd-parity unitary singlets generated by the operators $\bar\psi\gamma^\mu\psi$ and $\bar V^{\mu\nu}V_\nu-\bar V_\nu V^{\mu\nu}$ are physically connected through the gauge field $B^\mu$. Thus, a particle of the corresponding quantum numbers is not associated with a specific field, but is a joint physical attribute of all three fields.

Strong nucleonic binding forces also operate between the sets of oppositely charged fields $V(N=\pm2)$ and $\psi(N=\mp1)$. We therefore anticipate the existence of low-lying fermion excitations of nucleonic charge $\pm1$, with spins and parities $\frac{1}{2}^\pm$, $\frac{3}{2}^\pm$, $\cdots$. Spin-$\frac{1}{2}$ excitations, for example, are produced by the effect on the vacuum state of the non-Hermitian operators

$$\bar\psi_a\gamma_\mu V_b{}^\mu,\quad \bar V_a{}^\mu\gamma_\mu\psi_b.$$

With an additional $\gamma_5$ factor, the spin-$\frac{1}{2}$ states of opposite parity are produced. It is important to recognize that these excitations form nonuplets. The multiplicity is $3\times3=9$ since the unitary transformations of the $\psi$ and $V$ fields are still independent. A different kind of fermion multiplet structure is generated by operator products of the form (omitting spinor and vector indices)

$$[(\bar\psi_a\bar\psi_b+\bar\psi_b\bar\psi_a)\psi_c-\tfrac{1}{2}(\delta_{bc}\bar\psi_a+\delta_{ac}\bar\psi_b)\bar\psi\psi]V_d,$$

and their adjoints. The multiplicity of this set is $3\times15=45$. The interchange of $\psi$, $\bar\psi$ with $V$, $\bar V$ gives another set. One should also note the boson states generated by

$$(\bar\psi_a\bar\psi_b\pm\bar\psi_b\bar\psi_a)V_c$$

and their adjoints. The multiplicity is $3\times6=18$ or $3\times3=9$ according to the choice of sign.

The interaction between the Fermi field $\psi$ and the Bose field $V_\mu$ must now be introduced. Nucleonic charge conservation dictates the non-Hermitian combinations $\psi_a\psi_b\bar V_c$ and $\bar\psi_a\bar\psi_b V_c$. But no linear combination of these forms can be invariant under common unitary transformations of the $\psi$ and $V$ fields. The mechanism for the breakdown of unitary symmetry is thereby identified.

The violation of unitary symmetry is minimized if scalar products are formed between $\bar V$ and a $\psi$, $V$ and a $\bar\psi$, which leaves objects that have the transformation properties of a component of $\psi$ and $\bar\psi$. We shall designate the preferred direction in the three-dimensional unitary space as the third axis. It must be associated with an electrically neutral component of $\psi$. The resulting theory is invariant under $U_2$, a single two-dimensional unitary group of transformations on the $\psi$ and $V$ fields. The generators of the subgroup $SU_2$ are the total isotopic spin operators

$$T_a={}^{(\psi)}T_a+{}^{(V)}T_a,\quad a=1,2,3$$

and the multiplicative phase group is generated by the total hypercharge

$$Y={}^{(\psi)}Y+{}^{(V)}Y.$$

These operators are constants of the motion and the related currents are conserved.

The three components of $\bar\psi_a$ and $\bar V_a$ can now be given standard isotopic spin and hypercharge labels. Thus $\bar\psi_1$ and $\bar\psi_2$ are designated $Y=+1$, $T=\frac{1}{2}$, $T_3=+\frac{1}{2}$ and $-\frac{1}{2}$, respectively while $\bar\psi_3$ is assigned $Y=T=0$. The precise significance of these statements, together with the electrical charge labels, is given by commutation relations of the type

$$[\bar\psi,T_a]=t_a\bar\psi,\quad [\bar\psi,Y]=y\bar\psi,\quad [\bar\psi,Q]=q\bar\psi,$$

where

$$t_3=\tfrac{1}{2}\begin{pmatrix}+1&&\\&-1&\\&&0\end{pmatrix},\quad y=\begin{pmatrix}+1&&\\&+1&\\&&0\end{pmatrix},\quad q=\begin{pmatrix}+1&&\\&0&\\&&0\end{pmatrix}$$

and

$$t_1+it_2=\begin{pmatrix}\cdot&1&\cdot\\\cdot&\cdot&\cdot\\\cdot&\cdot&\cdot\end{pmatrix},\quad t_1-it_2=\begin{pmatrix}\cdot&\cdot&\cdot\\1&\cdot&\cdot\\\cdot&\cdot&\cdot\end{pmatrix}.$$

For completeness we record the matrices $u_a$ and $v_a$ that are similarly associated with the operators that are not constants of the motion,

$$U_a={}^{(\psi)}U_a+{}^{(V)}U_a,\qquad a=1,2.$$
$$V_a={}^{(\psi)}V_a+{}^{(V)}V_a,$$

They are given by

$$u_1+iu_2=\begin{bmatrix}\cdot&\cdot&\cdot\\ \cdot&\cdot&1\\ \cdot&\cdot&\cdot\end{bmatrix},\quad u_1-iu_2=\begin{bmatrix}\cdot&\cdot&\cdot\\ \cdot&\cdot&\cdot\\ \cdot&1&\cdot\end{bmatrix},$$

$$v_1+iv_2=\begin{bmatrix}\cdot&\cdot&\cdot\\ \cdot&\cdot&\cdot\\ 1&\cdot&\cdot\end{bmatrix},\quad v_1-iv_2=\begin{bmatrix}\cdot&\cdot&1\\ \cdot&\cdot&\cdot\\ \cdot&\cdot&\cdot\end{bmatrix}.$$

All these matrices comprise a three-dimensional representation of the $SU_3$ generator commutation relations. Note that the commutation relations for the field operators are of the form illustrated by

$$[V_\mu,T_a]=-V_\mu t_a=-t_a{}^{\mathrm{tr}}V_\mu$$

so that the negative transposed matrices appear. The latter constitute an inequivalent three-dimensional representation of the $SU_3$ commutation relations.

The structure of the $\psi-V$ coupling term is not yet completely specified. There is the choice of using a symmetrical or an antisymmetrical combination of $\psi_a$ and $\psi_b$, before one forms the scalar product with $\bar{V}_b$ and sets $a=3$. If the antisymmetrical combination is chosen, we get, effectively,

$$\psi_3\psi_1\bar{V}_1+\psi_3\psi_2\bar{V}_2,$$

and its adjoint, so that $\bar{V}_3$ and $V_3$ do not appear in this coupling. Such a theory is invariant under independent phase transformations of the $V_3$ field, which would imply the existence of another conservation law on the same dynamical level as isotopic spin and hypercharge conservation. Since there is no evidence for additional selection rules, we reject that possibility and choose the symmetrical combination. The space-time character of the vector coupling is then also fixed. If we use Hermitian fermion fields, the possible double nucleonic charged vector structures are

$$\psi_a\alpha^\mu\nu_{1,2}\psi_b,\quad \psi_a\gamma_5\alpha^\mu\nu_{1,2}\psi_b,$$

where $\nu_1$ and $\nu_2$ are the real, symmetrical nucleonic charge reflection matrices. In view of the symmetry of the real matrices $\alpha^\mu$, the antisymmetry of the real matrix $\gamma_5$, and Fermi-Dirac statistics, the two structures are, respectively, antisymmetrical and symmetrical in $a$ and $b$. This selects the pseudovector combination. The result of these considerations is the Hermitian coupling term

$$f'\tfrac{1}{2}[\psi_3 i\gamma_5\alpha^\mu\nu_1\psi V_\mu{}^{(1)}+\psi_3 i\gamma_5\alpha^\mu\nu_2\psi V_\mu{}^{(2)}]$$
$$=f'(1/2^{1/2})[\bar{\psi}_3\gamma_5\gamma^\mu\beta\bar{\psi}V_\mu+\psi_3\gamma_5\beta\gamma^\mu\psi\bar{V}_\mu],\quad (2)$$

where $\beta$ is the real, antisymmetrical matrix $i\gamma^0$, and

$$V_\mu{}^{(1)}=2^{-1/2}(V_\mu+\bar{V}_\mu),\quad V_\mu{}^{(2)}=2^{-1/2}i(V_\mu-\bar{V}_\mu).$$

The latter is the explicit diagonalization transformation for the nucleonic charge matrix

$$\nu_3=\begin{pmatrix}0&-i\\ i&0\end{pmatrix}.$$

The equivalence of the two forms also depends upon the specific identifications

$$\nu_1=\begin{pmatrix}1&\\ &-1\end{pmatrix},\quad \nu_2=\begin{pmatrix}&1\\ 1&\end{pmatrix}.$$

The description of $V_\mu$ as a pseudovector field is to some extent conventional but it is a very natural one in the framework of Hermitian fields, where space reflection is represented by the purely geometrical transformation (spatial coordinates are omitted)

$$P:\ \psi\to\beta\psi,\quad (V_0,V_k)\to(-V_0,V_k),$$
$$(B_0,B_k)\to(B_0,-B_k).$$

The repetition of this operation reverses the sign of $\psi$, while leaving the Bose fields unchanged. This can be expressed by the following property of the parity operator in relation to nucleonic charge, or angular momentum,

$$P^2=(-1)^N=(-1)^{2J}.$$

One can always combine $P$ with a nucleonic phase transformation such as $\exp[-\frac{1}{2}\pi iN]$, which obeys

$$(Pe^{-\frac{1}{2}\pi iN})^2=1.$$

The latter is represented in its effect on the non-Hermitian fields by

$$Pe^{-\frac{1}{2}\pi iN}:\ \psi\to\gamma^0\psi,\quad \bar{\psi}\to\bar{\psi}\gamma^0,\quad (V_0,V_k)\to(V_0,-V_k),$$
$$(B_0,B_k)\to(B_0,-B_k).$$

The theory possesses another discrete invariance operation in the combined reflection of nucleonic and hypercharge. It is represented by

$$\psi\to\bar{\psi}\gamma^0e^{-\pi it_2},\quad \bar{\psi}\gamma^0\to e^{+\pi it_2}\psi,$$
$$G:\ V_\mu\to\bar{V}_\mu e^{-\pi it_2},\quad \bar{V}_\mu\to e^{+\pi it_2}V_\mu,$$
$$B_\mu\to -B_\mu,$$

where the isotopic spin rotation is introduced to maintain the structure of the isotopic spin currents. The repetition of this operation, which we shall designate as hyperparity, reverses the sign of those field components that carry hypercharge and leaves the others unchanged. This is expressed by

$$G^2=(-1)^Y=(-1)^{2T}.$$

To complete the theory we must specify the dynamical properties of the $A_\mu$ and $Z_\mu$ fields and introduce the leptonic component of the Fermi field,

$$\chi=\psi(N=0).$$

The six elements of this field can be classified by a leptonic charge $L=\pm1$ in such a way that all values of $Q$ are represented for a given $L$,

$$\chi=\chi(L=\pm1;Q=0,\pm1,\mp1).$$

One may arbitrarily identify the physical realization of $L=Q$ with the muon, and of $L=-Q$ with the electron. We shall label these components of $\chi$ as $\chi_2$ and $\chi_3$, respectively. In accordance with the various electric charge assignments, the coupling with the electromagnetic field is given by

$$eA_\mu[-\bar\psi_1\gamma^\mu\psi_1-i(\bar V_1{}^{\mu\nu}V_{\nu1}-\bar V_{\nu1}V_1{}^{\mu\nu}) +\bar\chi_2\gamma^\mu\chi_2-\bar\chi_3\gamma^\mu\chi_3+i(\bar Z^{\mu\nu}Z_\nu-\bar Z_\nu Z^{\mu\nu})]. \quad (3)$$

At this dynamical level the internal symmetry groups are restricted to the phase transformations that define the charges $N$, $Y$, and $Q$. There is also the discrete operation of charge reflection,

$$C:\quad \begin{aligned} &\psi\leftrightarrow\bar\psi\gamma^0, && \chi\leftrightarrow\bar\chi\gamma^0,\\ &V_\mu\leftrightarrow\bar V_\mu, && Z_\mu\leftrightarrow(\pm)\bar Z_\mu,\\ &A_\mu\to-A_\mu, && B_\mu\to-B_\mu.\end{aligned}$$

The charge parity operator obeys

$$C^2=1.$$

The structure of the interaction term between the $Z$ field and the leptonic field is relatively well established in the form[7]

$$e'\bar Z_\mu[\bar\chi_1\gamma^\mu(1+i\gamma_5)\chi_2+\bar\chi_3\gamma^\mu(1-i\gamma_5)\chi_1] +e'Z_\mu[\bar\chi_2\gamma^\mu(1+i\gamma_5)\chi_1+\bar\chi_1\gamma^\mu(1-i\gamma_5)\chi_3]. \quad (4a)$$

This coupling term incorporates the connection between neutrino polarization and the electrical charge of the accompanying lepton that is expressed by the requirement of invariance under the substitution

$$\chi_1\to e^{i\lambda(i\gamma_5)}\chi_1,\quad \chi_2\to e^{i\lambda}\chi_2,\quad \chi_3\to e^{-i\lambda}\chi_3.$$

It also contains an assertion of dynamical equivalence between $\mu$ and $e$ in the sense of the charge reflection transformation

$$(1+i\gamma_5)\chi_2\leftrightarrow(1-i\gamma_5)\chi_3,\quad Z\leftrightarrow\bar Z.$$

Finally, the nonequivalence of the left-polarized neutrinos that accompany $\mu^+$ and $e^+$ is represented by the distinction between the field operators $\bar\chi_1$ and $\chi_1$. The only parity operation admitted by this coupling term is

$$CP:\quad \begin{aligned} &\chi\to i\bar\chi,\quad \bar\chi\to i\chi,\\ &(Z_0,Z_k)\leftrightarrow(-\bar Z_0,\bar Z_k),\\ &(A_0,A_k)\leftrightarrow(-A_0,A_k).\end{aligned}$$

We shall introduce an interaction between the $Z$ field and nucleonic charge-bearing fields on the basis of an analogy between $\chi_a$ and $\psi_a$, $a=1,2,3$, one that has been foreshadowed in the labelling of field components. The single electrically charged and the two uncharged components of $\psi$ are to be set into correspondence with the single uncharged and the two charged components of $\chi$. The substitution $\bar\chi_1\chi_2\to\bar\psi_1\psi_2$ preserves the charge property of the operator while $\bar\chi_1\chi_3\to\bar\psi_1\psi_3$ reflects it. Hence the latter must be accompanied by a $Z$-charge reflection. The result is the $\psi-Z$ coupling term

$$e'\bar Z_\mu[\bar\psi_1\gamma^\mu(1+i\gamma_5)\psi_2+\bar\psi_1\gamma^\mu(1-i\gamma_5)\psi_3] +e'Z_\mu[\bar\psi_2\gamma^\mu(1+i\gamma_5)\psi_1+\bar\psi_3\gamma^\mu(1-i\gamma_5)\psi_1]. \quad (4b)$$

The specific identifications of $\psi_2$ and $\psi_3$ relative to $\gamma_5$ is in tentative anticipation of the empirical requirements.

An interaction between the $V$ and $Z$ fields is suggested by the recognition that $-\bar\psi_2\gamma^\mu\psi_1$ and $-\bar\psi_3\gamma^\mu\psi_1$ are the fermion contributions to the current vectors associated with $T_1+iT_2$ and $V_1-iV_2$, respectively. The electromagnetic field interacts with the current of total charge $Q=T_3+\frac{1}{2}Y$. It is very natural to suppose that the charged partners of the electromagnetic field also interact with the total quantities of the indicated type. This supplies the coupling term

$$e'\bar Z_\mu[i(\bar V_1{}^{\mu\nu}V_{\nu2}-\bar V_{\nu1}V_2{}^{\mu\nu})+i(\bar V_1{}^{\mu\nu}V_{\nu3}-\bar V_{\nu1}V_3{}^{\mu\nu})] +e'Z_\mu[i(\bar V_2{}^{\mu\nu}V_{\nu1}-\bar V_{\nu2}V_1{}^{\mu\nu}) +i(\bar V_3{}^{\mu\nu}V_{\nu1}-\bar V_{\nu3}V_1{}^{\mu\nu})]. \quad (4c)$$

We shall include no boson counterparts to the $\gamma_5$ terms in (4b), although a basis for such terms could be found in the similarity between $(1\pm i\gamma_5)\psi$ and

$$V^{\mu\nu}\pm i(\epsilon V)^{\mu\nu};$$
$$(\epsilon V)^{\mu\nu}=\tfrac{1}{2}\epsilon^{\mu\nu\lambda\kappa}V_{\lambda\kappa}.$$

This hypothesis assigns the entire dynamical origin of the nonconservation of space parity to the fundamental fermion field. It must meet such experimental tests as those provided by the observed polarization properties in the nonleptonic decays of hyperons.

The complete interaction between the $Z$ field and the nucleonic charge bearing fields possesses the weak parity property

$$CP:\quad \begin{aligned} &\psi\to i\bar\psi,\quad \bar\psi\to i\psi,\\ &(V_0,V_k)\leftrightarrow(-\bar V_0,\bar V_k),\\ &(B_0,B_k)\leftrightarrow(-B_0,B_k).\end{aligned}$$

The repetition of this operation is described, for all fields, by

$$(CP)^2=(-1)^{N+L}=(-1)^{2J}.$$

The dynamical structure of the theory is contained in the four coupling terms (1), (2), (3), and (4a)–(4c).

### FURTHER SPECULATIONS

An alternative view of the $Z$ field and of leptonic properties is suggested by an attempt to extend the pattern of multiplicities that have been established for the fundamental spin $\frac{1}{2}$ field $\psi$ to the set of fundamental

[7] Apart from notational differences and a change in sign for $\gamma_5$, this is the coupling term (14) of Ref. 4.

unit spin fields $A$, $B$, $Z$, and $V$. We recall that $\psi$ has six charged components and six neutral components, with either sense of charge. Of the unit spin fields, the number of nucleonic charged components is six, being the fields $V$ and $\bar{V}$, and the number of electrically neutral fields is also six, comprising the fields $A$, $B$, and the four electrically uncharged components of $V$ and $\bar{V}$. But when we count the $N=0$ components $[A,B,Z,\bar{Z}]$ and the electrically charged components $[Z,\bar{Z},V_1,\bar{V}_1]$ we find each number to be only four. Should the occult power of multiplicity six be sufficiently impressive, we would be led to introduce another spin one field carrying $N=0$ and $Q=\pm1$, that is, a second $Z$ field!

We shall try to accommodate two $Z$ fields dynamically by using a vector form of coupling for one field and pseudovector coupling for the other. This is illustrated by the leptonic coupling term

$$2^{1/2}e'[Z_\mu\tfrac{1}{2}\chi\alpha^\mu t\chi+Z_\mu'\tfrac{1}{2}\chi\alpha^\mu i\gamma_5\{t_3,t\}\chi], \qquad (1')$$

which is written in terms of Hermitian fields. Here

$$Zt=Z_1t_1+Z_2t_2$$

and $t_{1,2,3}$ are the imaginary antisymmetrical unit isotopic matrices (they are the $SU_3$ matrices $2u_2$, $2v_2$, $2t_2$, respectively). The structure (1′) is left invariant under the neutrino parity reflection

$$\chi\rightarrow[-it_3+\gamma_5(1-t_3^2)]\chi,$$

combined with the interchange

$$Z_\mu\leftrightarrow Z_\mu'.$$

An alternative form employs the fields

$$W_\mu=2^{-1/2}(Z_\mu+Z_\mu'),\quad W_\mu'=2^{-1/2}(Z_\mu-Z_\mu'),$$

namely,

$$e'[W_\mu\tfrac{1}{2}\chi\alpha^\mu(1+i\gamma_5t_3)t(1+i\gamma_5t_3)\chi$$
$$+W_\mu'\tfrac{1}{2}\chi\alpha^\mu(1-i\gamma_5t_3)t(1-i\gamma_5t_3)\chi]. \qquad (2')$$

The transformation induced by neutrino parity reflection now appears as

$$W_\mu\rightarrow W_\mu,\quad W_\mu'\rightarrow -W_\mu'.$$

If the $W'$ term were absent (2′) would be the leptonic $Z$-field coupling (4a). But the same practical result appears if the particles of the $W'$ field are sufficiently more massive than the $W$ particles. This is analogous to the empirical mass relation between the two types of charged leptons, $\mu$ and $e$. One might hope that both kinds of mass splitting have a common dynamical origin, which would also identify the mechanism for the individual violation of $P$ and $C$ invariance, since the effect of either transformation is to interchange $W$ and $W'$.

We thus seek a field that is coupled to the $N=0$, $Q=\pm1$ members of the spin $\tfrac{1}{2}$ and spin 1 duodecuplets. In view of the unique role it must play, the field may be presumed to lie outside the established framework, which suggests that it is a spin 0 field. This field $\phi$ does not carry nucleonic charge and we infer that it is also electrically neutral. (Compare the $\sigma$ field of Ref. 4.) The proposed analogy between the (non-Hermitian) fields $W'$ and $\chi_2$ is realized by the coupling terms

$$-h\phi\bar{\chi}_2\chi_2-h'\tfrac{1}{2}\phi^2\bar{W}'^\mu W_\mu' \qquad (3')$$

in which $h$ and $h'$ are dimensionless coupling constants. The spinless field is a weak parity scalar. It also reverses sign under the charged lepton parity reflection

$$\chi\rightarrow[(1-t_3^2)+i\gamma_5t_3]\chi,$$
$$W_\mu\rightarrow W_\mu,\quad W_\mu'\rightarrow -W_\mu'.$$

In consequence of the large mass displacements it must produce, we cannot assume that (3′) is a weak interaction. An important aspect of the field $\phi$ is its vacuum expectation value $\phi_0$, a quantity that has the dimensions of mass and which is not required to vanish by its space-time transformation properties. If it is assumed that no mass constant is associated with $\phi$, the field equations imply the vanishing vacuum expectation value

$$\langle h\bar{\chi}_2\chi_2+h'\phi\bar{W}'^\mu W_\mu'\rangle=\phi_0F(\phi_0^2)=0.$$

It is conceivable that the stability of the vacuum requires that $\phi_0\neq0$. Its magnitude would then be given by the (presumably unique) root of

$$F(\phi_0^2)=0.$$

Either sign of $\phi_0$ can be chosen. But once a sign is adopted, one can no longer perform a charged lepton parity reflection. This implies that the muon, in particular, has acquired a mass, as the term $-h\phi_0\bar{\chi}_2\chi_2$ qualitatively suggests. The structure of the $h'$ coupling term also indicates that particles associated with the $W'$ and $\phi$ fields will receive substantial mass increments. It is interesting that the possibility of electron parity reflection ($\chi_3\rightarrow -i\gamma_5\chi_3$) would persist if only the $W$-lepton coupling were considered, but fails on including the $W'$ term. Thus, the interference of the two kinds of weak interaction imply a nonzero electron mass. It is not proposed that such higher order effects of the weak interactions produce the observed electron mass, but perhaps they supply the seed that flowers under the influence of electromagnetic interactions.

The same analogies and arguments used in constructing (4b) and (4c) give the vector and pseudovector forms that couple $Z$ and $Z'$ to the nucleonic charge-bearing fields. We have observed that a neutrino parity reflection, $\chi_1\rightarrow\gamma_5\chi_1$, induces the interchange of $Z$ and $Z'$ or the substitution $W\rightarrow W$, $W'\rightarrow -W'$. No transformation of $\psi$ and $V$ can reverse the latter sign change. Thus, the presence of two kinds of weak interactions generates both an electron and a neutrino mass, but only the former could be amplified through electromagnetic action. It is to be hoped that the new possibilities offered by the idea of the $W'$ field will stimulate more precise measurements of the neutrino mass and of lepton polarizations. The best available polarization measurements still allow a 10% leeway for the effect of the reversed polarizations characteristic of $W'$ coupling. That is, a $W'$ particle with only a little more than twice the mass of the $W$ particle is compatible with present observations.

It should be remarked that the new hypothetical domain of strong interactions described by the $W'$ and $\phi$ fields is quite insulated from the strongly interacting nuclear world. This solves the problem of giving the large muon mass a dynamical origin without implying strong nonelectromagnetic muon-nucleon interactions. A direct experimental confrontation of these ideas may require higher energies than are now available. Perhaps the most immediate possibility for an experimental test lies in the detection of anomalous electromagnetic production of muon pairs, proceeding through the charged $W'$ field and the strong coupling of the latter to muons through the intermediary of the $\phi$ field.

## SIMPLE CALCULATIONS

The intent of this section is to obtain a first orientation toward some of the physical implications of the theory by using very simple calculational methods. This includes the application of perturbation theory to interactions that are not weak. When such methods lead to gratifying numerical results, that must be regarded as a sign of physical processes at work to bring about unanticipated simplifications, which are exploited but not explained by the crude calculational techniques.

We begin with the highly symmetrical and physically degenerate situation that is created by the strong nucleonic charge binding forces, interaction (1), and try to estimate the effect of the symmetry-destroying coupling between the $\psi$ and $V$ fields, interaction (2).

### Baryons ($\frac{1}{2}^+$)

Nine degenerate states of unit nucleonic charge are represented by the vectors

$$\langle ab| \sim \langle\,|\bar{\psi}_a V_b$$

together with their adjoints

$$|ab\rangle \sim \psi_a \bar{V}_b|\,\rangle .$$

The notation indicates that the states transform under the two independent groups of $U_3$ transformations in the same manner as the operators acting on the invariant vacuum state. An explicit construction of the states would include functions of the unitary invariants formed from the two sets of field operators.

To have a nonvanishing matrix with respect to these states, a perturbation must conserve the nucleonic charges associated with each field. The $\psi-V$ coupling is first effective in the second order of $f'$. That perturbation is symbolized by

$$\begin{aligned} M_{f'}{}^{(2)} &= f'^2\psi_3\psi\bar{V}V\bar{\psi}\bar{\psi}_3 ,\\ &= f'^2\sum_{cd}\psi_3\psi_c\bar{\psi}_d\bar{\psi}_3\bar{V}_c V_d . \end{aligned}$$

In view of the product structure of the states and of the individual terms in this summation, the computation of the matrix with respect to the nonuplet of baryon states essentially reduces to

$$\langle V_b\bar{V}_c V_d\bar{V}_{b'}\rangle = \delta_{bb'}\delta_{cd}k_1 + \delta_{bc}\delta_{b'd}k_2 ,$$

according to the requirement of $V$ unitary invariance, together with

$$\begin{aligned} &\langle\bar{\psi}_a\psi_3\psi_c\bar{\psi}_d\bar{\psi}_3\psi_{a'}\rangle \\ &\quad = \delta_{aa'}(\delta_{cd}+\delta_{c3}\delta_{d3})l_1 + (\delta_{a3}\delta_{cd}\delta_{a'3}+\delta_{a3}\delta_{c3}\delta_{a'd} \\ &\qquad\qquad + \delta_{ac}\delta_{d3}\delta_{a'3}+\delta_{ac}\delta_{da'})l_2 , \end{aligned}$$

which also invokes the symmetry between $\psi_3$ and $\psi_c$, $\bar{\psi}_3$ and $\bar{\psi}_d$, in the fundamental interaction. On combining the factors, and omitting a term that displaces the whole multiplet, we obtain the following matrix form:

$$\begin{aligned} &\langle ab|M_{f'}{}^{(2)}|a'b'\rangle \\ &\quad = \delta_{aa'}\delta_{b3}\delta_{b'3}m_1 + \delta_{bb'}\delta_{a3}\delta_{a'3}m_2 \\ &\qquad\qquad + (\delta_{ab}\delta_{a'b'}+\delta_{ab}\delta_{a'3}\delta_{b'3}+\delta_{a'b'}\delta_{a3}\delta_{b3})m_3 . \end{aligned}$$

Various submatrices are supplied by the conservation laws of isotopic spin and hypercharge. Thus, the two states with $b=3$ and $a=1$, 2 are characterized by $Y=1$, $T=\frac{1}{2}$, and $T_3=+\frac{1}{2}$, $-\frac{1}{2}$, respectively. This set is identified with the nucleon

$$N^{+,0}:\ \langle 13|,\ \ \langle 23|,$$

and the corresponding mass displacement is

$$\langle M\rangle_N = m_1 .$$

The states with $a=3$ and $b=1$, 2 possess $Y=-1$, $T=\frac{1}{2}$ and $T_3=-\frac{1}{2}$, $\frac{1}{2}$, respectively. This is the cascade particle,

$$\Xi^{-,0}:\ \langle 31|,\ \ \langle 32|,$$

and

$$\langle M\rangle_\Xi = m_2 .$$

Three states with quantum numbers $Y=0$, $T=1$ and $T_3=1$, $-1$, 0, the $\Sigma$-particle, are, respectively,

$$\Sigma^{+,-,0}:\ \langle 12|,\ \ -\langle 21|,\ \ 2^{-1/2}(-\langle 11|+\langle 22|),$$

and

$$\langle M\rangle_\Sigma = 0 ,$$

which sets the mass origin for this calculation.

Finally, we consider the two states with $Y=T=0$,

$$2^{-1/2}(\langle 11|+\langle 22|),\ \ \langle 33|.$$

The submatrix of $M$ for these states is

$$\begin{pmatrix} 2m_3, & 2^{3/2}m_3 \\ 2^{3/2}m_3, & m_1+m_2+3m_3 \end{pmatrix},$$

where the second row and column refer to ⟨33| and its adjoint. We shall denote the two mass eigenvalues of this matrix by $\Lambda-\Sigma$ and $Y^0-\Sigma$, where $Y^0>\Lambda$. Accordingly,

$$Y^0+\Lambda-2\Sigma=m_1+m_2+5m_3,$$
$$-(Y^0-\Sigma)(\Sigma-\Lambda)=2m_3(m_1+m_2)-2m_3^2$$

and, in the same kind of notation,

$$-(\Sigma-N)=m_1,\quad \Xi-\Sigma=m_2.$$

Then

$$\tfrac{1}{5}(Y^0+\Lambda-N-\Xi)=m_3,$$

and the implied relation among the masses of the five particles $N$, $\Lambda$, $\Sigma$, $\Xi$, and $Y^0$ is contained in

$$[\tfrac{1}{5}(Y^0+\Lambda-N-\Xi)]^2=\tfrac{1}{5}(Y^0+\Lambda-N-\Xi)(N+\Xi-2\Sigma)+\tfrac{1}{2}(Y^0-\Sigma)(\Sigma-\Lambda).$$

This can be presented conveniently as the quadratic mass formula

$$(\tfrac{1}{5}M_b-\tfrac{7}{5}M_a)^2=2(\Sigma-\Lambda)M_a+M_a^2,\tag{5}$$

where

$$M_a=\tfrac{1}{2}(N+\Xi)-\tfrac{1}{4}(3\Lambda+\Sigma)$$

and

$$M_b=Y^0+2\Lambda-3\Sigma.$$

We shall add another physical requirement to the mass formula, which is in the nature of a stability condition. It is that $N$, the lowest mass of the multiplet, shall be stationary about a minimum value with respect to variations of the physical parameter represented by the mass $Y^0$, or equivalently that

$$dM_a/dM_b=0.$$

The consequences are the two relations

$$M_b-7M_a=0$$

and

$$M_a[M_a+2(\Sigma-\Lambda)]=0.$$

If the second factor were to vanish in the latter equation, implying that $M_b=-14(\Sigma-\Lambda)$, we should assert that

$$\Lambda-\Sigma=\frac{1}{11}(Y^0-\Lambda)>0,$$

in contradiction to the empirical order of the masses. We therefore conclude that both $M_a$ and $M_b$ vanish,

$$\tfrac{1}{2}(n+\Xi)=\tfrac{1}{4}(3\Lambda+\Sigma)\tag{6}$$

and

$$Y^0=3\Sigma-2\Lambda.\tag{7}$$

The latter result, in the form

$$\Sigma-\Lambda=\tfrac{1}{3}(Y^0-\Lambda)>0,$$

predicts the correct sequence of $\Lambda$ and $\Sigma$. It follows, incidentally, from

$$\frac{d^2M_a}{dM_b^2}=\frac{1}{25}\frac{1}{\Sigma-\Lambda}>0$$

that $N$ does assume a minimum value at the stationary point.

The mass relation (6) will be recognized as the Gell-Mann-Okubo mass formula,[8] which has been previously derived from an octuplet form of a unitary symmetry theory. It is well known to be accurately satisfied. Indeed, according to the mass values (in MeV)

$$N=939,\quad \Lambda=1115,\quad \Sigma=1193,\quad \Xi=1318,$$

we have

$$M_a=-6\text{ MeV},$$
$$=-\frac{1}{13}(\Sigma-\Lambda),$$

which represents an error of only 0.5 percent, in comparison with either term of $M_a$.

What name shall be assigned to the ninth baryon? According to (7), its mass should be approximately

$$Y^0\simeq 1349\text{ MeV}.$$

Is there a particle with $T=Y=0$ in the neighborhood of this mass? The obvious candidate is $Y_0^*$ at 1405 MeV. With this identification the latter is predicted to be a particle of the same spin and parity as the eight other baryons, $\frac{1}{2}^+$, which provides a crucial experimental test of this simple treatment. In other classifications[9] this particle has been assigned quantum numbers $\frac{1}{2}^-$ and $\frac{3}{2}^+$.

It may be that 56 MeV, or a 4% error, will seem an uncomfortably large discrepancy between theory and experiment for the mass of $Y_0^*$, in view of the high precision attained by the mass formula (6). But, remarkably enough, it is the deviation from the latter that is unacceptable since the experimental values imply a negative sign for the right-hand side of the quadratic mass formula (5).

$$2(\Sigma-\Lambda)M_a+M_a^2=-\left[\frac{5}{13}(\Sigma-\Lambda)\right]^2.$$

One would expect the difficulty to disappear when higher order perturbations are included. This raises the question whether small effects, of the magnitude suggested by the 6-MeV discrepancy in the mass formula (6), could suffice to produce the needed 56-MeV displacement in $Y^0$. This is not impossible since the stability condition for $N$ asserts conversely ($dY^0/dN=\infty$) the great sensitivity of $Y^0$ to small perturbations.

When higher order perturbations are included, the two-dimensional matrix for the states with $Y=T=0$ becomes a general real symmetrical matrix. This can be conveyed by two additional contributions; a multiple, $m_0$, of the unit matrix, and an added term $2m$, associated with the ⟨33| diagonal element. The mass formula (5)

[8] S. Okubo, Progr. Theoret. Phys. (Kyoto) **27**, 949 (1962), and Ref. 5.

[9] S. L. Glashow and A. H. Rosenfeld, Phys. Rev. Letters **10**, 192 (1963); R. E. Behrends and L. F. Landovitz, *ibid.* **11**, 296 (1963).

is retained, with the substitutions

$$M_a \to M_a + m + \tfrac{3}{4}m_0, \quad M_b \to M_b - 3m_0,$$
$$\Sigma - \Lambda \to \Sigma - \Lambda + m_0.$$

If $m$ and $m_0$ are regarded as fixed numbers, the stability condition would imply the vanishing of the quantities that replace $M_a$ and $M_b$, or that

$$m + \tfrac{3}{4}m_0 = 6 \text{ MeV},$$
$$Y^0 = 1349 + 3m_0.$$

With $m=0$ we obtain an upward displacement of $Y^0$ by 24 MeV, which is an appreciable fraction of the discrepancy. But, instead, the problem will be intensified by setting $m_0=0$.

We must recognize that $m$ is a dynamical parameter, subject to variation, in seeking the condition for stability. The latter now reads

$$\frac{1}{25}[M_b - 7(M_a+m)](1-7dm/dY^0) = (\Sigma - \Lambda + M_a + m)(dm/dY^0)$$

so that neither $M_b$ nor $M_a+m$ are required to vanish. How great a dynamical variation of $m$, and how large a deviation in value from 6 MeV is needed to account for the empirical data? An accurate estimate of the latter is obtained from

$$M_a + m = \frac{M_b^2}{50(\Sigma-\Lambda)+14M_b}.$$

On inserting the numerical values $M_a = -6$ MeV, $M_b = 56$ MeV, we obtain

$$m = 6.67 \text{ MeV}$$

and then

$$dm/dY^0 = 0.022.$$

This reassuring conclusion increases the plausibility of the claim that $Y_0^*$ is the ninth baryon.

In the approximation that ignores these corrections, the mass constants in the perturbation matrix are given by

$$m_3 = -\tfrac{1}{3}(m_1+m_2) = \tfrac{1}{2}(\Sigma-\Lambda)$$

and the two-dimensional submatrix for states with $Y=T=0$ becomes

$$(\Sigma-\Lambda)\begin{pmatrix} 1 & 2^{1/2} \\ 2^{1/2} & 0 \end{pmatrix}.$$

The two eigenvectors are

$$\Lambda: \quad 6^{-1/2}(\langle 11| + \langle 22| - 2\langle 33|)$$

and

$$Y_0^*: \quad 3^{-1/2}(\langle 11| + \langle 22| + \langle 33|).$$

All nine states can be conveniently displayed by writing $\langle ab|$ in a square array, with rows and columns labeled by $a(\bar{\psi})$ and $b(V)$, respectively, and using the particle symbol for the corresponding state,

$$\langle ab| = \begin{bmatrix} -2^{-1/2}\Sigma^0 + 6^{-1/2}\Lambda + 3^{-1/2}Y_0^* & \Sigma^+ & N^+ \\ -\Sigma^- & 2^{-1/2}\Sigma^0 + 6^{-1/2}\Lambda + 3^{-1/2}Y_0^* & N^0 \\ \Xi^- & \Xi^0 & -(\tfrac{2}{3})^{1/2}\Lambda + 3^{-1/2}Y_0^* \end{bmatrix}. \tag{8}$$

When higher order perturbations are included, in the manner just described, the two states become

$$\begin{aligned} \Lambda:& \quad (5.17)^{-1/2}(\langle 11| + \langle 22| - 1.78\langle 33|) \\ Y_0^*:& \quad (3.25)^{-1/2}(\langle 11| + \langle 22| + 1.12\langle 33|) \end{aligned} \tag{9}$$

and the diagonal elements of the square array are changed into

$$\begin{aligned} \langle 11| &= -2^{-1/2}\Sigma^0 + 0.44\Lambda + 0.55Y_0^*, \\ \langle 22| &= 2^{-1/2}\Sigma^0 + 0.44\Lambda + 0.55Y_0^*, \\ \langle 33| &= -0.78\Lambda + 0.62Y_0^*. \end{aligned} \tag{10}$$

### Mesons ($1^-$)

The known particles of this type will be tentatively related to the states generated by the operator products $\bar{\psi}_a\psi_b$, rather than the alternative $\bar{V}_aV_b$. With respect to such states the $\psi - V$ coupling term can be reduced to the structure symbolized by

$$M_{f'}^{(2)} = f'^2\psi_3\bar{\psi}\psi\bar{\psi}_3.$$

Within the unitary octuplet, the scalar $\psi\bar{\psi}$ can be effectively replaced by a constant. We shall adopt a more dynamical attitude toward this replacement, however, and conjecture that, in the evaluation of the perturbation matrix, all such operator unitary scalars are dominated by numbers, in the nature of vacuum expectation values. Then the states and the perturbation can be interpreted literally as

$$\langle ab| = \langle|\bar{\psi}_a\psi_b, \; |ab\rangle = \bar{\psi}_b\psi_a|\rangle, \quad M_{f'}^{(2)} = f'^2\bar{\psi}_3\psi_3, \tag{11}$$

apart from multiplicative constants, and the necessity of subtracting vacuum expectation values from the operators used in constructing the states. This viewpoint becomes significant when the unitary singlet state is added to form a set of nine such states. Although it is not assumed that the singlet state is degenerate with the other eight states[10] we do proceed to describe all the

[10] After completing this work I was made aware of a somewhat related contribution of S. Okubo, Phys. Letters 5, 165 (1963) which is a perturbation treatment of nine degenerate states. As our discussion will show, better accord with experiment is reached if one includes a displacement of the unperturbed singlet state relative to the octuplet of states. *Note added in proof.* See also F. Gürsey, T. D. Lee, and M. Nauenberg, Phys. Rev. (to be published).

states in terms of the set (11). This involves the complete neglect of $V$-field operators in the construction of the singlet state.

The structure of the perturbation matrix for the nine states is given in part by

$$\langle ab|M_{f'}{}^{(2)}|a'b'\rangle = f'^2\langle\bar\psi_a\psi_b\bar\psi_3\psi_3\bar\psi_{b'}\psi_{a'}\rangle$$
$$=\lambda(\delta_{a3}\delta_{a'3}\delta_{bb'}+\delta_{aa'}\delta_{b3}\delta_{b'3}).$$

The operators $\bar\psi$ and $\psi$ have been paired to form invariant vacuum expectation values, while recalling that such terms have been removed in constructing the states, and noting that the vacuum expectation value of the perturbation can be discarded since it contributes only a common displacement of the multiplet. The symmetry of the matrix under the substitution $a\leftrightarrow b$, $a'\leftrightarrow b'$ assures hyperparity conservation. There is another perturbation term, assigning a constant $\lambda'$ to the unitary singlet state, that represents the dynamical difference between the unitary singlet and octuplet. This contribution to the perturbation matrix is

$$\tfrac{1}{3}\lambda'\delta_{ab}\delta_{a'b'}.$$

Isotopic spin and hypercharge distinguish various submatrices. The states with $Y=1$, $T=\frac{1}{2}$, and $T_3=\frac{1}{2}$, $-\frac{1}{2}$ are

$$K^{*+,0}:\ \langle 13|,\ \langle 23|,$$

while the states

$$\bar K^{*-,0}:\ \langle 31|,\ \langle 32|$$

are characterized by $Y=-1$, $T=\frac{1}{2}$, and $T_3=-\frac{1}{2}$, $\frac{1}{2}$, respectively. For all these states we have

$$\langle M\rangle_{K^*}=\lambda.$$

The three states with $Y=0$, $T=1$ and $T_3=1,-1,0$ are

$$\rho^{+,-,0}:\ \langle 12|,\ \ -\langle 21|,\ \ 2^{-1/2}(-\langle 11|+\langle 22|)$$

and

$$\langle M\rangle_\rho=0.$$

There are two states with $Y=T=0$,

$$2^{-1/2}(\langle 11|+\langle 22|),\ \ \langle 33|.$$

The corresponding submatrix is

$$\begin{bmatrix} \frac{2}{3}\lambda', & \frac{2^{\frac{1}{2}}}{3}\lambda' \\ \frac{2^{\frac{1}{2}}}{3}\lambda', & \frac{1}{3}\lambda'+2\lambda \end{bmatrix}.$$

The two eigenvalues of this matrix will be denoted by $\omega-\rho$ and $\phi-\rho$, with $\phi>\omega$. Then we have

$$\phi+\omega-2\rho=\lambda'+2\lambda,$$
$$(\phi-\rho)(\omega-\rho)=\tfrac{4}{3}\lambda'\lambda,$$

where, in this notation

$$K^*-\rho=\lambda.$$

We obtain, successively,

$$\phi+\omega-2K^*=\lambda'$$

and the mass relation for odd parity, unit spin mesons:

$$(\phi-\rho)(\omega-\rho)=\tfrac{4}{3}(K^*-\rho)(\phi+\omega-2K^*). \qquad (12)$$

It is a well-known aspect of field spectral properties that the squared mass plays the same role for Bose fields as does the mass with Fermi fields. Hence the particle symbols in (12) are to be interpreted as squared masses.[11] The experimental values are (in MeV)

$$\rho^{1/2}=755\pm5,\quad \omega^{1/2}=\ 781\pm0.8,$$
$$K^{*1/2}=888\pm3,\quad \phi^{1/2}=1019\pm0.5,$$

or (in BeV$^2$)

$$\rho=0.570\pm0.008,\quad \omega=0.610\pm0.001,$$
$$K^*=0.789\pm0.005,\quad \phi=1.038\pm0.001.$$

Both sides of Eq. (12) are quite small, owing to the respective factors

$$\omega-\rho=0.040\pm0.009,$$
$$\phi+\omega-2K^*=0.070\pm0.010,$$

and they are equal within the experimental errors,

$$0.019\pm0.004=0.020\pm0.003.$$

These errors are largely associated with $\rho$. It is desirable, therefore, to isolate $\rho$ in the mass formula (12), as given by

$$(\rho+\tfrac{1}{6}(\phi+\omega)-\tfrac{4}{3}K^*)^2=K^{*2}-\omega\phi+K^*(\phi+\omega-2K^*)$$
$$+\frac{1}{36}(\phi+\omega-2K^*)^2. \qquad (13)$$

The effect of errors in $K^*$ is minimized on the right-hand side, in virtue of the approximate equality between $\phi+\omega$ and $2K^*$. Even more striking is the near cancellation of $K^{*2}$ and $\omega\phi$, according to

$$(\omega\phi)^{1/2}-K^*=0.007\pm0.005.$$

The square root of the right-hand side in (13) has the numerical value $0.211\pm0.001$. If we choose its sign to be negative, $\rho$ is predicted to be

$$\rho=0.566\pm0.007$$

or

$$\rho^{1/2}=752\pm5\ \text{MeV}.$$

The agreement with the measured value is amazing, particularly in view of the seemingly crude dynamical assumptions involved.

[11] A comment to this effect is generally ascribed to R. P. Feynman (unpublished).

Incidentally, a simplified version of (13), from which some very small terms are omitted is

$$K^*=\tfrac{1}{4}[3\rho+\tfrac{1}{2}(\phi+\omega)+3(\phi\omega)^{1/4}(\phi^{1/2}-\omega^{1/2})].$$

It is still fulfilled within experimental error.

The parameters of the perturbation matrix are related by

$$\frac{\lambda'}{\lambda}=\frac{\phi+\omega-2K^*}{K^*-\rho}\simeq\tfrac{1}{3}.$$

Thus, the two-dimensional submatrix is

$$\frac{2}{9}\lambda\begin{pmatrix}1, & 2^{-1/2}\\ 2^{-1/2}, & 19/2\end{pmatrix}$$

and its eigenvectors are

$$\begin{aligned}\omega:&\quad (2.013)^{-1/2}[\langle 11|+\langle 22|-0.116\langle 33|],\\ \phi:&\quad (1.003)^{-1/2}[0.058(\langle 11|+\langle 22|)+\langle 33|],\end{aligned}\tag{14}$$

which differ very little from $2^{-1/2}(\langle 11|+\langle 22|)$ and $\langle 33|$. Using the latter, the nine states are displayed in the square array

$$\langle ab|=\begin{bmatrix}-2^{-1/2}\rho^0+2^{-1/2}\omega, & \rho^+, & K^{*+}\\ -\rho^-, & 2^{-1/2}\rho^0+2^{-1/2}\omega, & K^{*0}\\ \bar{K}^{*-}, & \bar{K}^{*0}, & \phi\end{bmatrix},\tag{15}$$

where row and column are labeled by $a(\bar{\psi})$ and $b(\psi)$. The diagonal elements of the array are given somewhat more precisely by

$$\begin{aligned}\langle 11|&=-2^{-1/2}\rho+0.705\omega+0.058\phi,\\ \langle 22|&=\phantom{-}2^{-1/2}\rho+0.705\omega+0.058\phi,\\ \langle 33|&=\qquad\qquad -0.081\omega+0.997\phi.\end{aligned}\tag{16}$$

In order to exhibit the couplings among the baryons and the mesons we shall introduce phenomenological or secondary fields that possess the transformation properties of the corresponding particle states. Thus, we have a $\tfrac{1}{2}^+$ baryon spinor field

$$\Psi_{ab}\sim\bar{\psi}_a V_b,$$

with its adjoint

$$\bar{\Psi}_{ab}\sim\bar{V}_a\psi_b,$$

and a $1^-$ meson vector field

$$U_{ab}\sim\bar{\psi}_a\psi_b.$$

As an initial approximation we construct such couplings to be invariant under the two independent groups of $U_3$ transformations associated with the $\psi$ and $V$ primary fields (should this be called $W_3$ invariance?). The unitary structure required for the coupling of baryons and mesons is illustrated by

$$\mathfrak{L}_{U\Psi}=g_{U\Psi}\sum_{abc}U_{ab}\Psi_{bc}\bar{\Psi}_{ca}=g_{U\Psi}\,\mathrm{Tr}U\Psi\bar{\Psi}.$$

The irreducibility of the summation expresses the essential equivalence of all nine meson states. When written out, using particle symbols for the corresponding fields, this becomes, approximately,

$$\begin{aligned}\mathfrak{L}_{U\Psi}/g_{U\Psi}=2^{-1/2}\omega&\left[\bar{N}N+\bar{\Sigma}\Sigma+\tfrac{1}{3}\bar{\Lambda}\Lambda+\tfrac{2}{3}\bar{Y}Y+\frac{2^{\frac{1}{2}}}{3}(\bar{\Lambda}Y+\bar{Y}\Lambda)\right]+\phi\left[\bar{\Xi}\Xi+\tfrac{2}{3}\bar{\Lambda}\Lambda+\tfrac{1}{3}\bar{Y}Y-\frac{2^{\frac{1}{2}}}{3}(\bar{\Lambda}Y+\bar{Y}\Lambda)\right]\\ &+2^{-1/2}\rho[-\bar{N}2tN-\bar{\Sigma}t\Sigma+\bar{\Sigma}(3^{-1/2}\Lambda+(\tfrac{2}{3})^{1/2}Y)-(3^{-1/2}\bar{\Lambda}+(\tfrac{2}{3})^{1/2}\bar{Y})\bar{\Sigma}^g]\\ &+(\bar{N}K^*)(-(\tfrac{2}{3})^{1/2}\Lambda+3^{-1/2}Y)+(6^{-1/2}\bar{\Lambda}+3^{-1/2}\bar{Y})(\Xi K^*)+2^{1/2}\Sigma^g(\Xi tK^*)\\ &+(-(\tfrac{2}{3})^{1/2}\bar{\Lambda}+3^{-1/2}\bar{Y})(\bar{K}^*N)+(\bar{K}^*\bar{\Xi})(6^{-1/2}\Lambda+3^{-1/2}Y)-2^{1/2}\Sigma(\bar{K}^*t\bar{\Xi}).\end{aligned}\tag{17}$$

Here $t$ represents the appropriate isotopic spin matrices,

$$\rho t=\rho^0t_3+\rho^-2^{-1/2}(t_1-it_2)-\rho^+(t_1+it_2)$$

and similarly for $\Sigma t$, while $\Sigma^g$ and $\bar{\Sigma}^g$ indicate the effect of the hyperparity operation (omitting spin matrices):

$$\begin{aligned}\bar{\Sigma}^g&=e^{\pi it_2}\Sigma=(\Sigma^-,\Sigma^+,-\Sigma^0),\\ \Sigma^g&=e^{\pi it_2}\bar{\Sigma}=(\bar{\Sigma}^+,\bar{\Sigma}^-,-\bar{\Sigma}^0).\end{aligned}$$

It will be noted that $\omega$ is not coupled precisely to the total phenomenological current of nucleonic charge nor is $\rho$ coupled precisely to the total phenomenological current of isotopic spin.[12] But what is most significant at the moment is the essential absence of interaction between $\phi$ and the nucleon. This is consistent with the marked weakness of $\phi$ production relative to $\omega$ production in $\pi-N$ collisions,[13] and indicates that $1^-$ mesons have been assigned correctly to the $\bar{\psi}_a\psi_b$ states. The alternative assignment would reverse the roles of $N$ and $\Xi$. If the linear combinations (16) are used, the most important effect is the substitution $2^{-1/2}\omega\rightarrow 0.705\omega+0.058\phi$. Then the intrinsic ratio of $\phi$ to $\omega$ production in pion-nucleon interactions is presumably of the magnitude indicated by $(0.058/0.705)^2=0.69\times10^{-2}$. This could easily be increased by a factor $\sim1.5$ in view of the relatively large uncertainty in $\phi+\omega-2K^*$. There is an experimental upper limit to this ratio measured in the reaction $\pi^-+p\rightarrow\pi^-+p+\omega,\ \phi$. When corrected for phase-space factors, it is $0.016/0.55=2.9\times10^{-2}$.

[12] Compare J. J. Sakurai, Ann. Phys. (N. Y.) **11**, 1 (1960).

[13] Y. Y. Lee, W. Moebs, B. Rose, D. Sinclair, and J. Vander Velde, Phys. Rev. Letters **11**, 508 (1963); M. Abolins, R. Lander, W. Mehlhop, N. Xuong, and P. Yager, *ibid.* **11**, 381 (1963).

**Mesons ($0^-$)**

If one assumes that $0^-$ mesons are also related to $\bar{\psi}_a\psi_b$ states, the preceding discussion can be applied to these particles. States that describe particles in the rest system will be obtained by using $\gamma_5$ instead of the matrices required for $1^-$ particles, $\gamma_k$, $k=1, 2, 3$. The algebraic similarity of these matrices implies that the constant $\lambda$ associated with the second-order $\psi-V$ coupling is the same for both types of mesons. That is, when one follows the procedure of inserting vacuum expectation values for unitary and space-time scalars, the two calculations symbolized by

$$\lambda_1\delta_{kl}=f'^2\langle\bar{\psi}_1\gamma_k\psi_3\bar{\psi}_3\psi_3\bar{\psi}_3\gamma_l\psi_1\rangle$$
$$=f'^2\langle\psi_3\bar{\psi}_3\rangle\langle\psi_3\bar{\psi}_3\rangle\langle\bar{\psi}_1\gamma_k\gamma_l\psi_1\rangle$$

and

$$\lambda_0=f'^2\langle\bar{\psi}_1\gamma_5\psi_3\bar{\psi}_3\psi_3\bar{\psi}_3\gamma_5\psi_1\rangle$$
$$=f'^2\langle\psi_3\bar{\psi}_3\rangle\langle\psi_3\bar{\psi}_3\rangle\langle\bar{\psi}_1\gamma_5\gamma_5\psi_1\rangle\,,$$

will yield the same values for $\lambda_0$ and $\lambda_1$. The experimental $0^-$ meson masses are (in MeV)

$$\pi^{1/2}=138.04\pm0.05\,,\quad K^{1/2}=496.0\pm0.3\,,\quad \eta^{1/2}=548.5\pm0.6$$

and ($\mathrm{BeV}^2$)

$$\pi=0.01904\,,\quad K=0.2460+0.0005\,,\quad \eta=0.3001\pm0.0006.$$

Thus

$$\lambda_0=K-\pi=0.2270\pm0.0005 \tag{18}$$

is to be compared with

$$\lambda_1=K^*-\rho=0.219\pm0.009.$$

They are indeed equal, within experimental error.[14]

A ninth $0^-$ meson has not yet been identified. That presumably means that the unitary singlet is displaced far above the octuplet. In the limit of infinite singlet mass the G-M-O formula is obtained for the known mesons,

$$K=\tfrac{1}{4}(3\eta+\pi)\,.$$

This relation is obeyed moderately well; the right-hand side equals $0.2298\pm0.0005$, which is a discrepancy of 7%. It is significant that $K$ exceeds the value computed in this way. That enables the discrepancy to be accounted for by a ninth meson. We shall designate this unknown fourth type of $0^-$ meson as $\delta$. The mass relation

$$(\delta-\pi)(\eta-\pi)=\tfrac{4}{3}(K-\pi)(\delta+\eta-2K)\,,$$

or

$$\delta-\pi=\frac{(K-\pi)(2K-\eta-\pi)}{K-\frac{1}{4}(3\eta+\pi)} \tag{19}$$

predicts

$$\delta=2.45\pm0.1\,,\quad \delta^{1/2}=1560\pm30\ \mathrm{MeV}. \tag{20}$$

Without taking this number too seriously, it does seem reasonable to anticipate that the $0^-$ meson $\delta$ exists somewhere in the vicinity of 1.5 BeV.

The two parameters of the perturbation matrix for these particles are related by

$$\lambda_0'/\lambda_0=(\delta+\eta-2K)/(K-\pi)\simeq 10.$$

The two dimensional submatrix for $Y=T=0$ is

$$\frac{20}{3}\lambda\begin{pmatrix}1, & 2^{-1/2}\\ 2^{-1/2}, & \frac{4}{5}\end{pmatrix}$$

and the eigenvectors are

$$\begin{aligned}\eta:&\quad(4.66)^{-1/2}[\langle11|+\langle22|-1.63\langle33|]\,,\\ \delta:&\quad(3.51)^{-1/2}[\langle11|+\langle22|+1.23\langle33|]\,.\end{aligned} \tag{21}$$

The diagonal elements of a square array are given by

$$\begin{aligned}\langle11|&=-2^{-1/2}\pi^0+0.465\eta+0.533\delta\,,\\ \langle22|&=2^{-1/2}\pi^0+0.465\eta+0.533\delta\,,\\ \langle33|&=-0.755\eta+0.654\delta\,.\end{aligned} \tag{22}$$

A rough approximation to these numerical coefficients is used for simplicity in writing

$$\langle ab|=\begin{bmatrix}-2^{-1/2}\pi^0+\frac{1}{2}\eta+\frac{1}{2}\delta, & \pi^+, & K^+\\ -\pi^-, & 2^{-1/2}\pi^0+\frac{1}{2}\eta+\frac{1}{2}\delta, & K^0\\ \bar{K}^-, & \bar{K}^0, & -2^{-1/2}\eta+2^{-1/2}\delta\end{bmatrix}\,. \tag{23}$$

The same approximation will be used for the secondary pseudoscalar field that is associated with these particles,

$$\Phi_{ab}\sim\bar{\psi}_a\psi_b\,.$$

The unitary structure required for an initial approximation to the coupling between the $0^-$ mesons and the baryons is given by

$$\mathcal{L}_{\Phi\Psi}=g_{\Phi\Psi}\,\mathrm{Tr}\Phi\Psi\bar{\Psi}\,,$$

while couplings among the mesons are illustrated by

$$\mathcal{L}_{U\Phi^2}=g_{U\Phi^2}\,\mathrm{Tr}U\Phi\Phi$$

and

$$\mathcal{L}_{\Phi U^2}=g_{\Phi U^2}\,\mathrm{Tr}\Phi UU\,.$$

[14] This property of the data has been noticed by S. Coleman and S. L. Glashow, Phys. Rev. **134**, B671 (1964).

The baryon coupling term is

$$\begin{aligned}\mathcal{L}_{\Phi\Psi}/g_{\Phi\Psi}=\tfrac{1}{2}\eta\Big[&\bar{N}N+\bar{\Sigma}\Sigma-2^{1/2}\bar{\Xi}\Xi-\frac{2^{3/2}-1}{3}\bar{\Lambda}\Lambda+\frac{2-2^{1/2}}{3}\bar{Y}Y+\frac{2+2^{1/2}}{3}(\bar{\Lambda}Y+\bar{Y}\Lambda)\Big]\\ &+\tfrac{1}{2}\delta\Big[\bar{N}N+\bar{\Sigma}\Sigma+2^{1/2}\bar{\Xi}\Xi+\frac{2^{3/2}+1}{3}\bar{\Lambda}\Lambda+\frac{2+2^{1/2}}{3}\bar{Y}Y-\frac{2-2^{1/2}}{3}(\bar{\Lambda}Y+\bar{Y}\Lambda)\Big]\\ &+2^{-1/2}\pi[-\bar{N}2tN-\bar{\Sigma}t\Sigma+\bar{\Sigma}(3^{-1/2}\Lambda+(\tfrac{2}{3})^{1/2}Y)-(3^{-1/2}\bar{\Lambda}+(\tfrac{2}{3})^{1/2}\bar{Y})\Sigma^{g}]\\ &+(\bar{N}K)(-(\tfrac{2}{3})^{1/2}\Lambda+3^{-1/2}Y)+(6^{-1/2}\bar{\Lambda}+3^{-1/2}\bar{Y})(\Xi K)+2^{1/2}\Sigma^{g}(\Xi tK)\\ &+(-(\tfrac{2}{3})^{1/2}\bar{\Lambda}+3^{-1/2}\bar{Y})(\bar{K}N)+(\bar{K}\bar{\Xi})(6^{-1/2}\Lambda+3^{-1/2}Y)-2^{1/2}\Sigma(\bar{K}t\bar{\Xi}).\end{aligned}\tag{24}$$

The explicit space-time structure is obtained by extending the known pion-nucleon coupling, which we use in the pseudovector form

$$(f_{\pi N}/\mu_\pi)\Phi_\pi{}^\lambda\bar{\Psi}_N i\gamma_5\gamma_\lambda 2t\Psi_N.$$

Thus, the pion coupling between $Y_0^*$ and $\Sigma$ is given by

$$(\tfrac{2}{3})^{1/2}(f_{\pi N}/\mu_\pi)\Phi_\pi{}^\lambda[-\bar{\Psi}_\Sigma i\gamma_5\gamma_\lambda\Psi_Y+\bar{\Psi}_Y i\gamma_5\gamma_\lambda\Psi_{\Sigma^g}].$$

From this we calculate the single-pion emission contribution to the width of $Y_0^*$,

$$\Gamma_\pi(Y_0^*)=\frac{4}{3}\frac{f_{\pi N}^2}{4\pi}\frac{p_\pi^3}{\mu_\pi^2}=30\text{ MeV},\tag{25}$$

where $p_\pi$ is the pion momentum, and

$$f_{\pi N}^2/4\pi=0.1$$

has been used. This is somewhat smaller than the experimentally indicated width[15] of $\sim$50 MeV.

The complete absence of $\pi-\Xi$ coupling should be noted. It supports the identification of the $0^-$ mesons as $\bar{\psi}_a\psi_b$ states. Also of interest is the lack of $\bar{N}K\Sigma$ interaction, which is not altered by including $\pi$-mesons. However, the relation between $\Lambda$ and $\Sigma$ couplings should be sensitive to the effect of the symmetry-destroying coupling, particularly if it is amplified by the small pion mass.

The coupling between the vector field $U$ and the pseudoscalar field $\Phi$ can be given the following space-time structure:

$$\mathcal{L}_{U\Phi^2}=g_{U\Phi^2}\sum_{abc}U_{ab}{}^\lambda(1/2i)(\Phi_{bc}\Phi_{\lambda ca}-\Phi_{\lambda bc}\Phi_{ca}).$$

One should observe that the bilinear $\Phi$ factor is traceless and that terms with $a=b=c$ are not present. The unitary structure is written out as

$$\begin{aligned}\mathcal{L}_{U\Phi^2}/g_{U\Phi^2}=&(\phi-2^{-1/2}\omega)\bar{K}K+2^{-1/2}\rho(\bar{\pi}t\pi+\bar{K}2tK)\\ &+2^{1/2}\pi\bar{K}tK^*+2^{1/2}\bar{K}^*tK\pi+(2^{-1/2}-\tfrac{1}{2})\\ &\times(\delta\bar{K}K^*+\bar{K}^*K\delta)-(2^{-1/2}+\tfrac{1}{2})\\ &\times(\eta\bar{K}K^*+\bar{K}^*K\eta),\end{aligned}\tag{26}$$

[15] A general summary of experimental information is given by Matts Roos, Rev. Mod. Phys. 35, 314 (1963).

where

$$\bar{\pi}=-e^{\pi i t_2}\pi=(-\pi^-,-\pi^+,\pi^0).$$

This phenomenological interaction describes three known decay processes, $\rho\to\pi+\pi$, $K^*\to K+\pi$, $\phi\to K+\bar{K}$, and gives a first approximation to their rates in terms of a common coupling constant. We find that

$$\Gamma(\rho,K^*,\phi)=(1,\tfrac{3}{4},1)\frac{1}{3}\left(\frac{g^2}{4\pi}\right)\frac{p^3}{\mu^2},$$

where $p$ is the relative momentum of the decay products and $\mu$ is the unstable particle mass in each reaction. The theoretical values for the ratios are

$$\Gamma(K^*)/\Gamma(\rho)=1/3.5,\quad \Gamma_K(\phi)/\Gamma(\rho)=1/50.\tag{27}$$

The experimental results for the widths are (in MeV)

$$\Gamma(\rho)=120\pm10,\quad \Gamma(K^*)=50\pm10,\quad \Gamma(\phi)=3.1\pm1.0.$$

From the first of these we obtain

$$g_{U\Phi^2}{}^2/4\pi\sim5.$$

The widths predicted for $K^*$ and $\phi$ are then

$$\Gamma(K^*)=35\pm3,\quad \Gamma_K(\phi)=2.4\pm0.2.\tag{28}$$

The latter should not be compared with the measured total width until we have estimated the contribution of the alternative process $\phi\to\rho+\pi$.

This is described by the phenomenological coupling term

$$\mathcal{L}_{\Phi U^2}=g_{\Phi U^2}\sum_{abc}\Phi_{ab}\tfrac{1}{4}\epsilon_{\mu\nu\lambda\kappa}U_{bc}{}^{\mu\nu}U_{ca}{}^{\lambda\kappa},$$

which is symmetrical in the two $U$ factors. On using the approximate forms of the square arrays it is written out as

$$\begin{aligned}\mathcal{L}_{\Phi U^2}/g_{\Phi U^2}=&\tfrac{1}{2}\delta[\omega\omega+\bar{\rho}\rho+(1+2^{1/2})\bar{K}^*K^*+2^{1/2}\phi\phi]\\ &+\tfrac{1}{2}\eta[\omega\omega+\bar{\rho}\rho+(1-2^{1/2})\bar{K}^*K^*-2^{1/2}\phi\phi]\\ &-2^{1/2}\pi\bar{K}^*tK^*+2^{1/2}\bar{\pi}\rho\omega+2^{-1/2}\omega\bar{K}^*K\\ &+2^{-1/2}\bar{K}K^*\omega+\phi\bar{K}^*K+\bar{K}K^*\phi\\ &-2^{1/2}\rho\bar{K}^*tK-2^{1/2}\bar{K}tK^*\rho,\end{aligned}\tag{29}$$

which contains no $\pi\rho\phi$ coupling term. The more precise version obtained by the substitution (16),

$$2^{-1/2}\omega \rightarrow 0.705\omega+0.058\phi,$$

gives a small coupling[16] which implies an almost forbidden $\phi \rightarrow \rho+\pi$ decay.

The rate for the latter process depends upon the coupling constant $g_{\Phi U^2}$ which cannot be inferred from any similar decay rate. It can, however, be related to the partial width for $\omega \rightarrow \pi^+ + \pi^- + \pi^0$, if this reaction is viewed as proceeding through the stages $\omega \rightarrow \rho+\pi \rightarrow 3\pi$. We shall temporarily adopt a value quoted in the literature,[17] which asserts that

$$(\mu_\omega g_{\Phi U^2})^2/4\pi \sim 6.$$

With the definition

$$g_{\pi\rho\phi} = 0.116 g_{\Phi U^2}$$

we find that

$$\Gamma_\pi(\phi) = [(\mu_\phi g_{\pi\rho\phi})^2/4\pi](p^3/\mu^2)_\phi = 1.0 \text{ MeV}.$$

The theoretical branching ratio

$$\Gamma_\pi(\phi)/\Gamma_K(\phi) = 0.42$$

is consistent with a measurement[18] that gives $0.35 \pm 0.2$. The total width

$$\Gamma(\phi) = 3.4 \text{ MeV}$$

agrees well with the observed value.[19]

## L'ENVOI

To avoid a paper of excessive length, these elementary studies of particle phenomena will be continued in other publications. But we cannot refrain from a retrospective glance, for it is necessary to emphasize the distinction between the general dynamical theory that has been proposed and the specific implications that have been obtained. The latter also involve a dynamical conjecture, that the effect of the symmetry-destroying interaction can be discussed by perturbation theory, which requires an approximate phenomenological validity for the underlying symmetry group $W_3$. This need not be true. It is conceivable that the $\psi - V$ coupling term is sufficiently strong that its major effect is the establishment of a dynamical regime governed by the common symmetry group $SU_3$, while only secondarily destroying that symmetry. In this light, the crucial experimental spin-parity determination for $Y_0^*$ (1405 MeV) will test whether $W_3$ or $SU_3$ is the more realistic underlying symmetry group.[20] Should this trial of the $W_3$ hypothesis fail, such of its successes as the interpretation of the weak $\phi$ production in $\pi - N$ collisions will demand new explanations.

*Note added in proof.* It has been suggested [for example, I. S. Gerstein and K. T. Mahanthappa, Phys. Rev. Letters **12**, 570 (1964)] that the $\frac{3}{2}^+$ baryon resonances belong to a $3\times6=18$ dimensional representation of $W_3$. This presumably is not correct since the representation that contains protons and positive pions is the $3\times15=45$ dimensional one that is mentioned (for that purpose) in the text. An equivalent remark has been published by R. E. Cutkosky, Phys. Rev. Letters **12**, 530 (1964). The 45-dimensional $W_3$ representation contains $SU_3$ representations of dimensionality 10, 8, and 27. The experimental situation with regard to the 8- and 27-fold representations is unclear. There are some resonances below 2 BeV that may be $\frac{3}{2}^+$, but it is likely that most of these states are more massive. It seems that the mechanism which destroys $W_3$ symmetry is more effective in separating the various $SU_3$ multiplets than in splitting the individual ones.

If the separation of $SU_3$ multiplets in a $W_3$ representation is large enough, a new possibility appears, which has been noted by P. Freund and Y. Nambu, Phys. Rev. Letters **12**, 714 (1964). Fermions need not have a common parity. Thus, the known $\frac{3}{2}^-$ baryon resonances might belong to the same 45-dimensional $W_3$ representation as the tenfold $\frac{3}{2}^+$ resonances, and a $\frac{1}{2}^- Y_0^*$ (1405) could be the ninth baryon. With this interpretation, the pseudovector pion coupling (24) gives a width for $Y_0^*$ of the observed magnitude. A resonance that may belong to a 27-fold representation has been found [R. Alvarez, Z. Bar-Yam, W. Kern, D. Luckey, L. S. Osborne, S. Tazzari, and R. Fessel, Phys. Rev. Letters **12**, 710 (1964)]. Perhaps this particle will help to establish the larger pattern of $W_3$ symmetry.

A uniform description of nine $1^-$ mesons and nine $0^-$ mesons is given in the text. It unites unitary singlet and octuplet states while retaining a mass displacement between the multiplets. This procedure is successful for the $1^-$ mesons. It is somewhat justified by the small singlet-octuplet displacement implied by the proximity of the $\omega$ and $\rho$ masses. The situation is not so favorable for the $0^-$ mesons $\pi$, $K$, $\eta$, $\delta$ since the singlet must be considerably more massive than the octuplet. That should imply some difference in structure for the two types of states. One could attempt to represent this relative distortion by including an "overlap" factor $\alpha(\alpha^2<1)$ in the matrix element, of the symmetry destroying perturbation, that connects singlet and

[16] If the production processes $\pi+N \rightarrow \pi+N+(\omega,\phi)$ are dominated by the $\pi\rho(\omega,\phi)$ coupling, one again estimates the intrinsic ratio to be in the order of one percent.

[17] See, for example, G. Feinberg and H. S. Mani, Phys. Rev. Letters **11**, 448 (1963).

[18] P. L. Connolly, E. Hart, K. Lai, G. London, G. Moneti, *et al.*, Phys. Rev. Letters **10**, 371 (1963).

[19] N. Gelfand, D. Miller, M. Nussbaum, J. Ratau, J. Schultz, *et al.*, Phys. Rev. Letters **11**, 438 (1963).

[20] I am grateful to S. F. Tuan for reminding me of the possibility that the ninth baryon $Y^0$ is an unknown metastable particle. The $\pi+\Sigma$ threshold is centered at 1331 MeV. It is quite conceivable that higher order perturbations displace $Y^0$ downward from 1349 MeV below this nearby threshold. Such a particle would decay radiatively into $\Lambda$ and could be confused with $\Sigma^0$ in high-energy production experiments. This alternative would permit the more usual assignment $\frac{1}{2}^-$ for $Y_0^*$ (1405) but raises the problem of identifying the other members of the nonuplet to which $Y_0^*$ should belong. Or is this particle a member of the triplet generated by $\psi_a$?

octuplet states. The modified (mass)$^2$ formula is

$$(\eta-\pi)(\delta-\pi)=\tfrac{4}{3}(K-\pi)(\eta+\delta-2K)+8/9(1-\alpha^2)(K-\pi)^2$$

or

$$\delta-\pi=\frac{K-\pi}{K-\frac{1}{4}(3\eta+\pi)}\times[2K-\eta-\pi-\tfrac{2}{3}(1-\alpha^2)(K-\pi)]\geq\tfrac{4}{3}(K-\pi).$$

The lower limit, $\delta^{1/2}\geq 567$ MeV, is reached for $\alpha=0$. The maximum value $\delta^{1/2}\leq 1560$ MeV occurs for $|\alpha|=1$. The actual magnitude of $\alpha$ will be determined by considerations of dynamical stability. Thus, increasing $\alpha$ raises $\delta$, which then tends to decrease $\alpha$, the overlap factor. The $\delta$ mass should be some reasonable average of the two unattainable extremes: 567 MeV$<\delta^{1/2}<$1560 MeV.

A new meson that decays into $\pi^+ + \pi^- + \eta$ has been announced [G. R. Kalbfleisch, L. W. Alvarez, A. Barbaro-Galtieri, O. I. Dahl, P. Eberhard, *et al.*, Phys. Rev. Letters **12**, 527 (1964); M. Goldberg, M. Gundzik, S. Lichtman, J. Leitner, M. Primer *et al.*, *ibid.* **12**, 546 (1964)]. Its properties are consistent with the quantum numbers $T=Y=0$, $J^P=0^-$, and $G=+1$, which are those of the $\delta$ meson. The mass observed for the particle is 959±2 MeV. The value of the overlap parameter required to represent this mass is $|\alpha|=0.53$. If we accept the identification of the new meson with $\delta$, the ratio of the two perturbation parameters is reduced to $\lambda'/\lambda=3.2$, and the particle states with $T=Y=0$ become

$$\eta=(3.89)^{-1/2}[\langle 11|+\langle 22|-1.37\langle 33|],$$
$$\delta=(4.13)^{-1/2}[\langle 11|+\langle 22|+1.46\langle 33|].$$

This change improves the accuracy of the approximate square array, Eq. (23).

# Field Theory of Matter. II

JULIAN SCHWINGER*
*Harvard University, Cambridge, Massachusetts*
(Received 21 July 1964)

A qualitative dynamical interpretation is given for observed regularities of nonleptonic phenomena in strong, electromagnetic, and weak interactions.

## INTRODUCTION

THE strongly interacting particles exhibit various approximate regularities that appear to have rather special dynamical origins. Nonleptonic decays are governed by the $\Delta T=\frac{1}{2}$ rule but with appreciable $\Delta T=\frac{3}{2}$ admixture, which is not naturally explained as a symmetry property of the weak interactions. Electromagnetic mass differences show a dominant $\Delta T=1$ effect in the approximately equal spacing of $\Sigma^-$, $\Sigma^0$, and $\Sigma^+$, but $\Delta T=2$ must also be significant to account for the very nature of the pion mass spectrum ($\pi^+=\pi^-\neq\pi^0$). The baryon ($\frac{1}{2}$+) and baryon resonance ($\frac{3}{2}$+) multiplets are approximately equally spaced when they are regarded as $U$ or $V$ multiplets, in the sense of the three isotopic spins $T$, $U$, $V$ that characterize $SU_3$ symmetry. Thus, equal spacing for the $U=1$ baryon multiplet $N^0$, $\frac{1}{2}(3^{1/2}\Lambda+\Sigma^0)$, $\Xi^0$ or the $V=1$ multiplet $\Xi^-$, $\frac{1}{2}(3^{1/2}\Lambda-\Sigma^0)$, $N^+$ expresses the Gell-Mann–Okubo (GO) mass relation. It is a directly observable property of, say, the $U=\frac{3}{2}$ resonance multiplet $N^{*-}$, $Y^{*-}$, $\Xi^{*-}$, $\Omega^-$. This means that the mechanism predominantly responsible for the mass splittings obeys $\Delta U=1$ and $\Delta V=1$.

We shall give a qualitative dynamical interpretation of these regularities in the framework of a recently proposed field theory of matter.[1] In doing this we face another such question, but one that, for the moment, lacks a definite experimental basis. How does the breakdown of the underlying $W_3$ symmetry produce predominantly $SU_3$ symmetry?

## BROKEN SYMMETRIES

### $W_3$ and $SU_3$

The first in the hierarchy of interactions couples the gauge field $B$ to the nucleonic charge-bearing fields $\psi_a(N=\pm1)$ and $V_a(N=\pm2)$, $a=1, 2, 3$. That interaction is invariant with respect to the two independent internal symmetry groups $U_3(\psi)$ and $U_3(V)$. Such is the underlying symmetry $W_3$. It is broken by the interaction that exchanges nucleonic charge between $\psi$ and $V$ in accordance with the unitary structure $\bar{\psi}_3(\bar{\psi}V)$ and its adjoint. Although this is a strong interaction, it will be expedient to discuss its implications in the language of perturbation theory.

The low-lying states that one identifies with degenerate families of physical particles are created by combinations of fields with opposite signs of nucleonic charge, as in $\bar{\psi}_a\psi_b$, $\bar{\psi}_aV_b$, and $\bar{\psi}_a\bar{\psi}_b\psi_cV_d$. All these combinations are characterized by invariance under common phase transformations of $\psi$ and $V$,

$$\psi_a\rightarrow e^{i\alpha}\psi_a,\quad V_a\rightarrow e^{i\alpha}V_a,\quad a=1,2,3,$$

as distinguished from the phase transformation of nucleonic charge. Thus the total nucleonic charges associated with the $\psi$ and $V$ fields in these states are related by

$$\tfrac{1}{2}N(V)=-N(\psi)=N.$$

If we assume that all other types of states (such as the triplets generated by $\psi_a$ and $V_a$) are much more massive, the most important effect on the low-lying states, of the $W_3$ destroying interaction, comes from the part that is invariant under the common phase transformation. This is represented by the set of perturbation terms symbolized as

$$(\bar{\psi}_3\psi_3)^n[(\bar{\psi}V)(\bar{V}\psi)]^n\quad n=1,\cdots.$$

The product of $2n$ field operators at distinct space-time points that is implied by $(\bar{\psi}_3\psi_3)^n$ can be decomposed into irreducible parts by means of the vacuum expectation value. Thus, $(\bar{\psi}_3\psi_3)^n$ is additively represented by a number $\langle(\bar{\psi}_3\psi_3)^n\rangle$; irreducible operator pairs

$$[\bar{\psi}_3\psi_3]=\bar{\psi}_3\psi_3-\langle\bar{\psi}_3\psi_3\rangle,$$

multiplied by numbers which are the vacuum expectation values $\langle(\bar{\psi}_3\psi_3)^{n-1}\rangle$; and so forth. When all terms in the perturbation series are rearranged in this way, the resulting functional form is

$$f_0((\bar{\psi}V)(\bar{V}\psi))+[\bar{\psi}_3\psi_3]f_1((\bar{\psi}V)(\bar{V}\psi))+\cdots.\quad(1)$$

A vacuum expectation value becomes singular as two points coincide. If the symmetry destroying interaction is sufficiently localized in space and time, owing to the dominance of very massive states, the consecutive terms in the series (1), which involve successively fewer replacements of operator products by vacuum expectation values, may be of diminishing numerical importance. Under such circumstances, the primary effect of the mechanism that breaks $W_3$ symmetry is to introduce a dynamical regime of $SU_3$ symmetry, as the symmetry

* Supported in part by the U. S. Air Force Office of Scientific Research under Contract No. AF 49(638)-1380.

[1] Julian Schwinger, Phys. Rev. Letters **12**, 237 (1964) and Phys. Rev. **135**, B816 (1964). See also *Symmetry Principles at High Energy* (W. H. Freeman and Company, Inc., San Francisco, 1964).



Reprinted from *The Physical Review* **136**, B1821–B1824 (1964).

group that governs common[2] transformations of $\psi_a$ and $V_a$ in the coupling term $f_0((\bar{\psi}V)(\bar{V}\psi))$.

The same mechanism has the secondary effect of breaking down $SU_3$ symmetry. This aspect of the perturbation is dominated by the term with $[\bar{\psi}_3\psi_3]$ as a factor. The latter responds as a component of a three-dimensional Euclidean vector under the $SU_2$ subgroups that govern transformations in the 23 plane ($U$ isotopic spin) and the 31 plane ($V$ isotopic spin). Accordingly, matrix elements of this perturbation will obey just those selection rules, $\Delta U=1$, $\Delta V=1$, that are observed in the dominant part of the mass splitting that violates $SU_3$ symmetry. A relative measure of the higher perturbation terms can be had by comparing the 8 MeV excess in the mass of $\Lambda$, over the value given by the GO formula, with the average of the mass splittings between $\Sigma$ and $N$, $\Xi$ and $\Sigma$. This ratio is 1:23. If a similar ratio connects the first two terms of (1), the intervals between the $SU_3$ multiplets of a common $W_3$ representation should be several BeV. The $W_3$ concept will acquire an observational basis when one finds some evidence of this larger pattern among the particles.

The dynamical processes that result in broken $SU_3$ symmetry possess the residual symmetry that is described by the invariance group $U_2$, generated by the isotopic spin $T$ ($SU_2$) and the hypercharge $Y$.

### Electromagnetism and $SU_2$

The electric current vector is formed additively from the operator products $\bar{\psi}_1\psi_1$ and $\bar{V}_1V_1$ (omitting spinor and vector indices). A part of the current is proportional to the electromagnetic vector potential $A$. Electromagnetic mass displacements and the breakdown of $SU_2$ symmetry are produced by electromagnetic vacuum fluctuations through this induced current, together with the second order effect of the linearly coupled potential. The nature of the two mechanisms is symbolized by

$$(\bar{\psi}_1\psi_1+\bar{V}_1V_1)+(\bar{\psi}_1\psi_1+\bar{V}_1V_1)^2. \tag{2}$$

The operators $\bar{\psi}_1\psi_1$ and $\bar{V}_1V_1$ behave as components of three-dimensional Euclidean vectors with respect to $T$ isotopic spin transformations (12 plane). Therefore they have matrix elements obeying $\Delta T=1$. The products of these operators also possess matrix elements with $\Delta T=2$. But such operator products can be decomposed into irreducible parts through the introduction of the vacuum expectation value. A term such as $[\bar{\psi}_1\psi_1]\langle\bar{\psi}_1\psi_1\rangle$ is characterized by $\Delta T=1$. These vacuum expectation value contributions, and the $\Delta T=1$ effect, will dominate to the extent that the electromagnetic mechanism is localized. Thus, the long-range nature of electromagnetic action would seem to explain the comparative importance of $\Delta T=2$ effects, as measured by the 1:3 (mass)$^2$ ratio of $\pi^+-\pi^0$ and $K^0-K^+$.

### Weak Interactions and Hypercharge

The weak interactions of the strongly interacting particles are described by a coupling of the charged vector field $Z$ with currents of the form indicated by $\bar{\psi}_1\psi_2+\bar{V}_1V_2(\Delta T=1,\ \Delta Y=0)$ and $\bar{\psi}_1\psi_3+\bar{V}_1V_3$ ($\Delta T=\frac{1}{2}$, $|\Delta Y|=1$). The self-action of these currents through the intermediary of the $Z$ field contains a part symbolized by

$$(\bar{\psi}_2\psi_1+\bar{V}_2V_1)(\bar{\psi}_1\psi_3+\bar{V}_1V_3)$$

and its adjoint. This perturbation destroys the conservation of hypercharge ($|\Delta Y|=1$) and induces isotopic spin transitions with $\Delta T=\frac{3}{2}$ and $\frac{1}{2}$. The decomposition of the operator by means of vacuum expectation values gives the term

$$\bar{\psi}_2\langle\psi_1\bar{\psi}_1\rangle\psi_3+\bar{V}_2\langle V_1\bar{V}_1\rangle V_3 \tag{3}$$

and its adjoint, which is characterized by $|\Delta Y|=1$ and $\Delta T=\frac{1}{2}$. We conclude from the observed dominance of the latter effect that the current self-coupling is effectively localized, or that the $Z$-field excitations are quite massive. The relative importance of the $\Delta T=\frac{3}{2}$ effect is measured by the amplitude ratio for $K^+\to\pi^++\pi^0$ and $K_1{}^0\to\pi^++\pi^-$, which is 1:23.

We recall the detailed structure of the fermion current that is coupled to $\bar{Z}_\mu$,

$$\bar{\psi}_1\gamma^\mu(1+i\gamma_5)\psi_2+\bar{\psi}_1\gamma^\mu(1-i\gamma_5)\psi_3.$$

The opposite signs of $i\gamma_5$ in the two terms should be noted. The $\Delta T=\frac{1}{2}$ contribution derived from the second-order effect of this coupling contains the Green's function of the $Z$ field and of the $\psi_1$ field. When the product of the Green's functions is approximated by a four-dimensional delta function, the resulting localized interaction is proportional to

$$-(\tfrac{1}{8})\bar{\psi}_2\gamma^\mu(1+i\gamma_5)\gamma_\mu(1-i\gamma_5)\psi_2$$

with its adjoint, which equals

$$\bar{\psi}_2(1-i\gamma_5)\psi_3+\bar{\psi}_3(1+i\gamma_5)\psi_2. \tag{4}$$

The complete $\Delta T=\frac{1}{2}$, $CP$ invariant perturbation is obtained by adding the parity-preserving contribution of the $V$ field, which is represented by

$$\bar{V}_2{}^\mu V_{\mu3}+\bar{V}_3{}^\mu V_{\mu2}. \tag{5}$$

No general statement can be made about the relative magnitude of the parity-violating and parity-preserving effects, particularly since the strong interactions of the broken $SU_3$ symmetry scheme are of decisive influence on the observed phenomena.[3]

[2] The additional possibility of independent phase transformations for $\psi$ and $V$ is described by the nucleonic charge phase transformation and by the common phase transformtaion that leaves invariant the low-lying states.

[3] Indeed, the existence of these processes seems to involve the breakdown of $SU_3$ symmetry, beyond the mere role of supplying the necessary energy. Thus the weak parity-violating coupling between $K^*$ and $\pi$ that accounts for $K_1{}^0\to\pi+\pi$ (see Ref. 4)

**PHENOMENOLOGY**

In view of the serious technical obstacles to performing field theory calculations from first principles, it is useful to convert the characteristic ideas of this field theory of matter into a corresponding phenomenology. Such a program is suggested by the structure of the electromagnetic interaction. The electromagnetic potential $A_\mu$ is coupled to the electric current vector

$$j^\mu = e[-\bar{\psi}_1\gamma^\mu\psi_1 - i(\bar{V}_1^{\mu\nu}V_{1\nu} - \bar{V}_{1\nu}V_1^{\mu\nu})].$$

This operator generates meson states of spin-parity $1^-$, as represented by phenomenological fields $U_{11}^\mu$. Thus, one could attempt to describe all linear electromagnetic interactions as proceeding through the fields of $1^-$ mesons with suitable quantum numbers. The least massive of these, $\omega$ and $\rho^0$, might be of major importance in this description. Similarly, the vector and pseudovector currents that are coupled to the $Z$ field can be represented approximately by the phenomenological fields of known $1^-$ and $0^-$ mesons. These couplings would describe all such weak interactions.[4] The so-called Goldberger-Treiman relations are an immediate consequence of this point of view which is, in a sense, a return to the original idea of Yukawa.

The nonleptonic weak interaction (3), as clarified by (4) and (5), can be represented by suitable components of the phenomenological fields associated with $0^-$ and $0^+$ mesons. Particles of the latter type have not been identified with certainty,[5] but these excitations will exist somewhere in the mass spectrum. Such scalar fields will also represent the part of the perturbation (2) that produces $\Delta T=1$ electromagnetic mass displacements, as well as the portion of (1) that generates the mass splittings of broken $SU_3$ symmetry.[6] Scalar fields can also be used to describe the break-down of $W_3$ symmetry.

This is an outline of a phenomenological field theory, which gives quantitative expression to the general ideas that have been expressed here. It will be developed in another publication.

would vanish if the $K$ and $\pi$ masses were equal, owing to the transversality of the four-vector $K^*$ field. A similar remark applies to baryon decays. As we shall discuss elsewhere, simple models of parity-conserving decays depend upon coupling constant differences that would be zero were $SU_3$ symmetry not broken. [*Added in proof*. See J. Schwinger, Phys. Rev. Letters **13**, 355 and 500 (1964).]

[4] This idea is used to compute the absolute rate for the $\Delta T=\frac{3}{2}$ process $K^+\rightarrow\pi^++\pi^0$ in J. Schwinger, Phys. Rev. Letters **12**, 630 (1964).

[5] Some possibilities are discussed by S. Coleman and S. L. Glashow, Phys. Rev. **134**, B671 (1964).

[6] A perturbation that is a numerical multiple of a phenomenological boson field generates a displacement of that field, which is its nonvanishing vacuum expectation value. We propose the word *vacuon* for this property. (Despite its hybrid etymology, such a terminology seems preferable to the use of a biologicographic argot.) The vacuon concept is not new [J. Schwinger, Phys. Rev. **104**, 1164 (1956); Ann. Phys. (N. Y.) **2**, 407 (1957); A. Salam and J. Ward, Phys. Rev. Letters **5**, 390 (1960)], but its most effective application is that of Ref. 5. We have now connected thie phenomenological procedure with a fundamental dynamical theory.

*Notes added in proof*. (1) Some statements about the group $SU_3$ need clarification and correction. For each field, $\psi$ and $V$, the generators of $SU_3$ transformations are obtained from the $U_3$ generators by forming a traceless combination, with the aid of the corresponding nucleonic charge,

$$\psi:\ 'T_{ab}=T_{ab}-\tfrac{1}{3}\delta_{ab}(-N)$$

$$V:\ 'T_{ab}=T_{ab}-\tfrac{1}{3}\delta_{ab}(-\tfrac{1}{2}N).$$

The complete generators

$$T_{ab}=T_{ab}(\psi)+T_{ab}(V)$$

are related by

$$'T_{ab}=T_{ab}-\tfrac{1}{3}\delta_{ab}F,$$

where

$$F=-N(\psi)-\tfrac{1}{2}N(V)=T_{11}+T_{22}+T_{33}.$$

We shall call this quantity field charge. Unit field charge is assigned to $\bar{\psi}$ and $\bar{V}$,

$$[\bar{\psi},F]=\bar{\psi},\quad [\bar{V},F]=\bar{V},$$

while $\psi$ and $V$ carry the opposite sign of charge. Hypercharge and electrical charge are identified as

$$Y=T_{11}+T_{22},\quad Q=T_{11},$$

and

$$Q-\tfrac{1}{2}Y=\tfrac{1}{2}(T_{11}-T_{22})=T_3.$$

The charges $F$, $Y$, and $Q$ have integer eigenvalues. If an $SU_3$ hypercharge $'Y$ is defined by

$$'Y={}'T_{11}+{}'T_{22}=Y-\tfrac{2}{3}F,$$

we can give the electrical charge the alternative forms

$$Q=T_3+\tfrac{1}{2}Y=T_3+\tfrac{1}{2}{}'Y+\tfrac{1}{3}F.$$

[Field charge is evidently related to what has been called triality. See, for example, G. Baird and L. Biedenharn, *Symmetry Principles at High Energy* (W. H. Freeman and Company, San Francisco, 1964). Triality is only conserved modulo 3, however.]

The characteristic property of low-lying states that is noted in the text can now be restated as: Low-lying states carry zero field charge. For such states there is no distinction between $'T_{ab}$ and $T_{ab}$. In particular, $'Y=Y$. The $U_3$ matrices $t_{ab}$ and the $SU_3$ matrices $'t_{ab}$ that are associated with the fields differ, however. [I am indebted to A. Radkowski for a comment on this point.] Thus,

$$[\bar{\psi},T_{ab}]=t_{ab}\bar{\psi},\quad [\bar{\psi},{}'T_{ab}]={}'t_{ab}\bar{\psi},$$

where

$$(t_{ab})_{cd}=\delta_{ca}\delta_{bd}$$

and

$$'t_{ab}=t_{ab}-\tfrac{1}{3}\delta_{ab}.$$

The electric charge and hypercharge matrices are

$$q=t_{11}=\begin{pmatrix}1&&\\&0&\\&&0\end{pmatrix},\quad y=t_{11}+t_{22}=\begin{pmatrix}1&&\\&1&\\&&0\end{pmatrix}$$

and not

$$'t_{11}=q-\tfrac{1}{3}, \quad 't_{11}+'t_{22}=y-\tfrac{2}{3}.$$

The field charge $F$ is conserved, in addition to the operators $'T_{ab}$, in the dynamical regime that has been described as governed by $SU_3$ symmetry. Accordingly, the relevant symmetry group is $U_3$, rather than $SU_3$. Stable particles with unit field charge will exist in this idealization. In particular, the operators $\psi_a$ and $V_a$ generate triplets of fermions and bosons, respectively. When the full dynamical effect of the $\psi$–$V$ coupling is included, however, $F$ ceases to be conserved and the triplets become highly unstable, particularly if they are quite massive. Examples of possible decay modes are (field symbols designate triplet particles)

$$\psi_{1,2}\rightarrow\Lambda+\bar{K}, \quad \psi_3\rightarrow\Lambda+\eta$$

and

$$V_{1,2}\rightarrow\Lambda+\Xi, \quad V_3\rightarrow\Lambda+\Lambda.$$

Such resonant states may be observed eventually. It is the negative aspect of these predictions that needs immediate emphasis: Stable or long-lived triplets should not exist. [Compare T. D. Lee (to be published).]

(2) It is asserted in the text that the various symmetry-destroying mechanisms can be represented by suitable vacuons. While that description conveys the major effect of these perturbations one should not overlook its incompleteness. Thus, the operator $\bar{\psi}_3\psi_3 f_1((\bar{\psi}V)(\bar{V}\psi))$ is only approximately represented by a numerical multiple of $\bar{\psi}_3\psi_3$ or the corresponding scalar field. The incompleteness of the vacuon description is particularly significant for parity-conserving weak decays, since a simple vacuon treatment gives no effect. This is a consequence of detailed cancellations, which express the possibility of absorbing the weak scalar vacuon into the strong vacuon responsible for broken $SU_3$ symmetry. It is easy to underestimate the extent of the cancellation and just this happened in the recent notes that are cited in Ref. 3. Parity-conserving decays were there ascribed primarily to the breakdown of $SU_3$ coupling constant relations. But, when the relevant mechanism is represented by the vacuon that dominates the mass splittings of broken $SU_3$ symmetry, there is an additional weak contribution that cancels this source of parity conserving decays. Something similar occurs in the vacuon treatment of singlet-octuplet mixing. It is necessary to base the dynamical explanation of parity-conserving nonleptonic decays on the fact that the symmetry-destroying mechanisms implied by the theory cannot be represented precisely by the numerical displacement of scalar fields.

# Field Theory of Matter. IV

JULIAN SCHWINGER*
*Harvard University, Cambridge, Massachusetts*
(Received 5 April 1965)

The relativistic dynamics of $0^-$ and $1^-$ mesons in the idealization of $U_3$ symmetry is derived from the hypothesis that a compact group of transformations on fundamental fields induces a predominantly local and linear transformation of the phenomenological fields that are associated with particles. The physical picture of phenomenological fields as highly localized functions of fundamental fields implies that the interaction term of the phenomenological Lagrange function can have symmetry properties, expressed by invariance under the compact transformation group, that have no significance for the remainder of the Lagrange function, which describes the propagation of the physical excitations. It is verified that the meson interaction term derived by considering fundamental fermion fields is invariant under the parity-conserving group $U_6 \times U_6$. The implied connection between the $\rho\pi\pi$ and $\omega\rho\pi$ coupling constants is well satisfied. There is a brief discussion of the dynamics of fermion-particle triplets, from which it is shown that the invariance of the similarly derived interaction term implies the mass degeneracy of the singlet and octuplet of $1^-$ mesons, without relation to $0^-$ masses. The triplets are also used to illustrate the derivation of gauge- and relativistically invariant electromagnetic properties. The mass degeneracy of the nine $1^-$ mesons, and of nine $2^+$ mesons, can be inferred from the commutation properties of bilinear combinations of the fundamental field.

## INTRODUCTION

THE preceding paper of this series[1] describes a program for establishing contact between the fundamental fields $\psi_{\zeta a}(x)$ and $V_a^{\mu}(x)$, $V_a^{\mu\nu}(x)$, and the phenomenological fields that represent observed particles. This is accomplished by a technique of comparative kinematical transformation. Compact groups of kinematical transformations on the fundamental fields are exploited as a device for conveying dynamical information concerning the highly localized structure of the phenomenological fields. The hypothesis of completeness for stable and unstable particles permits a linear representation of the essentially localized transformations induced on the phenomenological fields. This implies a correspondence between group generators at the fundamental and the phenomenological levels. Each kind of generator is a quadratic function of the appropriate kind of field. The quadratic functions of the fundamental fields, as objects with various tensor transformation properties, are also represented linearly by the phenomenological fields of bosons with suitable spins and parities. Through the machinery of relativistic field theory, particularly the distinction and the relation between independent and dependent field components, these alternative phenomenological identifications of group generators serve to determine phenomenological field dynamics. In this paper the program will be illustrated by the dynamics of $0^-$ and $1^-$ bosons, in the idealization of $U_3$ symmetry. We shall also consider briefly the dynamics of spin-$\frac{1}{2}$ particle triplets. The extension to baryon interactions and the inclusion of unitary symmetry-breaking effects will be dealt with separately.

* Supported in part by the U. S. Air Force Office of Scientific Research under Contract No. A.F. 49(638)-1380.
[1] J. Schwinger, Phys. Rev. **135**, B816 (1964); **136**, B1821 (1964), which are referred to in the text as I and II, respectively. The third article is contained in the published account of the second *Coral Gables Conference on Symmetry Principles at High Energy* (W. H. Freeman and Company, Inc., San Francisco, 1964).

## MESON FIELDS

The discussion of III directed particular attention to the group of parity-preserving transformations on the twelve-component field $\psi_{\zeta a}(x)$. This group has the structure $U_6 \times U_6$. The corresponding generators can be displayed as

$$M_{AB}^{(\pm)}(x)=\psi_a^{\dagger}(x)\tfrac{1}{2}(1\pm\gamma^0)\frac{1}{2}\begin{pmatrix}1+\sigma_3 & \sigma_1+i\sigma_2\\ \sigma_1-i\sigma_2 & 1-\sigma_3\end{pmatrix}\psi_b(x),$$

where $A$, for example, is a sextuple-valued index that combines $a$ and the double-valued spin label. The equal-time commutation relations obeyed by these combinations are

$$[M_{AB}^{(+)}(x),M_{CD}^{(-)}(x')]=0$$

and

$$[M_{AB}^{(\pm)}(x),M_{CD}^{(\pm)}(x')] =\delta(\mathbf{x}-\mathbf{x}')\{\delta_{BC}M_{AD}^{(\pm)}(x)-\delta_{AD}M_{CB}^{(\pm)}(x)\}.$$

We also note that

$$(M_{AB}^{(\pm)})^{\dagger}=M_{BA}^{(\pm)}.$$

The transformation induced on the odd-parity objects

$$M_{AB}=-\psi_a^{\dagger}\gamma_5\tfrac{1}{2}(1+\gamma^0)\frac{1}{2}\begin{pmatrix}1+\sigma_3 & \sigma_1+i\sigma_2\\ \sigma_1-i\sigma_2 & 1-\sigma_3\end{pmatrix}\psi_b$$

is represented by the equal-time commutation relations

$$[M_{AB}(x),M_{CD}^{(+)}(x')]=\delta(\mathbf{x}-\mathbf{x}')\delta_{BC}M_{AD}(x),$$
$$[M_{AB}(x),M_{CD}^{(-)}(x')]=\delta(\mathbf{x}-\mathbf{x}')(-)\delta_{AD}M_{CB}(x).$$



Reprinted from *The Physical Review* **140**, B158–B163 (1965).

To these we add the commutation relations of

$$M_{AB}^{\dagger}=\psi_a^{\dagger}\tfrac{1}{2}(1+\gamma^0)\gamma_5\frac{1}{2}\begin{pmatrix}1+\sigma_3 & \sigma_1+i\sigma_2\\ \sigma_1-i\sigma_2 & 1-\sigma_3\end{pmatrix}\psi_b$$
$$=(M_{BA})^{\dagger},$$

namely,

$$[M_{AB}^{\dagger}(x),M_{CD}^{(+)}(x')]=\delta(\mathbf{x}-\mathbf{x}')(-)\delta_{AD}M_{CB}^{\dagger}(x),$$
$$[M_{AB}^{\dagger}(x),M_{CD}^{(-)}(x')]=\delta(\mathbf{x}-\mathbf{x}')\delta_{BC}M_{AD}^{\dagger}(x).$$

When completed by

$$[M_{AB}(x),M_{CD}^{\dagger}(x')]=\delta(\mathbf{x}-\mathbf{x}')\{\delta_{BC}M_{AD}^{(-)}(x)-\delta_{AD}M_{CB}^{(+)}(x)\},$$

we have also indicated the commutator structure of the group $U_{12}$.

The matrices that represent parity-preserving transformations, on the phenomenological fields of octuplet-plus-singlet families of $0^-$ and $1^-$ particles, commute with $A^0$. The latter is the antisymmetrical real matrix that specifies boson-field equal-time commutation relations. Accordingly, it is possible and convenient to diagonalize $A^0$. This is accomplished by introducing the non-Hermitian combinations

$$\phi^{\pm}=(\tfrac{1}{2}m_0)^{1/2}\phi\mp i(2m_0)^{-1/2}\phi^0$$

and

$$U_k^{\pm}=(2m_1)^{-1/2}U^0{}_k\pm i(\tfrac{1}{2}m_1)^{1/2}U_k,$$

which obey the commutation relations (all other commutators vanish)

$$[\phi_{ab}^+(x),\phi_{cd}^-(x')]=\delta(\mathbf{x}-\mathbf{x}')\delta_{bc}\delta_{ad},$$
$$[U_{kab}^+(x),U_{lcd}^-(x')]=\delta(\mathbf{x}-\mathbf{x}')\delta_{kl}\delta_{bc}\delta_{ad}.$$

The quantities $m_0$ and $m_1$ are mass scaling factors which could vary arbitrarily among the multiplet members, subject only to the symmetry requirements of the assumed dynamical level. Thus, with the assumption of $U_3$ symmetry, we might use a different mass factor for a singlet than for the members of an octuplet. We shall, in fact, employ a unique mass for all nine fields of a given spin so that $m_0$ and $m_1$ are just numbers. The necessity for this will be indicated later. It will also be discovered that $m_0=m_1$.

The 72 odd-parity field components comprised in the 36 non-Hermitian combinations $\phi_{ab}^+$, $U_{kab}^+$ can be arrayed in the matrix

$$M_{AB}=\frac{(\frac{1}{2}m)^{3/2}}{(-g)}\begin{pmatrix}\phi_{ab}^++U_{3ab}^+ & U_{1ab}^++iU_{2ab}^+\\ U_{1ab}^+-iU_{2ab}^+ & \phi_{ab}^+-U_{3ab}^+\end{pmatrix}$$

which is displayed more compactly as

$$M_{AB}=(\tfrac{1}{2}m)^{3/2}/(-g)(\phi_{ab}^++\sigma_k{}^TU_{kab}^+).$$

The adjoint matrix is

$$M_{AB}^{\dagger}=(M_{BA})^{\dagger}=(\tfrac{1}{2}m)^{3/2}/(-g)(\phi_{ab}^-+\sigma_k{}^TU_{kab}^-).$$

Here, $m$ is a mass parameter which can be identified conveniently with the common value of $m_0$ and $m_1$, while $g$ is an arbitrary dimensionless constant. Note that the commutation properties of the phenomenological fields are given by

$$[M_{AB}(x),M_{CD}^{\dagger}(x')]=\delta(\mathbf{x}-\mathbf{x}')\delta_{BC}\delta_{AD}(m^3/4g^2).$$

The use of a common notation indicates the linear correspondence that is being established between the phenomenological fields and bilinear combinations of the fundamental $\psi$ operators. This correspondence is expressed covariantly by

$$-g\bar{\psi}_a\gamma_5\psi_b\leftrightarrow m^{3/2}m_0{}^{1/2}\phi_{ab}$$
$$-g\bar{\psi}_k\gamma^{\mu}\psi_b\leftrightarrow m^{3/2}m_1U_{ab}{}^{\mu}$$
$$g\bar{\psi}_ai\gamma^{\mu}\gamma_5\psi_b\leftrightarrow m^{3/2}m_0{}^{-1/2}\phi_{ab}{}^{\mu}$$
$$g\bar{\psi}_a\sigma^{\mu\nu}\psi_b\leftrightarrow m^{3/2}m_1{}^{-1/2}U_{ab}{}^{\mu\nu},$$

where a distinction among the various mass constants is still retained.

The correspondence has another meaning. It presents the phenomenological fields as the basis for a particular representation of the group $U_6\times U_6$. The commutation relations between $M_{AB}$ and $M_{CD}^{(\pm)}$ imply that $M_{AB}$ transforms as the product of independent six-dimensional representations. This is symbolized by

$$M\sim 6_-{}^*\times 6_+,$$

and similarly

$$M^{\dagger}\sim 6_+{}^*\times 6_-,$$

while

$$M^{(+)}\sim 6_+{}^*\times 6_+,\quad M^{(-)}\sim 6_-{}^*\times 6_-.$$

Accordingly, the generators $M^{(\pm)}$ must be identified with suitable bilinear combinations of the phenomenological fields $M$ and $M^{\dagger}$. The required combinations are simply

$$M_{AB}^{(+)}=(4g^2/m^3)(M^{\dagger}M)_{AB}$$

and

$$M_{AB}^{(-)}=-(4g^2/m^3)(MM^{\dagger})_{AB},$$

which possess all the commutation properties required of $U_6\times U_6$ generators.

The linear correspondence between phenomenological fields and bilinear $\psi$ combinations also implies the following correspondence

$$M_{AB}^{(\pm)}\leftrightarrow-(m^{3/2}/4g)(\pm m^{1/2}S_{ab}+m_1{}^{1/2}U_{ab}{}^0-\sigma_k{}^Tm_0{}^{-1/2}\phi_{kab}\mp\sigma_k{}^Tm_1{}^{-1/2}U_{[k]ab}),$$

where

$$U_{[k]}=\tfrac{1}{2}\epsilon_{klm}U_{lm}$$

and

$$g\bar{\psi}_a\psi_b\leftrightarrow m^2S_{ab}.$$

The result is a local relation between the dependent and independent components of the fields associated with $0^-$ and $1^-$ particles. It is expressed by the matrix equations

$$-m_1^{1/2}U^0+\sigma_k{}^T m_0{}^{-1/2}\phi_k = (g/m^{3/2})[\phi^-+\sigma_k{}^TU_k{}^-,\ \phi^++\sigma_k{}^TU_k{}^+]$$

and

$$-m^{1/2}S+\sigma_k{}^Tm_1{}^{-1/2}U_{[k]} = (g/m^{3/2})\{\phi^-+\sigma_k{}^TU_k{}^-,\ \phi^++\sigma_k{}^TU_k{}^+\}.$$

These are written out somewhat more explicitly as

$$-m^{3/2}m_1{}^{1/2}U^0=g([\phi^-,\phi^+]+[U_k{}^-,U_k{}^+]),$$

$$m^{3/2}m_0{}^{-1/2}\phi_k=g([\phi^-,U_k{}^+]+[U_k{}^-,\phi^+] -i\epsilon_{klm}\{U_l{}^-,U_m{}^+\}),$$

$$m^{3/2}m_1{}^{-1/2}U_{[k]}=g(\{\phi^-,U_k{}^+\}+\{U_k{}^-,\phi^+\} -i\epsilon_{klm}[U_l{}^-,U_m{}^+]),$$

and

$$-m^2S=g(\{\phi^-,\phi^+\}+\{U_k{}^-,U_k{}^+\}).$$

## MESON DYNAMICS

The Lagrange function of octuplet plus singlet families of $0^-$ and $1^-$ particles is given by

$$\mathfrak{L}=\mathrm{tr}[-\phi^\mu\partial_\mu\phi+\tfrac{1}{2}\phi^\mu\phi_\mu-\tfrac{1}{2}\phi M_0{}^2\phi] +\mathrm{tr}[-\tfrac{1}{2}U^{\mu\nu}(\partial_\mu U_\nu-\partial_\nu U_\mu)+\tfrac{1}{4}U^{\mu\nu}U_{\mu\nu}-\tfrac{1}{2}U^\mu M_1{}^2U_\mu] +\mathfrak{L}_{\mathrm{int}}(\phi,U),$$

where $M_0{}^2$ and $M_1{}^2$ only distinguish between singlet and octuplet of the corresponding species. The constraint equations are

$$M_1{}^2U^0=-(\partial\mathfrak{L}_{\mathrm{int}}/\partial U^0)-\partial_kU^{0k},$$

$$\phi_k=-(\partial\mathfrak{L}_{\mathrm{int}}/\partial\phi_k)+\partial_k\phi,$$

$$U_{[k]}=-(\partial\mathfrak{L}_{\mathrm{int}}/\partial U_{[k]})+(\nabla\times\mathbf{U})_k,$$

and we identify the local parts of these relations with those established in the previous section. The resulting tentative construction of $\mathfrak{L}_{\mathrm{int}}$ is indicated by

$$\mathfrak{L}_{\mathrm{int}}=\mathrm{tr}[-U^0(M_1{}^2/m^2)\text{`}m^2U^0\text{'}-\phi_k(1/m)\text{`}m\phi_k\text{'} -U_{[k]}(1/m)\text{`}mU_{[k]}\text{'}],$$

where the quantities in quotation marks are the quadratic functions of the independent field variables, and a partial anticipation of the relations $m_0=m_1=m$ has been introduced, merely for typographical simplification.

We now invoke the requirement of relativistic invariance in order to obtain connections among the various constants that have been employed. To illustrate this procedure let us isolate the terms in $\mathfrak{L}_{\mathrm{int}}$ that are linear in the components of $\phi^\mu$. These terms are

$$(g/m^{3/2})\ \mathrm{tr}[m_0{}^{1/2}\phi^k\{m_1{}^{1/2}U^l,m_1{}^{-1/2}\,{}^*U_{kl}\} +m_0{}^{-1/2}\phi^0\{m_1{}^{1/2}U^k,m_1{}^{1/2}\,{}^*U_{0k}\}],$$

where we have introduced the dual tensor

$$^*U_{\mu\nu}=\tfrac{1}{2}\epsilon_{\mu\nu\lambda\kappa}U^{\lambda\kappa}.$$

Since the two contributions are parts of a single scalar they must possess the same coefficients for any combination of unitary components. The unitary singlets contribute to this interaction as well as the octuplets. It is now clear that all possible values for $m_0$ and $m_1$ must be identical, which we also adopt as the natural choice for $m$,

$$-m_0=m_1=m.$$

One should note that there is another term obtained in this way, which is not contained in the tentative form of $\mathfrak{L}_{\mathrm{int}}$. It is

$$-(g/m)\ \mathrm{tr}\phi^k\{U^0,{}^*U_{0k}\},$$

which is a product of three dependent field components. It is an example of a structure that is outside the simplified framework of local and linear transformations, but one which is demanded by the requirements of relativistic invariance.

Another such pair of terms, parts of a single relativistic scalar, is

$$-ig\ \mathrm{tr}[(M_1{}^2/m^2)U^0[\phi^0,\phi]+U_k[\phi^k,\phi]].$$

Evidently,

$$M_1(\text{octuplet})=m,$$

but no information is obtained in this way about $M_1$ (singlet), since $\mathrm{tr}[\phi^0,\phi]=0$. (See the last section, however.) There is no necessary relation between $m$ and the masses of the $0^-$ particles.

The outcome of these considerations is the following scalar interaction term of the Lagrange function,[2]

$$\mathfrak{L}_{\mathrm{int}}=ig\ \mathrm{tr}(\phi^\mu[U_\mu,\phi]+\tfrac{1}{2}U^{\mu\nu}[U_\mu,U_\nu]) -(g/m)\ \mathrm{tr}(\phi\tfrac{1}{4}\{^*U_{\mu\nu},U^{\mu\nu}\}+\phi^\nu\{U^\mu,{}^*U_{\mu\nu}\}) -i(g/m^2)\ \mathrm{tr}(\tfrac{1}{2}U^{\mu\nu}[\phi_\mu,\phi_\nu]+\tfrac{1}{3}U^{\mu\nu}U_{\nu\lambda}U_\mu{}^\lambda).$$

It is particularly interesting that this structure can also be presented as

$$\mathfrak{L}_{\mathrm{int}}=-((2g)^4/m^5)\ \mathrm{Tr}[M^{(+)}M^\dagger M-M^{(-)}MM^\dagger -\tfrac{1}{3}(M^{(+)})^3+\tfrac{1}{3}(M^{(-)})^3],$$

where Tr refers to the sextuple-valued indices that label the fields $M^{(\pm)}$, $M$, and $M^\dagger$. The latter form makes it evident that $\mathfrak{L}_{\mathrm{int}}$ is invariant under the transformations of the group $U_6\times U_6$. Let it be emphasized that we have *derived* this property of the interaction term[3] from our fundamental dynamical assumptions concerning localizability and completeness. It is entirely comprehensible that $\mathfrak{L}_{\mathrm{int}}$ should possess this invariance as a kinematical expression of the highly localized dynamical relation between phenomenological and fundamental

[2] These results were reported at the second Coral Gables Conference on Symmetry Principles at High Energy, January, 1965.

[3] Some authors have postulated that the interaction term in a Lagrange function referring to fundamental fields (translated from the original Quark) is invariant under a $U_6\times U_6$ transformation group, which is also regarded as imbedded in a larger noncompact group. See, for example, K. Bardakci, J. Cornwall, P. Freund, and B. Lee, Phys. Rev. Letters **14**, 48 (1965). In our view, such hypotheses are irrelevant to the emergence of the parity-conserving $U_6\times U_6$ group at the phenomenological level.

fields, without such transformations having the slightest relevance to the remainder of the phenomenological Lagrange function, which characterizes the propagation of the physical excitations.[4]

There is one difficulty, at least, with this result. It concerns the manner in which $U^\mu$ appears, through the implication for electromagnetic interactions. Let us recall that the $\psi$ contribution to the electric current vector is

$$j^\mu(x) = -e\bar{\psi}_1\gamma^\mu\psi_1(x) \to (e/g)m^2U_{11}{}^\mu(x).$$

The field equations deduced from the action principle by varying $U^\mu$ are

$$M_1{}^2U^\mu + \partial_\nu U^{\mu\nu} = ig[\phi,\phi^\mu] + ig[U_\nu,U^{\mu\nu}] - (g/m)\{{}^*U^{\mu\nu},\phi_\nu\},$$

and the consequence for the operator of total electrical charge is contained in

$$(M_1{}^2/g)\int(d\mathbf{x})U^0 = \int(d\mathbf{x})(i[\phi,\phi^0] + i[U_k,U^0{}_k] - (1/m)\{{}^*U^0{}_k,\phi_k\}).$$

The last term on the right-hand side is objectionable. In its absence, only octuplet terms appear and, on replacing $M_1$ with the constant $m$, we obtain just the total charge operator anticipated for $0^-$ and $1^-$ particles. It is true that the additional term vanishes in a first approximation, for which $\phi_k = \partial_k\phi$ and $\partial_k{}^*U^{0k} = 0$. But we should prefer that the required charge operator emerge precisely. This can be achieved by a redefinition of the fields $\phi_\mu$ and $U_{\mu\nu}$ within the framework of cubic interaction terms, to which our discussion is limited. The required transformation is given by

$$\phi^\mu \to \phi^\mu - (g/m)\{{}^*U^{\mu\nu},U_\nu\},$$

$$U_{\mu\nu} \to U_{\mu\nu} - (g/m)\{{}^*U_{\mu\nu},\phi\} - (g/m)^*(\partial_\mu\{U_\nu,\phi\} - \partial_\nu\{U_\mu,\phi\}),$$

and the new version of the interaction is

$$\mathcal{L}_{\text{int}} = ig\ \mathrm{tr}(\phi^\mu[U_\mu,\phi] + \tfrac{1}{2}U^{\mu\nu}[U_\mu,U_\nu]) - \tfrac{3}{2}(g/m)\ \mathrm{tr}(\phi^*U^{\mu\nu}U_{\mu\nu}) - i(g/m^2)\ \mathrm{tr}(\tfrac{1}{2}U^{\mu\nu}[\phi_\mu,\phi_\nu] + \tfrac{1}{3}U^{\mu\nu}U_{\nu\lambda}U_\mu{}^\lambda).$$

## COMPARISON WITH EXPERIMENT

Tests of the prediction that $U\phi\phi$, $\phi UU$, and $UUU$ couplings are all governed by a single coupling constant are available only for the first two, which are involved in various meson decays. In order to minimize the effects of symmetry-breaking interactions, which have not yet been included, we confine our attention to the particles $\pi$, $\rho$, and $\omega$. We shall also identify $m$ with the observed mass of the $\rho$ particle. This is based on the treatment of interactions that break $U_3$ symmetry, discussed in II, which associates the perturbation with the third axis of the unitary space. The phenomenological analysis of $\rho\pi\pi$ and $\omega\rho\pi$ couplings given in I (which also explained the suppression of $\phi\rho\pi$ coupling) employed the interaction terms

$$\mathcal{L}_{U\phi^2} = \tfrac{1}{2}g_{U\phi^2}i\ \mathrm{tr}\{U^\mu[\phi_\mu,\phi]\}$$

and

$$\mathcal{L}_{\phi U^2} = \tfrac{1}{2}g_{\phi U^2}\ \mathrm{tr}\{\phi\ {}^*U^{\mu\nu}U_{\mu\nu}\},$$

where

$$(g_{U\phi^2})^2/4\pi\sim 5, \quad (m_\omega g_{\phi U^2})^2/4\pi\sim 6.$$

The effective coupling constant $g_{\phi U^2}$ is identified with $-3(g/m)$, $m = m_\rho$, while $g_{U\phi^2} = -3g$. The latter connection can be verified by removing the interaction term $U^{\mu\nu}[\phi_\mu,\phi_\nu]$, either with a field transformation in the manner indicated before, or by a perturbation approximation appropriate to the decay $\rho\to\pi+\pi$. Thus, the two independent evaluations of $g$ agree closely,[2,5] and

$$g^2/4\pi\sim\tfrac{1}{2}.$$

## FERMION TRIPLETS

The inability to obtain information about the mass $M_1$ (singlet) can be traced back to the correspondence between the field $U_{aa}{}^\mu$ and the current $-\bar{\psi}_a\gamma^\mu\psi_a$, which is the current of $\psi$ field charge (as defined in II). The $0^-$ and $1^-$ mesons carry zero field charge. Accordingly, we now consider the simplest particle multiplet that possess nonzero field charge. This is the spin-$\frac{1}{2}$ unitary triplet field $\Psi_a(x)$, which is in correspondence with the fundamental Fermi field $\psi_a(x)$. The decomposition

$$\Psi_a = \tfrac{1}{2}(1+\gamma^0)\Psi_a + \tfrac{1}{2}(1-\gamma^0)\Psi_a$$

supplies the bases $\Psi_A{}^{(+)}$ and $\Psi_A{}^{(-)}$ for the representations $6_+$ and $6_-$ of the kinematical group $U_6\times U_6$. The phenomenological generators $M^{(\pm)}$ now acquire a fermion contribution which is given by

$$(M_{AB}{}^{(+)})_f = \Psi_A{}^{(+)\dagger}\Psi_B{}^{(+)}.$$

$$(M_{AB}{}^{(-)})_f = \Psi_A{}^{(-)\dagger}\Psi_B{}^{(-)}.$$

Through addition and substraction we obtain the equivalent unitary forms

$$(m^2/g)(-U^0 + (1/m)\sigma_k{}^T\phi_k)_f = \Psi^\dagger\Psi + (\Psi^\dagger\sigma_k\Psi)\sigma_k{}^T,$$

$$(m^2/g)(-S + (1/m)\sigma_k{}^TU_{[k]})_f = \Psi^\dagger\gamma^0\Psi + (\Psi^\dagger\gamma^0\sigma_\kappa\Psi)\sigma_k{}^T,$$

[4] This distinction, which requires that one take seriously the possibility of spatio-temporal description at subparticle dimensions, is lost completely in a global or $S$-matrix particle description. The physical incorrectness of $S$-matrix theories built on relativistically generalized $SU_6$ symmetry is beginning to be recognized (see the 29 March 1965 issue of Physical Review Letters).

[5] Some other discussions of the relation between the $\rho\pi\pi$ and $\omega\rho\pi$ coupling constants have appeared very recently: K. Bardakci, J. Cornwall, P. Freund, and B. Lee, Phys. Rev. Letters **14**, 264 (1965); B. Sakita and K. Wali, *ibid.* **14**, 404 (1965); I. Gerstein, *ibid.* **14**, 453 (1965). Also W. Rühl, CERN, February, 1965 (unpublished).

or

$$-m^2(U^0)_f = g\bar{\Psi}\gamma^0\Psi,$$
$$m(\phi_k)_f = g\bar{\Psi}i\gamma_k\gamma_5\Psi,$$

and

$$m(U_{kl})_f = g\bar{\Psi}\sigma_{kl}\Psi,$$
$$-m^2(S)_f = g\bar{\Psi}\Psi.$$

The part of $\mathfrak{L}_{\text{int}}$ that refers to the fermion triplets and the $0^-$, $1^-$ boson fields is then obtained in the tentative form

$$(\mathfrak{L}_{\text{int}})_f = g\,\text{tr}[U^0(M_1^2/m^2)\bar{\Psi}\gamma^0\Psi - (1/m)\phi^k\bar{\Psi}i\gamma_k\gamma_5\Psi - (1/2m)U^{kl}\bar{\Psi}\sigma_{kl}\Psi].$$

Now, these coupling terms can also be displayed as the $U_6\times U_6$ invariant structure

$$(\mathfrak{L}_{\text{int}})_f = -(2g/m)^2\,\text{Tr}[M^{(+)}\Psi^{(+)\dagger}\Psi^{(+)} + M^{(-)}\Psi^{(-)\dagger}\Psi^{(-)}],$$

provided only that

$$M_1 = m$$

is also valid for the unitary singlet combination of $1^-$ particles. Thus, it requires only a very weak extension of the invariance property of $\mathfrak{L}_{\text{int}}$ already established for mesons to reach the empirically valid conclusion that the octuplet and singlet of $1^-$ particles are mass degenerate in the dynamical idealization of $U_3$ symmetry. How different this dynamical argument is from the $SU_6$ assertion that the $0^-$ and the $1^-$ mesons constitute an initially mass degenerate multiplet, which is split by a spin-dependent perturbation.[6]

The relativistically invariant form of the triplet interaction term is

$$(\mathfrak{L}_{\text{int}})_f = g\,\text{tr}[-U^\mu\bar{\Psi}\gamma_\mu\Psi - (1/m)\phi^\mu\bar{\Psi}i\gamma_\mu\gamma_5\Psi - (1/2m)U^{\mu\nu}\bar{\Psi}\sigma_{\mu\nu}\Psi].$$

The additional terms demanded by relativistic invariance are

$$g\,\text{tr}[-U^k\bar{\Psi}\gamma_k\Psi + (1/m)\phi^0\bar{\Psi}i\gamma^0\gamma_5\Psi + (1/m)U^{0k}\bar{\Psi}i\gamma^0\gamma_k\Psi].$$

If one were to supplement these by the pseudoscalar interaction

$$-g\,\text{tr}(\phi\bar{\Psi}\gamma_5\Psi),$$

all the additional terms would be comprised in the $U_6\times U_6$ invariant contribution

$$(2g/m)^2\,\text{Tr}[M\bar{\Psi}^{(+)}\gamma_5\Psi^{(-)} + M^\dagger\bar{\Psi}^{(-)}\gamma_5\Psi^{(+)}].$$

The fermion triplets illustrate most simply the derivation of electromagnetic properties from the coupling

$$\mathfrak{L}_{\text{em}}{}^A = j^\mu A_\mu \to (e/g)m^2U_{11}{}^\mu A_\mu.$$

We give first the general verification of the gauge-invariant nature of this description. The field transformation

$$U_{11}{}^\mu \to U_{11}{}^\mu + (e/g)A^\mu,$$
$$F^{\mu\nu} \to F^{\mu\nu} - (e/g)U_{11}{}^{\mu\nu}$$

converts the electromagnetic coupling into the explicitly gauge-invariant form

$$\mathfrak{L}_{\text{em}}{}^F = -(e/g)\tfrac{1}{2}U_{11}{}^{\mu\nu}F_{\mu\nu}$$

while introducing, through the substitution $gU_{11}{}^\mu \to gU_{11}{}^\mu + eA^\mu$, the gauge-covariant combination $\partial_\mu - ieqA_\mu$ involving the appropriate charge matrix $q$.

If we simplify the dynamics of the $U$ field by retaining only the interaction with the $\Psi$ field, the field equations become

$$m^2U^\mu + \partial_\nu U^{\mu\nu} = -g\bar{\Psi}\gamma^\mu\Psi,$$
$$U_{\mu\nu} - (\partial_\mu U_\nu - \partial_\nu U_\mu) = (g/m)\bar{\Psi}\sigma_{\mu\nu}\Psi,$$

and therefore

$$(m^2-\partial^2)(e/g)U_{11}{}^\mu = -e[\bar{\Psi}_1\gamma^\mu\Psi_1 + (1/m)\partial_\nu\bar{\Psi}_1\sigma^{\mu\nu}\Psi_1].$$

This gives immediately a form-factor description of electromagnetic properties. In particular, the anomalous magnetic moment of a positively charged particle is obtained as two $\rho$ meson magnetons[7] (in the approximation that ignores the indirect contribution of other particles).

*Note added in proof.* The relations between dependent and independent field variables contain nonlocal contributions, which are demanded by the structure of phenomenological relativistic field theory. These include the linear space-derivative terms illustrated by

$$\phi_k = -(\partial\mathfrak{L}_{\text{int}}/\partial\phi_k) + \partial_k\phi.$$

It is important to recognize that such terms are implicit in the commutation properties of bilinear combinations of the $\psi$ field. This fact is the basis for another, and more fundamental proof that the nine $1^-$ mesons are mass degenerate in the $U_3$ dynamical idealization.

It has been known for some time[8] that the commutation properties of local bilinear combinations of Dirac fields, defined by a limiting procedure from initially distinct points, contain additional space-derivative terms which are overlooked in a formal calculation. (Such terms were not mentioned in the text precisely because they are nonlocal. A question on this point by Coleman served to elicit these additional remarks.) Consider, for example, the equal-time commutator

$$[\bar{\psi}_a(x)\gamma^0\psi_b(x), \bar{\psi}_c(x')\gamma_k\psi_d(x')] = \lim_{\epsilon\to 0}[\bar{\psi}_a(x)\gamma^0\psi_b(x), \bar{\psi}_c(x'-\tfrac{1}{2}\epsilon)\gamma_k\psi(x'+\tfrac{1}{2}\epsilon)]$$

in which the limit $\epsilon\to 0$ is performed symmetrically in

[6] F. Gürsey and L. Radicati, Phys. Rev. Letters 13, 173 (1964).

[7] A related statement can be found in P. Freund, Institute for Advanced Study, March, 1965 (unpublished).

[8] J. Schwinger, Phys. Rev. Letters 3, 296 (1959).

space. The result is

$$[\bar\psi_a(x)\gamma^0\psi_b(x),\bar\psi_c(x')\gamma_k\psi_d(x')]$$
$$=\delta(\mathbf{x}-\mathbf{x}')\{\delta_{bc}\bar\psi_a(x)\gamma_k\psi_d(x)-\delta_{ad}\bar\psi_c(x)\gamma_k\psi_b(x)\}$$
$$-\tfrac{1}{2}i\partial_k\delta(\mathbf{x}-\mathbf{x}')(\delta_{bc}K_{ad}+\delta_{ad}K_{cb}),$$

where

$$K_{ab}=\lim_{\epsilon\to 0}\tfrac{1}{3}i\langle\bar\psi_a(x)\boldsymbol{\epsilon}\cdot\boldsymbol{\gamma}\psi_b(x+\epsilon)\rangle$$

is assumed to be nonzero, in consequence of the singularity of the vacuum expectation value, and finite. In the idealization of $U_3$ dynamical symmetry, we have

$$K_{ab}=K\delta_{ab}$$

and the additional nonlocal term in this commutator becomes

$$-iK\delta_{ad}\delta_{bc}\partial_k\delta(\mathbf{x}-\mathbf{x}').$$

Similar commutators are

$$[\bar\psi_a(x)i\gamma_k\gamma_5\psi_b(x),\ \bar\psi_c(x')i\gamma^0\gamma_5\psi_d(x')]$$
$$=\delta(\mathbf{x}-\mathbf{x}')\{\delta_{bc}\bar\psi_a(x)\gamma_k\psi_d(x)-\delta_{ad}\bar\psi_c(x)\gamma_k\psi_b(x)\}$$
$$-iK\delta_{ad}\delta_{bc}\partial_k\delta(\mathbf{x}-\mathbf{x}')$$

and

$$[\bar\psi_a(x)\sigma_k\psi_b(x),\bar\psi_c(x')i\gamma^0\gamma_l\psi_d(x')]$$
$$=-\delta(\mathbf{x}-\mathbf{x}')\epsilon_{klm}\{\delta_{bc}\bar\psi_a(x)\gamma_m\psi_d(x)-\delta_{ad}\bar\psi_c(x)\gamma_m\psi_b(x)\}$$
$$+iK\delta_{ad}\delta_{bc}\epsilon_{klm}\partial_m\delta(\mathbf{x}-\mathbf{x}').$$

The transcription of these commutators into properties of phenomenological fields gives

$$m^2[U_{ab}{}^0(x),U_{kcd}(x')]$$
$$=\delta(\mathbf{x}-\mathbf{x}')g\{\delta_{ad}U_{kcb}(x)-\delta_{bc}U_{kad}(x)\}$$
$$-i(g^2/m^2)K\delta_{ad}\delta_{bc}\partial_k\delta(\mathbf{x}-\mathbf{x}')$$

and

$$[\phi_{kab}(x),\phi_{cd}{}^0(x')]$$
$$=\delta(\mathbf{x}-\mathbf{x}')g\{\delta_{ad}U_{kcb}(x)-\delta_{bc}U_{kad}(x)\}$$
$$-i(g^2/m^2)K\delta_{ad}\delta_{bc}\partial_k\delta(\mathbf{x}-\mathbf{x}'),$$
$$[U_{[k]ab}(x),U_{lcd}{}^0(x')]$$
$$=-\delta(\mathbf{x}-\mathbf{x}')\epsilon_{klm}g\{\delta_{ab}U_{mcb}(x)-\delta_{bc}U_{mad}(x)\}$$
$$+i(g^2/m^2)K\delta_{ad}\delta_{bc}\epsilon_{klm}\partial_m\delta(\mathbf{x}-\mathbf{x}').$$

These are statements about the dependence of $U^0$, $\phi_k$, and $U_{[k]}$ on $U^0{}_k$, $\phi$, and $U_l$, respectively. The appearance of the space derivative of the three-dimensional delta function implies that $U^0$, $\phi_k$, and $U_{[k]}$ additively contain $\partial_k U^{0k}$, $\partial_k\phi$, and $(\nabla\times U)_k$, respectively. The unit coefficients that occur in the last two relations are reproduced by the value

$$K=m^2/g^2.$$

We conclude that $m^2U_{ab}{}^0(x)$ contains the additive linear term $-\partial_k U_{ab}{}^{0k}(x)$. The comparison with the Lagrangian relation

$$M_1{}^2U^0=-(\partial\mathcal{L}_{\rm int}/\partial U^0)-\partial_k U^{0k}$$

shows that

$$M_1{}^2=m^2$$

is valid for all nine $1^-$ mesons, at the dynamical level of $U_3$ symmetry.

*Note added in proof*. These considerations can be extended to an octuplet plus singlet family of $2^+$ mesons. A correspondence exists between the phenomenological tensor fields $h^{\mu\nu}{}_{ab}$, ${}^\lambda\Gamma^{\mu\nu}{}_{ab}$ and bilinear combinations of the $\psi$ fields constructed with the aid of the Dirac matrices $\gamma^\mu$, $\sigma^{\mu\nu}$, and single coordinate derivatives. In particular, the commutation properties of $h^{kl}{}_{ab}$ and ${}^0\Gamma^{mn}{}_{ab}-\frac{1}{2}\partial^m h^{0n}{}_{ab}$ can be compared with those of $i\bar\psi_a(\gamma^k\partial^l+\gamma^l\partial^k)\psi_b$ and $i\bar\psi_a\sigma^{0n}\partial^m\psi_b-\frac{1}{3}\delta^{mn}i\bar\psi_a\sigma^{0k}\partial_k\psi_b$, where the coordinate derivatives act antisymmetrically on both fields. These phenomenological-field commutation relations depend explicitly upon the particle masses. In the idealization of $U_3$ dynamical symmetry, the comparison implies the mass degeneracy of octuplet and singlet for the $2^+$ mesons. This result is very gratifying in view of the experimental situation concerning the following particles: $f$(1.25 BeV, $T=0$), $A_2$(1.32 BeV, $T=1$), $K^*$(1.43 BeV, $T=\frac{1}{2}$, $Y=\pm1$), and $f'$(1.50 BeV, $T=0$), which seem to form just such a family of $2^+$ mesons, in striking analogy with the well-known $1^-$ mesons [see three contributions on spin-2 mesons in Phys. Rev. Letters, 16 August 1965]. The dynamical couplings of $2^+$ mesons should also have a universal character akin to those of $1^-$ mesons, since the former provide a basis for the representation of gravitational interactions analogous to the electromagnetic use of spin-1 mesons.

Finally, we must mention an important aspect of the comparative transformation program which was discussed in the lecture upon which III is based, but is not contained in the published version. The assumption that fundamental field transformations induce linear transformations on phenomenological fields forces one to restrict the compact group $U_{12}$ to the parity-preserving subgroup $U_6\times U_6$. But parity-changing transformations can be represented if nonlinear transformations are admitted. This leads to a further unveiling of phenomenological dynamics, including quartic couplings among the mesons and quadratic meson couplings with baryons, which are also specified by the universal coupling constant $g$.

JULIAN SCHWINGER

# Relativistic quantum field theory

*Nobel Lecture, December 11, 1965*

The relativistic quantum theory of fields was born some thirty-five years ago through the paternal efforts of Dirac, Heisenberg, Pauli and others. It was a somewhat retarded youngster, however, and first reached adolescence seventeen years later, an event which we are gathered here to celebrate. But it is the subsequent development and more mature phase of the subject that I wish to discuss briefly today.

I shall begin by describing to you the logical foundations of relativistic quantum field theory. No dry recital of lifeless «axioms» is intended but, rather, an outline of its organic growth and development as the synthesis of quantum mechanics with relativity. Indeed, relativistic quantum mechanics – the union of the complementarity principle of Bohr with the relativity principle of Einstein – *is* quantum field theory. I beg your indulgence for the mode of expression I must often use. Mathematics is the natural language of theoretical physics. It is the irreplaceable instrument for the penetration of realms of physical phenomena far beyond the ordinary experience upon which conventional language is based.

Improvements in the formal presentation of quantum mechanical principles, utilizing the concept of action, have been interesting by-products of work in quantum field theory. Both my efforts in this direction[1] and those of Feynman[2] (which began earlier) were based on a study of Dirac concerning the correspondence between the quantum transformation function and the classical action. We followed quite different paths, however, and two distinct formulations of quantum mechanics emerged which can be distinguished as differential and integral viewpoints.

In order to suggest the conceptual advantages of these formulations, I shall indicate how the differential version transcends the correspondence principle and incorporates, on the same footing, two different kinds of quantum dynamical variable. It is just these two types that are demanded empirically by the two known varieties of particle statistics. The familiar properties of the variables $q_k$, $p_k$, $k = 1 \cdots n$, of the conventional quantum system enable one to *derive* the form of the quantum action principle. It is a differential statement about

Reprinted from *Nobel Lectures – Physics, 1963–1970,* Elsevier, Amsterdam, 1972, pp. 140–152.

time transformation functions,

$$\delta\langle t_1|t_2\rangle = (i/\hbar)\langle t_1|\delta[\int_{t_2}^{t_1} dt L]|t_2\rangle \tag{1}$$

which is valid for a certain class of kinematical and dynamical variations. The quantum Lagrangian operator of this system can be given the very symmetrical form

$$L = \sum_{k=1}^{n} \frac{1}{4}\left(p_k \frac{dq_k}{dt} - q_k \frac{dp_k}{dt} + \frac{dq_k}{dt} p_k - \frac{dp_k}{dt} q_k\right) - H(q,p,t) \tag{2}$$

The symmetry is emphasized by collecting all the variables into the $2n$-component Hermitian vector $z(t)$ and writing

$$L = \frac{1}{4}\left(za\frac{dz}{dt} - \frac{dz}{dt}az\right) - H(z,t) \tag{3}$$

where $a$ is a real antisymmetrical matrix, which only connects the complementary pairs of variables.

The transformation function depends explicitly upon the choice of terminal states and implicitly upon the dynamical nature of the system. If the latter is held fixed, any alteration of the transformation function must refer to changes in the states, as given by

$$\delta\langle t_1| = (i/\hbar)\langle t_1|G_1 \qquad \delta|t_2\rangle = -(i/\hbar)\, G_2|t_2\rangle \tag{4}$$

where $G_1$ and $G_2$ are infinitesimal Hermitian operators constructed from dynamical variables of the system at the specified times. For a given dynamical system, then,

$$\delta[\int_{t_2}^{t_1} dt L] = G_1 - G_2 \tag{5}$$

which is the quantum principle of stationary action, or Hamilton's principle, since there is no reference on the right hand side to variations at intermediate times. The stationary action principle implies equations of motion for the dynamical variables and supplies explicit expressions for the infinitesimal operators $G_{1,2}$. The interpretation of these operators as generators of transformations on states, and on the dynamical variables, implies commutation relations. In this way, all quantum-dynamical aspects of the system are derived from a single dynamical principle. The specific form of the commutation relations obtained from the symmetrical treatment of the usual quantum system is given by the matrix statement

$$[z(t), z(t)] = i\hbar a^{-1} \tag{6}$$

Note particularly how the antisymmetry of the commutator matches the antisymmetry of the matrix *a*.

We may now ask whether this general form of Lagrangian operator,

$$L = \frac{1}{4}\left(xA\frac{dx}{dt} - \frac{dx}{dt}Ax\right) - H(x,t) \tag{7}$$

also describes other kinds of quantum systems, if the properties of the matrix $A$ and of the Hermitian variables $x$ are not initially assigned. There is one general restriction on the matrix $A$, however. It must be skew-Hermitian, as in the realization by the real, antisymmetrical matrix $a$. Only one other simple possibility then appears, that of an imaginary, symmetrical matrix. We write that kind of matrix as $i\alpha$, where $\alpha$ is real and symmetrical, and designate the corresponding variables collectively by $\zeta(t)$. The replacement of the antisymmetrical $a$ by the symmetrical $\alpha$ requires that the antisymmetrical commutators which characterize $z(t)$ be replaced by symmetrical anticommutators for $\varsigma(t)$, and indeed

$$\{\zeta(t), \zeta(t)\} = \hbar\,\alpha^{-1} \tag{8}$$

specifies the quantum nature of this second class of quantum variable. It has no classical analogue. The consistency of various aspects of the formalism requires only that the Lagrangian operator be an even function of this second type of quantum variable.

Time appears in quantum mechanics as a continuous parameter which represents an abstraction of the dynamical rôle of the measurement apparatus. The requirement of relativistic invariance invites the extension of this abstraction to include space and time coordinates. The implication that space-time localized measurements are a useful, if practically unrealizable idealization may be incorrect, but it is a grave error to dismiss the concept on the basis of *a priori* notions of measurability. Microscopic measurement has no meaning apart from a theory, and the idealized measurement concepts that are implicit in a particular theory must be accepted or rejected in accordance with the final success or failure of that theory to fulfill its avowed aims. Quantum field theory has failed no significant test, nor can any decisive confrontation be anticipated in the near future.

Classical mechanics is a determinate theory. Knowledge of the state at a given time permits precise prediction of the result of measuring any property of the system. In contrast, quantum mechanics is only statistically determinate. It is the probability of attaining a particular result on measuring any property of the system, not the outcome of an individual microscopic observation, that is predictable from knowledge of the state. But both theories are causal – a knowledge of the state at one time implies knowledge of the state at a later time. A quantum state is specified by particular values of an optimum set of compatible physical properties, which are in number related to the number of degrees of freedom of the system. In a relativistic theory, the concepts of «before» and «after» have no intrinsic meaning for regions that are in space-like relation. This implies that measurements individually associated with different regions in space-like relation are causally independent, or compatible. Such measurements can be combined in the complete specification of a state. But since there is no limit to the number of disjoint spatial regions that can be considered, a relativistic quantum system has an infinite number of degrees of freedom.

The latter statement, incidentally, contains an implicit appeal to a general property that the mathematics of physical theories must possess – the mathematical description of nature is not sensitive to modifications in physically irrelevant details. An infinite total spatial volume is an idealization of the finite volume defined by the macroscopic measurement apparatus. Arbitrarily small volume elements are idealizations of cells with linear dimensions far below the level of some least distance that is physically significant. Thus, it would be more accurate, conceptually, to assert that a relativistic quantum system has a number of degrees of freedom that is extravagantly large, but finite.

The distinctive features of relativistic quantum mechanics flow from the idea that each small element of three-dimensional space at a given time is physically independent of all other such volume elements. Let us label the various degrees of freedom explicitly – by a point of three-dimensional space (in a limiting sense), and by other quantities of finite multiplicity. The dynamical variables then appear as

$$\chi_{a,\mathbf{x}}(t) \equiv \chi_a(t = x^0, \mathbf{x}) \tag{9}$$

which are a finite number of Hermitian operator functions of space–time co-ordinates, or quantum fields. The dynamical independence of the individual volume elements is expressed by a corresponding additivity of the Lagrangian operator

$$L = \int (\mathrm{d}\mathbf{x})\, \mathscr{L} \tag{10}$$

where the Lagrange function $\mathscr{L}$ describes the dynamical situation in the infinitesimal neighborhood of a point. The characteristic time derivative or kinematical part of $L$ appears analogously in $\mathscr{L}$ in terms of the variables associated with the specified spatial point. The relativistic structure of the action principle is completed by demanding that it present the same form, independently of the particular partitioning of space-time into space and time. This is facilitated by the appearance of the action operator, the time integral of the Lagrangian, as the space-time integral of the Lagrange function. Accordingly, we require, as a sufficient condition, that the latter be a scalar function of its field variables, which implies that the known form of the time derivative term is supplemented by similar space derivative contributions. This is conveyed by

$$\mathscr{L} = \frac{1}{4}\left(\chi A^{\mu}\partial_{\mu}\chi - \partial_{\mu}\chi A^{\mu}\chi\right) - \mathscr{H}(\chi) \tag{11}$$

where the $A^{\mu}$ are a set of four finite skew-Hermitian matrices. A specific physical field is associated with submatrices of the $A^{\mu}$, which are real and antisymmetrical for a field $\varphi$ that obeys Bose-Einstein statistics, or imaginary and symmetrical for a field $\psi$ obeying Fermi-Dirac statistics. Finally, the boundaries of the four-dimensional integration region, formed by three-dimensional space at the terminal times, are described by the invariant concept of the space-like surface $\sigma$, a three-dimensional manifold such that every pair of points is in space-like relation. The ensuing invariant form of the action principle of relativistic quantum field theory is (we now use atomic units, in which $\hbar = c = 1$)

$$\delta\langle\sigma_1|\sigma_2\rangle = i\langle\sigma_1|\delta[\textstyle\int_{\sigma_2}^{\sigma_1}(\mathrm{d}x)\,\mathscr{L}]|\sigma_2\rangle \tag{12}$$

Relativity is a statement of equivalence within a class of descriptions associated with similar but different measurement apparatus. Space-time coordinates are an abstraction of the role that the measurement apparatus plays in defining a space-time frame of reference. The empirical fact, that all connected space-time locations and orientations of the measurement apparatus supply equivalent descriptions, is interpreted by the mathematical requirement of invariance under the group of proper orthochronous inhomogenous Lorentz transformations, applied to the continuous numerical coordinates. There is another numerical element in the quantum-mechanical description that has a measure of arbitrariness and expresses an aspect of relativity. I am referring to the quantum-mechanical use of complex numbers and of the mathematical equivalence of the two square roots of $-1$, $\pm$ i. What general property of any measurement apparatus is subject to our control, in principle, but offers only the choice of two alternatives? The answer is clear – a macroscopic material system can be constructed of matter, or of antimatter! But let us not conclude too hastily that a matter apparatus and an antimatter apparatus are completely equivalent. It is characteristic of quantum mechanics that the dividing line between apparatus and system under investigation can be drawn somewhat arbitrarily, as long as the measurement apparatus always possesses the classical aspects required for the unambiguous recording of an observation. To preserve this feature, the interchange of matter and antimatter must be made on the whole assemblage of macroscopic apparatus and microscopic system. Since the observational label of this duality is the algebraic sign of electric charge, the microscopic interchange must reverse the vector of electric current $j^\mu$, while maintaining the tensor $T^{\mu\nu}$ that gives the flux of energy and momentum. But this is just the effect of the coordinate transformation that reflects all four coordinates.

It is indeed true that the action principle does not retain its general form under either of the two transformations, the replacement of $i$ with $-i$, and the reflection of all coordinates, but does preserve it under their combined influence. In more detail, the effect of complex conjugation is equivalent to the reversal of operator multiplication, which distinguishes fields with the two types of statistics. The reflection of all coordinates, a proper transformation, can be generated by rotations in the attached Euclidean space obtained by introducing the imaginary time coordinate $x_4 = i\,x^0$. This transformation alters reality properties, distinguishing fields with integral and half-integral spin. The combination of the two transformations replaces the original Lagrange function

$$\mathcal{L}(\varphi_{\mathrm{int}}, \varphi_{1/2\,\mathrm{int}}, \psi_{\mathrm{int}}, \psi_{1/2\,\mathrm{int}}) \tag{13}$$

with

$$\mathcal{L}(\varphi_{\mathrm{int}}, i\varphi_{1/2\,\mathrm{int}}, i\psi_{\mathrm{int}}, \psi_{1/2\,\mathrm{int}}) \tag{14}$$

If only fields of the types $\varphi_{\mathrm{int}}$, $\Psi_{\frac{1}{2}\,\mathrm{int}}$ are considered, which is the empirical connection between spin and statistics, the action principle is unaltered in form. This invariance property of the action principle[3] expresses the relativity of matter and antimatter. That is the content of the so-called TCP theorem. The anomalous response of the field types $\varphi_{\frac{1}{2}\,\mathrm{int}}$, $\Psi_{\mathrm{int}}$ is also the basis for the theoretical rejection of these possibilities as contrary to general physical requirements of positiveness, namely, the positiveness of probability, and the positiveness of energy.

The concept of space-like surface is not limited to plane surfaces. According to the action principle, an infinitesimal deformation of the space-like surface on which a state is specified changes that state by

$$\delta\langle\sigma| = i\langle\sigma|\int d\sigma_\mu\, T^{\mu\nu}\delta x^\nu \tag{15}$$

which is the infinitely multiple relativistic generalization of the Schrödinger equation

$$\delta\langle t| = i\langle t|H(-\delta t) \tag{16}$$

This set of differential equations must obey integrability conditions, which are commutator statements about the elements of the tensor $T^{\mu\nu}$. Since rigid displacements and rotations can be produced from arbitrary local deformations, the operator expressions of the group properties of Lorentz transformations must be a consequence of these commutator conditions. Foremost among the latter are the equal-time commutators of the energy density $T^{00}$, which suffice to convey all aspects of relativistic invariance that are not of a three-dimensional nature. A system that is invariant under three-dimensional translations and rotations will be Lorentz invariant if, at equal times[4],

$$-i[T^{00}(x), T^{00}(x')] = -(T^{0k}(x) + T^{0k}(x'))\,\partial_k\,\delta(\mathbf{x}-\mathbf{x}') \tag{17}$$

This is a sufficient condition. Additional terms with higher derivatives of the delta function will occur, in general. But there is a distinguished class of physical system, which I shall call local, for which no further term appears. The phrase «local system» can be given a physical definition within the framework we have used or, alternatively, by viewing the commutator condition as a measurability statement about the property involved in the response of a system to a weak external gravitational field[5]. Only the external gravitational potential $g_{00}$ is relevant here. A physical system is local if the operators $T^{\mu\nu}$, which may be explicit functions of $g_{00}$ at the same time, do not depend upon time derivatives of $g_{00}$. The class of local systems is limited[6] to fields of spin 0, ½, 1. Such fields are distinguished by their physical simplicity in comparison with fields of higher spin. One may even question whether consistent relativistic quantum field theories can be constructed for non-local systems.

The energy density commutator condition is a very useful test of relativistic invariance. Only a month or so ago I employed it to examine whether a relativistic quantum field theory could be devised to describe magnetic as well as electric charge. Dirac pointed out many years ago that the existence of magnetic charge would imply a quantization of electric charge, in the sense that the product of two elementary charges, $eg/\hbar c$, could assume only certain values. According to Dirac, these values are any integer or half-integer. In recent years, the theoretical possibility of magnetic charge has been attacked from several directions. The most serious accusation is that the concept is in violation of Lorentz invariance. This is sometimes expressed in the language of field theory by the remark that no manifestly scalar Lagrange function can be constructed for a system composed of electromagnetic field and electric and magnetic charge-bearing fields. Now it is true that there is no relativistically invariant theory for arbitary $e$ and $g$, so that no formally invariant version could exist. Indeed, the unnecessary assumption that $\mathscr{L}$ is a scalar must be relinquished in favor of the more general possibilities that are compatible with the action principle. But the energy commutator condition can still be applied. I have been able to show that energy and momentum density operators can be exhibited which satisfy the commutator condition, together with the three-dimensional requirements, provided $eg/\hbar c$ possesses one of a discrete set of values. These values are integers, which is more restrictive than Dirac's quantization condition. Such general considerations shed no light on the empirical elusiveness of magnetic charge. They only emphasize that this novel theoretical possibility should not be dismissed lightly.

The physical systems that obey the commutator statement of locality do not include the gravitational field. But, this field, like the electromagnetic field, requires very special consideration. And these considerations make full use of the relativistic field concept. The dynamics of the electromagnetic field is characterized by invariance under gauge transformations, in which the phase of every charge-bearing field is altered arbitrarily, but continuously, at each space-time point while electromagnetic potentials are transformed inhomogeneously. The introduction of the gravitational field involves, not only the use of general coordinates and coordinate transformations, but the establishment at each point of an independent Lorentz frame. The gravitational field gauge transformations are produced by the arbitrary reorientation of these local coordinate systems at each point while gravitational potentials are transformed linearly and inhomogeneously. The formal extension of the action principle to include the gravitational field can be carried out[7], together with the verification of consistency conditions analogous to the energy density commutator condition[8]. To appreciate this tour de force, one must realize that the operator in the role of energy density is a function of the gravitational field, which is influenced by the energy density. Thus the object to be tested is only known implicitly. It also appears that the detailed specification of the spatial distribution of energy lacks physical significance when gravitational phenomena are important. Only integral quantities or equivalent asymptotic field properties are physically meaningful in that circumstance. It is in the further study of such boundary conditions that one may hope to comprehend the significance of the gravitational field as the physical mediator between the worlds of the microscopic and the macroscopic, the atom and the galaxy.

I have now spoken at some length about fields. But it is in the language of particles that observational material is presented. How are these concepts related? Let us turn for a moment to the early history of our subject. The quantized field appears initially as a device for describing arbitrary numbers of indistinguishable particles. It was defined as the creator or annihilator of a particle at the specified point of space and time. This picture changed somewhat as a consequence of the developments in quantum electrodynamics to which Feynman, Tomonaga, myself, and many others contributed. It began to be appreciated that the observed properties of so-called elementary particles are partly determined by the effect of interactions. The fields used in the dynamical description were then associated with noninteracting or bare particles, but there was still a direct correspondence with physical particles. The weakness of electromagnetic interactions, as measured by the small value of the fine structure constant $e^2/\hbar c$ is relevant here, for the same view-point failed disastrously when extended to strongly interacting nucleons and mesons. The resulting wide spread disillusionment with quantum field theory is an unhappy chapter in the history of high-energy theoretical physics, although it did serve to direct attention toward various useful phenomenological calculation techniques.

The great qualitative difference between weakly interacting and strongly interacting systems was impressed upon me by a particular consideration which I shall now sketch for you[9]. In the absence of interactions there is an immediate connection between the quantized Maxwell field and a physical particle of zero mass, the photon. The null mass of the photon is the particle transcription of a field property, that electromagnetism has no well-defined range but weakens geometrically. Now one of the most important interaction aspects of quantum electrodynamics is the phenomenon of vacuum polarization. A variable electromagnetic field induces secondary currents, even in the absence of actual particle creation. In particular, a localized charge creates a counter charge in its vicinity, which partially neutralizes the effect of the given charge at large distances. The implication that physical charges are weaker than bare charges by a universal factor is the basis for charge renormalization. But once the idea of a partial neutralization of charge is admitted one cannot exclude the possibility of total charge neutralization. This will occur if the interaction exceeds a certain strength such that an oppositely charged particle combination, of the same nature as the photon, becomes so tightly bound that the corresponding mass diminishes to zero. Under these circumstances no long-range fields would remain and the massless particle does not exist. We learn that the connection between the Maxwell field and the photon is not an *a priori* one, but involves a specific dynamical aspect, that electromagnetic interactions are weaker than the critical strength. It is a natural speculation that another such field exists which couples more strongly than the critical amount to nucleonic charge, the property carried by all heavy fermions. That hypothesis would explain the absolute stability of the proton, in analogy with the electromagnetic explanation of electron stability, without challenging the uniqueness of the photon.

A field operator is a localized excitation which, applied to the vacuum state, generates all possible energy-momentum, or equivalently, mass states that share the other distinguishing properties of the field. The products of field operators widen and ultimately exhaust the various classes of mass states. If an isolated mass value occurs in a particular product, the state is that of a stable particle with corresponding characteristics. Should a small neighborhood of a particular mass be emphasized, the situation is that of an unstable particle, with a proper lifetime which varies inversely as the mass width of the excitation. The quantitative properties of the stable and unstable particles that may be implied by a given dynamical field theory cannot be predicted with presently available calculation techniques. In these matters, to borrow a phrase of Ingmar Bergman, and St. Paul, we see through a glass, darkly. Yet, in the plausible qualitative inference that a substantial number of particles, stable and unstable, will exist for sufficiently strong interactions among a few fields lies the great promise of relativistic quantum field theory.

Experiment reveals an ever growing number and variety of unstable particles, which seem to differ in no essential way from the stable and long-lived particles with which they are grouped in tentative classification schemes. Surely one must hope that this bewildering complexity is the dynamical manifestation of a conceptually simpler substratum, which need not be directly meaningful on the observational level of particles. The relativistic field concept is a specific realization of this general groping toward a new conception of matter.

There is empirical evidence in favor of such simplification at a deeper dynamical level. Strongly interacting particles have been rather successfully classified with the aid of a particular internal symmetry group. It is the unitary group $SU_3$. The dimensionalities of particle multiplets that have been identified thus far are 1, 8, and 10. But the fundamental multiplet of dimensionality 3 is missing. It is difficult to believe in the physical significance of some transformation group without admitting the existence of objects that respond to the transformations of that group. Accordingly, I would describe the observed situation as follows. There are sets of fundamental fields that form triplets[10] with respect to the group $U_3$. The excitations produced by these fields are very massive and highly unstable. The low lying mass excitations of mesons and baryons are generated by products of the fundamental fields. If these fields are assigned spin $\frac{1}{2}$, as a specific model, it is sufficient to consider certain products of two and three fields to represent the general properties of mesons and baryons, respectively.

The cogency of this picture is emphasized by its role in clarifying a recent development in symmetry classification schemes. That is the provocative but somewhat mysterious suggestion that the internal symmetry group $SU_3$ be combined with space-time spin transformations to form the larger unitary group $SU_6$. This idea, with its relativistic generalizations, has had some striking numerical successes but there are severe conceptual problems in reconciling Lorentz invariance with any union of internal and space–time transformations, as long as one insists on immediate particle interpretation. The situation is different if one can refer to the space-time localizability that is the hallmark of the field concept[11]. Let us assume that the interactions among the fundamental fields are of such strength that field products at practically coincident points suffice to describe the excitation of the known relatively low lying particles. The resulting quasi-local structures are in some sense fields that are associated with the physical particles. I call these phenomenological fields, as distinguished from the fundamental fields which are the basic dynamical variables of the system. Linear transformations on the fundamental fields can simulate the effect of external probes, which may involve both unitary and spin degrees of freedom. If these external perturbations are sufficiently gentle, the structure of the particles will be maintained and the phenomenological field transformed linearly with indefinite multiplets. It is not implausible that the highly localized interactions among the phenomenological fields will exhibit a corresponding symmetry. Thus, combined spin and unitary transformations appear as a device for characterizing some gross features of the unknown inner field dynamics of physical particles, as it operates in the neighborhood of a specific point. But these transformations can have no general significance for the transfer of excitations from point to point, and only lesser symmetries will survive in the final particle description.

Phenomenological fields are the basic concept in formulating the practical calculation methods of strong interaction field theory. They serve to isolate the formidable problem of the dynamical origin of physical particles from the more immediate questions referring to their properties and interactions. In somewhat analogous circumstances, those of non-relativistic many-particle physics, the methods and viewpoint of quantum field theory[12] have been enormously successful. They have clarified the whole range of cooperative phenomena, while employing relatively simple approximation schemes. I believe that phenomenological relativistic quantum field theory has a similar future, and will replace the algorithms that were introduced during the period of revolt from field theory. But the intuition that serves so well in non-relativistic contexts does not exist for these new conditions. One has still to appreciate the precise rules of phenomenological relativistic field theory, which must supply a self-consistent description of the residual interactions, given that the strong fundamental interactions have operated to compose the various physical particles. And when this is done, how much shall we have learned, and how much will remain unknown, about the mechanism that builds matter from more primitive constituents? Are we not at this moment,

*... like stout Cortez when with eagle eyes*
*He star'd at the Pacific – and all his men*
*Look'd at each other with a wild surmise –*
*Silent, upon a peak in Darien.*

And now it only remains for me to say: Tack så mycket för uppmärksamheten.

1. Some references are: J. Schwinger, *Phys. Rev.*, 82 (1951) 914, 91 (1953) 713; *Proc. Natl. Acad. Sci. (U. S.)*, 46 (1960) 883.
2. R. Feynman, *Rev. Mod. Phys.*, 20 (1948) 36; R. Feynman and A. Hibbs, *Quantum Mechanics and Path Integrals*, McGraw-Hill, New York, 1965.
3. In the first two papers cited in Ref. 1 I have assumed space-reflection invariance and shown the equivalence between the spin-statistics relation and the invariance of the action principle under combined time reflection and complex conjugation. It was later remarked by Pauli that the separate hypothesis of space-reflection invariance

was unnecessary, W. Pauli, *Niels Bohr and the Development of Physics*, McGraw-Hill, New York, 1955.

4. J. Schwinger, *Phys. Rev.*, 127 (1962) 324.
5. J. Schwinger, *Phys. Rev.*, 130 (1963) 406.
6. J. Schwinger, *Phys. Rev.*, 130 (1963) 800.
7. J. Schwinger, *Phys. Rev.*, 130 (1963) 1253.
8. J. Schwinger, *Phys. Rev.*, 132 (1963) 1317.
9. J. Schwinger, *Phys. Rev.*, 125 (1962) 397; 128 (1962) 2425.
10. J. Schwinger, *Phys. Rev.*, 135 (1964) B 816; 136 (1964) B 1821.
11. J. Schwinger, *Second Coral Gables Conference on Symmetry Principles at High Energy*, Freeman, San Francisco, 1965; *Phys. Rev.*, 140 (1965) B 158.
12. The general theory is described by P. Martin and J. Schwinger, *Phys. Rev.*, 115 (1959) 1342.

# Particles and Sources

JULIAN SCHWINGER*
*Harvard University, Cambridge, Massachusetts*
(Received 6 July 1966)

It is proposed that the phenomenological theory of particles be based on the source concept, which is abstracted from the physical possibility of creating or annihilating any particle in a suitable collision. The source representation displays both the momentum and the space-time characteristics of particle behavior. Topics discussed include: spin and statistics, charge and the Euclidean postulate, massless particles, and $SU_3$ and spin. It is emphasized that the source description is logically independent of hypotheses concerning the fundamental nature of particles.

## INTRODUCTION

THE particle is presently the central concept in interpreting the raw data of high-energy experimental physics. The discovery of the meson $\eta^*$ (960 MeV) by studying mass correlations among five $\pi$ mesons[1] is a recent example of the experimental procedure that defines a particle primarily by energy and momentum characteristics. The latter aspect of particle behavior is of such obvious significance that many theoretical studies in particle phenomenology concentrate on it completely. But the particle has another, complementary, aspect—a degree of spatial localizability. Is it possible to give a useful phenomenological definition or characterization of particles that does not emphasize unduly either of these complementary facets of the particle concept?

Any particle can be created in a collision, given suitable partners before and after the impact to supply the appropriate values of the spin and other quantum numbers, together with enough energy to exceed the mass threshold. In identifying new particles it is a basic experimental principle that the specific reaction is not otherwise relevant. Then let us abstract from the physical presence of the additional particles involved in creating a given one, and consider them simply as the *source* of the physical properties that are carried by the created particle. The ability to give some localization in space and time to a creation act may be represented by a corresponding coordinate dependence of a mathematical source function, $S(x)$. The effectiveness of the source in supplying energy and momentum may be described by another mathematical source function, $S(p)$. The complementarity of these source aspects can then receive its customary quantum interpretation, as illustrated by

$$S(p)=\int (dx)e^{-ipx}S(x).$$

The source of a particular particle must also have the multiplicity necessary to represent its spin and those internal quantum numbers appropriate to the dynamical level of description that is used.

For simplicity, in the following we shall only consider a restricted time scale such that the possible instability of any particle is not significant. This restriction can always be removed. Particles of zero mass will receive special attention, and it is otherwise understood that the mass does not vanish.

## SPINLESS PARTICLES

As a first step in supplying a quantitative framework for the source concept consider a spinless particle of mass $m$, without internal quantum numbers. We begin with the vacuum state $|0\rangle$, and then let a weak source operate in some space-time region. The probability amplitude for the generation of a single particle in a state specified by a small momentum range about the momentum $\mathbf{p}$ will be written

$$\langle 1_p|0_-\rangle^S=\left[\frac{(d\mathbf{p})}{(2\pi)^3}\frac{1}{2p^0}\right]^{1/2} i\int (dx)e^{-ipx}S(x),$$
$$p^0=+(\mathbf{p}^2+m^2)^{1/2},$$

which is an invariant expression if $S(x)$ is transformed as a scalar. The subscript on the vacuum state indicates the time sense—this is the vacuum state before the source has acted. The probability amplitude for the inverse process appears as

$$\langle 0_+|1_p\rangle^S=\left[\frac{(d\mathbf{p})}{(2\pi)^3}\frac{1}{2p^0}\right]^{1/2} i\int (dx)e^{ipx}S(x),$$

which refers to the vacuum state after the source, now functioning as a sink, has acted to annihilate the particle. The two processes are related by the "$TCP$" operation of complex conjugation and space-time coordinate reflection. The factors of $i$ have been inserted to make these expressions consistent with the further restriction of $S(x)$ to be a real function. The latter property symbolizes the reciprocity between creation and annihilation mechanisms.

These postulated representations of the creation and annihilation aspects of a source can be united by considering the vacuum probability amplitude

$$\langle 0_+|0_-\rangle^S,$$

where

$$S=S_1+S_2$$

* Supported in part by the U. S. Air Force Office of Scientific Research under Contract No. A. F. 49(638)-1380.

[1] G. R. Kalbfleisch *et al.*, Phys. Rev. Letters **12**, 527 (1964); M. Goldberg *et al.*, *ibid.* **12**, 546 (1964).



Reprinted from *The Physical Review* **152**, 1219–1226 (1966).

and $S_2$, effectively localized in time prior to $S_1$, creates a particle which is subsequently annihilated by $S_1$. The notion of weak source, which remained unexplained before, means that only single-particle exchange between the sources is numerically significant. This is expressed by the probability-amplitude composition law

$$\langle 0_+|0_-\rangle^S \cong \langle 0_+|0_-\rangle^{S_1}\langle 0_+|0_-\rangle^{S_2}+\sum_p \langle 0_+|1_p\rangle^{S_1}\langle 1_p|0_-\rangle^{S_2}$$

$$\cong 1+\int \frac{(d\mathbf{p})}{(2\pi)^3}\frac{1}{2p^0}\int (dx)(dx')iS_1(x)\times e^{ip(x-x')}iS_2(x').$$

The second form involves a further, but temporary simplification, which is to retain only contributions that are linear in each partial source. A vacuum amplitude like $\langle 0_+|0_-\rangle^{S_2}$, which is restricted by the probability condition

$$1\cong|\langle 0_+|0_-\rangle^S|^2+\sum_p|\langle 1_p|0_-\rangle^S|^2,$$

should deviate from unity by terms that are quadratic functions of $S_2$. These effects are reinstated by writing $\langle 0_+|0_-\rangle^S$ as a functional of $S$, rather than of the constituent sources $S_1$ and $S_2$. That is accomplished by the expression

$$\langle 0_+|0_-\rangle^S\cong 1+\tfrac{1}{2}i\int (dx)(dx')S(x)\Delta_+(x-x')S(x'),$$

where

$$\Delta_+(x-x')=\Delta_+(x'-x)$$

$$=i\int \frac{(d\mathbf{p})}{(2\pi)^3}\frac{1}{2p^0}e^{ip(x-x')},\quad x^0>x^{0\prime}.$$

The function $\Delta_+(x)$ is invariant with respect to the transformations of the proper, orthochronous, homogeneous Lorentz group. The additional symmetry,

$$\Delta_+(-x)=\Delta_+(x),$$

is also conveyed by the statement of invariance under the attached Euclidean rotation group ($x_4=ix^0$).

The quadratic terms in $S_1$ and $S_2$ that appear in $\langle 0_+|0_-\rangle^S$ reproduce the structure of the product of $\langle 0_+|0_-\rangle^{S_1}$ and $\langle 0_+|0_-\rangle^{S_2}$. Furthermore, the property that

$$-\mathrm{Re}\, i\Delta_+(x-x')=\mathrm{Re}\int \frac{(d\mathbf{p})}{(2\pi)^3}\frac{1}{2p^0}e^{ip(x-x')},$$

for all $x-x'$, leads immediately to the necessary relations

$$|\langle 0_+|0_-\rangle^S|^2\cong 1-\sum_p|\langle 1_p|0_-\rangle^S|^2,$$

$$\cong 1-\sum_p|\langle 0_+|1_p\rangle^S|^2.$$

When the production source $S_2$ and the detection source $S_1$ are located within certain macroscopic volumes, the trajectory of the exchanged particle is correspondingly limited. All this defines a region associated with the specified particle. Outside that region other independent acts of creation and detection can be considered, and similarly represented. Thus, the restriction to weak sources, or single-particle states, is easily removed subject to the limitation that the various particles remain physically independent, in virtue of their spatial separation. This situation is described by multiplying the vacuum probability amplitudes associated with the various independent source pairs,

$$\langle 0_+|0_-\rangle^S=\prod\left[1+\tfrac{1}{2}i\int (dx)(dx')S(x)\Delta_+(x-x')S(x')\right].$$

To represent this as a property of a single source, uniting the several spatially isolated parts, we have only to write

$$\langle 0_+|0_-\rangle^S=\exp\left[\tfrac{1}{2}i\int (dx)(dx')S(x)\Delta_+(x-x')S(x')\right].$$

Here is the simplest example of our answer to the problem of giving a phenomenological particle representation in which both localization and momentum aspects receive proper attention. The corresponding momentum-space formula is

$$\langle 0_+|0_-\rangle^S=\exp\left[\tfrac{1}{2}i\int \frac{(dp)}{(2\pi)^4}S(-p)\frac{1}{p^2+m^2-i\epsilon}S(p)\right],$$

which involves the limiting process $\epsilon\to+0$.

It is a reasonable extrapolation to apply this result under conditions for which the interactions among the particles are still not significant but the particles need not be macroscopically isolated, so that microscopic quantum interference effects come into play. Consider again a production and detection source $S_2$ and $S_1$, respectively, and write

$$\langle 0_+|0_-\rangle^S=\langle 0_+|0_-\rangle^{S_1}\times\exp\left[i\int (dx)(dx')S_1(x)\Delta_+(x-x')S_2(x')\right]\times\langle 0_+|0_-\rangle^{S_2}$$

$$=\sum_{\{n\}}\langle 0_+|\{n\}\rangle^{S_1}\langle\{n\}|0_-\rangle^{S_2}.$$

To identify the individual multiparticle probability amplitudes, let us note that

$$i\int (dx)(dx')S_1(x)\Delta_+(x-x')S_2(x')=\sum_p iS_{1p}{}^*iS_{2p},$$

where

$$iS_p=\left[\frac{(d\mathbf{p})}{(2\pi)^3}\frac{1}{2p^0}\right]^{1/2}iS(p)$$

is just the quantity designated by $\langle 1_p|0_-\rangle^S$ in the discussion of a weak source. We now find that

$$\langle\{n\}|0_-\rangle^S=\langle 0_+|0_-\rangle^S\prod_p(iS_p)^{n_p}/[n_p!]^{1/2}$$

and

$$\langle 0_+|\{n\}\rangle^S=\langle 0_+|0_-\rangle^S\prod_p(iS_p^*)^{n_p}/[n_p!]^{1/2},$$

in which the possible values of each $n_p=0, 1, 2, \cdots$. Probability normalization conditions are satisfied in consequence of the property

$$|\langle 0_+|0_-\rangle^S|^2=\exp[-\sum_p|S_p|^2].$$

The significant conclusion is that the system under investigation is a Bose-Einstein (B.E.) ensemble of indistinguishable particles. It will be recognized that this characteristic has been introduced implicitly by regarding the source function as an ordinary numerical quantity. The latter assumption now emerges as the mathematical representation of B.E. statistics.

Even though an application to spinless particles is inappropriate, it can be appreciated that an analogous representation of Fermi-Dirac (F.D.) statistics demands that the source have such properties that $(S_p)^2=0$, for all $p$. This implies that the source functions, $S(x)$, of F.D. particles are totally anticommutative quantities, as realized by the elements of an exterior algebra.[2] The general correspondence between particle statistics and the mathematical nature of the source is thus expressed by

$$\text{B.E.:}\quad [S(x),S(x')]=0,$$
$$\text{F.D.:}\quad \{S(x),S(x')\}=0.$$

Let us also note here a consequence of the implicit reference to a source defined in a certain spatiotemporal region. Suppose the source is rigidly displaced, in the sense that

$$S(x)\to S(x+X).$$

Then

$$S_p\to e^{ipX}S_p$$

and

$$\langle\{n\}|0_-\rangle^S\to e^{iPX}\langle\{n\}|0_-\rangle^S,$$

which identifies the total energy and momentum of this state

$$P^\mu=\sum_p n_p p^\mu.$$

## PARTICLES WITH SPIN

The extension of the preceding discussion to include particles with spin can be done in a variety of ways for specific values. If we wish a uniformly applicable description, however, the unique source representation is the multispinor. We consider sources $S_{\zeta_1\cdots\zeta_n}(x)$, where each $\zeta_\kappa$ is a four-valued Dirac spin index, and $S$ has a definite symmetry pattern with respect to permutations of the $n$ spin indices. The totally symmetric choice suffices for most purposes, but other possibilities will be considered.

[2] There is a related discussion of exterior algebras in J. Schwinger, Proc. Natl. Acad. Sci. **48**, 603 (1962).

The general expression for the vacuum amplitude is taken to be

$$\langle 0_+|0_-\rangle^S=\exp\left[\tfrac{1}{2}i\int(dx)(dx')S(x)G_+(x-x')S(x')\right].$$

In order that this source representation be an invariant one, the matrix function $G_+(x)$ must have the transformation properties

$$G_+(x)=L^TG_+(lx)L,$$

where

$$L=\prod_{\kappa=1}^{n}L_\kappa(l)$$

and each $L_\kappa(l)$ produces the individual spin transformation that accompanies the proper orthochronous coordinate transformation

$$\bar{x}^\mu=l^\mu{}_\nu x^\nu.$$

We recall the transformation properties (the index $\kappa$ is omitted)

$$L^T\beta L=\beta,$$
$$L^T\alpha^\mu L=l^\mu{}_\nu\alpha^\nu,$$

where the real matrices $\alpha^\mu$ and $\beta$ are, respectively, symmetrical and antisymmetrical. In another notation,

$$-i\beta=\gamma^0,\quad \alpha^\mu=\gamma^0\gamma^\mu.$$

The matrix function $G_+(x-x')$ must be constructed by multiplying the invariant function $\Delta_+(x-x')$ with matrices that satisfy the covariance properties of $G_+$. The result should describe a particle of definite spin and parity in its rest coordinate system. As we shall verify, this is accomplished by

$$G_+(x-x')=\prod_{\kappa=1}^{n}[\gamma^0(m-\gamma^\mu(1/i)\partial_\mu)]_\kappa\Delta_+(x-x').$$

The individual factors here are antisymmetrical under transposition of the matrices, combined with the exexchange of $x$ with $x'$, and

$$G_+(x'-x)^T=(-1)^nG_+(x-x').$$

Since this operation effectively interchanges the two source factors in the vacuum amplitude we learn that even $n$ demands B.E. statistics and odd $n$, F.D. statistics. The polynomial factor of $G_+$ is also represented by

$$x^0>x^{0\prime}:\quad G_+(x-x')=i\int\frac{(d\mathbf{p})}{(2\pi)^3}\frac{1}{2p^0}\times e^{ip(x-x')}\prod_{\kappa=1}^{n}[\gamma^0(m-\gamma p)]_\kappa.$$

We proceed to identify single-particle states, labeled by momentum and spin, as implied by the effect of a weak production and detection source,

$$\sum_{p\lambda}\langle 0_+|1_{p\lambda}\rangle^{S_1}\langle 1_{p\lambda}|0_-\rangle^{S_2} = \int \frac{(d\mathbf{p})}{(2\pi)^3}\frac{1}{2p^0} iS_1(-p)\prod_\kappa [\gamma^0(m-\gamma p)]_\kappa S_2(p).$$

The individual Hermitian matrix combinations have the projection property

$$[\gamma^0(m-\gamma p)]^2 = 2p^0[\gamma^0(m-\gamma p)].$$

This permits us to write

$$\sum_\lambda \langle 0_+|1_{p\lambda}\rangle^{S_1}\langle 1_{p\lambda}|0_-\rangle^{S_2} = \sum_{\zeta_1\cdots\zeta_n} iS_1{}^{(-)}{}_{p\zeta_1\cdots\zeta_n} iS_2{}^{(+)}{}_{p\zeta_1\cdots\zeta_n},$$

where the two spinors are

$$S_p{}^{(+)} = \left[\frac{(d\mathbf{p})}{(2\pi)^3}(2p^0)^{n-1}\right]^{1/2}\prod_\kappa\left[\frac{\gamma^0(m-\gamma p)}{2p^0}\right]_\kappa S(p),$$

$$S_p{}^{(-)} = \left[\frac{(d\mathbf{p})}{(2\pi)^3}(2p^0)^{n-1}\right]^{1/2} S(-p)\prod_\kappa\left[\frac{\gamma^0(m-\gamma p)}{2p^0}\right]_\kappa.$$

In the rest system, $p^0=m$, the projection factor $\frac{1}{2}(1+\gamma^0)$ selects $\gamma^0=+1$ for each of the $n$ Dirac indices, and the source thus produces particles of definite parity. It is purely conventional to have all values of $\gamma^0$ be $+1$, and this can be altered by redefining the source function with suitable $\gamma_5$ factors.

Now let us consider a totally symmetrical spinor source. In the rest system the restriction to $\gamma_\kappa{}^0=+1$, $\kappa=1,\cdots n$, effectively reduces the source to a totally symmetrical function of two-valued spin indices, which can be identified with the eigenvalues of $\sigma_{3\kappa}$. A totally symmetric combination of $n$ spins has a definite spin angular momentum,

$$s=\tfrac{1}{2}n.$$

Thus, all possible half-integral spin values can be represented by a suitable odd integer $n$, and all possible integer spins can be represented by an even integer $n$, except $s=0$. For the latter one has to take $n=2$ and use an antisymmetrical spinor source. The relation between statistics and the even-odd nature of $n$ will be recognized as the connection between spin and statistics. Note that more complicated symmetry patterns give equivalent descriptions, to the extent that a definite spin appears in the rest system. Consider, for example, $n=3$ with the requirement of antisymmetry in a pair of Dirac indices. The latter contributes zero spin in the rest system and we have a possible description of an $s=\frac{1}{2}$ F.D. particle.

The individual single-particle amplitudes are easily identified. If we use a totally symmetric spinor, for example, the extreme values of the spin magnetic quantum number correspond to identical choices for the $n$ component spins, as suggested by

$$\langle 1_{ps}|0_-\rangle^S = iS_{p++\cdots}{}^{(+)}$$
$$\langle 0_+|1_{ps}\rangle^S = iS_{p++\cdots}{}^{(-)},$$

and the other states can be generated from these. There is a simple relation between the two spinors $S_p{}^{(\pm)}$:

$$S_p{}^{(-)} = S_p{}^{(+)*}.$$

This conveys the Hermitian nature of the matrices $\gamma^0(m-\gamma p)$, and the reality of the source $S(x)$. For B.E. particles, the physically necessary property

$$|\langle 0_+|0_-\rangle^S|^2 = \exp[-\sum_p S_p{}^{(+)*}S_p{}^{(+)}]<1$$

is deduced on remarking that ($n$ is even)

$$-\mathrm{Re}\,iG_+(x-x') = \mathrm{Re}\int\frac{(d\mathbf{p})}{(2\pi)^3}\frac{1}{2p^0} \times e^{ip(x-x')}\prod_\kappa[\gamma^0(m-\gamma p)]_\kappa,$$

which can be expressed as

$$-\mathrm{Re}\,i\int(dx)(dx')S(x)G_+(x-x')S(x') = \sum_p S_p{}^{(+)*}S_p{}^{(+)}>0.$$

An analogous positiveness statement exists for F.D. particles. Let $\epsilon(x-x')$ state the algebraic sign of $x^0-x^{0\prime}$ and, for $x^0-x^{0\prime}\neq 0$, consider the function ($n$ is odd)

$$-\mathrm{Re}\,i\epsilon(x-x')G_+(x-x') = \mathrm{Re}\int\frac{(d\mathbf{p})}{(2\pi)^3}\frac{1}{2p^0}e^{ip(x-x')}\prod_\kappa[\gamma^0(m-\gamma p)]_\kappa,$$

which we shall define by the right-hand side for all $x-x'$. Then, if $S_1(x)$ is a real *commutative* spinor, we have

$$-\mathrm{Re}\,i\int(dx)(dx')S_1(x)\epsilon(x-x')G_+(x-x')S_1(x') = \sum_p S_{1p}{}^{(+)*}S_{1p}{}^{(+)}>0.$$

## CHARGED PARTICLES

The specification of the mass of a particle (stable or unstable) defines it uniquely, within present experimental knowledge, except when the particle carries electrical or nucleonic charge. These are exactly conserved quantities. The *TCP* theorem then assures us of the existence of two particles with opposite charges, and identical masses. This situation can be represented by supplying the real source function with another index, upon which the appropriate charge acts as a matrix,

$$l=\begin{pmatrix}0 & -i\\ i & 0\end{pmatrix}.$$

The diagonalization of this matrix defines the complex sources of charged particles.

The introduction of the antisymmetrical charge matrix puts the connection between spin and statistics in a new light. It should be evident that the construction of $G_+(x-x')$ in the previous section was essentially uniquely determined by the various space-time requirements imposed upon it. The resulting structure possesses a symmetry that is specified completely by $n$, the number of Dirac spin indices. But now we have the option of inserting the Lorentz-invariant matrix $l$ as a factor in $G_+$, which would alter the symmetry and destroy the spin-statistics connection. It is the physical positiveness properties that enable one to reject this possibility, and retain the correlation between spin and statistics. That invites us to re-examine the physical basis of these positiveness assertions.

We cannot improve on the B.E. situation in which sources appear directly in probability statements. For F.D. particles, however, the positiveness property emerged as an algebraic observation without direct reference to F.D. sources, which are a more abstract concept. In this circumstance we must turn to the physical system with which the source is coupled. We shall be able to treat both statistics in a unified way, however. Let the real multicomponent source $S(x)$ be coupled to the Hermitian operators $\Psi(x)$, which we assume to transform contragradiently to $S(x)$ under Lorentz transformations. It is immaterial for the following discussion whether F.D. sources anticommute or commute with $\Psi(x)$. To have a uniform treatment of both statistics we assume commutativity, and take the source term in a phenomenological Lagrange function as

$$\mathcal{L}_{\text{source}}=\Psi(x)S(x)=S(x)\Psi(x).$$

The action principle supplies the differential statement

$$\delta_S\langle 0_+|0_-\rangle^S=i\langle 0_+|\int(dx)\delta S(x)\Psi(x)|0_-\rangle^S$$

and the repetition of this operation gives

$$\begin{aligned}\delta_S{}^2\langle 0_+|0_-\rangle^S&=-\langle 0_+|\int(dx)(dx')(\delta S(x)\Psi(x)\delta S(x')\Psi(x'))_+|0_-\rangle^S\\&=-\int(dx)(dx')(\delta S(x)\delta S(x'))_+\\&\qquad\times\langle 0_+|(\Psi(x)\Psi(x'))_+|0_-\rangle^S.\end{aligned}$$

We have not written out possible additional terms, which occur when some of the components of $\Psi(x)$ are explicit functions of $S(x)$ or its coordinate derivatives. The overt reference to statistics appears on writing

$$(\delta S(x)\delta S(x'))_+=\begin{cases}\text{B.E.:} & \delta S(x)\delta S(x')\\ \text{F.D.:} & \epsilon(x-x')\delta S(x)\delta S(x').\end{cases}$$

A comparison with the general source representation identifies the vacuum expectation values,

$$\langle(\Psi(x)\Psi(x'))_+\rangle=\begin{cases}\text{B.E.:} & -iG_+(x-x')\\ \text{F.D.:} & -i\epsilon(x-x')G_+(x-x'),\end{cases}$$

apart from possible additional delta-function terms, which do not contribute for $x\neq x'$. Now we observe that

$$2\operatorname{Re}\langle(\Psi(x)\Psi(x'))_+\rangle=\langle\{\Psi(x),\Psi(x')\}\rangle$$

is a positive matrix structure. The positiveness properties we have noted thus emerge as necessary quantum consequences of the coupling between source and physical system.

## EUCLIDEAN POSTULATE

It is a remarkable fact that all F.D. particles carry some kind of charge. The experimental proof of nonidentity between electron and muon neutrinos[3] confirms an early suggestion[4] that neutrinos would be no exception to that rule. A representation of this regularity is given by the following abstract Euclidean postulate: The vacuum probability amplitude must be transformable into the attached Euclidean space in such a way that the original time axis cannot be identified. We first illustrate this for B.E. particles.

The basic Euclidean transformations are

$$i(dx)\to(d_4x)=dx_1\cdots dx_4,$$

$$-i\Delta_+(x)\to\Delta_E(x)=\int(d_4p)\frac{e^{ip_\mu x_\mu}}{p^2+m^2},$$

$$p^2=p_\mu p_\mu>0.$$

The transformation of B.E. sources is taken to be

$$S(x)\to\exp[\tfrac{1}{4}\pi i(\prod_{\kappa=1}^{n}\gamma_\kappa{}^0+1)]S_E(x).$$

The product of the even number of $\gamma^0$ matrices is symmetrical, and

$$-S(x)[\prod_\kappa\gamma_\kappa{}^0]S(x)\to S_E(x)S_E(x).$$

For each spinor index $\kappa$, we define the matrices

$$\alpha_\mu=\exp[-\tfrac{1}{4}\pi i\prod\gamma_\kappa{}^0](-i)\gamma_\mu\exp[\tfrac{1}{4}\pi i\prod\gamma_\kappa{}^0],\quad\mu=1\cdots 4$$

which maintain the commutativity of those referring to different indices. The new set has the algebraic property

$$\tfrac{1}{2}\{\alpha_\mu,\alpha_\nu\}=\delta_{\mu\nu}.$$

Each of the $\alpha_\mu$ is imaginary and antisymmetrical, which places all of them on the same footing. The resulting Euclidean transformation of the vacuum amplitude is

$$\langle 0_+|0_-\rangle^S\to\exp\left[-\frac{1}{2}\int(d_4x)(d_4x')S_E(x)G_E(x-x')S_E(x')\right]$$

[3] G. Danby *et al.*, Phys. Rev. Letters **9**, 36 (1962).

[4] J. Schwinger, Ann. Phys. **2**, 407 (1957). See also K. Nishijima, Phys. Rev. **108**, 907 (1957). This is the introduction of a leptonic charge not related to neutrino helicity.

where

$$G_E(x-x')=\prod_{\kappa=1}^{n}(m-\alpha_\mu\partial_\mu)_\kappa\Delta_E(x-x').$$

This is a Euclidean form in which all coordinate axes are indeed indistinguishable.

There is an alternative version which is produced by the further transformation

$$S_E(x)\to\prod_\kappa[\exp(\tfrac{1}{4}\pi i(\alpha_5-1))]_\kappa S_E(x),$$

where

$$\alpha_5=\alpha_1\alpha_2\alpha_3\alpha_4=\alpha_5{}^*=\alpha_5{}^T.$$

Now the function

$$G_E(x-x')=\prod_{\kappa=1}^{n}(\alpha_5 m+i\alpha_\lambda\partial_\lambda)_\kappa\Delta_E(x-x')$$

is real. If the associated Euclidean sources are taken to be real, so also is this Euclidean transcription of the vacuum probability amplitude.

It is interesting to examine further the algebraic basis of the Euclidean transformation, at least in the simplest B.E. situation, $n=2$. Let $\alpha_\mu$, $\mu=1\cdots5$ be a set of anticommuting matrices of unit square such that the product $\alpha_1\cdots\alpha_5$ is the unit matrix. With the latter, the five $\alpha_\mu$ and the ten $\alpha_\mu\alpha_\nu$, $\mu<\nu$, comprise 16 independent matrices. If there be $n_s$ symmetrical matrices among the $\alpha_\mu$, and $n_a=5-n_s$ antisymmetrical ones, the decomposition of the 16 matrices into symmetrical and antisymmetrical members is counted as

$$N_s=1+n_s+n_sn_a=10-(n_s-3)^2,$$
$$N_a=n_a+\tfrac{1}{2}n_s(n_s-1)+\tfrac{1}{2}n_a(n_a-1)=6+(n_s-3)^2.$$

Two independent systems of this type comprise 256 matrices, which are equivalent to the full $16\times16$ matrix algebra if there are just $\frac{1}{2}\,16\times15$ antisymmetrical matrices. This requires that $N_sN_a=60$, or that

$$(n_s-3)^2=0,\,4.$$

The first possibility, $n_s=3$, $n_a=2$, is the one realized in the Minkowski metric where the three symmetrical matrices are $-i\gamma_k$, $k=1,2,3$, and the remaining two are the antisymmetrical matrices $\gamma^0$, $-i\gamma_5$. The second possibility contains two alternatives, $n_s=1$, $n_a=4$, or $n_s=5$, $n_a=0$. This is the Euclidean realization, in which the $\alpha_\mu$, $\mu=1\cdots4$, are antisymmetrical and $\alpha_5$ is symmetrical, or alternatively $\alpha_\mu'=i\alpha_\mu\alpha_5$, $\mu=1\cdots4$ and $\alpha_5'=\alpha_5$ are all symmetrical. Only in the Minkowski metric, however, is any one set of 16 matrices equivalent to a $4\times4$ matrix algebra. The latter contains six antisymmetrical matrices, whereas the Euclidean realizations require ten such matrices.

We now consider both statistics. In response to the Euclidean source transformation

$$S(x)\to RS_E(x),$$

the function $G_E$ emerges as

$$G_E(x-x')=-R^T[\textstyle\prod_\kappa\gamma_\kappa{}^0]R$$
$$\times\textstyle\prod_\kappa(m-R^{-1}(1/i)\gamma_\mu R\partial_\mu)_\kappa\Delta_E(x-x').$$

The matrix factor $-R^T[\prod_\kappa\gamma_\kappa{}^0]R$ must be a Euclidean scalar. It is symmetrical for B.E. particles, and antisymmetrical for F.D. particles. The matrices $R^{-1}(1/i)\gamma_\mu R$, $\mu=1\cdots4$, are required to have a common symmetry. We shall let $\alpha_\mu$, $\mu=1\cdots4$, specifically designate antisymmetrical matrices, and construct the alternative symmetrical matrices as $i\alpha_\mu\alpha_5$. The appropriate Euclidean scalar matrix must commute with the $\alpha_\mu$ and anticommute with the $i\alpha_\mu\alpha_5$. The latter requirement can be satisfied by the factor $\prod_\kappa\alpha_{5\kappa}$, and either choice requires the existence of a Euclidean scalar matrix that commutes with all the $\alpha_\mu$ and has a definite symmetry, as demanded by the statistics. The $\alpha_{\mu\kappa}$, $\kappa=1\cdots n$ generate an algebra of dimensionality $16^n$. If the latter is the full $4^n\times4^n$ matrix algebra, only the symmetrical unit matrix can commute with all elements. This is the B.E. situation. The antisymmetrical scalar matrix of F.D. statistics can be realized only by adjoining such a matrix, $l$, to the algebra generated by the $\alpha_\mu$. The necessity for this additional element is also evident in the remark that $4\times4$ matrices do not permit a representation of Euclidean symmetries.

A suitable Euclidean transformation matrix for F.D. statistics is

$$R=\exp[\tfrac{1}{4}\pi i(l\prod_{\kappa=1}^{n}\gamma_\kappa{}^0+1)]$$

and

$$G_E(x-x')=l\textstyle\prod_\kappa(m-\alpha_\mu\partial_\mu)_\kappa\Delta_E(x-x'),$$

which makes explicit the charge carried by every kind of F.D. particle. The alternative Euclidean representation

$$G_E(x-x')=l\textstyle\prod_\kappa(\alpha_5m+i\alpha_\mu\partial_\mu)_\kappa\Delta_E(x-x')$$

is produced by the transformation

$$S_E(x)\to\textstyle\prod_\kappa[\exp(\tfrac{1}{4}\pi i(\alpha_5-1))]_\kappa S_E(x).$$

The new $G_E$ function is imaginary. Nevertheless real Euclidean source functions imply a real Euclidean transcription of the vacuum probability amplitude, according to the definition of complex conjugation in an exterior algebra,[2]

$$(S(x)S(x'))^*=S(x')S(x)$$
$$=-S(x)S(x').$$

## MASSLESS PARTICLES

In order to deal with the special circumstances posed by particles of zero mass, we return to the stage of identifying single-particle states and consider, for $m=0$, the individual projection matrices

$$\gamma^0(-\gamma p)/2p^0=\tfrac{1}{2}(1-i\gamma_5\sigma_p).$$

Here $\sigma_p$ is the component of $\boldsymbol{\sigma}$ that is parallel to the momentum of the particle. The isolated occurrence of the $\gamma_5$ matrices demands a further specification of particle sources. It is sufficient to consider symmetrical spinors that are subject to the single-invariant constraint

$$\left[\left(\frac{1}{n}\sum_{\kappa=1}^{n} i\gamma_{5\kappa}\right)^2-1\right]S(x)=0.$$

The latter requires that each $i\gamma_{5\kappa}$ have the same eigenvalue, which becomes the common eigenvalue of $-\sigma_{p\kappa}$. This is the familiar statement that massless B.E. particles have only two helicity states, those for which the component of angular momentum parallel to the linear momentum equals $\pm s$, $s=\frac{1}{2}n$. The photon and graviton are represented by $n=2$ and $n=4$, respectively. Incidentally, the Euclidean version of the constraint is

$$\left[\left(\frac{1}{n}\sum_{\kappa} \alpha_{5\kappa}\right)^2-1\right]S_E(x)=0.$$

A massless particle obeying F.D. statistics can be described by a real source which is a symmetrical spinor obeying one of the two constraints:

$$\left[l\frac{1}{n}\sum_{\kappa=1}^{n} i\gamma_{5\kappa}-1\right]S(x)=0$$

or

$$\left[l\frac{1}{n}\sum_{\kappa=1}^{n} i\gamma_{5\kappa}+1\right]S(x)=0.$$

In either situation each of the $i\gamma_{5\kappa}$ has the same eigenvalue, which is the common eigenvalue of $-\sigma_{p\kappa}$, and is also equal to the eigenvalue of $l$, or $-l$, according to the nature of the particle. A given particle has two helicity states, of angular momentum $\pm s$, $s=\frac{1}{2}n$. The $e$ and $\mu$ neutrinos, presumed to be massless, are represented by the two alternatives available for $n=1$.

The Euclidean transcription of the F.D. constraints requires an additional source transformation in order to be compatible with the reality of Euclidean sources. This transformation involves the fermionic charge reflection matrix $r_l$, a real symmetrical matrix, which can be chosen as

$$r_l=\begin{pmatrix}1 & 0\\ 0 & -1\end{pmatrix}.$$

An alternative charge reflection matrix is $ir_l l$. With the further transformation

$$S_E(x)\to\exp(\tfrac{1}{4}\pi r_l l)S_E(x),$$

the constraint equations become

$$[r_l(1/n)\sum_{\kappa}\alpha_{5\kappa}\mp 1]S_E(x)=0$$

while leaving unaltered the imaginary structure

$$G_E(x-x')=l\prod\nolimits_\kappa(i\alpha_\mu\partial_\mu)_\kappa\Delta_E(x-x').$$

## $SU_3$ AND SPIN

The rapid increase in experimental information concerning massive, strongly interacting particles has brought a provisional answer to the long quest for a particle classification scheme. It is supplied by the multiplets of the internal symmetry group $SU_3$. Let the two types of three-valued unitary indices be designated by $a$ and $a^*$. The various unitary multiplets are labeled by symmetrical functions of indices $a_1, \cdots a_r$, and of indices $a_{r+1}^*, \cdots a_\nu^*$, which are irreducible with respect to contraction of an index of type $a$ with an $a^*$ index. Thus, the various particle sources can be indicated by

$$S_{\zeta_1\cdots\zeta_n;a_1\cdots a_r,a_{r+1}^*\cdots a_\nu^*}(x).$$

It should be emphasized that the unitary indices are a means of supplying the quantum numbers appropriate to a given particle, which may still need to be identified through its mass value, and that no presumption exists concerning the masses of the different particles which are united in a particular multiplet.

As we have already remarked in connection with the spin indices, there are alternative ways, of designating sources, that employ more complicated symmetry patterns. While an antisymmetrical pair of spin indices is affectively inert, an antisymmetrical combination of similar unitary indices is equivalent to a complex-conjugate index. In this way, other symmetry patterns can be reduced to the totally symmetric structures (zero spin is an exception, of course). Several different sources may appear in consequence of the reduction process. The possibility thus suggested of a more inclusive classification can be illustrated by considering symmetric functions of the combined indices

$$A=\zeta,a\quad A^*=\zeta,a^*.$$

The generalized source

$$S_{A_1\cdots A_r;A_{r+1}^*\cdots A_n^*}(x),$$

a symmetrical function of $A_1\cdots A_r$, and of $A_{r+1}^*\cdots A_n^*$, unites the sources of particles, with various spins and a common parity, that belong to several unitary multiplets. Thus, the source with $r=1$, $n=2$, $S_{\zeta_1a_1;\zeta_2a_2^*}$, describes spin 0 and spin 1 meson multiplets, each containing a unitary octuplet and singlet. This source can be applied to the well-established $0^-$ and $1^-$ mesons, both of which form octuplet + singlet families. The baryon multiplets comprised in $S_{\zeta_1a_1,\zeta_2a_2,\zeta_3a_3}$ have spin $\frac{1}{2}$ and $\frac{3}{2}$, forming a unitary octuplet and a decuplet, respectively. The latter is applicable to the known system of $\frac{1}{2}^+$ baryons and $\frac{3}{2}^+$ baryon resonances. These wider classifications have been achieved by a combinatorial union of spin and unitary indices. There is no reference to a continuous group of transformations on all the indices.

**CONCLUSION**

The source concept emerges as a valid and useful phenomenological particle description. It displays the complementary aspects of particle behavior, and supplies the connection between spin and statistics in a simple and direct way. The sources also serve as carriers of quantum numbers and give concrete expression to family relations among particles.

There is, in this description, no commitment to any specific view of particle structure. The extreme $S$-matrix attitude can be introduced by insisting that no source function $S(p)$ is defined for momentum values that do not obey $-p^2=m^2$. But if it is considered meaningful to define $S(p)$ over a wider momentum range, one has admitted a concept of matter more fundamental than that of the particle. It seems desirable to have a phenomenological description of particles which is logically independent of hypotheses concerning the deeper nature of these objects. Such hypotheses may then be suggested by the formal representation of observed regularities.

*Note added in proof.* The complete independence of the phenomenological source description from other formulations needs more emphasis. The text contains reference to the field-theoretic version of the *TCP* theorem, and to operator-based positiveness properties. Neither is required. The essential tool is the formal expression of completeness for the multi-particle states in the two forms

$$\sum_n \langle 0_-|n\rangle^S\langle n|0_-\rangle^S=1$$
$$\langle 0_-|n\rangle^S=\langle n|0_-\rangle^{S*},$$

and

$$\sum_n \langle 0_+|n\rangle^S\langle n|0_+\rangle^S=1$$
$$\langle n|0_+\rangle^S=\langle 0_+|n\rangle^{S*},$$

where $n$ symbolizes the whole set of occupation numbers. Using the multispinor representation, which unifies all spins and associated statistics, the consideration of a production and detection source gives

$$\langle n|0_-\rangle^S=\langle 0_+|0_-\rangle^S\prod_{p\lambda}(iS_{p\lambda})^{n_{p\lambda}}/(n_{p\lambda}!)^{1/2}$$
$$\langle 0_+|n\rangle^S=\langle 0_+|0_-\rangle^S\prod_{p\lambda}{}'(iS_{p\lambda}^*)^{n_{p\lambda}}/(n_{p\lambda}!)^{1/2}$$

where $\Pi$ and $\Pi'$ refer, respectively, to some standard multiplication order and its inverse. Here

$$S_{p\lambda}=\left[\frac{(d\mathbf{p})}{(2\pi)^3}\frac{(2m)^n}{2p^0}\right]^{1/2}u_{p\lambda}^*(\prod_\kappa\gamma_\kappa^0)S(p)$$

and

$$\prod_\kappa(m-\gamma p/2m)_\kappa=\sum_\lambda u_{p\lambda}u_{p\lambda}^*(\prod\gamma_\kappa^0),$$

with

$$u_{p\lambda}^*(\Pi\gamma_\kappa^0)u_{p\lambda'}=\delta_{\lambda\lambda'}.$$

It is the comparison of the two completeness expressions that indicates the necessity for the general rule

$$(S(x)S(x'))^*=S(x')S(x).$$

Then either form gives

$$1=|\langle 0_+|0_-\rangle^S|^2\exp[\sum_{p\lambda}S_{p\lambda}^*S_{p\lambda}].$$

The usual connection between spin and statistics assures the reality of the combination

$$S(x)(\Pi_\kappa\gamma_\kappa^0)S(x'),$$

which is the basis of the direct evaluation

$$|\langle 0_+|0_-\rangle^S|^2=\exp[-\sum_{p\lambda}S_{p\lambda}^*S_{p\lambda}].$$

An attempt to reverse the connection between spin and statistics by introducing an antisymmetrical matrix $q$ into the general source structure will now founder on the contradiction between the direct evaluation, which recognizes the indefinite nature of the $q$ spectrum, and the summation over all particle states, which involves only the magnitudes of the $q$ eigenvalues. Similarly, any reference to internal properties is limited to a symmetrical positive matrix which can always be transformed into the unit matrix. Here is the assurance that a charged particle has an oppositely charged counterpart of equal mass.

The source representation of the invariance operation designated as *TCP* combines space-time reflection with transposition—the reversal in multiplication order of all sources. The former is performed through the attached Euclidean group and leaves the vacuum amplitude unchanged, but at the expense of making half-integer spin sources imaginary. The actual sign reversal of a product of such sources is then compensated by the transposition of the sources, in virtue of the connection between spin and statistics.

We have built the source description on a principle of the unity of the source; sources that are effective in different space-time regions are constituents of one general source. Perhaps it should be pointed out that only the two usual statistics are admitted by this principle. In the consideration of a production and detection source that leads to the identification of the particle states, it is necessary to give an ordered form to the product of sources. This demands the existence of relations of the type

$$S(x')S(x)=\lambda S(x)S(x')$$

which exhibit the algebraic properties of the source function, and limit $\lambda$ to one of the alternatives, $\pm 1$.

# CHIRAL DYNAMICS

J. SCHWINGER*
*Harvard University, Cambridge, Massachusetts*

Received 6 April 1967

The cumbersome operator techniques and weak interaction orientation of current algebra are replaced by a non-operator method based on strong interaction phenomenology. Some new results include alternative possibilities of $\pi\pi$ scattering lengths, and a treatment of $\rho$ and $A_1$ decay widths that supports the interpretation of $A_1$ as the axial partner of $\rho$.

This note was stimulated by some recent work of Weinberg [1]. He has shown how the results of current-algebra can be easily reproduced by certain calculational rules used in conjunction with an appropriate Lagrange function. Current-algebra is still considered primary, however. I propose to further this simplification and clarification by eliminating all reference to current-algebra. The non-operator method that replaces it is the phenomenological source theory now under development [2]. For our present purposes, however, it suffices to think of a numerical effective Lagrange function, the coupling terms of which are directly applicable to the corresponding processes**.

We begin with the strongly interacting***, low energy pion-nucleon system and the empirical absence of S-wave scattering in the amplitude combination $2a_{\frac{1}{2}} + a_{\frac{1}{2}} \approx 0$. The corresponding Lagrange function is

$$\mathcal{L} = -\tfrac{1}{2}(\partial_\mu \Phi)^2 - \tfrac{1}{2} m_\pi^2 \Phi^2 - \bar{\Psi}(-\gamma^\mu \mathrm{i}\partial_\mu + M)\Psi + (f/m_\pi)\bar{\Psi}\mathrm{i}\gamma^\mu\gamma_5\tau\Psi \cdot \partial_\mu\Phi - (f_0/m_\pi)^2\,\bar{\Psi}\gamma^\mu\tau\Psi\cdot\Phi\times\partial_\mu\Phi,$$

where it is known that $f \approx 1.0$ and $f_0 \approx 0.8$. In addition to being invariant under the infinitesimal isotopic rotation

$$\Phi \to \Phi - \delta\omega \times \Phi, \qquad \Psi \to [1 + \mathrm{i}\tfrac{1}{2}\tau\cdot\delta\omega]\Psi,$$

this structure, apart from the pion mass term, is invariant under the infinitesimal transformation

$$\Phi \to \Phi + \delta\varphi, \qquad \Psi \to [1 + \mathrm{i}(f_0/m_\pi)^2\ \tau\cdot\Phi\times\delta\varphi]\Psi,$$

if one temporarily disregards more complicated $\pi$-N couplings than the ones considered. In the neighbourhood of $\Phi = 0$ the $\Psi$ transformations form a compact group. It has the structure of the four-dimensional Euclidean rotation group, or, if one introduces the parameters $\delta\omega_\pm = \delta\omega \pm (2f_0/m_\pi)\delta\varphi$, it is the chiral group $SU2\times SU2$. The group structure becomes exact ‡ if one replaces the Abelian transformation $\Phi \to \Phi + \delta\varphi$ with

$$\Phi \to \Phi + \delta\varphi + (f_0/m_\pi)^2\,[2\Phi\delta\varphi\cdot\Phi - \delta\varphi\,\Phi^2].$$

* Supported in part by the Air Force Office of Scientific Research.

** It is not meaningful to question the use of the coupling terms "in lowest order". That is the nature of a numerical effective Lagrange function, which gives a direct description of the phenomena.

*** Here is a fundamental difference in attitude from current-algebra. The latter uses concepts and numerical parameters derived from the analysis of weak interactions to predict some strong interaction effects. We reverse the logical sequence and restore the primary role of strong interactions.

‡ This is a new approach to chiral transformations. The $\Phi$ transformations are non-linear and thereby avoid reference to the non-physical $\sigma$ field required for their linear representation [3]. The $\Psi$ transformations do not contain $\gamma_5$ and have a convenient static limit.



Reprinted from *Physics Letters* **24B**, 473–476 (1967).

We can now return to $\mathcal{L}$ and ask for the more detailed form that would give precise validity to the infinitesimal response* $\delta\mathcal{L} = -m_\pi^2\Phi\cdot\delta\varphi$. It is

$$\mathcal{L} = -\tfrac{1}{2}[1+(f_0/m_\pi)^2\Phi^2]^{-2}(\partial_\mu\Phi)^2 - \tfrac{1}{2}(m_\pi^2/f_0)^2\log[1+(f_0/m_\pi)^2\Phi^2] - \bar\Psi(-\gamma^\mu i\partial_\mu + M)\Psi +$$
$$+[1+(f_0/m_\pi)^2\Phi^2]^{-1}\{(f/m_\pi)\bar\Psi i\gamma^\mu\gamma_5\tau\Psi\cdot\partial_\mu\Phi - (f_0/m_\pi)^2\bar\Psi\gamma^\mu\tau\Psi\cdot\Phi\times\partial_\mu\Phi\}.$$

This differs from a related formula of Weinberg only in the second term, which is derived from the pion mass term by the arbitrary requirement that $\mathcal{L}$ respond linearly to the $\delta\varphi$ transformation. There is an analogous requirement in Weinberg's work (PCAC), but it is applied to a different field**.

The predictions concerning low energy $\pi$-$\pi$ scattering are contained in***

$$\mathcal{L}_{\pi\pi} = (f_0/m_\pi)^2\Phi^2[(\partial_\mu\Phi)^2 + \tfrac{1}{4}m_\pi^2\Phi^2].$$

The implied scattering amplitudes are

$$m_\pi a_0 = \tfrac{9}{4}(f_0/4\pi) \approx 0.1, \qquad a_2 = -\tfrac{2}{3}a_0.$$

Had we omitted the second term‡ of $\mathcal{L}_{\pi\pi}$, these results would be

$$m_\pi a_0 = f_0^2/4\pi \approx 0.05, \qquad a_2 = -2a_0.$$

The extension of the infinitesimal parameters $\delta\omega$, $\delta\Phi$ to space-time functions gives

$$\delta\mathcal{L} + m_\pi^2\Phi\cdot\delta\varphi = -\partial_\mu\delta\omega\cdot[-\Phi\times\partial^\mu\Phi + \bar\Psi\gamma^\mu\tfrac{1}{2}\tau\Psi] - \partial_\mu\delta\varphi\cdot[\partial^\mu\Phi - (f/m_\pi)\bar\Psi i\gamma^\mu\gamma_5\tau\Psi],$$

in which we have retained only the simplest contributions‡‡. The equivalent coefficient of $-\tfrac{1}{2}\partial_\mu\delta\omega_+$ is

* The assumption of a linear response of $\mathcal{L}$ to the chiral transformation is a hypothesis of simplicity. It is the analogue of PCAC (see footnote ‡‡).

** Namely, the non-physical $\pi$-field of the $\sigma$-model.

*** The complete form of $\mathcal{L}$ is not required to infer $\mathcal{L}_{\pi\pi}$. It suffices to note that $\mathcal{L}_\pi$, comprising the quadratic $\pi$ field terms, has the following response to the infinitesimal chiral transformation,

$$\delta\mathcal{L}_\pi + m_\pi^2\Phi\cdot\delta\varphi = -2(f_0/m_\pi)^2[\delta\varphi\cdot\Phi(\partial_\mu\Phi)^2 + \tfrac{1}{2}m_\pi^2\delta\varphi\cdot\Phi\Phi^2].$$

This is just $-\delta\mathcal{L}_{\pi\pi}$, if only cubic terms in the $\pi$ field are retained.

‡ While chiral transformations are suggested by the observed properties of the $\pi$N system, the specific way that chiral symmetry is violated must be learned from the physical properties of the $\pi\pi$ system. The linear response of $\mathcal{L}$ is one simple possibility. Another one is that $-\tfrac{1}{2}m_\pi^2\Phi^2$ is the whole non-invariant term of $\mathcal{L}$.

‡‡ Some additional terms in the construction of the two vectors are given here

$$j^\mu = -\Phi\times\partial^\mu\Phi + \bar\Psi\gamma^\mu\tfrac{1}{2}\tau\Psi - (f/m_\pi)\bar\Psi i\gamma^\mu\gamma_5\tau\Psi\times\Phi,$$

$$(2f_0/m_\pi)k^\mu = \partial^\mu\Phi + (f_0/m_\pi)^2[2\Phi\,\Phi\cdot\partial^\mu\Phi - 3\Phi^2\partial^\mu\Phi] - (f/m_\pi)\bar\Psi i\gamma^\mu\gamma_5\tau\Psi + 2(f_0/m_\pi)^2\bar\Psi\gamma^\mu\tau\Psi\times\Phi.$$

Sufficient structure is now exhibited to confirm the exact geometrical behaviour of these isotopic and axial current vectors under constant $\delta\omega$, $\delta\varphi$ transformations,

$$\delta j^\mu = -\delta\omega\times j^\mu - (2f_0/m_\pi)\delta\varphi\times k^\mu, \qquad \delta k^\mu = -\delta\omega\times k^\mu - (2f_0/m_\pi)\delta\varphi\times j^\mu.$$

An analogous operator statement contains commutators of current components. The variation of $\mathcal{L}$ for space-time dependent $\delta\omega$, $\delta\varphi$ implies differential equations. They are

$$\partial_\mu j^\mu = 0, \qquad \partial_\mu k^\mu = (m_\pi/2f_0)m_\pi^2\Phi,$$

which alphabetize to CIC and PCAC.

$$-\Phi \times \partial^\mu \Phi + (m_\pi/2f_0)\partial^\mu \Phi + \bar{\Psi}\gamma^\mu \tfrac{1}{2}\tau\Psi - (f/f_0)\bar{\Psi} i\gamma^\mu \gamma_5 \tfrac{1}{2}\tau\Psi .$$

We now base the weak interaction theory for pions and nucleons on the assumption that leptons are coupled to the charged components of this vector *. The immediate implications include: the Feynman-Gell-Mann hypothesis connecting pion $\beta$-decay to nucleonic vector $\beta$-decay coupling; the Goldberger-Treiman relation between pion instability and the nucleon axial vector coupling; the nucleonic axial vector-vector ratio $-G_A/G_V = f/f_0 \approx 1.2$, which is the Adler-Weisberger relation, as interpreted by Tomozawa and Weinberg.

Although it involves a considerable extrapolation from low-energy phenomena, this description can be extended to the unit spin particles $\rho$ and $A_1$ by introducing corresponding non-Abelian gauge fields $U_\mu$ and $P_\mu$. (This probability is also mentioned by Weinberg). A simplified form of the Lagrange function ** is

$$\mathcal{L} = -[D_\mu \Phi + 2^{-\frac{1}{2}} m P_\mu]^2 - \tfrac{1}{2} m_\pi^2 \Phi^2 - \tfrac{1}{4}(\partial_\mu U_\nu - \partial_\nu U_\mu + g U_\mu \times U_\nu + g P_\mu \times P_\nu)^2 - \tfrac{1}{2} m^2 U_\mu^2 +$$

$$- \tfrac{1}{4}(D_\mu P_\nu - D_\nu P_\mu)^2 - \tfrac{1}{2} m^2 P_\mu^2 - \bar{\Psi}[\gamma^\mu(-i\partial_\mu - g\tfrac{1}{2}\tau \cdot U_\mu) + M]\Psi + 2(f/m_\pi)\bar{\Psi} i\gamma^\mu\gamma_5 \tau\Psi \cdot [D_\mu\Phi + 2^{-\frac{1}{2}} m P_\mu] +$$

$$- \tfrac{1}{2}(g/m)^2 \bar{\Psi}\gamma^\mu \tau\Psi \cdot \Phi \times [D_\mu \Phi + 2^{\frac{1}{2}} m P_\mu]$$

with $D_\mu = \partial_\mu + gU_\mu \times$. Here $m$ is the $\rho$ mass and $2^{-\frac{1}{2}}(m_\pi/m)\, g = f_0$. Also involved is the hypothesis that, when the $\rho$ dynamics is explicitly exhibited, S-wave $\pi$-N scattering is not a consequence of a direct $\pi$-N coupling. The absence of the latter is made evident by the transformation

$$P_\mu \to P_\mu - 2^{-\frac{1}{2}}(1/m) D_\mu \Phi ,$$

which gives

$$\mathcal{L} = -\tfrac{1}{2}(D_\mu\Phi)^2 - \tfrac{1}{2} m_\pi^2 \Phi^2 - \tfrac{1}{4}[\partial_\mu U_\nu - \partial_\nu U_\mu + gU_\mu \times U_\nu + gP_\mu \times P_\nu - 2^{-\frac{1}{2}}(g/m)(D_\mu\Phi \times P_\nu + P_\mu \times D_\nu \Phi) +$$

$$+ (g/2m^2) D_\mu \Phi \times D_\nu \Phi]^2 - \tfrac{1}{2} m^2 U_\mu^2 - \tfrac{1}{4}[D_\mu P_\nu - D_\nu P_\mu - 2^{-\frac{1}{2}}(1/m)(D_\mu D_\nu - D_\nu D_\mu) \times \Phi]^2 - \tfrac{1}{2}(2m^2) P_\mu^2 +$$

$$- \bar{\Psi}[\gamma^\mu(-i\partial_\mu - g\tfrac{1}{2}\tau \cdot U_\mu) + M]\Psi + (f/m_\pi)\bar{\Psi} i\gamma^\mu\gamma_5\tau\Psi \cdot [D_\mu\Phi + 2^{\frac{1}{2}} m P_\mu] - 2^{-\frac{1}{2}}(g^2/m)\bar{\Psi}\gamma^\mu\tau\Psi \cdot \Phi \times P_\mu .$$

We recognize here Weinberg's remarkable mass relation $m_{A_1} = 2^{\frac{1}{2}} m_\rho$, which is accurately fulfilled. The $\rho$-mediated static coupling between nucleon and pion is

$$-(1/m^2)\, g\bar{\Psi}\gamma^\mu \tfrac{1}{2}\tau\Psi \cdot g\Phi \times \partial_\mu \Phi = -(f_0/m_\pi)^2 \bar{\Psi}\gamma^\mu \tau\Psi \cdot \Phi \times \partial_\mu\Phi ,$$

as before. When the static $\pi$-$\pi$ interaction is constructed from this approach, it emerges as

$$\mathcal{L}_{\pi\pi} = \tfrac{1}{2}(1/m^2)(g\Phi \times \partial_\mu\Phi)^2 + (g^2/8m^2)[(\partial_\mu \Phi^2)^2 + m_\pi^2(\Phi^2)^2]$$

where the first term is contributed by $\rho$ exchange.

The coupling terms responsible for $\rho$ instability are

* The known chiral behaviour of the leptons in weak interactions is thereby extended to mesons and baryons. One need not speculate on the significance of the dual role played by chiral transformations in strong and weak interactions to draw weak interaction consequences from established strong interaction properties. I find some wry amusement in recalling that the paper which introduced the $\sigma$ field in order to realize $\pi$-field chiral transformations (footnote *** on page 473) also contains a totally unnoticed precursor of the universal V-A weak interaction theory. Unfortunately, the two sets of ideas remained unconnected then (1956-7).

** Various higher powers of $\Phi$ are omitted.

$$\mathcal{L}_{\rho\pi} = -gU^{\mu}\cdot\Phi\times\partial_{\mu}\Phi - (g/4m^2)(\partial^{\mu}U^{\nu} - \partial^{\nu}U^{\mu})\cdot\partial_{\mu}\Phi\times\partial_{\nu}\Phi \rightarrow -\tfrac{3}{4}gU^{\mu}\cdot\Phi\times\partial_{\mu}\Phi.$$

The value of $g$ implied by the observed $\rho$ width [note the factor $\frac{3}{4}$] is conveyed by * $2^{-\frac{1}{2}}(m_{\pi}/m)g \approx 1.0$, to be compared with $f_0 \approx 0.8$. The decay of $A_1$ is described by

$$\mathcal{L}_{A_1\rho\pi} = 2^{-\frac{3}{2}}(g/m)(\partial^{\mu}U^{\nu} - \partial^{\nu}U^{\mu})\cdot(\partial_{\mu}\Phi\times P_{\nu} + P_{\mu}\times\partial_{\nu}\Phi) +$$

$$+ 2^{-\frac{3}{2}}(g/m)(\partial^{\mu}U^{\nu} - \partial^{\nu}U^{\mu})\cdot\Phi\times(\partial_{\mu}P_{\nu} - \partial_{\nu}P_{\mu}) \rightarrow -2^{-\frac{1}{2}}gm\,U^{\mu}\times\Phi\cdot P_{\mu},$$

and the observed $A_1$ width is fitted with ** $2^{-\frac{1}{2}}(m_{\pi}/m)\,g \approx 0.7$. According to Weinberg, there is some difficulty about the $A_1$ width from the current-agebraic standpoint.

Non-Abelian gauge invariance is violated by tha mass terms of the unit spin particles. This gives

$$\delta\mathcal{L} + m_{\pi}^2\,\Phi\cdot\delta\varphi = -\partial_{\mu}\delta\omega\cdot(m^2/g)\,U^{\mu} - \partial_{\mu}\delta\varphi\cdot[D^{\mu}\Phi - 2^{\frac{1}{2}}mP^{\mu}]$$

from which we recover the previous weak interaction results, with the additional information that nucleon vector and axial vector form factors are governed by the $\rho$ and $A_1$ masses, respectively.

These methods can also be applied to the general meson-baryon system. That will be discussed in another communication.

I am greatly indebted to S. Weinberg for telling me about his work in advance of publication.

*References*

1. S. Weinberg, Phys. Rev. Letters 18 (1967) 188 and to be published.
2. J. Schwinger, Phys. Rev. 152 (1966) 1219 and to be published.
3. J. Schwinger, Ann. Phys. (N. Y.) 2 (1957) 407;
   M. Gell-Mann and M. Levy, Nuovo Cimento 16 (1960) 705.

* This number corresponds to $m_{\rho}$ = 760 MeV, $\Gamma_{\rho} \approx$ 130 MeV. Since properties of the $\rho$ are not well established, we note some other values: $\Gamma_{\rho}$ = 120 MeV, "$f_0$" = 0.94; $\Gamma_{\rho}$ = 100 MeV, "$f_0$" = 0.86. A somewhat more precise value of $f_0$ as implied by s-wave $\pi$N scattering, is $f_0 = 0.81 \pm 0.01$. This is a good place to point out that the phenomenological Lagrange function does not state that $2a_{\frac{3}{2}} + a_{\frac{1}{2}}$ is exactly zero at threshold. The predominantly p-wave coupling measured by $f$ also implies a small s-wave scattering contribution. This produces only a tiny correction to $f_0^2$ but gives

$$(2a_{\frac{3}{2}} + a_{\frac{1}{2}})/(a_{\frac{1}{2}} - a_{\frac{3}{2}}) = -\tfrac{1}{2}(m_{\pi}/M)(f/f_0)^2 = -0.12 \text{ or } \quad m_{\pi}(2a_{\frac{3}{2}} + a_{\frac{1}{2}}) \approx -0.03.$$

Direct fits of the scattering data agree with this, but there is some conflict with the indirect determinations based on dispersion relations.

** The qualitative consistency of these coupling constants lends support to the interpretation of $A_1$ as the axial partner of $\rho$.

* * * * *

# PARTIAL SYMMETRY*

Julian Schwinger
Harvard University, Cambridge, Massachusetts
(Received 17 April 1967)

Partial symmetry is proposed as the general basis for a phenomenological method that replaces current algebra. Its practical merit is displayed through a reinterpretation of combined internal and spin transformations. Among various results referring to coupling constants, magnetic moments, and electromagnetic decays, the following is noted:

$$-G_A/G_V = 2^{-1/2}\times 5/3 = 1.18.$$

I have suggested[1] that the operator techniques of current algebra[2] are not the most effective way to explore the utility of kinematical transformation groups as a means of conveying dynamical information. The phenomenological description of the low-energy $\pi N$ system by a numerical effective Lagrange function was the context used to introduce its replacement. The chiral group $SU(2)\otimes SU(2)$ appears in this example as a partial-symmetry group—that is, a set of transformations under which a significant portion of an effective Lagrange function, referring to specific physical circumstances, remains invariant. The notion of partial symmetry includes the current-algebraic approach, in a much simpler form. Equal-time commutation relations of group generators (currents), a purely kinematical concept, are implicit[3] in a dynamical framework that supplies equations of motion for the generators (current conservation or nonconservation equations). The search for approximate realizations of the commutators among a small set of particle states can be viewed as the quest for an artificial but relevant dynamical situation in which the group would have dynamical



Reprinted from *Physical Review Letters* **18**, 923–926 (1967).

significance. The partial-symmetry attitude emphasizes that one is revealing, not "explaining," particle relationships that ultimately must be assertions about the self-consistency of the dynamical mechanisms that govern the subparticle world.

In order to stress the practical advantages of the partial-symmetry method, we shall use it to reinterpret the much discussed group of transformations involving internal and spin degrees of freedom [usually SU(6) or U(6) ⊗ U(6), but we shall only consider a U(4) factor of U(4) ⊗ U(4)]. The particles principally involved are $\pi$, $\rho$, $\omega$, and $N$, $N^*$, although we shall later add $\Phi$ and the other members of the baryon octuplet and decuplet.

It is important to recognize that no change in the earlier chiral-group discussion is required by the inclusion of $N^*$. An elementary approach to the baryon chiral transformation is based on the $\pi$-$\rho$ coupling term

$$g\rho^{\mu}\cdot\partial_{\mu}\pi\times\pi.$$

Consider the chiral transformation $\pi \to \pi + \delta\pi + \cdots$, where $\delta\pi$ is a constant. If we retain only the latter, and not higher powers in the $\pi$ field, the change in the $\pi$-$\rho$ coupling induces a simple gauge transformation on the $\rho$ field,

$$\delta\rho_{\mu} = \partial_{\mu}[(g/m_{\rho}^{2})\pi\times\delta\pi].$$

The corresponding response of any isospin-bearing field is the appropriate chiral transformation. For a nucleon field,

$$\delta N = ig\times\tfrac{1}{2}\tau\cdot(g/m_{\rho}^{2})\pi\times\delta\pi N,$$

and comparison with the form of the $s$-wave $\pi N$ coupling, $(f_0/m_\pi)^2\bar{N}\gamma^{\mu}\tau N\cdot\partial_{\mu}\pi\times\pi$, gives the relation

$$\tfrac{1}{2}(g/m_{\rho})^2 = (f_0/m_{\pi})^2.$$

No mixing of the $N$ and $N^*$ fields appears, since the chiral transformations depend only upon isospin. This is in sharp distinction to the more usual $\gamma_5$ representation of chiral transformations, in which such mixing does occur.[4]

We shall describe the mesons by appropriate (pseudo)scalars, vectors, and antisymmetrical tensors, with the Lagrange function supplying the first-order differential connections, as in the following $\pi$ and $\rho$ contributions:

$$\mathcal{L}_{\pi} = -\pi^{\mu}\cdot\partial_{\mu}\pi + \tfrac{1}{2}\pi^{\mu}\cdot\pi_{\mu} - \tfrac{1}{2}m_{\pi}^{2}\pi^{2},$$

$$\mathcal{L}_{\rho} = -\tfrac{1}{2}\rho^{\mu\nu}\cdot(\partial_{\mu}\rho_{\nu} - \partial_{\nu}\rho_{\mu}) + \tfrac{1}{4}\rho^{\mu\nu}\cdot\rho_{\mu\nu} - \tfrac{1}{2}m_{\rho}^{2}\rho^{\mu}\cdot\rho_{\mu}.$$

The single meson interactions with the baryons involve the field components $\pi^{\mu}$, $\rho^{\mu}$, $\omega^{\mu}$, $\rho^{\mu\nu}$, and $\omega^{\mu\nu}$. Suppose we restrict our attention to arbitrarily small momentum exchanges between the mesons and baryons. This introduces two types of simplifications: When viewed in the rest frame of the baryons, only positive-parity meson-field components contribute to the interaction; the positive- and negative-parity meson components are uncoupled, in the absence of the coordinate derivative terms. The positive-parity field components can be assembled in the 4×4 Hermitian matrix [$H_{\rho 3} = \rho_{12}$, etc.]

$$M = (\vec{\pi} + \vec{\mathrm{H}}_{\rho}):\vec{\sigma}\tau + \vec{\mathrm{H}}_{\omega}\cdot\vec{\sigma} - m_{\rho}\rho^{0}\cdot\tau - m_{\omega}\omega^{0}1,$$

with appropriate scalar products in the three-dimensional isotopic and coordinate spaces. The quadratic terms in $\mathcal{L}$ that refer to these components are reproduced by $\mathrm{tr}(\frac{1}{4}M)^2$, together with a similar structure that contains $\vec{\pi} - \vec{\mathrm{H}}_{\rho}$. The trace is invariant under the group U(4) of transformations[5] on the $\vec{\sigma}$ and $\tau$ matrices in $M$.

Let us test a hypothesis of partial symmetry —that the meson-baryon coupling also possesses U(4) symmetry when the baryon rest-frame fields $N$ and $N^*$ are united in $\Psi_{A_1A_2A_3}$. The latter is totally symmetrical in the four-valued indices $A = \sigma, \tau$. Such a coupling is

$$(g/2m_{\rho})\Psi^{\dagger}\Big(\sum_{\alpha=1}^{3} M_{\alpha}\Big)\Psi,$$

where each $M_{\alpha}$ acts only on the corresponding indices. The scale factor

$$g/2m_{\rho} = 2^{-1/2}f_0/m_{\pi}$$

is established by the general relationship between $-g\rho^0$ and the density of isospin. The nucleon part of this interaction is[6]

$$2^{-1/2}(f_0/m_\pi)N^\dagger[(5/3)(\vec{\pi}+\vec{H}_\rho):\vec{\sigma}\tau + \vec{H}_\omega\cdot\vec{\sigma} - m_\rho \rho^0\cdot\tau - 3m_\omega \omega^0]N.$$

The implied $\pi N$ coupling constant $f$ is given by

$$f = 2^{-1/2}(5/3)f_0 = 1.18 f_0 = 0.95.$$

This is in striking agreement[7] with the indirect $\beta$-decay measurement

$$f/f_0 = -G_A/G_V = 1.18 \pm 0.02,$$

and quite close to the $\pi N$ scattering determination

$$f = 1.01 \pm 0.01.$$

Unlike the SU(6) treatment, so-called central masses of meson and baryon multiplets play no role in this discussion. Concerning $\pi NN^*$ coupling we shall only remark that, if the width of $N^*$ is compared with a rough calculation that neglects recoil effects, the corresponding value of $f \simeq 1.1$.

Electromagnetic properties are introduced by the following substitutions, which are designed to reproduce the charges of the baryons[8]:

$$\rho_\mu^{(0)} \to (e/g)A_\mu, \quad \omega_\mu \to \tfrac{1}{3}(e/g)(m_\rho/m_\omega)A_\mu.$$

The implied magnetic field coupling predicts the total moments

$$\mu_p = \frac{5}{3}\frac{e}{2m_\rho} + \frac{1}{3}\frac{e}{2m_\omega}, \quad \mu_n = -\frac{5}{3}\frac{e}{2m_\rho} + \frac{1}{3}\frac{e}{2m_\omega}.$$

If the mass difference between $\omega$ and $\rho^0$ is ignored, we recover the famous ratio $-\mu_p/\mu_n = \frac{3}{2}$. An assumed mass difference of 25 MeV gives the ratio 1.48. The experimental value is 1.46. The absolute moment predictions are about 15% too small.

The extension of the internal unitary space to three dimensions introduces the other members of the baryon multiplets. We keep our U(4) notation and write successively, $\Psi_{A_1A_2,\sigma 3}$, $\Psi_{A,\sigma 3,\sigma' 3}$, and $\Psi_{\sigma 3,\sigma' 3,\sigma'' 3}$. SU(3) dynamical symmetry is not assumed. Nevertheless, the coupling of $M$ to the various U(4) indices must be universal to reproduce the isospin dependence of electric charge. Consequently, the $\pi$-baryon couplings are just those that appear in the literature parametrized as $d/f = \alpha/(1-\alpha) = \frac{3}{2}$.

The even-parity $\Phi$ field components are assembled in the $2\times 2$ matrix

$$M_\Phi = \vec{H}_\Phi\cdot\vec{\sigma} - m_\Phi \Phi^0 1.$$

A corresponding U(2) partial-symmetry hypothesis gives the typical meson-baryon coupling term

$$(g/2m_\rho)\Psi^\dagger\Big(\sum_{\alpha=1} M_\alpha\Big)\Psi + (g_\Phi/2m_\Phi)\Psi^\dagger M_{\Phi 3}\Psi,$$

which here refers to a $\Psi$ with a single 3 index, representing the particles with $Y=0$. The hypercharge component of electric charge is reproduced with the substitution

$$\Phi_\mu \to -\tfrac{2}{3}(e/g_\Phi)A_\mu.$$

Some magnetic moments obtained in this way are

$$\mu_\Lambda = -\frac{2}{3}\frac{e}{2m_\Phi},$$

$$\mu_{\Sigma^\pm} = \pm\frac{4}{3}\frac{e}{2m_\rho} + \mu_{\Sigma^0}, \quad \mu_{\Sigma^0} = \frac{4}{9}\frac{e}{2m_\omega} + \frac{2}{9}\frac{e}{2m_\Phi}.$$

The ratios

$$\frac{\mu_\Lambda}{\mu_n} = \frac{(2/m_\Phi)}{(5/m_\rho)-(1/m_\omega)},$$

$$\frac{\mu_{\Sigma^+}}{\mu_p} = \frac{(4/m_\rho)+\frac{4}{3}(1/m_\omega)+\frac{2}{3}(1/m_\Phi)}{(5/m_\rho)+(1/m_\omega)}$$

predict $\mu_\Lambda = -0.706$ and $\mu_{\Sigma^+} = 2.71$, in nucleon magnetons. The current experimental values are $-0.73 \pm 0.16$ and $2.3 \pm 0.6$, respectively.

Finally, here are some remarks about the pion coupling of the unit-spin mesons. Without attempting to fully implement an invariant structure, we recognize in $\mathcal{L}_\rho + \mathcal{L}_\omega$, $m_\omega \cong m_\rho$, the odd-parity field combinations that are arranged in the matrices

$$\rho_k\cdot\sigma_k\tau + \omega_k\sigma_k, \quad \rho_k^0\cdot\sigma_k\tau + \omega_k^0\sigma_k.$$

Now consider the $\rho^0$ and $\vec{\pi}$ coupling terms in

$$(g/2m_\rho)\tfrac{1}{4}i\,\mathrm{tr}\,\{(\rho_k^0\cdot\sigma_k\tau+\omega_k^0\sigma_k)[M,\rho_l\cdot\sigma_l\tau+\omega_l\sigma_l]\} = g\rho_k^0\cdot\rho^0\times\rho_k + (g/m_\rho)\epsilon_{klm}(\omega_k^0\rho_l+\rho_k^0\omega_l)\cdot\pi_m.$$

The relativistic form of the latter is [$^*\omega_{12} = \omega_{03}$, etc]

$$\mathcal{L}_{\omega\rho\pi} = (g/m_\rho)(^*\omega^{\mu\nu}\rho_\nu + {}^*\rho^{\mu\nu}\omega_\nu)\cdot\pi_\mu .$$

We present two electromagnetic tests of this prediction. The substitution $\rho_\nu^{(0)} \to (e/g)A_\nu$ gives, effectively,

$$\mathcal{L}_{\omega\gamma\pi} = -(e/m_\rho)^*\omega^{\mu\nu}F_{\mu\nu}\pi^{(0)},$$

and the width anticipated for the decay $\omega\to\gamma+\pi$ is 0.96 MeV. The present experimental value is $1.2\pm 0.2$ MeV. The further electromagnetic substitution, $^*\omega^{\mu\nu} \to \frac{1}{3}(e/g)(m_\rho/m_\omega)^*F^{\mu\nu}$, produces

$$\mathcal{L}_{\gamma\gamma\pi} = -\tfrac{1}{3}(e^2/g)(1/m_\omega)^*F^{\mu\nu}F_{\mu\nu}\pi^{(0)}.$$

The width predicted for $\pi^0\to\gamma+\gamma$ is 7.4 eV. The measured value is $7.4\pm 1.5$ eV.

---

*Publication assisted by the U. S. Air Force Office of Scientific Research.

[1]Julian Schwinger, to be published.

[2]Murray Gell-Mann, Physics 1, 63 (1964).

[3]J. Schwinger, Phys. Rev. 130, 406 (1963). There is a fuller discussion in my contribution to Theoretical Physics (International Atomic Energy Agency, Vienna, 1963). Cf. Secs. 6 and 8. It seems that this point has had to be rediscovered: M. Veltman, Phys. Rev. Letters 17, 553 (1966); M. Nauenberg, Phys. Rev. 154, 1455 (1967).

[4]Except in a rest frame of the particles. But such a rest frame is useless for the discussion of $\gamma_5$ transformations.

[5]To completely realize $U(4)\otimes U(4)$ transformations, other field components are required which would make asertions about $\eta$ and $\eta^*$ couplings, for example, but we shall not write them out.

[6]It suffices to consider the proton state with $S_z=\frac{1}{2}$ and to represent it unsymmetrically by $\Psi_{\sigma_1+,\sigma_2+,\sigma_3-}$, where the first two spins are in a triplet state. Then $\langle\sum_\alpha \sigma_\alpha{}^z\tau_\alpha{}^3\rangle = (\vec{S}_{12}-\vec{S}_3)\cdot\vec{S}/\frac{3}{4} = 5/3$, since $\vec{S}_{12}\cdot\vec{S}=1$ and $\vec{S}_3\cdot\vec{S}=-\frac{1}{4}$.

[7]The classic SU(6) value for $-G_A/G_V$ is 5/3. The same result is obtained from current-commutator matrix elements that are restricted to the baryon octuplet and decuplet. This has produced much discussion of more elaborate mixing schemes. The new situation seems to reflect the different physical interpretation given to chiral transformations. Such a transformation on $N$ generates $N+\pi$, not $N$ and $N^*$ with orbital motion.

[8]Whenever confusion of indices may occur, (0) is used to designate neutral particles.

# Gauge Fields, Sources, and Electromagnetic Masses

JULIAN SCHWINGER*
*Harvard University, Cambridge, Massachusetts*
(Received 17 July 1967)

The hypothesis of strong-interaction gauge fields, with non-Abelian gauge invariance broken only by the $1^-$ particle mass terms, gives a natural source theory setting for the introduction of electromagnetic effects. The electromagnetic potential vector appears as a compensating field in the mass terms of the neutral $1^-$ particles. The resulting electromagnetic self-action is used to discuss mass displacements. The pion electromagnetic mass is computed in a number of ways—by direct calculation of various processes and by chiral methods, in two variants. The relationship of these approaches is established. A phenomenological modification of the chiral evaluation gives perfect agreement with the observed value. It is found, however, that the $(m_\pi/m_\rho)^2$ terms, which are neglected in this method, are not very small. Baryon electromagnetic mass splittings are described by a simple adaptation of gross mass-spectrum empirics. Agreement with the data is excellent.

THE great utility of phenomenological gauge-field descriptions of the $1^-$ and $1^+$ mesons in connection with unitary and chiral transformations has become evident recently.[1] This raises again the question of the relative role of the photon gauge field. Previous discussions have used the language of quantum field theory.[2] The phenomenological orientation associated with the source concept[3] has produced a new situation, however. We shall give a simple solution of the problem. With its aid, we discuss in some detail the electromagnetic contributions to the mass difference between charged and neutral pions. There is also a brief treatment of baryon electromagnetic mass splittings.

## GAUGE INVARIANCE I

Let us consider the isotopic spin gauge field associated with $\rho$. An illustrative phenomenological Lagrange function is that of the $\rho+\pi$ system,

$$\mathcal{L}=-\tfrac{1}{2}(D_\mu\pi)^2-\tfrac{1}{2}m_\pi^2\pi^2-\tfrac{1}{4}(\rho_{\mu\nu})^2-\tfrac{1}{2}m_\rho^2(\rho_\mu)^2,$$

where

$$D_\mu=\partial_\mu+g\rho_\mu\times=\partial_\mu-igt\cdot\rho_\mu$$

and

$$\rho_{\mu\nu}=\partial_\mu\rho_\nu-\partial_\nu\rho_\mu+g\rho_\mu\times\rho_\nu.$$

This Lagrange function is invariant under the infinitesimal isotopic gauge transformation

$$\delta\pi=-\delta\omega\times\pi,$$
$$\delta\rho_\mu=-\delta\omega\times\rho_\mu+(1/g)\partial_\mu\delta\omega,$$

with the exception of the $\rho$ mass term,

$$\delta\mathcal{L}=-(m_\rho^2/g)\rho^\mu\cdot\partial_\mu\delta\omega.$$

This implies, incidentally, that

$$\partial_\mu\rho^\mu=0.$$

When $\delta\omega$ is directed along the third isotopic spin axis, the gauge transformation of $\rho_{\mu3}$ is simply

$$\delta g\rho_{\mu3}=\partial_\mu\delta\omega.$$

We now recognize the possibility of realizing complete invariance under this transformation, through the compensating effect of the electromagnetic gauge transformation

$$\delta e_0A_\mu=\partial_\mu\delta\omega,$$

provided the $\rho$ mass term is generalized to

$$-\tfrac{1}{2}m_\rho^2(\rho_{\mu1,2})^2-\tfrac{1}{2}m_\rho^2[\rho_{\mu3}-(e_0/g)A_\mu]^2.$$

It is an old idea to couple the photon directly to the unit spin mesons, as in $m_\rho^2(e_0/g)\rho^\mu{}_3A_\mu$, but questions of gauge invariance have produced uneasiness with this recipe[4] [rightly so, since the $(A_\mu)^2$ term is omitted]. In contrast, gauge invariance is our guide.

## THE PARTICLES

The direct coupling between $\rho^\mu{}_3$ and $A^\mu$ means that these are not the fields associated with the particles $\rho^0$ and photon ($\gamma$). The relevant part of the Lagrange function is

$$\mathcal{L}_{\gamma\rho^0}=-\tfrac{1}{4}(F_{\mu\nu})^2-\tfrac{1}{4}(\rho_{\mu\nu3})^2-\tfrac{1}{4}m_\rho^2[\rho_{\mu3}-(e_0/g)A_\mu]^2,$$

with

$$F_{\mu\nu}=\partial_\mu A_\nu-\partial_\nu A_\mu,\quad \rho_{\mu\nu3}=\partial_\mu\rho_{\nu3}-\partial_\nu\rho_{\mu3}.$$

The diagonalization transformation,[5]

$$\rho_{\mu3}=[1+(e_0/g)^2]^{-1/2}[\rho_\mu{}^{(0)}+(e_0/g)\gamma_\mu],$$
$$A_\mu=[1+(e_0/g)^2]^{-1/2}[\gamma_\mu-(e_0/g)\rho_\mu{}^{(0)}],$$

---

* Supported in part by the Air Force Office of Scientific Research.
[1] (a) S. Weinberg, Phys. Rev. Letters **18**, 188 (1967); (b) J. Schwinger, Phys. Letters **24B**, 473 (1967).
[2] J. Schwinger, Rev. Mod. Phys. **36**, 609 (1964).
[3] J. Schwinger, Phys. Rev. **152**, 1219 (1966); **158**, 1391 (1967).
[4] For a recent quantum field theory discussion, see N. Kroll, T. D. Lee, and B. Zumino, Phys. Rev. **157**, 1376 (1967); also T. D. Lee, S. Weinberg, and B. Zumino, Phys. Rev. Letters **18**, 1029 (1967).
[5] These transformations also appear in Ref. 2. Something similar occurs in Appendix B of the first paper mentioned in Ref. 4. I became aware of this Appendix only after the present paper was written, when the published article was distributed. Although it does not have the background of the non-Abelian gauge treatment of $\rho$, one can recognize in Eq. (B8) the counterpart of our mass-term prescription.



Reprinted from *The Physical Review* **165**, 1714–1721 (1968).

gives

$$\mathcal{L}_{\gamma\rho^0}=-\tfrac{1}{4}(\gamma_{\mu\nu})^2-\tfrac{1}{4}(\rho^{(0)}{}_{\mu\nu})^2-\tfrac{1}{2}m_{\rho^0}{}^2(\rho^{(0)}{}_\mu)^2,$$
$$m_{\rho^0}{}^2=[1+(e_0/g)^2]m_\rho{}^2.$$

If we include a photon-source term, written as $e_0J^\mu A_\mu$, the latter becomes

$$e_0J^\mu A_\mu=eJ^\mu[\gamma_\mu-(e_0/g)\rho_\mu{}^{(0)}],$$

where

$$e=e_0[1+(e_0/g)^2]^{-1/2}$$

is identified as the physical unit of charge. This is also evident from the relation

$$g\rho_{\mu3}=e\gamma_\mu+(g^2-e^2)^{1/2}\rho_\mu{}^{(0)}.$$

To *first* order in $e$, and in terms of the physical $\rho$ field $(\rho_{\mu1,2},\rho_\mu{}^{(0)})$, we have

$$D_\mu=\partial_\mu-igt\cdot\rho_\mu-iet_3\gamma_\mu,$$

which exhibits $et_3$ as the absolute charge matrix of unit isotopic spin.

The simple diagonalization transformation gives a first indication of $\rho$-meson electromagnetic mass splittings

$$m_{\rho^0}{}^2=[1-(e/g)^2]^{-1}m_\rho{}^2,$$

while maintaining the masslessness of the photon. It is interesting to apply perturbation methods to these questions. We do not distinguish here between $e_0$ and $e$. The coupling term $(em_\rho{}^2/g)\rho^\mu{}_3A_\mu$ displays $(em_\rho{}^2/g)\rho^\mu{}_3$ as an effective photon source. The self-coupling of that source supplies an additional term in the action $w$:

$$\delta w_\rho=\tfrac{1}{2}(em_\rho{}^2/g)^2\int(dx)(dx')\rho^\mu{}_3(x)D_+(x-x')\rho_{\mu3}(x').$$

The insertion of the unperturbed $\rho$ field gives

$$\int(dx')D_+(x-x')\rho_{\mu3}(x')=-(1/m_\rho{}^2)\rho_{\mu3}(x),$$

which implies the mass displacement

$$\delta m_{\rho^0}{}^2\simeq(e/g)^2m_\rho{}^2.$$

The same coupling term displays $(em_\rho{}^2/g)A_\mu$ as an effective $\rho^0$ source. That produces the additional action term

$$\delta w_A=\tfrac{1}{2}(em_\rho{}^2/g)^2\int(dx)(dx')A^\mu(x)\left(g_{\mu\nu}-\frac{1}{m_\rho{}^2}\partial_\mu\partial_\nu\right)\times\Delta_+(x-x')_\rho A^\nu(x').$$

It is not gauge-invariant. But the gauge variance is cancelled by that of the quadratic $A$ term in $\mathcal{L}$, $-\tfrac{1}{2}(em_\rho/g)^2(A_\mu)^2$. The use of the Lorentz gauge, $\partial_\mu A^\mu=0$, and reference to the unperturbed photon field,

$$\int(dx')\Delta_+(x-x')_\rho A^\nu(x')=(1/m_\rho{}^2)A^\nu(x),$$

results in complete cancellation of the quadratic $A$ terms.

The electromagnetic vector potential can also be eliminated, in the manner of the perturbation discussion, but without approximation. The electromagnetic field equations are

$$-\partial^2A_\mu=(m_\rho{}^2e_0/g)[\rho_{\mu3}-(e_0/g)A_\mu].$$

They are solved by

$$A_\mu(x)=(m_\rho{}^2e_0/g)\int(dx')D_+'(x-x')\rho_{\mu3}(x'),$$

where $D_+'$ is the propagation function associated with the mass $(e_0/g)m_\rho$. [Reader: Do not too hastily conclude that this is a predicted photon mass.] The electromagnetic terms in $w$ are

$$w_e=\int(dx)\{-\tfrac{1}{2}(\partial_\mu A_\nu)^2-\tfrac{1}{2}m_\rho{}^2[-2(e_0/g)\rho^\mu{}_3A_\mu+(e_0/g)^2(A_\mu)^2]\}$$
$$=\tfrac{1}{2}(m_\rho{}^2e_0/g)^2\int(dx)(dx')\rho^\mu{}_3(x)D_+'(x-x')\rho_{\mu3}(x').$$

The implied field equations for $\rho^\mu{}_3$, referring to noninteracting particles, are then

$$-\partial^2\rho^\mu{}_3(x)+\partial^\mu\partial_\nu\rho^\nu{}_3(x)+m_\rho{}^2\rho^\mu{}_3(x)=(m_\rho{}^2e_0/g)^2\int(dx')D_+'(x-x')\rho^\mu{}_3(x').$$

The condition for natural oscillations with unit spin is

$$p^2+m_\rho{}^2=(m_\rho{}^2e_0/g)^2[p^2+(e_0/g)^2m_\rho{}^2]^{-1}$$

or

$$p^2\{p^2+[1+(e_0/g)^2]m_\rho{}^2\}=0.$$

Here again are the masses of the photon and of $\rho^0$. Both are in evidence since $\rho^\mu{}_3$ is a mixture of the fields of the two particles. If we approximate $D_+'$ by $D_+$ we lose the ability to describe the photon and must restrict application to $\rho^0$.

Let us use the latter framework to construct Green's function for $\rho^0$. The momentum form of the field equations, including a source term, is

$$[p^2+m_\rho{}^2-([e_0/g]m_\rho{}^2)^2(1/p^2)]\rho^\mu{}_3(p)-p^\mu p_\nu\rho^\nu{}_3(p)=J^\mu{}_3(p).$$

This differs from the usual version only through the appearance of an effective, momentum-dependent mass. Accordingly,

$$G_{\mu\nu}(p)_{\rho^0}=[p^2+m_\rho{}^2-([e_0/g]m_\rho{}^2)^2(1/p^2)]^{-1}\times\{g_{\mu\nu}+[m_\rho{}^2-([e_0/g]m_\rho{}^2)^2(1/p^2)]^{-1}p_\mu p_\nu\}.$$

We are particularly interested in the charge-dependent part of this function. It is

$$G_{\mu\nu}(p)_{\rho}{}' - G_{\mu\nu}(p)_{\rho} = \delta G_{\mu\nu}(p)_{\rho}$$

$$\simeq \left(\frac{e}{g}\right)^2 \frac{1}{p^2}\left(\frac{m_\rho^2}{p^2+m_\rho^2}\right)^2\left[g_{\mu\nu}+\frac{p_\mu p_\nu}{m_\rho^2}\right] + \left(\frac{e}{g}\right)^2 \frac{1}{p^2}\frac{1}{p^2+m_\rho^2} p_\mu p_\nu .$$

This can also be obtained through the perturbation solution

$$\delta G_{\mu\nu}(p)_\rho = G_\mu{}^\lambda(p)_\rho[(e/g)m_\rho^2]^2(1/p^2)G_{\lambda\nu}(p)_\rho$$

$$= \left(\frac{e}{g}m_\rho^2\right)^2 \frac{1}{p^2}\frac{1}{(p^2+m_\rho^2)^2} \times\left[g_{\mu\nu}+\frac{2}{m_\rho^2}p_\mu p_\nu + \frac{p^2}{m_\rho^2}\frac{1}{m_\rho^2}p_\mu p_\nu\right].$$

To the extent that $\delta G_{\mu\nu}(p)_\rho$ enters in gauge-invariant combinations, it is equivalent to

$$g^2\delta G_{\mu\nu}(p)_\rho = g_{\mu\nu}(e^2/p^2)[m_\rho^2/(p^2+m_\rho^2)]^2,$$

which can be interpreted as photon exchange, modified by the $\rho$ form factor, $m_\rho^2/(p^2+m_\rho^2)$, at both emission and absorption.

## SOURCE THEORY

Before discussing electromagnetic masses, it is necessary to recognize the relation provided by source theory[3] between processes involving two real particles and processes that refer to a virtual particle. The vacuum amplitude that represents an arbitrary number of noninteracting particles, of any type, is of the form

$$\langle 0_+|0_-\rangle^S = \exp\left[i\tfrac{1}{2}\int (dx)(dx')S(x)G_+(x-x')S(x')\right].$$

If two partial sources $S_1$ and $S_2$ are made explicit, their multiplicative contribution appears as

$$\exp\left[i\int (dx)(dx')S_1(x)G_+(x-x')S_2(x') + i\int (dx)S_1(x)\chi(x) + i\int (dx)\chi(x)S_2(x)\right],$$

where $\chi(x)$ is the field

$$\chi(x) = \int (dx)G_+(x-x')S(x').$$

The vacuum amplitude terms that are bilinear in $S_1$ and $S_2$ are combined in

$$-\int (dx)(dx')S_1(x)[\chi(x)\chi(x') - iG_+(x-x')]S_2(x').$$

Accordingly, the two types of processes are related by the correspondence

$$\chi(x)\chi(x') \to -iG_+(x-x').$$

## $\pi$ ELECTROMAGNETIC MASS

The electromagnetic modification of the $\rho^0$ propagation function introduces new processes associated with $\rho$ exchange. They are superimposed upon the phenomenological framework which already incorporates the physics of strong interactions. This implies changes of phenomenological parameters, including particle masses. The simplest example is the displacement of the charged pion mass, arising from the $\rho$-$\pi$ coupling mechanism. The relevant Lagrangian terms are

$$\mathcal{L}_{\pi\rho} = g\rho^\mu(\partial_\mu\pi\times\phi - \pi\times\partial_\mu\phi) - \tfrac{1}{2}g^2(\rho_\mu\times\phi)^2,$$

in which we have distinguished between the $\pi$ field of interest, designated as $\phi$, and the $\pi$ field that will describe an exchanged particle. We shall also use simplifications associated with the small (mass)$^2$ ratio $(m_\pi/m_\rho)^2 \simeq 1/30$ and therefore discard the $\partial_\mu\phi$ term. It would contribute as $(\partial_\mu\phi)^2 \to -m_\pi^2\phi^2$.

The electromagnetically induced $\rho$-exchange contribution of the first term in $\mathcal{L}_{\pi\rho}$ is

$$\delta w_\pi^{(1)} = \tfrac{1}{2}g^2\int (dx)(dx')(\partial_\mu\pi\times\phi)_3(x)\delta G^{\mu\nu}(x-x')_\rho \times(\partial_\nu\pi\times\phi)_3(x') \to \tfrac{1}{2}g^2(1/i)\int (dx)(dx')\phi(x)\cdot\phi(x') \times\delta G^{\mu\nu}(x-x')_\rho\partial_\mu\partial_\nu'\Delta_+(x-x')_\pi,$$

where the second step introduces a virtual pion. In the latter form, $\phi$ refers to charged pions only. On adding the direct contribution of the second term in $\mathcal{L}_{\pi\rho}$, while continuing to regard $(m_\pi/m_\rho)^2$ as very small, we get

$$\delta w_\pi' = -\tfrac{1}{2}g^2\int dx(\phi(x))^2(1/i) \times\int \frac{(dp)}{(2\pi)^4}\delta G^{\mu\nu}(p)_\rho\left[g_{\mu\nu}-\frac{p_\mu p_\nu}{p^2}\right].$$

This identifies the $(\text{mass})^2$ displacement contribution $(m_{\pi^\pm}^2 - m_{\pi^0}^2)$:

$$\delta m_\pi^{2\prime} = (1/i)\int \frac{(dp)}{(2\pi)^4} g^2 \delta G^{\mu\nu}(p)_\rho \left[g_{\mu\nu} - \frac{p_\mu p_\nu}{p^2}\right]$$

$$= 3e^2(1/i)\int \frac{(dp)}{(2\pi)^4}\frac{1}{p^2}\left(\frac{m_\rho^2}{p^2+m_\rho^2}\right)^2.$$

In Euclidean spherical coordinates,

$$(1/i)(dp) \to \pi^2 p^2 dp^2$$

and

$$\delta m_\pi^{2\prime} = (3\alpha/4\pi)m_\rho^2.$$

This contribution to $\delta m_\pi$ is

$$\delta m_\pi' = 3.6 \text{ MeV},$$

which is a substantial fraction ($\sim 80\%$) of the observed 4.6 MeV.

There has been much attention recently to the role of the $\pi$ meson in chiral transformations and the related significance of $A_1$(1080 MeV) as the axial partner of $\rho$. Chiral invariance is violated by the $\pi$ mass term. An independent approach to the pion electromagnetic mass is based on computing the additional violation of chiral invariance that is produced by

$$w_e = \tfrac{1}{2}(m_\rho^2 e/g)^2 \int (dx)(dx')\rho^\mu{}_3(x) D_+(x-x')\rho_{\mu 3}(x').$$

The response to infinitesimal homogeneous chiral transformations is given by

$$\delta\rho_\mu = -(2g/m_A)\delta\varphi \times (A_\mu),$$
$$\delta(A_\mu) = -(2g/m_A)\delta\varphi \times \rho_\mu,$$

where

$$m_A = 2^{1/2} m_\rho$$

and[6]

$$(A_\mu) = A_{1\mu} + (1/m_A)D_\mu \pi$$

combine the field of the $1^+$ $A_1$ particle with that of the $0^-$ pion. Accordingly,

$$\delta_\varphi w_e = (m_\rho^2 e/g)^2 (2g/m_A)\int (dx)(dx')$$
$$\times [(A^\mu)\times\delta\varphi]_3(x) D_+(x-x')\rho_{\mu 3}(x')$$

and

$$\delta_\varphi^2 w_e = 2e^2 m_\rho^2 \int (dx)(dx') D_+(x-x')\{[(A^\mu)\times\delta\varphi]_3(x)$$
$$\times [(A_\mu)\times\delta\varphi]_3(x') + [(\rho^\mu \times \delta\varphi)\times\delta\varphi]_3(x)\rho_{\mu 3}(x')\}.$$

[6] The notation differs slightly from the Physics Letter of Ref. 1(b), where $-P_\mu$ is used for $(A_\mu)$ and $A_{1\mu}$.

We emphasize again that we are now considering processes that are not contemplated in the strong interaction phenomenology. The consideration of a virtual $A_1$, $\pi$, or $\rho$ meson gives

$$\delta_\varphi^2 w_e = 2e^2 m_\rho^2 (\delta\varphi)^2 i \int (dx)(dx') D_+(x-x')$$
$$\times [G^\mu{}_\mu(x-x')_\rho - G^\mu{}_\mu(x-x')_{(A)}],$$

where $\delta\varphi$ here refers only to charged components, and meson Green's functions for any one isotopic component are used. This structure is identified as the chiral response of the additional charged pion term

$$\delta w_\pi = -e^2 m_\rho^2 \int (dx)(\phi(x))^2 (1/i)$$
$$\times \int \frac{(dp)}{(2\pi)^4} D_+(p)[G^\mu{}_\mu(p)_\rho - G^\mu{}_\mu(p)_{(A)}]$$

and gives

$$\delta m_\pi^2 = 2e^2 m_\rho^2 (1/i)\int \frac{(dp)}{(2\pi)^4}\frac{1}{p^2}[G^\mu{}_\mu(p)_\rho - G^\mu{}_\mu(p)_{(A)}].$$

Here

$$G_{\mu\nu}(p)_{(A)} = \frac{g_{\mu\nu} + (1/m_A^2)p_\mu p_\nu}{p^2+m_A^2} + \frac{(1/m_A^2)p_\mu p_\nu}{p^2+m_\pi^2},$$

and $(m_\pi^2 \ll m_\rho^2)$

$$G^\mu{}_\mu(p)_{(A)} = 3/(p^2+m_A^2) + 1/m_A^2.$$

That combines with

$$G^\mu{}_\mu(p)_\rho = 3/(p^2+m_\rho^2) + 1/m_\rho^2$$

to give $(m_A^2 = 2m_\rho^2)$

$$\delta m_\pi^2 = 6e^2 m_\rho^2 (1/i)\int \frac{(dp)}{(2\pi)^4}\frac{1}{p^2}\left[\frac{1}{p^2+m_\rho^2} - \frac{1}{p^2+m_A^2}\right]$$
$$= (3\alpha/2\pi)(\ln 2) m_\rho^2.$$

This prediction is[7]

$$\delta m_\pi = 5.0 \text{ MeV}.$$

There is another chirality-based derivation of this result that brings dynamics more into evidence. It uses $\delta_\varphi w_e$ rather than $\delta_\varphi^2 w_e$. We write

$$\delta_\varphi w_e = (m_\rho^2 e^2/g)\int (dx)(dx')[(\partial^\mu \pi + g\rho^\mu \times \phi)\times\delta\varphi]_3(x)$$
$$\times D_+(x-x')\rho_{\mu 3}(x') + (m_\rho^2 e/g)^2 (2g/m_A)$$
$$\times \int (dx)(dx')[A_1{}^\mu \times \delta\varphi]_3(x) D_+(x-x')\rho_{\mu 3}(x'),$$

[7] We have reproduced the result of a current-algebra derivation by T. Das, G. Guralnick, V. Mathur, F. Low, and J. Young, Phys. Rev. Letters 18, 759 (1967).

which exhibits several new processes induced by the electromagnetic interaction. Only one of these is self-contained in its contribution to a structure that can be recognized as the chiral response of a charged pion mass term. It is

$$-e^2m_\rho^2(1/i)\int(dx)(dx')\phi(x)\cdot\delta\varphi D_+(x-x')G^\mu{}_\mu(x-x')_\rho.$$

For the others, we use relevant strong-interaction couplings.[1b] They are

$$\mathcal{L}_{\text{int}}=g\rho^\mu\cdot\partial_\mu\pi\times\phi+(g/m_A)\tfrac{1}{2}A_1{}^{\mu\nu}\times\phi\cdot\rho_{\mu\nu}.$$

The $\rho$-$\pi$ coupling term gives the contribution

$$e^2m_\rho^2(1/i)\int(dx)\cdots(dx'')\phi(x)\cdot\delta\varphi D_+(x'-x'')\times G^{\mu\nu}(x-x'')_\rho\partial_\mu\partial_\nu{}'\Delta_+(x-x')_\pi,$$

which combines with the simple $\rho$-exchange contribution to produce the following partial evaluation of the (mass)$^2$ displacement:

$$e^2m_\rho^2(1/i)\int\frac{(dp)}{(2\pi)^4}\frac{1}{p^2}G^{\mu\nu}(p)_\rho\left[g_{\mu\nu}-\frac{p_\mu p_\nu}{p^2}\right]=3e^2m_\rho^2(1/i)\int\frac{(dp)}{(2\pi)^4}\frac{1}{p^2}\frac{1}{p^2+m_\rho^2}.$$

The complete result obtained by adding the effect of $A_1+\rho$ exchange is

$$\delta m_\pi^2=3e^2m_\rho^2(1/i)\int\frac{(dp)}{(2\pi)^4}\frac{1}{p^2+m_\rho^2}\left[\frac{1}{p^2}-\frac{1}{p^2+m_A^2}\right]=6e^2(m_\rho^2)^2(1/i)\int\frac{(dp)}{(2\pi)^4}\frac{1}{p^2}\frac{1}{p^2+m_\rho^2}\frac{1}{p^2+m_A^2},$$

which is equivalent to the first chiral evaluation. This is indirect evidence for the consistency of the dynamical scheme with chirality requirements.

The chirality calculations seem to suggest that the particle $A_1$ is essential to a pion electromagnetic mass calculation. And yet, a not unsatisfactory result was obtained from the $\pi+\rho$ system alone. We shall implement the idea that the consideration of $A_1$ is significant, but not fundamental, by exhibiting the physical processes that relate the two distinct calculational schemes.

We have already referred to the $A_1\rho\pi$ coupling, which arises from the following Lagrangian term ($\partial_\mu\phi$ contributions are omitted):

$$-\tfrac{1}{4}[A_{1\mu\nu}+(g/m_A)\rho_{\mu\nu}\times\phi]^2.$$

Thus, a more complete account of this interaction is

$$\mathcal{L}_{A_1\rho\pi}=(g'/m_A)\tfrac{1}{2}A_1{}^{\mu\nu}\times\phi\cdot\rho_{\mu\nu}-(g'/m_A)^2\tfrac{1}{4}(\rho_{\mu\nu}\times\phi)^2,$$

where we have written $g'$ rather than $g$ in order to incorporate form-factor effects that are not considered in the simple Lagrange function. Their presence is indicated by the observed $A_1$ width, which requires[1b]

$$g'\simeq g/1.2.$$

We now consider electromagnetic modifications in the exchange of $\rho+A_1$ and of $\rho$ alone, corresponding to the two terms of $\mathcal{L}_{A_1\rho\pi}$. This gives

$$\delta w_\pi''=(g'/2m_A{}^2)(1/2i)\int(dx)(dx')\phi(x)\cdot\phi(x')\times\delta G_{\mu\nu,\lambda\kappa}(x-x')_\rho G_{\mu\nu,\lambda\kappa}(x-x')_{A_1}-(g'/2m_A)^2(1/i)\times\int(dx)(\phi(x))^2\delta G_{\mu\nu,\mu\nu}(0)_\rho$$

$$\cong(g'/2m_A)^2\int(dx)(\phi(x))^2(1/i)\int\frac{(dp)}{(2\pi)^4}\times[\tfrac{1}{2}\delta G_{\mu\nu,\lambda\kappa}(p)_\rho G_{\mu\nu,\lambda\kappa}(p)_{A_1}-\delta G_{\mu\nu,\mu\nu}(p)_\rho].$$

Here again $\phi(x)$ refers to charged $\pi$ mesons, in the approximation $m_\pi^2\ll m_\rho^2$. Also

$$G_{\mu\nu,\lambda\kappa}(p)_{A_1}=f_{\mu\nu,\lambda\kappa}[1/(p^2+m_A{}^2)]$$

and

$$\delta G_{\mu\nu,\lambda\kappa}(p)_\rho=f_{\mu\nu,\lambda\kappa}(e/g)^2(1/p^2)[m_\rho^2/(p^2+m_\rho^2)]^2,$$

where

$$f_{\mu\nu,\lambda\kappa}=p_\mu p_\lambda g_{\nu\kappa}-p_\nu p_\lambda g_{\mu\kappa}+p_\nu p_\kappa g_{\mu\lambda}-p_\mu p_\kappa g_{\nu\lambda}.$$

We note that

$$\tfrac{1}{4}(f_{\mu\nu,\lambda\kappa})^2=3(p^2)^2$$

and

$$\tfrac{1}{2}f_{\mu\nu,\mu\nu}=3p^2.$$

Hence, this contribution to $\delta m_\pi^2$ is

$$\delta m_\pi^{2\prime\prime}=3(g'/m_A)^2(e/g)^2(1/i)\times\int\frac{(dp)}{(2\pi)^4}\left(\frac{m_\rho^2}{p^2+m_\rho^2}\right)^2\left[1-\frac{p^2}{p^2+m_A{}^2}\right]=3(g'/g)^2e^2(1/i)\int\frac{(dp)}{(2\pi)^4}\left(\frac{m_\rho^2}{p^2+m_\rho^2}\right)^2\frac{1}{p^2+m_A{}^2}.$$

Indeed, if $g'=g$, the two physical contributions to $\delta m_\pi^2$ combine, according to

$$\frac{1}{(p^2+m_\rho^2)^2}\left[\frac{1}{p^2}+\frac{1}{p^2+m_A^2}\right]=2\frac{1}{p^2}\frac{1}{p^2+m_\rho^2}\frac{1}{p^2+m_A^2},$$

to reproduce the result of the chirality calculations. But if we retain the empirical distinction between $g'$ and $g$, the second contribution is

$$\delta m_\pi^{2\prime\prime}=(3\alpha/4\pi)(g'/g)^2(2\ln 2-1)m_\rho^2,$$

or

$$\delta m_\pi''=1.4/(1.2)^2=1.0\text{ MeV}.$$

Now the predicted value is

$$\delta m_\pi=3.6+1.0=4.6\text{ MeV},$$

which is embarrassingly accurate.

The deceptiveness of this agreement is emphasized by examining the heretofore ignored $(m_\pi/m_\rho)^2$ corrections. While that is outside the framework of the chirality methods, it is a straightforward calculational question in our dynamical approach. We consider the principal mechanism of $\rho+\pi$ exchange. It is easily seen that $\delta m_\pi^{2\prime}$ is modified into

$$\delta m_\pi^{2\prime}=(1/i)\int\frac{(dp)}{(2\pi)^4}g^2\delta G^{\mu\nu}(p)_\rho \times\left[g_{\mu\nu}-\frac{(p-2P)_\mu(p-2P)_\nu}{(p-P)^2+m_\pi^2}\right],$$

where

$$P^2+m_\pi^2=0.$$

The longitudinal terms in $\delta G^{\mu\nu}(p)_\rho$ still give no contribution, since

$$p^\mu p^\nu[g_{\mu\nu}-(p-2P)_\mu(p-2P)_\nu/(p^2-2pP)]=2p^\mu P_\mu$$

vanishes on integration. Accordingly,

$$\delta m_\pi^{2\prime}=e^2(1/i)\int\frac{(dp)}{(2\pi)^4}\frac{1}{p^2}\frac{1}{(p^2+m_\rho^2)^2} \times\left[3+\frac{2pP+4m_\pi^2}{p^2-2pP}\right],$$

where the factor in brackets can also be written as

$$3+(p^2+4m_\pi^2)[1/(p^2-2pP)]-1.$$

The result of averaging over the four-dimensional Euclidean angle between $p$ and $P$, with the aid of

$$\frac{2}{\pi}\int_0^\pi\sin^2\vartheta d\vartheta\frac{1}{1+t^2-2t\cos\vartheta}=1,$$

is given by

$$\left\langle\frac{1}{p^2-2pP}\right\rangle=\frac{1}{p^2}\frac{2}{(1+4m_\pi^2/p^2)^{1/2}+1}.$$

Therefore,

$$\delta m_\pi^{2\prime}=\frac{\alpha}{4\pi}\int_0\frac{dx}{(x+1)^2}\left[3+2\frac{1+\lambda x}{(1+\lambda x)^{1/2}+1}-1\right],$$

where

$$x=m_\rho^2/p^2,\quad\lambda=(2m_\pi/m_\rho)^2.$$

An asymptotic evaluation based on the smallness of $\lambda$ gives

$$\delta m_\pi^{2\prime}\cong(3\alpha/4\pi)m_\rho^2[1+(m_\pi^2/m_\rho^2)(\ln(m_\rho^2/m_\pi^2)+\tfrac{1}{2})].$$

Owing to the logarithmic factor, this is a 13% correction. It raises the computed value of $\delta m_\pi$ by somewhat less than $\frac{1}{2}$ MeV.

There are processes that contribute only to the $(m_\pi/m_\rho)^2$ corrections. The best established of these is the $\omega\rho\pi$ coupling[8]

$$(g/m_\rho)({}^*\omega^{\mu\nu}\rho_\nu+{}^*\rho^{\mu\nu}\omega_\nu)\cdot\partial_\mu\pi.$$

The electromagnetic modification of $\rho^0$ exchange gives the following action term for neutral pions:

$$2(g/m_\rho)^2\int(dx)(dx')\partial_\mu\phi_3(x){}^*\omega^{\mu\lambda}(x)\delta G_{\lambda\kappa}(x-x')_\rho \times\partial_\nu\phi_3(x'){}^*\omega^{\nu\kappa}(x').$$

There is no contribution from the longitudinal part of $\delta G_{\lambda\kappa}$, and the latter is effectively proportional to $g_{\lambda\kappa}$. The consideration of a virtual $\omega$ meson ($m_\omega\simeq m_\rho$) then implies, approximately, the neutral pion term

$$-3(em_\rho)^2\int(dx)(\partial_\mu\phi_3)^2(1/i)\int\frac{(dp)}{(2\pi)^4}\frac{1}{(p^2+m_\rho^2)^3} =-\frac{3\alpha}{4\pi}\frac{1}{2}\int(dx)(\partial_\mu\phi_3)^2.$$

A redefinition of the $\pi^0$ field, or the equivalence $(\partial_\mu\phi)^2\to -m_\pi^2\phi^2$, identifies this as producing a $\pi^0$ mass decrease. Thus $\delta m_\pi^2$ is further increased by $(3\alpha/4\pi)m_\pi^2$, which raises the discrepancy to 16%.

Another relevant mechanism is the $A_2\rho\pi$ coupling. But we shall terminate this discussion here, with the general remark that to achieve better than ~10% accuracy in computing the pion electromagnetic mass splitting seems to require detailed reference to fairly

[8] J. Schwinger, Phys. Rev. Letters **18**, 923 (1967).

high-energy phenomena. There may include multiple exchange modifications of the processes that give the bulk of the electromagnetic mass effect. A particularly interesting subset of the latter can be described as electromagnetic modification of $\rho^\pm$ exchange, with these modifications produced within the coupled $\rho$ system by the primary mechanism of electromagnetically modified $\rho^0$ exchange.

## GAUGE INVARIANCE II

The possibility of realizing electromagnetic gauge invariance by compensation between the electromagnetic potential $A_\mu$ and a neutral meson gauge field has been illustrated by the $T=1$ field $\rho_\mu$. We now include the $T=0$ fields $\omega_\mu$ and $\phi_\mu$. The only problem is the relative coupling strengths of the various mesons. For simplicity, we place $\omega$ entirely in the $U_2$ subspace of the three-dimensional unitary space. Then, following the suggestion[8] that mass is a significant factor, we consider the $3\times3$ matrix

$$\tfrac{1}{2}(t_{11}-t_{22})m_\rho\rho_{\mu3}+\tfrac{1}{2}(t_{11}+t_{22})m_\omega\omega_\mu+2^{-1/2}t_{33}m_\phi\phi_\mu,$$

where $t_{aa}$ has unit entry in the $a$th row and column and is zero elsewhere. The above matrix is such that the trace of its square equals

$$\tfrac{1}{2}[m_\rho^2(\rho_{\mu3})^2+m_\omega^2(\omega_\mu)^2+m_\phi^2(\phi_\mu)^2].$$

We assume this matrix to be the significant structure in coupling the vector fields to a given system, with the $U_3$ matrices $T_{aa}$ of that system replacing the elementary matrices $t_{aa}$. It is then required that the linear interaction of $\rho_\mu$, $\omega_\mu$, and $\phi_\mu$ with $A_\mu$ reproduce the coupling of $A_\mu$ to the electric charge, as represented by the $SU_3$ matrix

$$T_{11}-\tfrac{1}{3}(T_{11}+T_{22}+T_{33})=\tfrac{1}{2}(T_{11}-T_{22})+\tfrac{1}{6}(T_{11}+T_{22})-\tfrac{1}{3}T_{33}.$$

Using the normalization already established in the $\rho$ discussion, we infer the following gauge-invariant structure of the neutral gauge-field mass terms:

$$-\tfrac{1}{2}m_\rho^2[\rho_{\mu3}-(e_0/g)A_\mu]^2-\tfrac{1}{2}m_\omega^2[\omega_\mu-\tfrac{1}{3}(e_0/g)(m_\rho/m_\omega)A_\mu]^2-\tfrac{1}{2}m_\phi^2[\phi_\mu+\tfrac{1}{3}2^{1/2}(e_0/g)(m_\rho/m_\phi)A_\mu]^2.$$

We shall not discuss the new diagonalization problem or the altered relation between $e_0$ and $e$. Let us ignore that distinction and write the linear coupling between $A_\mu$ and the neutral mesons as

$$(m_\rho^2e/g)A^\mu V_\mu^{(0)},$$

with

$$V_\mu^{(0)}=\rho_{\mu3}+\tfrac{1}{3}(m_\omega/m_\rho)\omega_\mu-\tfrac{1}{3}2^{1/2}(m_\phi/m_\rho)\phi_\mu.$$

Viewed as a photon source, the self-couping of the latter is described by

$$w_e=\tfrac{1}{2}(m_\rho^2e/g)^2\int(dx)(dx')V^{(0)\mu}(x)D_+(x-x')V^{(0)}{}_\mu(x').$$

The following application of this electromagnetic term falls quite short of the quantitative ambition of the preceding section.

## BARYON ELECTROMAGNETIC MASSES

We give a brief and highly empirical discussion of baryon electromagnetic mass splittings. This is patterned after a recent observation[9] concerning the gross mass spectrum of the baryons. It was noted that

$$M=M_0(1+\vartheta H_3),$$

where $M_0$ varies from octuplet to decuplet but is fixed within each multiplet, while $\vartheta$ is a universal constant,[10]

$$\vartheta=0.119.$$

The quantum number $H_3$ is defined additively on the unitary indices in $\psi_{abc}$, with the basic values: $+1$, for a single 3 index; $-1$, for an antisymmetrical 12 pair; 0 otherwise. From the viewpoint of the three isotopic spins that characterize $SU_3$ symmetry, $T$(12 plane), $U$(23 plane), $V$(31 plane), the symmetry-breaking effect of $H_3$ is described by $\Delta T=0$, $\Delta U=1$, $\Delta V=1$, which is known[11] to produce Gell-Mann–Okubo mass relations. The $H_3$ mass formula also supplies specific connections among the arbitrary constants of such relations. If $\Lambda$ is omitted, one has the following correspondences with hypercharge:

$$8(N,\Sigma,\Xi):\quad H_3=1-\tfrac{3}{2}Y-\tfrac{1}{2}Y^2,$$
$$10:\quad H_3=1-Y.$$

The electromagnetic coupling term contains $\Delta T=2$ ($\rho\rho$) and $\Delta T=1$ ($\rho\omega,\rho\phi$) contributions. The square of the electric charge

$$Q=T_3+\tfrac{1}{2}Y=T_{11}-\tfrac{1}{3}\sum_{a=1}^{3}T_{aa}$$

also has that character. We propose to represent the $\Delta T=2$ electromagnetic term as a multiple of $Q^2$, and treat the residual $\Delta T=1$ component by analogy with $H_3$. Thus, we define $H_1$ additively in its action on $\psi_{abc}$, with the basic values: $+1$ for a single 1 index; $-1$ for an antisymmetrical 23 pair; 0, otherwise. For the dec-

[9] J. Schwinger, Phys. Rev. Letters **18**, 797 (1967).

[10] As a curiousity, we remark that $m=m_\rho(1+\tfrac{3}{2}\vartheta H_3)$, $m_\rho=755$ MeV, where $H_3$ is defined on $\psi_{ab}{}^*$, gives quite a good account of the masses of $K^*(H_3=1)$ and $\phi(H_3=2$, if it is assumed that $\varphi$ is very closely represented by $\psi_{33}{}^*)$.

[11] J. Schwinger, Phys. Rev. **136B**, 1821 (1964). There is a brief discussion here of the use of meson fields to represent electromagnetic and weak interactions.

uplet we have, simply,

$$10:\quad H_1=1+Q,$$

while

$$N:\quad H_1=1+Q,$$
$$\Sigma:\quad H_1=\tfrac{1}{2}+\tfrac{3}{2}Q,$$
$$\Xi:\quad H_1=1+2Q.$$

These are represented collectively by

$$8(N,\Sigma,\Xi):\quad H_1=\tfrac{1}{2}+\tfrac{1}{2}Y^2+(\tfrac{3}{2}-\tfrac{1}{2}Y)Q.$$

Our proposed formula for the electromagnetic splitting of an isotopic multiplet with central mass $M_0$ is

$$M=M_0[1-\epsilon(H_1-\langle H_1\rangle)]+\lambda(Q^2-\langle Q^2\rangle),$$

where $\langle\ \ \rangle$ refers to a multiplet average. We note that

$$8:\quad \langle H_1\rangle=1-\tfrac{1}{2}H_3,$$
$$10:\quad \langle H_1\rangle=\tfrac{3}{2}-\tfrac{1}{2}H_3.$$

The constants $\epsilon$ and $\lambda$, assumed to be universal, can be determined from the properties of the $\Sigma$ multiplet,[12]

$$\Sigma^- -\Sigma^0=4.88\pm0.06\text{ MeV},$$
$$\Sigma^- -\Sigma^+=7.97\pm0.11\text{ MeV}.$$

An acceptable fit is given by

$$3\epsilon M(\Sigma)=8.0\text{ MeV},$$
$$\lambda+4.0\text{ MeV}=4.8\text{ MeV},$$

or

$$\lambda=0.8\text{ MeV},\quad \epsilon=2.2(4)\times10^{-3}.$$

The nucleon splitting is now predicted:

$$N^0-N^+=-0.8+(8/3)(939/1192)=1.3\text{ MeV},$$

in agreement with the observed 1.29 MeV. For the cascade particle we anticipate

$$\Xi^- -\Xi^0=0.8+2(8/3)(1318/1192)=6.7\text{ MeV},$$

which is compatible with the measured 6.5±0.2 MeV. It is interesting that the Coleman-Glashow electromagnetic mass formula[13]

$$\Sigma^- -\Sigma^+=\Xi^- -\Xi^0+N^0-N^+$$

now appears as a consequence of the gross structure mass relation

$$3\Sigma=2\Xi+N$$

or

$$\Sigma-N=2(\Xi-\Sigma).$$

This connection is built into the $H_3$ mass formula, and is accurately obeyed.

The comparison with the present crude data on decuplet electromagnetic mass splittings is as follows:

$$\Xi^{*-}-\Xi^{*0}=0.8+(8/3)(1532/1192)=4.2\text{ MeV}$$
Exp: 4.9±2.2 MeV,

$$Y_1^{*-}-Y_1^{*+}=2(8/3)(1385/1192)=6.2\text{ MeV},$$
Exp: 5.8±3.0 MeV,

$$N^{*-}-N^{*++}=-3(0.8)+3(8/3)(1238/1192)=5.9\text{ MeV}$$
Exp: 7.9±6.8 MeV,

$$N^{*0}-N^{*++}=-4(0.8)+2(8/3)(1238/1192)=2.3\text{ MeV}$$
Exp: 0.5±0.9 MeV.

The agreement is satisfactory, with the possible exception of the last entry.

The $\Delta T=1$ component of electromagnetic mass splitting is compared in strength with the $SU_3$ symmetry-breaking mechanism by the ratio

$$\epsilon/\vartheta=1.9\times10^{-2}.$$

When one contrasts mass intervals, rather than the universal parameters, the ratio varies within a unitary multiplet. In the uniformly spaced decuplet, for example, the ratio of the $H_1$ coefficient to the $H_3$ coefficient changes from $1.9\times10^{-2}$ to $2.4\times10^{-2}$. An analogous comparison is produced within the $0^-$ meson spectrum by using the (mass)$^2$ difference $\pi^+-\pi^0$ to remove the $\Delta T=2$ part of $K^0-K^+$, giving $(K-\pi)^0-(K-\pi)^+$, and dividing this by the $SU_3$ (mass)$^2$ splitting, $K-\pi$,

$$[(K-\pi)^0-(K-\pi)^+]/(K-\pi)=2.3\times10^{-2}.$$

It would seem that the same mechanisms are at work to shape the meson mass spectrum.

---

[12] All experimental mass values are taken from A. Rosenfeld *et al.*, Rev. Mod. Phys. **39**, 1 (1967).

[13] S. Coleman and S. Glashow, Phys. Rev. Letters **6**, 423 (1961).

**Gauge Fields, Sources, and Electromagnetic Masses,** JULIAN SCHWINGER [Phys. Rev. **165**, 1714 (1968)]. The following *Note added in proof* was omitted in the printing:

(1) The discussions in the text are simplified by using the electromagnetic action term

$$w_e=\tfrac{1}{2}(m_\rho{}^2e/g)^2\int(dx)(dx')\rho_3{}^\mu(x)D_{\mu\nu}(x-x')\rho_3{}^\nu(x'),$$

where the property

$$\partial_\mu D^{\mu\nu}(x-x')=0$$

guarantees the conserved nature of the photon source described by $\rho_3{}^\mu$. The momentum form of $D_{\mu\nu}$ is

$$D_{\mu\nu}(p)=(p^2-i\epsilon)^{-1}(g_{\mu\nu}-p_\mu p_\nu/p^2).$$

As an example, we now have

$$g^2\delta G_{\mu\nu}(p)_\rho=e^2[m_\rho{}^2/(p^2+m_\rho{}^2)]^2D_{\mu\nu}(p),$$

while the chiral computation of $\delta m_\pi{}^2$ becomes

$$\delta m_\pi{}^2=2e^2m_\rho{}^2(1/i)\int\frac{(dp)}{(2\pi)^4}D_{\mu\nu}(p)[G^{\mu\nu}(p)_\rho-G^{\mu\nu}(p)_{(A)}],$$

which no longer requires the detailed compensation of longitudinal terms in the two Green's functions. Other applications of $D_{\mu\nu}$ occur in J. Schwinger, Phys. Rev. Letters **19**, 1154 (1967), which is later referred to as PMFF.

(2) As an alternative to introducing the empirical coupling constant $g'$ in $A_1\rho\pi$ coupling, one can add the coupling term

$$-\gamma(g/m_A)\rho^{\mu\nu}\cdot A_{1\mu}\times\partial_\nu\phi,$$

as mentioned in PMFF. This gives no effect in the limit $(m_\pi/m_\rho)^2=0$, and leaves the predicted mass splitting for that circumstance at 5.0 MeV.

(3) In examining finite $(m_\pi/m_\rho)^2$ corrections we considered $\rho+\pi$ exchange, but not $\rho+(A_1+\pi)$ exchange. Had we done so we would have found a logarithmically divergent result. This is discussed in PMFF from the view point of chiral transformations. The remedy proposed there is the recognition of a phenomenological form factor connecting $\rho^0$ and the photon, which is effective also in producing finite electromagnetic modifications of nuclear $\beta$ decay. There is a related proposal of T. D. Lee (to be published) who suggests that all photon couplings pass through an intermediate neutral meson. This is done, presumably, since Lee's field algebra method, with its unacknowledged ambivalence between phenomenological and fundamental descriptions, has no other way to introduce an effective photon form factor. The avowedly phenomenological source theory is not so limited and allows one to consider the more plausible possibility that the $\rho\gamma$ form factor describes the exchange of known particles, rather than a new one. A reexamination of the $\pi$ electromagnetic mass problem suggests that the significant masses in the form factor are $\sim 5m_\rho$ [Tung-Mow Yan (to be published)].

(4) General confirmation of the $\rho\gamma$ coupling description comes from the clashing beam experiments of Auslander *et al.* [Phys. Letters **25B**, 433 (1967)], who clearly see the intermediary $\rho^0$ meson in $e^++e^-(\rightarrow\gamma\rightarrow\rho^0\rightarrow)\pi^++\pi^-$. They express their results through the resonance factor

$$km_\rho{}^4/[(p^2+m_\rho{}^2)^2+m_\rho{}^2\Gamma^2],$$

$$k=0.59\pm0.15,\quad \Gamma=93\pm15\text{ MeV},$$

and also state the maximum cross section to be $1.2\pm0.2$ $\mu$b. The theoretical value of the latter is

$$\sigma=4\pi(1/137)^2(g^2/4\pi)^{-1}(m_\rho\Gamma)^{-1},$$

which assumes no relation between $g$ and $\Gamma$. If the width is fixed at 93 MeV we infer

$$g^2/4\pi=3.1\pm0.5,$$

to be compared with

$$g^2/4\pi=3.4\pm0.2,$$

which is deduced from the strong-interaction relation $2^{-1/2}(g/m_\rho)=f_0/m_\pi$ and the experimental $s$-wave $\pi N$ scattering parameter $f_0=0.84\pm0.03$. The effective coupling constant of $\rho\rightarrow\pi+\pi$ decay,

$$\hat{g}=(\tfrac{3}{4}+\tfrac{1}{4}\gamma)g$$

determines $\Gamma$ and $k=(\hat{g}/g)^2$. The $k$ measurements imply

$$\hat{g}/g=0.77\pm0.10,\quad \gamma=0.1\pm0.4,$$

and similar results are obtained from $\Gamma$. There is new experimental evidence for the $A_1$ particle [R. Juhala *et al.*, Phys Rev. Letters **19**, 1335 (1967)]. The measured width of $120\pm15$ MeV indicates that $\gamma=0.5\pm0.1$. A value of $\gamma\sim0.4$ seems compatible with the various experiments.

Reprinted from *The Physical Review* **167**, 1546 (1968).

# Sources and Magnetic Charge

JULIAN SCHWINGER
*Institute for Theoretical Physics,* State University of New York, Stony Brook, New York*
(Received 8 April 1968)

A beginning is made on a phenomenological reconstruction of the theory of magnetic charge. The concept is introduced by reference to a new kind of photon source. It is shown that photon exchange between different source types is relativistically invariant. The space-time generalization of this coupling involves an arbitrary vector. The only way to remove a corresponding arbitrariness of physical predictions is to recognize the localization of charge and impose a charge quantization condition. The consideration of particles that carry both kinds of charge loosens the charge restrictions. The great strength of magnetic attraction indicated by $g^2/4\pi = 4(137)$ suggests that ordinary matter is a magnetically neutral composite of magnetically charged particles that carry fractional electric charge. There is a brief discussion of such a magnetic model of strongly interacting particles, which makes contact with empirical classification schemes. Additional remarks on notation, and on the general nature of the source description, are appended.

THE concept of magnetic charge has great theoretical appeal, since it provides a beautiful explanation of the observed quantization of electric charge.[1] It has received some recent attention from the standpoint of operator quantum field theory.[2] Owing to the singular nature of operator field products, the resulting formalism, which makes liberal use of limiting processes, has a delicate and tentative aspect. This must diminish considerably the impact of the assertion that magnetic charge is a physical possibility, and certainly makes the operator theory ill-suited for quantitative application. It is our intention here to use the phenomenological and nonoperator approach of source theory[3] to provide a new foundation for the idea of magnetic charge, and to develop its implications sufficiently that one recognizes the existence of phenomenological charge quantization.

*Sources.* We first review the description of photon emission and absorption by sources, $J^\mu(x)$, which are introduced as idealizations of realistic mechanisms. Complete processes of emission and absorption are contained in the vacuum amplitude

$$\langle 0_+|0_-\rangle^J = \exp\left[i\frac{1}{2}\int (dx)(dx')J^\mu(x)D_+(x-x')J_\mu(x')\right], \quad (1)$$

where

$$D_+(x-x') = D_+(x'-x),$$

$$x^0 > x^{0\prime}:\quad D_+(x-x') = i\int d\omega_k e^{ik(x-x')}, \quad (2)$$

and

$$d\omega_k = \frac{(d\mathbf{k})}{(2\pi)^3}\frac{1}{2k^0}, \quad k^0 = |\mathbf{k}|. \quad (3)$$

The vectorial source must obey the conservation condition

$$\partial_\mu J^\mu(x) = 0. \quad (4)$$

The consideration of a causal arrangement, with emission source $J_2^\mu$ and detection source $J_1^\mu$, gives

$$\langle 0_+|0_-\rangle^J = \langle 0_+|0_-\rangle^{J_1} \times \exp\left[\int d\omega_k iJ_1^\mu(-k)g_{\mu\nu}iJ_2^\nu(k)\right]\langle 0_+|0_-\rangle^{J_2}, \quad (5)$$

where

$$J^\mu(k) = \int (dx)e^{-ikx}J^\mu(x) \quad (6)$$

and

$$k_\mu J^\mu(k) = 0. \quad (7)$$

Polarization vectors are introduced by writing

$$g^{\mu\nu} = \sum_{\lambda=1,2} e_{k\lambda}{}^\mu e_{k\lambda}{}^\nu + (k^\mu\bar{k}^\nu + \bar{k}^\mu k^\nu)/(k\bar{k}), \quad (8)$$

in which $\bar{k}^\mu$ is obtained from $k^\mu$ by reflecting the spatial components, in some coordinate frame. In that frame, the two $e_{k\lambda}{}^\mu$, $\lambda = 1,2$, are unit orthogonal spatial vectors that are perpendicular to the photon momentum $\mathbf{k}$. As a result

$$\int d\omega_k iJ_1^\mu(-k)g_{\mu\nu}J_2^\nu(k) = \int d\omega_k \sum_\lambda i\mathbf{J}_1(-k)\cdot\mathbf{e}_{k\lambda}\mathbf{e}_{k\lambda}\cdot i\mathbf{J}_2(k), \quad (9)$$

and the subsequent analysis into multiparticle states, through the causal decomposition

$$\langle 0_+|0_-\rangle^J = \sum_{\{n\}} \langle 0_+|\{n\}\rangle^{J_1}\langle\{n\}|0_-\rangle^{J_2}, \quad (10)$$

yields

$$\langle\{n\}|0_-\rangle^J = \langle 0_+|0_-\rangle^J \prod_{k\lambda} (iJ_{k\lambda})^{n_{k\lambda}}/(n_{k\lambda}!)^{1/2},$$

$$\langle 0_+|\{n\}\rangle^J = \langle 0_+|0_-\rangle^J \prod_{k\lambda} (iJ_{k\lambda}^*)^{n_{k\lambda}}/(n_{k\lambda}!)^{1/2}, \quad (11)$$

* Permanent address: Harvard University, Cambridge, Mass.
[1] P. A. M. Dirac, Proc. Roy. Soc. (London) **A133**, 60 (1931); Phys. Rev. **74**, 817 (1948).
[2] J. Schwinger, Phys. Rev. **144**, 1087 (1966).
[3] J. Schwinger, Phys. Rev. **152**, 1219 (1966); **158**, 1391 (1967).



Reprinted from *The Physical Review* 173, 1536–1544 (1968).

where

$$J_{k\lambda}=(d\omega_k)^{1/2}\mathbf{e}_{k\lambda}\cdot\mathbf{J}(k). \tag{12}$$

Given one pair of polarization vectors $\mathbf{e}_{k\lambda}$, another pair is produced by rotation about $\mathbf{k}$ through 90°, as expressed by

$$^*\mathbf{e}_{k\lambda}=(\mathbf{k}/k^0)\times\mathbf{e}_{k\lambda}. \tag{13}$$

Let us suppose that a different type of photon source exists, designated by $^*J^\mu(x)$, such that the effective component for the emission of the photon labeled $k\lambda$ is not along $\mathbf{e}_{k\lambda}$, but is in the perpendicular direction specified by $^*\mathbf{e}_{k\lambda}$:

$$^*J_{k\lambda}=(d\omega_k)^{1/2}\,{}^*\mathbf{e}_{k\lambda}\cdot{}^*\mathbf{J}(k). \tag{14}$$

The two types of sources are not intrinsically distinguishable through the act of photon exchange since

$$\sum_\lambda \mathbf{e}_{k\lambda}\mathbf{e}_{k\lambda}=\sum_\lambda {}^*\mathbf{e}_{k\lambda}\,{}^*\mathbf{e}_{k\lambda}, \tag{15}$$

but they can be contrasted by exchanging a photon between the two varieties. If we arrange that $^*J^\mu$ emits and $J^\mu$ absorbs, the vacuum amplitude is

$$\langle 0_+|0_-\rangle^{J,{}^*J}=\langle 0_+|0_-\rangle^J \times\exp\left[\int d\omega_k\sum_\lambda i\mathbf{J}(-k)\cdot\mathbf{e}_{k\lambda}\,{}^*\mathbf{e}_{k\lambda}\cdot i\,{}^*\mathbf{J}(k)\right] \times\langle 0_+|0_-\rangle^{{}^*J}, \tag{16}$$

where

$$\sum_\lambda \mathbf{e}_{k\lambda}\,{}^*\mathbf{e}_{k\lambda}\cdot{}^*\mathbf{J}(k)=\sum_\lambda \mathbf{e}_{k\lambda}\mathbf{e}_{k\lambda}\cdot{}^*\mathbf{J}(k)\times\mathbf{k}/k^0 ={}^*\mathbf{J}(k)\times\mathbf{k}/k^0, \tag{17}$$

since $^*\mathbf{J}\times\mathbf{k}$ only has components along the two directions represented by $\mathbf{e}_{k\lambda}$.

*Invariance.* It is now vital to recognize that, despite its three-dimensional appearance,

$$\mathbf{J}(-k)\times{}^*\mathbf{J}(k)\cdot\mathbf{k}/k^0 \tag{18}$$

is a Lorentz scalar, so that the description of photon-mediated coupling between different source types is Lorentz-invariant. This property is a consequence of the conserved nature of the sources,

$$\mathbf{k}\cdot\mathbf{J}(k)=k^0J^0(k),\quad \mathbf{k}\cdot{}^*\mathbf{J}(k)=k^0\,{}^*J^0(k), \tag{19}$$

which is an aspect of the masslessness of the photon,

$$k^0=|\mathbf{k}|. \tag{20}$$

We examine the response to the infinitesimal Lorentz transformation indicated by

$$\delta\mathbf{J}=\delta\mathbf{v}J^0=\delta\mathbf{v}(\mathbf{k}/k^0)\cdot\mathbf{J}, \tag{21}$$

which, with a similar equation for $^*\mathbf{J}$, gives

$$\delta[\mathbf{J}\times{}^*\mathbf{J}\cdot\mathbf{k}/k^0]=-[(\mathbf{k}/k^0)\times(\delta\mathbf{v}\times\mathbf{k}/k^0)]\cdot\mathbf{J}\times{}^*\mathbf{J} +\delta(\mathbf{k}/k^0)\cdot\mathbf{J}\times{}^*\mathbf{J}. \tag{22}$$

But, since

$$\delta\mathbf{k}=\delta\mathbf{v}k^0,\quad \delta k^0=\delta\mathbf{v}\cdot\mathbf{k}, \tag{23}$$

we have

$$\delta(\mathbf{k}/k^0)=\delta\mathbf{v}-(\mathbf{k}/k^0)\delta\mathbf{v}\cdot\mathbf{k}/k^0 =(\mathbf{k}/k^0)\times(\delta\mathbf{v}\times\mathbf{k}/k^0), \tag{24}$$

invoking explicitly the zero mass of the photon, and the statement is verified.

*Space-Time Description.* The logical connection between the photon and the long-range Coulomb interaction of static charges is introduced by performing a space-time extrapolation of the source coupling that is first established by considering causal arrangements involving propagating photons. Preparatory to carrying out such a space-time generalization in the present situation, we observe that

$$\mathbf{J}(-k)\times{}^*\mathbf{J}(k)\cdot\mathbf{k}/k^0=\epsilon^{\mu\nu\lambda\kappa}J_\mu(-k)f_\nu(k)ik_\lambda\,{}^*J_\kappa(k), \tag{25}$$

where $f_\nu(k)$ is *any* vector that obeys

$$ik^\nu f_\nu(k)=1 \tag{26}$$

and $\epsilon^{\mu\nu\lambda\kappa}$ is the totally antisymmetrical symbol normalized by

$$\epsilon^{0123}=+1. \tag{27}$$

The coupling under discussion can be regarded as operating between $J^\mu$ and the effective source

$$J^\mu(k)|_{\text{eff}}=\epsilon^{\mu\nu\lambda\kappa}f_\nu(k)ik_\lambda\,{}^*J_\kappa(k), \tag{28}$$

which correctly obeys

$$k_\mu J^\mu(k)|_{\text{eff}}=0. \tag{29}$$

The space-time description of this effective source is

$$J^\mu(x)|_{\text{eff}}=\epsilon^{\mu\nu\lambda\kappa}\int(dx')f_\nu(x-x')\partial_\lambda'\,{}^*J_\kappa(x'), \tag{30}$$

where

$$\partial_\nu f^\nu(x-x')=\delta(x-x'). \tag{31}$$

A space-time transcription of the vacuum amplitude is now at hand. We write it as

$$\langle 0_+|0_-\rangle^{J,{}^*J}=\exp[iW(J,{}^*J)], \tag{32}$$

with

$$W(J,{}^*J)=\frac{1}{2}\int(dx)(dx')J^\mu(x)D_+(x-x')J_\mu(x') +\frac{1}{2}\int(dx)(dx')\,{}^*J^\mu(x)D_+(x-x')\,{}^*J_\mu(x') +\int(dx)(dx')(dx'')J^\mu(x)\epsilon_{\mu\nu\lambda\kappa}f^\nu(x-x') \times D_+(x'-x'')\partial''^\lambda\,{}^*J^\kappa(x''), \tag{33}$$

which makes use of the equivalence

$$\int (dx')D_+(x-x')f^\nu(x'-x'') = \int (dx')f^\nu(x-x')D_+(x'-x''). \quad (34)$$

*Fields.* Auxiliary quantities—fields—are introduced as measures of the effects experienced by additional weak sources. This is conveyed by the differential expression

$$\delta W(J,{}^*J) = \int (dx)[\delta J^\mu(x)A_\mu(x)+\delta\,{}^*J^\mu(x)B_\mu(x)], \quad (35)$$

where

$$A_\mu(x) = \partial_\mu\lambda(x)+\int (dx')D_+(x-x')J_\mu(x') + \int (dx')(dx'')f^\nu(x-x')D_+(x'-x'') \times{}^*(\partial_\mu''\,{}^*J_\nu(x'')-\partial_\nu''\,{}^*J_\mu(x'')) \quad (36)$$

and

$$B_\mu(x) = \partial_\mu\,{}^*\lambda(x)+\int (dx')D_+(x-x')\,{}^*J_\mu(x') + \int (dx')(dx'')f^\nu(x'-x)D_+(x'-x'') \times{}^*(\partial_\mu''J_\nu(x'')-\partial_\nu''J_\mu(x'')). \quad (37)$$

We have used a notation for dual tensor, in accordance with the definition

$${}^*A^{\mu\nu} = \tfrac{1}{2}\epsilon^{\mu\nu\lambda\kappa}A_{\lambda\kappa}. \quad (38)$$

The arbitrary gauge functions $\lambda(x)$, ${}^*\lambda(x)$ appear in consequence of the conservation restrictions on the sources.

A particularly advantageous gauge choice is

$$\lambda(x) = -\int (dx')(dx'')f^\nu(x-x')D_+(x'-x'')J_\nu(x''),$$
$${}^*\lambda(x) = \int (dx')(dx'')f^\nu(x'-x)D_+(x'-x'')\,{}^*J_\nu(x''), \quad (39)$$

for, then,

$$A_\mu(x) = -\int (dx')f^\nu(x-x')F_{\mu\nu}(x'),$$
$$B_\mu(x) = \int (dx')f^\nu(x'-x)\,{}^*F_{\mu\nu}(x'), \quad (40)$$

where

$$F_{\mu\nu}(x) = \int (dx')D_+(x-x')[\partial_\mu'J_\nu(x')-\partial_\nu'J_\mu(x') - {}^*(\partial_\mu'\,{}^*J_\nu(x')-\partial_\nu'\,{}^*J_\mu(x'))], \quad (41)$$

and (${}^{**}A_{\mu\nu} = -A_{\mu\nu}$),

$${}^*F_{\mu\nu}(x) = \int (dx')D_+(x-x')[\partial_\mu'\,{}^*J_\nu(x')-\partial_\nu'\,{}^*J_\mu(x') + {}^*(\partial_\mu'J_\nu(x')-\partial_\nu'J_\mu(x'))]. \quad (42)$$

The identity

$$\partial_\nu\,{}^*(\partial^\mu A^\nu-\partial^\nu A^\mu) = \epsilon^{\mu\nu\lambda\kappa}\partial_\nu\partial_\lambda A_\kappa = 0 \quad (43)$$

leads immediately to

$$\partial_\nu F^{\mu\nu}(x) = J^\mu(x), \qquad \partial_\nu\,{}^*F^{\mu\nu}(x) = {}^*J^\mu(x), \quad (44)$$

which justifies the description of the two source varieties as electric and magnetic currents. The relations

$$\partial_\mu F_{\nu\lambda}+\partial_\nu F_{\lambda\mu}+\partial_\lambda F_{\mu\nu} = \epsilon_{\mu\nu\lambda\kappa}\partial_\alpha\,{}^*F^{\alpha\kappa} = \epsilon_{\mu\nu\lambda\kappa}\,{}^*J^\kappa \quad (45)$$

and

$$\partial_\mu\,{}^*F_{\nu\lambda}+\partial_\nu\,{}^*F_{\lambda\mu}+\partial_\lambda\,{}^*F_{\mu\nu} = -\epsilon_{\mu\nu\lambda\kappa}J^\kappa \quad (46)$$

are used in deriving

$$\partial_\mu A_\nu(x)-\partial_\nu A_\mu(x) = F_{\mu\nu}(x) - {}^*\Big[\int (dx')(f_\mu(x-x')\,{}^*J_\nu(x') - f_\nu(x-x')\,{}^*J_\mu(x'))\Big],$$
$$\partial_\mu B_\nu(x)-\partial_\nu B_\mu(x) = {}^*F_{\mu\nu}(x) - {}^*\Big[\int (dx')(f_\mu(x'-x)J_\nu(x') - f_\nu(x'-x)J_\mu(x'))\Big]. \quad (47)$$

*Action.* In view of the linear relation between fields and sources, the quantity $W$ is given by

$$W = \frac{1}{2}\int (dx)[J^\mu(x)A_\mu(x)+{}^*J^\mu(x)B_\mu(x)]. \quad (48)$$

The identities

$$\int (dx)J^\mu A_\mu = \int (dx)\tfrac{1}{2}F^{\mu\nu}(\partial_\mu A_\nu-\partial_\nu A_\mu),$$
$$\int (dx)\,{}^*J^\mu B_\mu = \int (dx)\tfrac{1}{2}\,{}^*F^{\mu\nu}(\partial_\mu B_\nu-\partial_\nu B_\mu), \quad (49)$$

and

$$\int (dx)\tfrac{1}{2}F^{\mu\nu}(\partial_\mu A_\nu - \partial_\nu A_\mu) = \int (dx)\tfrac{1}{2}F^{\mu\nu}F_{\mu\nu} + \int (dx)\,{}^*J^\mu B_\mu\,,$$

$$\int (dx)\tfrac{1}{2}{}^*F^{\mu\nu}(\partial_\mu B_\nu - \partial_\nu B_\mu) = \int (dx)\tfrac{1}{2}{}^*F^{\mu\nu}\,{}^*F_{\mu\nu} + \int (dx) J^\mu A_\mu\,, \tag{50}$$

enable one to rewrite $W$ in equivalent forms that contain only fields. Thus,

$$W = \int (dx)[\tfrac{1}{2}F^{\mu\nu}(\partial_\mu A_\nu - \partial_\nu A_\mu) - \tfrac{1}{4}F^{\mu\nu}F_{\mu\nu}]$$
$$= \int (dx)[\tfrac{1}{2}\,{}^*F^{\mu\nu}(\partial_\mu B_\nu - \partial_\nu B_\mu) - \tfrac{1}{4}\,{}^*F^{\mu\nu}\,{}^*F_{\mu\nu}]\,, \tag{51}$$

from which we infer the additional equivalent forms

$$W = \int (dx)[J^\mu A_\mu + {}^*J^\mu B_\mu - \tfrac{1}{2}F^{\mu\nu}(\partial_\mu A_\nu - \partial_\nu A_\mu) + \tfrac{1}{4}F^{\mu\nu}F_{\mu\nu}] \tag{52}$$

and

$$W = \int (dx)[J^\mu A_\mu + {}^*J^\mu B_\mu - \tfrac{1}{2}\,{}^*F^{\mu\nu}(\partial_\mu B_\nu - \partial_\nu B_\mu) + \tfrac{1}{4}\,{}^*F^{\mu\nu}\,{}^*F_{\mu\nu}]\,. \tag{53}$$

The last two structures have a significant property. Considered as a functional of appropriate fields, they are stationary with respect to variations of those fields. This gives the fundamental quantity $W$ the character of an action. Indeed, the consequences of the stationary requirement comprise all the field equations necessary to eliminate the fields and express $W$ in terms of the sources. For the first of these forms, the variables are $A_\mu$ and $F^{\mu\nu}$ with $B_\mu$ regarded as a functional of $F_{\mu\nu}$ that is given by

$$B_\mu(x) = \int (dx') f^\nu(x'-x)\,{}^*F_{\mu\nu}(x')\,. \tag{54}$$

The variational equations are

$$\partial_\nu F^{\mu\nu}(x) = J^\mu(x)\,,$$
$$\partial_\mu A_\nu(x) - \partial_\nu A_\mu(x) = F_{\mu\nu}(x) - {}^*\Big[\int (dx')\big(f_\mu(x-x')\,{}^*J_\nu(x') - f_\nu(x-x')\,{}^*J_\mu(x')\big)\Big]\,. \tag{55}$$

On differentiating the dual of the last equation, we get

$$\partial_\nu\,{}^*F^{\mu\nu}(x) = {}^*J^\mu(x)\,, \tag{56}$$

and enough information is available to reconstruct all the fields, within the gauge arbitrariness of $A_\mu$. The second action form uses $B_\mu$ and ${}^*F^{\mu\nu}$ as variables, with

$$A_\mu(x) = -\int (dx') f^\nu(x-x') F_{\mu\nu}(x')\,. \tag{57}$$

The discussion is entirely analogous.

*Charge Quantization*. The action principles restate the differential dependence of $W$ upon the sources:

$$\delta W = \int (dx)[\delta J^\mu A_\mu + \delta\,{}^*J^\mu B_\mu]\,. \tag{58}$$

To simplify the discussion let us hold the magnetic source fixed and choose

$$\delta J^\mu(x) = \partial_\nu \delta M^{\mu\nu}(x) = -\partial_\nu \delta M^{\nu\mu}(x)\,, \tag{59}$$

which satisfies identically the constraint

$$\partial_\mu \delta J^\mu(x) = 0\,. \tag{60}$$

This gives

$$\delta W = \int (dx)\tfrac{1}{2}\delta M^{\mu\nu}(x)[\partial_\mu A_\nu(x) - \partial_\nu A_\mu(x)]$$
$$= \int (dx)\tfrac{1}{2}\delta M^{\mu\nu}(x) F_{\mu\nu}(x) - \int (dx)(dx')\delta\,{}^*M^{\mu\nu}(x) f_\mu(x-x')\,{}^*J_\nu(x')\,. \tag{61}$$

The fundamental problem posed by the space-time extrapolation from the initial photon-mediated causal circumstances is now evident. There is an explicit dependence upon the physically undetermined function $f_\mu(x-x')$. The permissible restriction to the class of functions that only connect points in spacelike relation would remove that dependence when $\delta M_{\mu\nu}$ and ${}^*J_\nu$ are causally separated, but otherwise is of no avail.

One avenue of escape exists—that the $f_\mu$-dependent part of $\delta W$ is not continuously variable but is restricted to integer multiples of $2\pi(\hbar)$, for then the arbitrariness disappears from $\exp(i\delta W)$, which is the physically significant quantity. This possibility cannot be realized unless there is a discrete element intrinsic to the nature of photon sources. Sources are introduced to represent the common features of analogous mechanisms and to abstract from the vagaries of individual mechanisms. In this situation, we must conclude that charge is discretely localized. The idealizations involved in introducing sources cannot violate that general physical law if a consistent theory is to be constructed.

A point-charge representation of independent electron and magnetic sources is indicated by

$$J^\mu(x)=\sum_e e\int dx_e{}^\mu\delta(x-x_e)\,,$$
$$*J^\mu(x)=\sum_g g\int dx_g{}^\mu\delta(x-x_g)\,, \qquad (62)$$

where $x_e{}^\mu$ and $x_g{}^\mu$ are the coordinates of charge-bearing points. Since sources are present neither initially nor finally in the situation described by the vacuum amplitude, it may be supposed that focal points exist from which positive and negative charges appear and at which they eventually disappear. (Or, one can recognize that, in a description limited to a finite space-time region, charge can be introduced from the outside through the boundaries and then be subsequently withdrawn.) A variation in the paths of the electric charges, for example, leads to

$$\delta W=\sum_e e\delta\int dx_e{}^\mu A_\mu(x_e)$$
$$=\sum_e e\int \tfrac{1}{2}d\sigma_e{}^{\mu\nu}(\partial_\mu A_\nu-\partial_\nu A_\mu)(x_e) \qquad (63)$$

or

$$\delta W=\sum_e e\int \tfrac{1}{2}d\sigma_e{}^{\mu\nu}F_{\mu\nu}(x_e)$$
$$-\sum_{eg} eg\int d\,{}^*\sigma_e{}^{\mu\nu}f_\mu(x_e-x_g)dx_{g\nu}, \qquad (64)$$

where the two-dimensional surfaces are bounded by the initial and varied paths. The latter form is not restricted to small path changes. With respect to integration over the variable $x=x_e-x_g$, the product $d\,{}^*\sigma_e{}^{\mu\nu}dx_{g\nu}$ defines a three-dimensional directed surface area,

$$d\,{}^*\sigma_e{}^{\mu\nu}dx_{g\nu}=d\sigma^\mu. \qquad (65)$$

Just such a surface occurs in expressing a consequence of the differential equation

$$\partial_\mu f^\mu(x)=\delta(x)\,, \qquad (66)$$

namely,

$$\int d\sigma_\mu f^\mu(x)=1\,, \qquad (67)$$

but this is a closed surface that surrounds the origin. We must now restrict $f_\mu$ to a class of functions such that $\int d\sigma_\mu f^\mu$ assumes only discrete values, for any integration surface. This indicates that the support domain for such a function on a regular surface enclosing the origin must be limited to a finite number of points.

If these points are sufficiently continuous in moving to neighboring surfaces, we can picture a number of filaments drawn out from the origin. Let there be $\nu$ of these, and let $r_\alpha$, $\alpha=1,\cdots,\nu$ be the contribution of any portion of a closed surface that contains only the $\alpha$th point, where

$$\sum_{\alpha=1}^{\nu} r_\alpha=1. \qquad (68)$$

We also recognize the possibility that a surface encounters the $\alpha$th point on its boundary, and assign the integration value $\frac{1}{2}r_\alpha$ to this arrangement. The uniqueness of $\exp(i\delta W)$ then requires, for any pair of electric and magnetic charges, that

$$eg\tfrac{1}{2}r_\alpha=2\pi n_\alpha\,, \qquad (69)$$

where $n_\alpha$ is an integer. This is the strongest condition, the analogous one involving $egr_\alpha$ being satisfied as a consequence. On summing over the $\nu$ points, we reach the charge quantization relation

$$\tfrac{1}{2}eg=2\pi n\,, \qquad (70)$$

with

$$n=\sum_{\alpha=1}^{\nu} n_\alpha. \qquad (71)$$

The individual weights $r_\alpha$ are necessarily restricted to the rational form

$$r_\alpha=n_\alpha/n. \qquad (72)$$

If the $\nu$ points are all equivalent and $r_\alpha=1/\nu$, we have

$$n=n_\alpha\nu\,, \qquad (73)$$

which describes the situation where the minimum value assumed by a nonzero charge product $eg/4\pi$ is the integer $\nu$.

We have now achieved the effective elimination of the arbitrary elements in $\delta W$, which can be replaced by

$$\delta W=\sum_e e\int \tfrac{1}{2}d\sigma_e{}^{\mu\nu}F_{\mu\nu}(x_e). \qquad (74)$$

But this expression is no longer the change of a quantity $W$, and the question of integrability arises. Consider, for a particular electric charge, a continuous deformation of paths that returns to the initial one and thereby defines a surface that encloses a three-dimensional volume. If there is to be no change in $\exp(iW)$, it is necessary that

$$e\oint \tfrac{1}{2}d\sigma_{\mu\nu}F^{\mu\nu}=e\int d\sigma_\mu\partial_\nu{}^*F^{\mu\nu}=e\int d\sigma_\mu{}^*J^\mu=2\pi n. \qquad (75)$$

An individual magnetic charge contributes the value $g$ to this three-dimensional volume integration if it is in the interior, and $\frac{1}{2}g$ if it is on the surface of the volume.

Thus, the integrability requirement leads back to

$$\tfrac{1}{2}eg=2\pi n\,, \tag{76}$$

the general charge quantization condition.

*Dual Charged Particles*. The preceding discussion assumed that electric and magnetic charges are associated with different particles, which brings us to the interesting possibility that magnetically charged particles also carry electric charge. In this situation, a displacement of source particles produce a variation of both electric and magnetic sources, and we must write

$$\delta W=\sum_a\left[e_a\delta\int dx_a{}^\mu A_\mu(x_a)+g_a\delta\int dx_a{}^\mu B_\mu(x_a)\right], \tag{77}$$

where $a$ is a particle label. This leads to

$$\delta W=\sum_a\left[e_a\int \tfrac{1}{2}d\sigma_a{}^{\mu\nu}F_{\mu\nu}(x_a)+g_a\int\tfrac{1}{2}d\sigma_a{}^{\mu\nu}\,{}^*F_{\mu\nu}(x_a)\right]$$
$$-\sum_{ab}\left[e_ag_b\int d\,{}^*\sigma_a{}^{\mu\nu}f_\mu(x_a-x_b)dx_{b\nu}\right.$$
$$\left.+g_ae_b\int d\,{}^*\sigma_a{}^{\mu\nu}f_\mu(x_b-x_a)dx_{b\nu}\right]. \tag{78}$$

Consider first the integrability condition that is required when the $f_\mu$-dependent terms in $\delta W$ are effectively eliminated. It is $({}^{**}F_{\mu\nu}=-F_{\mu\nu})$

$$e_a\int d\sigma_\mu{}^*J^\mu-g_a\int d\sigma_\mu J^\mu=2\pi n\,, \tag{79}$$

and the presence of another particle on the surface of the three-dimensional volume gives the new charge quantization condition

$$\tfrac{1}{2}(e_ag_b-g_ae_b)=2\pi n_{ab}. \tag{80}$$

Before examining it, we must note the existence of a conflict with the independent quantization statement involving $f_\mu$, unless

$$f_\mu(x-x')=-f_\mu(x'-x)\,, \tag{81}$$

which is compatible with the differential equation obeyed by the function. When $f_\mu(x-x')$ was introduced, the points $x$ and $x'$ referred to distinct electric and magnetic charge regions, respectively. In the present situation, with particles carrying both electric and magnetic charges, an additional symmetry property is required. It implies that every filament of $f_\mu(x)$ has its image, or that $\nu$ is even. The integer $n$ of the charge quantization condition must also be even.

The charge quantization situation changes significantly on considering particles with both kinds of charge. This can be appreciated by examining the specific possibility that $e_a/g_a$ is a fixed constant independent of $a$; there is *no* restriction on the individual products $e_ag_b$. But no conflict with the earlier discussion exists. If we consider a group of particles which are magnetically neutral as a whole, $\sum g_a=0$, their total electric charge, $e=\sum e_a$, satisfies the previous charge quantization condition:

$$\tfrac{1}{2}eg_b=2\pi n\,,\quad n=\sum_a n_{ab}. \tag{82}$$

In the special circumstance we have described, the total electric charge is also zero. As another example, if there are particles with a common magnetic charge $g$ but different electric charges, $e_1\neq e_2$, the new condition asserts that

$$\tfrac{1}{2}(e_1-e_2)g=2\pi n\,, \tag{83}$$

but does not limit the individual electric charges.

*Composite Particles*. The formal symmetry between electric and magnetic charge is violated in nature by the great disparity between the charge units. If we use the smallest even integer to relate them, we conclude from

$$e^2/4\pi=1/137 \tag{84}$$

that

$$g^2/4\pi=4(137)\,. \tag{85}$$

The enormously strong forces of attraction that must operate between positive and negative charges are consistent with the fact that the units of matter thus far observed are magnetically neutral. This opens the possibility that ordinary matter may be composed of magnetically charged constituents, which carry electric charges of values different from those characteristic of magnetically neutral matter.[4] We now discuss briefly a model of matter based on this idea, which has some contact with empirical classification schemes.

Consider a set of particles with two choices of magnetic charge, $-g_0$, $(N-1)g_0$, and two analogous choices of electric charge, $-e_0$, $(N-1)e_0$. The integer $N$ may be $2,3,\cdots$. The various charge quantization conditions are all satisfied if

$$Ne_0g_0/4\pi=2\,, \tag{86}$$

which assumes that the smallest even integer is realized. If that is also true for the charge unit $e$ of magnetically neutral particles,

$$eg_0/4\pi=2\,, \tag{87}$$

we conclude that

$$e_0=e/N\,. \tag{88}$$

The individual electric charges of the magnetically charged particles are thus $-1/N$ and $(N-1)/N$ in units of $e$. They differ by a unit charge.

[4] This approach to composite structure has nothing in common with attempts to describe observed particles in terms of the nonrelativistic behavior of weakly interacting constituents.

The minimum number of constituents required to produce a neutral combination, without using antiparticles, is $N$. Let us assume that the pattern of magnetic charge required for neutrality, $-g_0$ repeated $N-1$ times with $(N-1)g_0$ occurring once, is duplicated in the electric charges exhibited by each of the $N$ particles, namely, $-e_0=-(1/N)e$ is repeated $N-1$ times and $(N-1)e_0=[(N-1)/N]e$ occurs once. The outcome is a set of magnetically neutral particles displaying integer electric charges that range from $N-1$ to $-1$. To the extent that electric charge and other unspecified properties are of secondary dynamical importance, this set, or some subset of it, may be recognizable as a particle multiplet with a broken mass spectrum. The analogous multiplet that is constructed completely from antiparticles is a different one, if $N>2$. That is an empirical characteristic of baryons, which are Fermi-Dirac particles. If we assume that all magnetically charged particles are fermions, it is necessary that $N$ be odd. The first possibility is $N=3$.

The resulting baryon model composes these particles from three constituents, of magnetic charge $2g_0$, $-g_0$, $-g_0$, each with three choices of electric charge $\frac{2}{3}$, $-\frac{1}{3}$, $-\frac{1}{3}$, in units of $e$. This pattern of fractional electric charges is familiar in an empirical model, based on the symmetry group $SU_3$, which was introduced[5] without reference to the magnetic charge concept[6] that makes fractional electric charge understandable and physically acceptable. The consequences of the simplest meson model, comprising a magnetically charged electric triplet and an antiparticle triplet, are also familiar empirically in various nonuplet realizations.

*Note added in proof.* These remarks are made in response to a referee's comments.

(1) "The notation $D_+$ for the function defined in (2) is unfortunate, since $D_+(x-x')$ is universally used for the integral in (2) for *all* $x^0-x^{0\prime}$." It is also stated that $D_F$, $D_c$, or $D_{1R}$ are the usual notations for the symmetric function defined in (2).

In fact, there has been no such historical consensus concerning the positive frequency function, notations such as $D^{(+)}$ being not uncommon. Since 1949, I have consistently used the symbol $D_+$ to designate a function of definition equivalent to (2). Apart from a different choice of factors and principal symbol, this has also been Feynman's usage. An advantage over notations such as $D_F$ and $D_c$ (the symbol $D_{1R}$ is unfamiliar to me) is the uniformity with which one represents the alternative boundary conditions of outgoing waves ($D_+$) or incoming waves ($D_-$).

(2) "... for the case of point sources (62), the action (48) diverges due to Coulomb field singularities calling into question the theory based on (48)."

Presumably the *action* expression that is meant is (52) or (53). This would be a serious charge indeed if source theory were based on the field concept. It is not. The discussion in the text began with known results for distributed sources. It was eventually recognized that localization of charge must be introduced to reach a consistent theory of electric and magnetic charge. At this point I should have reconstructed the theory to deal with the new circumstance. I did not do so in the full knowledge that nothing I intended to discuss would be changed thereby. But I do concede that a more complete discussion is instructive for, in contrast with operator field theory, self-action holds no terrors for the sourcerer.

The theory starts from a description of the exchange of photons between distinct, casually arranged sources, which is in no way dependent upon the details of the charge distributions. On considerlng point electric charges and placing them all on the same footing we arrive at the proper time structure

$$W=\frac{1}{2}\sum_{a\neq b} e_a e_b \int ds ds' v_a^{\mu}(s) v_{\mu b}(s') \times D_+[x_a(s)-x_b(s')]+\sum_a W_a,$$

where $v^\mu=dx^\mu/ds$, and $W_a$ characterizes a single charge. The principle of source unity requires that $W_a$ have the same structure as the mutual coupling terms. But we must also note that the individual charged particles will have been described already, under physical conditions of non-interaction. This is represented by the action term $-m_a\int ds_a$, where $m_a$ is the observed mass of the particle. It is the essence of the phenomenological source theory that physical parameters, once identified under restricted physical conditions, do not alter their meaning when more general circumstances are examined. Hence we cannot include the electromagnetic self-action associated with

$$\mathrm{Re}D_+(x-x')=(1/4\pi)\delta[(x-x')^2],$$

for it would change the already correctly assigned mass $m_a$. This is a simple example of mass *normalization*. One can say that the observed mass already includes any inertial electromagnetic effect, which should not be counted twice. With this attitude we clearly separate particle phenomenology from speculations about inner particle structure. (That was also the intention of relativistic renormalization theory, as it was originally formulated in 1947, but the methods then used were too cumbersome and this simple idea was displaced by the regulators and counter terms that came into fashion). We conclude that

$$W_a=\frac{1}{2}e_a^2\int ds ds' v_a^{\mu}(s) v_{\mu a}(s') i\, \mathrm{Im}D_+[x_a(s)-x_a(s')],$$

which is exactly what is needed to give a consistent

[5] M. Gell-Mann, Phys. Letters 8, 214 (1964).
[6] The magnetic model of matter will be elaborated elsewhere.

quantum mechanical account of emission and absorption of photons by a single accelerated charge.

Now that photon sources are endowed with the more specific point structure, it is important to be remined that sources are never objects of explicit physical interest. The source is an instrument—a practical calculational tool on the one hand and, on the other, a device for initiating theory at a particularly simple level by utilizing idealizations of realistic particle behavior. Such idealizations may not violate general physical laws, however, and, conversely, provide a means for apprehending the existence of such laws when they appear as necessary restrictions on the nature of sources. That was our attitude in deducing, from the hypothesized existence of both electric and magnetic sources, the localization and quantization of charge.

We add a remark on the intrinsic symmetry between electricity and magnetism, which is not limited to the discrete substitution

$$F_{\mu\nu}\rightarrow {}^*F_{\mu\nu}, \qquad {}^*F_{\mu\nu}\rightarrow -F_{\mu\nu},$$
$$J_\mu\rightarrow {}^*J_\mu, \qquad {}^*J_\mu\rightarrow -J_\mu,$$

but is more fully expressed by the rotation

$$J_\mu\rightarrow J_\mu\cos\vartheta+{}^*J_\mu\sin\vartheta$$
$${}^*J_\mu\rightarrow -J_\mu\sin\vartheta+{}^*J_\mu\cos\vartheta,$$

with analogous equations for $F_{\mu\nu}$, ${}^*F_{\mu\nu}$. This property underlies the statement that no charge quantization would exist if all particles had the same $g/e$ ratio for, by a suitable rotation in the $e$, $g$ space, the situation is reduced to that of pure electricity. The charge rotational symmetry is inherent in the photon exchange definition of electric and magnetic sources since it is equivalent to a rotation of all polarization vectors. In order that the theory erected on this foundation [Eq. (33)] maintain that symmetry, it is necessary that $f_\mu(x-x')$ have the symmetry property given in Eq. (81). (Of course, the $e$, $g$ rotational symmetry is only partial since the distinction between electricity and magnetism is an absolute one in the real world. That implies the existence of nonelectromagnetic but charge-dependent interactions of which the so-called weak interactions are a known example.) We can now recognize that the self-action of particles carrying both electric and magnetic charges introduces nothing new since $e_a{}^2$ is replaced by the invariant combinaton $e_a{}^2+g_a{}^2$ while no $e_ag_a$ term can appear.

The unsymmetrical treatment of mutual and self actions is admittedly awkward when the source description is translated into field language. But no new physics is involved and elaborate formal devices are unnecessary. It suffices to proceed as in the text, with the understanding that the real parts of self-action terms are to be struck out when the fields are eliminated and attention returned to the sources. Nevertheless, in the interests of completeness we shall indicate that this mental process could be realized by a well-defined mathematical procedure. For simplicity we consider only electrically charged particles. The point charge form of $W$, which contains no real self-action terms, is unaltered if $D_+(x-x')$ is replaced by $D_+(x-x'+\lambda)$, where $\lambda^\mu\rightarrow 0$ through space-like values. But, for any finite $\lambda^\mu$, the real self-action terms exist and could be added and substracted to give the form

$$W(\lambda)=\frac{1}{2}\int(dx)J^\mu(x)A_\mu(x+\lambda)-w_{\text{self}}(\lambda),$$

where $A_\mu(x)$ retains its original meaning in terms of the total source $J_\mu(x)$. In the following we shall understand that a symmetrization between $\lambda^\mu$ and $-\lambda^\mu$ is used. Then

$$W(\lambda)=\frac{1}{4}\int(dx)F^{\mu\nu}(x)F_{\mu\nu}(x+\lambda)-w_{\text{self}}(\lambda)$$
$$=\int(dx)[J^\mu(x)A_\mu(x+\lambda)-\tfrac{1}{4}F^{\mu\nu}(x)F_{\mu\nu}(x+\lambda)]$$
$$-w_{\text{self}}(\lambda),$$

and the latter form has the stationary action property:

$$\delta_A W(\lambda)=\int(dx)\delta A_\mu(x+\lambda)[J^\mu(x)-\partial_\nu F^{\mu\nu}(x)]=0,$$

since ($\lambda^\mu\rightarrow -\lambda^\mu$)

$$\int(dx)\delta A_\mu(x)\partial_\nu F^{\mu\nu}(x+\lambda)\rightarrow\int(dx)\delta A_\mu(x)\partial_\nu F^{\mu\nu}(x-\lambda)$$
$$=\int(dx)\delta A_\mu(x+\lambda)\partial_\nu F^{\mu\nu}(x).$$

Analogous $\lambda$-generalizations of Eqs. (52) and (53) give the action expressions for the field descriptions of sources composed of dual charged particles.

Finally, a word about the charge quantization condition for dual charged particles when the $\lambda$ process is used. If unphysical elements are not to appear during the limiting operation, a charge quantization condition must hold for almost all $\lambda^\mu$. It is

$$(e_ag_b-e_bg_a)\frac{1}{2}\left(\int_{\sigma(\lambda)}+\int_{\sigma(-\lambda)}\right)d\sigma_\mu f^\mu(x)=2\pi n,$$

where $\sigma(\lambda)$ is obtained from the arbitrary three-dimensional surface $\sigma$ by the rigid displacement $\lambda^\mu$, and the necessary symmetrization between $\lambda^\mu$ and $-\lambda^\mu$ is made explicit. The critical situation occurs when one of the even number of filaments comprising $f^\mu(x)$ pierces $\sigma(\lambda)$ once but does not intersect $\sigma(-\lambda)$. This is not an exceptional possibility, but refers to a $\lambda$-domain of nonzero measure. The implied quantization condition is

$$\tfrac{1}{2}(e_a g_b - e_b g_a) = 2\pi n_{ab}, \quad n_{ab} \text{ even}.$$

This way of obtaining Eq. (80) differs from that of the text where the factor of $\frac{1}{2}$ derives from an exceptional geometrical situation, although the definition for that circumstance clearly depends on the same limiting process we have just described.

# A Magnetic Model of Matter

## A speculation probes deep within the structure of nuclear particles and predicts a new form of matter.

Julian Schwinger

*And now we might add something concerning a certain most subtle Spirit, which pervades and lies hid in all gross bodies.*

—Newton

Although electromagnetic phenomena are the best understood of all nature's manifestations, there are still great mysteries in this and related areas. Here are four of them:

1) The fundamental electromagnetic equations of Maxwell show an intrinsic symmetry between electric and magnetic quantities which, incidentally, is unique to the four dimensions of space and time. Yet no magnetic counterpart to electric charge is known experimentally. How can one account for this, while retaining the qualitative idea of reciprocity between electric charge and magnetic charge?

2) The unit of electric charge is universal. It is observed, with fantastic precision, to be identical on all charged particles, despite wide variations in their other characteristics. What unknown general principle is at work?

3) A new periodic table is coming into being through the artificial creation of subnuclear particles and their tentative grouping into families. Two approximate but significant properties have been recognized, isotopic spin and hypercharge, which serve also to specify the electric charge of the particle. What is the dynamical meaning of these properties that are related to but distinct from electric charge? In addition, more inclusive classification schemes have been proposed, which lend themselves to the interpretation that nuclear particles have constituents with fractional electric charges. In view of the strict regularity noted in 2), how can such models have physical significance?

4) The behavior of all particles with respect to strong, electromagnetic, and weak interactions has seemed consist-

The author is Higgins Professor of Physics at Harvard University, Cambridge, Massachusetts. This article is based on lectures delivered over several years, most recently at Lindau, Germany, in July 1968.

Reprinted from *Science* 165, 757–761 (1969).

ent with a general symmetry property in which the interchange of left and right, symbolized by *P*(parity), is combined with the exchange of positive and negative charge, *C*. But recently, phenomena have been observed that indicate a weak violation of this *CP* symmetry. What dynamical mechanism is responsible?

I shall put forward a speculative hypothesis, which has in its favor only one argument—that it does connect and give tentative answers to all these questions. However wide of the truth this hypothesis may be, it can serve to bring into better focus the nature of the quest for order and understanding that underlies the activity of the high-energy physicist.

It began long ago when Dirac (*1*) pointed out that, according to the quantum laws of atomic physics, the existence of a magnetic charge would lead to a quantization of electric charge in which only integral multiples of a fundamental unit could occur. I have never seriously doubted that here was the missing general principle referred to in 2). And Dirac himself noted the basis for the reconciliation called for in 1). The law of reciprocal electric and magnetic charge quantization is such that the unit of magnetic charge, deduced from the known unit of electric charge, is quite large. It should be very difficult to separate opposite magnetic charges in what is normally magnetically neutral matter. Thus, through the unquestioned quantitative asymmetry between electric and magnetic charge, their qualitative relationship might be upheld.

What is new is the proposed contact with the mysteries noted under 3) and 4), which are testimonials to the advance of experimental science. This appears if one considers the properties of particles that carry both electric and magnetic charges. Now the conditions of charge quantization are less stringent, and fractional electric charge becomes a physical possibility, consonant with the integral charges necessarily carried by magnetically neutral particles. Such dual-charged particles supply a physical realization for the constituents used in the empirical models of the so-called hadrons (*2*), which are the strongly interacting nuclear particles. Furthermore, in the introduction of particles with definite ratios between electric and magnetic charges, a mechanism for *CP* violation has made its appearance. Electric and magnetic charges, like electric and magnetic fields, behave oppositely under spatial reflection, whereas the equations of electromagnetism are symmetrical between positive and negative charges, when both types are considered together. If the dual-charged particles, and their antiparticles, realize a certain ratio between electric and magnetic charge, but not its negative value, the rule of *CP* invariance is broken. This physical model requires further elaboration to explain, rather paradoxically, why the observed violation of *CP* invariance is so remarkably weak. The same refinement is also relevant in the establishment of the detailed correspondence with the empirical mass spectrum that relates to the meaning of isotopic spin and hypercharge. We now turn from this brief survey to more specific but elementary discussions of the various items.

## Maxwell's Equations

The form of these equations for the electric field **E** and the magnetic field **H**, in which $c$ is the speed of light,

$$\nabla \times \mathbf{H} - \frac{1}{c}\frac{\partial}{\partial t}\mathbf{E} = \frac{4\pi}{c}\mathbf{j}_e$$
$$\nabla \cdot \mathbf{E} = 4\pi\rho_e$$
$$-\nabla \times \mathbf{E} - \frac{1}{c}\frac{\partial}{\partial t}\mathbf{H} = \frac{4\pi}{c}\mathbf{j}_m$$
$$\nabla \cdot \mathbf{H} = 4\pi\rho_m$$

makes evident the symmetry

$$\mathbf{E} \rightarrow \mathbf{H},\ \mathbf{H} \rightarrow -\mathbf{E} \qquad \rho_e \rightarrow \rho_m,\ \rho_m \rightarrow -\rho_e$$

with the electric and magnetic currents, $\mathbf{j}_e$ and $\mathbf{j}_m$, following the pattern of the charge densities $\rho_e$ and $\rho_m$. This is a particular example of the invariance expressed by the rotation through the arbitrary angle $\theta$.

$$\mathbf{E}' = \mathbf{E}\cos\theta + \mathbf{H}\sin\theta$$
$$\mathbf{H}' = -\mathbf{E}\sin\theta + \mathbf{H}\cos\theta$$
$$\rho_e' = \rho_e\cos\theta + \rho_m\sin\theta$$
$$\rho_m' = -\rho_e\sin\theta + \rho_m\cos\theta$$

In purely electromagnetic considerations, the observed absence of magnetic charge is equally well described as the coexistence of electric and magnetic charge in the universal ratio indicated by

$$\rho_m/\rho_e = \tan\theta$$

We also note that the following combinations formed from electric charges $e_1, e_2$ and magnetic charges $g_1, g_2$ are invariant under the redefinitions produced by the rotation through the angle $\theta$:

$$e_1e_2 + g_1g_2,\quad e_1g_2 - e_2g_1$$

## Charge Quantization

Here is an elementary argument in support of the existence of charge quantization (*3*). Consider the nonrelativistic behavior of a particle with mass $m$, carrying electric charge $e_1$ and magnetic charge $g_1$, which moves with velocity **v** in the field of a stationary body that possesses charges $e_2$ and $g_2$. There is an equivalent description for the relative motion of two particles with arbitrary masses. The equation of motion is

$$m\frac{d\mathbf{v}}{dt} = e_1\left(\mathbf{E} + \frac{1}{c}\mathbf{v}\times\mathbf{H}\right) + g_1\left(\mathbf{H} - \frac{1}{c}\mathbf{v}\times\mathbf{E}\right)$$

where the following forms of the field strengths at the point with vector **r**, of magnitude $r$,

$$\mathbf{E} = e_2\frac{\mathbf{r}}{r^3},\quad \mathbf{H} = g_2\frac{\mathbf{r}}{r^3}$$

assign the origin of coordinates to the position of the stationary body. The explicit statement

$$m\frac{d\mathbf{v}}{dt} = (e_1e_2 + g_1g_2)\frac{\mathbf{r}}{r^3} + (e_1g_2 - e_2g_1)\frac{1}{c}\mathbf{v}\times\frac{\mathbf{r}}{r^3}$$

involves just the invariant charge combinations that were noted. The associated moment equation is

$$\mathbf{r} \times m \frac{d\mathbf{v}}{dt} = (e_1 g_2 - e_2 g_1) \frac{1}{c} \frac{\mathbf{r} \times (\mathbf{v} \times \mathbf{r})}{r^3}$$
$$= (e_1 g_2 - e_2 g_1) \frac{1}{c} \frac{d}{dt} \frac{\mathbf{r}}{r}$$

and we recognize the conserved angular momentum vector

$$\mathbf{J} = \mathbf{r} \times m\mathbf{v} - (e_1 g_2 - e_2 g_1) \frac{1}{c} \frac{\mathbf{r}}{r}$$

The quantization of the component of this angular momentum along the connecting line of the particles then gives the charge quantization condition ($2\pi\hbar$ is Planck's constant)

$$(e_1 g_2 - e_2 g_1)/\hbar c = \nu$$

where $\nu$ is an integer. The exclusion of (integer + ½) values, which were admitted by Dirac, seems plausible in this purely orbital situation, but it requires a rather subtle argument in support. Equally subtle is the suggestion that, if there are dual-charged particles, rather than just electrically charged particles and magnetically charged particles, the integer $\nu$ must be even (*4*).

I shall try only to indicate what is involved in these arguments through the following consideration on behalf of integer quantization. Since matter is normally magnetically neutral, any purely magnetically charged particle, for example, has an oppositely charged counterpart somewhere. If one is interested in an electric charge $e$, at the point $\mathbf{r}$, which is in the neighborhood of the magnetic charge $g$ at the origin, it should not be necessary to refer to the compensating charge $-g$ at the point $\mathbf{R}$, if the latter is sufficiently remote from the origin. But, on examining the additional electromagnetic angular momentum of this system, which is

$$-(eg/c)\left[\frac{\mathbf{r}}{r} - \frac{\mathbf{r}-\mathbf{R}}{|\mathbf{r}-\mathbf{R}|}\right]$$

we see that the angular momentum associated with the charge $-g$ does not vanish as this particle recedes to infinity but contributes an additive constant. The total angular momentum of the three-particle system is integral, as we confirm by noting that the electromagnetic angular momentum vanishes when $\mathbf{R} \to 0$ and the magnetic charge is neutralized. On shifting our viewpoint between the physically equivalent two-particle system and the three-particle system with an infinitely remote compensating charge, paradoxical transitions between (integer + ½) and integer values of the angular momentum will be avoided if $eg/\hbar c$ is restricted to integral values.

It can be useful to regard

$$-\frac{1}{c}(e_1 g_2 - e_2 g_1)$$

as the radial component of a spin angular momentum vector $\mathbf{S}$ (*5*)

$$-\frac{1}{c}(e_1 g_2 - e_2 g_1) = \mathbf{S} \cdot \mathbf{r}/r$$

The complete spin vector is introduced by defining the momentum $\mathbf{p}$

$$m\mathbf{v} = \mathbf{p} + \mathbf{S} \times \mathbf{r}/r^2$$

which gives

$$\mathbf{J} = \mathbf{r} \times \mathbf{p} + \mathbf{S} = \mathbf{L} + \mathbf{S}$$

The properties of $\mathbf{p}$ and $\mathbf{S}$ are indeed those suggested by this familiar combination involving the orbital angular momentum vector **L**. On introducing the radial momentum $p_r$ according to

$$\mathbf{p} = \frac{\mathbf{r}}{r} p_r + \frac{\mathbf{L} \times \mathbf{r}}{r^2}$$

we infer the kinetic energy

$$T = \tfrac{1}{2} m\mathbf{v}^2 = \frac{1}{2m}\left( p_r^2 + \frac{\mathbf{J}^2 - (\mathbf{J}\cdot\mathbf{r}/r)^2}{r^2} \right)$$

where

$$\mathbf{J} \cdot \frac{\mathbf{r}}{r} = \mathbf{S} \cdot \frac{\mathbf{r}}{r} = -\nu\hbar$$

The total angular momentum spectrum is, correspondingly

$$\mathbf{J}^2 = j(j+1)\hbar^2, \; j = |\nu|, \; |\nu|+1, \; \ldots$$

In the present experimental situation, with only electric charge known, the consideration of a hypothetical magnetic charge $g$ gives the electric charge quantization condition

$$eg/\hbar c = 2n$$

in which the evenness of the integer $\nu = 2n$ is assumed. From the observed unit of electric charge, as measured by

$$e^2/\hbar c \simeq 1/137$$

we deduce a unit of magnetic charge, on choosing $n = 1$

$$g_0^2/\hbar c \simeq 4(137)$$

Forces between magnetic charges are superstrong, in comparison with the strong nuclear forces for which coupling constants are $\sim 10$. The above electric-charge quantization condition also governs the total electric charge of a magnetically neutral aggregate of dual-charged particles. Let $e_a, g_a$ denote the various dual-charge assignments, which obey

$$\sum_a g_a = 0 \quad , \quad \sum_a e_a = e$$

Allowing for the possibility that the smallest magnetic charge $g_0$ resides on a particle with electric charge $e_0$, we conclude from

$$\sum_a (e_a g_0 - e_0 g_a)/\hbar c = \sum_a 2n_a$$

that

$$eg_0/\hbar c = 2n$$

The importance of the latter remark is that the electric charge on a dual-charged particle need not be an integral multiple of the charge unit. Let us imagine a situation in which all particles are dually charged with a universal charge ratio

$$g/e = \tan\,\theta$$

Then the integer in any charge-quantization condition vanishes, and no further restrictions appear. Of course, as the use of the angle $\theta$ signals, this situation is just a charge-rotated version of pure electric charges, where no charge quantization exists. But it emphasizes the weakening of charge quantization that the consideration of dual-charged particles entails. Suppose there are two kinds of dual-charged particles with a common magnetic charge $g_0$, but different electric charges, $e_1$ and $e_2$. The charge quantization condition is

$$(e_1 - e_2)\, g_0/\hbar c = 2n$$

which asserts that $e_1 - e_2$ is a multiple

of the charge unit but does not determine the individual charges. Alternatively, by forming a composite of the particle having charges $e_1, g_0$ with the antiparticle of charges $-e_2, -g_0$, we produce a magnetically neutral particle with electric charge $e_1 - e_2$, which must submit to normal charge quantization. If there are dual-charged particles with magnetic charges different from $g_0$, they must carry integral multiples of this smallest value to be consistent with the reciprocal quantization enforced on magnetic charge by the electrical charge unit $e$. If we compare the charges $e_3, 2g_0$, for example, with $e_1, g_0$, the charge quantization condition asserts that

$$(e_3 - 2e_1)\, g_0/\hbar c = 2n$$

The conclusion that $e_3 - 2e_1$ is a multiple of $e$ is again equivalent to the production of a magnetically neutral composite whose electric charge must have $e$ as a unit. In this way it is seen that the electric charge on a dual-charged particle with magnetic charge $g_0$ provides a new charge unit that is distinct from the known unit $e$.

## Hadron Models

We are being led to a picture in which hadronic matter is viewed as a magnetically neutral composite of dual-charged particles that are based electrically upon a new unit of charge. Such a picture must have enough variety to account for the two different kinds of hadrons: mesons, which are Bose-Einstein particles, and baryons, which are Fermi-Dirac particles. In perhaps the simplest kind of model, all dual-charged particles are alike, at least with regard to statistics, which must be Fermi-Dirac if baryons are to be built from them. It would not do to have only one value of magnetic charge, for then magnetically neutral composites could be produced in only one way, namely, by the combination of particle and antiparticle. That would only manufacture mesons. But it is enough to have just two different values of magnetic charge, which we take to be $2g_0$ and $-g_0$. Now a magnetically neutral composite is also formed from three constituents, of magnetic charges $2g_0, -g_0, -g_0$, and this is a Fermi-Dirac particle. It is satisfactory that this pattern of magnetic charge is unsymmetrical, in contrast with the meson pattern illustrated by $g_0, -g_0$, for it means that the antibaryon, with constituent magnetic charges $-2g_0, g_0, g_0$, is a fundamentally different particle. Magnetic charge thus supplies an interpretation for the empirical property of nucleonic charge. The latter could be identified, for example, with the total magnetic charge on doubly charged constituents measured in units of $2g_0$.

As in the unification of neutron and proton into the nucleon, which was the first use of isotopic spin, it is natural to regard the three values of magnetic charge as three choices available to the fundamental dual-charged particle. And the heuristic power of the theoretical reciprocity between electric and magnetic charges becomes apparent in the suggestion that the electric charge of this fundamental particle has the same threefold option: $2e_0, -e_0, -e_0$. On equating the nonvanishing difference of these charge values to the known unit $e$, we identify the new charge unit

$$e_0 = \tfrac{1}{3}e$$

The pattern of fractional electric charges, $\tfrac{2}{3}, -\tfrac{1}{3}, -\tfrac{1}{3}$, is just the one used in the empirical models. It has now been traced back to the qualitative symmetry between electric and magnetic charge, and the requirement of magnetic neutrality. We should note the consistency of the hypothesis that the electric-charge pattern is independent of the magnetic charge, which again states that $2e_0$ differs from $-e_0$ by an integer. Incidentally, the relation between the electric-charge unit of dual-charged particles and the unit of pure electric charge has its magnetic analog in

$$g_0 = \tfrac{1}{3}g$$

The unit of pure magnetic charge has a magnitude given by

$$g^2/\hbar c \simeq 36(137)$$

We come now to a very important question. What name shall we give to the fundamental dual-charged particle (6)? The particle ending *-on* is obligatory. As evidenced by the use of the provisional phrase "dual-charged particle," the basic aspect that should be commemorated in the name is the dualistic or dyadic character of the charge that the particle bears. There are various short Greek and Latin combining forms that could be applied: *bi-*, *di-*, *duo-*, *dyo-*, as well as longer words such as *dyadikos-*, of two. Dyadikon surely has a ring to it. But being mindful that mesotron became shortened to meson, I believe that dyon is a better choice. The symbol D will not often lead to confusion with deuterium, particularly if we add labels that indicate the electric and magnetic charges, for which we use $e$ and $g$ as units. Thus ${}^{2/3}\mathrm{D}^{-1/3}$ is the dyon with electric charge $-\tfrac{1}{3}e = -e_0$ and magnetic charge $\tfrac{2}{3}g = 2g_0$.

What is the mass of a dyon? Let us be clear about this; any estimate is sheer guesswork. We do not have the wit to connect the known properties of the composites—hadrons—with the unknown properties of the constituents—dyons. The interaction strength far surpasses anything for which such skill exists. But a beginning must be made. Consider the nonrelativistic behavior of two widely separated dyons, of common mass $M_D$, that are combined in a hydrogenlike structure. The energy expression is

$$H = \frac{1}{2m}\left[p_r^2 + \frac{J^2 - (\nu\hbar)^2}{r^2}\right] + (g_1g_2 + e_1e_2)/r$$

where $m$ is the reduced mass, $\tfrac{1}{2}M_D$. Hydrogen energy levels depend only upon the principal quantum number $n = n_r + l + 1$, where $l$ is here given by

$$l(l+1) = j(j+1) - \nu^2$$

As an initial approximation, let us ignore the fine structure of order $eg_0/\hbar c \sim 1$, and the hyperfine structure of order $e^2/\hbar c \simeq 1/137$. The appearance of the formulas will also be simplified by the adoption of atomic units for which $\hbar = c = 1$. With the specific choice

$$g_1 g_2 = -g_0^2$$

the Bohr formula supplies the total mass, or better, the squared mass as

$$M^2 = (2M_D)^2 \left[1 - \frac{1}{4}\frac{g_0^4}{n^2}\right]$$

This result is valid only when the second term is small compared to unity, corresponding to very large quantum numbers, $n \gg n_0$, where

$$n_0 = \tfrac{1}{2}\, g_0^2 \simeq 2(137)$$

But, *faute de mieux*, let us abandon caution and extrapolate down to zero mass! That is reached at $n = n_0$. The neighboring states identified by $n = n_0 + k$, where $k = 1, 2, \ldots$, are approximately represented by

$$M^2 = (2M_D)^2(2/n_0)k$$

This formula can be compared with empirical meson mass spectra (*7*). Simple but accurate representations of the mass splittings within the known families of $9 = 3 \times 3$ particles enable one to remove this mass structure, and the resulting squared masses are proportional to an integer with actual values of 0, 1, 2, or 3. The scale is supplied by the mass of the $\rho$-meson, which gives the identification

$$(2M_D)^2 = \tfrac{1}{2}\, n_0 m_\rho^2$$

The specific value of $n_0$ noted above refers to the individual magnetic charge magnitude $g_0$. If $2g_0$ were considered, $n_0$ would be four times larger. We shall use a weighted mean of these values, which effectively equates $n_0$ to 4(137), and then

$$M_D \sim (137/2)^{1/2}\, m_\rho \sim \text{6 billion electron volts}$$

I would not risk more than three groschen on the likelihood of this estimate, but at least it is an optimistic one, in relation to current accelerator plans.

Let us return to what was termed the fine and hyperfine structure of the mass spectrum. The hyperfine structure is an electric-charge dependence which causes the interaction strength to vary by a fraction

$$\sim e^2/g_0^2$$

The corresponding change in squared mass is

$$\sim M_D^2\, e^2/g_0^2 \sim (1/137)\, m_\rho^2$$

That is indeed the magnitude observed for the charge dependence as illustrated by the K-mesons

$$m_{K^0}^2 - m_{K^\pm}^2 \simeq 0.7 \times 10^{-2}\, m_\rho^2$$

but the sense of the splitting is opposite to what one would expect from the simple mechanism considered. The fine structure is represented by a variation in $n$ or $k$ that is of the order of unity. Such is the qualitative empirical situation, as illustrated by a comparison of the K-meson, which belongs to the $k = 0$ nonuplet, with the $\rho$-meson, a member of the $k = 1$ nonuplet:

$$m_K^2 \simeq 0.4\, m_\rho^2$$

But the quantitative details are wrong. Only the value of the electric charge would seem to be relevant, whereas observed mass spectra are labeled by and give meaning to the properties of isotopic spin and hypercharge. Something is missing.

## Charge Exchange

Are other known phenomena omitted in this dynamical scheme? Although conventional electromagnetic and strong interactions have been given a generalized electromagnetic interpretation, there is no reference to the so-called weak interactions. This type of interaction can be viewed as a mechanism of electric-charge exchange among members of the same particle family, including the lepton family (L) of electron, muon, and neutrino. It is possible, but not necessary, to regard this charge exchange as proceeding through the intermediary of an unknown, heavy, charged boson, as has been proposed several times under different names. The commonly employed symbol is W(weak), which we use in writing some typical particle reactions

$$L^{+} \leftrightarrow L^{0} + W^{+}, \quad {}^{-1/3}D^{2/3} \leftrightarrow {}^{-1/3}D^{-1/3} + W^{+}$$

There is a striking analogy with electromagnetic emission and absorption of photons (which motivated my own early speculations in this direction) since the observed interactions are essentially vectorial in character and have a certain universality in strength. In more detail, these weak interactions are known to be *CP*-conserving, but *C*- and *P*-violating, in a way that depends upon the sign of the electric charge that is exchanged. It would seem that they destroy the charge rotational invariance of the Maxwell equations and thereby help to establish the absolute distinction between electric and magnetic charge.

I find it natural to imagine a magnetic analog of these processes, with a correspondingly stronger coupling, that could be mediated by a boson of unit magnetic charge, S (strong). A typical magnetic charge-exchange process for the dyon is

$${}^{2/3}D^{-1/3} \leftrightarrow {}^{-1/3}D^{-1/3} + {}^{+}S$$

It is entirely possible that both S and W are fictitious and should be understood only as a shorthand for the direct exchange of charge between pairs of particles of various types. We shall not dwell on the conceivable magnetic counterparts of leptons, except to wonder if the neutral neutrino(s) could be common to both families. The mechanism represented by the magnetic particle S produces a rapid exchange of magnetic charge among the dyons that constitute a hadron. It may be that the result is a very short time scale averaging out of the magnetic charge on an individual dyon. Indeed, that is what is suggested by a naive view of the empirical baryon situation. The pattern of low-lying multiplets is correctly represented if we unite a three-valued electric label with a two-valued spin index and consider only totally symmetrical arrangements of three such index pairs as though three constituents were in a symmetrical orbital state and obeyed Bose-Einstein statistics. Conflict with the physical Fermi-Dirac statistics of dyons is avoided if we recognize the additional three-valued magnetic labels and combine them in a totally antisymmetric arrangement. This implies that each of the three magnetic assignments is equally probable for an individual dyon, thus giving an average magnetic charge of zero.

A mechanism for magnetic charge exchange is also indicated if large violations of *CP* invariance are to be avoided. Particularly relevant is the remarkable precision with which it is known that the neutron does not have an electric dipole moment (*8*). The associated length is measured to be $\lesssim 10^{-22}$ centimeter, in contrast with $\sim 10^{-14}$ centimeter for the magnetic dipole moment. We compare this situation with an elementary model in which a dyon, considered to be a particle with spin of ½, possesses intrinsic magnetic and electric dipole moments proportional to its spin vector and its electric and magnetic charge, respectively. Such models have often been applied to nucleon magnetic moments, with the paradoxical result that the constituents have masses smaller than the nucleon mass, rather than the much larger values that are required physically. This is an artifact of an overly naive nonrelativistic attitude, however, and it is removed if the magnetic (and electric) energy is incorporated as an added term in the relativistic expression for the squared mass of the composite particle. Then it is the mass of the composite particle that sets the scale, and one empirical factor of the order of unity gives a reasonable account of neutron and proton magnetic moments. The analogous electric dipole moment is proportional to the sum of dyon magnetic charges multiplied by spin vectors. Since the total magnetic charge is zero, the electric dipole moment would vanish if all three dyons were dynamically the same and therefore had identical average spin vectors. But surely the dyon of magnetic charge ⅔ is in a different environment than a dyon of magnetic charge −⅓ and should have a somewhat different average spin; this would lead to an unacceptable electric dipole moment, unless a mechanism restores the equivalence of all dyons by a rapid exchange of magnetic charge which effectively destroys the correlation between spin and magnetic charge.

The same mechanism for magnetic charge exchange will tend to suppress those effects of order $eg_0/\hbar c$ that were called fine structure. The exchange mechanism itself produces mass splittings, however. Among the consequences of these couplings is a displacement in the masses of the individual dyons. There is a plausible expression for the exchange interaction that produces a mass splitting of a threefold electric multiplet into a doublet and a singlet, which gives an elementary account of the empirical properties of isotopic spin and hypercharge. These considerations are too quantitative and too uncertain to merit further comment here. Suffice it to say that the general outlines of a mechanism have appeared, which may meet the challenge posed by the regularities observed in the properties of hadronic strong, electromagnetic, and weak interactions.

## Summary

A conceivable dynamical interpretation of the subnuclear world has been erected on the basis of the speculative but theoretically well-founded hypothesis that electric and magnetic charge can reside on a single particle. I hope that these suggestive, if inadequate, arguments will be sufficiently persuasive to encourage a determined experimental quest for the portal to this unknown new world of matter, for

*Nothing is too wonderful to be true, if it be consistent with the laws of nature, and in such things as these, experiment is the best test of such consistency.*

—Faraday

### References and Notes


1. P. A. M. Dirac, *Proc. Roy. Soc. London Ser. A* **133**, 60 (1931); *Phys. Rev.* **74**, 817 (1948).
2. The term hadron has been introduced in opposition to lepton, which designates particles, other than the photon and graviton, that do not have strong interactions. Lepton was well chosen since the Greek combining form *lepto-* includes "small, weak" among its meanings. But, unfortunately, the meanings of *hadro-* are limited to "ripe, thick," which this is, a bit.
3. The argument was presented for particles with either electric or magnetic charge in *Symmetry Principles at High Energy*, A. Perlmutter, J. Wojtaszek, G. Sudarshan, B. Kursunoglu, Eds. (Freeman, San Francisco, 1966). Also noted there is the charge-quantization condition for dual-charged particles, under the assumption that particles carry both charges $e,g$ and $e,-g$. I was not yet ready to face the apparently strong violation of *CP* invariance that occurs if only one of these particles exists.
4. Both arguments are presented in the language of the new theory of sources [J. Schwinger, *Phys. Rev.* **173**, 1536 (1968)]. A discussion that employed the more conventional and more cumbersome methods of operator field theory was given some time ago by T.-M. Yan [thesis, Harvard University (1968)]. It has been duplicated recently by D. Zwanziger [*Phys. Rev.* **176**, 1489 (1968)], although this author does not recognize that two different factors of 2 are involved.
5. Comments in this direction were made by A. Goldhaber [*Phys. Rev.* **140**, B1407 (1965)]. I came to this approach from the opposite direction, by asking how the spin of a particle could be removed in favor of its helicity, the spin component along the momentum direction. The mathematical problem is the same, with position and momentum vectors interchanged.
6. Unfortunately, the field of choice is not free of prior incursions. In the interests of an obscure literary reference that celebrates the empirical aspect of triadism, an untraditional and unmellisonant term was introduced and has found favor in some circles. I prefer to respect tradition and, more important, to emphasize the theoretical basis of the otherwise mysterious empirical characteristics.
7. J. Schwinger, *Phys. Rev. Lett.* **20**, 516 (1968).
8. W. Dress, J. Baird, P. Miller, N. Ramsey, *Phys. Rev.* **170**, 1200 (1968); C. Schull and R. Nathans, *Phys. Rev. Lett.* **19**, 384 (1967).

# THEORY OF SOURCES

J. SCHWINGER*
Department of Physics, Harvard University,
Cambridge, Mass., United States of America

**Abstract**

THEORY OF SOURCES. 1. Introduction; 2. Source theory; 3. Application of the source theory to quantum electrodynamics; 4. Strong interactions; 5. Conclusion.

## 1. INTRODUCTION

I would like to propose a new theory of particles. That may seem like a rather special topic but the particles I have in mind include the photon, which gives one a new approach to quantum electrodynamics. They include the graviton so that we can also deal with gravitational phenomena and the elusive, if uninteresting, quantum corrections to it. Also, though it has not been considered, the methods I am going to outline should be helpful in connection with the non-relativistic problems of many-particle physics, so that in a sense, perhaps, we come close to that feeling of universality which has been the inspiration for this meeting.

However, despite the generality of the theory and its variety of applications, it remains true that it is high-energy physics and the problems of high-energy physics that mark, so to speak, the stimulus for the construction of this theory and so I want first to outline what I see as the difficulties of the conventional approaches and, since the conventional approaches to high-energy physics form a continuum, I will try to polarize this by giving what I consider two extreme viewpoints within which antipodes my own approach will be located.

The two extreme positions, in my opinion, are operator field theory and S-matrix theory. Operator field theory is very familiar; its foundations are the basic ideas of quantum mechanics and relativity. That in fact will be characteristic of all theories we will discuss, but operator field theory is characterized by the assumption that a detailed space-time description of phenomena is possible and, corresponding to that, the fundamental dynamical variables in terms of which this mechanical dynamical system is characterized, are the field-operator functions of space and time. The specific formal operator field theory that I am describing here is one which is based fundamentally upon Hamiltonian or Lagrangian formulations and, as such, a specific Lagrangian is made use of (a function of the fundamental dynamical variables) and so there is a specific dynamics or, as Professor Salam refers to it, a distinguished dynamics. In other words, a specific form of a Hamiltonian or Lagrangian characterizes by hypothesis a particular system of fields as the fundamental dynamical variables. The idea of local action is that there is a Lagrange function and that operator products at the same point are the fundamental structures out of which the

* For the compilation of this paper the author is grateful for notes taken by Mr. R. White.



Reprinted from *Contemporary Physics* (Trieste Symposium 1968), IAEA, Vienna, 1969, pp. 59–83.

dynamics is constructed. The particle is then a secondary, derived concept from this point of view. The particle is a stable or quasi-stable excitation of the fields, that is, as in any quantum mechanical system, one looks for the quasi-stationary states and finds them appropriately in terms of the dynamical variables which, by themselves, do not necessarily have any immediate physical interpretation. Since experimentally it is the particle that we deal with and not this fundamental description in terms of fields, to make contact with experiment it is necessary to carry out a transition from the initial fundamental field theory to the phenomenological or the physical particles. That leads to the concept of renormalization because we know that a field operator for an interacting system does not create simply a particle, it is just a way of producing a localized excitation and it will generate every combination of particles with the same set of quantum numbers. If there is a particle with that set of quantum numbers it will generate it, but only for a fraction of the time and so, if one is interested in that particle, then the process of renormalization must be carried out, which is to change the emphasis from the field which generates the particle for only part of the time to the particles supposed always to be in existence.

Renormalization is more fundamental than it is usually considered in connection with perturbation theory which is just a particular mathematical technique, an important but limited one, and it is very familiar that, when one proceeds to convert this general formulation into specific concrete terms, one usually falls back on perturbation theory, and then divergences appear. I leave open the question whether the original theory is mathematically consistent and whether it is divergent or not, but certainly divergences do appear as soon as perturbation theory is used. It is very familiar that in electrodynamics all the divergences are compressed into a very few renormalization constants and therefore the actual physical predictions of the theory are entirely finite and I need not remind you in what excellent accord they are with experiment. But when we come to strongly interacting systems, which is, of course, of principal interest, then it is simply impractical to carry out the transition from the field to the particle point of view. For weakly interacting systems such as one meets in electrodynamics it is usually taken for granted that the transformation from field to particle is not a large one, that is to say, that the fields really are rather closely associated with the particles of interest, and for that reason perhaps perturbation theory is roughly relevant, and at least the logical connection between the particle and field is not very remote. But it seems very plausible that for strongly interacting systems we have no way of even being sure that we have the correct connections between particles and fields; in other words, specific hypotheses of the particle structure are required. We must assume, for example, that it takes the products of, say, three fundamental fields or two fundamental fields perhaps, to make this or that baryon or meson; this is a hypothesis and it is upon these hypotheses that the further development of the theory is based. In other words, I make here the rather serious objection that in order to be able to talk about the physically interesting phenomena at all one must begin with a speculation about how these particles are formed. And so I have the very logical, the very strong logical, objection to this formulation (quite apart from its practical difficulties) that it mixes together phenomenology in the sense of how the observed properties of the particles are correlated with an unwarranted speculation (which is possibly correct, but the probability is small) about the actual nature of the particles.

Contrasting with this is the second point of view, that of S-matrix theory, which arose not unnaturally as a reaction to field theory and its difficulties. Again, it is based upon the faithful servants of quantum mechanics and relativity, but here we have the complementary point of view; the description is now embedded in momentum space and space time is made use of not at all or perhaps only in a sort of macroscopic correlation limit. The point of view is not an operator one but rather the S-matrix is taken as fundamental and the labels of states are the variables of various particles. In principle, since one is not allowed to use a time description, it is presumably only stable particles with which one is concerned, those that have no time scale attached to them. The principles that have entered into this theory are basically unitarity plus analyticity. These are very general principles, and certainly no specific dynamics in the analogous sense of the Lagrangian or Hamiltonian field theory is used, but one hopes, that some requirement of self-consistency between the various particles will eventually distinguish our world from the infinity of possible other worlds. This is an interesting idea; I have no way of judging how successful it actually has been. But in a way also I would draw an objection to this, that again, in order to be able to discuss phenomenologically the properties of the particles, there is built into this procedure a hypothesis, namely that the particles have no underlying structure. The self-consistency or bootstrap idea is the notion that the observed particles are fundamental, that they create each other and that there is no deeper level in terms of which the particles are to be understood or explained. And that again is a hypothesis. It may be correct but I assume that the probability of that is also small. In any event, such a hypothesis should not, if possible, intervene in the day-by-day task of correlating data and arriving at a phenomenological theory in terms of the useful correlation and description of phenomena.

## 2. SOURCE THEORY

It is on the basis of these two extreme positions that we come to an intermediate position which is the idea of source theory. It is based again upon quantum mechanics and relativity. It is an intermediate position in the sense that while it is formulated in space-time (it recognizes that measurements are space-time operations), nevertheless it leaves open the question of to what extent an extrapolation down to a point is physically meaningful. In short, it is left open to test experimentally how far down in space and time the possibility of a detailed description can be pushed. Of course, it also makes use of momentum description (that we can hardly avoid), but it leaves open the question of how far throughout the totality of momentum space this is to be applied; that is to say, it is a more experimentally motivated theory, it does not take for granted that infinite momentum space is available from the very beginning, it is an extrapolation upward in momentum, downward in space and time, always testing through experimental comparison whether these extrapolations are in fact valid.

The basic notion is the source and the source is introduced as an attempt to incorporate into the mathematical formulation, as much as possible, what the practising experimental physicist does and to try to find a correlation between actual experimental practice and mathematical theory. We try to abstract what is relevant and to put that into the formulation of the theory. The source is introduced as an abstraction of those realistic collision processes which in fact one either can, or for unstable particles must, make use of in order to generate the particles in which one is interested. Perhaps this theory has arisen at this time as a natural reaction to the totally new state of affairs. We are familiar with the earlier stages of the development of quantum mechanics and relativity in which we were concerned only with a few stable particles. Now the situation is entirely different. The particles we are concerned with are almost all unstable. And instead of regarding this as an aberration I choose to regard it as a definition of particle, that in general, if we are interested in the particle, we must create it to study it. This is certainly true of the high-energy particles we are concerned with; they are not normally found but must be created. The creation of the particles of interest is the first stage of an experiment and in the final stage the particle must be detected and, in a general sense, that is the annihilation of a particle. The detection methods are invariably the transformation of that particle into other forms of matter more amenable to experimentation, the detection process is the destruction of the particular particle of interest. The purpose of the source theory is rather two-fold; firstly the sources, since they are associated with particle creation and annihilation acts, serve as identifying instruments for the particle in question when it enters into the theoretical framework. Secondly, the source of a particular particle unites all of those dynamical properties that various mechanisms for creating that particular particle have in common. However, it does not embody the individual peculiarities of this or that mechanism. But nevertheless, when the whole theory is constructed, it must be verified that the various mechanisms which can create a particle indeed have in common those starting points, those hypotheses of a notion of the source. So in that sense it is a very useful way of talking about dynamics because it incorporates from the very beginning a survey of all dynamical processes and then it begins to put in the details step by step.

What are the constructive principles that we will make use of in the formulation of this theory? They are basically two. Firstly causality, which is the idea that in fact the experimentalist must create the particles and then he must let time elapse as the particles get near each other and interact and then more time must elapse as the particles separate and are eventually detected. This idea of being able to control the space-time arrangement, to be able to make sure that one act precedes another act and so on, is the basic idea of any physics and it occurs here as a creative principle in the sense that we take initial causal arrangements which are so arranged because we can produce them through our experimental knowledge or abilities and use this as a process of extrapolation, always testing, of course, whether the extrapolation and the new physical phenomena that it predicts are indeed in accord with experiment. The second principle is the idea of space-time uniformity, by which I mean that when a particle is created by one source and then travels and is detected by another source, the initial coupling inferred from a certain causal arrangement is supposed to be more general than that original arrangement. The sources can be moved around and the coupling will still continue to be meaningful and so give logical connections between phenomena in which particles actually do exist as real particles and other phenomena in which they are not actually present, in which perhaps not enough energy occurs to actually create the particles. I also mean that sources occupying different regions of space are not independent physical phenomena, that they are simply different manifestations of the same underlying laws of physics.

The development of the theory starting from these constructive principles begins initially with the idea of particle as fundamental and the creation and detection of the particles, so to speak, as unanalysed acts but upon which is superimposed the self-consistency requirement. Out of this emerges, as a derived concept, a field theory; it is a phenomenological field theory because the fields are attached to the observed particles, and it is most interestingly a non-operator field theory. There are field equations in this theory but there are no commutation relations and no operators. Most of the complicated mathematical difficulties of the initial operator field theory attest to the enormously difficult concept of operators which are functions of points and operators multiplied together at points.

It is a phenomenological theory but it is a physical one which describes how couplings occur through the transport or the exchange of particles. As such it is non-local. The word phenomenological as I use it here, I think, does not have the same associations for me that it did for Professor Heisenberg. I regard this as a phenomenological theory in the sense that we are dealing with the actual phenomena, but it is a creative theory in the sense that different phenomena are connected by fundamental principles. The procedures used, however, are flexible and reflect the complexity of the physical problem and the amount of information available. If one talks about weakly interacting systems, and I think of electrodynamics or gravitational phenomena perhaps, then one begins with a rather specific primitive interaction, as I call it, by which I mean that there is essentially no detail about the inner structure of this primitive interaction; it is represented in fact by a local product of fields. By the repetition of processes in which particles are exchanged among sources one elaborates the dynamics.

The dynamics is not stated once and for all, but it grows organically by considering more and more complicated processes, as a result of which one comes closer and closer to the actual dynamics which is, of course, non-local. For example, beginning with the simplest properties of photons and electrons, one can, out of such a primitive interaction which refers to an experimentally conceivable, if idealized, situation, reconstruct quantum electrodynamics (as has, of course, been obtained first by operator field theories with all the well-known and verified experimental consequences) without any reference to either renormalization or divergences, and in the language of a numerical field theory. If one begins with the graviton (which is a rather more hypothetical beginning, but one can adduce rather strong arguments for the existence of the graviton indirectly) one obtains a reconstruction and a quantum extension of gravitational dynamics — a quantum extension because one is now enabled to discuss, as I think one could not really before, some of the quantum corrections analogous to electrodynamic corrections to gravitational phenomena. It is a purely conceptional problem, of no practical interest, but it has been a vexing question for very many years; I would like to think that it is now solved in principle. When we turn, however, to more interesting strongly interacting systems then, of course, we have no possibility of thinking that we know a simple dynamical mechanism (as we do in electrodynamics) out of which all the complicated machinery of interacting systems really evolves. We have to recognize the fact that there is an unlimited number of particles with all kinds of complicated interactions and we must begin in a very modest way. And so I suggest that we first turn to low-energy phenomenology, the known properties in the relatively low-energy domain of pions, nucleons, mesons, and so on, and try to codify what is known about these particles in terms of the notion of partial symmetry.

It is well known that in high-energy physics there are many conservation laws, but they are approximate. There are symmetries or conservation laws valid only in certain limited senses and one must, of course, in any formulation use them as a basis for codification or systematization of the phenomena. But I suggest doing this by means of creating a primitive interaction out of which the more elaborate theory will evolve. The important thing is that it be done entirely at the phenomenological level and no hypotheses about inner particle structure be involved. From there one tries to extrapolate to higher energies; one does this by beginning with the local couplings and using them as primitive interactions from which the non-local phenomena actually emerge. When we turn to very high energies, and this is really a totally untouched subject from this point of view, there is a natural analogy that I think I ought to point to. I like to think that what we are doing in this field (as we have done throughout all physics) is observing a certain kind of physical system (molecule, atom, nucleus, whatever it may be), initially beginning from the outside, observing only certain general features. As we move up in energy, we come closer, we explore more and more details. For example, consider the stage of nuclear physics in which at first we recognize only charges and magnetic dipole moments. We come closer and we begin to learn something about electric quadrupole moments. We come closer still and we get higher and higher moments. I would like to suggest that there is an analogy with high-energy physics in a sense that, as we go to higher energies we begin to recognize particles of higher and higher spins. The interactions which these higher and higher spin particles carry are, mathematically speaking, more and more singular just as multipole fields are more and more singular. But we know very well what the story is with multipole fields and multipole expansions; they are useful, if only a few are involved (low energy). If we get very close to the system or if we get inside it, the multipole expansion is useless; in fact, the series of multipole moments diverges. It is simply a way of writing the actually well-defined interaction of the physical systems which have a structure, and when we penetrate seriously into the structure the multipole expansion is useless. And so I suggest that we are meeting the same situation again; the present particles, the few low-energy low-spin particles we presently know about are, so to speak, analogous to low multipole moments of the more fundamental interaction. As we move up in energy we explore closer and closer and the multipole expansion will be useful for a while; eventually it will fail and must be replaced by the more fundamental dynamics in which no longer the particles which we observe, but rather the fundamental structures out of which they are constructed, are the instruments. But I submit to you that we have no right to think that we can guess this more fundamental interaction without the clues which will be obtained by moving ever inward, ever upward in energy, correlating the data until we have enough information to actually carry out that quantum jump Professor Heisenberg describes. The jump is usually not from an infinitely remote position but from as close as possible to the final goal. It can only be made by finding a formalism which does not have incorporated into it a hypothesis about the more fundamental structure but does enable one to make contact with it in the sense that it is flexible enough in its framework, and that is what I like to think the theory of sources is.

I have said before that, although the phenomenological theory of sources is designed to correlate data and to be as independent of speculations about inner structure as possible, nothing prevents one from speculating independently. What goes on in the innards of the particles we observe may be a manifestation of an entirely new form of matter, not merely fractionally electrically charged but also magnetically charged, and perhaps out of these incredibly strong but, amazingly enough, electromagnetic forces, the actual nature of strongly interacting systems is to be understood. This is a speculation and I have absolutely no idea of how to bring into contact this speculation at the one extreme and the evolving phenomenology of the actually observed particles at the other. But it indicates the eventual goal.

Now we come to the discussion of the formalism of the source theory. Figure 1 represents a realistic collision process whereby particular particles of interest, indicated with heavy lines, are either created or annihilated. The source theory recognizes that if we are concerned only with the particle and if a collision is there simply to produce it or detect it, everything else that enters into the collision serves only the purpose of supplying the properties of that particle. So we abstract from everything and we indicate that, as far as what we are interested in is concerned, a particle emerges or disappears within a more or less limited space-time region. (We have some ability to control this – there is some control over the various momenta that the particle will have.) There is a source in space and time describing the effectiveness of carrying out some kind of space-time localization – that is certainly possible – and, to some extent, the localization in energy and momentum. Certainly we can do that, but we leave open the question of how far it can be pushed. To the extent of our concern with that particle and only that particle we begin with the vacuum. In other words, we disregard everything else and recognize then that the source operates. It is a realistic process, but for our present purposes it simply creates or annihilates a particle. Consider

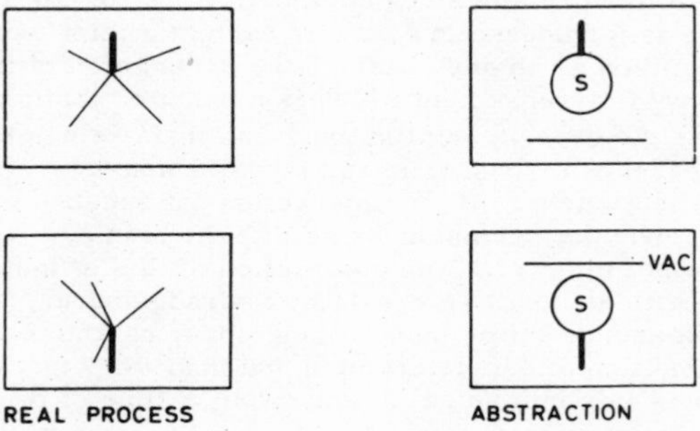


FIG. 1. Collision process.

a particular case, a very simple one, a spin zero stable particle, with no properties other than mechanical, and a weak source which at most can create a single particle which is specified by a momentum p. The subscripts on the zero which represents the vacuum state simply indicate the causal arrangement, the time sequence. We are talking about the creation of the particle and so initially (before the source operates) we have a vacuum. We then create the particle and for this there is a quantum mechanical probability amplitude. The quantum mechanical probability will be proportional to a certain invariant element of momentum space — so this enters as a kinematical factor and the rest is simply the definition of the source. (I use the symbol k to refer to spin zero specifically.)

Vacuum and one particle states are not orthogonal??

$$\langle 1_p | 0_- \rangle^k = i\sqrt{d\omega_p}\, K(p) \qquad d\omega_p = \frac{1}{(2\pi)^3}\frac{d\vec{p}}{p^0} .$$

But the notion of quantum mechanical complementarity is that the effectiveness of the source in the momentum language is complementary in a four-dimensional sense to the effectiveness of the source in terms of the description of space-time localizability.

$$K(p) = \int (dx)\, e^{-ipx} K(x)$$

In fact very little of the ideas of quantum mechanics is really needed beyond this and some of the general composition properties of probability amplitudes. Finally, there is the reversed process, the annihilation of the particle and the fact that the source is now being used as a sink; instead of emitting the momentum p it absorbs the momentum p.

$$\langle 0_+ | 1_p \rangle^k = i\sqrt{d\omega_p}\, K(-p)$$

The appearance of the minus sign involves no hypothesis about time reflection or anything of the sort. One will recognize later from the way in which it is actually used that if we make use of any other assumption we violate translational invariance in space and time. So in fact it is simply space-time uniformity which demands that. This is how the definition of the source is introduced in a simple context where there is only a weak source. We can then go on and consider the description of the first and simplest experimental arrangement which is a beam of particles.

As the experimentalist produces them, a number of particles may be present at a given time and they are non-interacting, not because we mentally switched off any interaction but because we have arranged the experimental circumstances so that the particles are far enough apart and do not interact. The abstraction of this is indicated in Fig. 2. We begin with nothing, we create by a strong source, which is effectively just a weak source acting many times, those particles we are interested in on the condition of non-interaction and then we detect them again. We may graduate this into an act of emission, a time of propagation and an act of detection and we put together the two basic acts of the emission and absorption as this is generalized to a strong source and the generalization is in terms of an exponential structure.

$$\langle 0_+ | 0_- \rangle = e^{iW}$$

I think it will be recognized that the exponential structure is the familiar way of describing a number of physically independent processes; we are just multiplying them and if there is an unlimited number of them, an exponential sum arises.

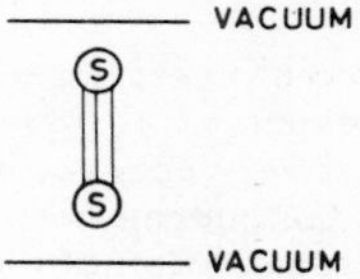


FIG. 2. Non-interacting particles.

In fact, the structure of W just puts together in a space-time language these individual acts. There is a reference to two sources; one will act individually as an emission source, the other as a detection source. But they are all part of one general mechanism and all put together into the general framework of source. W is symmetrical because it is a quadratic form and so the matrix that occurs between states is effectively symmetrical. In the causal arrangement with which we begin we infer what this propagation function is; it has the form

$$W = \frac{1}{i} \int (dx)\,(dx')\; K(x)\,\Delta_+(x - x')\,K(x')$$

$$\Delta_+(x - x') = \Delta_+(x' - x)$$

and

$$\Delta_+(x - x') = i \int d\omega_p\, e^{ip(x-x')} \quad x^0 > x^{0\prime} \quad p^2 = -m^2$$

which, you may recognize, puts together the square root of the invariant momentum amplitude and the plane wave that probes the emission source to pick out the effectiveness of emitting a certain particle, and the second square root factor and the plane wave that probes the detection source to measure its effect in detecting the particle. These two factors come together which is the origin of what is known as a propagation function, but the word propagation as such has entered nowhere. All that has been used is the fact that the particle that leaves the emission source is the same particle that hits the detection source; nothing has happened to it. And so one arrives in a very elementary kinematic way at the structure of a very familiar function starting out from these quite simple quantum mechanical relativistic notions. I mentioned an important physical requirement.

This vacuum-to-vacuum probability amplitude describes the whole process. We begin with the vacuum, we end with the vacuum and we describe complete acts of emission and absorption. But this quantity by itself has a physical significance; it is a vacuum-to-vacuum probability amplitude and its absolute square expresses the physical probability that if we begin with the vacuum we should end with the vacuum despite the intervention, the disturbance, produced by the sources. And as such this must be a number less than one.

$$|\langle 0_+|0_-\rangle^k|^2 = \exp\left(-\int d\omega_p |K(p)|^2\right) \leq 1$$

This particular structure meets all those requirements. One instantly checks that it has an exponential form, it is certainly less than or equal to one and it is qualitatively correct in that it is less than one by a measure of the extent to which the source does create a particle and the exponential is just a familiar representation of the multi-particle creations that may go on.

One may simply erect the whole theory of non-interacting particles of any spin on the same basis.

## 3. APPLICATION OF THE SOURCE THEORY TO QUANTUM ELECTRODYNAMICS

Let me turn to some interesting special features which occur for familiar particles. Consider the photon. To represent a unit spin we must use a vector, and so the coupling will necessarily be of the form

$$W = \frac{1}{i}\int (dx)(dx')\, J^\mu(x)\, D_+(x-x')\, J_\mu(x')$$

$$|\langle 0_+|0_-\rangle|^2 = \exp\left[-\int d\omega_p J^\mu(p)^* J_\mu(p)\right]$$

The exponential quadratic structure is just the statement of the basic act of the emission, propagation and absorption acts endlessly iterated without interaction. The propagation function is here called $D_+$ just as a reminder that we are talking of massless particles, the vector is the statement that we are interested in spin one. Now suppose we compute the vacuum persistence probability. We shall have exactly the same structure as for spin zero but now it contains a reference to the vectorial nature of the source and the indefinite matrix of space-time. Now we have something which by itself is certainly not less than one. If the vectorial source is unlimited, the exponent could be positive or negative and so we shall have a violation of quantum mechanical positive definiteness which I take to be a very basic requirement in the structure of the theory. It is necessary therefore to restrict the source so that the time component never exceeds the space component. It is also necessary to do that to have a proper account of spin; a necessary and in fact sufficient restriction is that the scalar, which can be constructed in the only possible way given the ingredients available, be equal to zero

$$p_\mu J^\mu(p) = 0$$

or, more familiarly in space and time, $\partial_\mu J^\mu(x) = 0$. This must hold and then in fact we satisfy all quantum mechanical restrictions and we correctly come back to just those modes of excitation which are the two transverse polarizations of the photon. The important thing, however, is that we arrive at this condition on sources out of general quantum mechanical restrictions. We therefore have a general property of sources for photons which must be characteristic of all possible mechanisms. If the source has a restriction, then every possible realization of the source must have that restriction and so we are uncovering a general law of physics. We know what it is — it is the conservation of charge. We are here reversing the more familiar discussion, the existence of the photon is now taken as the basis of the inference that a property like charge which satisfies conservation laws must exist in nature. To illustrate the fact that this is not merely phenomenology, that we are not just describing the fact that there are photons, I mention the constructive principles of, in particular, the notion of space-time extrapolation. We begin with sources in time-like, or actually, for the photon, in null arrangement — but let us say time-like, it is the more general situation — and we end up with a coupling with the quantity $W = JDJ$. Nothing in that space-time structure tells us that the component sources have to be in any particular arrangement. Therefore give it a wider validity, let it be meaningful also if the source is made up of two component parts, for example, which are in space-like relation (Fig. 3). These sources may in particular be considered in the limit where they vary very slowly with time, so we are, so to speak, roughly dealing with the situation where two static charge and current distributions are existing simultaneously and are interacting with each other. If we look at the structure of W, we will find that under those conditions of quasi-static sources it reduces to a single time integral of a quantity which is clearly identified from the connection between probability amplitudes or wave functions, as the energy that this system possesses under those particular quasi-static conditions.

$$\langle 0_+ | 0_- \rangle^J = \exp\left(-i \int dx^0 E\right)$$

where energy E is

$$E = \frac{1}{i} \int d\vec{x}\, d\vec{x}' \, \frac{J^0 J^0 - \vec{J} \cdot \vec{J}}{4\pi |\vec{x} - \vec{x}'|}$$

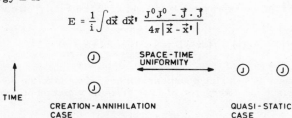

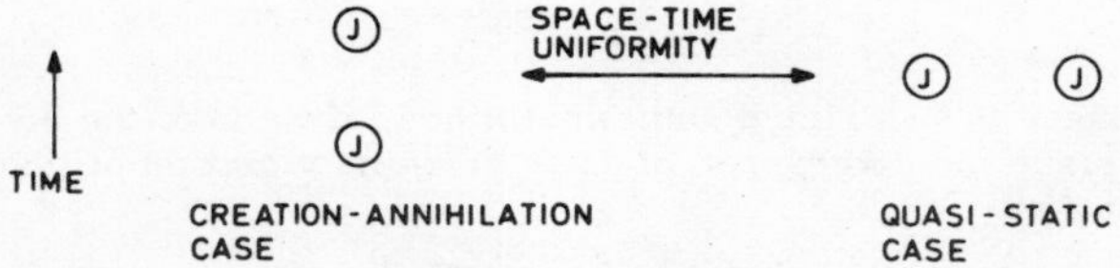


FIG. 3. Source made up of two component parts in space-like relation.

You will recognize the Coulomb and the Ampèrian laws of interaction of static or quasi-static charge distributions, derived here not from Maxwell's equations – we have not seen them yet – but from this notion of space-time extrapolation beginning with the properties of the photon as basic, representing them by couplings between sources, mediated by the photon, giving that coupling a wider space-time significance, extrapolating it here in space-time arrangements and arriving at the interaction laws of Coulomb and Ampère. We may express that a little more generally in the following way. We have begun with action at a distance, we may say the sources are coupled at remote space-time intervals; we could, however, ask whether that could be replaced by a local action – which of course is the introduction of the field concept except that, since we are not dealing with operators but numbers, it is a numerical field – and introduce the field just as one does it in classical physics. We will, so to speak, add a test source into a pre-existing arrangement of sources, and we will add a very weak source $\delta J^{\mu}$. That source, of course, must by itself – since it is superimposed on those already existing – satisfy the necessary restrictions that any photon source has to obey; it must be a vector which is divergenceless:

$$\partial_{\mu} \delta J^{\mu} = 0$$

If we ask how W alters, it will be linear in this new test source and there will be a quantity which measures the effect on that source of the pre-existing sources:

$$\delta W = \int (dx)\, \delta J^{\mu} A_{\mu}(x)$$

It is, of course, a vector potential, and we recognize immediately that the vector potential so defined is not uniquely determined, because $\delta J^{\mu}$ is not arbitrary but subject always to the conservation condition. Therefore, A is arbitrary to the extent of an additional gradient which will have no effect on integration by virtue of the divergenceless condition of $\delta J^{\mu}$. In short, we recognize that vector potentials are in principle subject to arbitrary gauge transformations.

$$A_{\mu}(x) \to A_{\mu}(x) + \partial_{\mu} \lambda(x)$$

What is A? From the expression W we can read it directly, and we find

$$A_{\mu}(x) = \int (dx')\, D_{+}(x - x')\, J_{\mu}(x') + \partial_{\mu}\lambda(x)$$

The arbitrariness is restricted somewhat since, if we take the divergence of $A_{\mu}$, then, since the divergence of $J_{\mu}$ is zero (no violation of that is thinkable), we find

$$\partial_{\mu} A^{\mu} = \partial^{2} \lambda(x)$$

and since the propagation function satisfies

$$-\partial^2 D_+(x - x') = \delta(x - x')$$

as an expression of the fact that energy and momentum are related by the d'Alembertian equation, we find, upon introducing $F_{\mu\nu} = \partial_\mu A_\nu - \partial_\nu A_\mu$, the Maxwell field equations:

$$\partial_\nu F^{\mu\nu} = J^\mu$$

derived from the starting point of photons and sources.

Once we have a local description, fields being introduced as derived or secondary quantities, we want to recognize the fundamental significance of the quantity W which I have so far refrained from giving a name; action is the name. Let us recognize various reformulations of this structure W. We begin with action at a distance; source, propagation function, source.

$$W = \frac{1}{2}\int (dx)\, J^\mu D_+ J_\mu = \frac{1}{2}\int (dx)\, J^\mu A_\mu = \frac{1}{4}\int (dx) F^{\mu\nu} F_{\mu\nu}$$

All these are equivalent by virtue of the interrelations of the definitions of the fields. Also

$$W = \int (dx)\,[J^\mu A_\mu + \mathscr{L}]$$

where

$$\mathscr{L} = -\frac{1}{4} F^{\mu\nu} F_{\mu\nu} = -\frac{1}{2} F^{\mu\nu}(\partial_\mu A_\nu - \partial_\nu A_\mu) + \frac{1}{4} F^{\mu\nu} F_{\mu\nu}$$

The point is the following. Think how W depends upon the variables; think of J, the source and the fields as being independent. Suspend your information about the connection that actually exists. Then the variation of W will consist first of the variation of J times A and then there will be the additional terms that come from varying the fields wherever they are. If only the vector potential is used we get

$$\delta W = \int (dx)\,[\delta J A + \delta_A(\mathscr{L} + JA)]$$

But remember the definition of vector potential; it is the expression of the way in which the quantity W depends upon J and so

$$\delta_A(\mathscr{L} + JA) = 0$$

In other words, we derive a principle of stationary action. So we have an action formulation, at least in the limited physical circumstances of the existence of a local Lagrange function, that is to say of a local function of fields, which, however, we must expect to be eventually superseded by non-local interactions, by non-local Lagrangians as the dynamics develops. How does the dynamics develop? It develops by first extending the source concept. Let us think in terms of the electron. Our starting point is what I now call a simple source (Fig. 4). I use the symbol $\eta$ for electron source, but I ask you to remember what this circle with the symbol in it stands for. It is an idealization of the realistic collision process in which electrons are generated. And how are they generated? Either by perhaps a beta decay process or more simply by a collision in which an initially slow electron is made an energetic electron by some kind of collision act. In other words, the process of creating the particle is inevitably abrupt and we know that the one dynamical feature of electrons and the electric charge they carry is that an accelerated charge radiates photons.

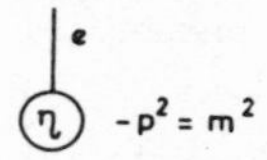

FIG. 4. A simple source.

In short, the same physical act that creates an electron by some kind of collision process unavoidably can and will create an electron and a photon. We can avoid that only by a control of energy and momentum impossible in practice. The point is particularly acute here because the photon has no mass, and photons of arbitrarily low energy exist, but nevertheless I point to this as a general feature. A collision that can create a particle of certain quantum numbers can in principle (given enough energy and momentum) create any other combination of particles with the same over-all quantum numbers if indeed those particles have knowledge of each other dynamically. If they do not, then nothing of this kind happens. But the photon is intimately related to the electric charge that the electron carries, and so, while in principle it would be possible to separate the two acts of the creation of an electron and an electron and a photon, in practice it would be madness because we would then sever the dynamical connection that actually exists between the photon and the electron. And so therefore we are led to extend the concept of source, saying that under those conditions in which more energy is available under the same set of quantum numbers, the electron <u>and</u> the photon can be created. On this qualitative basis one can erect the whole of quantum electrodynamics. If we say this, if we introduce the idea of extended source, then we have a new coupling. We have now a source that creates two particles which can be detected by <u>their</u> respective sources. So we now have cubic couplings among sources. We do not want to regard a simple source and an extended source as different entities; we want to say that the only thing that is different about them is the purely accidental feature of how much energy is available. If enough energy is available to create a photon, it will be created. If it is not available it will not be created. But nothing else distinguishes the sources and so we try to unify them.

Consider Fig. 5 which suggests the unity of the sources. An electron source has a particle travelling to it. This expresses the fact that it is going to detect that particle, therefore the lower extended electron source should also have a particle travelling from it. But that particle, of course, is not a real particle. We are now extending the particle concept off the mass shell, as the language goes, which simply means, we are taking the same mathematical structure and continuing to give it a meaning in its space-time form. But it no longer has the same meaning in terms of particle structure. The point is that only by this symmetry could we then consider another arrangement in which one source is simple, detecting a real electron, and the other source becomes extended, emitting an electron and photon. I have added symbols like A and $\psi$ to express the fact that the coupling is after all in terms of the idea of more or less local action and is going to be expressed not in terms of the sources, which may be infinitely remote, but rather in terms of the fields, which indicate how the action of the sources is brought into this more or less limited space-time region. This is not, however, an operator field theory although fields <u>are</u> involved because we are talking about particles with certain kinematical properties

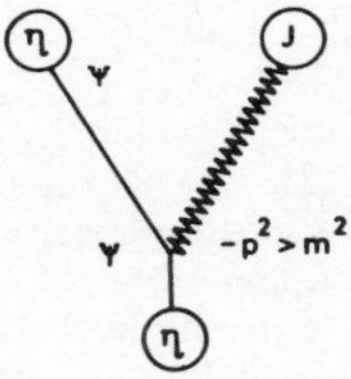


FIG. 5. Diagram showing the unity of sources.

and they had always the same expression. But the mechanism of dealing with these fields and their associated particles is entirely different. A particular example is

$$\mathscr{L} = \ldots + e\bar{\psi}\gamma^\mu\psi A_\mu$$

$$\mathscr{J}^\mu = e\bar{\psi}\gamma^\mu\psi \qquad \partial_\mu \mathscr{J}^\mu = 0$$

This is a cubic coupling with appropriate vectorial properties and so on, introduced in order to express the idea of the extended source. But at the same time we are now realizing for the first time, in terms of an explicit dynamical mechanism, an example of a general photon source. $\mathscr{J}^\mu$ is an electron current but it is now a specific dynamical mechanism. Here is the simplest example of self-consistency between the general framework and the specific realization. So here, for example, is a primitive interaction; it is not the end of the theory, it is the beginning.

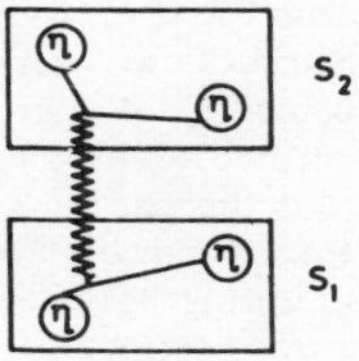


FIG. 6. Extended sources.

Once we have introduced cubic couplings, we infer logically the existence of quartic, etc. In other words, we now have more and more complicated phenomena that necessarily are called into being. Consider the extended source, $S_1$ (Fig. 6). What does it do? It radiates an electron which is detected and then, out of that limited space-time region, $S_1$, a photon occurs. In other words, here we have a more complicated realization of a photon source. Now it has inner structure, but out of it comes a photon and it must be represented as a photon source by the same general machinery as any other photon source. Also shown is its inverse, $S_2$.

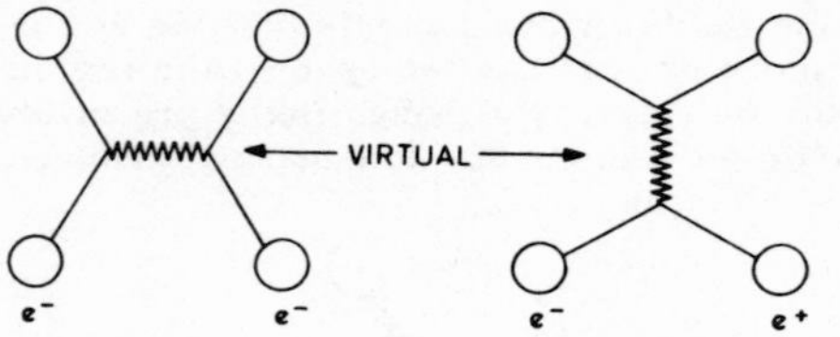


FIG. 7. Scattering skeletons.

These extended sources can now be applied by a process of space-time extrapolation to other arrangements. Figure 7 shows two such arrangements. I ask you to appreciate that the topology of the connections here is the same. We have just moved the sources around. But as we do that we now have an arrangement in which the four electron sources are used to emit or absorb real electrons. We have the process of electron-electron scattering, in which the photon is now virtual, merely by the process of coupling and applying the same space-time arrangement in other forms. The second possibility has instead of a space-like photon a time-like photon. It is the same arrangement and it is the familiar process of electron-positron virtual annihilation as part of the contribution to scattering. What we arrive at in this way, I would have called scattering skeletons. They will be further corrected and improved at later stages. As we have more complicated couplings, we have more complicated possibilities for sources, more complicated interactions and still more complicated possibilities for sources.

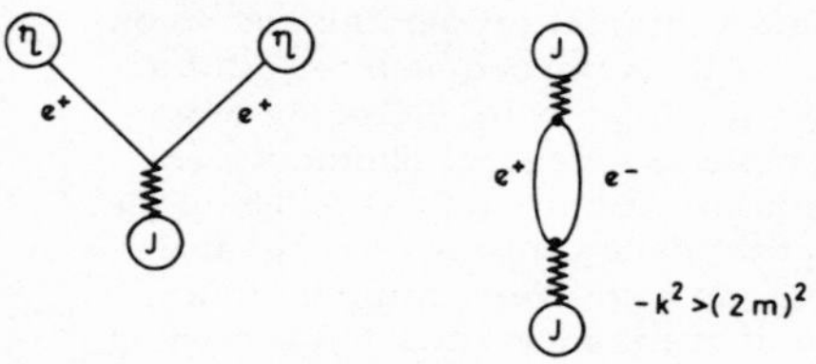


FIG. 8. Creation of electron-positron pairs.

Now there is a next stage, in which we recognize that where single-particle exchange could occur, given enough energy and momentum, two particles could also occur. Consider again the primitive interaction which is a coupling between two electron sources and a photon source. One application of it is to say, let the photon source emit enough mass; it is then a virtual photon, so that it can create an electron-positron pair (see Fig. 8). If we consider the coupling between photon sources, starting out from our initial point of view in which we allowed only enough energy and momentum (or mass) to emit a photon and then detect it, if these sources are given a wider meaning so that they emit more energy or more mass corresponding to the pair threshold, this pair, so to speak, details the mechanism; the virtual photon field creates the electron-positron pair which, after perhaps a macroscopic interval, recombines to a virtual photon which is detected by the source. This means, therefore, that the two sources are now coupled together by a more elaborate function which represents not only the possibility of a single photon exchange but this and yet more complicated possibilities. Its very familiar momentum form which refers only to the existence of the photon must now be extended to describe a wider range of physical circumstances, not by modifying the initial description of the photon, because that remains intact – it is our phenomenological starting point – but by recognizing the existence of new modes of excitation if the energy and momentum are available to excite them. And since these modes of excitation simply represent the propagation from one space-time region to another of a certain amount of energy-momentum or mass, it looks just like (and the whole formalism automatically supplies this) a new spectrum of particles with a variable mass in which all the dynamics is comprised in a weight function which must be real and positive and describes the particular phenomena involved.

$$D_+(k) = \frac{1}{k^2 - i\epsilon} \rightarrow \bar{D}(k) = \frac{1}{k^2 - i\epsilon} + \int dM^2 \frac{a(M^2)}{k^2 + M^2 - i\epsilon}$$

In other words, there is no such thing as renormalization because our starting point is not a more fundamental field but the particle itself and what we are doing is not changing the initial description of the photon but widening the experimental circumstances. For example, if we actually compute the function $a(M^2)$ for the simplest process, we find (I indicate only the limit where the virtual mass involved is large compared to the electron mass) it is the fine structure constant divided by $m^2$ as the only available mass that occurs (it could be no other from purely dimensional considerations). This leads to a perfectly convergent integral for $\bar{D}(k)$. Suppose that there actually did exist – they do not – massless charged particles. Then you might recognize that this integral over virtual masses will have a rather singular structure and in fact it would influence the Coulomb law. It would then be altered in the characteristic way

$$\frac{1}{r} \rightarrow \frac{1}{r^{1+\alpha A}} \qquad r > \frac{1}{m}$$

associated with the possibility of creating pairs of oppositely charged massless particles, had they existed. Coulomb would never have discovered his law if indeed massless charged particles had existed. I mention this to indicate the fact that the formalism we are discussing could equally well have begun with a source appropriate to gravitons, supposed to be massless particles of spin two. Instead of a vector source it would be a tensor source and the whole theory would evolve in the same way. It is perhaps of interest to mention that, after all, for gravitons massless charged particles do exist because charge here means gravitating charge, that is mass, and all particles carry mass in a physical sense. I do not mean rest mass but mass. And so, for example, a graviton source can also emit pairs of gravitons which are massless gravitating particles and can emit pairs of photons and the formalism can be and has been worked out; what goes on is very easily understood on dimensional grounds – instead of the dimensionless fine structure constant we have the dimensional gravitational constant and the square of the wave number of the virtual particles must occur to keep the dimensions. And in fact here is the corresponding asymptotic form, very interestingly, Newton's potential is changed, but it is changed not by an altered power but by an additional cubic interaction.

$$\frac{1}{r} \rightarrow \frac{1}{r} + c\,\frac{G}{r^3} \qquad r > \frac{1}{m}$$

$$\sqrt{G} \sim 10^{-32}\ \text{cm}$$

A student of mine has computed what this constant is; it is known for graviton pairs and for photon pairs. I remind you of the fact that the basic unit of length which the phenomenological gravitation constant introduces is the incredibly small number $10^{-32}$ cm. This modification could be regarded as saying that we do not really know what distances are to within an uncertainty of plus or minus that amount. It is obviously of no physical practical interest, but I mention this really to indicate that we now have the possibility of answering these conceptually interesting quantum questions concerned with gravitational phenomena. We have a theory which can cover the whole range from the super-microscopic to the galaxies in the sense of recapturing the Einstein field equations, and so on.

Now, continuing with this development, we see, once we have recognized that as we ascend the scale of energy, more and more complicated multiparticle exchange processes can occur, that the photon propagation function $D_+$ must be replaced by $\bar{D}_+$ to describe the fact that if enough energy is available we will have an electron-positron pair instead of a photon and more complicated things. This must occur everywhere because, after all, the propagation function refers to idealized sources and all sources have at least that property and then more besides, which represents their peculiar individual features. Suppose, for example, that we consider the primitive interaction itself and ask what happens as we move up the scale of energy or mass that the photon source emits. If enough energy is available, the virtual photon will convert to a pair and then reconvert again to finally produce the electron-positron pair that is actually detected. This is in fact exactly that mechanism of pair exchange which is the origin of the modification of the propagation action. Wherever the photon field occurs, related to its source by the propagation function D, it must be replaced by $\bar{D}$. Of course, it must be done also for the electron fields and their sources. Let me not discuss this. But more than this occurs. Because once we have created an electron-positron pair, it not only interacts by re-annihilation but it interacts by normal Coulomb scattering, so that we have the scheme of Fig. 9, where A is localized near the source J and the

$\psi$'s are localized in a macroscopically different space-time interval. So we recognize the existence of a non-local coupling between the $\psi$'s and A which occurs by the propagation of an excitation, represented by a propagation function with a continuous mass spectrum. It is just like a particle, only it is a whole spectrum of particles. And so one infers in a

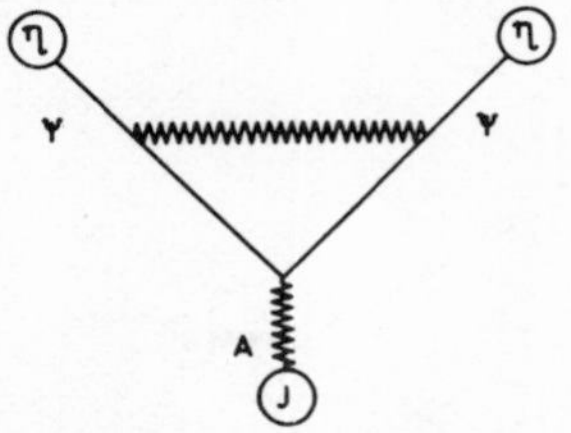


FIG. 9. Interaction of electron-positron pair by normal Coulomb scattering.

purely intuitive way the existence of spectral forms (more familiarly, dispersion relations) which in fact represent what is here the modification of the electromagnetic properties. The vector potential and the electron fields are no longer coupled locally but non-locally. And in addition to an electric form factor, new processes come into existence; the current of the anomalous magnetic moment of familiar value of $\alpha/2\pi$ appears here automatically, simply by keeping in mind the causal and physical arrangement that one has described. I beg you to remember again that we have arrived at all these things by a numerical and not an operator field theory.

Now, continuing this process of evolution, we see that we have here a theory which I think is in close accord with what the physicist actually does. He does not begin with the final theory and draw mathematical consequences, he elaborates the theory, he describes a certain class of phenomena and then he extrapolates and sees how far it works. Then, if it does not work, he discovers new phenomena and so on. The final theory is never in hand, when it is in hand the subject has become uninteresting.

I have mentioned, for example, the phenomena of photon-photon scattering. How do we arrive at all these processes? We do not consider what is actually of interest. In the case of the anomalous magnetic moment of the electron, we consider a causal arrangement in which the sources we are actually interested in are involved, but are involved in a way in which real particles occur. And so also in photon-photon scattering we are interested in the coupling among four photon sources which emit only photons, but consider an arrangement in which four photon sources are involved in a process in which real electrons occur. Figure 10 is an extended photon source. It creates a pair. Then these pairs propagate at macroscopic intervals and they are scattered in such a way that they reconvene and they annihilate and are detected by that extended photon source. This is a particular coupling and we can describe it completely, in fact, very simply, because the kinematic theory is so restricted that this process occurs in one and only one way. There is nothing to integrate over; it is an elementary kinematical procedure, and by studying it we arrive at a coupling among the four photon sources referring to this arrangement. Now we try to put it in terms of space and time and extrapolate it. It still gives us information about how four photon sources are coupled under different arrangements. What this amounts to is that one derives in an intuitive way the existence of a double spectral form, and in fact it is the Mandelstam representation with the rules of Cutkosky found by considering this physical process.

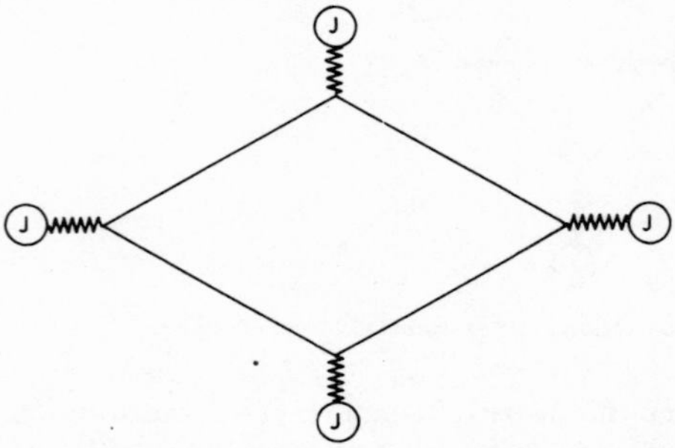


FIG. 10. An extended photon source.

One could go on in this way and still get more complicated processes and so on, in the sense of recognizing new processes produced by new multiparticle exchanges. Among these multiparticle exchange systems are bound systems, say positronium, where one must consider not only the exchanges of single photons once or twice but endlessly. So the next stage is a consideration not of a finite number of particle exchanges but the iteration of this process. One expresses that iteration by writing down integral equations, and one is naturally led in terms of the space-time picture to multiparticle Green's functions and all the rest of the phenomenological description which has been expressed in that language.

## 4. STRONG INTERACTIONS

We come now to what started all this, i.e. strong interactions. First we must look for unifying principles and I point to the idea of partial symmetry. Partial symmetry has various applications; consider partial symmetry on masses. I remind you of the Lagrange function of the $\pi$ field as a spin-zero field, of the Lagrange function of the $\rho$ regarded as a spin-one field and I ask what would the world be if one were to omit the mass term of the $\pi$ (one can do this only in the fairly explicit language of an action or a Lagrangian formulation in which individual terms can be considered).

$$\pi : \qquad \mathscr{L} = -\tfrac{1}{2}(\partial_\mu \pi)^2 - \tfrac{1}{2} m_\pi^2 \pi^2$$

$$\rho : \qquad \mathscr{L} = -\tfrac{1}{4}(\rho_{\mu\nu})^2 - \tfrac{1}{2} m_\rho^2 (\rho_\mu)^2$$

$$\rho_{\mu\nu} = \partial_\mu \rho_\nu - \partial_\nu \rho_\mu$$

If the pion mass were zero, then only derivatives of the field would occur and there would be invariance under displacements by constants which I will call $\delta\phi$. If the $\rho$ mass term were absent, then one gets the photon and one has invariance under gauge transformation. But in fact this Lagrange function is not invariant and the actual response to an infinitesimal displacement of the $\pi$ field and to an infinitesimal gauge transformation on the $\rho$ field is

$$\delta\mathscr{L} = -m_\pi^2 \pi \cdot \delta\phi - \frac{m_\rho^2}{g} \rho^\mu \partial_\mu \delta\omega$$

We want to start with phenomenology. Let us talk about $\pi$N and the related particle system. We may analyse it, and the brief discussion of photon-photon scattering and the connection with the Mandelstam representation is an indication that we can, in fact, here also consider multiparticle exchanges and in particular those few processes which can be regarded as produced by single-particle exchange, such as the exchange of one nucleon between pion - nucleon combinations or the exchange of a $\rho$ between nucleon pairs or pion pairs, elementary mechanisms which will give rise to (but certainly not the whole story of) pion-nucleon scattering. My point is, to make a start, we think of these single-particle exchanges as the analogy of primitive interactions, not the final theory, but as the basis from which by iterations the complete dynamics will be evolved. So consider then these single-particle exchanges and think of them as primitive interactions, represented by local products of fields, for example, the NN$\pi$, NN$\rho$, $\rho\pi\pi$, and so on. Here are familiar structures for them now in the language of this phenomenological field theory:

$$\frac{f}{4\pi} \bar{N}\gamma^\mu \gamma_5 \tau N \partial_\mu \pi \qquad g\bar{N}\gamma^\mu \tfrac{1}{2}\tau N\rho_\mu \qquad g' \rho^\mu \partial_\mu \pi \times \pi$$

This is simply an incorporation of the various quantum numbers for local structure, but now beyond this we have to add some hypotheses. One possible hypothesis is that the primitive interactions, not the whole theory, obey the partial symmetries that I have noted; and so, for example, that the primitive interactions are invariant under displacements of the $\rho$ field, which is just a way of saying, as is familiar, that the $\rho$ field must interact with the current of isotopic spin. Also, if we go on and include all carriers of isotopic spin including the $\rho$ itself so that we have self-couplings of the $\rho$, then we have introduced in this simple way the idea that the $\rho$ dynamics is that of a non-Abelian gauge theory, that is, $\rho$ coupled to itself in a way that corresponds to gauge transformations, but gauge transformations of a non-Abelian kind. We then arrive at the following transformation properties for the $\rho$, $\pi$, and N fields:

$$\delta\rho_\mu = -\delta\omega \times \rho_\mu + \frac{1}{g}\partial_\mu \delta\omega \qquad \delta\pi = \delta\phi$$

$$g\rho^\mu \cdot \partial_\mu(\pi \times \delta\phi) \Rightarrow \delta\rho_\mu = \frac{g}{m_\rho^2}\partial_\mu(\pi \times \delta\phi)$$

Consequently,

$$\delta N = i \frac{g^2}{2m_\rho^2} \tau \cdot \pi \times \delta\phi \, N$$

The form of the $\rho$ transformation follows simply from the fact that $\rho$ is a gauge field and carries isotopic spin. The displacement of the $\pi$ field implies that the $\rho\pi$ coupling is not invariant and since it changes in a manner exactly like a $\rho$ gauge transformation, it may be encountered by a change in $\rho$ gauge. In view of this transformation the $\rho$N coupling is not invariant, but this may be countered by a redefinition of the nucleon field, which is a set of three transformations because of isotopic spin multiplicity. Then there are isotopic spin transformations. We put all these together and we recognize that we have a group of six parameters which show the structure of the group as that of the four-dimensional Euclidean group which factors in a familiar way into two three-dimensional Euclidean groups or the equivalent SU(2) groups; the parameters which characterize these individual SU(2) groups are fixed, there is no freedom here. These transformations are unique, the group structures uniquely defined and the parameters are given by

$$\delta\omega_\pm = \delta\omega \pm \frac{\sqrt{2}\,g}{m_\rho} \delta\phi$$

What we are recognizing here is, at the phenomenological strong interaction level, the existence of chiral groups, because the $\delta\phi$ as a $\pi$ displacement has opposite parity from $\delta\omega$ as a purely inert mechanical rotation. We now set up a new hypothesis which seems to be hardly avoidable. We have recognized that instead of isotopic spin we have two isotopic spin groups. We should then extend the fields to be carriers of them, replace $\rho$ by two $\rho$ fields, which are given by

$$\rho_\pm = \rho \pm a \qquad \rho \sim 1^- \qquad a \sim 1^+$$

with

$$\delta(\rho \pm a) = -\delta\omega_+ \times (\rho \pm a) + \frac{1}{g} \partial \delta\omega_\pm$$

$$\delta a_\mu = \frac{\sqrt{2}}{m_\rho} \partial_\mu \delta\phi(x) + \text{homogeneous transformations}$$

Then we have the possibility of insisting even further on the invariance under these groups of the primitive interactions. For example, here is the $\pi$ Lagrange function:

$$\mathscr{L} = -\left(\partial_\mu \pi - \frac{m_\rho}{\sqrt{2}} a_\mu\right)^2 - \tfrac{1}{4}(\rho_{\mu\nu})^2$$

$$- \tfrac{1}{4}(a_{\mu\nu})^2 - \tfrac{1}{2} m_\rho^2 (\rho_\mu^2 + a_\mu^2)$$

This is not the whole Lagrangian, it is a simplified part which keeps only quadratic terms. Here we have partial invariance, partial symmetry, under the non-Abelian chiral groups. This is a definite structure. We see, however, that $\pi$ and a are mixed together; we must carry out a diagonalization process. Here it is:

$$a_\mu \rightarrow A_{,\mu} + \frac{1}{\sqrt{2}\, m_\rho} \partial_\mu \pi$$

The Lagrangian function of the few particles that are being considered so far then reduces to the correct and previously written $\pi$ and $\rho$ Lagrange function,

$$\mathcal{L} = \mathcal{L}_\pi + \mathcal{L}_\rho - \tfrac{1}{4}(A_{,\mu\nu})^2 - \tfrac{1}{2}\, 2m_\rho^2 A_{,\mu}^{\;2}$$

and since everything is completely fixed, we have now predicted the mass of $A_1$. It is a well-known prediction first found by Weinberg in quite different ways which involve hypotheses about high-energy behaviour rather than the low-energy behaviour exploited here. It is this famous $\sqrt{2}$ ratio which seems very well obeyed in nature. I think there is hardly any doubt any longer about the physical existence of $A_1$.

$$\frac{m_{A_1}}{m_\rho} = \sqrt{2}$$

Now, continuing with the dynamics evolved in this way, we have a much more complicated scheme. a is mixed with $\pi$ and so there are more complicated couplings between $\rho$ and $\pi$'s. Just to describe what the situation is, we find a new coupling constant effective in the actual decay of $\rho$ into two $\pi$'s. It turns out to be three-quarters of the initial coupling constant g. Let us make a temporary assumption of $\rho$ dominance. I tried to emphasize in the beginning that we are talking about primitive interactions which are single-particle exchanges, and it is not a necessary assumption. If indeed we had a more detailed analysis of the experimental data, we could separate that part of the S-wave pion-nucleon low-energy scattering which is produced by $\rho$ exchange from that part which is not. At the moment this analysis does not yet exist and it is conventionally assumed that it is dominated by $\rho$ exchange. It must be nearly dominated. Let us assume that it is. It is a familiar assumption but it is merely an approximate one for our purposes. It then gives rise to a definite value of

$$\frac{g^2}{4\pi} = 3.4 \pm 0.3$$

coming from the known S-wave $\pi$N scattering. If we use this known value of g and use the effective coupling constant of 3/4 we predict a $\rho$ width of 95 MeV. We also predict, of course, the effective coupling constant of $\rho\pi\pi$ decay $g'/g = 3/4$. The colliding beam experiments which have been done in Novosibirsk have shown us very directly that the $\rho$ meson is an intermediary between electron-positron colliding partners and charged pion pairs, and the preliminary data have given a value of the width $\Gamma = 93 \pm 15$. Their derived value of $g'/g$ is $0.77 \pm 0.10$. These numbers are in striking agreement, I think, with these anticipations from the simple chiral group. The simplest chiral invariant theory also predicts a width of $A_1$. It is too large from the most recent clear-cut measurements made on the neutral $A_1$. There are various possibilities. I mention an interesting one here, which has to do with the particle we have not mentioned yet, the sigma meson, which now seems to be known experimentally as a very broad S-wave resonance somewhere around 700 MeV close to the $\rho$ mass. There is the theoretical possibility that the observed decay of $A_1$, which is after all into three pions, can be a process of $\rho\pi$ or $\sigma\pi$ decay and one can exhibit models in which that decay shows destructive interference. Through this interference between the rather sharp $\rho\pi$ decay and the very broad $\sigma\pi$ background, one can produce, depending on the model, a fairly substantial reduction of the $A_1$ width, without in any way, of course, affecting the $\rho$. I do not think one can use this to account entirely for the observed width, but for the rest we can use further chiral invariant interactions which are analogous to magnetic moment, like interactions, Pauli terms which can always be added independently of the particular invariant structure and can be introduced to reduce the $A_1$ width. This increases the $\rho$ width and then we must use a small value of g. So the final story on this will only be available when one has analysed all the relevant experimental data including in particular the question of what is the value of g as inferred from only the $\rho$ exchange in pion-nucleon S-wave scattering, without this simplifying assumption.

I mention merely some additional topics to which these notions could be applied; for example, the possibility of introducing electromagnetic coupling of the strongly interacting particles by simply insisting upon gauge invariance, balancing the gauge invariance of the vector potential against the gauge invariance of the $\rho$ field (neutral $\rho$ field as well as $\omega$ and so on) which gives one a very immediate derivation of the now familiar coupling of vector potentials to fields in which the field plays the role of the current. The same procedures are, of course, used for weak interactions in particular, since the phenomenology of strong interactions have already given us a chiral group.

The known chiral behaviour of strong interactions can then be used immediately to make predictions about the chiral behaviour anticipated for weak interaction. Here we are completely reversing the logical order made use of in current algebra which begins with hypotheses about weak interactions and turns them about to make predictions about strong interactions. I think we must agree that logically, in the sense of the natural order of things, the strong interactions dominate the physics and not the other way round. I also mentioned the fact that the primitive interactions I have written down, $\rho\pi\pi$ and so on, are not the end of the story, they are the starting point, and out of this one must consider multiparticle exchanges in far more realistic dynamics in which requirements such as unitarity will automatically come out from the causal structure and use them, so to speak, to interpolate between the resonances. This naturally raises the question whether in fact, when we have interpolated and come to the position of a new resonance, the strong interaction dynamics already available could not be used to predict new, but of course already recognized, resonances; that is, the scheme has some aspect of self-consistency by interpolating from one resonance to others where we could not insist that we get back to the one we started with. These ideas of self-consistency have been exploited before and I anticipate that the general structure will not be basically different, but nothing has been done so far.

I mention the fact that the ideas of partial symmetry, of considering a restricted domain of physics or of omitting temporarily certain parts of Lagrange functions as in higher symmetries, can be used in other ways, in particular if we consider physical processes where no momentum is exchanged. We can recognize already within the framework of the non-interacting particles that a wider group opens up. It is the group which has been suggested long before as a means of uniting spins and internal degrees of freedom. If we transfer the symmetry so recognized, as a testing hypothesis to interactions, we then find some of the very successful predictions previously obtained from SU(6) and one new prediction: the ratio of the axial vector to the vector coupling in weak decay, $G_A/G_V = 5/3\sqrt{2} = 1.18$. This is a very striking number indeed and represents, I think, one of the as yet unchallenged successes of this particular point of view.

## 5. CONCLUSION

I have tried to indicate that here is a new and a very physical, intuitive and practical approach not only in high-energy physics but in the whole range of physical phenomena.

Source theory is not an engineer's theory with fixed calculational rules, not a mathematician's theory with emphasis on irrelevant subtleties but a physicist's theory which recognizes the evolutionary growth of experimental knowledge and helps pave the way for an eventual more fundamental theory by not confusing phenomenology with speculation.

# How Massive Is the $W$ Particle?

Julian Schwinger
*University of California at Los Angeles, Los Angeles, California 90024*
(Received 8 June 1972)

It is suggested that the mass of the hypothetical unit-spin charged particle mediating the weak interactions is ~53 BeV.

Slowly, attention has been drawn to the idea[1] that the weak and electromagnetic interactions are, not merely analogous phenomena, but aspects of a unified dynamical mechanism. Some time ago, I applied this concept to estimate the mass of the hypothetical charged carriers of the weak interaction. That work was reported at a Columbia University colloquium in 1967, but remained unpublished. Recently, I became aware that other authors[2] had subsequently used similar reasoning, but had arrived at a different estimate. This note is devoted to an outline of my quite elementary considerations, and also isolates the hypotheses responsible for the discordant answers.

The leptons will be described in the manner introduced in the 1957 paper cited in Ref. 1. The conserved leptonic charge $L=\pm 1$ distinguishes particles of equal electric charge, $\mu^+(L=+1)$, $e^+(L=-1)$, which is the basis for asserting that different neutrinos accompany $\mu$ and $e$: $\mu^+\bar{\nu}$ $(L=-1)$, $e^+\nu(L=+1)$. For a given leptonic charge, there is an electric-charge triplet; $L=+1: \mu^+, \nu, e^-$. The matrices acting in the electric-charge space are the $3\times 3$ antisymmetrical isotopic-spin matrices $t_a$, $a=1,2,3$, which are completed by the six symmetrical products $\{t_a, t_b\}$; the electric-charge matrix is identified as $t_3$. The chiral charge-bearing currents observed in the leptonic processes can be presented as $2j^\mu_{12}$, $2j^\mu_{21}$, in which

$$j^\mu = \tfrac{1}{2}\psi\gamma^0\gamma^\mu T\psi\,,$$

$$T = \frac{1+i\gamma_5 t_3}{2}\, 2^{1/2} t\, \frac{1+i\gamma_5 t_3}{2} = 2^{-3/2}(t + i\gamma_5\{t, t_3\})\,, \qquad (1)$$

and

$$t_{12} = t_1 + it_2\,,\quad t_{21} = t_1 - it_2\,. \qquad (2)$$

Thus, the phenomenological coupling responsible for the decay $\mu^+ \to e^+ + \nu_L + \nu_R$ is

$$2^{-1/2}G\, 2j^\mu_{12} 2j_{\mu 21}\,,\quad Gm_{\text{prot}}^2 = 1.0\times 10^{-5}\,. \qquad (3)$$

The particular normalization adopted for $T_{12}$, $T_{21}$ is such that $U_2$-group commutation relations are obeyed,

$$[T_{12}, T_{21}] = T_{11} - T_{22} = 2T_{11} - (T_{11}+T_{22})\,, \qquad (4)$$

where

$$T_{11} = t_3 \qquad (5)$$

is the electric-charge matrix, and

$$T_{11} + T_{22} = i\gamma_5 + \tfrac{3}{2} t_3(1 - i\gamma_5 t_3) \qquad (6)$$

commutes with the other $T$ matrices. (It reduces to $i\gamma_5$ when multiplying $T_{12}$, $T_{21}$.) This is analogous to the $U_2$ subgroup of strong interactions, where $T_{11}=Q$ and $T_{11}+T_{22}=Y$.

The electromagnetic field $A^\mu$, which couples to electric charge, is now extended to the set of fields $A^\mu_{ab}$, $A^\mu_{11}=A^\mu$, through the hypothesis of $U_2$-invariant coupling[3]:

$$e\sum_{ab} A^\mu_{ab} j_{\mu ba}\,,\quad e^2/4\pi = \alpha\,. \qquad (7)$$

That symmetry is broken by the introduction of a mass, $m_W$, for the charged particles represented by $A^\mu_{12}$, $A^\mu_{21}$. The resulting quasilocal coupling of the associated currents is then

$$\frac{1}{2}\frac{e^2}{m_W{}^2}\sum_{1,2} j^\mu_{ab} j_{\mu ba} = \frac{e^2}{m_W{}^2} j^\mu_{12} j_{\mu 21}\,, \qquad (8)$$

and comparison with (3) gives



Reprinted from *Physical Review* D7, 908–909 (1973).

$$\frac{e^2}{m_W{}^2} = 2^{3/2}G\,. \tag{9}$$

This predicts the mass

$$m_W = \left(\frac{4\pi\alpha}{2^{3/2}G}\right)^{1/2} = 53 \text{ BeV}\,. \tag{10}$$

The authors cited in Ref. 2 have concentrated on the $SU_2$ subgroup resembling isotopic spin and, in effect, replaced (7) with

$$e\sum_a A_a^\mu j_{\mu a}, \quad a = 1, 2, 3\,, \tag{11}$$

where $j_3$ is the isovector part of the electric current. (There is also an isoscalar term.) The implied coupling of the charge-bearing currents, produced by the exchange of a particle of mass $m_W'$, is then

$$\frac{1}{2}\frac{e^2}{m_W'^2}\sum_{a=1,2} j_a^\mu j_{\mu a} = \frac{1}{2}\frac{e^2}{m_W'^2} j_{12}^\mu j_{\mu 21}\,. \tag{12}$$

This gives the relation

$$\frac{1}{2}\frac{e^2}{m_W'^2} = 2^{3/2}G \tag{13}$$

and the mass

$$m_W' = 2^{-1/2}m_W = 37 \text{ BeV}\,. \tag{14}$$

Since the decision between the rival partial-symmetry groups will not be an immediate one, there may be time to improve these estimates by incorporating coupling-constant symmetry violations induced by the splitting of the mass spectrum. That will also involve $m_Z$, the mass of the particle represented by the field $A_{22}^\mu$. (In the 1957 paper, the heavy charged particles were symbolized by $Z$, but custom now dictates otherwise.) We shall only remark that the simplest hypothesis for $U_2$-symmetry breaking, as expressed by the mass term

$$-\sum_{ab} A_{ab}^\mu f_b A_{\mu ba}\,, \qquad f_1 = 0, \quad f_2 = m_W{}^2\,, \tag{15}$$

suggests that

$$m_Z = 2^{1/2}m_W = 75 \text{ BeV}\,, \tag{16}$$

but any value can be accepted.

The theory described by Weinberg uses two coupling constants, $g$ and $g'$. The situation with $g = g'$ is that of $U_2$-symmetry, while $g' \gg g$ is required for the $SU_2$ hypothesis. The $g = g'$ version of Weinberg's theory differs in several respects from the phenomenological $U_2$-theory presented here. Apart from the omission of muons, presumably to avoid notational complications, there is a specific hypothesis concerning the dynamical origin of the boson and lepton masses (it already appears in the 1957 paper), which is designed to produce a renormalizable operator field theory. Yet, the examples of the compensation between various processes that result in good high-energy behavior[4] do not seem to involve this additional hypothesis, but rather depend on the non-Abelian gauge structure of the theory. From the viewpoint of a phenomenological theory, which has no reference to renormalization, it becomes an open question whether one has need for hypotheses concerning the dynamical origin of experimental masses, at least for the practical questions of the anticipatable future. Source-theory techniques should be useful in exploring this and related problems.

---

[1]J. Schwinger, Ann. Phys. (N.Y.) 2, 407 (1957); S. Weinberg, Phys. Rev. Letters 27, 1688 (1971); also see Physics Today 25, (No. 4), 17 (1972).

[2]J. Schechter and Y. Ueda, Phys. Rev. D 2, 736 (1970); T. D. Lee, Phys. Rev. Letters 26, 801 (1971).

[3]Another investigation of the unification of the electromagnetic field with the fields of electrically charged particles has been reported [J. Schwinger, Rev. Mod. Phys. 36, 609 (1964)]. It is noted there that the charged unit-spin particles acquire an additional magnetic moment, as expressed by the gyromagnetic ratio $g = 2$.

[4]The matrix notation used here simplifies the consideration of classes of processes. Thus, for lepton collisions that create bosons, $l + l \to A + A$, specifically, longitudinally polarized heavy bosons (examples are $\nu + \nu \to W + W$, $e + e \to W + W$, $e + \nu \to W + Z$), there are two mechanisms, symbolized by $l + l \to A + (l) + A$ and $l + l \to (A) \to A + A$, where parentheses indicate a virtual particle. The lepton exchange process has a simple high-energy limit, growing with energy, that involves the commutator of the $T_{ab}$ matrices. The boson-exchange process, which is dominated by a magnetic-moment-like coupling, contains an analogous commutator of the boson field matrix $A_{ab}$. In the limit that all phenomenological particle masses in propagation functions are neglected, there is exact cancellation between the two mechanisms.

# Classical Radiation of Accelerated Electrons. II. A Quantum Viewpoint*

Julian Schwinger
*University of California, Los Angeles, California 90024*

(Received 27 November 1972)

The known classical radiation spectrum of a high-energy charged particle in a homogeneous magnetic field is rederived. The method applies, and illuminates, an exact (to order $\alpha$) expression for the inverse propagation function of a spinless particle in a homogeneous field. An erratum list for paper I is appended.

For a long time I have wanted to reexamine a classic situation of classical electrodynamics, that of high-energy charged particles radiating in a homogeneous magnetic field, from the modern quantum viewpoint that employs the machinery of propagation (Green's) functions. Since the electromagnetic and relativistic aspects of the problem are quite transparent, the comparison should be instructive in giving the more abstract quantum procedure a concrete interpretation in a particular instance. And, as an added bonus, the necessary ability to treat motion in magnetic fields that goes beyond the lowest orders in a perturbative expansion should be helpful in answering questions about very strong fields, to which recent astrophysical speculations have directed attention. This paper is devoted to describing one such procedure, and applying it to rederive (for a spin-0 particle) the known classical radiation result.[1] Another method is indicated in a separate paper of Yildiz. A subsequent joint paper will contain the analogous spin-$\frac{1}{2}$ calculation, and a discussion of the anomalous magnetic moment in strong fields.

The language and methodology of source theory[2] will be used (which should not seriously impede readers who are untutored in this art). The initial action expression of spin-0 charged particles with mass $m$,

$$\int(dx)[K(x)\phi(x) - \tfrac{1}{2}\phi(x)(\Pi^2 + m^2)\phi(x)] ,$$
$$\Pi = (1/i)\partial - eqA , \qquad (1)$$

is supplemented by the action contribution associated with the exchange of one virtual photon [cf. Eq. (4-14.2) of PSF II[2]],

$$-\tfrac{1}{2}\int(dx)(dx')\phi(x)M(x,x')\phi(x') . \qquad (2)$$

Here, written in a symbolic notation, we have

Reprinted from *Physical Review* D7, 1696–1701 (1973).

$$M = ie^2 \int \frac{(dk)}{(2\pi)^4}(2\Pi - k)\frac{1}{k^2}\frac{1}{(\Pi - k)^2 + m^2}(2\Pi - k) + \text{c.t.}\ , \tag{3}$$

where the contact term (c.t.) is a linear function of $\Pi^2$ that is designed to satisfy the normalization conditions. These require that $M$ and its first derivative with respect to $\Pi^2$ vanish, in the null-field situation, at $\Pi^2 + m^2 = 0$. The stationary action principle, applied to the sum of (1) and (2), yields the field solution that is conveyed symbolically by

$$\phi = \bar{\Delta} K\ , \tag{4}$$

where

$$\bar{\Delta}^{-1} = \Pi^2 + m^2 + M \tag{5}$$

is the inverse of the modified (to order $\alpha$) propagation function.

The exact evaluation of $M$ for an applied homogeneous electromagnetic field has two basic ingredients. The first is the exponential representation of the particle and photon propagation functions,

$$\frac{1}{(\Pi - k)^2 + m^2 - i\epsilon} = i\int_0^\infty ds_1 e^{-is_1[(\Pi - k)^2 + m^2]}\ ,$$
$$\frac{1}{k^2 - i\epsilon} = i\int_0^\infty ds_2 e^{-is_2 k^2}\ , \tag{6}$$

together with their product:

$$\frac{1}{k^2}\frac{1}{(\Pi - k)^2 + m^2} = -\int_0^\infty ds\, s\int_0^1 du\, e^{-ism^2u}e^{-isH}. \tag{7}$$

In the latter we have used the parametrization

$$s_1 = su, \quad s_2 = s(1-u) \tag{8}$$

and introduced the "Hamiltonian"

$$H = u(\Pi - k)^2 + (1-u)k^2 = (k - u\Pi)^2 + u(1-u)\Pi^2\ . \tag{9}$$

The second one is the replacement of the $k$ integration by an algebraic procedure associated with the vector $\xi_\mu$ that is complementary to $k_\mu$,

$$[\xi_\mu, k_\nu] = i g_{\mu\nu}\ . \tag{10}$$

Then, on using the four-dimensional transformation functions (primes to designate eigenvalues are omitted)

$$\langle \xi | k\rangle = \frac{1}{(2\pi)^2}e^{i\xi k}\ ,$$
$$\langle k | \xi\rangle = \frac{1}{(2\pi)^2}e^{-i\xi k}\ , \tag{11}$$

we have

$$\langle \xi = 0 | f(k) | \xi = 0\rangle = \int \langle \xi = 0 | k\rangle (dk) f(k) \langle k | \xi = 0\rangle$$
$$= \int \frac{(dk)}{(2\pi)^4} f(k)\ . \tag{12}$$

The two devices transform $M$ into

$$M = -ie^2 \int ds\, s\, du\, e^{-is\,m^2u}\langle (2\Pi - k)e^{-isH}(2\Pi - k)\rangle + \text{c.t.}\ , \tag{13}$$

where the expectation value refers to the $\xi = 0$ state.

The "time" development described by $H$ is made explicit by introducing quantities such as

$$\xi(s) = e^{isH}\xi e^{-isH}\ , \tag{14}$$

which obey the equations of motion

$$\frac{d\xi(s)}{ds} = \frac{1}{i}[\xi(s), H]\ . \tag{15}$$

The full set of equations of motion is

$$\frac{dk(s)}{ds} = 0\ ,$$
$$\frac{d\xi(s)}{ds} = 2[k - u\Pi(s)]\ , \tag{16}$$
$$\frac{d\Pi(s)}{ds} = 2ueqF[\Pi(s) - k]\ ,$$

where the last equation applies the commutator

$$[\Pi_\mu, \Pi_\nu] = [-i\partial_\mu - eqA_\mu, -i\partial_\nu - eqA_\nu]$$
$$= ieqF_{\mu\nu}\ . \tag{17}$$

The simplicity of the homogeneous field situation is the linearity of the equations of motion, which permits their exact solution.[3] Thus, writing the last equation of (16) as

$$\frac{d}{ds}[e^{-2ueqFs}\Pi(s)] = \frac{d}{ds}[e^{-2ueqFs}k]\ , \tag{18}$$

we get

$$\Pi(s) = e^{2ueqFs}\Pi + (1 - e^{2ueqFs})k\ . \tag{19}$$

This is followed by the integration of the $\xi$ equation:

$$\xi(s) = \xi + 2ks - \frac{e^{2ueqFs} - 1}{eqF}\Pi$$
$$-2uks + \frac{e^{2ueqFs} - 1}{eqF}k\ , \tag{20}$$

or

$$eqF[\xi(s) - \xi] = Dk - A\Pi\ , \tag{21}$$

where

$$A = e^{2ueqFs} - 1 \tag{22}$$

and

$$D=e^{2ueqFs}-1+2(1-u)eqFs\ . \qquad (23)$$

The solution of the equations of motion is used to rewrite the expectation value in (13) as

$$\begin{aligned}\langle(2\Pi-k)e^{-isH}(2\Pi-k)\rangle &=\langle e^{-isH}[2\Pi(s)-k](2\Pi-k)\rangle\\ &=\langle e^{-isH}\rangle 4\Pi(1+A^T)\Pi-\langle e^{-isH}k\rangle 2(2+A+2A^T)\Pi\\ &\quad+\langle e^{-isH}k(1+2A)k\rangle\ ,\end{aligned} \qquad (24)$$

where the transposed form of $A$ is

$$A^T=e^{-2ueqFs}-1\ , \qquad (25)$$

according to the antisymmetry of $F_{\mu\nu}$. The expectation values with one and two additional factors of $k$ are reduced to the basic expectation value, $\langle e^{-isH}\rangle$, as follows. We first note that

$$\begin{aligned}0&=\langle[\xi,e^{-isH}]\rangle\\ &=\langle e^{-isH}[\xi(s)-\xi]\rangle\ ,\end{aligned} \qquad (26)$$

which implies [Eq. (21)]

$$\langle e^{-isH}k\rangle=\langle e^{-isH}\rangle\frac{A}{D}\Pi\ . \qquad (27)$$

Similarly,

$$\begin{aligned}0&=\langle[\xi_\mu,[\xi_\nu,e^{-isH}]]\rangle\\ &=\langle e^{-isH}[\xi_\mu(s)\xi_\nu(s)-\xi_\mu(s)\xi_\nu-\xi_\nu(s)\xi_\mu+\xi_\mu\xi_\nu]\rangle\\ &=\langle e^{-isH}[\xi_\mu(s)-\xi_\mu][\xi_\nu(s)-\xi_\nu]\rangle+\langle e^{-isH}[\xi_\mu,\xi_\nu(s)]\rangle\end{aligned} \qquad (28)$$

leads to

$$\begin{aligned}0&=\langle e^{-isH}(Dk-A\Pi)_\mu(Dk-A\Pi)_\nu\rangle\\ &\quad+\langle e^{-isH}\rangle i(eqFD^T)_{\mu\nu}\end{aligned} \qquad (29)$$

and then

$$\langle e^{-isH}k_\mu k_\nu\rangle=\langle e^{-isH}\rangle\left[\left(\frac{A}{D}\Pi\right)_\nu\left(\frac{A}{D}\Pi\right)_\mu-i\left(\frac{eqF}{D}\right)_{\mu\nu}\right]. \qquad (30)$$

One can verify that the right-hand side matches the left-hand side in its symmetry in $\mu$ and $\nu$. The implied algebraic property is

$$AA^T+D+D^T=0\ . \qquad (31)$$

It is confirmed by noting, first, that

$$AA^T+A+A^T=0 \qquad (32)$$

and, then, that $D-A$ is an antisymmetrical matrix.

The material for the main task, the evaluation of $\langle e^{-isH}\rangle$, is now at hand. We construct a differential equation,

$$\begin{aligned}i\frac{\partial}{\partial s}\langle e^{-isH}\rangle&=\langle e^{-isH}H\rangle\\ &=\langle e^{-isH}\rangle u\Pi^2-\langle e^{-isH}k\rangle 2u\Pi+\langle e^{-isH}k^2\rangle\ ,\end{aligned} \qquad (33)$$

from which, with the aid of Eqs. (27) and (30), it immediately follows that

$$\begin{aligned}i\frac{\partial}{\partial s}\ln\langle e^{-isH}\rangle&=u\Pi^2-2u\Pi\frac{A^T}{D^T}\Pi\\ &\quad+\Pi\frac{A^T}{D^T}\frac{A}{D}\Pi-i\,\mathrm{tr}\left(\frac{eqF}{D}\right)\ ,\end{aligned} \qquad (34)$$

where the trace refers only to the vector indices. In order to have a symmetrical matrix in the $\Pi$ quadratic form, we rewrite this structure as

$$u\Pi^2+\Pi\left[\frac{A^T}{D^T}\frac{A}{D}-u\left(\frac{A}{D}+\frac{A^T}{D^T}\right)\right]\Pi-ieq\,\mathrm{tr}\left(F\frac{uA+1}{D}\right)\ , \qquad (35)$$

which makes use of the commutator (17). Now,

$$\frac{A^T}{D^T}\frac{A}{D}-u\left(\frac{A}{D}+\frac{A^T}{D^T}\right)=-\frac{1}{2eqF}\left(\frac{\partial D/\partial s}{D}-\frac{\partial D^T/\partial s}{D^T}\right), \qquad (36)$$

since

$$\begin{aligned}\frac{\partial D}{\partial s}&=2eqF(uA+1)\ ,\\ \frac{\partial D^T}{\partial s}&=-2eqF(uA^T+1)\ ,\end{aligned} \qquad (37)$$

while $A$ and $D$ obey (31), in which $A$ and $A^T$ are commutative. Accordingly, (35) becomes

$$u\Pi^2+\Pi\left(-\frac{1}{2eqF}\right)\frac{\partial}{\partial s}\ln\left(-\frac{D}{D^T}\right)\Pi-\tfrac{1}{2}i\frac{\partial}{\partial s}\mathrm{tr}\ln D\ , \qquad (38)$$

and the integrated version is

$$\begin{aligned}\langle e^{-isH}\rangle&=C\frac{1}{[\det(D/2eqF)]^{1/2}}\\ &\quad\times\exp\left\{-i\left[su\Pi^2-\Pi\frac{1}{2eqF}\ln\left(-\frac{D}{D^T}\right)\Pi\right]\right\}.\end{aligned} \qquad (39)$$

The $F$-dependent factor is inserted into the determinant in order to simplify the form of the integration constant $C$.

To evaluate $C$, we consider the limit of small $s$, where

$$\begin{aligned}\frac{D}{2eqF}&=s+u^2eqFs^2+\cdots,\\ -\frac{D}{D^T}&=1+2u^2eqFs+\cdots\ .\end{aligned} \qquad (40)$$

Then, (39) exhibits the dominant behavior

$$\langle e^{-isH}\rangle \sim C\frac{1}{s^2}\ , \tag{41}$$

in view of the four-dimensional nature of the determinant. The singularity at $s=0$ arises from the large values of $k$ that are increasingly demanded, as $s\to 0$, by complimentarity with $\xi=0$. Accordingly, the limiting structure is given by the elementary $k$ integral

$$\int\frac{(dk)}{(2\pi)^4}e^{-isk^2}=\frac{1}{(4\pi)^2}\frac{1}{is^2}\ , \tag{42}$$

and

$$C=-\frac{i}{(4\pi)^2}\ . \tag{43}$$

We present the result as

$$\langle e^{-isH}\rangle=-\frac{i}{(4\pi)^2}\frac{e^{-isu(1-u)\Pi^2}}{s^2}\left(\det\frac{2eqFs}{D}\right)^{1/2}\times\exp\left\{i\Pi\left[\frac{1}{2eqF}\ln\left(-\frac{D}{D^T}\right)-u^2s\right]\Pi\right\}\ , \tag{44}$$

which is so written that the last two factors approach unity as $F\to 0$. The remainder, the structure of $\langle e^{-isH}\rangle$ for $F=0$, is immediately evident from the second version of $H$ in Eq. (9) and the integral (42).

We now have before us all the ingredients to construct $M$ as the double parametric integral of Eq. (13). It is, however, not necessary to display $M$ in detail in order to make the principal application of this paper – the derivation of the classical radiation spectrum. Since the properties of the real charged particle are essentially characterized by $\Pi^2+m^2=0$, we only need $M$ for this circumstance. And, since radiative decay is the question of interest, it is only the imaginary part of $M$ that is required. There is, furthermore, a simplification associated with the concentration on classical radiation. To appreciate it, let us note that in the classical limit the $k$ integral of $e^{-isH}$ should be dominated by the point of stationary phase,

$$\frac{\delta H}{\partial k}=2(k-u\Pi)=0\ . \tag{45}$$

The value of $k$ thus selected, $k=u\Pi$, is not that of a real photon, in general,

$$-k^2=-u^2\Pi^2=u^2m^2\ , \tag{46}$$

but becomes so if $u$ is sufficiently small. In this circumstance, we can express the energy of the radiated photon, $k^0=\omega$, relative to the energy of the particle, $\Pi^0=E$, by

$$\frac{\omega}{E}=u\ll 1\ , \tag{47}$$

which is evidently a classical restriction. With this identification, the $u$ integral of Im $M$ becomes a spectral integral for the radiation.

There is yet another simplification associated with the restriction to high particle energy,

$$E\gg m\ . \tag{48}$$

We first remark that the periodicity of motion in the magnetic field $\vec{H}$ has its representation in the exponential function $e^{2ueqFs}$, where the nonzero eigenvalues of $F$ are $\pm iH$. [Recall that $\operatorname{tr}F^2=-F^{\mu\nu}F_{\mu\nu}=2(\vec{E}^2-\vec{H}^2)$.] This gives the identification

$$2eHus=\omega_0\tau\ , \tag{49}$$

where $\tau$ is a time variable and $\omega_0$ is the rotational frequency. According to the classical equations of motion $[\dot{\vec{p}}=e\vec{v}\times\vec{H}]$, the high-energy form of $\omega_0$ is

$$\omega_0=\frac{eH}{E}\ , \tag{50}$$

and one can write (49) as

$$2Eus=\tau\ . \tag{51}$$

Now, the point about high energies is this. Only a small fraction of the orbit, $\sim m/E$, is involved in classical radiation toward a particular direction. We therefore expect that the dominant contributions will come from values of $us$ such that

$$2eHus\sim\frac{m}{E}\ll 1\ . \tag{52}$$

Under these circumstances the logarithmic function in (44) has the leading terms

$$\frac{1}{2eqF}\ln\left(-\frac{D}{D^T}\right)\simeq u^2s+\tfrac{1}{3}u^4(1-u)^2(eqF)^2s^3\ , \tag{53}$$

which are comparable, since

$$\Pi\frac{1}{2eqF}\ln\left(-\frac{D}{D^T}\right)\Pi\simeq-m^2u^2s-\tfrac{1}{3}u^4(1-u)^2(eHE)^2s^3=-m^2u^2s[1+\tfrac{1}{3}(1-u)^2(eHusE/m)^2]\ . \tag{54}$$

The evaluation used here,

$$\Pi(eqF)^2\Pi\simeq-(eHE)^2\ , \tag{55}$$

assumes zero momentum parallel to the magnetic field, confining the motion to the plane perpendicular to the field. In the strict classical limit under consideration, we should also replace $1-u$ by unity,[4] in (54). As for the determinant of (Eq. 44), the expansion

$$\frac{D}{2eqFs}=1+u^2eqFs+\tfrac{2}{3}u^3(eqFs)^2+\cdots \tag{56}$$

and the evaluation

$$\begin{aligned}\left(\det\frac{D}{2eqFs}\right)^2&=\det\frac{D}{2eqFs}\left(\frac{D}{2eqFs}\right)^T\\ &\simeq\det[1+\tfrac{4}{3}u^3(eqFs)^2]\\ &=1-\tfrac{2}{3}u(2eHus)^2\end{aligned} \tag{57}$$

show that the determinant reduces to unity in the classical high-energy limit, where both $u$ and $2eHus$ are small quantities.

The terms of (24) that have one or two additional factors of $k$ clearly become relatively negligible in the classical limit (as one can verify). In the high-energy limit, we also have

$$A^T\simeq -2ueqFs+2(eqFus)^2\,, \tag{58}$$

where the antisymmetrical $F$ term, which introduces the commutator $[\Pi,\Pi]$, is a negligible quantum correction. Accordingly,

$$\begin{aligned}\frac{\langle(2\Pi-k)e^{-isH}(2\Pi-k)\rangle}{\langle e^{-isH}\rangle}&\simeq-4[m^2+2(eHEus)^2]\\ &=-4E^2[(m^2/E^2)+\tfrac{1}{2}(\omega_0\tau)^2]\,,\end{aligned} \tag{59}$$

where the last version begins the process of introducing the classical time variable $\tau$, for comparison with the known result. The other basic combination is (54), which now reads

$$\Pi\frac{1}{2eqF}\ln\left(-\frac{D}{D^T}\right)\Pi\simeq-\omega\tau[\tfrac{1}{2}(m^2/E^2)+\tfrac{1}{24}(\omega_0\tau)^2]\,. \tag{60}$$

Putting together the limiting forms of the various parts of $M$ gives

$$M\to\frac{\alpha}{\pi}E\int d\omega\int_0^\infty\frac{d\tau}{\tau}\left\{\left(\frac{m^2}{E^2}+\tfrac{1}{2}\omega_0^2\tau^2\right)\exp\left[-i\omega\left(\frac{1}{2}\frac{m^2}{E^2}\tau+\tfrac{1}{24}\omega_0^2\tau^3\right)\right]-\frac{m^2}{E^2}\exp\left(-i\omega\frac{1}{2}\frac{m^2}{E^2}\tau\right)\right\}, \tag{61}$$

which now incorporates the contact term that is required to make $M$ vanish in the absence of the magnetic field ($\omega_0=0$). What is needed for the description of radiative decay is

$$-\frac{1}{E}\mathrm{Im}M=\int d\omega\frac{1}{\omega}P(\omega)\,, \tag{62}$$

where

$$\begin{aligned}P(\omega)&=\frac{\alpha}{\pi}\omega\left[\int_0^\infty d\tau\left(\frac{m^2}{E^2}+\tfrac{1}{2}\omega_0^2\tau^2\right)\frac{\sin\omega[\frac{1}{2}(m^2/E^2)\tau+\frac{1}{24}\omega_0^2\tau^3]}{\tau}-\tfrac{1}{2}\pi\frac{m^2}{E^2}\right]\\ &=\frac{\alpha}{\pi}\omega\frac{m^2}{E^2}\left[\int_0^\infty dx(1+2x^2)\frac{\sin\frac{3}{2}\xi(x+\frac{1}{3}x^3)}{x}-\tfrac{1}{2}\pi\right];\end{aligned} \tag{63}$$

the last form introduces the variables

$$x=\tfrac{1}{2}\omega_0\tau\frac{E}{m}\,,\qquad \xi=\tfrac{2}{3}\frac{\omega}{\omega_0}\left(\frac{m}{E}\right)^3. \tag{64}$$

The physical identification of $P(\omega)$ follows on writing the inverse propagation function (omitting $\mathrm{Re}M$) as

$$\vec{\Pi}^2+m^2-\left[E+\tfrac{1}{2}i\left(-\frac{1}{E}\mathrm{Im}M\right)\right]^2, \tag{65}$$

which displays (62) as the damping constant of the system. Therefore $\omega^{-1}P(\omega)$ is the probability per unit time for radiation into a unit $\omega$ interval, and $P(\omega)$ is the spectral distribution of the radiated power. The results stated in Eq. (63) coincide with the classically derived ones contained in Eqs. (II-5) and (II-7) of paper I.

## APPENDIX

Paper I seems to have escaped proofreading, since it contains a number of rather obvious typographical errors. Among these are the following:

(1) In the first of the three equations of (I.30), read

$$\frac{d\Omega}{4\pi}\ \text{for}\ \frac{d\Omega}{4\omega}.$$

(2) In Eq. (I.44), read $\sin^2\theta\cos^2\phi$ for $\sin^2\theta\cos^2\theta$.

(3) In the sixth line after Eq. (II.9), read (II.2) for (II.7).

(4) For the fractional power occurring in the denominator of the unnumbered equation preceding (II.18), read $\frac{5}{2}$ instead of $\frac{7}{3}$.

(5) For the fractional power appearing in the denominator at the end of Eq. (II.19), read $\frac{1}{2}$ instead of $\frac{1}{3}$.

(6) In Eq. (II.20), insert

$$\frac{\Gamma(\frac{1}{3})}{2}\left(\frac{\omega}{2\omega_c}\right)^{2/3} \text{ instead of } \frac{\Gamma(\frac{2}{3})}{2}\left(\frac{\omega}{2\omega_c}\right)^{1/3} .$$

(7) In Eq. (II.37), read $\frac{\pi^2}{8}$ instead of $\frac{\pi^2}{4}$.

(8) In Eq. (III.31), read $\frac{n}{2n_c}$ instead of $\frac{n}{n_c}$.

(9) The denominator of Eq. (III.32) should contain $\pi^3$ instead of $\pi^2$.

---

*Work supported in part by the National Science Foundation.

[1]J. Schwinger, Phys. Rev. 75, 1912 (1949), referred to as paper I.

[2]A systematic development of this new approach to particle theory is described in J. Schwinger, *Particles, Sources, and Fields I* (Addison-Wesley, Reading, Mass., 1970) and Particles, Sources, and Fields II (to be published).

[3]This procedure resembles that introduced in an earlier paper [J. Schwinger, Phys. Rev. 82, 664 (1951)], but is here applied to the system of charged particle and photon.

[4]Retaining $u$ here gives the essence of the first quantum correction [J. Schwinger, Proc. Nat. Acad. Sci. U. S. 40, 132 (1954), and the Russian literature cited in *Synchrotron Radiation*, A. A. Sokolov and I. M. Ternov (Pergamon, New York, 1968)].

# How to Avoid $\Delta Y = 1$ Neutral Currents*

Julian Schwinger
*University of California at Los Angeles, Los Angeles, California 90024*
(Received 5 March 1973)

The problem is posed of exhibiting a mechanism that avoids $\Delta Y = 1$ neutral currents without invoking experimentally unknown types of particles. The proposed solution rejects the Cabibbo rotation in favor of a mixing, between two types of unit-spin mesons, that is produced by the $SU_3$-symmetry-breaking interaction. One quantitative prediction that is well satisfied is the identity of the strong-interaction coupling constants appearing in $\pi^{\pm}$ decay and in $\rho^0 \rightarrow e^+ + e^-$.

Unified theories of electromagnetic and weak interactions generally face a problem with hadronic neutral currents that change hypercharge. Such currents are strikingly suppressed in nature, but are usually implied by the Cabibbo rotation that introduces the $\Delta Y = 1$ charged currents. This has led to several suggestions, of varying degrees of charm, which are uniformly couched in the language of hypothetical subnuclear constituents.[1] The number of the latter has thereby been increased, from three, to four, five, seven, .... The phenomenological orientation of source theory[2] invites a more conservative attempt. Can one exhibit a mechanism for avoiding unwanted neutral currents that refers only to experimentally recognized types of particles? This note sketches an affirmative answer.

First we must review the archetypal treatment of the leptons.[3] These particles are grouped into leptonic charge triplets,[4] $L=+1$: $\mu^+, \nu, e^-$, and the chiral charge-bearing currents represented by

$$j^{\mu}_{ab} = \tfrac{1}{2}\psi\gamma^0\gamma^{\mu}T_{ab}\psi\,, \quad ab = 12, 21\,. \tag{1}$$

Here we have introduced the antisymmetrical matrices

$$T_{ab} = \tfrac{1}{2}(t_{ab} + i\gamma_5\{t_3, t_{ab}\}), \quad ab = 12, 21 \tag{2}$$

where[5]

$$\sqrt{2}\,t_{12} = t_1 + it_2\,, \quad \sqrt{2}\,t_{21} = t_1 - it_2\,, \tag{3}$$

and the $t_a$, $a = 1, 2, 3$ are the $3\times 3$ imaginary, antisymmetrical matrices of unit isotopic spin. The $T$ matrices obey the commutation relations of the

Reprinted from *Physical Review* D8, 960–964 (1973).

group $U_2$, as illustrated by

$$[T_{12}, T_{21}] = T_{11} - T_{22}, \tag{4}$$

in which

$$T_{11} = t_3 \tag{5}$$

is the electric charge matrix and

$$T_{11} + T_{22} = i\gamma_5 + \tfrac{3}{2} t_3 (1 - i\gamma_5 t_3) \tag{6}$$

commutes with all the $T$ matrices. The simplest dynamical hypothesis introduces a $U_2$-invariant coupling with four-vector fields $A^\mu_{ab}$, $a, b = 1, 2$:

$$e \sum_{a,b=1,2} A^\mu_{ab} j_{\mu ba}, \tag{7}$$

where the association between particles and fields is $A_{11} \leftrightarrow \gamma$, $A_{12,21} \leftrightarrow W^\pm$, $A_{22} \leftrightarrow Z$. The $U_2$ invariance is a partial symmetry that is broken by the large masses assigned to the particles $W$ and $Z$ (as discussed in Ref. 3).

The coupling between the vector fields $A^\mu_{ab}$ and the hadrons is pictured as proceeding through the intermediary of fields associated with the nonuplets of $1^+(0^-)$ and $1^-$ particles. The fields of such particles are conveniently represented by spinors of the second rank,

$$\psi_{\zeta\zeta'} \sim \psi_\zeta \psi_{\zeta'}, \quad \zeta, \zeta' = 1, \ldots, 4. \tag{8}$$

We emphasize that the factorization into individual spinors indicated here is purely symbolic, designed to facilitate contact with the leptonic structures; it carries *no* implication concerning the compositeness of the particles. Nonuplets of vector and axial-vector fields are thus symbolized by

$$v^\mu_{\alpha\beta} \sim \tfrac{1}{2}\psi\gamma^0\gamma^\mu t_{\alpha\beta}\psi,$$
$$a^\mu_{\alpha\beta} \sim \tfrac{1}{2}\psi\gamma^0\gamma^\mu i\gamma_5 t_{\alpha\beta}\nu\psi, \quad \alpha, \beta = 1, 2, 3 \tag{9}$$

in which the $t_{\alpha\beta}$ are antisymmetrical matrices obeying $U_3$ group commutation relations,

$$[t_{\alpha\beta}, t_{\gamma\delta}] = \delta_{\beta\gamma} t_{\alpha\delta} - \delta_{\alpha\delta} t_{\gamma\beta}, \tag{10}$$

together with

$$\sum_{\alpha=1}^{3} t_{\alpha\alpha} = \nu, \tag{11}$$

and $\nu$ is a $2\times 2$ antisymmetrical imaginary matrix representing a unit nucleonic charge. The explicit structure of the $t_{\alpha\beta}$ is given by

$$t_{\alpha\beta} = \tfrac{1}{2}(1+\nu)\tau_{\alpha\beta} - \tfrac{1}{2}(1-\nu)\tau_{\beta\alpha}, \tag{12}$$

where the $\tau_{\alpha\beta}$ are the elementary $3\times 3$ matrices with a single unit entry in the $\alpha$ row, $\beta$ column. They have the multiplication property

$$\tau_{\alpha\beta}\tau_{\gamma\delta} = \delta_{\beta\gamma}\tau_{\alpha\delta}. \tag{13}$$

The two symbolic fields appearing in the construction (9) are associated with opposite nucleonic charge, which is the counterpart of the more usual procedure where the two unitary indices $\alpha$, $\beta$ are assigned to inequivalent, complex conjugate representations. In connection with nucleonic charge, we note the relation between the generators $T_{\alpha\beta}$ of the group $U_3$, where

$$N = \tfrac{1}{3}\sum_{\alpha=1}^{3} T_{\alpha\alpha} \tag{14}$$

represents nucleonic charge, and those of the reduced group $SU_3$ ($T'_{\alpha\beta}$) from which the nucleonic charge concept has been deleted. It is

$$T'_{\alpha\beta} = T_{\alpha\beta} - \delta_{\alpha\beta} N. \tag{15}$$

Also, the $U_2$ strong-interaction group having generators that combine isotopic spin and hypercharge, electric charge in particular, is identified with the corresponding subgroup of $SU_3$, so that

$$Q = T'_{11} = T_{11} - N. \tag{16}$$

As an aspect of this relation we expect that $1^-$ fields coupled to the photon are of the form

$$v^\mu_{11} - \tfrac{1}{3}\sum_{\alpha=1}^{3} v^\mu_{\alpha\alpha} \sim \tfrac{1}{2}\psi\gamma^0\gamma^\mu(t_{11} - \tfrac{1}{3}\nu)\psi. \tag{17}$$

The problem in constructing hadronic couplings on the leptonic model begins with the choice of charge-bearing currents analogous to (1, 2), since the electric charge axis of the unitary space could be combined with either of the other two axes. Leaving that decision open for the moment, we introduce analogs of the matrices (2) as ($\beta = 2$ or 3)

$$\begin{aligned} T_{12} &= \tfrac{1}{2}(t_{1\beta} + i\gamma_5\{t_{11}, t_{1\beta}\}) \\ &= \tfrac{1}{2}(1 + i\gamma_5\nu)t_{1\beta}, \\ T_{21} &= \tfrac{1}{2}(t_{\beta 1} + i\gamma_5\{t_{11}, t_{\beta 1}\}) \\ &= \tfrac{1}{2}(1 + i\gamma_5\nu)t_{\beta 1}, \end{aligned} \tag{18}$$

which uses the equivalence illustrated by

$$\begin{aligned} \{t_{11}, t_{1\beta}\} &= \tfrac{1}{2}(1+\nu)\tau_{1\beta} + (1-\nu)\tau_{\beta 1} \\ &= \nu t_{1\beta}. \end{aligned} \tag{19}$$

Now we evaluate the commutator

$$\begin{aligned} [T_{12}, T_{21}] &= \tfrac{1}{2}(1 + i\gamma_5\nu)(t_{11} - t_{\beta\beta}) \\ &= T_{11} - T_{22}, \end{aligned} \tag{20}$$

where

$$\begin{aligned} T_{11} &= t_{11} - \tfrac{1}{3}\nu, \\ T_{22} &= \tfrac{1}{2}(1 - i\gamma_5\nu)t_{11} + \tfrac{1}{2}(1 + i\gamma_5\nu)t_{\beta\beta} - \tfrac{1}{3}\nu, \end{aligned} \tag{21}$$

and

$$T_{11} + T_{22} = (1 - i\gamma_5\nu)t_{11} + \tfrac{1}{2}(1 + i\gamma_5\nu)(t_{11} + t_{\beta\beta}) - \tfrac{2}{3}\nu. \tag{22}$$

The latter matrix commutes with all the $T$ matrices, specifically, because

$$[t_{1\beta}, t_{11}+t_{\beta\beta}]=[t_{\beta 1}, t_{11}+t_{\beta\beta}] \\ =0 \tag{23}$$

and

$$(1-i\gamma_5\nu)(1+i\gamma_5\nu)=0\,. \tag{24}$$

Note that all four of the currents obtained in this way can be represented as linear combinations of the $v$ and $a$ fields.

The conventional response given to the problem of choosing between the unitary axes 2 and 3 is that the weak interactions select a particular direction in the 23 plane, one that is inclined at a fairly small angle $\theta_C \sim 0.2$ relative to the second axis (Cabibbo rotation). But there is another possibility. Perhaps nature utilizes both axes, with the respective currents constructed as different combinations of $1^{\pm}$ fields, which combinations also reflect the mixing action of the strong $SU_3$-symmetry-breaking interactions. To explore this idea, we place an appropriate superscript on the hadronic fields to distinguish the choice of $\beta=2$ or 3, and write out the $U_2$-invariant coupling, apart from a common factor, as

$$A_{11}[(v_{11}-\tfrac{1}{3}\sum v_{\alpha\alpha})^{(2)}+(v_{11}-\tfrac{1}{3}\sum v_{\alpha\alpha})^{(3)}]+A_{21}[\tfrac{1}{2}(v_{12}+a_{12})^{(2)}+\tfrac{1}{2}(v_{13}+a_{13})^{(3)}] \\ +A_{12}[\tfrac{1}{2}(v_{21}+a_{21})^{(2)}+\tfrac{1}{2}(v_{31}+a_{31})^{(3)}]+A_{22}[\tfrac{1}{2}(v_{11}-a_{11})^{(2)}+\tfrac{1}{2}(v_{22}+a_{22})^{(2)}-\tfrac{1}{3}\sum v_{\alpha\alpha}^{(2)} \\ +\tfrac{1}{2}(v_{11}-a_{11})^{(3)}+\tfrac{1}{2}(v_{33}+a_{33})^{(3)}-\tfrac{1}{3}\sum v_{\alpha\alpha}^{(3)}], \tag{25}$$

where vector indices are suppressed.

As the simplest realization of the fields $v^{(2)}, a^{(2)}$ and $v^{(3)}, a^{(3)}$, we consider just two sets of particle fields $v, a$ and $v', a'$, which are mixed together by the strong symmetry-breaking interaction. A crude picture of the latter will be based on the situation encountered in the well-established $1^-$ nonuplet, where $\rho$ and $\omega$ are approximately degenerate in mass, while $K^*$ and $\phi$ are displaced upward. Thus, field components with one or two 3-indices are perturbed, and mixed:

$$v_{13}^{(2)}=v_{13}\cos\theta_3-v'_{13}\sin\theta_3,\quad v_{13}^{(3)}=v_{13}\sin\theta_3+v'_{13}\cos\theta_3, \\ v_{33}^{(2)}=v_{33}\cos\theta_{33}-v'_{33}\sin\theta_{33},\quad v_{33}^{(3)}=v_{33}\sin\theta_{33}+v'_{33}\cos\theta_{33}, \tag{26}$$

while the other components are identified directly with distinct particle fields, as illustrated by

$$v_{11}^{(2)}=v_{11},\quad v_{11}^{(3)}=v'_{11},\quad v_{12}^{(2)}=v_{12}\,. \tag{27}$$

Our rough view of strong interaction effects also assumes the same mixing angles for $1^+$ and $1^-$ fields. To avoid unessential complications, we do not include the mixing described by $\theta_{33}$ in stating the resulting form of the coupling (and write $\theta_3=\theta$),

$$A_{11}[v_{11}-\tfrac{1}{3}\sum v_{\alpha\alpha}+v'_{11}-\tfrac{1}{3}\sum v'_{\alpha\alpha}]+A_{21}[\tfrac{1}{2}(v_{12}+a_{12})+\tfrac{1}{2}(v_{13}+a_{13})\sin\theta+\tfrac{1}{2}(v'_{13}+a'_{13})\cos\theta] \\ +A_{12}[\tfrac{1}{2}(v_{21}+a_{21})+\tfrac{1}{2}(v_{31}+a_{31})\sin\theta+\tfrac{1}{2}(v'_{31}+a'_{31})\cos\theta] \\ +A_{22}[\tfrac{1}{2}(v_{11}-a_{11})+\tfrac{1}{2}(v_{22}+a_{22})-\tfrac{1}{3}\sum v_{\alpha\alpha}+\tfrac{1}{2}(v'_{11}-a'_{11})+\tfrac{1}{2}(v'_{33}+a'_{33})-\tfrac{1}{3}\sum v'_{\alpha\alpha}]\,. \tag{28}$$

The two kinds of hadronic fields, $v, a$ and $v', a'$, have been distinguished through their (somewhat mixed) roles in mediating the weak interactions. Now we add a final assumption concerning their disparate roles in strong interactions. It is that the fields $v', a'$ are only slightly coupled to the quasistable hadrons that are of interest in weak-interaction measurements. If we ignore that coupling completely, we can effectively strike out the primed fields in (28). The outcome is a coupling with the charged bosons $W^{\pm}$ that differs from the result of a Cabibbo rotation ($\theta \simeq \theta_C$) only in the absence of the factor $\cos\theta_C \cong 0.98$, and a coupling with the neutral boson $Z$ that contains no hypercharge transitions (in contrast, the Cabibbo rotation would introduce the field combination $v_{23}+a_{23}+v_{32}+a_{32}$, for which $\Delta Y=1$). This is the principal consequence of our investigation.

The factor $\cos\theta_C$ is ranked as one of the minor successes of the Cabibbo theory. One would be more concerned about its absence were there not larger discrepancies outstanding in the usual theory. We cite in particular the ~10% difference between the (Goldberger-Treiman) prediction of 84 MeV for the pion decay constant and the observed value of 94 MeV. It is worth recognizing, then, that the general viewpoint advocated here can accommodate such deviations from the special model just discussed. The modifications to be introduced are twofold: The fields $v, a$ and $v', a'$ are replaced by linear combinations of different particle fields of the respective types; the assumption that the primed fields are completely uncoupled from the low-lying hadrons is removed.

If the primed fields give an effective baryon contribution in the $\gamma$ coupling that is $\sim +2\%$ of the unprimed value, the result is essentially equivalent to introducing $\cos\theta_C$ in the $\Delta Y = 0$ part of the $W$ coupling with baryons.[6] There are two possibilities here: The various primed particles are coupled with normal strength to the familiar hadrons but there is almost complete destructive interference among the contributions of all the particles of this type for small momentum transfer; or, every primed particle is subnormally coupled to the usual baryons (is there a connection with a recent cosmic ray observation[7] of a particle that decays into hadrons with an anomalously long lifetime?). Next, let us recall that the charge-bearing components of the $a$ field associated with the lightest nonuplets have the following meaning in terms of fields attached to the particles $A_1$ and $\pi$:

$$a = A_1 + \frac{1}{m_A}\partial\pi, \quad m_A = \sqrt{2}\, m_\rho . \tag{29}$$

Now suppose that, both in the photon and $W$ couplings, the effective baryon contribution is reduced by ~10% through the mediation of more massive (unprimed) families of unit spin particles. The consequence is an increase of ~10% in the $\pi$-$W$ coupling, in comparison with the usual theory. So far, this is simply data fitting. But there is an important implication here for the value of the coupling constant $g$ associated with the unit-spin particles; it is best inferred through the behavior of $\pi$, which is an associate member of the lightest particle family, rather than by reference to baryons, the properties of which are also influenced by other families of such particles.

To express the last point more quantitatively, let us supply (28) with the factor

$$\sqrt{2}\,\frac{e}{g}\, m_\rho^{\,2} \cos\theta_C . \tag{30}$$

We have introduced $\cos\theta_C$ to compensate for the additional contribution of the primed particles to the electromagnetic interactions of the baryons. The factor of $\sqrt{2}$ refers to the field identification, expressed for the usual $1^-$ particles, by

$$v_{12} = \rho^+, \quad v_{21} = \rho^-, \quad v_{11} - v_{22} = \sqrt{2}\,\rho^0 , \tag{31}$$

having in mind that

$$v_{11} = \tfrac{1}{2}(v_{11} - v_{22}) + \tfrac{1}{2}(v_{11} + v_{22}) . \tag{32}$$

The constant $g$ is not to be equated to the constant $g_\rho$ that characterizes the lightest mesons, since the heavier mesons are supposed to produce a reduction by the factor $\frac{84}{94}$:

$$g = (\tfrac{84}{94}) g_\rho . \tag{33}$$

Now let us exhibit the electromagnetic coupling specifically associated with $\rho^0$:

$$(e/g)\, m_\rho^{\,2} \cos\theta_C\, A\rho^0 , \tag{34}$$

and the $W$ coupling of the pion:

$$2^{-1/2} e \cos\theta_C \left[W^-\left(\frac{m_\rho}{\sqrt{2}\,g}\right)\partial\pi^+ + W^+\left(\frac{m_\rho}{\sqrt{2}\,g}\right)\partial\pi^-\right]. \tag{35}$$

The latter displays the conventional $\pi$ coupling constant for weak interactions,

$$\frac{m_\rho}{\sqrt{2}\,g} = 94 \text{ MeV}, \tag{36}$$

from which it follows that

$$\frac{g^2}{4\pi} = 2.7 . \tag{37}$$

Apart from the factor $\cos\theta_C$, the same coupling constant appears in the electromagnetic interaction (34), which predicts the rate for the decay $\rho^0 \to e^+ + e^-$. A recent colliding-beam measurement[8] gives

$$\frac{1}{4\pi}\left(\frac{g}{\cos\theta_C}\right)^2 = 2.8 \pm 0.16 , \tag{38}$$

or

$$\frac{g^2}{4\pi} = 2.7 \pm 0.15 , \tag{39}$$

which is quite satisfactory agreement.

The value deduced for $g_\rho$ is measured by

$$\frac{g_\rho^{\,2}}{4\pi} = 3.4 . \tag{40}$$

It is interesting that a very similar number is obtained from the identification of low-energy $s$-wave $\pi N$ scattering with the consequence of $\rho$ exchange.[9] Of course, we must now reconcile this agreement with our picture in which $\pi$ and $N$ interact, not only through the exchange of $\rho$, but also by means of more massive particles of the same type. Since we are, at the moment, quite free to adjust unknown coupling constants and masses, there is no immediate difficulty here. As an illustration, let us suppose that the same constant $g$ appears in coupling the various mesons to the photon and to $W$. In addition, imagine that the ~10% contribution associated with the heavier mesons is dominated by one particle family with a squared mass ~3 and a hadronic coupling constant $\sim\frac{1}{3}$ relative to the lightest mesons. Then the additional contribution associated with the heavier mesons in $\pi N$ coupling, where squared coupling constants appear, will be only several percent.

Another application where the distinction between $g$ and $g_\rho$ can be significant is in the calculation of magnetic moments. We follow the discussion of Ref. 9, Sec. 3.8, where it is remarked that the predicted magnitudes of magnetic moments are some 15% too low, as illustrated by

$$\mu_p - \mu_n = \frac{5}{3}\frac{e}{m_\rho} = 4.1\,\mu_N\,. \tag{41}$$

The factor

$$\frac{g_\rho}{g}\cos\theta_C = 1.10 \tag{42}$$

describes a 10% increase associated with the lightest $1^-$ mesons. Presumably, the primed particles will produce a small additional increase, while the heavier particles of the same type as $\rho^0$, $\omega$, $\phi$ will make a negative contribution. To the extent that the magnetic moments associated with heavier mesons have a corresponding inverse mass factor, a substantial fraction of the 10% increase could remain.

The last remarks are characteristic of the situation produced by this approach to electromagnetic and weak interactions. The quantitative predictive power of the theory is limited, but the qualitative situation seems to be improved. The application of these ideas to electromagnetic mass splittings and nonleptonic decays will be deferred to another publication.

---

*Work supported in part by the National Science Foundation.

[1]A survey of such models is given by B. Zumino, Lectures at Cargèse Summer Institute, CERN report, 1972 (unpublished).

[2]A systematic development of source theory will be found in J. Schwinger, *Particles, Sources and Fields* (Addison-Wesley, Reading, Mass.), Vol. I (1970) and Vol. II (1973).

[3]It is also reviewed briefly in J. Schwinger, Phys. Rev. D **7**, 908 (1973).

[4]The first use of such a classification is in E. Konopinski and H. Mahmoud, Phys. Rev. **92**, 1045 (1953). However, the independent introduction of this concept of leptonic charge [J. Schwinger, Ann. Phys. (N.Y.) **2**, 407 (1957)] owed nothing to the earlier work but was a reaction to the then new situation of parity nonconservation in which helicity had become the only discrete quantum number generally attributed to the neutrino. The denial of the latter view was the later vindicated prediction of two neutrinos. See also K. Nishijima, *Fundamental Particles* (Benjamin, New York, 1963), Sec. 7-8.

[5]Note the change in definition from that used in Ref. 3.

[6]One should also note that, at the accuracy level of 1%, electromagnetic differences between $\rho^0$ and $\rho^\pm$ may be significant.

[7]K. Niu, E. Mikumo, and Y. Maeda, Prog. Theor. Phys. **46**, 1644 (1971).

[8]As reported by the Particle Data Group, Phys. Lett. **39B**, 1 (1972).

[9]See, for example, J. Schwinger, *Particles and Sources* (Gordon and Breach, New York, 1969), Sec. 3.3.

# A Report on Quantum Electrodynamics

*Julian Schwinger*

My assignment is to trace the development of quantum electrodynamics. This topic is intended to be an example of the general theme of the symposium – the 'Development of the Physicist's Conception of Nature'. But, as Dirac pointed out, the phrase 'the physicist's conception' implies a degree of unanimity that rarely exists during the period of active development of a subject. Only when the material is finally frozen in the textbooks can one speak of 'the physicist's conception'. At any interesting moment during the period of development there are discordant viewpoints of individual physicists. What is the present status of quantum electrodynamics, the modern version of which is now some twenty-five years of age? Perhaps I can emphasize that the development phase is not yet terminated by telling you that I have just finished writing a book – *Particles, Sources, and Fields*, Vol. II – in which the major quantitative accomplishments of modern quantum electrodynamics have been derived, not as they were done historically, but by a conceptually and computationally simpler method. It is through this method, source theory, that I believe future generations will learn quantum electrodynamics from their textbooks.

There is nothing unusual in the displacement of a particular historical attitude toward a subject by a conceptually sounder and simpler one. For example, none of us has learned classical mechanics in the strict spirit of Newton, but much more as Euler, many years later, recast and clarified the subject. I believe that the time has come for quantum electrodynamics to be so recast and clarified.

My programme for today has two parts, of unequal length. First of all, being conscious of the historical orientation of this symposium, I shall not launch into an exposition of the new viewpoint. Rather, I shall trace the highlights in the historical development of quantum electrodynamics in order to underline, to emphasize, the conceptual difficulties in the conventional approach. In other words, I shall not be concerned today with the quantitative successes of quantum electrodynamics. They are far too familiar to merit discussion here. My attention is focused entirely on conceptual difficulties, as I see them. Perhaps I should say now that, while the difficulties of electrodynamics are on stage today, they were not the initial stimulus for the development of the different viewpoint of source theory. Rather that came from the situation in strong interaction physics.

Thanks to the continuing development of high energy accelerators, we are now acquainted with a bewildering variety of particles, which run the whole gamut of stability characteristics. And we have learned to group some of these particles into

Reprinted from *The Physicist's Conception of Nature*, edited by J. Mehra, Reidel, Dordrecht, 1973, pp. 413–426.

families, multiplets and supermultiplets, comprising particles with related dynamical characteristics, which particles can individually range from absolutely stable to long-lived to strongly unstable. We are being told that stable and unstable particles are not intrinsically different, and require a uniform kind of description. This is the situation that, to me, seemed to demand a new point of view, one that was more physically motivated, in which the practices of the experimentalist took precedence over the *a priori* definitions of the theorist. While the need for something new is most evident in strong interaction physics, electrodynamics provides the most familiar and most dynamically elementary situation in which to illustrate the ideas.

My first topic, then, is an historically oriented discussion of electrodynamics pointing toward *source theory*. The second topic, about which I shall say only a few words, is the suggestion that quantum electrodynamics, far from being an almost closed and exhausted subject, which I suppose is the prevalent opinion, may yet have a stunningly dramatic impact on the theories of strong and weak interactions. I shall list this exciting potentiality under the heading of *magnetic charge*. Perhaps I should remark here that neither of these two sets of ideas is totally new; there are several years of history behind them. Yet, relative to the time scale that has characterized this symposium, they may pass as reasonably fresh proposals.

Essentially everything I shall discuss today has its origin in one or another work of Dirac. That will not astonish you. It is unnecessary, of course, but nevertheless I remind you that *quantum* electrodynamics began in the famous paper of 1927 [appropriately, this is the first paper in the collection, *Selected Papers on Quantum Electrodynamics*, Dover, 1958]. Here Dirac first extended the methods of quantum mechanics to the electromagnetic field. We need not be concerned now with the technical point that the quantization was applied to radiation field oscillators. The fact remains that, through the extension of quantum mechanics to electromagnetism, the electromagnetic field was promoted from a classical quantity to a quantum mechanical operator – an operator field. A cloud of formalism later obscured it all, but what Dirac did is simply this. If you think of Maxwell's equations as the equation of one photon, as Schrödinger's equation is that of one electron, the quantization to the electromagnetic field became the first example of second quantization, so-called, which is a procedure for the compact description of a multi-particle system. In this initial example, the particles obeyed Bose–Einstein statistics. Inevitably, and it is only surprising that Dirac himself did not do it, the same procedure was extended to particles obeying Fermi-Dirac statistics, with the essential difference that anti-commutation relations replace commutation relations, to express the characteristic Pauli exclusion principle. At that time, in 1928, we had a situation in which multi-particle systems obeying Bose or Fermi statistics could be described in terms of operator functions of space and time – the quantized $\psi$ field for electrons, the quantized $A$ field for photons. And these operator fields had clear and immediate physical interpretations in terms of the emission and absorption, the creation and annihilation, of the physical particles.

The year 1928 is more significant for another event – the introduction of the Dirac relativistic equation for the electron. It was first viewed as a single particle equation, which ran into the difficulty of the physically meaningless negative energy solutions. The answer proposed by Dirac in 1930 was to exploit the exclusion principle by filling up all the 'negative energy' states to form the physical vacuum state. One electron added to this infinite sea would be stable against spontaneous photon emission, as befitted a physical particle. And it was recognized that the removal of a 'negative energy' particle produced an excitation that acted physically as a particle of the same mass but opposite electric charge – the positron. Thus, the only way out of the difficulties posed by the Dirac equation was to admit from the beginning that a multi-particle system was being described. But then, why not initially adopt a second quantized description, with the operator field unifying the electron and positron as two alternative states of a single particle? With this formalism the vacuum becomes again a physically reasonable state with no particles in evidence. The picture of an infinite sea of negative energy electrons is now best regarded as an historical curiosity, and forgotten. Unfortunately, this episode, and discussions of vacuum fluctuations, seem to have left people with the impression that the vacuum, the physical state of nothingness (under controlled physical circumstances), is actually the scene of wild activity.

I shall now try to describe the resulting situation as it appeared toward the end of the third decade of our century. It was the almost inexorable outcome of a combination of ingredients. They were: the classical equations of Maxwell, the laws of quantum mechanics, the extension from the commutators of Bose statistics to the anticommutators of Fermi statistics, and, the Dirac equation with its electron–positron interpretation. The resulting mathematical formalism was clear cut. The real problem was to understand physically what had happened, and that problem would occupy a generation of physicists.

The mathematical situation is illustrated by writing down a system of field equations. They are, ($\hbar = c = 1$),

$$[\gamma^\mu (1/i\, \partial_\mu - eqA_\mu(x)) + m]\, \psi(x) = 0,$$
$$F_{\mu\nu}(x) = \partial_\mu A_\nu(x) - \partial_\nu A_\mu(x),$$
$$\partial_\nu F^{\mu\nu}(x) = \tfrac{1}{2}\psi(x)\, \gamma^0 \gamma^\mu eq\psi(x).$$

These equations only deviate from what was customary then, or indeed now, by emphasizing the symmetry between positive and negative charge – between electron and positron – through the introduction of the charge matrix $q$,

$$q = \begin{pmatrix} 0 & -i \\ i & 0 \end{pmatrix}.$$

This matrix acts on a two-valued index which, in addition to the four-valued spinor index, is carried by the field $\psi(x)$. The more usual formalism diagonalizes $q$, which has eigenvalues $+1$ and $-1$, and replaces the eight component $\psi$ by the pair of four-component eigenvector fields $\psi$ and $\psi^*$. In the formalism used here, the symmetry between electron and positron is expressed by the invariance of the system of field equations under the simple transformation

$$C: \begin{array}{l} \psi(x) \to r_q \psi(x) \\ A_\mu(x) \to - A_\mu(x) \\ F_{\mu\nu}(x) \to - F_{\mu\nu}(x), \end{array}$$

where

$$r_q = \begin{pmatrix} 1 & 0 \\ 0 & -1 \end{pmatrix}$$

is the charge reflection matrix,

$$r_q^{-1}\, q\, r_q = - q .$$

The field equations are not the whole story, of course, They are supplemented by equal time commutation relations, or, to be precise, commutation relations for the fields $A$, $F$, anticommutation relations for the components of $\psi$. I shall not write these out explicitly.

In putting together these various equations, the physical meaning of the symbols had seemed to be clear. The charge and mass of the electron (positron) is $e$ and $m$ respectively, and $\psi(x)$ symbolizes the creation and annihilation of an electron (positron), while $A$ and $F$ represent the creation and annihilation of a photon. Is it true? Not at all. Through the innocent process of combining these equations into a non-linear coupled system, the physical meanings of all the symbols have changed. They no longer refer to the physical particles, but rather to a deeper level of description.

As a rough indication of this state of affairs, suppose we begin with an excitation of $\psi$, presumably referring to the presence of an electron (positron). But then there is an electric current flowing, which creates an accompanying electromagnetic field. And this field reacts on the particle field $\psi$ to alter whatever initial excitation we assumed. In short, we have a coupled system, and the description of one physical electron is not to be written down *a priori*, but emerges only after solving a complicated dynamical problem. The situation is similar with the photon. An assumed photon field $A$ will excite the electron-positron field $\psi$, and the associated electric current acts to alter the initially assumed electromagnetic field. In short, $\psi$ and $A$ have become abstract dynamical variables, with the aid of which one constructs the physical states, but they in themselves do not have an elementary physical interpretation. In particular, the constants appearing in the field equations, $e$ and $m$, are not identifiable as the physical charge and mass of the electron (we now write them as $e_0$, $m_0$).

The question naturally arises as to what kind of direct physical interpretation can one give to the operator fields $\psi(x)$ and $A(x)$? The answer is given in terms of the vacuum state. I recall that, for us, the vacuum is the state in which no particles exist. It carries no physical properties; it is structureless and uniform. I emphasize this by saying that the vacuum is not only the state of minimum energy, it is the state of *zero* energy, zero momentum, zero angular momentum, zero charge, zero whatever.

Physical properties, structure, come into existence only when we disturb the vacuum, when we excite it. What shall we say about the excitation symbolized by $\psi(x)\,|\,\text{vac.}\rangle$?

The reference to the specific space-time point $x$ means that this is a localized excitation, with the complementary property that arbitrary amounts of energy and momentum are available. The spinor nature of the field implies an angular momentum of $\frac{1}{2}$, relative to the point $x$. And the field $\psi$ injects a unit magnitude of charge. Evidently, one physical realization of these properties is an electron (by which we henceforth mean electron-positron). But an appropriate combination of one electron and one photon, or indeed several photons, can also qualify, to which we add one electron and one electron-positron pair, and so forth. In short, $\psi(x)\,|\text{vac.}\rangle$ is a superposition of all multi-particle states that can realize the particular circumstances that are symbolized by the field $\psi$. We indicate this situation by writing

$$\psi\,|\text{vac.}\rangle = a\,|1\text{ el.}\rangle + \sum a'\,|1\text{ el. }1\text{ ph.}\rangle + \sum a''\,|1\text{ el. }1\text{ ph. }1\text{ ph.}\rangle + \\ + \sum a'''\,|1\text{ el. }1\text{ el. }1\text{ el.}\rangle + \cdots,$$

where the various coefficients are probability amplitudes for the respective kinds of particle states. The analogous discussion of the vector field $A$ gives

$$A\,|\text{vac.}\rangle = b\,|1\text{ ph.}\rangle + \sum b'\,|1\text{ el. }1\text{ el.}\rangle + \cdots.$$

In practice, the spectral properties of these excitations are more conveniently handled in terms of numerical functions that describe the correlations between different excitations localized at various spacetime points. Such a function is the vacuum expectation value

$$\langle\text{vac.}|\,\psi(x)\,\psi(x')\,|\text{vac.}\rangle \equiv \langle\psi(x)\,\psi(x')\rangle.$$

Through its translationally invariant dependence on the variable $x-x'$, the energy-momentum characteristics of the various excitations are displayed. Even more convenient, for technical reasons, is the time ordered product of fields, symbolized by $(\psi(x)\,\psi(x'))_+$, where the operator associated with the later time value stands to the left. In the particular example of the anticommuting field $\psi$ that is now under discussion, it is advisable to counter the discontinuous appearance of a minus sign by an antisymmetrical function of the time difference, $\varepsilon(x-x')$. Then, with some additional factors that are useful, we come to the definition of the propagation function or Green's function, which is the effective mathematical instrument of quantum field theory,

$$G(x, x') = i\langle(\psi(x)\,\psi(x'))_+\rangle\,\gamma^0\varepsilon(x-x').$$

This example refers to a Fermi–Dirac field. There is an analogous definition for the propagation function $D(x, x')$ associated with the Bose–Einstein electromagnetic field, although there are special features accompanying the gauge invariance characteristic of the Maxwell field.

These propagation functions obey infinite systems of linear inhomogeneous differential equations, obtained by adjoining analogous time ordered correlation functions involving increasing numbers of additional fields. As the first example of this system, we have the following, produced by applying the Dirac differential field equation to $G(x, x')$:

$$\left(\gamma \frac{1}{i} \partial + m_0\right) G(x, x') - e_0 q \gamma i \langle (A(x) \psi(x) \psi(x'))_+ \rangle \gamma^0 \varepsilon(x - x') = \delta(x - x').$$

The inhomogeneous term originates in the discontinuity of $G(x, x')$ at $x^0 = x^{0\prime}$, thereby introducing $\delta(x^0 - x^{0\prime})$, which discontinuity is measured by the equal-time anticommutator of the fields, a unit matrix multiple of $\delta(\mathbf{x} - \mathbf{x}')$. The additional factors appearing in the definition of $G(x, x')$ are such as to yield just the four-dimensional delta function. Thus, the infinite set of linear, numerical equations combines the non-linear operator field equations with the basic (anti-) commutation relations. And, through the reference to the vacuum state in the definitions, there are boundary conditions to select the appropriate solution of this equation system.

Now, imagine (!) that we have solved this infinite system and have $G(x, x')$, along with $D(x, x')$, before us. These functions will display the various multiparticle states comprising $\psi \mid \text{vac.}\rangle$ and $A \mid \text{vac.}\rangle$ in the form of additive contributions to the complete propagation function, as in

$$G(x, x') = AG_{\text{el.}}(x - x') + \cdots,$$
$$D(x, x') = BD_{\text{ph.}}(x - x') + \cdots.$$

Here $G_{\text{el.}}$ and $D_{\text{ph.}}$ are the propagation functions of the respective physical particles, the weight factors $A \sim a^2$ and $B \sim b^2$ are relative probabilities for the formation of the single particle states by the excitations $\psi(x)$ and $A(x)$, and the dots stand for the propagation functions that describe all the multi-particle excitations that are also implied by these elementary excitations.

Clearly, then, the propagation functions $G$ and $D$ do not describe the physical particles directly. They contain much more information, but of a kind that is rarely of immediate interest, in the nature of the background that accompanies any specific reaction that is under scrutiny. There is no difficulty, however, in isolating the parts of $G$ and $D$ that refer to single-particle propagation. Corresponding to the experimental requirement of following a particle for a sufficient time to measure its energy-momentum characteristics, its mass, it suffices to consider a kind of asymptotic form for these functions, such that the continuous mass spectrum of multi-particle states will lead to a suppression of those contributions. Nevertheless, the outcome of this procedure still contains a reference to the special mechanism used to create the particles, in the weight factors $A$ and $B$. Accordingly, one must also remove these factors by a process of *renormalization*:

$$G_{\text{renorm.}} = \frac{1}{A} G, \qquad D_{\text{renorm.}} = \frac{1}{B} D.$$

Accompanying this change of scale is the replacement of the original parameters $m_0$ and $e_0$ by the physical mass and charge, $m$ and $e$. The latter is particularly simple since the comparison of alternative and equivalent ways of expressing the coupling between two charges:

$$e_0^2 D = e^2 D_{\text{renorm.}},$$

gives

$$e^2 = e_0^2 B.$$

Renormalization, then, is the process of transferring attention from the underlying dynamical variables with which the theory begins to the physical level at which the observed particles are in evidence. It is not a concept to which there are intrinsic conceptual objections, and indeed must occur in any theory that contains structural assumptions about the constitution of the physical particles, in which abstract, underlying, dynamical variables appear.

Only now that I have described the fundamental significance of renormalization will I refer to the fact that is usually emphasized about this procedure. When one solves the coupled system of differential equations for the time-ordered field correlation functions by successive substitution, as a power series in $e_0^2$, the results contain divergent integrals. Very small distances or, equivalently, very high momenta make a disproportionately large contribution to the results, with the consequence that solutions to the coupled equations do not really exist – at least, not in the sense of a perturbative expansion. Whether this is merely a deficiency in the elementary mathematical approach, which could disappear in a more sophisticated scheme, or, an intrinsic failure of the physical model, I do not know. Nor am I very sure that it is particularly important to select one of the choices, since it must remain true that phenomena at very small distances play a substantial role in the whole story. What is important is that the renormalized equations, on the other hand, do have solutions – finite solutions, with numerical consequences that are in overwhelming agreement with experiment.

Here is something worth understanding. Somehow, the process of renormalization has removed the unbalanced reference to very high momenta that the unrenormalized equations display. The renormalized equations only make use of physics that is reasonably well established. In contrast, the unrenormalized equations critically involve phenomena in regions where we cannot pretend to know the physics. By what right do we, as physicists, claim that the laws of physics, even within the narrow domain of electrodynamics, will forever remain the same as we extrapolate to arbitrarily high energies? All experience suggests that new and unexpected phenomena will come into play. The resulting logical situation is very interesting. From the same system of equations emerge two very different attitudes. The unrenormalized description constitutes a model of the dynamical structure of the physical particles, which is sensitive to details at distances where we have no particular reason to believe in the correctness of the physics – an implicit speculation about inner structure – while the renormalized description removes these unwarranted speculations and concentrates on the reasonably known physics that is germane to the behaviour of the particles.

This way of putting the matter can hardly fail to raise the question whether we have to proceed in this tortuous manner of introducing physically extraneous hypotheses, only to delete these at the end in order to get physically meaningful results. Clearly there would be a great improvement, both conceptually and computationally, if we could identify and remove the speculative hypotheses that are implicit in the unrenormalized equations, thereby working much more at the phenomenological level. For electrodynamics, this may be just a *tour de force*, since it is not claimed that new numerical results will appear in this way. But, for strong interaction physics the issue is crucial. If we are caught in a formalism that has built into it implicit hypotheses, inner structural assumptions that have small chance of being correct, we shall have grave difficulties in fighting through to the correct theory. I believe that we need a more flexible kind of theory, one that can incorporate experimental results, and extrapolate them in a reasonable manner without falling into the trap of the wholesale extrapolation that infringes on unexplored areas where surprises are sure to await. I am well aware that this kind of flexible approach is anathema to the mathematically oriented [State your axioms! What are the calculational rules?], but I continue to hope that it has great appeal to the true physicist (Where are you?).

Having said all this, let us return to electrodynamics and try to identify the implicit hypothesis, the one that has introduced speculative structural assumptions. I say that it is the introduction of operator fields. Do not misunderstand me. The *field* concept is unavoidable, barring some totally new approach to the space-time continuum, which is not being advocated. But *operator* fields? That is another matter. An operator such as $\psi(x)$ is defined by the totality, or at least a sufficiently large class, of its matrix elements. And the overwhelming proportion of these refer to energies and momenta that are far outside experimental experience. Unavoidably, then, an operator field theory makes reference to phenomena in experimentally unexplored regions. It is simple enough to identify the innocent use of field operators as the culprit. But, what to replace it with?

Usually, during the development of a particular subject, several competing but equivalent formalisms are available. The example of matrix and wave mechanics naturally comes to mind. Yet one of these may be specially suited for the transition to the next levels of description. Who, in the mid-nineteenth century could have suspected the significance of the otherwise not very useful Hamiltonian formalism in the much later developments of statistical mechanics and quantum mechanics? It is important that these variant approaches exist since, in seeking a new theory, one of them may sufficiently narrow the mental gap that needs to be traversed to make this journey feasible. The human mind is not adapted to a quantum jump in ideas. A small step for mankind is all that one can reasonable expect.

I mention this in order to recall that other formalisms for quantum electrodynamics were in existence. During the 25 year period of quantum electrodynamical development, there was great formal progress in the manner of presenting the laws of quantum mechanics, all of which had its inspiration in a paper of Dirac. This paper (which is No. 26 in the collection, *Selected Papers on Quantum Electrodynamics*, Dover, 1958) discussed for the first time the significance of the Lagrangian in quantum mechanics.

I have always been puzzled that it took so long to do this, but a faint glimmering of the reason appeared when I reread this paper recently and noticed that even Dirac himself thought that the action principle required the use of coordinates and velocities rather than coordinates and momenta, despite the existence of the classical action expression

$$W = \int_{t_2}^{t_1} \mathrm{d}t \left[ \sum_k p_k \frac{\mathrm{d}q_k}{\mathrm{d}t} - H(p, q) \right].$$

[Incidentally, this same hang-up seems to persist in recent articles claiming that the quantum action principle is inapplicable to curved spaces.] Eventually, these ideas led to Lagrangian or action formulations of quantum mechanics, appearing in two distinct but related forms, which I distinguish as differential and integral. The latter, spearheaded by Feynman, has had all the press coverage, but I continue to believe that the differential viewpoint is more general, more elegant, more useful, and more tied to the historical line of development as the quantum transcription of Hamilton's action principle.

The quantum action principle is a variational statement about the transformation function that connects states at different times, or on different space-like surfaces. Its bare expression, leaving out all the labels, is

$$\delta \langle \, | \, \rangle = i \langle | \, \delta W \, | \rangle,$$

where $W$ is an action operator which, for particle mechanics, has just the classical form, while, for field mechanics, it is illustrated by the Maxwell–Dirac expression

$$W = \int (\mathrm{d}x) \{ -\tfrac{1}{2} F^{\mu\nu} (\partial_\mu A_\nu - \partial_\nu A_\mu) + \tfrac{1}{4} F^{\mu\nu} F_{\mu\nu} - \tfrac{1}{2} \psi \gamma^0 [\gamma^\mu (1/i \, \partial_\mu - e_0 q A_\mu) + m_0] \psi \}.$$

The classical context of action principles is incomplete, and somewhat misleading, since the *operator* field variations have individual operator character, specifically, distinguishing F. D. fields ($\psi$) from B. E. fields ($A$, $F$). (A fairly detailed discussion of these two types can be found in *Quantum Kinematics and Dynamics*, Benjamin, 1970.) The quantum action principle supplies the field equations (equations of motion) as in the classical model and also the (anti-) commutation relations that are characteristically quantum mechanical. This approach emphasized the indivisible unity of the quantum principles.

The action formulation facilitates the consideration of a more general situation in which the system of interest is externally driven, perturbed. A sufficiently simple yet general form is obtained by adding to the action a linear functional of the fields, as in

$$\int (\mathrm{d}x) \, [A^\mu (x) \, J_\mu (x) + \psi (x) \, \gamma^0 \eta (x)],$$

where the $J_\mu(x)$ are arbitrary commuting numbers and the components of $\eta(x)$ are arbitrary anticommuting numbers. The inhomogeneous form of the resulting field equations,

$$[\gamma^{\mu}(1/i\,\partial_{\mu} - e_0 q A_{\mu}(x)) + m_0]\,\psi(x) = \eta(x),$$
$$\partial_{\mu}F^{\mu\nu}(x) - \tfrac{1}{2}\psi(x)\,\gamma^0 e_0 q \gamma^{\mu}\psi(x) = J^{\mu}(x),$$

supplies the designation of these external quantities as sources. There are various uses of these external couplings, which are an idealized version of the interventions that constitute measurements on the system. The commutation relations of the respective fields are implicit in the source terms of the field equations, for example. And, with the possibility of exciting arbitrary states through the agency of the sources, it suffices to choose only the vacuum state in considering the over-all development of the system. Thus, the basic quantity that contains all physical information about the system is the vacuum amplitude

$$\langle \text{vac.} \mid \text{vac.}\rangle^{\eta J} \equiv \langle 0_+ \mid 0_-\rangle^{\eta J},$$

where $+$ and $-$ are causal labels designating times that follow or precede the region where the sources are manipulated.

The expansion of the vacuum amplitude functional in powers of the sources defines an infinite sequence of functions; they are the totality of Green's functions previously introduced as time-ordered field correlation functions:

$$\langle 0_+ \mid 0_-\rangle^{\eta J} = 1 + i\tfrac{1}{2}\int (\mathrm{d}x)\,(\mathrm{d}x')\,\eta(x)\,\gamma^0 G(x, x')\,\eta(x') + i\tfrac{1}{2}\int (\mathrm{d}x)\,(\mathrm{d}x')\,J^{\mu}(x)\,D(x, x')_{\mu\nu}\,J^{\nu}(x') + \cdots.$$

Corresponding to this functional union of all the propagation functions, the infinite sequence of dynamical equations for these functions is replaced by a small number of functional equations. The use of the action principle gives:

$$\delta\langle 0_+ \mid 0_-\rangle^{\eta J} = i\langle 0_+| \int (\mathrm{d}x)\,[A^{\mu}(x)\,\delta J_{\mu}(x) + \psi(x)\,\gamma^0\delta\eta(x)]\,|0_-\rangle^{\eta J},$$

or

$$\frac{1}{i}\frac{\delta}{\delta J_{\mu}(x)}\langle 0_+ \mid 0_-\rangle^{\eta J} = \langle 0_+|\,A^{\mu}(x)\,|0_-\rangle^{\eta J}$$

$$\frac{1}{i}\frac{\delta}{\delta\eta(x)\,\gamma_0}\langle 0_+ \mid 0_-\rangle^{\eta J} = \langle 0_+|\,\psi(x)\,|0_-\rangle^{\eta J},$$

and then the Dirac operator equation, for example, becomes

$$\left\{\left[\gamma^{\mu}\left(\frac{1}{i}\,\partial_{\mu} - e_0 q\,\frac{1}{i}\,\frac{\delta}{\delta J^{\mu}(x)}\right) + m_0\right]\frac{1}{i}\,\frac{\delta}{\delta\eta(x)\,\gamma^0} - \eta(x)\right\}\langle 0_+ \mid 0_-\rangle^{\eta J} = 0.$$

Successive functional differentiation of this equation now yields one sequence of the coupled Green's function equations, the other following from the analogous treatment of the Maxwell operator field equations.

The functional equations can be given a variety of other forms with different emphases. We remark, for example, that the Dirac functional equation can be expressed as

$$\left\{i\left[\eta(x), W\left(\frac{1}{i}\frac{\delta}{\delta\eta\gamma^0}, \frac{1}{i}\frac{\delta}{\delta J}\right)\right] - \eta(x)\right\}\langle 0_+ \mid 0_-\rangle^{\eta J} = 0,$$

or

$$e^{iW}\eta(x)\, e^{-iW}\langle 0_+ \mid 0_-\rangle^{\eta J} = 0,$$

where $W$ is constructed from the operator action by the functional substitutions. The formal (!) solution of this and the analogous Maxwell system is given in terms of functional delta functions by

$$\langle 0_+ \mid 0_-\rangle^{\eta J} = e^{iW[(1/i)(\delta/\delta\eta\gamma^0),\,(1/i)(\delta/\delta J)]}\delta[\eta]\,\delta[J].$$

Then the functional Fourier construction

$$\delta[\eta]\,\delta[J] = \int [\mathrm{d}\psi]\,[\mathrm{d}A]\; e^{i\int(\psi\gamma^0\eta + JA)}$$

supplies the explicit integral expression (to within a constant factor),

$$\langle 0_+ \mid 0_-\rangle^{\eta J} = \int [\mathrm{d}\psi]\,[\mathrm{d}A]\; e^{i\{\int(\psi\gamma^0\eta + JA) + W[\psi, A]\}}.$$

It continues to surprise me that so many people seem to accept this formal statement as a satisfactory *starting* point of a theory.

Another procedure is of more interest for our present purposes. Write

$$\langle 0_+ \mid 0_-\rangle^{\eta J} = e^{i\{\int(\psi\gamma^0\eta + JA) + W[\psi A]\}},$$

where $\psi(x)$ and $A(x)$ are now *numerical* fields, numbers of the same types as $\eta(x)$ and $J(x)$, and $W[\psi A]$ is only implicitly defined through a stationary requirement:

$$\frac{\delta}{\delta\psi\gamma^0}W + \eta = 0, \qquad \frac{\delta W}{\delta A} + J = 0.$$

In consequence of the stationary property, we have

$$\frac{1}{i}\frac{\delta}{\delta\eta\gamma^0}\langle 0_+ \mid 0_-\rangle^{\eta J} = \psi\,\langle 0_+ \mid 0_-\rangle^{\eta J},$$

$$\frac{1}{i}\frac{\delta}{\delta J}\langle 0_+ \mid 0_-\rangle^{\eta J} = A\,\langle 0_+ \mid 0_-\rangle^{\eta J},$$

and the Dirac functional equation, for example, becomes

$$\left[\gamma\left(\frac{1}{i}\partial - e_0 q\left(A + \frac{1}{i}\frac{\delta}{\delta J}\right)\right) + m\right]\psi = \eta;$$

we recall again that $\psi(x)$, $A(x)$ are numerical fields. Here, then, is a formulation completely equivalent to the original one in terms of operator fields, commutation relations, and all the rest, but now expressed in the language of numerical sources, numerical fields. And it is this formulation that has the flexibility to permit a new beginning, a fresh, more physical approach to particle theory. The sources were initially tied to the operators $\psi$ and $A$ which describe elementary, multi-particle excitations. Why not abandon the whole operator framework and define the sources *ab initio* in terms of the excitation of single, physical particles? This is the starting point of source theory, and if any of this discussion seems moderately reasonable, I invite you to follow the systematic evolution of this phenomenological attitude in the two existing volumes of *Particles, Sources, and Fields*, Addison-Wesley, Vol. I (1970), Vol. II (1973).

The concept of magnetic charge in quantum mechanics was introduced by Dirac in 1931 and further developed in 1948. He showed that the existence of magnetic charge would provide an elementary and beautiful explanation of the empirical quantization of electric charge, which is otherwise a mysterious regularity of nature. This still remains a most compelling argument in favour of a property that, thus far, has stubbornly refused to surface experimentally. Other facets of this intriguing possibility come to light when one considers the more general situation of particles that carry both electric charge ($e$) and magnetic charge ($g$). First, let me make the following irritating statement. As far as electrodynamics is concerned, there is no observational difference between our world, in which magnetic charge does not exist, and a hypothetical world in which all atomic particles carry both charges, in a manner that is invariant under the rotation

$$e' = e\cos\theta - g\sin\theta$$
$$g' = g\sin\theta + e\cos\theta\,,$$

and satisfy the weak charge requirement

$$\frac{e^2 + g^2}{\hbar c} < 1\,.$$

This equivalence follows immediately from the charge quantization condition, referring to any two particles with charges $e_1$, $g_1$, and $e_2$, $g_2$, namely

$$\frac{e_1 g_2 - e_2 g_1}{\hbar c} = \text{integer}\,,$$

for, under the stated charge restrictions, the integer can only be zero. Accordingly, if one were to mark, in a two-dimensional $e$–$g$ space, points corresponding to various kinds of particles, these points would lie on a single straight line passing through the origin. It then suffices to specialize the arbitary $e$–$g$ coordinate system by defining this absolute line to be the axis of electric charge; all particles will now have zero magnetic charge. Note, however, that these charges need not be equally spaced, as they are empirically. For that situation to appear, there must exist other particles that violate the weak charge restriction and do not lie on the line of pure electric charge. These are magnetically (and possibly also electrically) charged particles. The charge quantization condition tells us that, in contrast with the customary electric charge magnitude,

$$\frac{e^2}{\hbar c} = \frac{1}{137},$$

such magnetic charges are measured by, essentially,

$$\frac{g^2}{\hbar c} = 137.$$

It is the very large magnitude of this unit that suggests the connection with the strongly interacting particles, the hadrons. Of course, hadrons are not magnetically charged. But neither are atoms electrically charged. In short, the emerging picture is of hadrons as magnetically neutral composites of entities that carry magnetic charge and, necessarily, also electric charge. I have christened these hypothetical particles: dyons. (An aside to any indignant member of the audience who is thinking – you warned us about introducing speculations concerning inner structure, and now you indulge in this outrageous speculation! My warning was directed against the implicit, unrecognized speculation. One must not confuse structural hypothesis and phenomenology. These are two distinct approaches which, hopefully, will eventually merge in a more fundamental theory.) And now a wholly new situation appears. The charge quantization condition is much less restrictive for dyons, permitting *fractional* charges, that is, fractions of the units of pure electric and magnetic charge. Here, unexpectedly, is contact with the empirical 'quark' model of hadrons, which uses the electric charge pattern, 2/3, −1/3, −1/3. For a dyon this pattern extends to magnetic charge as well, and is reasonably inferred from considerations of magnetic neutrality and particle statistics. (A semi-popular account is presented in *Science* **165**, 757 (1969).)

This is the basis for thinking that, underlying the mysterious strong interactions, is another and new manifestation of electromagnetism. As for the weak interactions, let me talk only about the superweak one, that is *CP* violation, which has seemed so unrelated to the rest of physics. The point, quite simply, is that dyons have a built-in mechanism for *CP* violation. The operation of charge reflection, *C*, interchanges particles and anti-particles, producing the substitution $e,\ g \rightarrow -e,\ -g$. Space reflection, *P*, acts oppositely on electric and magnetic charges, as it does on electric and magnetic fields. Thus, the combined CP effect is $e,\ g \rightarrow -e,\ g$; for some particular fractional charge assignments, it is 2/3, 2/3 → −2/3, 2/3, or −1/3, 2/3 → 1/3, 2/3. But the resulting charge assignments are not included in the list of particles: (2/3, −1/3, −1/3), (2/3, −1/3, −1/3), or of antiparticles: (−2/3, 1/3, 1/3), (−2/3, 1/3, 1/3), and *CP* is not an invariance transformation of the system. There is more to the story, which connects it with the ordinary weak interactions, but I refer you to the *Science* article for a short account of that.

How beautiful it would be if the logically sound concepts of magnetic charge and dyons should prove to be at the heart of the subnuclear world! But I leave you with this sobering remark. We have heard so much about the importance of beauty in physical theory. No doubt a correct theory will be beautiful (a cynic will say that our concept of beauty would evolve to make it so), but a merely beautiful theory has small chance of being correct. In short, beauty, as a criterion for validity, is necessary but not sufficient.

# Photon Propagation Function: Spectral Analysis of Its Asymptotic Form

**(electrodynamics/source theory/high energy/spectral forms/field theory)**

JULIAN SCHWINGER

University of California, Los Angeles, Calif. 90024

*Contributed by Julian Schwinger, June 3, 1974*

**ABSTRACT** The physical attitudes of source theory, displacing those of renormalized, perturbative, operator field theory, are used in a simple discussion of the asymptotic behavior of the photon propagation function. A guiding principle is the elementary consistency requirement that, under circumstances where a physical parameter cannot be accurately measured, no sensitivity to its precise value can enter the description of those circumstances. The mathematical tool is the spectral representation of the propagation function, supplemented by an equivalent phase representation. The Gell-Mann–Low equation is recovered, but with their function now interpreted physically as the spectral weight function. A crude inequality is established for the latter, which helps in interpolating between the initial rising behavior and the ultimate zero at infinite mass. There is a brief discussion of the aggressive source theory viewpoint that denies the existence of a "bare charge".

The asymptotic, high momentum behavior of the photon propagation function has been extensively discussed, but invariably with the methods and preconceptions of renormalized perturbative, operator field theory. This note represents another contribution to the ongoing program of displacing operator field theory by the more physically motivated and computationally simpler source theory (1).

Our starting point is the spectral representation of the photon propagation function $D(k)$, as it enters in the coupling between two charges of individual strength $e$:

$$e^2D(k) = \frac{e^2}{k^2}\frac{1}{1 - k^2\int dM^2 \frac{a(M^2)}{k^2+M^2}}. \quad [1]$$

Here, the factor $1/k^2$ refers to the physical photon, while the last factor gives an account of vacuum polarization effects, processes in which an excitation of variable mass $M$ is set up by the coupling of the photon with charged particles. The spectral weight function $a(M^2)$ is real and positive; it can be written as

$$a(M^2) = \frac{e^2}{M^2}s\left[\frac{M^2}{m^2}, e^2\right], \quad [2]$$

where the function called $s$ is dimensionless. The explicit factor of $e^2$ appears as a matter of convenience, and does not prejudice the dependence on the coupling constant $e^2$ of this quantitative measure of the dynamics. The mass $m$ is either to be interpreted literally as the electron mass, in the idealization of pure electrodynamics, or, in a more sketchy representation of the realistic situation, is a typical mass of the charged particles that are dynamically significant at the particular level of excitation under consideration. We can now present [1] as

$$k^2e^2D(k) = \frac{e^2}{1 - k^2e^2\int \frac{dM^2}{M^2}\frac{s(M^2/m^2, e^2)}{k^2+M^2}}$$

$$= \frac{1}{\frac{1}{e^2} - k^2\int \frac{dM^2}{M^2}\frac{s}{k^2+M^2}}. \quad [3]$$

We are interested in the behavior of [3] for such large momenta that

$$|k^2| \gg m^2. \quad [4]$$

Let us introduce an intermediate mass $\lambda$:

$$|k^2| \gg \lambda^2 \gg m^2, \quad [5]$$

and then proceed to decompose the last denominator in [3] at the spectral mass $\lambda$,

$$\frac{1}{e^2} - k^2\int_{\sim m^2}^{\lambda^2}\frac{dM^2}{M^2}\frac{s}{k^2+M^2} - k^2\int_{\lambda^2}^{\infty}\frac{dM^2}{M^2}\frac{s}{k^2+M^2}$$

$$\cong \frac{1}{e^2} - \int_{\sim m^2}^{\lambda^2}\frac{dM^2}{M^2}s - k^2\int_{\lambda^2}^{\infty}\frac{dM^2}{M^2}\frac{s}{k^2+M^2}. \quad [6]$$

The symbol $\sim m^2$ is here just a qualitative reminder of the effective threshold for the spectral integral, and the approximation of the last form is an application of the inequality $|k^2| \gg m^2$. If we now define the quantity

$$\frac{1}{e_\lambda^2} = \frac{1}{e^2} - \int_{\sim m^2}^{\lambda^2}\frac{dM^2}{M^2}s; \qquad e_\lambda^2 = \frac{e^2}{1 - e^2\int_{\sim m^2}^{\lambda^2}\frac{dM^2}{M^2}s}, \quad [7]$$

we realize an alternative, asymptotic presentation of [3] in the analogous form

$$k^2e^2D(k) \cong \frac{e_\lambda^2}{1 - k^2e_\lambda^2\int_{\lambda^2}^{\infty}\frac{dM^2}{M^2}\frac{s}{k^2+M^2}}. \quad [8]$$

All explicit reference to spectral masses less than $\lambda$ has disappeared from [8], with a corresponding introduction of an effective coupling constant $e_\lambda^2 > e^2$. It is now natural to rewrite the spectral weight function $s$ by using $\lambda$ as the reference mass for the spectral variable $M$, in place of $m$, while replacing $e^2$ by the coupling constant $e_\lambda^2$:

$$s\left[\frac{M^2}{m^2}, e^2\right] \equiv \sigma\left[\frac{M^2}{\lambda^2}, e_\lambda^2, \frac{m^2}{\lambda^2}\right]. \quad [9]$$



Reprinted from the *Proceedings of the National Academy of Sciences U.S.A.* 71, 3024–3027 (1974).

The ability to do this depends, of course, on the existence of an inverse to the relation [7], yielding $e^2$ as a unique function of $e_\lambda^2$. The third dimensionless variable that enters the function $\sigma$ is small in value,

$$m^2/\lambda^2 \ll 1. \qquad [10]$$

It would seem to be an essential characteristic of a reasonable theory that we should be able to neglect this small number, thereby replacing [9] with

$$s\left[\frac{M^2}{m^2}, e^2\right] \cong \sigma\left[\frac{M^2}{\lambda^2}, e_\lambda^2\right]. \qquad [11]$$

What is involved here can be stated as a principle of physical consistency: that, under circumstances where a physical parameter cannot be accurately measured, no sensitivity to its precise value can enter the description of those circumstances. The parameter is the particle mass $m$, and the circumstances are expressed by $|k^2| \gg m^2$. The latter implies that insufficient time-distance is available to carry out the energy-momentum measurement necessary to determine $m$ with any precision. One should not hastily conclude that we have set $m = 0$, for this quantity is still present in the structure of $e_\lambda^2$ (Eq. [7]).

The relation [11] must hold for arbitrary $\lambda$, consistent with $\lambda^2 \gg m^2$. Accordingly, for a value of $M$ such that

$$M^2 \gg m^2, \qquad [12]$$

we can choose $\lambda = M$ and deduce the following:

$$s\left[\frac{M^2}{m^2}, e^2\right] \cong \sigma(1, e_M^2) \equiv \sigma(e_M^2), \qquad [13]$$

or

$$s\left[\frac{M^2}{m^2}, e^2\right] = \sigma\left[\frac{e^2}{1 - e^2 \int_{\sim m^2}^{M^2} \frac{dM'^2}{M'^2} s\left[\frac{M'^2}{m^2}, e^2\right]}\right], \qquad [14]$$

which is a functional equation for the spectral weight function $s$.

Although we have foresworn the use of perturbation theory in the context of renormalized operator theory, it is appropriate here to mention the situation in which $\sigma(e_M^2)$ can be usefully expanded in a power series in $e_M^2$,

$$\sigma(e_M^2) = \sigma_0 + e_M^2\,\sigma_1 + (e_M^2)^2\,\sigma_2 + \ldots, \qquad [15]$$

with $e_M^2$ itself appearing as a power series in $e^2$:

$$e_M^2 = e^2 + (e^2)^2 \int_{\sim m^2}^{M^2} \frac{dM'^2}{M'^2} s\left[\frac{M'^2}{m^2}, e^2\right] + \ldots$$
$$\cong e^2 + (e^2)^2\,\sigma_0 \log\frac{M^2}{\sim m^2} + \ldots. \qquad [16]$$

In the last version, the symbol $\sim m^2$ is used to suggest the magnitude of the additive constant in the leading term of the asymptotic form of the spectral integral. The resulting expansion is

$$s\left[\frac{M^2}{m^2}, e^2\right] \cong \sigma_0 + e^2\,\sigma_1 + (e^2)^2\left[\sigma_0\sigma_1 \log\frac{M^2}{\sim m^2} + \sigma_2\right] + \ldots. \qquad [17]$$

That the first two terms of the power series are asymptotically independent of $M^2$ is in accord with, and provides an elementary explanation for, the results of explicit calculations. For spin $^1/_2$ charged particles, these are given by ($\alpha = e^2/4\pi$)

$$e^2\sigma_0 = \frac{\alpha}{3\pi}, \quad (e^2)^2\sigma_1 = \frac{\alpha^2}{4\pi^2}. \qquad [18]$$

We also learn, without effort, the leading logarithmic $M^2$ dependence of the next power of $\alpha$, as exhibited in

$$M^2 a(M^2) \cong \frac{\alpha}{3\pi} + \frac{\alpha^2}{4\pi^2} + \frac{\alpha^3}{12\pi^3}\left(\log\frac{M^2}{m^2} + \text{const.}\right). \qquad [19]$$

To discuss further the functional equation of [14], we note that

$$\frac{1}{e_M^2} = \frac{1}{e^2} - \int_{\sim m^2}^{M^2} \frac{dM'^2}{M'^2} s\left[\frac{M'^2}{m^2}, e^2\right] \qquad [20]$$

obeys

$$-M^2 \frac{d}{dM^2}\frac{1}{e_M^2} = s\left[\frac{M^2}{m^2}, e^2\right] = \sigma(e_M^2). \qquad [21]$$

This is integrated as

$$\log\frac{M^2}{\sim m^2} = \int_{e^2}^{e_M^2} \frac{dx}{x^2\sigma(x)}, \qquad [22]$$

where $\sim m^2$ here generalizes its limited meaning in [16] to be the value of $M^2$ at which the asymptotic form for $(1/e_M^2) - (1/e^2)$ extrapolates to zero. Alternatively, one can write [22] as*

$$\log\frac{M^2}{m^2} = \int_{\sim e^2}^{e_M^2} \frac{dx}{x^2\sigma(x)}, \qquad [23]$$

in which $\sim e^2$ stands for the function of $e^2$ to which the asymptotic form for $e_M^2$ reduces on extrapolating $M^2$ down to $m^2$. Thus, given a non-negative function $\sigma(x)$, the quantity $e_M^2$ can be found as a function of $M^2/\sim m^2$ and $e^2$ from [22], or of $M^2/m^2$ and $\sim e^2$ from [23], after which $s(M^2/m^2, e^2)$ is constructed as $\sigma(e_M^2)$. We also observe here that the existence of the more general form [11], involving the arbitrary parameter $\lambda$, can now be verified. On subtracting [22] or [23] from a similar equation, with $M^2$ replaced by $\lambda^2$, we get

$$\log\frac{M^2}{\lambda^2} = \int_{e_\lambda^2}^{e_M^2} \frac{dx}{x^2\sigma(x)}. \qquad [24]$$

Hence $e_M^2$, and $s(M^2/m^2, e^2)$, can be exhibited as functions of the variables $M^2/\lambda^2$ and $e_\lambda^2$. The earlier cited functional forms are maintained, with $\sim m^2$ and $e^2$, or $m^2$ and $\sim e^2$, replaced by $\lambda^2$ and $e_\lambda^2$.

The only general property of $\sigma(x)$ to which reference has been made is positiveness. There is, however, one additional important requirement. It refers to the ultimate asymptotic limit, $M^2 \to \infty$. First, we insert the obvious remark that this limit can have little to do with pure electrodynamics, but must involve the totality of physics, with its presently unknown aspects playing the dominant role. Then we comment that the denominator of [3] is a monotonically decreasing

* Although the variables are different (but see later), this equation is of the Gell-Mann–Low form, with their function $\psi(x)$ identified as $x^2\sigma(x)$. This simple connection with the spectral weight function has escaped notice for two decades.

function of $k^2$,

$$\frac{d}{dk^2}\left[1 - k^2e^2\int \frac{dM^2}{M^2}\frac{s}{k^2+M^2}\right] = -e^2\int dM^2 \frac{s}{(k^2+M^2)^2} < 0. \quad [25]$$

Accordingly, if $D(k)$ is to possess no space-like singularities, which are beyond the pale, that denominator, equal to unity at $k^2 = 0$, must still be non-negative at $k^2 = \infty$, or

$$e^2\int_{\sim m^2}^{\infty} \frac{dM^2}{M^2} s\left[\frac{M^2}{m^2}, e^2\right] \leq 1. \quad [26]$$

The immediate inference from the mere existence of the left side is that

$$M^2 \to \infty: \qquad s\left[\frac{M^2}{m^2}, e^2\right] \to 0, \quad [27]$$

vanishing more rapidly than $1/\log M^2$. In other words, the limiting value of $e_M{}^2$:

$$e_\infty{}^2 = \frac{e^2}{1 - e^2\int_{\sim m^2}^{\infty}\frac{dM^2}{M^2}s}, \; e^2 < e_\infty{}^2 \leq \infty, \quad [28]$$

is such that

$$\sigma(e_\infty{}^2) = 0. \quad [29]$$

The same elementary conclusion also follows from Eq. [23], for example, with $M^2 \to \infty$, and, indeed, that has been its exclusive basis to this point. Whether the upper integration limit, $e_\infty{}^2$, is finite or infinite, the right side of the cited equation cannot match the infinite value of the left side if $\sigma(x)$ is everywhere greater than zero. Hence, $\sigma(x)$ must vanish suitably at $x = e_\infty{}^2$.

There is a large gap between the initial rising behavior of the function $\sigma(x)$ that is given in Eqs. [15] and [18]:

$$x \ll 1: \qquad \sigma(x) = \sigma_0 + \sigma_1 x + \ldots; \qquad \sigma_0, \sigma_1 > 0, \quad [30]$$

and the ultimate descent to zero. But one can, at least, provide some assurance that $\sigma(x)$ must eventually cease its increase and begin its decrease toward zero. It is stated by the following very crude inequality:

$$\sigma(x) < 1/x. \quad [31]$$

To derive this, we now regard $-k^2 = z$ as a complex variable (that $-k^2$ is the real limit of a complex variable is implicit in the spectral representation), and consider the function ($e_\infty{}^2 < \infty$)

$$F(z) = \frac{e_\infty{}^2}{e^2}\left[1 + ze^2\int_0^\infty \frac{dM^2}{M^2}\frac{s}{M^2 - z}\right], \quad [32]$$

which approaches unity as $z \to \infty$. The new physical input is the assertion that the positive function $s(M^2/m^2, e^2)$ is, in fact, greater than zero for all finite $M^2 > 0$. The validity of this contention for $M^2 < \sim m^2$ depends upon the possibility of constructing excitations that only involve photons (scattering of light by light). We first remark that the bounded function $F(z)$, which is also presented as

$$F(z) = 1 + e_\infty{}^2\int_0^\infty dM^2 \frac{s}{M^2 - z}, \quad [33]$$

can have no complex zeros, $z = x + iy$, $y \neq 0$, since the imaginary part of the implied equation:

$$y\int_0^\infty dM^2 \frac{s}{(M^2 - x)^2 + y^2} = 0, \quad [34]$$

cannot be satisfied. As to zeros on the real axis, we note that the limiting form of [34] for positive $x$ is $s = 0$, $M^2 = x$, which is excluded by our physical assertion; there is obviously no zero of [33] on the negative real axis. Therefore, the singularities of $F(z)$ are totally comprised in the branch line extended from the origin along the entire positive real axis. The branch of $\log F(z)$ that vanishes at infinity also has only this branch line of singularities, whence it follows that

$$\log F(z) = \frac{1}{\pi}\int_0^\infty dM^2 \frac{\phi(M^2)}{M^2 - z}. \quad [35]$$

The real weight function appearing here, $\phi(M^2) = \phi(M^2/m^2, e^2)$, must vanish at infinity, and also at the origin, since $F(z)$ is bounded there.

The equivalence thus established,

$$\frac{e_\infty{}^2}{e^2}\left[1 + ze^2\int_0^\infty \frac{dM^2}{M^2}\frac{s}{M^2 - z}\right] = \exp\left[\frac{1}{\pi}\int_0^\infty dM^2 \frac{\phi}{M^2 - z}\right], \quad [36]$$

can also be presented as

$$1 + ze^2\int_0^\infty \frac{dM^2}{M^2}\frac{s}{M^2 - z} = \exp\left[z\frac{1}{\pi}\int_0^\infty \frac{dM^2}{M^2}\frac{\phi}{M^2 - z}\right], \quad [37]$$

which uses the $z = 0$ relation

$$e_\infty{}^2/e^2 = \exp\left[\frac{1}{\pi}\int_0^\infty (dM^2/M^2)\phi\right]. \quad [38]$$

The restriction $e_\infty{}^2 < \infty$ can be removed in this form. Now let $z \to M^2 + i0$ in [37]. The real and imaginary parts of the resulting equation are

$$1 - M^2e^2P\int_0^\infty \frac{dM'^2}{M'^2}\frac{s'}{M^2 - M'^2} = \exp\left[-M^2\frac{1}{\pi}P\int_0^\infty \frac{dM'^2}{M'^2}\frac{\phi'}{M^2 - M'^2}\right]\cos\phi \quad [39]$$

and

$$\pi e^2 s = \exp\left[-M^2\frac{1}{\pi}P\int_0^\infty \frac{dM'^2}{M'^2}\frac{\phi'}{M^2 - M'^2}\right]\sin\phi, \quad [40]$$

which uses the Cauchy principal value for the integrals. The latter relation shows that the positive character of $s$, for all $0 < M^2 < \infty$, also applies to $\sin\phi$ and, therefore,

$$0 < \phi(M^2/m^2, e^2) < \pi \quad [41]$$

in that open interval. Equation [40] exhibits $s$ in terms of the phase function $\phi$; its inverse is

$$\tan\phi = \frac{\pi e^2 s}{1 - M^2e^2P\int_0^\infty \frac{dM'^2}{M'^2}\frac{s'}{M^2 - M'^2}}. \quad [42]$$

We have now provided the proton propagation function with the alternative spectral representation

$$k^2e^2D(k) = e^2 \exp\left[k^2 \frac{1}{\pi}\int_0^\infty \frac{dM^2}{M^2}\frac{\phi(M^2/m^2, e^2)}{k^2+M^2}\right]. \quad [43]$$

The path that began at Eq. [4] can be followed again to produce the asymptotic form

$$k^2e^2D(k) \cong e_\lambda^2 \exp\left[k^2 \frac{1}{\pi}\int_{\lambda^2}^\infty \frac{dM^2}{M^2}\frac{\phi}{k^2+M^2}\right], \quad [44]$$

with

$$e_\lambda^2 = e^2 \exp\left[\frac{1}{\pi}\int_0^{\lambda^2}(dM^2/M^2)\phi\right]. \quad [45]$$

Then, on writing

$$\phi(M^2/m^2, e^2) \cong \Phi(M^2/\lambda^2, e_\lambda^2) = \Phi(e_M^2), \quad [46]$$

we deduce that

$$\frac{M^2 \dfrac{d}{dM^2} e_M^2}{e_M^2} = \frac{1}{\pi}\Phi(e_M^2), \quad [47]$$

or

$$\log\frac{M^2}{\sim m^2} = \pi\int_{e^2}^{e_M^2}\frac{dx}{x\Phi(x)}, \quad [48]$$

together with the variant form that is analogous to [23].

The identity that is thus suggested,

$$x\sigma(x) = \frac{1}{\pi}\Phi(x), \quad [49]$$

hinges upon the asymptotic equivalence of the two definitions of $e_M^2$. As a first bit of evidence in this direction, we remark that the two asymptotic forms for $k^2e^2D(k)$, Eqs. [8] and [44], agree in equating their respective definitions of $e_\lambda^2$ with the value obtained on extrapolation to $k^2 = 0$. And, from another direction, let us insert a large space-like value of $k^2 = \lambda^2 > 0$ in the two propagation function forms, [3] and [43]. The respective use of the logarithmically valid approximation

$$\frac{\lambda^2}{\lambda^2+M^2} \sim \begin{cases} 1: & M^2 < \lambda^2 \\ 0: & M^2 > \lambda^2, \end{cases} \quad [50]$$

then yields

$$k^2e^2D(k)\big|_{k^2=\lambda^2} \sim e_\lambda^2 \quad [51]$$

for both representations. Here also is the contact with the meaning ascribed to $e_\lambda^2$ in the Gell-Mann–Low formulation. The inequality stated in [31] now follows from the general property $\Phi(e_M^2) = \phi(M^2/m^2, e^2) < \pi$.

If, somehow, one were supplied with the function $\sigma(x)$, the location of its first positive zero (the only one of physical interest) would determine $e_\infty$, the quantity that plays the role of the "bare charge" of renormalized operator field theory. This speculative cornerstone of operator field theory is removed in an aggressive interpretation of phenomenologically oriented source theory, through the insistence that no such finite limit exists:

$$e_\infty^2 = \infty. \quad [52]$$

Since the question thus raised is likely to remain moot for some time to come, we add only the following observation. The consequence of [52] that emerges from [28],

$$e^2\int_{\sim m^2}^\infty \frac{dM^2}{M^2}s\left[\frac{M^2}{m^2}, e^2\right] = 1, \quad [53]$$

would seem to be an eigenvalue equation for $e^2$. Yet the analogous consequence inferred from [38],

$$\exp\left[-\frac{1}{\pi}\int_{\sim m^2}^\infty \frac{dM^2}{M^2}\phi\left(\frac{M^2}{m^2}, e^2\right)\right] = 0, \quad [54]$$

only asserts that $\phi$ either approaches a non-zero limit as $M^2 \to \infty$, or approaches zero essentially no more rapidly than

$$M^2 \to \infty: \quad \phi(M^2) \sim \frac{1}{\log M^2 \log\log M^2}. \quad [55]$$

Indeed, the relation [39] shows how [53] follows from [54] as $M^2 \to \infty$, when $s$ is regarded as a function to be computed from $\phi$. It is plausible that the additional parameter that would be introduced through a finite limit for $\phi(M^2)$ is avoided by having $\phi(M^2)$ vanish in the logarithmically slow manner indicated in [55], which, incidentally, corresponds to the asymptotic behavior

$$x \to \infty: \Phi(x) \sim (1/x)e^{-x}. \quad [56]$$

There is no immediate sign here of an eigenvalue condition on $e^2$, but it would be hard to avoid the inference that the very particular asymptotic behavior of $\phi(M^2)$ could only be realized within a quite circumscribed dynamical scheme.

This work was supported in part by the National Science Foundation.

1. Schwinger, J. (1970 and 1973) *Particles, Sources, and Fields* (Addison-Wesley, Reading, Mass.), Vols. I and II.

# Source Theory Viewpoints in Deep Inelastic Scattering

**(electron scattering/high energy/spectral forms/electromagnetic form factors/ photon cross-sections)**

JULIAN SCHWINGER

University of California, Los Angeles, Calif. 90024

*Contributed by Julian Schwinger, October 10, 1974*

**ABSTRACT** The phenomena of deep inelastic electron scattering on single nucleons, viewed as virtual photon absorption, are reexamined with the nonspeculative, phenomenological attitude of source theory. The use of a double spectral representation for forward scattering, and of experimental inputs from the low energy resonance region and the high energy real photon diffraction region, lead naturally to the general observed characteristics of deep inelastic scattering. A reasonably successful description of deviations from simple scaling behavior is also presented.

No experiment in the recent history of high energy physics has had more impact on the theoretical community at large than the deep inelastic scattering experiment of the MIT-SLAC collaboration (1). Very high energy electrons are inelastically scattered off individual nucleons, resulting in the production of various nucleonic excited states, or resonances. It is found that, with increasing inelasticity, this resonance structure very quickly blends into a smooth pattern that shows a remarkably simple, scaling dependence upon the two independent variables of the experiment, which measure the energy and invariant momentum transfer to the nucleonic system. It was the emergence of this scaling behavior that set off an orgy of speculative model building and abstract theorizing,* which has raged unchecked until quite recent experiments on hadronic production in electron-positron collisions dealt a body blow to the confident (but discordant) predictions that accompanied the various speculative viewpoints. Perhaps the time is now propitious for a reassessment of the situation, one that focuses more on correlating experimental facts and less on the urge to speculate about the ultimate constituents of matter. The systematic evolution of particle physics that is based on the epistemological attitude of the last sentence is known as source theory (3). Although it arose in response to the continuing crisis in high energy physics, the major attention for some time has been given to honing its blade on the whetstone of electrodynamics. Here we begin to wield this weapon in the arena for which it was forged.

An inelastic transition of the electron, with the space-like momentum change $q^\mu$, creates an electromagnetic field $F^{\mu\nu}(q)$ that interacts with the nucleon, of initial time-like momentum $p^\mu$. The experiments with which we are concerned work with unpolarized nucleons, and do not detect individual hadronic components of the final state. A useful approach to the evaluation of the desired total transition rate is through the consideration of forward scattering, by the nucleon, of the "photon" of momentum $q$. The probability for the persistence of this two-particle state, of individual momenta $q$ and $p$, is diminished below unity by just the required total transition probability. A convenient gauge invariant expression for the forward scattering probability amplitude is given by

$$1 + iVd\omega_p 4e^2m^2\left\{-\frac{1}{2}F^{\mu\nu}(-q)F_{\mu\nu}(q)H_1(q^2, qp) - \frac{1}{2}F^{\mu\nu}(-q)(g_{\nu\lambda} + m^{-2}p_\nu p_\lambda)F^{\lambda\kappa}(q) \times (g_{\kappa\mu} + m^{-2}p_\kappa p_\mu)H_2(q^2, qp)\right\}, \quad [1]$$

where $V$ represents the space-time interaction volume, $e$ is the electric charge quantum in rationalized units, $m$ is the nucleon mass, and $H_{1,2}(q^2, qp)$ are two functions of the scalar variables that can be formed from the vectors $q$ and $p$. The latter are constructed with the aid of the metric tensor $g_{\mu\nu}$, which is such that $p^2 = -m^2$, $q^2 > 0$. Since the nucleon is unpolarized, it is completely characterized by its momentum $p$, which appears in the invariant momentum space measure

$$d\omega_p = \frac{(d\mathbf{p})}{(2\pi)^3}\frac{1}{2p^0}, \qquad p^0 = (\mathbf{p}^2 + m^2)^{1/2}, \quad [2]$$

and in the second of the two possible gauge invariant combinations through the projection tensor $g_{\mu\nu} + m^{-2}p_\mu p_\nu$, which selects vectors orthogonal to $p$. The functions $H_{1,2}$ also reflect, in their even dependence upon the variable $qp$, the symmetry between $q$ and $-q$ of the field structure in [1].

The expression of the fields in terms of the vector potential $A$ converts [1] into

$$1 + iVd\omega_p 4e^2A^\mu(-q)A^\nu(q)[T_{1\mu\nu}H_1 + T_{2\mu\nu}H_2], \quad [3]$$

where the two symmetrical tensors are

$$T_{1\mu\nu} = m^2(q_\mu q_\nu - q^2g_{\mu\nu}), \qquad q^\mu T_{1\mu\nu} = 0, \quad [4]$$

* For an instructive exposition of the predominant viewpoints in these matters, together with some citations of the literature, see ref. 2.



Reprinted from the *Proceedings of the National Academy of Sciences U.S.A.* **72,** 1–5 (1975).

and

$$T_{2\mu\nu} = q^2 p_\mu p_\nu - qp(q_\mu p_\nu + p_\mu q_\nu) + (qp)^2 g_{\mu\nu} + m^2(q^2 g_{\mu\nu} - q_\mu q_\nu), \qquad q^\mu T_{2\mu\nu} = p^\mu T_{2\mu\nu} = 0. \quad [5]$$

It is also useful to consider the two distinct polarizations, in the Lorentz gauge, $qA(q) = 0$. For longitudinal polarization ($L$), $A^\mu$ is proportional to a unit time-like vector lying in the $q$-$p$ plane,

$$A^\mu(q) = \frac{q^2 p^\mu - qp\, q^\mu}{[q^2(m^2q^2 + (qp)^2]^{1/2}} A(q), \quad [6]$$

and [3] becomes

$$L{:}\ 1 + iV d\omega_p 4e^2 A(-q)A(q)m^2q^2H_1, \quad [7]$$

while transverse polarization ($T$) is similarly represented by a unit space-like vector, orthogonal to both $q$ and $p$, which leads to

$$T{:}\ 1 + iV d\omega_p 4e^2 A(-q)A(q)m^2[-q^2H_1 + (q^2 + \nu^2)H_2]. \quad [8]$$

Here, we have introduced the symbol

$$\nu = -qp/m, \quad [9]$$

which gives the electron energy loss in the rest frame of the initial nucleon (laboratory system).

The inferred persistence probabilities, illustrated by

$$L{:}\ 1 - V\, d\omega_p\, 8e^2\, A(-q)\, A(q)\, m^2q^2\, Im\, H_1, \quad [10]$$

display the total transition probabilities in the decrement below unity. These probabilities can be expressed as cross-sections, through division by the interaction volume $V$, and by the relative flux for the collision, namely,†

$$d\omega_p\, A(-q)\, A(q)\, 4[m^2q^2 + (qp)^2]^{1/2}. \quad [11]$$

This yields ($\alpha = e^2/4\pi$)

$$\sigma_L = (8\pi\alpha/m\nu)[1 + (q/\nu)^2]^{-1/2} m^2q^2\, Im\, H_1 \quad [12]$$

and

$$\sigma_T = (8\pi\alpha/m\nu)[1 + (q/\nu)^2]^{-1/2} m^2[(q^2 + \nu^2) \times Im\, H_2 - q^2\, Im\, H_1], \quad [13]$$

both of which must be positive quantities. The cross-section forms make apparent that the $H$ functions have the dimensions of inverse momentum to the fourth power. We also note that

$$\frac{\sigma_L}{\sigma_L + \sigma_T} = \frac{q^2}{q^2 + \nu^2} \frac{Im H_1}{Im H_2}, \quad [14]$$

and that on setting $q^2 = 0$, only $\sigma_T$ survives to become the total photon cross-section

$$\sigma_\gamma = (8\pi\alpha/m^2)\, m^3\nu\, Im\, H_2(q^2 = 0, -qp = m\nu). \quad [15]$$

An important experimental fact is that, with increasing photon energy, the latter approaches an essentially constant limit, the same for proton and neutron, which is represented by

$$\nu/m \gg 1{:}\ m^3\nu\, Im\, H_2(0, m\nu) \cong 1.2. \quad [16]$$

Also of interest is the manner of approach of the two cross-sections, $\sigma_{\gamma n}$ and $\sigma_{\gamma p}$, which can be represented approximately by

$$\nu/m \gg 1{:}\ \sigma_{\gamma n}/\sigma_{\gamma p} \cong 1 - \frac{1}{4}(m/\nu)^{1/2}. \quad [17]$$

We shall now exhibit double spectral forms for the $H$ functions. It is characteristic of source theory (3) that such spectral forms are inferred by first considering the causal propagation through the system of various excitations. Here, there are two independent inputs associated with the momentum combinations $p \pm q$, both of which are initially time-like. The resulting double spectral forms are

$$H_{1,2}(q^2,qp) = \int \frac{dM_+^2}{M_+^2}\frac{dM_-^2}{M_-^2} \times \frac{2h_{1,2}(M_+^2, M_-^2)}{[(p+q)^2 + M_+^2][(p-q)^2 + M_-^2]}, \quad [18]$$

where $-i0$ is understood in each denominator, and the crossing symmetry between $q$ and $-q$ is expressed by the symmetry of the real, dimensionless weight functions $h_{1,2}$ in the two quadratic mass variables. The process of space-time extrapolation that leads to [18] could be accompanied by an extrapolation of the mass spectrum that supports the weight functions. Rather then relying on specific dynamical models, we shall accommodate our views in this matter to the requirements of experiment. Comparison with experiment is also the basis for omitting possible single spectral forms, which, in this situation, involve the combination

$$\frac{1}{(p+q)^2 + M^2} + \frac{1}{(p-q)^2 + M^2} = 2\frac{M^2 - m^2 + q^2}{[(p+q)^2 + M^2][(p-q)^2 + M^2]}. \quad [19]$$

Apart from the $q^2$ term in the numerator, this could be incorporated into the double spectral form. The evidence for the absence of the additional $q^2$ factor will come from experimental results on form factors.‡

Having in mind that the momentum $q$ is absorbed,

† Some authors use a flux definition in which the square root of Eq. [11] is replaced by $-qp - 1/2\, q^2$.

‡ Since it is somewhat aside from our main concern, we only remark on the fact that, in application to the forward Compton scattering amplitude for real photons (Eq. [15]):

$$f = \frac{\alpha}{2m}(2m\nu)^2 H_2(q^2 = 0, -qp = m\nu), \qquad \sigma_\gamma = \frac{4\pi}{\nu} Imf,$$

the double spectral form can be reduced to a single spectral form, with an additive constant. This combination is such that use of the simplest information supplied by the elastic form factors (Eq. [31]) leads automatically to the correct Thompson amplitude, $-\alpha/m$, for low energy photon scattering by the proton.

not emitted, by the nucleon, we infer from [**18**] that

$$\frac{1}{\pi} Im\, H_{1,2} = \int \frac{dM_+^2}{M_+^2}\frac{dM_-^2}{M_-^2}\frac{\delta[(p+q)^2+M_+^2]}{(p-q)^2+M_-^2} \times 2h_{1,2}(M_+^2,M_-^2) = \int \frac{dM_+^2}{M_+^2}\frac{dM_-^2}{M_-^2} \times \frac{\delta(q^2-2m\nu-m^2+M_+^2)}{q^2+\frac{1}{2}(M_+^2+M_-^2)-m^2} h_{1,2}(M_+^2, M_-^2). \quad [20]$$

An alternative presentation of the latter form is obtained by writing

$$\frac{1}{q^2+\frac{1}{2}(M_+^2+M_-^2)-m^2} = \int_0^\infty \frac{d\zeta}{M_+^2} \times \exp\left[-\frac{q^2}{M_+^2}\zeta\right]\exp\left[-\frac{M_+^2+M_-^2-2m^2}{2M_+^2}\zeta\right] \quad [21]$$

The following definition of new dimensionless functions

$$h_{1,2}(\zeta,m^2/M_+^2) = \pi\int\frac{dM_-^2}{M_-^2} \times \exp\left[-\frac{M_+^2+M_-^2-2m^2}{2M_+^2}\zeta\right]h_{1,2}(M_+^2,M_-^2), \quad [22]$$

then gives

$$Im\, H_{1,2} = \frac{1}{(M_+^2)^2}\int_0^\infty d\zeta \exp\left[-(q^2/M_+^2)\zeta\right] \times h_{1,2}(\zeta,m^2/M_+^2)\big|_{M_+^2=m^2+2m\nu-q^2} \quad [23]$$

It should be remarked that the mass $m$ of the argument $m^2/M_+^2$ not only refers to the nucleon mass, but also characterizes, in order of magnitude, the resonance region. What is being underscored here is the physical hypothesis that, at the level of excitation realized in these experiments, there is, as yet, no significant dependence upon any larger, hypothetical, mass parameter, which would signal the onset of entirely new physical phenomena.

Elastic scattering receives a direct discussion in terms of the two form factors $F_{1,2}(q^2)$ that enter the Dirac matrix current combination

$$\gamma^\mu F_1 - (1/2m)\sigma^{\mu\nu}\, i\, q_\nu F_2. \quad [24]$$

As defined here, $F_1(q^2 = 0)$ equals the nucleon electric charge (1 or 0, for proton and neutron, respectively), while $F_2(q^2 = 0)$ gives the appropriate anomalous magnetic moment, in nucleon magnetons. The results of this standard calculation can be stated in terms of the longitudinal and transverse cross-sections:

$$\sigma_{L,T} = (8\pi\alpha/m\nu)[1+(q/\nu)^2]^{-1/2}\delta\left(\frac{q^2-2m\nu}{m^2}\right) \times \pi\left\{L:\ G_E^2,\ T:\ \frac{q^2}{4m^2}G_M^2\right\}, \quad [25]$$

where

$$G_E = F_1 - (q^2/4m^2)F_2, \qquad G_M = F_1 + F_2, \quad [26]$$

which is to say that

$$M_+^2 \sim m^2:\ m^2q^2\frac{1}{\pi}ImH_{1,2} = \delta\left(\frac{M_+^2}{m^2}-1\right) \times \left\{G_E^2,\ \frac{G_E^2+(q^2/4m^2)G_M^2}{1+(q^2/4m^2)}\right\}. \quad [27]$$

The experimental data on the various elastic form factors, $G_{E,M}$, are all roughly represented by the so-called dipole function,

$$(1+q^2/m_0^2)^{-2}, \qquad m_0 = 0.9\, m. \quad [28]$$

In particular, the coefficients of this function in $G_M$, for proton and neutron, are in the ratio of the respective magnetic moments, which, in magnitude, is approximately 3:2.

For comparison with the constructions of Eqs. [**20–23**], we introduce the known excitation spectrum, where the nucleon mass is isolated below the threshold of the continuum. Accordingly, for $M_+^2 \sim m^2$, we have

$$h_{1,2}(M_+^2,M_-^2) = \delta[(M_+^2/m^2)-1]\left\{h_{1,2}\,\delta[(M_-^2/m^2)-1] + h_{1,2}(M_-^2/m^2)\right\},$$
$$h_{1,2}(\zeta,\ m^2/M_+^2) = \pi\,\delta[(M_+^2/m^2)-1]\,h_{1,2}(\zeta), \quad [29]$$

where

$$h_{1,2}(\zeta) = h_{1,2} + \int_{>m^2}\frac{dM_-^2}{M_-^2}\exp\left[-\frac{M_-^2-m^2}{2m^2}\zeta\right] \times h_{1,2}(M_-^2/m^2). \quad [30]$$

This yields

$$\left\{G_E^2,\ \frac{G_E^2+(q^2/4m^2)G_M^2}{1+(q^2/4m^2)}\right\} = h_{1,2} + q^2 \times \int\frac{dM_-^2}{M_-^2}\frac{h_{1,2}(M_-^2/m^2)}{q^2+\frac{1}{2}(M_-^2-m^2)} = \int_0^\infty d\zeta\exp[-(q^2/m^2)\zeta]h_{1,2}'(\zeta), \quad [31]$$

and we note that the identification of an average (1/2) $(M_-^2 - m^2)$ with $m_0^2$ yields $M_- = 1.6\ m = 1.5$ GeV, which is well within the resonance region. Among other inferences from these equations, the constants $h_{1,2}$ are identified, by setting $q^2 = 0$, as the squared nucleon charge, 1 or 0, for proton and neutron (see footnote‡). The introduction of the derivative function of $\zeta$ in the last form of [**31**] involves the following property:

$$h_{1,2}(\zeta = 0) = h_{1,2} + \int\frac{dM_-^2}{M_-^2}h_{1,2}(M_-^2/m^2) = 0, \quad [32]$$

which is the first of a set of "superconvergence relations" that express the vanishing of the left side in Eq. [**31**] as $q^2/m^2 \to \infty$. Indeed, to reproduce the $(m^2/q^2)^4$ behavior of the left side for high momenta it is necessary that

$$\zeta \ll 1: \ h_{1,2}'(\zeta) \sim \zeta^3, \qquad [33]$$

with the attendant consequences inferred from [30]. It is also this strong decrease of the left side of [31] for large $q^2/m^2$ that argues against the presence of a single spectral form, with its implied constant multiple of $q^2$ on the right-hand side of that equation, since such a term would be difficult to compensate.

The latter remark refers to elastic form factors, of course, but the weight of evidence about inelastic form factors is that they behave in much the same manner for sufficiently large $q^2$, which we regard as reasonable justification for the complete omission of the single spectral form in favor of the double spectral form. That the shape of the form factors is essentially universal for sufficiently large $q^2$ indicates that $h_{1,2}(\zeta, m^2/M_+^2)$ generally contains a factor $\bar{h}_{1,2}(\zeta)$, with the characteristics indicated in Eq. [33]. And, with decreasing $m^2/M_+^2$, the delta function spike of [29] must give way to an increasingly smooth dependence on $m^2/M_+^2$. Indeed, one could reasonably anticipate that

$$m^2/M_+^2 \to 0: \ h_{1,2}(\zeta, m^2/M_+^2) \to \bar{h}_{1,2}(\zeta). \qquad [34]$$

There is, however, a qualification of this statement that originates in the experimental data for $q^2 = 0$.

Let us compare the observed high energy behavior of the photo cross-section, as stated in [16], but numerically rounded off to unity, for simplicity, with the construction of [23]:

$$M_+^2/m^2 \gg 1: \ \int_0^\infty d\zeta h_2(\zeta, m^2/M_+^2) \cong 2M_+^2/m^2. \qquad [35]$$

Obviously, we cannot omit the $m^2/M_+^2$ dependence of the left side, or, if we were to do so, setting $M_+^2/m^2 = \infty$ for consistency, we would conclude that $\int d\zeta \bar{h}_2(\zeta)$ does not converge at the upper limit. The inference is that [34] holds with the additional proviso that $(m^2/M_+^2)\zeta \ll 1$, whereas, in the opposite situation, $h_2(\zeta, m^2/M_+^2)$ behaves something like

$$(m^2/M_+^2)\zeta \gtrsim 1: \ h_2(\zeta, m^2/M_+^2) \cong \exp[-(m^2/2M_+^2)\zeta]. \qquad [36]$$

If, for definiteness, we accept the particular exponential function [36], we can add to Eqs. [33] and [34] the further information that $[(m^2/2M_+^2)\zeta \ll 1]$

$$\zeta \gg 1: \ \bar{h}_2(\zeta) \cong 1. \qquad [37]$$

Some evidence in favor of [36] and [37] will be adduced later.

The term "deep inelastic scattering" refers to the regime in which both $2\nu/m$ and $q^2/m^2$ are large, in such a way that the ratio

$$\omega = 2m\nu/q^2 \qquad [38]$$

has any value in excess of unity. The essential experimental observations about this region are the following. For both proton and neutron, the cross-section ratio, $\sigma_L/\sigma$, $\sigma = \sigma_L + \sigma_T$, is quite small, within relatively large experimental errors; the scaling behavior that is characteristic of the region is expressed by

$$\sigma = \frac{4\pi\alpha}{q^2} f(\omega), \qquad [39]$$

where $f(\omega)$ approaches a constant $\sim 1$ for sufficiently large $\omega$, and vanishes as $\omega \to 1$ in a manner not inconsistent with that of $(\omega - 1)^3$; the ratio $f_n(\omega)/f_p(\omega)$ decreases from unity at large $\omega$ to a somewhat uncertain limit as $\omega \to 1$.

We first remark that

$$q^2/\nu^2 = (2/\omega)^2 \ (m^2/q^2) \qquad [40]$$

is small for the circumstances of interest. Accordingly, if $Im\ H_1$ and $Im\ H_2$ have similar behaviors in the deep inelastic regions, thereby extending the similarities in the resonance region that are noted in Eq. [27], the cross-section ratio (Eq. [14])

$$\frac{\sigma_L}{\sigma} \cong \frac{q^2}{\nu^2} \frac{Im H_1}{Im H_2} \qquad [41]$$

will indeed be small, and should continue to diminish with increasing values of $q^2/m^2$. Then we note, using Eqs. [12] and [13], together with the smallness of $q^2/\nu^2$, that

$$\frac{q^2}{4\pi\alpha}\sigma \cong 2m\nu\, q^2\, Im\ H_2 = \frac{\omega}{(\omega - 1)^2} \times \int_0^\infty d\zeta \exp[-\zeta/(\omega - 1)] h_2(\zeta, m^2/M_+^2). \qquad [42]$$

The last step applies Eq. [23], with the observation that

$$M_+^2 = m^2 + 2m\nu - q^2 \cong (\omega - 1)q^2. \qquad [43]$$

It is at once apparent that the essential independence of the function $h_2$ on $m^2/M_+^2 \ll 1$, for the finite $\zeta$ that is realized by the exponential factor in [42], leads immediately to the scaling behavior recorded in [39], with the identification

$$f(\omega) = \frac{\omega}{(\omega - 1)^2} \int_0^\infty d\zeta \exp[-\zeta/(\omega - 1)]\bar{h}_2(\zeta) = \frac{\omega}{\omega - 1} \int_0^\infty d\zeta \exp[-\zeta/(\omega - 1)]\bar{h}_2'(\zeta). \qquad [44]$$

The second of these forms is conveniently used, first for large $\omega$, where

$$\omega \gg 1: \ f(\omega) = \int_0^\infty d\zeta \bar{h}_2'(\zeta) \cong 1, \qquad [45]$$

which employs and vindicates [37], and then as $\omega \to 1$, where small values of $\zeta$ dominate and [33] is applied:

$$\omega \to 1: \ f(\omega) \sim \frac{1}{\omega - 1} \times \int_0^\infty d\zeta^{\exp[-\zeta\ (\omega-1)]} \zeta^3 \sim (\omega - 1)^3. \qquad [46]$$

Some additional consequences depend explicitly upon the exponential function of [**36**], which can be approximately joined with the exponential function exhibited in [**23**] to produce

$$\exp\left[-(q^2 + {}^1/_2 m^2/M_+^2)\zeta\right], \qquad [47]$$

replacing $\exp[-\zeta/(\omega - 1)]$. Thus, in application to deep inelastic scattering events for which $2m\nu \gg q^2 \gg (1/2)\,m^2$, one could reasonably extrapolate the experimental results of [**17**] by the substitution

$$\frac{1}{4}\left(\frac{m}{\nu}\right)^{1/2} \cong \frac{1}{2}\left(\frac{\frac{1}{2}m^2}{M_+^2}\right)^{1/2} \rightarrow \frac{1}{2}\left(\frac{q^2}{M_+^2}\right)^{1/2} \cong \frac{1}{2}\frac{1}{(\omega - 1)^{1/2}}, \qquad [48]$$

thereby leading to

$$\omega \gg 1:\ \sigma_n/\sigma_p \cong 1 - \frac{1}{2}\frac{1}{(\omega - 1)^{1/2}}. \qquad [49]$$

Indeed, a not unsatisfactory fit to the data for $\omega > 2$ is obtained in this way. The introduction of the function [**47**] also changes the large $\omega$ evaluation of [**45**] into (the first of the two forms in [**44**] is used here)

$$\omega \gg 1:\ f(\omega) \rightarrow \frac{1}{\omega}\int_0^\infty d\zeta\, \exp\left[-\left(q^2 + \frac{1}{2}m^2\Big/\omega q^2\right)\zeta\right] = \frac{q^2}{q^2 + \frac{1}{2}m^2}. \qquad [50]$$

This gives an account of deviations from scaling behavior for small values of $q^2 \sim (1/2)\,m^2$, one that is in qualitative accord with the trend of the data.§ More generally, if [**47**] is used, and the approximations of [**42**] and [**43**] avoided, one replaces [**42**] by

$$\frac{q^2}{4\pi\alpha}\,\sigma \cong \left(1 + \frac{q^2}{\nu^2}\right)^{1/2} \frac{2m\nu}{2m\nu + \frac{3}{2}m^2}\, \frac{q^2}{q^2 + \frac{1}{2}m^2}\, f(\omega_S), \qquad [51]$$

where

$$\omega_S = \frac{2m\nu + \frac{3}{2}m^2}{q^2 + \frac{1}{2}m^2}, \qquad [52]$$

which combines a description of scaling deviations with the suggestion that $\omega_S$ is a better scaling variable than $\omega$. Combinations similar to $\omega_S$ have already been used, to some advantage, in widening the scaling region (**1**). Here we also find that the introduction of $\omega_S$ in [**49**] extends the range of an acceptable fit to the data down to $\omega_S > 1.5$. We also remark, as a purely empirical observation, that the replacement of $\omega_S - 1$ by $\omega_S - (3/4)$, which does not conflict with the large $\omega$ status of [**49**], gives a quite respectable account of the whole range of present data: $20 > \omega > 1.14$. If the empirical formula is taken seriously, it predicts a null limit for $\sigma_n/\sigma_p$ as the scaling variable approaches unity.

We have now seen that the general characteristics of deep inelastic scattering emerge as a reasonable interpolation between the known properties of the low energy resonance region and of the high energy diffractive region. And in doing this we have shunned the widespread practice of hanging such phenomenological correlations on the scaffolding of some speculative dynamical model. It is thereby emphasized that whatever understanding has been achieved cannot be adduced as evidence in favor of a particular model.

There is one significant point that remains to be mentioned. The slow exponential decrease of $h_2(\zeta, m^2/M_+^2)$ that is detailed in Eq. [**36**] is not what one might have expected from [**22**] on the basis of the physical spectrum. It appears that, under the high energy circumstances of diffractive scattering, the dominant support of the double spectral weight function occurs for $M_+^2 + M_-^2 - 2m^2 \sim m^2$, implying large *negative* values of $M_-^2$. While we have alluded briefly to the possibility of an extrapolation of the spectrum, the danger of serious conceptual problems posed by negative $M^2$ requires further investigation.

**Note added in proof.** A joint public announcement (Nov. 16, 1974) by the Stanford Linear Accelerator Center and the Brookhaven National Laboratory discloses the discovery of a new unit spin, neutral particle that decays, with a very long life-time, into hadrons, electron-positron and $\mu$-pairs. The electromagnetic leptonic decay rates are comparable to those of the known unit spin, neutral particles, but the hadronic modes are greatly suppressed. It is interesting that a family of particles with just these characteristics had been anticipated in the recent development (which was also motivated by source theory philosophy) of a new unified theory of electromagnetic and weak interactions, one that was intended to supply a less speculative interpretation for the experimental absence of certain neutral currents [Schwinger, J. (1973) *Phys. Rev.* **D8**, 960–964]. With this identification of the new particle, a substantial fraction of the observed hadronic decay rate is accounted for by electromagnetic mixing effects.

I am indebted to Wu-Yang Tsai, Kimball Milton, and Lester DeRaad for helpful comments on the manuscript. This work was supported in part by the National Science Foundation.

§ We recall here the hypothesis that the experiments under discussion do not reach such high excitation levels, as represented by a characteristic mass $M_0 \gg m$, that new phenomena come into play. The significant presence of an additional variable in [**23**], namely, $M_+^2/M_0^2 \cong (\omega - 1)(q^2/M_0^2)$, would show itself through a deviation from scaling behavior for *large* values of $q^2$.

1. Friedman, J. I. & Kendall, H. W. (1972) in *Annual Reviews of Nuclear Science* (Annual Reviews, Inc., Palo Alto, Calif.), Vol. 22, pp. 203–254.
2. Feynman, R. P. (1972) *Photon–Hadron Interactions* (W. A. Benjamin, Reading, Mass.).
3. Schwinger, J. (1970 and 1973) *Particles, Sources, and Fields* (Addison-Wesley, Reading, Mass.), Vols. I and II.

# Magnetic charge and the charge quantization condition*

Julian Schwinger
*University of California, Los Angeles, California 90024*
(Received 18 August 1975)

Two viewpoints concerning magnetic charge are distinguished: that of Dirac, which is unsymmetrical, and the symmetrical one, which embodies invariance under charge rotation. It is pointed out that the latter is not in conflict with the empirical asymmetry between electric and magnetic charge. The discussion is based on an action principle that uses field strengths and the vector potential $A$ as independent variables; a second vector potential $B$ is defined nonlocally in terms of the field strengths. This nonlocality is described by an arbitrary vector function $f^\nu(y)$, subject only to the restriction $\partial_\nu f^\nu(y) = \delta(y)$ and the additional requirement of oddness, in the symmetrical formulation. The charge quantization conditions for a pair of idealized charges, $a$ and $b$, are inferred by examining the dependence of the action $W$ on the choice of the arbitrary mathematical function $f$, and requiring the uniqueness of $\exp[iW]$. For the unsymmetrical viewpoint the half-integer condition of Dirac is obtained, $e_a g_b/4\pi = \frac{1}{2}n$, while the symmetrical formulation requires the integer condition $(e_a g_b - e_b g_a)/4\pi = n$. The Dirac injunction, "a string must never pass through a charged particle," is criticized as unnecessarily restrictive, owing to its origin in a classical action context. As simplified by a restriction to small momentum transfers, permitting the neglect of form-factor and vacuum-polarization effects, the dynamics of a realistic system of two spin-$\frac{1}{2}$ dyons is shown to involve the same interaction structure used in the idealized discussion.

## INTRODUCTION

Interest in magnetic charge has revived recently,[1] in contexts that are characterized by such terms as non-Abelian gauge fields, broken symmetry, color, . . . . It is not my intention to comment on this class of speculations. I wish only to review and attempt to clarify earlier remarks[2] on the troublesome charge quantization condition since they are sometimes misunderstood and because my own views have appreciably altered.

The original work of Dirac[3] pertained to particles with electric charge ($e$) in the presence of magnetic charge ($g$) and inferred the charge quantization condition ($\hbar = c = 1$, rational units)

$$eg/4\pi = \tfrac{1}{2}n, \quad n = 0, \pm 1, \pm 2, \ldots . \tag{1}$$

This statement incorporates the usual concept of an absolute distinction, not subject to continuous variation, between electric and magnetic charge. However, the general form of the Maxwell equations,

$$\begin{aligned} \partial_\nu F^{\mu\nu} &= J_e^\mu, \\ \partial_\nu {}^*F^{\mu\nu} &= J_m^\mu, \\ {}^*F^{\mu\nu} &= \tfrac{1}{2}\epsilon^{\mu\nu\kappa\lambda} F_{\kappa\lambda}, \end{aligned} \tag{2}$$

admits the one-parameter rotation group described by

$$\begin{aligned} J_e &\to \cos\phi J_e + \sin\phi J_m, \\ J_m &\to -\sin\phi J_e + \cos\phi J_m, \\ F &\to \cos\phi F + \sin\phi\, {}^*F, \end{aligned} \tag{3}$$

as supplemented by the statement produced with the aid of the iteration property of the dual:

$$^{**}F = -F. \tag{4}$$

The corresponding invariant form of the charge quantization for two particles, $a$ and $b$, is

$$(e_a g_b - e_b g_a)/4\pi = n, \tag{5}$$

where, for the moment, it suffices to say that $n$ assumes discrete values, including zero, with $n_0 \sim 1$ the magnitude of the smallest nonzero possibility. It is remarkable that this symmetrical viewpoint does *not* conflict with the empirical asymmetry between electric and magnetic charge.

To see this we make an invariant distinction between small charges, which are such that

$$(e_a{}^2 + g_a{}^2)/4\pi < n_0, \tag{6}$$

and large charges, which obey

$$(e_a{}^2 + g_a{}^2)/4\pi \geq n_0. \tag{7}$$

(The known unit of pure electric charge is comfortably small, $e^2/4\pi = \alpha \ll 1$.) Now apply the charge quantization condition to a pair of small charges, and note that

$$\left|\frac{e_a g_b - e_b g_a}{4\pi}\right| \leq \left(\frac{e_a{}^2 + g_a{}^2}{4\pi}\right)^{1/2}\left(\frac{e_b{}^2 + g_b{}^2}{4\pi}\right)^{1/2} < n_0, \tag{8}$$

from which it follows that the left-hand side vanishes. Hence, if only small charges are admitted, all possible two-dimensional points with coordinates $e_a, g_a$ must occupy a single line. And, by a conventional choice of coordinate system, this absolute line can be made the axis of pure electric



Reprinted from *Physical Review* **D12**, 3105–3111 (1975).

charge, thereby reducing to zero all magnetic charges. In this viewpoint, then, what is special about the world thus far disclosed by experiment is simply that no large charges have yet been produced. Incidentally, the geometrical inequality contained in Eq. (8) informs us that the minimum strength of a large charge, specifically, one that does not lie on the line of small charges, is actually given by

$$(e_a^2+g_a^2)/4\pi > n_0^2/\alpha \gg 1. \tag{9}$$

In the following discussion we shall consider both the unsymmetrical and the symmetrical viewpoints, although it is the latter, with its possibility of dyon fractional charges and the associated dynamical model of hadrons, that I support.

## ACTION

An unsymmetrical, and provisional, but convenient action expression that yields the pair of Maxwell equations is

$$W=\int(dx)[J_e^\mu A_\mu+J_m^\mu B_\mu-\tfrac{1}{2}F^{\mu\nu}(\partial_\mu A_\nu-\partial_\nu A_\mu)+\tfrac{1}{4}F^{\mu\nu}F_{\mu\nu}], \tag{10}$$

where $A$ and $F$ are subject to independent variations and $B$ is defined as

$$B_\mu(x)=\int(dx')\,{}^*F_{\mu\nu}(x')f^\nu(x'-x)+\partial_\mu\lambda_m(x), \tag{11}$$

in which $\lambda_m$ is arbitrary and $f^\nu(y)$ obeys

$$\partial_\nu f^\nu(y)=\delta(y). \tag{12}$$

We shall also introduce

$${}^*f^\nu(y)=-f^\nu(-y),\quad \partial_\nu{}^*f^\nu(y)=\delta(y), \tag{13}$$

so that (11) is presented alternatively as

$$B_\mu(x)=-\int(dx')\,{}^*f^\nu(x-x')\,{}^*F_{\mu\nu}(x')+\partial_\mu\lambda_m(x). \tag{14}$$

The stationary requirement on $A$ variations directly yields the first Maxwell set, while $F$ variations produce the relation

$$F_{\mu\nu}(x)=(\partial_\mu A_\nu-\partial_\nu A_\mu)(x)+\int(dx')\,{}^*[f_\mu(x-x')J_{m\nu}(x')-f_\nu(x-x')J_{m\mu}(x')], \tag{15}$$

the dual of which,

$${}^*F_{\mu\nu}(x)={}^*(\partial_\mu A_\nu-\partial_\nu A_\mu)(x)-\int(dx')[f_\mu(x-x')J_{m\nu}(x')-f_\nu(x-x')J_{m\mu}(x')], \tag{16}$$

leads to the second Maxwell set. Another consequence of Eq. (15) is the construction

$$A_\mu(x)=-\int(dx')f^\nu(x-x')F_{\mu\nu}(x')+\partial_\mu\lambda_e(x), \tag{17}$$

the analog of (14).

The analogy between (14) and (17) is not an actual symmetry in the sense of the charge rotation (3) unless

$${}^*f^\nu(y)=f^\nu(y)=-f^\nu(-y). \tag{18}$$

Then, despite its unsymmetrical appearance, the action expression (10) is invariant under the charge rotation. To verify this, consider the infinitesimal rotation

$$\delta J_e=\delta\phi J_m,\quad \delta J_m=-\delta\phi J_e,\quad \delta F=\delta\phi\,{}^*F,\quad \delta A=\delta\phi B, \tag{19}$$

which is completed by

$$\begin{aligned}\delta B_\mu(x)&=\delta\phi\int(dx')\,{}^*f^\nu(x-x')F_{\mu\nu}(x')\\&=\delta\phi\left[-A_\mu(x)+\int(dx')({}^*f^\nu-f^\nu)(x-x')F_{\mu\nu}(x')\right],\end{aligned} \tag{20}$$

where gradient terms have been omitted. The responses of the individual pieces of $W$ are

$$\delta\int(dx)(J_eA+J_mB)=\delta\phi\int(dx)(dx')J_m^\mu(x)({}^*f^\nu-f^\nu)(x-x')F_{\mu\nu}(x'), \tag{21}$$

$$\delta\int(dx)(-\tfrac{1}{2})F^{\mu\nu}(\partial_\mu A_\nu-\partial_\nu A_\mu)=-\delta\phi\int(dx)(J_mA+J_eB), \tag{22}$$

and

$$\begin{aligned}\delta\int(dx)(\tfrac{1}{4})F^{\mu\nu}F_{\mu\nu}&=\delta\phi\left[2\int(dx)J_mA-\int(dx)(dx')J_m^\mu(x)({}^*f^\nu-f^\nu)(x-x')F_{\mu\nu}(x')\right]\\&=\delta\phi\left[2\int(dx)J_eB+\int(dx)(dx')J_e^\mu(x)({}^*f^\nu-f^\nu)(x-x')\,{}^*F_{\mu\nu}(x')\right],\end{aligned} \tag{23}$$

where the last form involves the analog of (15),

$$^*F_{\mu\nu}(x) = (\partial_\mu B_\nu - \partial_\nu B_\mu)(x) - \int (dx')\, ^*[^*f_\mu(x-x')J_{e\nu}(x') - {}^*f_\nu(x-x')J_{e\mu}(x')]. \quad (24)$$

The average of the alternatives in Eq. (23) combines with the other contributions to give

$$\delta W = \delta\phi\tfrac{1}{2}\int (dx)(dx')(^*f^\nu - f^\nu)(x-x')[J_m^\mu(x)F_{\mu\nu}(x') + J_e^\mu(x)\, ^*F_{\mu\nu}(x')], \quad (25)$$

which indeed vanishes under the circumstances of Eq. (18).

The action expression (10) was characterized as provisional, for it cannot be applied as it stands to point charges, which constitute the only possibility for a consistent theory of electric and magnetic charges. The *mathematical* difficulty is, of course, the singular nature of the self-action for an individual charge. A simple device to alleviate the problem is based on the following replacement, which is applied to all the terms of (10):

$$\int (dx)G(x)H(x) \to \int (dx)G(x)H(x\pm\lambda) = \int (dx)G(x\pm\lambda)H(x). \quad (26)$$

Here, $\lambda$ is an arbitrary spacelike displacement that eventually tends to zero, and $\pm\lambda$ indicates the equally weighted average of the two possibilities. The formal action property of $W \to W(\lambda)$ is not affected by this replacement, the field equations being retained intact,[4] and the now finite self-actions of the charges, more precisely, their real parts, can be deleted without ambiguity. (The precise philosophy that underlies this incision need not detain us.) The only consequence in an equation like (25) is conveyed by the replacement $x - x' \to x - x' \pm \lambda$.

### CHARGE QUANTIZATION

The basic realization of a function $f$ that obeys Eq. (12) is given by

$$f_\mu(y) = \int_0^\infty d\xi_\mu \delta(y-\xi), \quad (27)$$

where the $\xi$ integration follows some path from the origin to infinity. Different paths produce different functions and we shall indicate specific choices by the superscripts (1) and (2). The difference of two such functions is

$$f_\mu^{(1)}(y) - f_\mu^{(2)}(y) = \left(\int^{(1)} - \int^{(2)}\right) d\xi_\mu \delta(y-\xi)$$
$$= -\int d\sigma_{\mu\nu}(\partial/\partial\xi_\nu)\delta(y-\xi)$$
$$= \partial^\nu \int d\sigma_{\mu\nu}\delta(y-\xi), \quad (28)$$

where the surface integral extends over the area bounded by the two paths. The analogous statements for functions that obey the symmetry restriction of (18) are

$$f_\mu(y) = \int_0^\infty d\xi_\mu \tfrac{1}{2}[\delta(y-\xi) - \delta(y+\xi)], \quad (29)$$

and

$$f_\mu^{(1)}(y) - f_\mu^{(2)}(y) = \partial^\nu \int d\sigma_{\mu\nu}\tfrac{1}{2}[\delta(y-\xi) + \delta(y+\xi)]. \quad (30)$$

In the following, differences of $f$ functions are written as

$$\delta f_\mu(y) = \partial^\nu m_{\mu\nu}(y), \quad (31)$$

which, despite the use of the $\delta$ symbol, is not limited to small changes; the surface integral constructions of the antisymmetrical tensor $m_{\mu\nu}(y)$ are exhibited in Eqs. (28) and (30).

The explicit dependence of $W(\lambda)$ on the $f$ function is contained in the $J_m B$ term. A change in the choice of $f$ induces

$$\delta W(\lambda) = -\int (dx)(dx')\, ^*F_{\mu\nu}(x)\delta f^\mu(x-x'\pm\lambda)J_m^\nu(x'), \quad (32)$$

which is not restricted to small changes since all $f$ dependence is concentrated in $\delta f$. The introduction of (31) converts $\delta W(\lambda)$ into

$$\delta W(\lambda) = \tfrac{1}{2}\int (dx)(dx')(\partial_\mu\, ^*F_{\nu\kappa} + \partial_\nu\, ^*F_{\kappa\mu} + \partial_\kappa\, ^*F_{\mu\nu})(x) \times m^{\mu\kappa}(x-x'\pm\lambda)J_m^\nu(x')$$
$$= -\tfrac{1}{2}\int (dx)(dx')\epsilon_{\mu\nu\kappa\sigma}J_e^\sigma(x) m^{\mu\kappa}(x-x'\pm\lambda)J_m^\nu(x')$$
$$= -\int (dx)(dx')J_e^\mu(x)\, ^*m_{\mu\nu}(x-x'\pm\lambda)J_m^\nu(x'). \quad (33)$$

We now insert a point-charge realization of these currents

$$J_e^\mu(x) = \sum_a e_a \int_{-\infty}^\infty ds \frac{dx_a^\mu(s)}{ds}\delta(x - x_a(s)),$$
$$J_m^\nu(x') = \sum_b g_b \int_{-\infty}^\infty ds' \frac{dx_b^\nu(s')}{ds'}\delta(x - x_b(s')), \quad (34)$$

and get

$$\delta W(\lambda) = -\sum_{a,b} e_a g_b \int dx_a^\mu dx_b^\nu \, {}^*m_{\mu\nu}(x_a - x_b \pm \lambda)$$
$$= -\sum_{a,b} e_a g_b \int \tfrac{1}{2}\, {}^*d\sigma_{ab}^{\mu\nu} m_{\mu\nu}(x_a - x_b \pm \lambda), \tag{35}$$

where

$$d\sigma_{ab}^{\mu\nu} = dx_a^\mu dx_b^\nu - dx_a^\nu dx_b^\mu . \tag{36}$$

In the symmetrical formulation, for which $m_{\mu\nu}(-y) = m_{\mu\nu}(y)$, $\delta W(\lambda)$ can be written in terms of the rotationally invariant combination of charges:

$$\delta W(\lambda) = -\tfrac{1}{2}\sum_{a,b} (e_a g_b - e_b g_a) \int \tfrac{1}{2}\, {}^*d\sigma_{ab}^{\mu\nu} m_{\mu\nu}(x_a - x_b \pm \lambda). \tag{37}$$

We first examine the unsymmetrical viewpoint, where (28) is appropriate,

$$\delta W(\lambda) = -\sum_{a,b} e_a g_b \int \tfrac{1}{2}\, {}^*d\sigma_{ab}^{\mu\nu} d\sigma_{\mu\nu} \delta(x_a - x_b \pm \lambda - \xi). \tag{38}$$

The product of orthogonal area elements defines a four-dimensional element of volume for the variable $x_a - x_b \pm \lambda - \xi$, where we now make a particular choice of $+\lambda$ or $-\lambda$. That variable may, or may not, pass through the origin in the course of the integration, correspondingly yielding 1, or 0, as the integral of the $\delta$ function. Consider, for definiteness, this situation. The two paths being compared are straight lines in the 12 plane, deviating by the angle $\theta$. The excursions of the variable $\xi$ in this plane are measured by the area element $d\sigma_{12}$. Perpendicular to the latter is ${}^*d\sigma_{ab}^{12} = d\sigma_{ab}^{30}$, and this integration over the appropriate $\delta$-function factors picks out the situation(s) where $x_a^0 - x_b^0 = 0$, $x_a^3 - x_b^3 = 0$, assuming, for simplicity only, that $\lambda$ lies in the 12 plane. At each occurrence of this situation, should the two-dimensional vector $(x_a - x_b \pm \lambda)_{1,2}$ lie within the triangle of vertex angle $\theta$, the complete integral would equal unity; otherwise, it would equal zero. As for $\pm\lambda$, one must also recognize the possibility that, for one of these choices, the integral equals unity while the other yields zero. Thus, the basic values of the four-dimensional integral are $1, \frac{1}{2}, 0$, with the possibility of $\frac{1}{2}$ dependent on the finiteness of the $\lambda$ displacement. On singling out a particular $ab$ pair, the uniqueness of $W(\lambda)$, mod$2\pi$, corresponding to the physical significance of $\exp[iW]$, requires that

$$e_a g_b(1, \tfrac{1}{2}) = 2\pi n, \tag{39}$$

with $n$ an integer. If the special situation producing $\frac{1}{2}$ is taken seriously, the quantization condition reads

$$e_a g_b / 4\pi = n; \tag{40}$$

if this special situation is disregarded, we get

$$e_a g_b / 4\pi = \tfrac{1}{2} n. \tag{41}$$

In the past, I have insisted that no anomalies should be tolerated during the limiting process $\lambda \to 0$, an attitude which results in the integer quantization condition (40). Yet, it was always apparent that this injunction might be unnecessarily strict, since the theory can be completely free of nonphysical elements only when $\lambda$ attains its null limiting value. Thus, it would suffice that the circumstances for which the $\lambda$ displacement is significant be a set of measure approaching zero in that limit. In this more permissive view, to which, in keeping with the times, I now subscribe, the Dirac quantization condition (1) is correct for the unsymmetrical formulation.

In the symmetrical formulation, where $\delta W(\lambda)$ is given by (37) and $m_{\mu\nu}(y)$ by (30), the presence of two disjoint areas, each with the weight factor $\frac{1}{2}$, replaces (39), for $\lambda = 0$, with

$$\tfrac{1}{2}(e_a g_b - e_b g_a) = 2\pi n \tag{42}$$

or

$$(e_a g_b - e_b g_a)/4\pi = n . \tag{43}$$

This is (5) with $n$ restricted to integer values, so that $n_0 = 1$. (The use of the strict construction of the limit $\lambda \to 0$, as described in the preceding paragraph, would further narrow $n$ to the even integers.)

## STRINGS

The original Dirac formulation has been prominent in recent developments, which makes it desirable to point out, again, the unnecessarily restrictive nature of the dictum "a string must never pass through a charged particle." To set the scene for this criticism, consider the part of the action (10) that depends explicitly on the coordinates of the point charges, as contained in the current constructions of Eq. (34). The response to a variation of these coordinates is

$$\delta W = \sum_a \delta \int dx_a^\mu [e_a A_\mu(x_a) + g_a B_\mu(x_a)]$$
$$= \sum_a \int \tfrac{1}{2} d\sigma_a^{\mu\nu} [e_a(\partial_\mu A_\nu - \partial_\nu A_\mu)(x_a) + g_a(\partial_\mu B_\nu - \partial_\nu B_\mu)(x_a)], \tag{44}$$

where

$$d\sigma_a^{\mu\nu} = \delta x_a^\mu dx_a^\nu - \delta x_a^\nu dx_a^\mu . \tag{45}$$

On introducing (15) and (24), this becomes

$$\delta W = \sum_a \int \tfrac{1}{2} d\sigma_a^{\mu\nu}[e_a F_{\mu\nu}(x_a) + g_a {}^*F_{\mu\nu}(x_a)]$$
$$- \sum_{a,b} e_a g_b \int d\sigma_{ab}^{\mu} f_\mu(x_a - x_b)$$
$$+ \sum_{ab} g_a e_b \int d\sigma_{ab}^{\mu}\, {}^*f_\mu(x_a - x_b), \qquad (46)$$

and here

$$d\sigma_{ab}^{\mu} = {}^*d\sigma_a^{\mu\nu} dx_{b\nu} \qquad (47)$$

is a directed three-dimensional element.

These variations are relevant to the derivation of equations of motion for the charged particles, where only field strengths should intervene. To achieve this, Dirac insisted that the additional $f$-dependent terms in (46) must vanish. Thus, if no point $\xi$ along the integration contour of (27) is allowed to coincide with a value assumed by $x_a - x_b$, these unwanted terms will disappear. But suppose such a coincidence does occur? Consider first the unsymmetrical situation in which particle $a$ is electrically charged, and we examine this contribution to $\delta W$ of a magnetically charged particle $b$:

$$-e_a g_b \int d\sigma_{ab}^{\mu} d\xi_\mu \delta(x_a - x_b - \xi). \qquad (48)$$

The integration element for the variable $x_a - x_b - \xi$, $d\sigma_{ab}^{\mu} d\xi_\mu$, is an infinitesimal four-dimensional volume. If, in the course of the integration, $x_a - x_b$ and $\xi$ become equal, the change in $W$ will be finite, of magnitude $e_a g_b = 2\pi n$, according to (41), and $\exp[iW]$ remains unaffected. In the symmetrical formulation, where ${}^*f_\mu = f_\mu$, the last terms of (46) read

$$-\sum_{a,b} (e_a g_b - e_b g_a) \int d\sigma_{ab}^{\mu} f_\mu(x_a - x_b), \qquad (49)$$

and the nonzero contribution associated with a particular pair of particles is similarly measured by $\frac{1}{2}(e_a g_b - e_b g_a) = 2\pi n$, as given in Eq. (42). Evidently, the same charge quantization conditions, expressing the uniqueness of $\exp[iW]$, should be and are encountered whether one examines the change of the $f$ function for fixed-particle trajectories or varies the trajectories for a given $f$ function.

## REALISTIC SYSTEM

The discussion thus far given is an idealized one, abstracted from the special dynamical features of particular types of particles. Accordingly, it would be instructive to indicate, at least, how the charge quantization condition enforces the consistency of the theory for a realistic system. To that end, but minimizing the difficulty of a fully general discussion, we consider the interaction of two spin-$\frac{1}{2}$ dyons, under conditions of sufficiently small momentum transfer that physical effects associated with vacuum polarization and charge form factors can be omitted. The motion of the individual dyons in a given electromagnetic field is described by Green's functions that obey

$$\left[\gamma^\mu\left(\frac{1}{i}\partial'_\mu - eA_\mu(x') - gB_\mu(x')\right) + m\right] G_+^{A,B}(x', x'') = \delta(x' - x''), \qquad (50)$$

where a charge matrix of eigenvalues $\pm 1$ is implicit in $e$ and $g$. In a matrix notation, this equation is written

$$HG_+^{A,B} = 1 \qquad (51)$$

or

$$G_+^{A,B} = i\int_0^\infty ds\, e^{-isH}, \qquad (52)$$

with

$$H = \gamma(p - eA - gB) + m. \qquad (53)$$

The analogy between $H$ and a Hamiltonian is used in deriving the proper-time equation of motion

$$\frac{dx^\mu}{ds} = \frac{1}{i}[x^\mu, H] = \gamma^\mu. \qquad (54)$$

We also exploit the significance of the matrix of $\exp(-is_1 H)$ as a transformation function,

$$\langle x'|e^{-is_1 H}|x''\rangle = \langle x's_1|x''0\rangle, \qquad (55)$$

and apply the differential action principle[5]

$$\delta\langle x's_1|x''0\rangle = -i\left\langle x's_1\right|\int_0^{s_1} ds\, \delta H\left|x''0\right\rangle \qquad (56)$$

to changes of the vector potentials,

$$\delta\langle x's_1|x''0\rangle = i\left\langle x's_1\right|\int_0^{s_1} ds\left(e\frac{dx}{ds}\delta A + g\frac{dx}{ds}\delta B\right)\left|x''0\right\rangle. \qquad (57)$$

The latter is also presented in functional notation as

$$\frac{1}{i}\frac{\delta}{\delta A_\mu(\xi)}\langle x's_1|x''0\rangle = \left\langle x's_1\right|\int_0^{s_1} ds\, e\frac{dx^\mu}{ds}\delta(\xi - x(s))\left|x''0\right\rangle \qquad (58)$$

and

$$\frac{1}{i}\frac{\delta}{\delta B_\mu(\xi)}\langle x's_1|x''0\rangle = \left\langle x's_1\right|\int_0^{s_1} ds\, g\frac{dx^\mu}{ds}\delta(\xi - x(s))\left|x''0\right\rangle. \qquad (59)$$

The current expressions appearing here, which resemble (34), are incomplete, however, for they are not conserved. Thus

$$\frac{\partial}{\partial \xi^\mu}\int_0^{s_1} ds\, e\frac{dx^\mu}{ds}\,\delta(\xi - x(s))$$
$$= -e[\delta(\xi - x(s_1)) - \delta(\xi - x(0))] \quad (60)$$

and

$$\frac{\partial}{\partial \xi^\mu}\frac{1}{i}\frac{\delta}{\delta A_\mu(\xi)}\langle x's_1|x''0\rangle$$
$$= -e[\delta(\xi - x') - \delta(\xi - x'')]\langle x's_1|x''0\rangle, \quad (61)$$

which relates the nonconservation to the gauge variance of the transformation function, and of the Green's function. What is missing is an electromagnetic model of the source, describing the history of the conserved charges before they apparently appear at $x''$, and after they seem to disappear at $x'$. This is effectively introduced by considering the gauge-invariant combination

$$\langle x's_1|x''0\rangle_{\text{inv}} = \exp[i\phi(x')]\langle x's_1|x''0\rangle \exp[-i\phi(x'')], \quad (62)$$

where

$$\phi(x') = \int (d\xi)[eA_\mu(\xi) + gB_\mu(\xi)]f^\mu(\xi - x'), \quad (63)$$

and analogously for $x''$. The $f$ function appearing here obeys (12) and can be exhibited, similarly to (27) but with notational changes, as

$$f^\mu(\xi - x') = \int_{s_1}^{\infty} ds\frac{dx^\mu}{ds}\,\delta(\xi - x(s)), \quad x(s_1) = x', \quad (64)$$

along with[6]

$$-f^\mu(\xi - x'') = \int_{-\infty}^{0} ds\frac{dx^\mu}{ds}\,\delta(\xi - x(s)), \quad x(0) = x''; \quad (65)$$

in these expressions $x(s)$ is numerically valued and the two paths are arbitrary. When the gauge-invariant transformation function is used, Eqs. (58) and (59) retain their form, but the proper time parameter now ranges from $-\infty$ to $+\infty$. The following remarks refer to the gauge-invariant Green's functions that are constructed as

$$G_+^{A,B}(x', x'')_{\text{inv}} = i\int_0^\infty ds_1\langle x's_1|x''0\rangle_{\text{inv}}. \quad (66)$$

The Green's function that describes the situation of interacting dyons $a$ and $b$ is presented symbolically in[7]

$$G_{ab} = \exp[iW_{ab}]G_{+,\text{inv}}^{A,B}|_a G_{+,\text{inv}}^{A,B}|_b, \quad (67)$$

where $W_{ab}$ is the part of the action expression of the preceding sections associated with two distinct current distributions, $J_{e,m}|_a$ and $J_{e,m}|_b$, and

$$J_e(\xi) \to \frac{1}{i}\frac{\delta}{\delta A(\xi)}, \quad J_m(\xi) \to \frac{1}{i}\frac{\delta}{\delta B(\xi)}; \quad (68)$$

the fields $A(\xi)$, $B(\xi)$ are set equal to zero after the differentiations. The effect of these differentiations is to introduce $\exp[iW_{ab}]$ into the over-all matrix element as an ordered operator, with the functional derivatives replaced by particle operators, in accordance with the gauge-invariant extensions of Eqs. (58) and (59). The current expressions thus obtained have just the form of (34), and their operator aspect can be removed through the evaluation of the matrix element as an infinite product of infinitesimal transformation matrix elements, yielding a functional integral form. The result is an interaction factor having precisely the structure used in the preceding discussion of the charge quantization condition. Of course, one would want to supplement this highly formal discussion with an explicit verification, which inevitably involves an approximation scheme. The most immediate one is a high-energy eikonal approximation[8] to the individual Green's functions in (67), which effectively results in the arbitrary particle paths of the functional integration being replaced by straight-line motion. The details are left to the reader.

*Added note.* At long last there is experimental evidence for the existence of magnetic charge. P. B. Price, E. K. Shirk, W. Z. Osborne, and L. S. Pinsky [Phys. Rev. Lett. 35, 487 (1975)] have detected a very heavily ionizing particle that has all the characteristics of a particle with magnetic charge $g = 137e$, or in rationalized units, $eg/4\pi = 1$. That is the smallest magnetic charge permissible in the symmetrical formulation, and twice the magnetic unit of the unsymmetrical Dirac version. While this does not prove the validity of the symmetrical viewpoint, as the discovery of magnetic charge $g = \frac{1}{2}(137e)$ would have disproved it, it does lend considerable support to the symmetrical formulation and encourages the serious study of the dyon model of hadronic phenomena. Since the smallest magnetic charge resides on a dyon, the observed particle should also carry a fractional electric charge, $\frac{1}{3}e$ or $\frac{2}{3}e$ in magnitude. That cannot be verified from the present data, but might be tested when such a particle is observed coming to rest.

*Work supported in part by the National Science Foundation.

[1]For example, Y. Nambu, Phys. Rev. D 10, 4262 (1974); G.'t Hooft, Nucl. Phys. B79, 276 (1974).

[2]J. Schwinger, Phys. Rev. 144, 1087 (1966); 173, 1536 (1968).

[3]P. A. M. Dirac, Phys. Rev. 74, 817 (1948); Proc. R. Soc. Lond. A133, 60 (1931).

[4]Compare Ref. 3.

[5]The concepts used here are fully developed, for simple physical contexts, in J. Schwinger, *Quantum Kinematics and Dynamics* (Benjamin, Reading, Mass., 1970).

[6]The alternative procedure, in which a compensating charge runs from $x''$ to $x'$, is less convenient here.

[7]This is the immediate generalization of Eq. (5-2.12) in J. Schwinger, *Particles, Sources and Fields* (Addison-Wesley, Reading, Mass., 1973), Vol. II. See also Chap. 7 of Ref. 5.

[8]Such techniques are discussed by H. M. Fried, *Functional Methods and Models in Quantum Field Theory* (MIT Press, Cambridge, Mass., 1972).

# CASIMIR EFFECT IN SOURCE THEORY*

JULIAN SCHWINGER
University of California, Los Angeles, Calif., U.S.A.

ABSTRACT

The theory of the Casimir effect, including its temperature dependence, is rederived by source theoretic methods, which do not employ the concept of zero point energy.

The Casimir effect is an observable [1] non-classical electromagnetic force of attraction between two parallel conducting plates. Its theory is usually derived [2] by comparing the divergent quantum zero point energies of two states: the true vacuum, and the perturbed 'vacuum' of the region defined by the boundaries of the conductors. Here is a challenge to source theory, [3] since that phenomenologically oriented method makes no reference to quantum oscillators and their zero point energy. In sketching our response [4] we use the simplification of a spinless photon with scalar source J(x), and scalar field A(x) that vanishes on the conductor surfaces; an inserted polarization multiplicity factor of two in the final expression for the force will suffice to produce the electromagnetic result.

The vacuum persistence probability amplitude is ($\hbar = c = 1$)

$$\langle 0_+ | 0_- \rangle^J = \exp[iW(J)],$$

$$W(J) = \frac{1}{2} \int (dx)(dx')\, J(x)\, D(x-x')\, J(x'),$$

where the translationally invariant propagation function is

$$D(x-x') = \int \frac{(d\underset{\sim}{p})}{(2\pi)^3}\, e^{i\underset{\sim}{p}\cdot(\underset{\sim}{r}-\underset{\sim}{r}')}\, \frac{i}{2E}\, e^{-iE|t-t'|}, \quad E = |\underset{\sim}{p}|.$$

The relation between field and source,

$$A(x) = \int (dx')\, D(x-x')\, J(x'),$$

can also be presented inversely as

---

* Support in part by the National Science Foundation.



Reprinted from *Letters in Mathematical Physics* **1**, 43–47 (1975).

$$J(x) = \int (dx')\, D^{-1}(x-x')\, A(x'),$$

with

$$D^{-1}(x-x') = \int \frac{(d\underset{\sim}{p})}{(2\pi)^3}\, e^{i\underset{\sim}{p}\cdot(\underset{\sim}{r}-\underset{\sim}{r}')} \left(E^2 + \frac{\partial^2}{\partial t^2}\right) \delta(t-t').$$

In the presence of conducting surfaces, which are introduced for a macroscopic time interval T and then withdrawn, the propagation function loses its spatial translational invariance, $D(x-x') \to D(x, x')$, $D^{-1}(x-x') \to D^{-1}(x, x')$, and the new functions depend upon parameters that specify the conductor configuration. An infinitesimal change of these parameters induces (at least) the modification

$$\delta W(J) = \frac{1}{2} \int (dx)(dx')\, J(x)\, \delta D(x, x')\, J(x')$$
$$= -\frac{1}{2} \int (dx)\, (dx')\, A(x)\, \delta D^{-1}(x, x')\, A(x').$$

The fractional change in the vacuum amplitude, $i\delta W(J)$, is compared with the description of two photon processes, $\frac{1}{2}\left[i \int (dx)\, J(x)\, A(x)\right]^2$, to yield the effective two photon source

$$iJ(x)J(x')\Big|_{eff} = -\, \delta D^{-1}(x, x').$$

The additional physical processes thus represented will also act within W(J) itself to give the new term

$$\delta W_0 = \frac{1}{2} \int (dx)(dx')\, D(x, x')\, J(x')J(x)\Big|_{eff}$$
$$= \frac{1}{2}\, i \int (dx)(dx')\, D(x, x')\, \delta D^{-1}(x', x).$$

On making the time variables explicit ($t-t' = \tau$), this reads

$$\delta W_0 = -\, \delta\varepsilon T,$$

with

$$\delta\varepsilon = -\frac{1}{2}\, i \int (d\underset{\sim}{r})(d\underset{\sim}{r}')\, d\tau\, D(\underset{\sim}{r}, \underset{\sim}{r}', \tau)\, \delta D^{-1}(\underset{\sim}{r}', \underset{\sim}{r}, -\tau),$$

where transient effects are omitted in comparison with the contribution of the long time interval T. The presence in the vacuum amplitude of the phase factor $\exp[-i\delta\varepsilon T]$ identifies $\delta\varepsilon$ as an energy shift.

The following configuration is considered. Conducting plates that define the total region are placed at $z = -L$ and $+L$, where L is a macroscopic distance. In addition, there are conducting plates at $z = 0$ and $z = a$. When examining the effect of an infinitesimal change in a, only the two regions on opposite sides of the plate at $z = a$ are relevant; the domain $z < 0$ can be ignored. The functions D and $D^{-1}$ are given for the region $0 < z < a$ by

$$D_a(x, x') = \int \frac{(d\underset{\sim}{p}_\perp)}{(2\pi)^2} e^{i\underset{\sim}{p}_\perp \cdot (\underset{\sim}{r} - \underset{\sim}{r}')_\perp} \Sigma_n \mu_n(z)\, \mu_n(z') \frac{i}{2E} e^{-iE|t-t'|}$$

$$D_a^{-1}(x, x') = \int \frac{(d\underset{\sim}{p}_\perp)}{(2\pi)^2} e^{i\underset{\sim}{p}_\perp \cdot (\underset{\sim}{r} - \underset{\sim}{r}')_\perp} \Sigma_n \mu_n(z)\, \mu_n(z') \left(E^2 + \frac{\partial^2}{\partial t^2}\right)\delta(t-t'),$$

where $\mu_n(z)$ are the normalized eigenfunctions $(2/a)^{1/2} \sin(n\pi z/a)$,

$$E^2 = \underset{\sim}{p}_\perp^{\,2} + (n\pi/a)^2,$$

and $\perp$ indicates the transverse xy plane. The orthogonality of the eigenfunctions restricts the contribution of $\delta D_a^{-1}$ to the effect of $\delta E^2$:

$$\delta\varepsilon_a = \frac{1}{2} A \int \frac{(d\underset{\sim}{p}_\perp)}{(2\pi)^2} \Sigma_n\ \delta E^2 \frac{1}{2E} e^{-iE\tau}\Big|_{\tau\to 0},$$

where A is the macroscopic area of the plates, and the limit $\tau \to 0$, as demanded by $\delta(\tau)$, is deferred until both contributions to the energy shift are added. After writing $\delta E^2 = -2(\delta a/a)(n\pi/a)^2$, recognizing that $(d\underset{\sim}{p}_\perp)/(2\pi)^2 \to (1/4\pi)\, dE^2$, and performing the E integration for fixed n, we find that

$$\delta\varepsilon_a/A = -\frac{1}{4\pi}\frac{\delta a}{a}\Sigma_n\left(\frac{n\pi}{a}\right)^2 \frac{e^{-i(n\pi/a)\tau}}{i\tau}$$

$$= \frac{1}{4\pi}\frac{\delta a}{a}\frac{1}{i\tau}\frac{d^2}{d\tau^2}\frac{1}{1 - e^{-i(\pi/a)\tau}}.$$

The analogous energy shift associated with the region $a < z < L$ is obtained by the substitution $a \to L-a$, $\delta a \to -\delta a$, or, since L is very large,

$$\delta\varepsilon_{L-a}/A = -\frac{1}{4\pi}\delta a \frac{1}{i\tau}\frac{d^2}{d\tau^2}\frac{1}{\pi i\tau}.$$

This gives the complete energy shift, or, more significantly, the force per unit area, as

$$F = -\frac{1}{A}\frac{\partial\varepsilon}{\partial a} = \frac{1}{8\pi a}\frac{1}{\tau}\frac{d^2}{d\tau^2}\left(\cot\frac{\pi}{2a}\tau - \frac{2a}{\pi\tau}\right)\Big|_{\tau\to 0}.$$

The introduction of the expansion

$$\cot x - \frac{1}{x} = -\frac{1}{3}x - \frac{1}{45}x^3 \ldots$$

then yields

$$F = -\frac{\pi^2}{480}\frac{1}{a^4},$$

which, multiplied by 2 for the two polarizations associated with $n \neq 0$, is the known electromagnetic result.

To include finite temperature effects, we make the following substitution [5] in D,

$$e^{-iE\tau} \to e^{-iE\tau} + \frac{2}{e^{\beta E} - 1} \cos E\tau ,$$

without altering $D^{-1}$. The effect on $\delta\varepsilon_a$ is conveyed by the replacement

$$\frac{e^{-i(n\pi/a)\tau}}{i\tau} \to \frac{e^{-i(n\pi/a)\tau}}{i\tau} - \frac{2}{\beta} \log \left(1 - e^{-(n\pi\beta/a)}\right) ,$$

where $\tau$ has been set equal to zero in the thermal term. The additional thermal contribution to F is

$$F_T = - \frac{\pi}{2} \frac{1}{\beta a^3} \Sigma_n \, n^2 \log \left(1 - e^{-(n\pi\beta/a)}\right) - \frac{\pi^2}{90} \frac{1}{\beta^4} ,$$

which employs the integral

$$- \int_0^\infty dx \, x^2 \log(1 - e^{-x}) = \frac{\pi^4}{45} .$$

Under the high temperature circumstances expressed by $\pi\beta/a << 1$, the use of the Euler-Maclaurin sum formula gives

$$F_T = - \frac{\pi}{2} \frac{kT}{a^3} \left[ \frac{1}{36} + 2 \sum_{m=2}^{\infty} (-1)^m \frac{(2m-4)!}{(2m)!} B_m \right] ,$$

where the Bernoulli numbers appearing here are illustrated by: $B_2 = 1/30$, $B_3 = 1/42$. The rapid convergence of this series is indicated by the first few successive partial sums referring to the quantity in brackets and beginning with 1/36: 0.02778, 0.03056, 0.03042, 0.03046. On the other hand, the introduction of the integral representation for (2m-4)!, but with 4 replaced by 2-s, s > 1, enables the series to be summed, and then analytically continued to evaluate the bracket of $F_T$ as

$$\frac{\zeta(3)}{4\pi^2} = 0.030448.$$

The final statement,

$$kT >> \pi/a: \quad F_T = - \frac{\zeta(3)}{8\pi} \frac{kT}{a^3} ,$$

when multiplied by 2, coincides with the known electromagnetic result.

REFERENCES

1. Sparnaay, M.J., Physica 24, 751 (1958).
2. Casimir, H.B.G., Proc. Kon. Ned. Akad. Wetenschap 51, 793 (1948).

For a survey of later developments, focused around the notion of zero point energy, see T.H. Boyer, Annals of Physics 56, 474 (1970).

3. Schwinger, J., Particles, Sources and Fields, Addison-Wesley, Reading, Mass., 1970 and 1973.
4. It is also a response to the appeal of C.M. Hargreaves, Proc. Kon. Ned. Akad. Wetenschap B68, 231 (1965) -- 'it may yet be desirable that the whole general theory be reexamined and perhaps set up anew.'
5. See, for example, Equation (2-2.48) of Ref. 3, with the Planck distribution introduced for the particle numbers.

# Deep inelastic scattering of leptons*

**(electrodynamics/strong interactions/weak interactions/high energy/polarization)**

JULIAN SCHWINGER

Department of Physics, University of California, Los Angeles, Calif. 90024

*Contributed by Julian Schwinger, July 23, 1976*

**ABSTRACT The description of deep inelastic scattering as an extrapolation between resonance production and diffractive scattering is extended to give successful quantitative representations for the structure functions of unpolarized electron scattering on protons and neutrons, and of neutrino, antineutrino scattering on nucleons. The same ideas also supply a prediction for the polarization asymmetry in deep inelastic electron–proton scattering (paper no. 2, to be published).**

Two previous articles (1, 2) gave a source theoretic analysis of deep inelastic scattering of electrons on nucleons. It was concluded that the general characteristics of these phenomena emerge as a reasonable extrapolation between the known properties of resonance production and of diffractive scattering, an observation that provides no support for the naive models of inner structure that now seem to be popularly accepted as reality. We press the point further in this note by producing a successful representation for the detailed shape of the scaling function in unpolarized electron scattering. And, in a new turn, the application of the same ideas to deep inelastic scattering of neutrinos and antineutrinos yields a satisfactory account of the known facts in this related domain of phenomena. Flushed with these triumphs, we then venture into still unexplored terrain by proposing a provisional expression for the electron–proton polarization asymmetry that will soon be probed experimentally in deep inelastic scattering (paper no. 2 of this series).

The formalism presented in the earlier papers, which are respectively cited as *I* (ref. 1) and *II* (ref. 2) will be modified somewhat in the course of these developments. Thus, the discussion in *I* of the high energy photo cross section was simplified by replacing a numerical factor $\cong 1.2$ with unity. That situation is improved by introducing the mass parameter of the dipole form factor representation, $m_0 = 0.9\ m = 0.84$ GeV, so that [*I*15] now reads

$$\sigma_\gamma = (8\pi\alpha/m_0^2)m_0^2 m\nu \operatorname{Im} H_2(q^2 = 0, -qp = m\nu),$$
$$8\pi\alpha/m_0^2 \cong 100\ \mu b, \qquad [1]$$

and [*I*16] becomes

$$\nu/m \gg 1:\ m_0^2 m\nu \operatorname{Im} H_2(0, m\nu) \cong 1. \qquad [2]$$

We also record here approximate representations of the high energy behavior of the proton and neutron cross sections,

$$\sigma_{\gamma p} \sim 100\ \mu b(1 + 0.6\ \nu_{\text{GeV}}^{-1/2}),$$
$$\sigma_{\gamma n} \sim 100\ \mu b(1 + 0.4\ \nu_{\text{GeV}}^{-1/2}). \qquad [3]$$

With this change, Eq. *I*35 appears as

$$M_+^2/m^2 \gg 1:\ \int_0^\infty d\zeta h_2(\zeta, m^2/M_+^2) \cong 2M_+^2/m_0^2, \qquad [4]$$

thereby replacing [*I*36] with

$$(m^2/M_+^2)\zeta \gtrsim 1:\quad h_2(\zeta, m^2/M_+^2) \cong \exp[-(m_0^2/2M_+^2)\zeta]. \qquad [5]$$

As a consequence, the improved scaling variable is altered to

$$\omega_s = \frac{2m\nu + m^2 + \frac{1}{2}m_0^2}{q^2 + \frac{1}{2}m_0^2} = \frac{2m\nu + 1.24}{q^2 + 0.36}, \qquad [6]$$

with corresponding changes in the two denominators of [*I*51].

The empirical universality in the large $q$ dependence of nucleon and resonance form factors suggests an identification of the asymptotic function $\bar{h}_2(\zeta)$ of [*I*44] with the function $h_2(\zeta)$ of [*I*31], at least for small values of $\zeta$. This is accomplished by the substitution

$$q^2/m^2 \longrightarrow (\omega_s - 1)^{-1}, \qquad [7]$$

with an implied restriction to $\omega_s$ values that are not large. Applied to the proton, of magnetic moment 2.79 magnetons, we get the provisional form

$$f_p(\omega_s) \cong \frac{\omega_s}{\omega_s - 1}\,\frac{1 + \frac{1}{4}(2.79)^2(\omega_s - 1)^{-1}}{1 + \frac{1}{4}(\omega_s - 1)^{-1}}\,\frac{1}{[1 + 1.2(\omega_s - 1)^{-1}]^4}$$
$$= \frac{\omega_s(\omega_s - 1)^3}{(\omega_s + 0.2)^4}\,\frac{\omega_s + 0.95}{\omega_s - 0.75}. \qquad [8]$$

This function behaves as $(\omega_s - 1)^3$, as $\omega_s \to 1$, and reaches unity for $\omega_s \to \infty$, which is qualitatively satisfactory. But the approach to unity for large $\omega_s$ is given by $1 - (2.1/\omega_s)$, which is not the behavior demanded by the asymptotic form of the photo cross section. According to Eq. 3 that is $1 + 0.6\ \nu_{\text{GeV}}^{-1/2}$, where

$$\nu_{\text{GeV}}^{-1/2} = \left[\frac{2m^2}{(0.94)\ 2m\nu}\right]^{1/2}$$
$$= \frac{2}{(0.9)(0.94)^{1/2}}\left[\frac{\frac{1}{2}m_0^2}{2m\nu}\right]^{1/2} \longrightarrow \frac{2.3}{(\omega_s - 3/4)^{1/2}} \qquad [9]$$

and the constant $-\frac{3}{4}$ appears rather arbitrarily at this stage since the domain of applicability is large $\omega_s$. The resulting function

$$1 + 1.4(\omega_s - 3/4)^{-1/2}, \qquad [10]$$

will be taken seriously down to $\omega_s = 3$, which is $\nu_{\text{GeV}} = 0.43$, or $M_+ = 1.30$ GeV, not far above the major resonance at 1.24 GeV. For $\omega_s < 3$ we simply return this function to unity at $\omega_s = 1$, in a continuous manner,

$$\omega_s < 3:\quad 1 + 1.15(\omega_s - 1) - 0.34(\omega_s - 1)^2. \qquad [11]$$

Then, as the simplest union of the information supplied by the resonance region and by the diffraction region, we take just the product of the two functions, [8] and [**10, 11**],

$$f_p(\omega_s) = \frac{\omega_s(\omega_s - 1)^3}{(\omega_s + 0.2)^4}\,\frac{\omega_s + 0.95}{\omega_s - 0.75} \times \begin{cases} 1 + 1.4(\omega_s - 0.75)^{-1/2} & \omega_s > 3 \\ 1 + 1.15(\omega_s - 1) - 0.34(\omega_s - 1)^2 & \omega_s < 3 \end{cases}. \qquad [12]$$

* This is paper no. 1 in a series. The second, final, paper will appear in a later issue of the PROCEEDINGS.



Reprinted from the *Proceedings of the National Academy of Sciences U.S.A.* 73, 3351–3354 (1976).

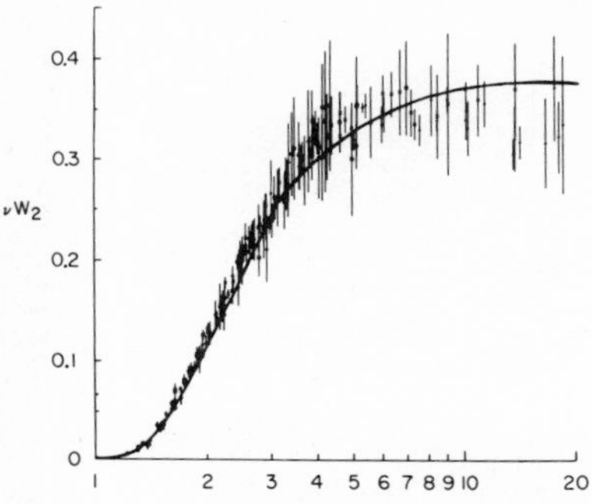

FIG. 1. Comparison between experiment and theory, according to Eq. 12, where $\nu W_2 = (1/\pi) f_p(\omega')$.

The comparison with experiment (3) is shown in Fig. 1, where the axis labeled $\nu W_2$, in the conventional parlance, is $(1/\pi)$ $f_p(\omega')$ and $\omega'$ is not exactly our scaling variable, but

$$\omega' = \frac{2m\nu + m^2}{q^2} = \omega_s + \frac{\frac{1}{2} m_0^2}{q^2}(\omega_s - 1) \qquad [13]$$

The agreement is seen to be quite satisfactory†.

The difference between the proton and neutron asymptotic photo cross sections is measured by

$$0.2\ \nu_{\mathrm{GeV}}^{-1/2} \longrightarrow 0.5(\omega_s - 3/4)^{-1/2}. \qquad [14]$$

This suggests the relation

$$f_p(\omega_s) - f_n(\omega_s) = \frac{1}{2}(\omega_s - 3/4)^{-1/2} f_p(\omega_s), \qquad [15]$$

which has already appeared in *I* as

$$f_n(\omega_s)/f_p(\omega_s) = 1 - \frac{1}{2}(\omega_s - 3/4)^{-1/2}. \qquad [16]$$

Indeed, it was this anticipation that motivated the additive constant in [**9**]. It is interesting that the same combination has also emerged from the resonance region, through the substitution

$$1 + \frac{1}{4}(q^2/m^2) \longrightarrow (\omega_s - 3/4)/(\omega_s - 1). \qquad [17]$$

It was remarked in *I* that [**16**] gave a "quite respectable account" of all available data‡. That is displayed in Fig. 3.

A source theoretic discussion of neutrino (antineutrino)-nucleon deep inelastic scattering (by which is meant the respective processes $\nu + N \rightarrow \mu^- + \text{any}$, $\bar\nu + N \rightarrow \mu^+ + \text{any}$) is presented in ref. 7. In contrast with the electromagnetic coupling to the nucleon current that is symbolized by $V_3 + S$ (isovector + isoscalar), the neutrino interaction involves the parity-violating, charge-changing combinations $(V_1 \pm iV_2) + (A_1 \pm iA_2)$, where $A_{1,2}$ represents the axial vector current. The presence of parity violation requires a third tensor,

$$T_3^{\mu\nu} = -q^2 i\epsilon^{\mu\nu\kappa\lambda} q_\kappa p_\lambda, \qquad [18]$$

and its associated function $H_3(q^2, qp)$. The form factor combinations that respectively determine $H_2$ and $H_3$ for elastic scattering are

$$\frac{G^2_{EV} + (q^2/4m^2)G^2_{MV}}{1 + (q^2/4m^2)} + G_A^2, \qquad G_A G_{MV}, \qquad [19]$$

where the $V$ subscript means that the charge and magnetic moment are proton–neutron differences, 1 and 4.7, respectively. The axial vector form factor is taken to be

$$G_A = -\frac{1.24}{[1 + (q^2/m_1^2)]^2}, \qquad \frac{1}{m_1^2} = 0.9\frac{1}{m^2}, \qquad [20]$$

in which the choice of $m_1 = 0.99$ GeV is consistent with the experimental determination (8) of $0.95 \pm 0.12$ GeV. In the limit

† After this work was finished, new data for larger $\omega$ values became available (4). Within the somewhat uncertain systematic errors, these results seem consistent with the anticipated slow decrease, for increasing $\omega$, to a limiting value in the neighborhood of $1/\pi = 0.32$. We also note that the approach to scaling exhibited in Fig. 18 of ref. 4 is well represented by the function $q^2/(q^2 + 0.36)$, as exhibited in Fig. 2.

‡ The paper cited in ref. 4 contains new measurements for small values of $x' = 1/\omega'$; these also are well represented by Eq. **16**.

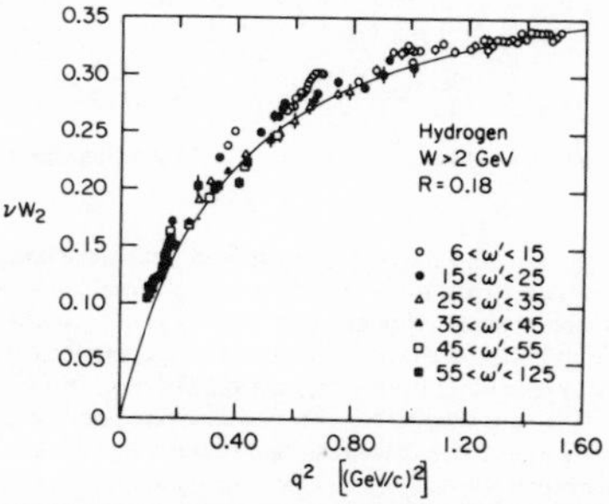


FIG. 2. Comparison between observed scaling deviations of $\nu W_2$ and theoretical function $q^2/(q^2 + 0.36)$, with factor 0.42. Here $W = M_+$, $R = \sigma_L/\sigma_T$.

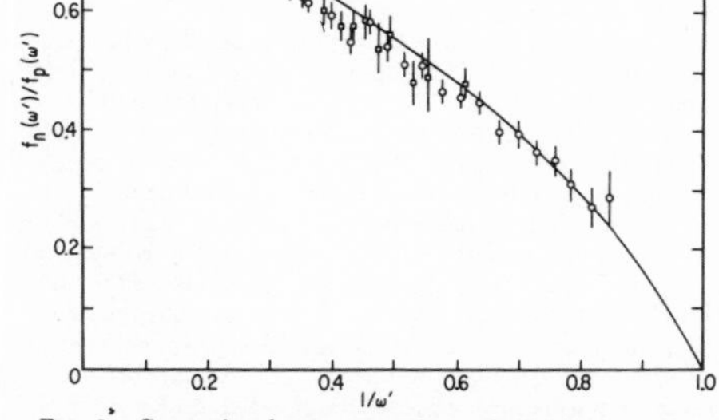

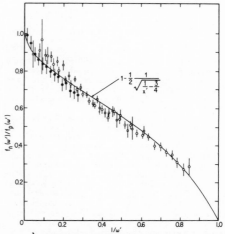

FIG. 3. Comparison between experimental values of $f_n(\omega')/f_p(\omega')$ and Eq. **16**. The symbol $x' = 1/\omega'$. Data are from: ●, Stein *et al.* (4); ○, Bodek *et al.* (5); and □, Poucher *et al.* (6).

of deep inelastic scattering, the total cross sections for incident energy $E$ are ($G = 1.02 \times 10^{-5}\, m^{-2}$)

$$\sigma^{\nu,\bar{\nu}} = \frac{2G^2}{3\pi^2} mE \int_0^1 dx[f_2(x) \mp x f_3(x)] \qquad [21]$$

where the functions $f_a(x) = f_a(\omega)$, $x = 1/\omega$, must obey

$$|2xf_3(x)| \leq f_2(x). \qquad [22]$$

We first note that the preponderance of neutrino experiments refer to an approximately equal mixture of neutrons and protons, which average appears as a trace in isotopic-spin space. Accordingly, the quadratic combination of interactions involved in $H_2$, and its scaling limit $f_2(x)$, are symbolized by $V_1{}^2 + V_2{}^2 + A_1{}^2 + A_2{}^2$, as compared with the electromagnetic analogue, $V_3{}^2 + S^2$, where the values associated with all $V_k{}^2$, $k = 1,2,3$, are the same. If we now remark that the equality of neutron and proton electromagnetic cross sections as $x \to 0$ is expressed by the effective vanishing of the isoscalar current in this limit, it is inferred that the isovector part of the neutrino function $f_2(x)$ approaches 2 as $x \to 0$. The substitution $q^2/m^2 \to x/(1-x)$ applied to [19] then yields the provisional identifications

$$f_2(x) \cong 2\frac{(1-x)^3}{(1+0.2\,x)^4}\left[\frac{1+4.52\,x}{1-0.75\,x} + g^2\right],$$

$$g = 1.24\left[\frac{1+0.2\,x}{1-0.1\,x}\right]^2 \qquad [23]$$

and

$$-\frac{2xf_3(x)}{f_2(x)} = 9.4\,xg\left[\frac{1+4.52\,x}{1-0.75\,x} + g^2\right]^{-1}. \qquad [24]$$

The latter function vanishes proportionately to $x$ as $x \to 0$, which is a statement about the approach to equality of the differential neutrino and antineutrino cross sections contained in [21]. But it is an $x^{1/2}$ behavior that might be expected in this diffractive limit. Hence, beginning (somewhat arbitrarily) at $x = 0.45$ and proceeding to smaller $x$ values, we multiply [24] by $(0.45/x)^{1/2}$, so that

$$-\frac{2xf_3(x)}{f_2(x)} = 9.4\,xg\left[\frac{1+4.52\,x}{1-0.75\,x} + g^2\right]^{-1} \begin{cases} x > 0.45: & 1 \\ x < 0.45: & (0.45/x)^{1/2} \end{cases}. \qquad [25]$$

The resulting function never exceeds unity (within a tolerance of less than 1%), in accordance with [22], while remaining everywhere greater than 0.88 in the interval $0.9 > x > 0.15$, and greater than 0.95 in the interval $0.8 > x > 0.25$. The effective near equality of $-2x\, f_3(x)$ and $f_2(x)$ except for small $x$ is just what is observed experimentally (9).

The diffractive modification of [23] for small $x$ is harder to specify. We first note that the approach of the V part to its limit is given by $2(1 + 1.5\,x)$. This is an approach from above, as compared with that encountered in Eq. 8 for the proton electromagnetic function and a diffraction factor is not needed here to amend the qualitative situation. Then, according to [23] the ratio of A and V contributions attains the limit $(1.24)^2$ as $x \to 0$. It is implausible that the low energy number 1.24 should determine the high energy limit; an asymptotic descent to unity would be a natural hypothesis. But, lacking any clear guide for this modification, and considering that the experimental values for small $x$ are particularly sensitive to uncertainties associated with scaling deviations, we leave this question for future study and retain [23] as it stands for our immediate purposes. (However, see Figs. 4 and 5.)

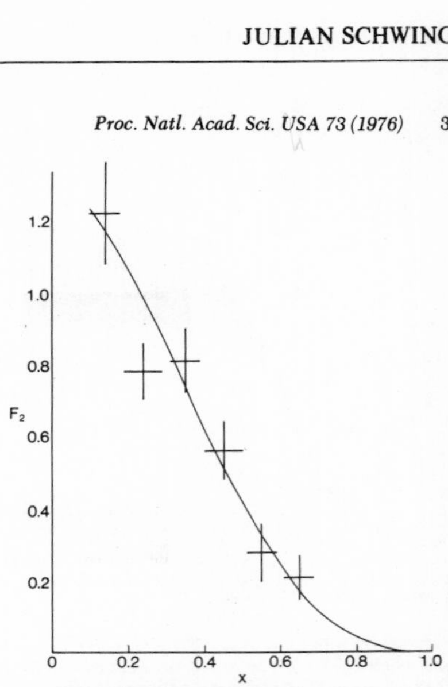


FIG. 4. Comparison between experimental values of $F_2(x) = (1/\pi)f_2(x)$ and Eqs. 23 and 33.

The total cross sections of [21] involve two integrals that are now assigned the theoretical values

$$\int_0^1 dx f_2(x) = 1.81, \qquad -\int_0^1 dx 2xf_3(x) = 1.49 \qquad [26]$$

which are in the ratio 0.82. The resulting prediction of the cross-section ratio,

$$\sigma^{\bar{\nu}}/\sigma^{\nu} = 0.42, \qquad [27]$$

compares favorably with a recent analysis (10) of CERN (European Center for Nuclear Research) data, selected to be in the scaling region, that gave

$$\sigma^{\bar{\nu}}/\sigma^{\nu} = 0.43 \pm 0.04. \qquad [28]$$

The same analysis also provided average values of $x$ for neutrino and antineutrino distributions:

$$\langle x\rangle^{\nu} = 0.26 \pm 0.02, \qquad \langle x\rangle^{\bar{\nu}} = 0.25 \pm 0.02. \qquad [29]$$

In view of the close similarity of $f_2(x)$ and $-2x\, f_3(x)$, these numbers can be compared approximately with the prediction

$$\int_0^1 dx\, x\, f_2(x) \Big/ \int_0^1 dx\, f_2(x) = 0.24. \qquad [30]$$

As for absolute values, we cite a recent high energy measurement (11) of the total cross sections that yielded

$$\sigma^{\nu} + \sigma^{\bar{\nu}} = (1.11 \pm 0.12)E_{\text{GeV}} \times 10^{-38}\ \text{cm}^2, \qquad [31]$$

or

$$\int_0^1 dx f_2(x) = 1.7 \pm 0.2, \qquad [32]$$

quite consistent with the prediction in [26]. Finally, the detailed shapes of $f_2(x)$ and $-2x\, f_3(x)$ are compared with the experimental results of ref. 9, as recalculated in ref. 10. These are presented in Figs. 4 and 5, where $F_2(x) = (1/\pi)f_2(x)$ and $xF_3(x) = (1/\pi)[-2x\, f_3(x)]$. The region $x < 0.1$ is omitted to avoid reference to measurements that can be significantly affected by systematic errors. Here we do adopt the suggestion that the

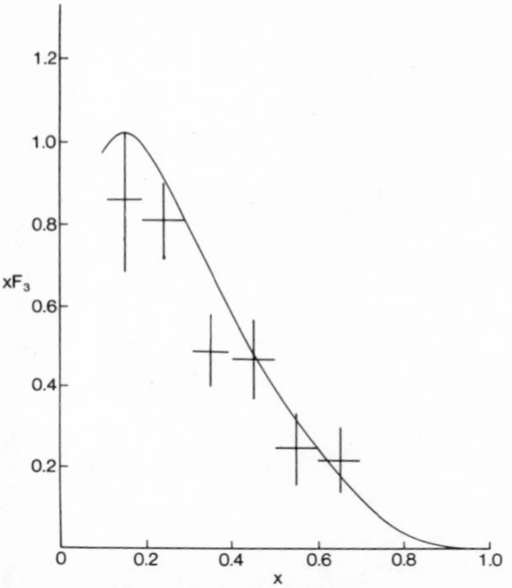


FIG. 5. Comparison between experimental values of $xF_3(x) = (1/\pi)[-2x\, f_3(x)]$ and Eqs. **23**, **25**, and **33**.

number 1.24 should become unity as $x \rightarrow 0$, which is realized by the simple diffractive form

$$1.24 \longrightarrow 1 + 0.24\, x^{1/2}. \qquad [\mathbf{33}]$$

The experimental data are seen to be reasonably well reproduced.

After this work had been completed, experimental data on neutrino–proton scattering became available (ref. 12). Our discussion refers to the equally weighted average of proton and neutron. Concerning the ratio $\sigma_p{}^\nu/\sigma_n{}^\nu$ as a function of $x$, we remark that it should decrease from unity at $x = 0$ in a diffractive manner (involving $x^{1/2}$) and, presumably, approach zero as $x \rightarrow 1$, since elastic $\nu p$ scattering vanishes. The simplest representation of these requirements is $\sigma_p{}^\nu/\sigma_n{}^\nu = 1 - x^{1/2}$ and thus we convert $\sigma^\nu$ of Eq. **21** into $\sigma_p{}^\nu$ by multiplication with $(1 - x^{1/2})/(1 - 0.5\, x^{1/2})$. The resulting function is in satisfactory accord with the data presented in Fig. 3 of ref. 12, in contrast with the curve exhibited there. This is illustrated by the comparison between the implied value of $\langle x \rangle_{\nu p} = 0.21$ and the experimental value, 0.22, cited in ref. 10.

There is also new evidence on scaling deviations for small $q^2$ and very large $\omega$ values, $\leq 1000$ (13). When the $\nu W_2$ values presented in Table I of ref. 13 are multiplied by $1 + (0.36/q^2)$, using mean values of $1/q^2$ for rather wide intervals, the results are entirely compatible with an approach for increasing $\omega$ to the limit $1/\pi = 0.32$.

I am indebted to Kimball Milton, Lester DeRaad, and Wu-Yang Tsai for conversations on their work and for assistance in preparing the figures.

1. Schwinger, J. (1975) "Source theory viewpoints in deep inelastic scattering," *Proc. Natl. Acad. Sci. USA* **72**, 1–5.
2. Schwinger, J. (1975) "Source theory discussion of deep inelastic scattering with polarized particles," *Proc. Natl. Acad. Sci. USA* **72**, 1559–1563.
3. Friedman, J. I. & Kendall, H. W. (1972) *Annu. Rev. Nucl. Sci.* **22**, 203–254.
4. Stein, S., Atwood, W. B., Bloom, E. D., Cottrell, R. L. A., DeStaebler, H., Jordan, C. L., Piel, H. G., Prescott, C. Y., Siemann, R. & Taylor, R. E. (1975) "Electron scattering at 4° with energies of 4.5–20 GeV," *Phys. Rev. D* **12**, 1884–1919.
5. Bodek, A., Dubin, D. L., Elias, J. E., Friedman, J. I., Kendall, H. W., Poucher, J. S., Riordan, E. M., Sogard, M. R., Coward, D. H. & Sherden, D. J. (1974) *The Ratio of Deep-Inelastic e-n to e-p Cross Sections in the Threshold Region*, Publication 1399 (Stanford Linear Accelerator Center, Stanford, Calif.).
6. Poucher, J. S., Breidenbach, M., Ditzler, R., Friedman, J. I., Kendall, H. W., Bloom, E. D., Cottrell, R. L. A., Coward, D. H., DeStaebler, H., Jordan, C. L., Piel, H. & Taylor, R. E. (1974) "High energy single-arm inelastic e-p and e-d scattering at 6 and 10°," *Phys. Rev. Lett.* **32**, 118–121.
7. DeRaad, L. L., Milton, K. A. & Tsai, W.-Y. (1975) "Deep-inelastic neutrino scattering: A double spectral form viewpoint," *Phys. Rev. D* **12**, 3747–3757, and (1976) *Phys. Rev. D* **13**, 3166 (Erratum).
8. Mann, W. A., Mehtani, U., Musgrave, B., Oren, Y., Schreiner, P. A., Singer, R., Yuga, H., Ammar, R., Barish, S., Cho, Y., Derrick, M., Engelmann, R. & Hyman, L. G. (1973) "Study of the reaction $\nu + n \rightarrow \mu^- + p$," *Phys. Rev. Lett.* **31**, 844–847.
9. Gargamelle Neutrino Collaboration (1975) "Experimental study of structure functions and sum rules in charge-changing interactions of neutrinos and antineutrinos on nucleons," *Nucl. Phys. B* **85**, 269–288.
10. Perkins, D. H. (1975) in *Proceedings of the 1975 International Symposium on Lepton and Photon Interactions at High Energies*, ed. Kirk, W. T. (Stanford Linear Accelerator Center, Stanford, Calif.), pp. 571–603.
11. Barish, B. C., Bartlett, J. F., Buchholz, D., Humphrey, T., Merritt, F. S., Sciulli, F. J., Stulle, L., Shields, D., Suter, H., Fisk, E. & Krafczyk, G. (1975) "Measurement of neutrino and antineutrino cross sections at high energy," *Phys. Rev. Lett.* **35**, 1316–1320.
12. FNAL–Michigan Neutrino Collaboration (1976) "Experimental study of inclusive deep inelastic neutrino-proton scattering," *Phys. Rev. Lett.* **36**, 639–642.
13. Chicago–Harvard–Illinois–Oxford Muon Collaboration (1976) "Measurement of nucleon structure function in muon scattering at 147 GeV/c," *Phys. Rev. Lett.* **37**, 4–7.